Aviation Meteorology

Aviation Meteorology

Navale Pandharinath

M.Sc. (Maths), M.Sc. (Statistics)

Director (Retd.),

India Meteorological Department,

Met. Centre, Hyderabad.

BSP BS Publications

A unit of **BSP Books Pvt. Ltd.**

4-4-309, Giriraj Lane, Sultan Bazar,

Hyderabad - 500 095

Phone : 040 - 23445605, 23445688

Published by :

BS Publications

A unit of **BSP Books Pvt., Ltd.**

4-4-309, Giriraj Lane, Sultan Bazar,
Hyderabad - 500 095
Phone : 040 - 23445605, 23445688
e-mail : info@bspbooks.net

ISBN : 978-93-85433-05-4 (HB)

Dedicated to

My wife

Smt. Sumati P. Navale

and to

My parents

Late Smt. Narasubai R. Navale

Late Shri. Rukmaji Rao N Rao Navale

Residents of Kalyani, Tq. Yellareddy,

Nizamabad, Dist., A.P.

Navale Pandharinath

एयर वाईस मार्शल (डॉ) अजित त्यागी
मौसम विज्ञान विभाग के महानिदेशक

Air Vice Marshal (Dr.) Ajit Tyagi

Director General of Meteorology &
Permanent Representative of India with W.M.O.

भारत सरकार
भारत मौसम विज्ञान विभाग
मौसम भवन, लोदी रोड,
नई दिल्ली–110 003

Government of India
India Meteorological Department
Mausam Bhavan, Lodi Road,
New Delhi - 110 003

Date: 16-11-2011

Foreword

I am pleased to write the foreword for the book entitled "Aviation meteorology".

Meteorology immensely contributed to the development of Aviation Industry, particularly in the safety of aircraft in flight. Meteorological support is essential in all three phases of aircraft flight – Take-off, cruising and Landing. The weather phenomena like poor visibility, height of low cloud base, wind shear, gusty winds, squalls dust/sand storms, thunderstorms, microbursts, tornadoes, hailstorm, CAT, Jet streams, tropical cyclones, frontal depressions, volcanic dust, snow falls etc, and altimetry (QNH, QFE) that play vitol roll in the various phases of aircraft flight. Till recently 75% of aviation disasters were attributed to the various weather phenomena referred above. There are a few books on aviation meteorology covering all these topics. The present book is systematically developed the subject and covers almost all weather phenomena that is sensitive to aircraft flights. It covers the complete latest syllabus of CPL & ATPL courses prescribed by DGCA. The book contains a good number objective questions with answers which will be very helpful not only to the trainee pilots but also to the students of meteorology in general.

I hope this book would serve as textbook for the pilots training courses that are taught at various aviation academies not only in India but also abroad and also in universities. This book will also serve as textbook/reference book for the students at post graduate courses offered in atmospheric science, meteorology, oceanography, environmental science and allied subjects taught at several Indian universities and abroad.

(Ajit Tyagi)

Preface

The objective of this book is to provide a compact course material on meteorology for CPL and ATPL or any advanced course of flying training. I had the experience of teaching aviation meteorology for different advanced flying courses, refreshers at Indian Airlines, Central Training Establishment, Hyderabad. After retirement as director IMD, i taught aviation meteorology at Andhra Pradesh Aviation Academy Begumpet and Flytech Aviation Academy Nadirgul Hyderabad. During this period i had gone through a number of books pertaining to aviation meteorology and ICAO, WMO publications and DGCA Circulars (Civil Aviation Requirements).

The science of aviation and meteorology both grew hand-in-hand. The later contributed a lot for the growth of aviation industry, particularly in the safety of aircraft inflight. Meteorological support is required in all three phases of aircraft flight, namely take-off/ climbout, cruising, landing and taxing phases. More than 75% of aviation disasters were linked to the weather phenomenon in all three phases of flight. The weather phenomena that plays vitol role in take-off and landing disasters include poor visibility, height of low cloud base, windshear, sudden gustry winds, squalls, thunderstorms, microbursts, downbursts, hailstorms, sand/dust storms, heavy precipitation, frontal storms, snowfalls and inundation of runways. All these topics are covered adequately. Though the number of disasters were few at cruising level, yet a majority accidents at cruising altitudes were associated with cumulonimbus clouds, CAT, Jetstreams and associated windshear, severe icing, marked mountain waves, tropical storms, volcanic dust. These topics are also covered adequately.

An airoplane moves in a medium of atmosphere which consists of various gases and aerosols. The lowest layer of air surrounding the earth which is called troposphere is the abode of all weather phenomena and a medium for all odds and evens in flying and inseperable for aircraft movements. An aircraft flying through the atmosphere is more vulnerable to the weather as compared to the other transports on land and sea. The first few chapters are dealt with solar energy, heat balance, atmospheric structure, physical laws, pressure, altimetry, QFE, QNH, pressure altitude, density altitude, water in the atmosphere, clouds, adiabatic process, lapse rates, stability and instability of environment surface wind, vertical motion of wind, visibility, fog and associated aviation hazards, air masses and fronts, frontal depressions and associated flying hazards are presented. A few chapters are dealt with tropical cyclones, ITCZ, Monsoons, Indian summer monsoon, general circulation of atmosphere-three cell model, zonal index with oscillation, global pressure, temperature and wind distribution described briefly.

According to DGCA Aircraft Rules 1937, the grant or renewal or variation of Flight Crew licences in various categories like SPL, PPL, GPL, CPL, ATPL etc. Candidates have to pass the relevant examinations (like Navigation, Technical General, Radio aids, Meteorology). The details of syllabus of meteorology for different categories of examinations is given. For other subjects can be had in the DGCA website http : // dgca.gov.in, for RTR-A from the website of Ministry of Communications http: //.india.com. The present volume covers the whole range of syllabus of all categories of pilot courses prescribed by DGCA, particularly CPL, ATPL the ultimate courses. As per the examination system of DGCA, which consist of multiple choice (objective) type questions (except performance) which are to be answered on OMR (Optical Mask Reader) sheets. Keeping this point in view relevant questions with answers (about 1300) have been provided at the end of each chapter, and a miscellaneous question bank at the end of this book.

As said earlier this volume is primarily meant for professional training of pilots of various categories, however it is equally useful for students studying courses in meteorology at college and university level.

The last few chapters deal with meteorological services for civil aviation and covers Area Forecast, Route Forecast, Aerodrome Warnings, briefing, display and documentation. World Area Forecast described sufficiently, which covers grid point data in digital and numerical form, prognostic charts of significant weather, SIGMET information,

VOLMET broad cast. Meteorological observations consists of units of measurement, observatory set up, Automatic weather observing system. Brief description of weather satellites, IMD Meteorological Data Processing system are given. Climatology for Aviators consists of brief description of climatic features in respect of all international aviation routes.

At the end of the volume, Aviation weather reports and Forecasts codes have been given which were evolved by WMO in consultation with ICAO. The current codes used by IMD as given in 9th edition of IMD codes presented the code forms of METAR, SPECI, CODAR, WINTEM, (reports) and TAF, ARFOR, ROFOR (forecasts) are presented. Great care has been taken to explain each term of the code. This is one of the major important feature of this book, which is very essential to individual pilots, ground training instructors because of its practical day to day use.

I will be happy to receive any useful comments or constructive suggestions to improve the book and pointing any printing mistakes.

- *Author*

Acknowledgement

In writing this book I have referred to a number of books and publications of WMO, ICAO, IMD, DGCA, AAI, UK Aeronautical Information Circulars, Oxford Aviation services Ltd. Handbook of Aviation, Meteorology HMSO etc. I owe my sincere thanks to all these institutions, authors of books and respective publishers. I express my thanks to Dr. R.R. Kelkar (Retd.) DGM, IMD and AVM Dr. Ajit Tyagi Director General Meteorology IMD for their encouragement and guidance in writing this book.

I profusely thank to M/S Anil Shah, Nikhil Shah B.S Publication for their patronage and keen interest in publishing this book and also books on meteorology in India. Thanks are also due to the staff B.S Publications, specially S/Shri. K.S. Raju, Naresh, Laxminarayana Jakkani, Kamesh, Smt. D. Prabha, Smt. S. Sandhya for their untiring help in bringing out this book.

Any suggestions for the improvement of the book from aviation academies/Pilots/Ground instructors or meteorologists shall be acknowledged with thanks.

- Author

Contents

Chapter 33

Chapter 34

Syllabus

As per the DGCA, civil aviation requirement series B, part II, section 7, date 16th June 2011

I. Syllabus for oral examination for issue of student Pilot Licence (SPL) for Aeroplanes, Microlight Aircraft, Gliders and Balloons.

Aviation meteorology

1. Elementary knowledge of atmosphere and its properties.
2. Basic knowledge of temperature, pressure and density and their relationship
3. (i) Elementary knowledge of relationship between pressure and wind

   (ii) Elementary knowledge of (a) variation of wind with height, (b) sea breeze and land breeze
4. Elementary knowledge of different types of clouds and precipitation. Basic understanding of hazards associated with certain types of clouds.
5. Elementary knowledge of the terms-visibility, fog, mist and haze
6. Elementary knowledge of variation of pressure with height, and the Q codes – QNH, QFE and QNE etc.
7. Basic understanding of METAR, SPECI and aerodrome warnings and their importance in aviation
8. Elementary knowledge of Anemometer, Aneroid Barometers, Wind sock etc.

II. The updated syllabus for the examination of Issue of Commercial Pilots Licence and Instrument Rating aeroplanes and helicopters. (updated – 16th June 2011)

1. Aviation Meteorology

1.1 The Atmosphere

(a) Composition, extent, vertical division

(b) Temperature

- vertical distribution of temperature
- transfer of heat; solar and terrestrial radiation, conduction, convection, advection and turbulence

- lapse rate, stability and instability
- development of inversions, types of inversions
- temperature near the earth's surface, surface effects, diurnal
- variation, effect of clouds, effect of wind

(c) Atmospheric pressure

- barometric pressure, isobars
- pressure variation with height, contours (isohypses)
- reduction of pressure to mean sea level QFF
- Surface low/upper-air low, surface high/upper-air high precipitation

(d) Atmospheric density : Interrelationship of pressure, temperature and density

(e) International Standard Atmosphere (ISA)

(f) Altimetry

- Pressure altitude, true altitude
- height, altitude, flight level
- altimeter setting: QNH, QFE, 1013.25 hPa
- effect of accelerated airflow due to topography

1.2 Wind

(a) Definition and measurement

(b) Primary cause of wind

- Primary cause of wind, pressure gradient, coriolis force, gradient wind
- relationship between isobars and wind

(c) General circulation

- general circulation around the globe

(d) Turbulence

- Turbulence and gustiness, types of turbulence
- origin and location of turbulence

(e) Variation of wind with height

- variation of wind in the friction layer

(f) Local winds: Anabatic and katabatic winds, land and sea breezes, venturi effects

(g) Standing waves

- origin of standing waves

1.3 Thermodynamics

(a) Humidity

- water vapour in the atmosphere
- temperature / dew point, mixing ratio, relative humidity

1.4 Clouds and Fog

(a) Clouds formation and description

- cloud types, cloud classification
- influence of inversions on cloud development

(b) Fog, mist, haze

- radiation fog
- advection fog
- steaming fog
- frontal fog
- orographic fog

1.5 Precipitation

(a) Development of precipitation

- development of precipitation
- types of precipitation
- type of precipitation, relationship with cloud types

1.6 Airmasses and Fronts

(a) Types of airmasses

- description, factors, affecting the properties of an airmass

- classification of airmasses, modifications of airmasses, areas of origin.

(b) Fronts

- boundaries between airmasses (fronts), general situation, geographic differentiation
- warm front, associated clouds and weather
- cold front, associated clouds and weather
- warm sector, associated clouds and weather
- weather behind the cold front
- occlusions, associated clouds and weather
- stationary front, associated clouds and weather
- movement of fronts and pressure systems, life cycle

1.7 Pressure Systems

(a) Location of the principal pressure areas

(b) Anticyclone; Anticyclones, types, general properties, cold and warm anticyclones, ridges and wedges, subsidence

(c) Non frontal depressions

- thermal, orographic – and secondary depressions, cold air pools, trough

(d) Tropical revolving storms

1.8 Climatology

(a) Typical weather situations in mid-latitudes

- westerly waves
- high pressure area
- uniform pressure pattern

(b) Local seasonal weather and wind

- SE & NE Monsoon, Pre-Monsoon, Northwesters, Kalbaisakhi
- Western Disturbance

1.9 Flight Hazards

(a) Icing

- weather conditions for ice accretion, topographical effects

(b) Turbulence

- effects on flight, avoidance

(c) Windshear

- definition of windshear
- weather conditions for windshear
- effects on flight

(d) Thunderstorms

- structure of thunderstorms, squall lines, life history, storm cells, electricity in the atmosphere, static charges
- Conditions for and process of development, forecast, location, type specification
- Thunderstorm avoidance, ground/airborne radar, storm scope
- development and effect of down bursts
- development of lightning discharge and effect of lightning strike on aircraft and flight execution

(e) Low and high level inversions: Influence on aircraft performance

(f) Hazards in mountainous areas

- Influence of terrain on clouds and precipitation, frontal passage
- vertical movements, mountain waves, windshear, turbulence, ice accretion
- development and effect of valley inversions

(g) Visibility reducing phenomena

- reduction of visibility caused by mist, smoke, dust, sand and precipitation
- reduction of visibility caused by low drifting and blowing snow

1.10 Meteorological Information

(a) Observation

- On the ground – surface wind, visibility and runway visual range, transmissometers; clouds – type, amount, height of base and tops, movement; weather – including all types of precipitation, air temperature, relative humidity, dew point, atmospheric pressure

- aircraft observations and reporting, data link systems, PIREPS

(b) Weather Charts

- significant weather charts
- surface charts
- upper air charts
- symbols and signs on analysed and prognostic charts

(c) Information for Flight Planning

- Aeronautical codes; METAR, TAF, SPECI, SIGMET, SNOWTAM, runway report
- Meteorological broadcasts for aviation: VOLMET, ATIS, HFVOLMET, ACARS
- Content and use of pre-flight meteorological documents
- Meteorological briefing and advice
- measuring and warning systems for low level windshear, inversion
- Special meteorological warnings

Syllabus for ATPL includes CPL syllabus and satellite meteorology

Satellite Meteorology (Radiation) Types of satellites

(iii) Satellite measurement of

(i) temperature profile vertically upwards
(ii) water vapour distribution in the atmosphere
(iii) cloud (top) temperatures
(iv) ocean temperatures and currents

(ii) Types of Images

(i) Visible and IR
(ii) Interpretation of satellite images

III. Syllabus for Examination for issue of Private Pilots licence for Aeroplanes and Pilots licence for Microlight Aircraft – Gliders and Balloons. (updated – 16th June 2011)

3. Aviation Meteorology

3.1 The Atmosphere

- Composition and structure
- vertical divisions
- Pressure, density and temperature
- barometric pressure, isobars
- changes of pressure, density and temperature with altitude
- altimetry terminology
- diurnal variation of temperature
- adiabatic process
- temperature lapse rate
- stability and instability

3.2 Humidity and Precipitation

- water vapour in the atmosphere
- vapour pressure
- dew point and relative humidity
- condensation and vaporization
- precipitation

3.3 Pressure and Wind

- high and low pressure areas
- motion of the atmosphere, pressure gradient
- vertical and horizontal motion, convergence, divergence
- surface and geostrophic wind
- effect of wind gradient and windshear on take-off and landing
- relationship between isobars and wind, Buys Ballot's law
- turbulence and gustiness
- local winds, fahn, land and sea breeze

3.4 Clouds formation

- Cooling by advection, radiation and adiabatic expansion – cloud types
- convection clouds
- orographic clouds
- stratiform and cumulus clouds
- flying conditions in each cloud type

3.5 Fog, Mist, Haze

- radiation, advection, frontal, freezing fog
- formation, and dispersal
- reduction of visibility due to mist, snow, smoke, dust and sand
- assessment of probability of reduced visibility
- hazards in flight due to low visibility, horizontal and vertical

3.6 Air masses

- description of and factors affecting the properties of air masses
- classification of air masses, region of origin modification of air masses during their movement.
- development of low and high pressure system
- weather associated with pressure systems

3.7 Ice accretion

- conditions conducive to ice formation
- effects of hoar frost, rime ice, clear ice
- effects of icing on aeroplane performance
- precautions and avoidance of icing conditions.
- powerplant icing
- precautions, prevention and clearance of induction and carburetor icing

3.8 Thunderstorms

- formation – air mass, frontal, orographic

- conditions required
- development process recognition of favourable conditions for formation
- hazards for aeroplanes
- effects of lightning and severe turbulence
- avoidance of flight in the vicinity of thunderstorms

3.9 Flight over hilly or mountainous areas

- hazards
- influence of terrain on atmospheric processes
- mountain waves, windshear, turbulence, vertical movement, rotor effects, valley winds

3.10 Indian Climatology

- general seasonal circulation in the troposphere
- local seasonal weather and winds
- tropopause influence on aircraft performance
- effect of ozone, radioactivity
- development and effect of valley inversions
- reduction of visibility caused by low drifting and blowing snow

3.11 Altimetry

- Operational aspects of pressure settings
- pressure altitude, density altitude
- height, altitude, flight level
- ICAO standard atmosphere

  QNH, QFE standard setting
- Transition altitude, layer and level

3.12 The meteorological organization

- aerodrome meteorological offices
- aeronautical meteorological stations
- forecasting service
- meteorological services at aerodromes
- availability of periodic weather forecasts

3.13 Weather analysis and forecasting

- weather charts, symbols signs
- significant weather charts
- prognostic charts for general aviation

3.14 Weather Information for Flight Planning

- reports and forecasts for departure, en-route, destination and alternate(s)
- interpretation of coded information METAR, TAF
- availability of ground reports for surface wind, windshear, visibility

3.15 Meteorological broadcasts for aviation

- VOLMET, ATIS, SIGMET

CHAPTER 1

The Sun

The naked eye stars in the sky is called celestial sphere. The sun is one of the normal sized star in the milky way galaxy. The galaxies are the building blocks of the universe. There are about 10 billion (10^9) galaxies in the universe. The milky way galaxy is rotating around its nucleus in accordance with law of gravitation. The distance of the sun from the galactic centre is about 30,000 light years. (1 light year = distance travelled by light in one year. Speed of the light = 3×10^8 m/s; 1lt yr = $365.25 \times 24 \times 60 \times 60 \times 3 \times 10^8$ m/s $\simeq 9.5 \times 10^{12}$ km). The link between the distant stars and the earth is the electromagnetic radiation emitted by the stars. The radiation emitted by the stars enables us to detect its source of direction, its composition, temperature and its velocity towards the earth or away from it (Doppler effect).

Big Bang theory: All matter in the universe initially was packed in a compact superdense agglomeration. It was hurled in all directions with a cataclysmic explosion (called Big Bang). Since then the mass is moving away from that nuclear centre and the universe is expanding. The recession of the galactic mass depended on the initial velocity, higher the velocity of matter farther is its recession.

***Solar System*:** The sun is at one of the foci of the ellipse. All planets, including the earth are revolving in elliptical orbits around the sun. The solar system as a whole is rotating and moving in the milkyway galaxy with a velocity of about 400 km/s; which is the absolute velocity of the sun. The origin of the solar system is very complex and still obscure.

***The Sun*:** The mass of the sun is about 1.99×10^{30} kg, its radius is about 6.97×10^8 m (or diameter = 14×10^5 km). The mass of the earth is about 6×10^{24} kg, its radius about 6.4×10^6 m. The gravitational attraction of the sun is about 274 m/s^2, while the gravitational attraction of the earth is about 9.81 m/s^2. About 99.87% mass of the solar system is contained in the sun (see Fig. 1.1). The mean earth-sun distance is 150×10^6 km.

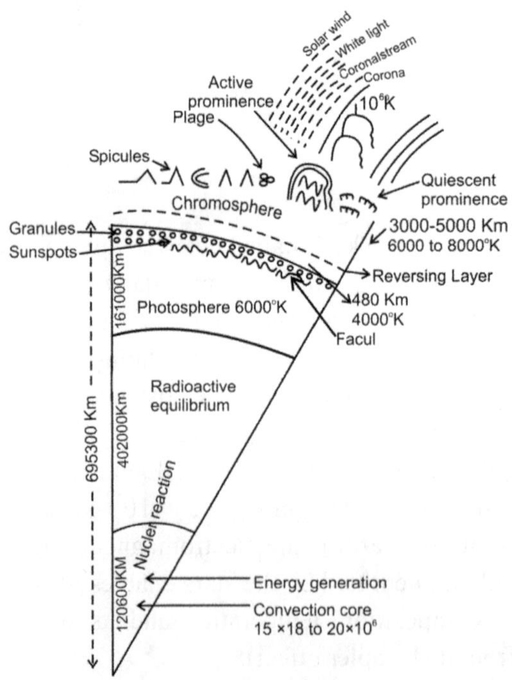

Fig. 1.1 Schematic representation of structure and activities of the sun with the exaggerated photosphere, chromosphere.

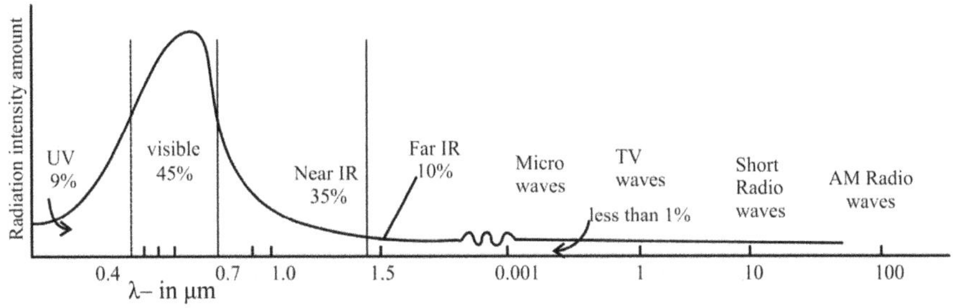

Fig. 1.2 Sun's electromagnetic spectrum, not to scale.

***Photosphere*:** The visible surface of the sun is called photosphere, its temperature is about 6000 °K. The photosphere consists of hot gases under high pressure. Photosphere produces radiations of the visible part of the solar spectrum, but the energy is produced in the deeper central region of the sun's sphere. The surface of the sun changes constantly without any rhythm. It appears as mottled or granular due to the boiling motion of the gases. The granules are visible to naked eye. Besides granules the photosphere shows bright regions called faculae and dark (cold) regions called sunspots. Sunspots are regions of strong magnetic field.

***Chromosphere*:** During the total solar eclipse a reddish ring of light observed around the sun's photosphere, which is called chromosphere or the colour sphere of the sun. It consists of hydrogen and helium gases at low pressure, but has temperature about 20,000 °K. Fibrilles and spicules are observed in the chromosphere. In chromosphere temperature rises slowly from 4400 °K to about 1.5 to 2.0 million °K. Between photosphere and chromosphere there is a cooler gas layer with temperature about 4000 °K, thickness about 500 km. This is called reversing layer. Above this layer temperature rises.

***Corona*:** Above chromosphere lies corona, which spreads far out into space with temperature 1.5 to 2.0 million °K. During total solar eclipse corona shows spectacular light of halo. Solar winds originate in corona, whose speed exceed 100 km/s and the plasma particles of corona exceed the escape velocity of the sun's atmosphere. This is called solar wind and it depends on the solar activity.

***Solar Activity*:** when the frequency of observable features of the sun (such as sunspots, solar wind, solar flares) are more than normal, then sun is said to be active or "disturbed", but when they are absent the sun is said to be "quiet". As seen from the sunspots the solar activity has about

11 years cycle. Sunspots begin to appear at latitude 35° on both hemispheres of the sun, this is sunspot minima phase. As the cycle advances, the number of spots increase in size and numbers and spread to the solar equator, which is sunspot maxima phase. Subsequently these spots decrease gradually and disappear. Again a new cycle begins with sunspots reappearing at latitude 35°. Sunspots are regions of cooler temperature and lower energy. The other observable features where energy emission is more than normal are called active regions. A facula appears before sunspot and persists even after the visible appearance of the spot. Faculae are also observed at high latitudes of the sun in the absence of sunspots.

Increased solar activity is seen in chromosphere in association with the sunspot groups in photosphere. Large bright regions are called plages.

Sometimes clouds of bright gas is seen projecting outwards from chromosphere, which are called prominences. When rapid changes are observed in the neighborhood of prominences there appears chromospheric bursts, coronal rays and holes.

The magnetic field in sunspots observed to exceed thousand times as compared to the magnetic field outside the sunspots. The polarity of the magnetic field in sunspot area changes in 22 years, which is twice the period of sunspot cycle. There does not appear to be a clear cut theory of solar activity. No sunspots observed during 1645 to 1715 (70 years) and there does not appear 11 year sunspot cycle before 1645. These facts were brought to the light by E. Maunder, the superintendent of Greenwich observatory in 1890. The 70 year period (1645-1715) of sunspot minima or complete absence of it, is called Maunder Minima.

***Cosmic Rays*:** The high energy particles of the galactic origin (outside the solar system) are called cosmic rays. The solar activity influences the intensity of cosmic rays that are falling on the earth. Cosmic rays propagate along the galactic magnetic field and are deviated by the earth's magnetic field (Moon has no magnetic field and has no radiation belts around it). During quiet sun, the solar wind is weak, and as a result magnetosphere shrinks and allows the cosmic rays to penetrate to the earth.

The sun's thermal radiation remains constant and is not affected by the solar activity but it effects short wave radiation (λ – less than 0.1 μm).

The effect of sun's gravity on earth's biosphere is insignificant compared to the gravity of the earth itself.

***Central Core*:** The central core of the sun has temperature 15 to 20 million °K. Above this convective core there is a radiative equilibrium layer with temperature ranging 10 million °K to 6000 °K. The convective core and radiative equilibrium region is characterized by nuclear reactions, which is the source region of solar radiative energy.

***Solar Flares*:** The solar flares are the result of gigantic nuclear explosions in the central core. Solar flares are the most spectacular activity of the charomosphere, but short lived (short period of life). There may be hundred odd flares daily but a few large flares may occur in a year. The large flares emit electromagnetic radiation and high energy charged particles of plasma. These particles may have velocity about 10 km/s. The Plasma particles consists of 91% ionized helium, which on reaching the earth interacts with atmosphere and earth's magnetic field. It causes ionospheric storm (high frequency radio blackout) and magnetic disturbances. The Solar wind associated with flare may have velocities 300-800 km/s.

***The Hetrogeneous Rotation of the Sun*:** Unlike the earth (which has all particles same angular velocity), the gaseous globe of the sun rotates a full turn at its equator in 25 days, and at its pole in 35 days. This rotation is called differential. The angular velocity of the sun's rotation also changes with depth, however the convection of the sun, its magnetic field and differential rotation are all related and interact with one another. The north pole of the sun is seen on the earth between 7 June to 7 December and the south pole of the sun is seen 8 December to 6 June.

Sun's Electromagnetic Spectrum

The temperature of sun's gases in photosphere has more than 6000 °K, which produce electrically charged particles with different speeds. As a result of their collision, electromagnetic waves of different wavelengths are emitted by the sun. These waves constitute the electromagnetic spectrum of the sun (see Fig. 1.2) which extends from gamma rays ($\lambda = 10^{-13}$) to radio waves ($\lambda = 10^{3} m$). The total range of λ is 10^{16} m.

About 99% of the solar radiation lies between $\lambda = 0.15$ to 4.0 μm, the distribution is as follows:

(i) About 9% of this lies in UV-radiation $\lambda = 0.15\mu m$ to 0.38 μm. This causes photochemical effects, bleaching and sunburn etc.

(ii) About 44% of this lies in visible portion, λ = 0.38 μm to 0.7 μm (that is violet to red).

(iii) A bout 46% lies in infrared region, λ = 0.7 μm to 2.3 μm. This causes radiant heat with some chemical effects.

All these radiations travel with a speed of 3×10^8 m/s in vacuum.

According to Wien's law

$$\lambda_{max} = \frac{\text{constant}}{T} = \frac{2897\mu mK}{T}$$

or

$$\lambda_{max} \simeq \frac{3000\ \mu mK}{T}$$

Sun emits energy at an average temperature of 6000 °K, while earth emits at an average surface temperature of 15 °C or 288 °K. It follows that, for sun $\lambda_{max} = \frac{3000\ \mu m K}{6000\ K} = 0.5\ \mu m$, that is the sun emits maximum radiation energy at wave length $\lambda_{max} = 0.5\ \mu m$.

For the earth

$$\lambda_{max} = \frac{3000\ \mu m\ K}{288\ K}$$

$$\simeq 10\ \mu m \qquad \text{(taking } 288^\circ \text{ as } 300^\circ\text{K)}$$

That is, the earth emits maximum amount of radiant energy at wavelength λ_{max} = 10 μm.

Because of these facts the terrestrial radiation is called long wave radiation and solar radiation is called shortwave radiation.

It is estimated that in all sun radiates each minute about 56×10^{26} cal of energy.

***Solar constant* (S):** The amount of solar radiation (energy) incident on unit area in unit time on a surface held at right angle to the solar beam at the outer boundary of the atmosphere is called solar constant and is given by

$$S = 2.0 \text{ ly/min or } 1359 \text{ W/m}^2,$$

where 1 ly (one Langley) = 1 cal/cm^2

The total solar energy intercepted by the earth in unit time = $\pi R_e^2 S$

where R_e = Radius of the earth.

This energy if spread uniformly over the whole of the earth, then the amount of solar energy received on unit area of the earth in unit time at the top of the atmosphere (Q_s) is given by

$$Q_s = \frac{\pi R_e^2 S}{4 \pi R_e^2} = \frac{S}{4} = 0.5 \text{ ly / min}$$

$$= 263 \text{ kly/year} \qquad \text{(kly = kilo Langley)}$$

$$= 339.75 \text{ W/m}^2$$

On an average the mean cloud coverage of the earth is slightly more than 50 %. Under this condition the distribution of insolation (incoming solar radiation) is given in Fig. 1.3.

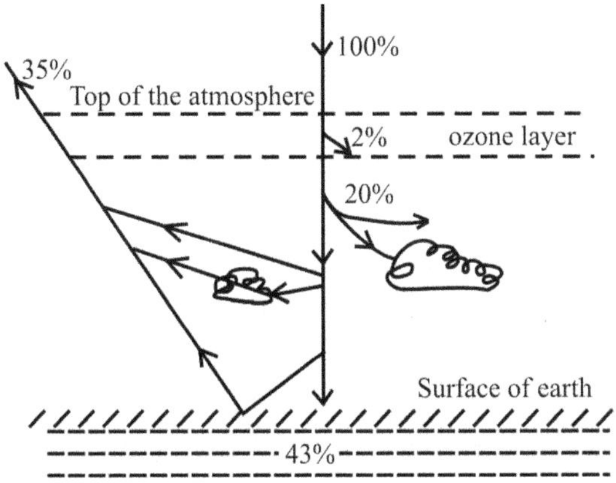

Fig. 1.3 Disposition of insolation in earth atmosphere.

Reflection and back scattering 35% (≃ 92 kly/year)

Absorption by ozone 2 % (≃ 5 kly/year)

Absorption by clouds, water vapour, dust etc 20 % (53 kly /year)

Absorption by the earths surface 43 % (113 kly / year)

The earth-atmosphere together absorb 65 % (171 kly / year)

Radiative Balance and Horizontal Transport of Heat

Past available records show that the global average of earth's surface temperature is about 15 °C and it is virtually constant. It means the amount of heat energy that is received by the earth (mainly in the form of shortwave radiation is equal to the amount of heat energy that is lost by the earth (by way of long wave IR-radiation). On the whole the earth atmosphere system radiates back nearly the same amount of energy that it is receiving from the sun. However the latitudinal distribution of incoming solar radiation to the earth and the outgoing radiation from the earth are different (that is they are not equal, see Fig. 1.4).

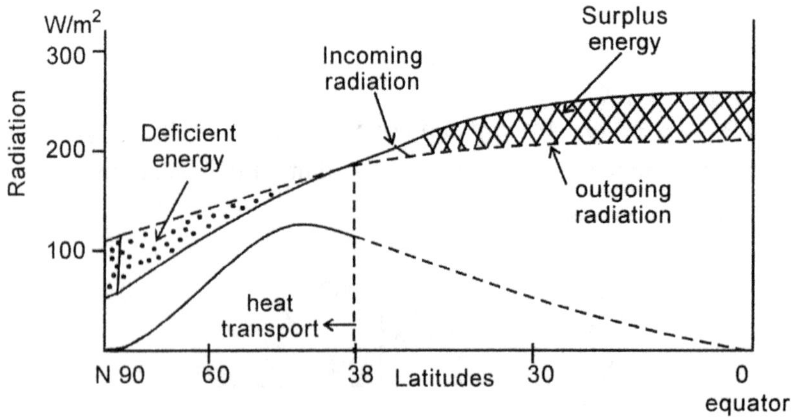

Fig. 1.4 Mean radiation balance northern hemisphere, solid line incoming solar radiation, dotted line outgoing earth radiation.

The Fig. 1.4 shows that in the equatorial region more energy is absorbed than that emitted by the earth. The earth-atmosphere system shows that there is surplus energy between latitudes 0°-35° (area hatched with cross lines), while more energy is emitted by the earth to space than received between latitudes 35° to 90°, indicating deficient energy in these latitudes (area hatched with dots). Since the earth as a whole is neither becoming warmer nor cooler shows that the excess energy absorbed in low latitudes is transported to higher latitudes, (otherwise equatorial regions would have much higher temperatures and mid and higher latitudinal regions would have been cooled to **very** low temperatures). This transfer of heat from lower to higher latitudes is carried by the action of wind and ocean currents. It has been calculated that ocean currents carry about 30% required heat transport from equator to polewards.

Questions

1. The temperature of the suns photospheric gases is about
   (a) 600°k (b) 6000°k (c) 60000°k (d) None
2. The global average temperature of the earth is about
   (a) 10°C (b) 15°C (c) 20°C (d) None
3. The sun emits maximum radiation energy at wavelength(λ):
   (a) 0.5 μm (b) 1 μm (c) 10 μm (d) None
4. The earth emits maximum radiation energy at wave length(λ)
   (a) 1 μm (b) 10 μm (c) 20 μm (d) None
5. Most of the insolation lies in:
   (a) Short wave radiation (b) Long wave radiation
   (c) UV radiation (d) None
6. Mean cloud coverage of the earth is
   (a) Slightly less than 25%
   (b) Slightly more than 50%
   (c) Slightly less than 75%
   (d) Slightly more than 90%
7. Most of the earth-atmosphere radiation lies in
   (a) Short wave radiation (b) Long wave radiation
   (c) UV radiation (d) None
8. In equatorial region (between latitudes 0–35^{0}):
   (a) More energy absorbed than emitted back to space
   (b) Less energy absorbed than emitted back to space
   (c) Energy absorbed is equal to the energy emitted back to space
   (d) None
9. In middle and higher latitudes (between 35–90^{0})
   (a) More energy absorbed than emitted back to space
   (b) Less energy absorbed than emitted back to space
   (c) Energy absorbed is equal to energy emitted back to space
   (d) None

10. At the surface (and in lower troposphere) heat transport takes place:
    (a) From poles to equator ward
    (b) From equator to pole wards
    (c) No heat transport takes place
    (d) None above

11. In troposphere at about 8 km. altitude level (and above)
    (a) Equatorial regions are warmer than polar regions
    (b) Equatorial regions are cooler than Polar Regions
    (c) Equatorial region and Polar Regions have the same temperature
    (d) None above

12. At the surface level
    (a) Equatorial regions are warmer than polar regions
    (b) Equatorial regions are cooler than polar regions
    (c) Equatorial regions and polar regions have the same temperature
    (d) None above

Answers

1.	(b)	2.	(b)	3.	(a)	4.	(b)
5.	(a)	6.	(b)	7.	(b)	8.	(a)
9.	(b)	10.	(b)	11.	(b)	12.	(a)

CHAPTER 2

The Motion of the Earth

The shape of the earth is roughly a sphere. The earth has two motions – (a) Rotation (b) Revolution.

The earth rotates about its axis through the poles (from west to east) with an equatorial speed of 464 m/s or 1669 km /hr, or angular speed of rotation is about 7.29×10^{-5} radians / sec or 0.26244 radians per hour. One complete rotation about its axis takes 24 hours. The poles are at rest with respect to the axis of earth, while all particles (objects) on the surface of the earth rotate roughly in circular paths. The particles on the earth's surface are moving fastest at the equator with speed about 1670 km/hr or 464 m/s, while they are moving in circular path at a speed of 835 km/hr at lat 60° and more slowly near the poles (in circular paths). The axis of the earth is inclined at an angle of $23 \frac{1}{2}^{\circ}$ to its elliptical orbit. This rotation causes day and night and diurnal variations of weather.

The rotating earth revolves eastwards round the sun in an elliptical orbit (with the sun at one of its Foci) with a speed (orbital speed) of about 1.073×10^{5} km/hr or 29770 m/s or 29.8 km/s. One revolution around the sun takes about 365 ¼ days. This revolution causes seasons.

Celestial Sphere

An imaginary sphere of an extremely large radius with its centre located at the centre of the earth is called celestial sphere. All heavenly bodies, the sun, the planets and the stars appear to be located on the inside surface of the celestial sphere; which has celestial poles and celestial equator (see Fig 2.1). The earth's equatorial plane that cuts the celestial sphere in a great circle is called celestial equator. The apparent path of the sun with respect to the celestial sphere is called ecliptic. The earth's axis makes an angle of about 66 ½ ° to the ecliptic. Solar system rotations are anti-clockwise when observed/viewed from the North Celestial Pole.

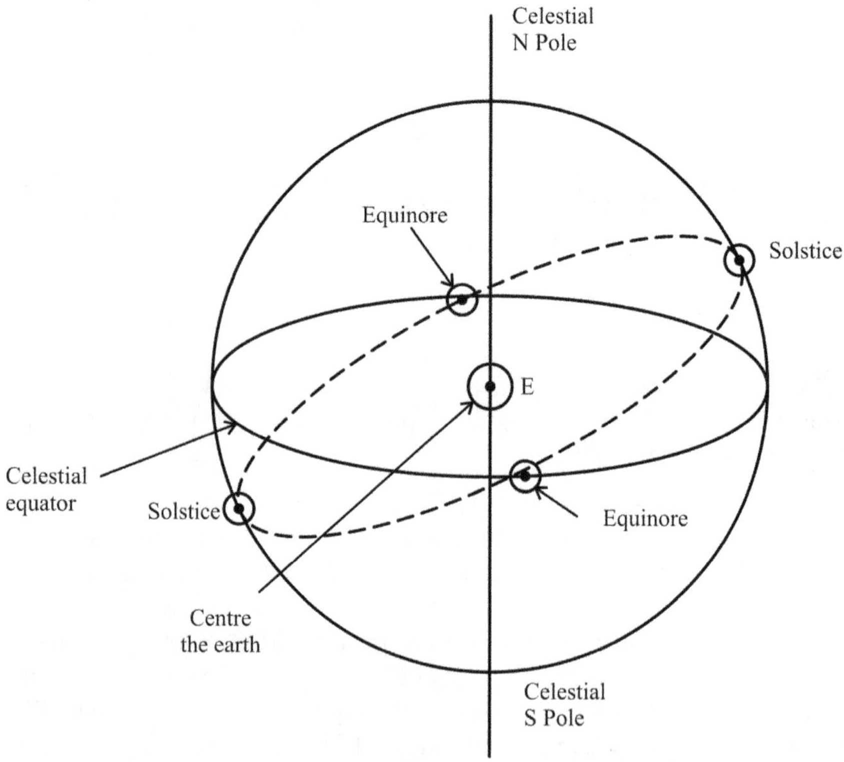

Fig. 2.1 Celestial sphere, equinoxes and solstices.

Equinoxes

When the sun apparently crosses the celestial equator (Ecliptic intersects the celestial equator), all the places on the earth experience equal (length) day and nights (that is the sun will be over the equator) is called Equinoxes. Every year equinoxes occur on March 21 and September 23.

Solstices

When the sun moves over to the Tropic of Cancer (Lat 23 ½ ° N farthest north near Aphelion) when the northern hemisphere will be tilted towards the sun, it is called summer solstice, June 21, which is associated with the longest day and shortest night in the Northern hemisphere or Mid summer in the Northern hemisphere. When the sun moves over to the Tropic of Capricorn (Lat 23 ½ ° S, farthest south, nearest to the sun or perihelion) it is called winter solstice, Dec 22, which is associated with longest day and shortest night in the southern hemisphere or Mid summer in the southern hemisphere. The two points on the ecliptic which are farthest from the celestial equator are solstices (see Fig. 2.1).

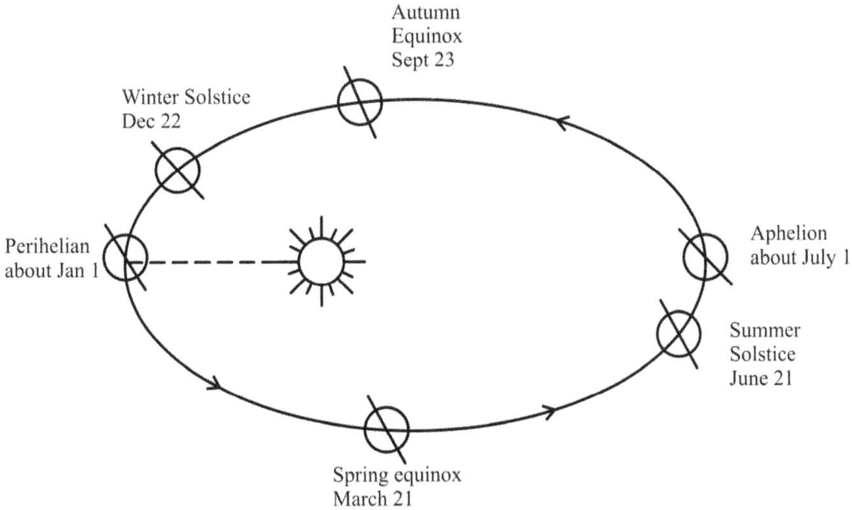

Fig. 2.2

Sidereal Day

The time period required for any point on the earth's surface for turning through an angle of 360° is called sidereal day. Sidereal day is determined by observing the rotation of earth with respect to a fixed star instead of sun.

1 sidereal day = 23 hrs 56 min and 4.09 seconds.

In scientific studies sidereal day is taken as standard.

Solar Day

The time period (t_i) between two successive crossing of the sun, across the meridian (over head) varies from day to day throughout the year. The average of this interval t,

$$\text{where } t = \left(\sum_{ti=1}^{365} t_i\right) / 365 \text{ is called mean solar day} = 24 \text{ hrs.}$$

Thus a sidereal day is shorter than the mean solar day by 3 minutes and 56 seconds.

$$1 \text{ Second} = \frac{1}{86164.09} \text{ sidereal day} = \frac{1}{86400} \text{ mean solar day}$$

Perihelion and Aphelion

According to Kepler's law, the earth is orbiting around the sun in an elliptical orbit with the sun being at one of the foci. The earth is nearest to the sun near about 1st January every year and farthest from the sun near about 1st July. The nearest position of the sun is called perihelion (earth-sun distance = 147×10^6 km) and the farthest position of the sun is called Aphelion (earth-sun distance = 152×10^6 km).

It may be noted that the sun is nearest to the earth during winter and farthest in summer. Thus earth receives more intense solar radiation during winter (of northern hemisphere) as compared to the summer (of northern hemisphere).

2.1 The Changing lengths of Day and Night Time

We know that the axis of the earth is tilted by about an angle of $23\frac{1}{2}^{\circ}$ to the plane of the earth's orbit around the sun. The region between Tropic of cancer (Latitude $23\frac{1}{2}^{\circ}$ N) to Tropic of Capricorn (Lat $23\frac{1}{2}^{\circ}$ S) is called Tropic. The region between Tropic of Cancer to Arctic circle (lat $66\frac{1}{2}^{\circ}$ N) and Tropic of Capricorn to Antarctic circle (Lat $66\frac{1}{2}^{\circ}$S) are called sub-tropics. The regions between Arctic circle to North pole and between Antarctic circle to South pole are called frigid zones or polar regions. The length of the Day (Night) varies with latitudes, the approximate periods are given in Table 2.1.

Table 2.1 Length of the Day or Night

Lat N/S	0° = equator	17°	41°	49°	63°	$66\frac{1}{2}°$	$67\frac{1}{2}°$	90°
Maximum length of Day (Night) hours	12	13	15	16	20	24	1 month	6 months

The regions which have continuous day light (24 hrs) [from lat $66\frac{1}{2}°$ to 90°] are generally called "the land of midnight sun". The day light varies from one day (24 hrs) at Arctic (or Antarctic) Circle to six months at the summer poles. Similarly the night time varies from one day (24 hrs) at Arctic (or Antarctic) circle to six months at the winter pole. As the year progresses the apparent path of the sun varies from place to place and also the length of day (or night) throughout the year except at equinoxes. This variation of the length of day or night causes seasonal changes.

At summer solstice (June 21), the sun is overhead (at meridian) at Tropic of Cancer. As we proceed northwards from Tropic of Cancer to North pole, the length of the day increases (or night decreases), and wihin Arctic circle (up to north pole) zone all places have 24 hrs day (while within Antarctic circle, all places have 24 hrs night). North Pole will be at the middle of six month's period continuous day (while the South pole lies in the middle of six months period continuous night/darkness).

At equinoxes (spring equinox March 21, Autumn equinox September 23) all places on the earth experience 12 hrs Day and 12 hrs Night.

At winter solstice (Dec 22) the sun will be overhead (at meridian) at Tropic of Capricorn. As we proceed southwards (from Tropic of Capricorn to South Pole) the length of the Day increases (while Night decreases) and within Antarctic Circle zone all places will have 24 hrs light (while all places within Arctic Circle will have 24 hrs night / darkness). South pole will be at the middle of six months period of continuous Day (and north pole will be at the middle of six month period of Night / darkness)

In summer hemisphere (that is when the sun is at Tropic of Cancer or Tropic of Capricorn) when the sun continuously shines for 24 hrs day (within Arctic circle or Antarctic circle) the apparent motion of the sun will be along the horizon (instead of sun rising in the east and setting in the west).

It may be noted that at any place, summer is warmer than winter due to direct sunshine and longer length of day.

2.2 Precession and Tides

In addition to the two motions referred to above, the earth has two other minor motions:

(a) Precession and

(b) Tides

(a) ***Precession*:** The movement of the earth about its polar axis is not steady. It wobbles on its axis solely like that of a dying top. The earth is not a perfect sphere but an oblate sphere. It has small bulge in the equatorial region (polar radius of the earth is about 6357 and equatorial radius 6378 km). This bulge is responsible for the wobbling in the earth's axis of rotation. The period of wobbling is about 26,000 years (cycle). Because of this wobbling, equinoxes are continually changing places in the earth's orbit. This effect is called precession of the equinoxes or simply precession.

(b) ***Tides*:** The cause of tides is complicated but the main factor is the gravitational pull of the moon and to a lesser extent the sun on the ocean waters of the earth. The rise and fall of the ocean waters due to the gravitational pull of the moon and the sun is called astronomical tides.

In coastal areas it is a common observation that the water of the sea rises twice in a day (roughly about every 12 hours) which covers a portion of the beach. This is called high tide. About 6 hours later the sea water recedes and goes down below normal level and thus a large portion of the beach is free from water. This is called low tide. In (24 hours) a day two high tides and two low tides are observed at any particular place (beach) in which the moon orbits around the earth.

The mass of the earth ($\simeq 5.98 \times 10^{24}$ kg) is about 81 times the mass of the moon and the common centre of mass (M_G) is about 1700 km below the earth's surface.

Consider the Fig. 2.3. The ocean water surface at A facing the moon will be attracted (pulled) by the moon. This causes high tide at A (as shown in Fig, which is exaggerated). Simultaneously at the same time at B on the opposite side of the earth a high tide occurs, due to the centrifugal force. As the waters pile up at A and B, correspondingly sea surface water at C and D (which are located on the perpendicular bisector of the line AB) falls (recedes/lowers). This results into low tides at the points C and D (while there are high tides at the points A and B). The positions of two high tides (and simultaneous two low tides) change their positions in accordance with the movement of the moon around the earth. As a result any point in the oceans below experiences two high tides (and two low tides) during the orbital period of the moon which is 24 hrs, 50 minutes.

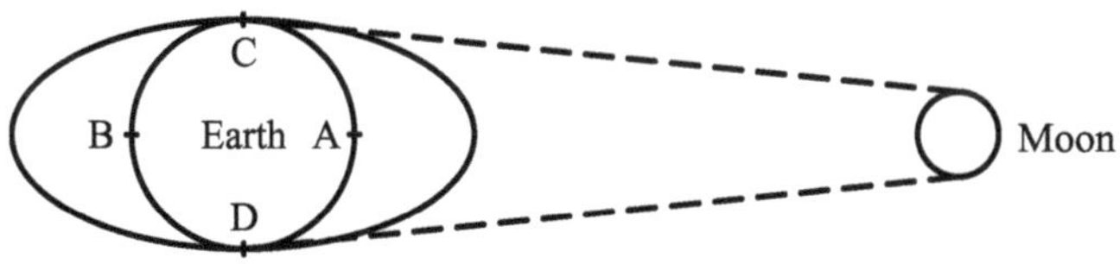

Fig. 2.3

In spite of the sun being much more massive than the moon, the gravitational attraction on the earth is small because of the very long distance. According to calculations, the solar tide is about $5/11^{th}$ of the lunar tide. At time of full moon/ new moon, the tidal force of the sun is added up to the tidal force of the moon. The high tides at new moon and full moon are termed spring tides (which is $1 + 5/11 = 16/11$).

At half moon (phase) the gravitational attraction of the sun acts at right angle to the gravitational attraction of the moon on earth. The corresponding high tides are below the average and these are termed Neap tides (which is $1–5/11 = 6/11$). Thus the ratio of a spring tide (16/11) and of Neap tide (6/11) is 16/11 : 6/11 or 8 :3.

It must be remembered the orbital path of the moon around the earth is elliptic and the earth is being at one of its focii. Consequently when the moon is closest (at perigee) and farthest (at apogee) the high tides are high and simultaneously the low tides are lowest (much lower than average).

It must be noted that the rise and fall of ocean waters at open sea due to tides is about one meter, but the height of the tides at coastal areas may go up to (15 to 16 feet) 5 meters. It is interesting to know that the entire territory of Moscow daily rises and falls about one and half $\left(1^{1/2}\right)$ feet from an average level. This rise and fall are due to daily ebbs and flows caused by the gravitational attraction of the moon and the sun on earth.

2.3 Meaning and Definitions of Some Terms

***Shape of the Earth*:** An oblate spheroid.

A Great Circle: The largest possible circle that can be described on the earth's surface or a circle on the earth's surface that cuts the earth into equal parts.

***Nautical mile (nm)*:** 1'(minute) of the arc of Great Circle is a nautical mile or 1' arc of meridian at the centre of curvature.

A standard nm = 1852 meters.

***Equator*:** It is a great circle whose plane at right angles (90°) to the earth's polar axis which passes through the centre of Great Circle.

***Meridians of longitude*:** They are semi-great circles joining the North and South poles of the earth.

***Prime Meriden*:** The meridian through the Greenwich (observatory) is called the prime meridian. It is numbered 0° (zero degrees). It is from this meridian all longitudes are measured. Longitude(s) is measured East and West of Greenwich up to 180°.

***Small circle*:** Any circle described on the surface of the earth other than great circle.

***Parallels of latitude*:** They are small circles which are parallel to the equator.

***Latitude*:** It is the arc of a meridian located between the equator and the parallel of latitude on which it is situated. It is measured North and South of the equator up to 90°.

***Kilometer*:** 1/10000 of the distance measured from the equator to the pole along a meridian.

1 km = 0.5 nm or 10000km = 5400 nm

***Rhumb lines*:** The lines drawn on the earth's surface which have a constant true direction.

Statue mile (Sm)

$$1\text{sm} = 3 \times 220 \times 8 = 5280 \text{ ft}$$

3 ft = 1 yd; 220 yds = 1 furlong

8 furlongs = 1 sm

1 nm = 1.15 sm = 1.85 km

1m = 3.28 feet

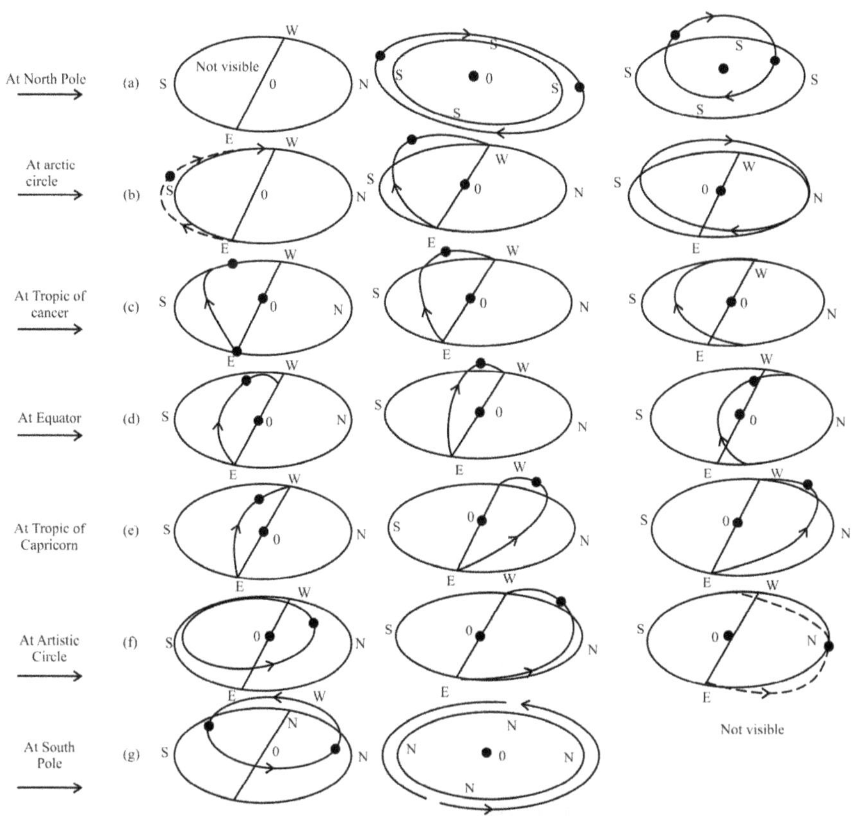

Apparent paths of the sun at about December 22 — Fig 2.6

Apparent paths of the sum at equinoxes, March 21 and Sept 23 — Fig 2.5

Apparent paths of the sun at about June 21 — Fig 2.4

The apparent paths of the sun between sunrise and sunset at various places at about June 21, equinoxes, and at about December 22 are respectively shown in Fig. 2.4, 2.5 & 2.6.

Time specification to classify different parts of the day

1.	Early morning	0400-0600 hrs
2.	Morning	0600-1200 hrs
3.	Forenoon	0800-1200 hrs
4.	Around noon	1100-1300 hrs
5.	After noon	1300-1600 hrs
6.	Evening	1600-2000 hrs
7.	Early night	2000-2300 hrs
8.	Around Midnight	2300-0100 hrs
9.	Late night	0100-0400 hrs

Questions

The motion of the earth

1. The earth rotates about its axis:
   (a) Though the poles from east to west
   (b) Through the poles from west to east
   (c) Through east-west at equator and from north to south
   (d) Through east-west at equator and from south to north
2. The particles on the surface of the earth move fastest
   (a) At the poles (b) At middle latitudes
   (c) At the equator (d) None
3. The polar axis of the earth is inclined to its revolving elliptical orbit by
   (a) About 10° (b) About $23\frac{1}{2}°$
   (c) About $66\frac{1}{2}°$ (d) About 90°
4. One complete rotation of the earth about its axis takes place in:
   (a) One year (b) One mouth (c) 24 hours (d) None

5. One complete revolution of the earth around the sun takes place in

(a) About one year ($365\frac{1}{4}$ days) (b) About one month

(c) About one day (d) None

6. An imaginary sphere of extremely large radius with its centre located at the centre the earth is called:

(a) Celestial equator (b) Celestial pole

(c) Celestial sphere (d) Astronomical tide

7. The earth's axis through poles makes an angle to the ecliptic by

(a) $23\frac{1}{2}$ (b) $66\frac{1}{2}^{\circ}$

(c) 90° (d) 0°

8. The apparent path of the sun with respect to the celestial sphere is called

(a) Celestial pole (b) Celestial equator

(c) Ecliptic (d) None

9. When the apparent movement of the sun crosses the celestial equator, all the places on the earth experience:

(a) Equal length of the day (or night)

(b) 24 hrs day

(c) 24 hrs night

(d) None

10. When the apparent movement of the sun moves over to Tropic Cancer it is called

(a) Equinox (b) Summer solstice

(c) Winter solstice (d) Ecliptic

11. The earth is orbiting around the sun in an elliptical orbit with the sun being at one of its foci. At about 1st January it (earth) is nearest to the sun. It is called:

(a) Perihelion (b) Aphelion

(c) Winter solstice (d) Both (a) and (c)

12. At about 1st July when the sun is farthest from the earth, it is called
   (a) Perihelion (b) Aphelion
   (c) Winter solstice (d) None

13. Earth receives more intense solar radiation during
   (a) Summer (b) Winter (c) Equinoxes (d) None

14. When the sun is at Tropic of Cancer, all places from Arctic circle to north pole experience:
   (a) Day, more than 24 hrs
   (b) Day, less then 24 hrs
   (c) Night, more than 24 hrs
   (d) None

15. When the sun is at Tropic of Capricorn, all places from Antarctic circle to south pole experience:
   (a) Day, more than 24 hrs (b) Day less than 24 hrs
   (c) Night, more than 24 hrs (d) None

16. The movement of the earth about its polar axis is not steady. It wobbles on its axis slowly, like that of a dying top. It is called:
   (a) Rotation (b) Revolution (c) Precession (d) Tides

17. The rise and fall of ocean waters twice in a day in coastal areas (high tides, low tides) is caused by the gravitational attraction of:
   (a) The moon (b) The sun
   (c) Both (a) and (b) (d) The earth

18. The orbital period of the moon around the earth is
   (a) 12 hrs (b) 24 hrs
   (c) 24 hrs, 50 minutes (d) None

19. The largest possible circle that can be described on the earth's surface is called:
   (a) A great circle (b) A small circle
   (c) Rhumbline (d) None

20. The arc of a meridian which makes an angle of 1' (minute) at the centre of curvature is called
   (a) Kilometer (b) Statue mile
   (c) Nautical mile (d) None

21. $\frac{1}{10,000}$ of the distance measured from equator to the pole along a meridian is called

(a) Kilometer (b) Statue mile

(c) Nautical mile (d) None

22. When the apparent movement of the sun moves over to Tropic of Capricorn it is called:

(a) Equinox (b) Summer solstice

(c) Winter solstice (d) None

Answers

1.	(b)	2.	(c)	3.	(b)	4.	(c)
5.	(a)	6.	(c)	7.	(b)	8.	(c)
9.	(a)	10.	(b)	11.	(d)	12.	(b)
13.	(b)	14.	(a)	15.	(a)	16.	(c)
17.	(c)	18.	(c)	19.	(a)	20.	(c)
21.	(a)	22.	(c)				

CHAPTER 3

The Atmosphere

The envelope of air surrounding the earth to great heights is called atmosphere. The atmospheric air is a mixture of gases and contains the particles of liquid, solid (which are suspended in air). These are bound to the earth by the gravitational attraction of the earth. Atmospheric gases obey the ideal gas laws and for practical purposes treated as fluid.

3.1 Composition of the Atmosphere

The composition of the dry atmosphere has:

Constituent	by volume	by wt.	Total weight × 10^{17} kg
Nitrogen (N_2)	78%	75.5%	38.65
Oxygen (O_2)	21%	23.1%	11.84
Argon (An)	0.93%	1.29%	0.66

The other minor gases include Neon, Helium, Methane, Krepton, Hydrogen, Xenon, Carbondioxide, Ozone and water vapour. The last three gases are variable both in time and space. This composition of atmospheric gases are practically in the same proportion up to an altitude of about 80-90 km. The non-gaseous constituents are dust, smoke, salt particles from sea spry and water particles, which are all variable.

More than 50% of the mass of atmosphere lies below an altitude of 5.5 km and 98% of mass below 30 km altitude. The lowest one kilometer of atmosphere contains about 10% of the mass of atmosphere. All biological and human activities except aircraft flights are confined to this lowest one kilometer, which is called Planetary Boundary Layer (PBL).

The mass of the atmosphere is about 5.6×10^{18} kg. The mass of the oceans water is about 1.4×10^{21} kg. The density of dry air at surface (msl) is 1.225 kg/m^3 which reduces to 50% (0.6125 kg/m^3) at 6 km altitude, 25% (0.30625 kg/m^3) at 12 km, at 18 km altitude to 10% (0.1225 kg/m^3) and at 30 km altitude to about 0.013 kg/m^3 (about 1%).

On an average at msl nitrogen exerts a pressure of about 760 hPa, oxygen 240 hPa and water vapour about 10 hPa.

Atmospheric mass (gases, liquid and solid particles together) is treated as fluid and it obeys gas laws. The main properties of atmospheric mass are:

(i) Molecular mobility; (ii) capacity for expansion and compression with adiabatic heating or cooling

3.2 Atmospheric Heat Process

Heat is a form of energy, which produces sensation of warmth in us.

The degree of hotness of a body is called temperature. Heat flows from one body (at higher temperature) to another body (at lower temperature) till they are at equal temperature. The measurement of temperature is made by thermometers. Heat is transferred in the atmosphere in five ways:

(i) conduction, (ii) convection, (iii) radiation (iv) advection and (v) condensation.

Conduction: The transfer of heat energy from hot surface to the adjacent cooler surface without the movement of molecules is called conduction. This process is important in transfer of heat very close the ground surface.

Note: air is a poor constructor of heat.

Convection: The transfer of heat energy with the movement of molecules and applies to fluids (liquids and gases). This process is important in atmosphere.

Radiation: The transfer of energy in the form of electromagnetic waves is called radiation. In this case no medium is required. All objects in the universe radiate as long as its temperature is more than 0 °K.

Advection: The transfer of heat energy from one area to another, through horizontal wind motion is called advection (i.e., by the movement of air masses).

Condensation: In atmosphere precipitation in the form of rain or snow releases latent heat, which warms the atmosphere.

The important methods of heat transfer from the earth to the\ atmosphere are: (i) convection (ii) Advection, (iii) latent heat of condensation of water vapour (which was transported upwards). Sun is the main source of heat energy for the earth and atmosphere, which is received in the form of radiation (insolation). The sun's rays do not heat the air in the atmosphere but heat the solid and liquid particles suspended in it. Because of this about 85% of the insolation hits the earth's surface, while only about 15% heats the atmospheric (aerosols) particles. The surfaces of the earth, which are heated by the suns' rays, pass heat energy to the air film in contact with the surface by way of conduction. By convection warm air moves up and it is replaced by cooler air. The transfer of heat from the surface of the earth to the atmospheric air takes place by conduction and convection, **but** it is called re-radiation or back radiation. Thus earth acts as the secondary source of heat for the lower atmosphere. In meteorology the incoming solar radiation is called insolation (also shortwave radiation) and the back radiation from the earth to space is called Terrestrial radiation (or long wave radiation or **IR** radiation). The absorption of solar energy at the surface of the earth depends on the angle of incidence of the sun's rays or inclination of the sun and the materials of the earth on which it falls. Different types of surfaces absorb differently on earth and create uneven heating (such as sea water, black soil, sandy soil, green trees etc.) which in turn results in horizontal motion of air. The unequal heating of the surface of the earth is the cause for Cell pattern (Hadley Cell, Ferrel Cell) of atmospheric circulation.

3.3 The Vertical Structure of the Atmosphere based on Temperature

In 1962, world meteorological organization (WMO) decided to divide the atmosphere into four regions (strata) based on temperature change with altitude. They are:

(i) Troposphere, (ii) Stratosphere, (iii) Mesosphere, (iv) Thermosphere. The salient features of these layers are given below.

Troposphere: The lowest layer of the atmosphere adjacent to the earth's surface is called troposphere. The altitude of this layer is about 16-18 Km in the equatorial regions (temperature at the top – 75 °C), about 11 Km in middle latitude (lat 30°), (temperature at the top –60 °C) and about 8 km at the poles (with temperature at top – 50 °C).

All weather systems, clouds practically confined to this layer. In this layer (mostly) temperature decreases with increasing height. The troposphere contains about 75% of the mass of the atmosphere and virtually all water vapour of the atmosphere. The average lapse rate (rate of change of temperature with height) is about 6.5 °C/km. The upper boundary of the troposphere is called tropopause, pause means break.

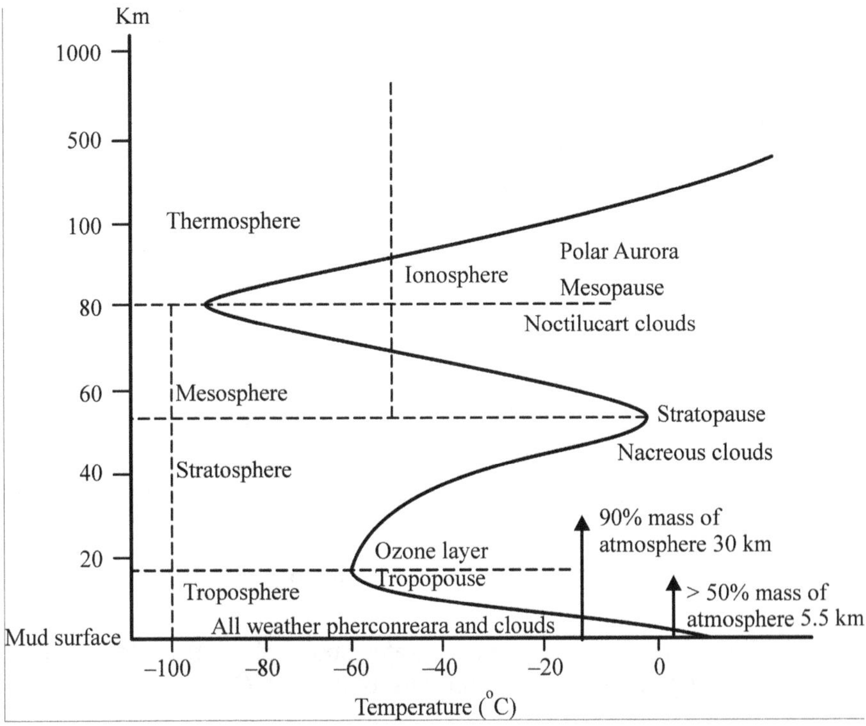

Fig. 3.1 Structure of the Atmosphere.

Tropopause

***Definition*:** The tropopause is the lowest level at which the lapse rate (of temperature) decreases to less than or equal to 2 °C/km at least for a layer of 2 km and above does not exceed this.

Tropopause is not a continuous surface. In middle latitudes two tropopauses are found. Lower one with tropical characteristics and the other with extra tropical characteristics. In between these two tropopauses lies the subtropical jet stream (around lat 30°). There is a sharp rise in temperature above tropical tropopause but in subtropical tropopause there is slight fall.

Troposphere is warmed up by the underlying earth surface and hence instability exists.

Stratosphere

The layer above tropopause is called stratosphere, which extends up to an altitude of 50 km, where its temperature is above 0 °C. There is an isothermal layer (lapse rate zero) above tropopause up to 20 km altitude and thereafter temperature rises (inversion) generally up to about 32 km altitude and thence rises rapidly and equals to the earths surface temperature (positive) at its top.

This layer has large concentration of ozone between altitudes 18-35 km (60000-115000 ft) with a maximum density at about 25 km (80000 ft). The increase of temperature in stratosphere is attributed to the presence of ozone, which is a green house gas. Ozone absorbs harmful solar UV-radiation (ultra-violet radiation) in the wavelengths 220 nm to 290 nm (1 nm = 10^{-9} m). UV-radiation induces skin cancer, damages eye and suppresses immune system in human beings. It effects the productivity of aquatic and terrestrial ecosystem. 1% decline in ozone concentration in stratosphere may result in 3% increase in the incidence of skin cancer among humans. Ozone layer absorbs about 2% of insolation (incoming solar radiation).

Stratosphere is stable, because it has cold temperature at its base and warm temperature at the top.

In equatorial region of stratosphere biennial wind oscillations are observed.

The boundary surface which separates stratosphere from Mesosphere is called stratopause (where temperature is about 0° C).

Mesosphere

The layer above stratopause is called Mesosphere, which extends up to 80 km altitude. In this layer temperature falls above stratopause and attains the lowest temperature about – 95 °C at 80 km altitude. Noctilucent clouds belong to Mesosphere, which are observed in higher latitudes during summer. The top of Mesosphere layer is called Mesopause.

Troposphere, Stratosphere and Mesosphere together is called **Homosphere** because the ratio of the constituents of the air in this layer (region) is practically constant (except ozone, carbon-dioxide, and water vapour which are variable).

Thermosphere

The region above Mesopause is called thermosphere which extends to great height (700-1000 km altitude). Thermosphere is also called hetrosphere because the composition of the ratio of constituent gases is heterogeneous. The main gases stratify according to their molecular weights. The temperature rises above Mesopause to about 1000 to 1200 °C at about 400 km altitude and the density of air falls to 3×10^{-12} kg/m^3 (very thin density) and pressure 10^{-8} mm of mercury.

The lowest layer contains oxygen and Nitrogen molecules (up to 500 km) and thereafter hydrogen (up to 1000 km). There is no boundary to the thermosphere. The upper part of the thermosphere merges with the interplanetary gas with temperature about 2000 °K which is called Exosphere. The thermospheric gases are found to be mostly in atomic state due to photo-dissociation by the insolation.

Ionosphere

The lower thermosphere and upper Mesosphere contains gases mostly in ionized state and hence it is called Ionosphere. The characteristic property of these ionized gases is that it reflects radio waves, which helps in long wave radio communication. Ionosphere is divided into D-region (50-90 km altitude), E-region (also called Kenelly Heaviside layer) 90-140 km altitude and F-region 140-500 km attitude. F-region is further divided into F_1 region 140-250 km altitude (also called Appleton layer) and F_2 region 250-500 km altitude.

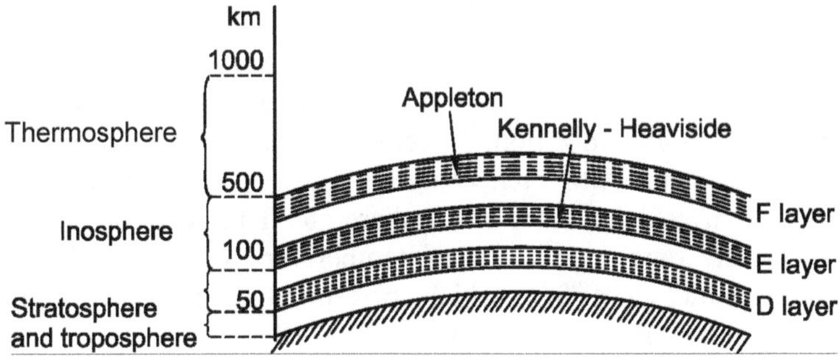

Fig. 3.2 Ionosphere layers.

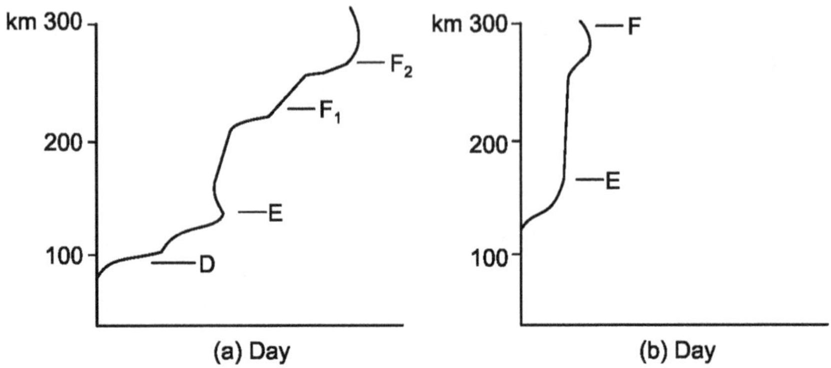

Fig. 3.3 Ionosphere diurnal effects.

D-region (50-90 km altitude) reflects low frequency radio waves but absorbs medium and high frequency radio waves. This region disappears during nights (in the absence of solar radiation). In this region air density is more than electron density.

E-region (90-140 km altitude) strongly reflects medium and high freqeuncy radio waves. This region weakens during might, but does not disappear. However during polar nights, E-region disappears. In this region electron density is (10^5/ cm^3) and air density is less.

F_1-region (140-250 km altitude) is important for the fact that it propagates medium and high frequency radio waves.

F_2-region (250-500 km altitude) is important for long distance radio communication. When the sun is low and during night, F_1-region merges with F_2-region.

In F-region electron density is high (10^6/cm^3) and air density is low.

3.4 International Standard Atmosphere (ISA)

International Civil Aviation Organisation (ICAO) ISA or simply ISA specifications were based on the statistical global average of temperature and pressure. Standardization is required for calibration of aircraft pressure instruments/Altimeter, design and testing of aircraft, jet engine performance, average state of atmosphere etc. This reference basis is to specify the average lapse rate of temperature (i.e., variation/fall of temperature with height) from which corresponding variations of pressure and density (using ideal gas low, $p = \rho RT$) are derived. The ISA specification (also called ICAO ISA specifications) are:

(i) Atmosphere is completely dry,

(ii) at mean sea level (msl) temperature 15 °C,

(iii) at msl atmospheric pressure 1013.25 hPa,

(iv) at msl air density 1.225 kg/m^3,

(v) lapse rate 1.98 °C/1000 ft or 6.5 °C/km) up to a standard tropopause height of 11 km (36090 ft) and remains constant (isothermal) at – 56.5 °C from 11 to 20 km (36090 ft to 65600 ft). Above 20 km temperature rises (inversion) by 1 °C/km or

0.3 °C/1000 ft up to 32 km (105000 ft). The layer from surface to 11 km is called troposphere. The layer 11 to 32 km is called stratosphere. These features are shown in Fig 3.4. It may be noted that ISA is defined only up to an altitude of 32 km and beyond that it is not defined because up to this altitude all aircraft flights are confined. However after the invention of high altitude jet aircrafts, a slight modification to ISA is made.

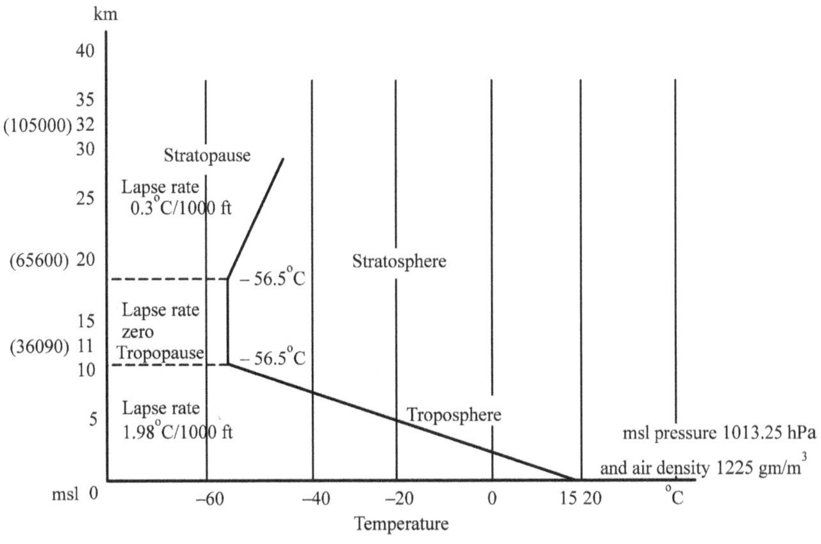

Fig. 3.4

3.5 Jet standard Atmosphere (JSA)

In aviation a slightly modified standard atmosphere is used called Jet Standard Atmosphere (JSA). This has been created to suite the Jet engine performance graph, where lower temperatures less than – 56.5 °C required. In JSA, there is no tropopause and hence temperature continue to fall with height. The lapse of ISA 1.98 °C/1000 ft changed to 2 °C/1000 ft in JSA and other specifications remain the same except tropopause which is absent.

It must be noted that both ISA and JSA temperatures are hypothetical and differ from actual day-to-day atmospheric (real world) environmental temperatures, which are obtained by daily radiosonde ascents. Thus actual temperature at a particular level of the atmosphere may be warmer or colder than the ISA/JSA values. This will be clear with following examples.

Example

A radiosonde observation on a particular day recorded a temperature of – 33 °C at 20000 ft. What is the deviation from ISA / JSA?

Solution

Comparison with ISA

In this case

lapse rate is 1.98 °C/1000 ft

The ISA temperature at 20000 ft would be

$$\left(+15-20000\times\frac{1.98}{1000}\right)^{\circ}\mathrm{C}$$

$$= 15 - 39.6 = -24.6\ ^{\circ}\mathrm{C}$$

It is given actual temperature was – 33 °C at 20000 ft. The actual temperature at 20000 ft is colder than ISA by 8.4°

∴ Actual temperature at 20000 ft on that day

$$= (\mathrm{ISA} - 8.4)\ ^{\circ}\mathrm{C}$$

Comparison with JSA

In this case the lapse rate is 2 °C/1000 ft

The JSA temperature at 20000 ft would be $\left(+15-20000\times\frac{2}{1000}\right)^{\circ}\mathrm{C}$

$$= (15 - 40)\ ^{\circ}\mathrm{C}$$
$$= -25\ ^{\circ}\mathrm{C}$$

It is given actual temperature at 20000 ft was – 33 °C, which is colder than JSA by 8 °C.

∴ Actual temperature at 20000 ft on that day = (JSA – 8) °C

Example

On a particular day at Srinagar, the radiosonde observation recorded (i) – 65 °C at 46090 ft altitude and (ii) at 66600 ft was – 75 °C. What is the deviation from ISA and JSA?

Solution

Case (**i)**

ISA temperature at 46090 ft is – 56.5 °C

(because it is isothermal from 36090 ft to 65600 ft at – 56.5 °C)

Actual temperature on that day at 46090 ft altitude level is colder than ISA by 8.5 °C or Actual temperature at 46090 ft = (ISA – 8.5 °C

JSA deviation: In this case lapse rate is 2 °C / 1000 ft

JSA temperature at 46090 ft altitude would be

$$\left(+15-\frac{2}{1000}\times 46090\right)\ ^{\circ}\text{C}=\left(+15-92.18\right)\ ^{\circ}\text{C}$$

$$\simeq -77\ ^{\circ}\text{C}$$

Actual temperature was warmer than JSA by 12 °C i.e., actual temperature at 46090 ft altitude = (JSA + 12) °C

Case **(ii)**

(a) ISA temperature at 65600 ft is – 56.5 °C and lapse rate inversion above this level is 0.3 °C/1000 ft

∴ ISA temperature at 66600 ft attitude would be

$$-56.5 + 0.3 = -56.2\ ^{\circ}\text{C}$$

It is given actual temperature at 66600 ft altitude on that day was –75° C i.e., actual temperature at 66600 ft was colder than ISA by 18.8 °C

∴ Actual temperature at 66600 ft altitude = (ISA – 18.8 °C)

(b) In JSA temperature lapse rate is 2 °C/1000 ft and there is no tropopause.

JSA temperature at 66600 ft would be

$$(+15-\frac{2}{1000}\times 66600)\ ^{\circ}\text{C}$$

$$15-133.2 \text{ or } -118.2\ ^{\circ}\text{C}$$

∴ Actual temperature at 66600 ft = (JSA + 43.2 °C)

Questions

1. The atmospheric gases and particles of liquid and solid (suspended in air) are bound to the earth, by:
   (a) Centrifugal force (b) Coriolis force
   (c) Earth's gravitational force (d) None
2. The mass of the atmosphere is treated as fluid, which obeys:
   (a) Boyle's law (b) Charles law
   (c) Gay-lussacs law (d) All above (a) (b) & (c)
3. The composition of the dry atmosphere up to an altitude about 80 km contains oxygen and nitrogen in a constant ratio by volume:
   (a) 21:78 (b) 78:21
   (c) 50:50 (d) None above
4. More than 50% of the mass of atmosphere lies below an altitude of:
   (a) 1 km (b) 5.5 km
   (c) 11.0 km (d) 30 km
5. About 98% of the mass of atmosphere lies below an altitude of:
   (a) 1 km (b) 5.5 km
   (c) 11.0 km (d) 30 km
6. The lowest one kilometer of the atmosphere contains about:
   (a) 10% of the total mass of atmosphere
   (b) 50% of the mass of the atmosphere
   (c) 90% of the mass of the atmosphere
   (d) None
7. The density of dry air at msl is about:
   (a) 1225 gm/m^3 (b) 1000 gm/m^3
   (c) 100 gm/m^3 (d) 10 gm/m^3
8. The density of dry air at an altitude 6 km drops as compared to the msl to:
   (a) 10% (b) 25%
   (c) 50% (d) None above

9. The density of dry air at an altitude of 12 km drops as compared to the msl to:
   (a) 10% (b) 25%
   (c) 50% (d) None above

10. The density of dry air at an altitude of 30 km drops as compared to the msl to about:
   (a) 10% (b) 25% (c) 50% (d) 1%

11. On an average at msl nitrogen exerts a pressure of about:
   (a) 760 mm of Hg (b) 760 hPa
   (c) 1013 hPa (d) None

12. On an average at msl oxygen exerts a pressure of about:
   (a) 1000 hPa (b) 240 hPa
   (c) 10 hPa (d) None

13. On an average at msl water vapour exerts a pressure of about:
   (a) 1000 hPa (b) 240 hPa
   (c) 10 hPa (d) None

14. The main properties of atmospheric mass molecules:
   (a) Mobility
   (b) Capacity for expansion
   (c) Compression
   (d) All above (a) (b) & (c)

15. Heat is transferred in the atmosphere by:
   (a) Conduction, convection (b) Radiation
   (c) Advection, condensation (d) All above (a) (b) & (c)

16. Air is a poor conductor of heat. Heat is transferred from earth to atmosphere mainly by:
   (a) Conduction
   (b) Convection
   (c) Advection, latent heat of condensation
   (d) both (b) and (c)

17. The sun's rays (insolation) heats the atmospheric particles (aerosols) by:
    (a) 10% (b) 15%
    (c) 85% (d) 100%

18. For the lower atmosphere the main source of heat is:
    (a) Insolation
    (b) Terrestrial radiation
    (c) Space radiation, cosmic radiation
    (d) All above

19. Horizontal motion of air (called wind) results from:
    (a) Equal absorption of insolation by all types surfaces
    (b) Differential absorption of insolation by the surfaces
    (c) Non-absorption of insolation by different surfaces
    (d) None above

20. Unequal heating of the surface of the earth by insolation is the cause of:
    (a) Tropical cell circulation
    (b) Ferrel mid latitude circulation
    (c) Polar cell circulation
    (d) Above all three meridional circulations

21. WMO divided the vertical atmosphere into four regions/stratas based on:
    (a) Temperature change with altitude
    (b) Pressure change with altitude
    (c) Density change with altitude
    (d) None

22. The lowest layer of the atmosphere close to earth's surface is called:
    (a) Mesosphere (b) Stratosphere
    (c) Troposphere (d) Thermosphere

23. Troposphere is mainly heated by:
    (a) Insolation (b) Terrestrial radiation
    (c) Cosmic radiation (d) None

24. The height and temperature of the tropopause in equatorial region is about:
   (a) 8-10 km, – 50° C (b) 11 km, – 60 °C
   (c) 16-18 km, – 75 °C (d) None

25. The height and temperature of mid-latitude tropopause is given by
   (a) 81 cm, – 50 °C (b) 11 km, – 60 °C
   (c) 16-18 km, – 75 °C (d) None

26. The height and temperature of polar latitude tropopause is given by
   (a) 8 km, – 50 °C (b) 11 m, – 60 °C
   (c) 16-18 km, – 75 °C (d) None

27. Tropapause is mainly described as:
   (a) The boundary layer between the troposphere and the stratosphere
   (b) The layer where temperature increases with height
   (c) The layer where the lapse rate is 2 °C/km or less
   (d) All of the above

28. The Stratosphere is described as the layer:
   (a) Above Mesopause where temperature increases with height
   (b) Above tropopause where temperature increases with height
   (c) Above stratopause where temperature increase with height
   (d) Below mesopause where gases are found in ionized state.

29. Ionosphere is described as the layer:
   (a) Where the atmospheric gases are found in ionized state
   (b) Where the atmospheric gases reflect the radio waves
   (c) Where the electron density is more and air density is less
   (d) All above.

30. Thermosphere is described as the layer:
   (a) Where the atmospheric gases composition has same proportion or homogeneous
   (b) The layer above mesopause where the composition of the air constituents are heterogeneous
   (c) Where the gases stratify according their molecular weights
   (d) Both (b) and (c)

31. All weather phenomena and clouds except mother of pearl are confined to:
    (a) Mesosphere
    (b) Planetary boundary layer
    (c) Tropopause
    (d) Troposphere
32. Ozone concentration layer and nacreous clouds are found in:
    (a) Troposphere (b) Stratosphere
    (c) Mesosphere (d) Thermosphere
33. In general, temperature falls with increasing height in troposphere because of this Troposphere is:
    (a) Stable (b) Unstable
    (c) Neutral (d) None above
34. In general, temperature rises with increasing height in stratosphere, consequently stratosphere presents:
    (a) Instability and weather
    (b) Stability, equatorial biennial wind oscillation
    (c) Neutrality and noctilncent clouds
    (d) None above
35. large concentration of ozone is found in:
    (a) troposphere between altitudes 8 – 15 km
    (b) stratosphere between altitudes is 18 – 35 km
    (c) mesosphere between altitudes 60 – 70 km
    (d) none above
36. Atmospheric ozone layer absorbs harmful radiation from the sun:
    (a) UV-radiation (b) IR-radiation
    (c) Microwave Radiation (d) Visible Radiation
37. Sun's UV-Radiation causes:
    (a) Skin cancer, damages eye
    (b) Supresss immune system in humanbeings
    (c) It effects the productivity of aquatic and terrestrial ecosystem
    (d) All above

38. Ionosphere extends between altitudes 50-500 km Ionosphere:
   (a) Consists of mostly ionized gases
   (b) Reflects radio waves
   (c) Helps in long wave radio communication
   (d) All above

39. In ISA the mean sea level temperature is taken as;
   (a) 0°C (b) 15°C
   (c) 27°C (d) none

40. In ISA the msl atmospheric pressure is taken as;
   (a) 1000 hPa (b) 760 mm Hg
   (c) 1013.25 hPa (d) None

41. In ISA the msl air density of atmosphere is taken as:
   (a) 1225 kg/m^3 (b) 1.225 kg/m^3
   (c) 1.225 kg/cm^3 (d) None above

42. Is ISA the tropopause height is:
   (a) 8 km (b) 11 km
   (c) 16 km (d) None above

43. In ISA temperature lapse rate in troposphere is:
   (a) 1.98°C / 1000 ft.(or 6.5°C / km (b) 2 °C / 1000 ft.
   (c) 1°C / km (d) None

44. In ISA, the temperature lapse rate between 11 and 20 km altitude (36090 ft to 65600 ft) is
   (a) 0°C / km (b) 1°C / km
   (c) 1.98°C /1000 ft (d) None above

45. In ISA, between 20 to 32 km altitude, temperature:
   (a) increases at the rate of 1 °C/km (0.3 °C/1000 ft)
   (b) decreases at the rate of 1.98°C / 1000ft
   (c) remain constant at – 56.5°C
   (d) none

46. In ISA, temperature at an altitude 6 km would be:
   (a) 39 °C (b) – 24 °C
   (c) – 39 °C (d) None

47. In ISA, temperature at altitude 10 km is:
    (a) – 65 °C (b) – 56.5 °C
    (c) – 50 °C (d) None above

48. In ISA, temperature at an altitude of 18 km would be
    (a) – 102 °C (b) – 56.5 °C
    (c) – 117 °C (d) None above

49. In ISA, temperature at an altitude of 24 km would be:
    (a) – 56.5 °C (b) – 52.5 °C
    (c) – 50.0 °C (d) None above

50. In ISA, temperature at an altitude of 30 km is:
    (a) – 46.5 °C (b) – 56.5 °C
    (c) – 66.5 °C (d) None above

51. On a particular day the radiosonde ascent at Hyderabad at altitude 30,000 ft had temperature – 40 °C. What was the ISA/JSA deviation?

I (a) ISA + 4.4 °C (b) ISA – 4.4 °C
  (c) ISA + 19.4 °C (d) None above

II (a) JSA + 5 C° (b) ISA – 5 C°
  (c) JSA – 20 °C (d) None above

52. On a particular day at New Delhi the radiosonde temperature (real temperature) at 26 km altitude was – 60 °C. What was the deviation from ISA.
    (a) ISA – 9.5 °C (b) ISA + 9.5 °C
    (c) ISA + 24.5 °C (d) None above

Answers

1. (c)	2. (d)	3. (a)	4. (b)
5. (d)	6. (a)	7. (a)	8. (c)
9. (b)	10. (d)	11. (b)	12. (b)
13. (c)	14. (d)	15. (d)	16. (d)
17. (b)	18. (b)	19. (b)	20. (d)

21.	(a)	22.	(c)	23.	(b)	24.	(c)
25.	(b)	26.	(a)	27.	(a)	28.	(b)
29.	(d)	30.	(d)	31.	(d)	32.	(b)
33.	(b)	34.	(b)	35.	(b)	36.	(a)
37.	(d)	38.	(d)	39.	(b)	40.	(c)
41.	(b)	42.	(b)	43.	(a)	44.	(a)
45.	(a)	46.	(b)	47.	(c)	48.	(b)
49.	(b)	50.	(a)	51.	(a)	52.	(a)

CHAPTER 4

Atmospheric Pressure

According to Kinetic theory of gases, the molecules of the atmosphere are perpetually (or constantly) moving at random and bombarding anything in their path and also colliding with one another. The bombardment of molecules on any unit area constitutes atmospheric pressure. Man is living at the bottom of the ocean of atmosphere. The weight of the atmospheric molecules from the top of the atmosphere to the bottom in a unit column (at sea level) is called atmospheric pressure. Atmospheric pressure is one of the three basic state variables, the other two are temperature and specific volume (or volume).

Pressure is defined as force per unit area.

$$\text{Pressure} = \frac{\text{Force}}{\text{Area}}$$

Pressure is a scalar quantity.

Unit of Atmospheric Pressure

At the surface of the earth, the atmospheric air exerts a pressure of about 10^5 Newtons/m^2 or 10^5 Pascal (or 14.7 lb/sq. inch), which is called one bar or one atmosphere (Atm). In meteorology one thousandth of a bar, called millibar or hectopascal (h Pa), is widely used.

$$1 \text{ h Pa} = 100 \text{ Pa}$$

$$1 \text{ bar} = 10^5 \text{ N/m}^2 = 10^5 \text{ Pa} = 1 \text{ Atm}$$

More accurately $1 \text{ Atm} = 1.01325 \times 10^5 \text{ Pa}$

$$= 1013.25 \text{ h Pa}$$

$$1 \text{ h Pa} = 1 \text{ millibar (mb)} = 10^2 \text{ N/m}^2 = 100 \text{ Pa.}$$

The Italian scientist Torricelli showed that the atmospheric air pressure supported the weight of the mercury (Hg) column of about 30 inches in a long glass tube when inverted in a dish of mercury (see Fig. 4.1). Thus the atmospheric pressure at any point is numerically equal to the weight of the mercury column.

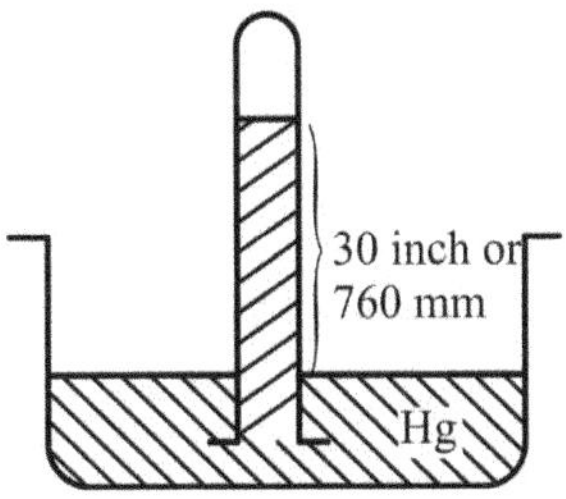

Fig. 4.1 Atmospheric pressure as the weight of mercury column.

Since the length of the mercury column varies with temperature and gravity, the atmospheric pressure is defined under standard conditions of temperature (0 °C or 273 °K) and gravity g = 9.80665 m/sec^2.

Thus $1 \text{ Atm} = 1013.25 \text{ hPa}$

$$= 760 \text{ mm of Hg column weight}$$

$$= 29.92 \text{ inches of Hg}$$

$$1 \text{ mm of Hg column weight} = 1.333224 \text{ hPa}$$

$$1 \text{ inch Hg} = 33.8639 \text{ hPa (or } 1 \text{ hPa} = 0.02953 \text{ inch Hg)}$$

This is called standard mm of Hg.

The instrument that is used for measuring the atmospheric pressure is called barometer. In Greek ‘baros’ meaning weight and ‘metron’ meaning measure.

Atmospheric pressure decreases with increasing altitude, about 1 hPa for every rise of 8.5 m in lower levels (near msl), but this fall is not linear, it is exponential. At any place atmospheric pressure is greatest at the surface of the earth. Variations in atmospheric pressure are caused by insolation, variations of atmospheric aerosols, differential heating of ground surfaces etc. On global scale at msl, pressure varies from less than 990 hPa to more than 1030 hPa.

The greatest sea level pressure recorded was 1083.8 hPa at Agata, in Saiberia on 31st December 1968, while the lowest sea level pressure recorded was 870 hPa (by dropsonde) in the Eye of Typhoon Tip (lat 16.7° N, long 137.8° E) in Pacific on 12th October 1979.

Atmospheric wind (horizontal air flow) depends on the pressure difference between two neighboring places and blows (in horizontal direction) from High pressure region to Low pressure region like water flow from higher region to lower. Wind flow (air circulation) ultimately results in the variation in weather phenomena. For this reason, the variation of pressure is of great importance in aviation. The relation between pressure and height is used in the construction of altimeters.

Static Pressure: the ambian atmospheric pressure at any station is called Barometric pressure or Static pressure when the air is calm (or at rest). The motion of air molecules or aerosols is entirely at random and which exert a uniform pressure in all directions. This is called Static pressure or Barometric pressure. In case of aircraft, it is measured or sensed so that it is not affected by turbulence. For this purpose a set of small holes are located at a point on the aircraft. This sensing point is called static source which is located by the side of fuselage (or on the side tube which is projecting into the air stream).

Dynamic Pressure: When a body/an aircraft moves through air it displaces surrounding air which causes compression of air on the facing edge of the body (aircraft). This compressing air pressure is called dynamic pressure. Dynamic pressure increases with the increasing speed of aircraft.

OR

When air is in motion, this will add as additional pressure on surface against the wind flow. This is called dynamic pressure or wind pressure. This type of pressure has application in the design of pitot tubes of air speed indicators.

Pitot Pressure: The pressure on the forward facing surface of body (aircraft) is the sum of static pressure and dynamic pressure. This total pressure is called pitot pressure.

Wind velocity and pressure relation at various exposed heights are given below.

Height in meters	Horizontal wind velocity in kmph	Horizontal pressure kg/m^2
(At surface) 0	80	40
3	96	58
6	108	73

Cabin Pressurization: As we go up into the atmosphere, pressure decreases, which implies decrease of oxygen content and causes trauma of decreased pressure on the body, namely breathlessness. It also causes several impacts on human body (like in the case tornado eye movement over an area).

Cabin pressurization was devised to overcome such problems. It is a method to maintain cabin pressure at a level lower than that of actual flight level. This will rush the air into the cabin to provide sufficient oxygen that will allow the people and crew to breath normally. For cabin pressurization the cabin structure is made strong to withstand the normal twisting and flexing forces of flight. Doors, windows, control cable openings are sealed. A source of compressed air for pressurization and ventilation made to allow large quantities of air to flow. It was designed for proper regulation of temperature, pressure, rate of change of pressure change and rate of outflow of air from the cabin.

4.1 Pressure Variation with Altitude

Pressure variation with increasing altitude is neither uniform nor linear but is exponential. However, on an average, the pressure lapse rate are:

(i) at the surface level (msl) pressure falls 1 hPa for a rise of about 27 ft,

(ii) at 20,000 ft level pressure falls 1 hPa for a rise of about 50 ft and

(iii) at 40,000 ft level pressure falls 1 hPa for a rise of about 100 ft.

The approximate pressure lapse rate in ISA, at different levels is given below.

No.	Level	1SA static pressure	Pressure lapse rate
1.	At msl	1013.25 hPa	27 ft/hPa
2.	At 10,000 ft	697 hPa	34 ft/hPa
3.	At 20,000 ft	466 hPa	47 ft/hPa
4.	At 40,000 ft	186 hPa	96 ft/hPa
5.	At 60,000 ft	72 hPa	263 ft/hPa

Note: Pressure lapse is constant in ISA

4.1.1 Pressure Variations

We know from gas law $p = \rho RT$, variation of pressure with height is greatly influenced by temperature. Warm air causes pressure fall slowly with height (decrease of pressure lapse rate), while cold air causes pressure fall rapidly with height (increase of pressure lapse rate). The hydrostatic equation is $dp = -g\rho d_z$ or $d_z = -\frac{dp}{g\rho}$ i.e., $d_z = -\frac{RT}{g\rho}$ dp (using gas equation)

If p is expressed in hPa, $g = 9.81$ m/s^2, P in kg/m^3, then

$R = 2.87 \times 10^2$ for dry air

i.e., $$dz = \frac{2.87 \times 10^2}{9.81} . \frac{Tdp}{P}$$

$$dz = -29.25$$

If we take dp = 1 hPa then

$$dz = -29.25\ I_p \text{ meters or } dz = 96\ \frac{T}{P} \text{ feet}$$

4.2 Barometric Pressure Corrections for Standard Conditions

In meteorology, in general, atmospheric pressure means static pressure, that is barometric pressure. For the purpose of comparable pressure readings at different places three corrections are made:

(i) Index correction,

(ii) temperature correction and

(iii) gravity correction

Index correction is made to eliminate error due to calibration, residual gas in the vacuum above mercury column and bending of light (refraction), capillarity effect.

Pressure readings made at room temperature which has to be reduced to some standard temperature 0 °C.

Pressure depends on g (acceleration due to gravity) which varies with latitudes and altitudes. Barometers are calibrated to standard $g = 9.80665\ m/s^2$, consequently barometer readings have to be reduced to the 'g' of that place.

4.3 Types of Mercury Barometers

At meteorological observatories/stations mostly Fortin or Kew pattern barometers are used.

In Fortin barometer the mercury level in the cistern has to be raised or lowered to just touch a fixed ivory pointer before taking readings.

In Kew pattern barometer, the cistern is fixed and hence no adjustment of mercury level required. IMD uses this type of barometers in observatories.

Aneroid barometer: Aneroid barometer essentially consists of:

(i) a closed metal chamber, which is completely or partly evacuated and

(ii) a strong spring system which prevents the collapse of chamber due to external atmospheric pressure. An equilibrium maintains between the forces due to the spring and that of external pressure. Aneroid chamber may be made of steel or beryllium copper which have elastic properties. The chamber itself can act as spring. Aneroid barometer is calibrated against mercury barometer which is accurate. Aneroid barometer is less accurate but it is compact and portable. The aneroid instrument has to be read in the same position either vertical or horizontal in which state it was calibrated. The instrument has to be slightly tapped before taking reading.

Reduction of Pressure to Standard Level

The observed pressure at a station is called *station level pressure*.

For comparison of pressure at various stations situated at different altitudes, it has to be reduced to a standard level. For all practical purposes, it is reduced to mean sea level (msl) and called *mean sea level pressure*.

In case of stations located above msl, we have to add the pressure of the unit area air column between the station and msl, to the station level pressure. It is assumed the temperature to be constant (isothermal) between station and msl (that is the msl temperature same as the station level temperature).

In case of station located below msl, we have to deduct the pressure of unit area air column between msl and station, from the station level pressure, assuming isothermal state. Meteorologists throughout the world use msl pressure for synoptic pressure charts. In Greek "syn" means together and "opsis" means looking. Thus synopsis stands for looking together. Pressure observations are taken at standard times simultaneously all over the world reduced to msl and then plotted on the charts. These charts are analysed by drawing isobars (equal pressure lines) at the interval of 2 hPa. The isobars thus drawn indicate Low pressure, High pressure regions, which gives a first indication about the weather.

4.4 Pressure Patterns

Pressure is one of the weather elements. Horizontal pressure destribution at msl has great bearing on weather. Some pressure patterns are shown in Fig 4.2. Isobars are found to be closed, elongated or straight.

The closed isobars, if indicate that pressure is decreasing outward, then it is called High pressure area (or region) or simply *High*. If the closed isobars indicate that pressure is increasing outward of the centre, it is called *Low* pressure area (or region) or simply Low. If the isobars are in the shape of V and pressure decreasing outward, the line passing though the vertices of V is called *ridgeline* or wedge. If the isobars are in the shape of V and pressure increasing outward, the line passing through the vertices of V is called *trough line*. The region between two Highs and two Lows is called *Col region*.

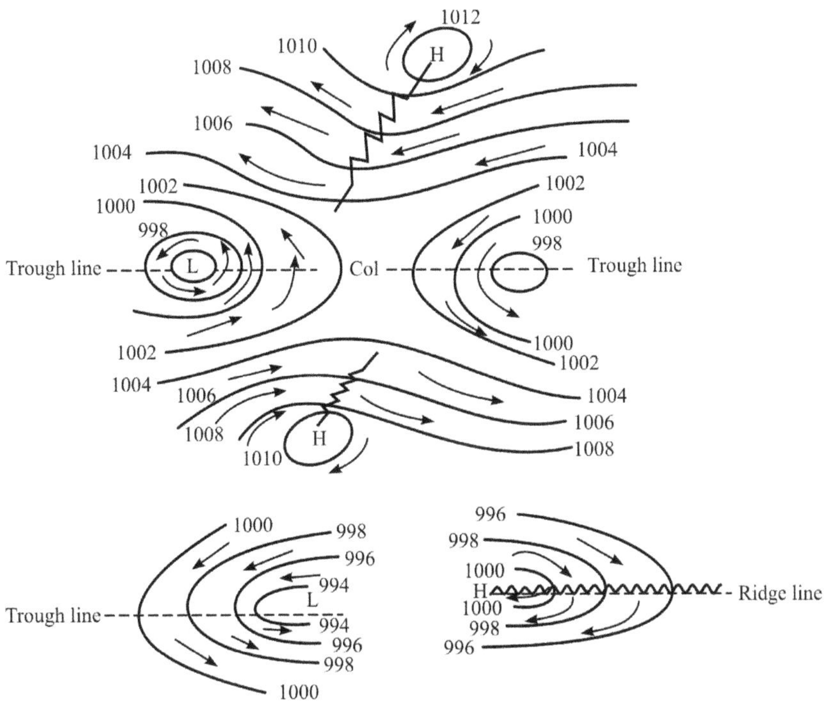

Fig. 4.2 Pressure Pattern in isobars.

4.5 Mean Sea Level (msl) Pressure Distribution

The horizontal distribution of msl pressure shows the following semi permanent features (see Fig 4.3). Low pressure regions are found along equator, called equatorial Low or Doldrums. There are high pressure regions along latitude belts of 25° to 35° North(N) and South(S). These are called sub tropical high pressure belts. Again low pressure belts are observed between latitudes 60 to 70° N and S, which are called sub-polar low pressure belts. In polar regions both around N pole and S pole, High pressure regions are found, which are called polar Highs (see Fig. 4.3).

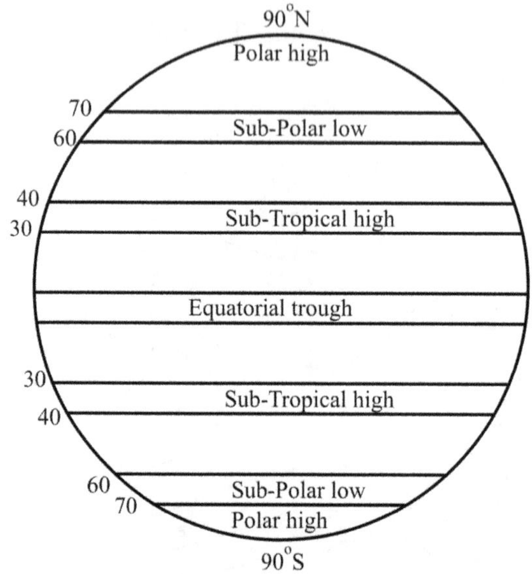

Fig. 4.3 msl Pressure distribution semi-permanent features.

4.6 Pressure Gradient

Horizontal rate of change of pressure for unit distance is called pressure gradient. It is measured perpendicular to the isobars from High to Low. In Fig. 4.4, the pressure gradient along AB is higher than along CD. If the isobars are closely spaced, the gradient is said to be steep or strong but if they are sparsely spaced (a few far apart) the gradient is said to be weak or flat. Wind strength depends on the pressure gradient, closer the isobars stronger the wind and vice versa.

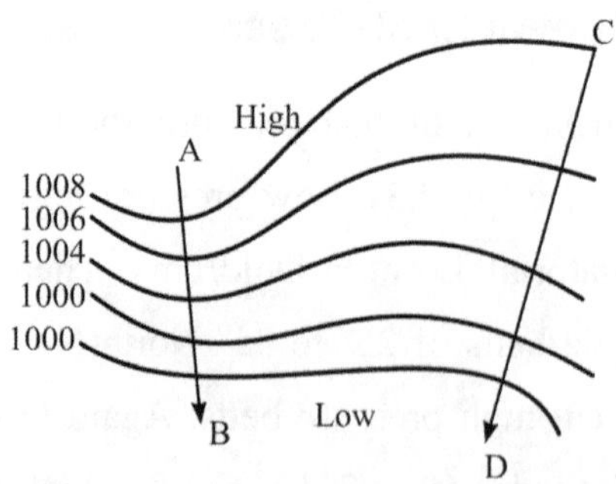

Fig. 4.4 Different pressure gradients.

4.7 Pressure Tendency

Pressure change during the preceding three hours of observations (synoptic observations are taken every three hours of standard time 00, 03, 06, 09, 12, 18, 21, 24/00 UTC) are plotted on a weather map. This gives the pressure tendency at a glance, the areas of pressure falling or rising. The distribution of pressure changes are brought out by drawing lines of equal tendency, which are called isallobars. The maximum fall or rise at the centre, which are denoted by is isallobaric Low or High. These areas indicate pressure field deepening (or intensifying) or filling.

4.8 The Semidiurnal Variation of Pressure

At any given place the atmosphere will be ever changing. This change may be periodic (regular) or non-periodic (irregular).

Periodic variations are regular but they have different periodicity. Of all periodicity 12 hours (half a day) is of importance. This is called semi-diurnal variation of pressure. As the earth rotates day and nights are formed. Heating of atmosphere during day time (due to insolation) and cooling at night (due to nocturnal cooling) causes rhythmic expansion and contraction, which results in pressure oscillations. Further it is assumed atmosphere itself possess 12 hrs period of natural oscillation. Because of this resonance, amplitude oscillations magnify. Consequently a double sinusoidal pressure wave travels around the earth along with the sun. It is observed a pressure maxima occurs at about 1000 hrs and 2200 hrs local meantime (LMT) and a pressure minima occurs at about 0400 hrs and 1600 hrs LMT (see Fig 4.5).

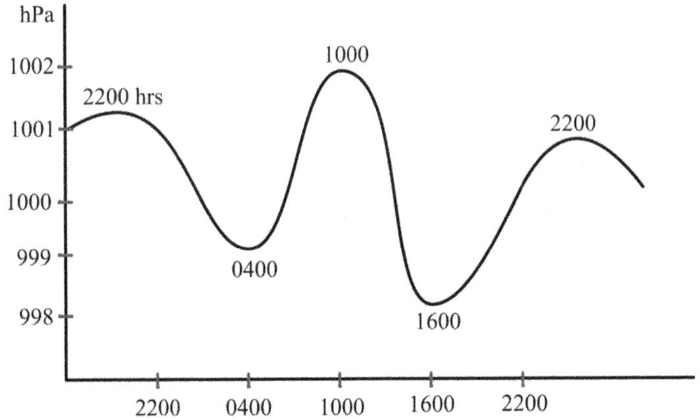

Fig 4.5 Semidiurnal variation of pressure.

The atmospheric air is a mixture of gases. Each gas exerts its own pressure called its partial pressure. The partial pressure exerted by an individual gas is proportional to its number of molecules present in the given volume of air (mixture). Thus the atmospheric pressure is the sum of partial pressures of all the gases present in the atmosphere including water vapour.

Questions

1. Every physical process of weather is the result of /accompanied by:
   (a) Movement of air (b) Pressure differential
   (c) Heat exchange (d) None above
2. What causes variations in altimeter setting between weather reporting points?
   (a) Unequal heating of the earth's surface
   (b) Variation of terrain elevation
   (c) Coriolis force
   (d) All above
3. What is meant by the term 'dew point'?
   (a) The temperature at which condensation and evaporation are equal
   (b) The temperature at which dew will always form
   (c) The temperature to which air must be cooled to become saturated
   (d) All above
4. The amount of water vapour which air can hold depends on the:
   (a) Dew point (b) Air temperature
   (c) Stability of the air (d) None above
5. Clouds, fog or dew will always form when:
   (a) Water vapour condenses
   (b) Water vapour is present
   (c) RH reaches 100%
   (d) All above

6. What are the processes by which moisture is added to unsaturated air?
   (a) Evaporation and sublimation
   (b) Heating and condensation
   (c) Supersaturation and evaporation
   (d) None above
7. Which is true regarding actual temperature and dew point temperature spread?

   The temperature spread:
   (a) Decrease as the relative humidity decreases
   (b) Decrease as the relative humidity increases
   (c) Increase as the relative humidity increases
   (d) None above
8. Moisture is added to a parcel of air by:
   (a) Sublimation and condensation
   (b) Evaporation and condensation
   (c) Evaporation and sublimation
   (d) All above
9. How much colder than standard temperature is the forecast temperature at 9000 ft, as indicated in the following excerpt from the winds and temperature aloft forecast?

Ft 6000	9000
0737-03	1043-09

   (a) 3 °C (b) 9 °C
   (c) 6°C (d) None above
10. Clouds, fog or dew will always form when:
    (a) Water vapour condenses
    (b) Water Vapour is present
    (c) The temperature and dew point are equal
    (d) None

11. To which meteorological condition does the term "dew point" refer?
    (a) The temperature to which air must be cooled to become saturated
    (b) The temperature at which condensation and evaporation are Equal
    (c) The temperature at which dew will always form
    (d) All above

12. The amount of water vapour that air can hold mostly depends on:
    (a) Relative humidity (b) Air temperature
    (c) Stability of air (d) None above

13. If an altimeter reads aerodrome elevation at ground level, then the sub-scale must have been set to:
    (a) QFE (b) QNH
    (c) QNE (d) None above

14. Density of moist air is:
    (a) Higher than dry air (b) Lower than dry air
    (c) Equal to dry air (d) None above

15. Under the same conditions of pressure and temperature, moist air is:
    (a) Lighter than dry air (b) Heavier than dry air
    (c) Has the same density as dry air (d) None above

16. The density of air:
    (a) Varies directly as pressure and temperature
    (b) Varies directly as pressure and inversely as temperature
    (c) Varies inversely as pressure and directly as temperature
    (d) Varies inversely as pressure and temperature

17. In troposphere, the density of air:
    (a) Increases with height (b) Decreases with height
    (c) Remains constant (d) None above

18. The major constituent of air by weight is:
    (a) Oxygen (b) Nitrogen
    (c) Carbondioxide (d) Water vapour

19. An adiabatic process is one in which heat content of the system:
    (a) Increases (b) Decreases
    (c) Remains constant (d) None above

Answers

1.	(c)	2.	(a)	3.	(c)	4.	(b)
5.	(c)	6.	(a)	7.	(b)	8.	(a)
9.	(c)	10.	(c)	11.	(a)	12.	(b)
13.	(b)	14.	(b)	15.	(a)	16.	(b
17.	(b)	18.	(b)	19.	(c)		

CHAPTER 5

Altimetry

For the safety of flight, the height and speed of aircraft are important which are provided to the pilot by ambient atmospheric pressure through (the instruments) pitot and static pressure systems.

Pressure-height relation is the basis of altimeter calibration. Pressure altimeters are aneroid barometers whose scale is graduated to read height/altitude instead of pressure. Altimeter uses static pressure to measure aircraft altitude or rate of change of altitude. ICAO altitudes are calibrated to read in ISA. The scale is linear.

Static pressure in an aircraft altimeter is sensed through a set of small holes located at a point on the aircraft which is not affected by turbulence. The sensing point is called static source. When the aircraft moves forward, the dynamic pressure is added to the static pressure. The pressure measured on the forward facing surfaces of an aircraft is the sum of static and dynamic pressures, which is called pitot pressure. Pitot and static pressure is supplied to the air data instruments through a system of tubes called pitot/static system.

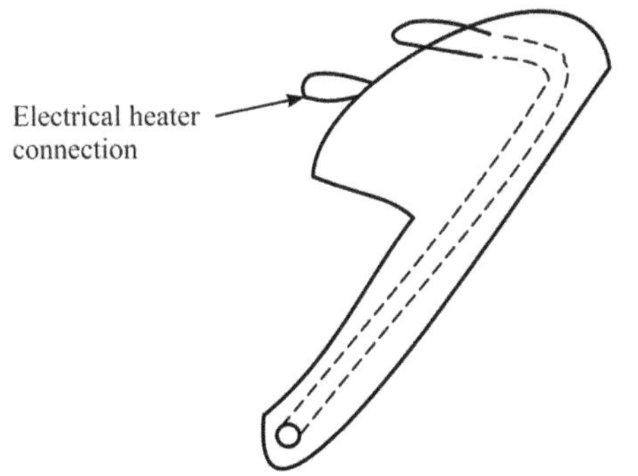

Fig. 5.1 Pitot head.

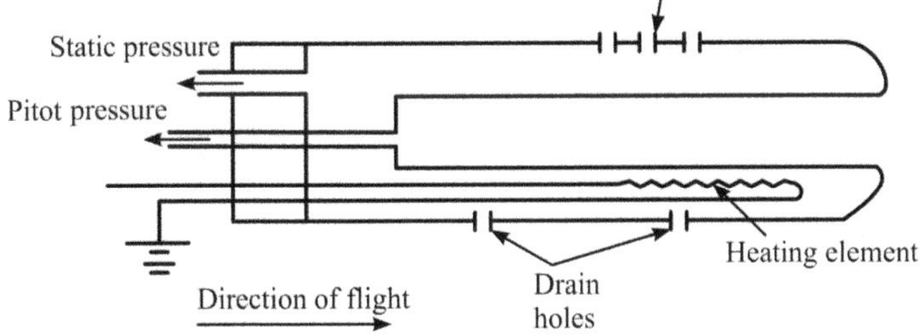

Fig. 5.2 Pitot /Static head.

5.1 The Pressure Altimeter

The pressure altimeter indicates the height of aircraft above a given datum pressure. It is an aneroid barometer calibrated (on the basis of ICAO ISA specifications) to read pressure in terms of height.

In ISA, atmospheric pressure or static pressure decreases with increasing height at a constant lapse rate, which is given below (Table 5.1).

Table 5.1 Rate of pressure change with altitude.

Height in ICAO ISA	Static Pressure hPa	Height change for 1 hPa Static Pressure Change
1. At msl	1013.25	27 ft
2. At 10,000 ft (FL 100)	697	34 ft
3. At 20,000 ft (FL 200)	466	47 ft
4. At 40,000 ft (FL 400)	186	96 ft
5. At 60,000 ft (FL 600)	72	263 ft

5.1.1 The Simple Altimeter

The schematic diagram of a simple pressure altimeter is shown in Fig. 5.3. An aneroid capsule is mounted in a sealed case which is connected to a static source. Increase in static pressure causes compression of capsule or decrease in static pressure causes expansion of the capsule. The capsule is connected to a restraining force of a leaf spring. The expansion or contraction of the capsule is transmitted to pointer through a lever, which will rotate over a calibrated card scale.

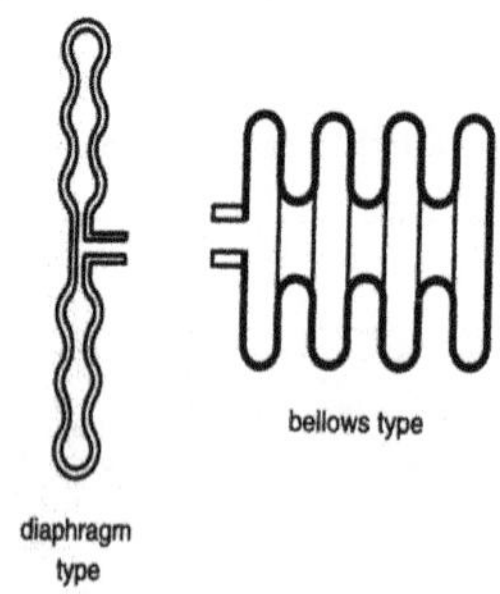

Aneroid capsule types.

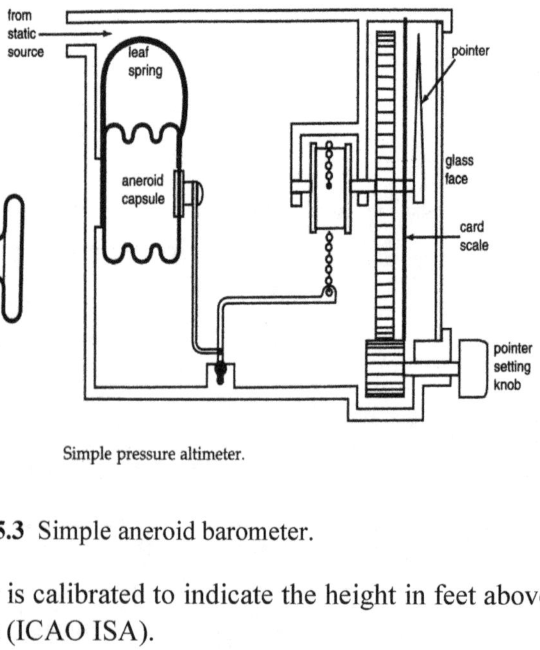

Simple pressure altimeter.

Fig. 5.3 Simple aneroid barometer.

The pressure altimeter is calibrated to indicate the height in feet above the datum of 1013.25 hPa (ICAO ISA).

The different rate of pressure change with altitude (non-linear distribution of atmospheric pressure) is achieved by careful design of the capsule manufacture.

It must be remembered that the altimeter is calibrated at ISA msl temperature of +15 °C, pressure 1013.25 hPa. As the aircraft climbs, the air temperature falls (which is not governed by ISA) and hence capsule deviates from ISA msl temperature of 15 °C. In sensitive altimeter, temperature deviation is compensated by a bimetallic strip introduced between capsule and the lever shaft.

In a sensitive altimeter there are three points (like the hands of a clock) for display of height (see Fig. 5.4).

One revolution of long hand pointer indicates height change of 1000 ft (each sub-division of 100 ft). In Fig. 5.4 it is indicating 500 ft.

One revolution of medium hand pointer indicates 10,000 ft (each sub-division 1000 ft). In Fig. 5.4 it is indicating 4000 ft.

The smallest pointer in one revolution covers 100,000 ft (each sub-division of 10,000 ft). In Fig. 5.4 it is indicating 20,000 ft. Thus Fig. 5.4 completely reads 24500 ft.

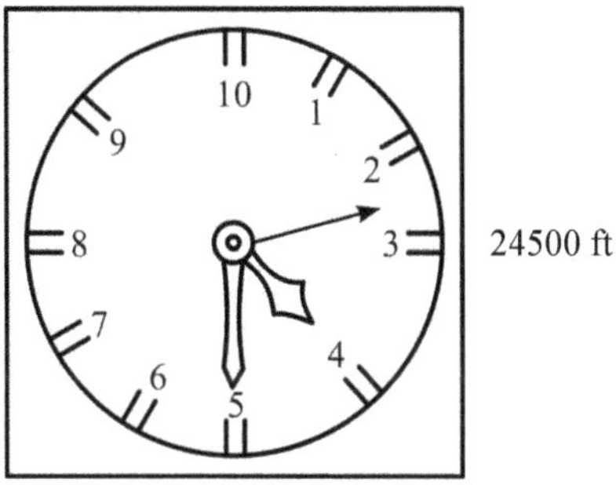

Fig. 5.4 Sensitive altimeter display

An aircraft altimeter consists of a main circular scale calibrated in hundreds, thousands of feet and a sub-scale is calibrated in hPa. The sub-scale is operated by a knob to set any pressure value as the base/datum. Corresponding to this base/datum value of pressure, the main scale will indicate altitude. This is called sub-scale setting. Sub-scale setting is made by QFE or QNH.

***QFE*:** Barometric pressure at aerodrome elevation or at runway threshold.

***QNH*:** QFE pressure (Barometric pressure at aerodrome elevation) reduced to (msl) mean sea level under ICAO conditions.

***QFE Altimeter Setting*:** When the sub-scale of altimeter set to QFE, the altimeter main scale reads zero when the aircraft is on the aerodrome (runway). This is called zero setting. The main scale will read the height of the aircraft above aerodrome if it is in the air.

***QNH Altimeter Setting*:** When the sub-scale of altimeter set to QNH, the altimeter main scale will read the elevation of the aerodrome when the aircraft is on the runway and it will read the elevation of the aircraft when it is in the air.

Note: QFE, QNH are useful only in the vicinity of an air field and for clearance of terrain obstructions near the airfield. However these settings are not suitable for cross-country flights.

5.2 Standard Altimeter Setting (SAS)

Standard altimeter setting is the mean sea level (msl) pressure in ICAO ISA is 1013.25 hPa ($\simeq$ 1013.2 hPa) or 29.92 inch Hg. Altitudes indicated in SAS setting are known as pressure altitudes (Z_p) or simply indicated altitudes.

For a cross country flights Air Traffic Control (ATC) allots cruising levels to aircrafts in Flight Levels (FL), which refer to pressure altitudes.

In practice a pressure altitude of 3000 ft is denoted as FL 30; FL 100 stands for 10000 ft; FL 300 stands for 30000 ft. Flight Levels are used for aircraft crusing at higher levels and they are usually separated by at least 500 ft or 1000 ft.

e.g. FL 60, FL 65, FL 70 etc., or F100, F110, F120 etc.

Note: Pressure at msl and temperature, humidity at various (FL) levels rarely correspond to ICAO ISA conditions. Thus indicated altitude (Z_p) differs from true altitude (z).

Altitude correction or "alticor": The correction required to be applied to the indicated altitude (z_p) to get true (real) altitude (z) is called "alticor". Alticor is required by a pilot while flying in mountainous terrain or to clear obstructions (with known elevation such as an airfield).

5.3 Some Terms and Q-code used in Aviation

Height: Vertical distance of an object/a point/ a level measured above a specified datum (with known elevation such as airfield).

When QFE is set on an altimeter it will indicate height of aircraft above that airfield.

Altitude: Vertical distance of an object/ a point/ a level above mean sea level (altitude that is indicated by an altimeter with QNH set).

Elevation: The vertical distance of an object/ a point/ a level on or affixed to the surface of the earth, measured from mean sea level.

Q-code

QFE: Barometric pressure reading at aerodrome elevation (or at runway threshold)

QNH: The QFE reduced to mean sea level under ICAO ISA conditions.

QNE: When 1013 hPa set on the altimeter subscale, it will read an altitude value on landing, is known as QNE value.

QFF: The QFE pressure reduced to mean sea level using station temperature, which is isothermal lapse rate between station and mean sea level (msl). This is used in synoptic charts.

QAN: Surface wind

QNT: Maximum speed of gust

QAB: (Surface) visibility

QBB: Amount, type of cloud and height

QAM: Weather

QAO: Upper wind at specified level

We have learnt about altimeter and its setting by sub scale.

5.4 Altimeter Setting

It concerned with the phase of the aircraft flight, namely take-off, cruising, landing and terrain clearance, vertical separation in controlled airspace.

***Transition Altitude*:** This is the highest altitude below which an aircraft will always fly on local QNH, that is below which the vertical position of an aircraft is controlled with reference to the height above aerodrome.

***Transition level*:** This is the lowest flight level above which the aircraft will always fly on standard QNH of 1013.25 hPa or the lowest flight level available for use above transition altitude.

Transition layer: The airspace in between transition altitude and the transition level.

Note: Pressure altimeter is essentially useful in terrain clearance, traffic separation for safe vertical separation and performance calculations of safety and efficiency.

(a) Altimeter will always read the vertical distance above the pressure level set.

(b) The lowest forecast msl pressure in an area is regional (local) QNH.

Terrain clearance: Height clearance above the obstacle, as shown in Fig. 5.5. Terrain clearance required for the safety of flight.

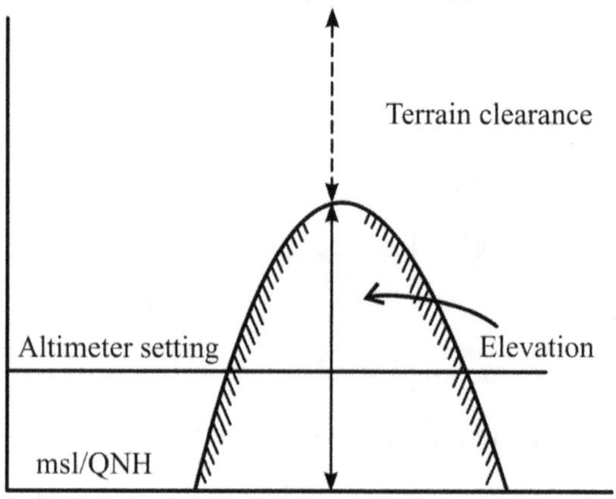

Fig. 5.5 Terrain Clearance.

Minimum Flight Level: For the safety of flight, a pilot should know the elevation of the highest ground/ obstacle along the cruising sector route, the minimum terrain clearance and QNH. The Fig. 5.6 explains the minimum safe flight level.

An aeroplane is not permitted to fly below a specified minimum altitude (above the ground and terrain obstacle) from safety point of view.

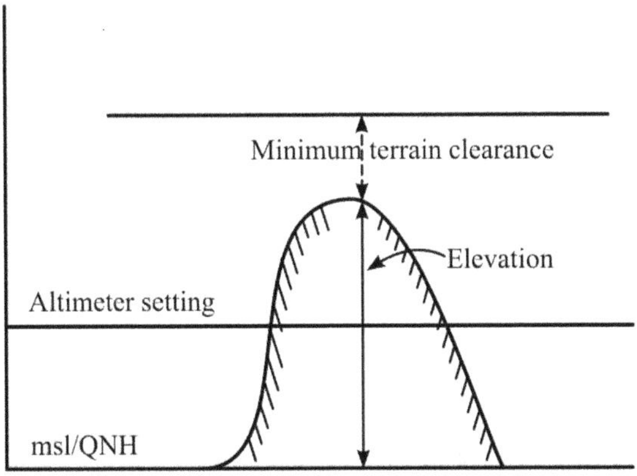

Fig. 5.6 Minimum Flight level.

This specified minimum flight en-route altitude has ATC considerations like controlled airspace, danger zones and semi-circular cruising altitude regulations as well. The minimum safe en-route altitude (MEA) is based on the elevation of the highest point along the route. The MEA is determined by adding increment an elevation to the elevation of the highest point in the en-route segment. The increment is given below.

Highest point Elevation en-route	Increment
5000 ft and below	1500 ft
over 5000 ft up to 10,000 ft	2000 ft
over 10,000 ft	10% of elevation + 1000 ft

5.5 Transition Altitude (TA) Computation

$$TA = E + 1500 \text{ ft} \quad \text{or} \quad E + H + 1000 \text{ ft}$$

where E = Elevation of the airfield

H = Height of the highest obstruction within 25 nautical miles around the aerodrome.

Transition attitude thus calculated is rounded off to the next 100 ft and the higher of the two values is taken.

5.6 Transition Level

Transition Level = TA + Pressure correction

+ Temperature correction

where

$$\text{Temperature correction} = (\text{TA} + \text{Pressure correction})\left(\frac{T_{ISA} - T_{min}}{T_{min}}\right)$$

T_{ISA} = ICAO ISA temperature at aerodrome level

T_{min} = Lowest surface minimum temperature

Transition level is rounded off to the next 500 ft.

If $T_{min} > T_{ISA}$, no temperature correction is applied.

5.7 Standard Procedure of Altimeter Setting

ICAO ISA standard QNH is 1013.25 hPa. Before take off, the altimeter is set to the current local QNH (altimeter setting for the aerodrome).

During climb, below transition altitude, the aircraft remains on this local QNH setting. After crossing the transition altitude, the aircraft altimeter has to be set to standard QNH 1013.25 hPa. This changeover should be completed within transition layer. After this, the aircraft remains at the flight level allotted (cruising level) on the same standard setting.

To assess terrain clearance latest and nearest QNH has to be used. Before initial approach for landing at the destination aerodrome, the transition level has to be obtained from ATC. On crossing the transition level, the altimeter setting must be changed to local QNH. A pilot can use local QFE setting instead of local QNH setting.

Thumb rule: Indicated altitude on the altimeter will be correct only as long as the airfield pressure set on the sub-scale remains the same. On landing back at the same airfield, the aircraft altimeter will read High (Hi, over read) if the pressure has fallen, it will read low (lo, under read) if the pressure has risen as compared to the pressure set before take-off.

In a similar way the aircraft altimeter will read High (Hi, more) at the destination airfield where the pressure QNH is Low and read Low (lo, less) at the destination airfield where the pressure QNH is High as compared to the pressure set at the take-off air field.

Thus we have the following thumb rule.

QNH	Altimeter Setting	True Altitude (Z)	Indicated Attitude (Z_p)
Hi	Lo	Hi	Lo
	OR		
Lo	Hi	Lo	Hi

5.8 Altimeter Temperature Error Correction

Pressure altimeters are calibrated under ICAO ISA conditions. Any deviation from ICAO ISA temperature, altimeter will show wrong readings. The error is proportional to the difference between the actual and ICAO ISA temperature. (The vertical distance of the aircraft is above altimeter setting datum, that is height above touch down).

The error is approximately 4 ft / 1000 ft for every 1°C temperature deviation. The following examples illustrate Thumb rule and temperature error correction.

Note: Pressure altitude is the height in 1SA above 1013.2 hPa (pressure level) at which the pressure equals to that of an object/ an aircraft/ a point / a level which is under consideration. In other words when the altimeter sub-scale is set to 1013.2 hPa, the altimeter will read the pressure altitude.

Ex: Given local QNH 1006 hPa and the aircraft is on the aerodrome ground, whose elevation is 40 ft. Find the pressure altitude of the aircraft.

***Sol*:** It is given local QNH is 1006 hPa. This means the ISA datum 1013.2 hPa level is below sea level (theoretically)

The pressure deviation from ISA is

about 7 hPa (1013.2 – 1006 = 7 hPa)

At msl 1 hPa pressure falls for every rise of 27 ft.

This means msl is 27 × 7 = 189 ft above ICAO datum

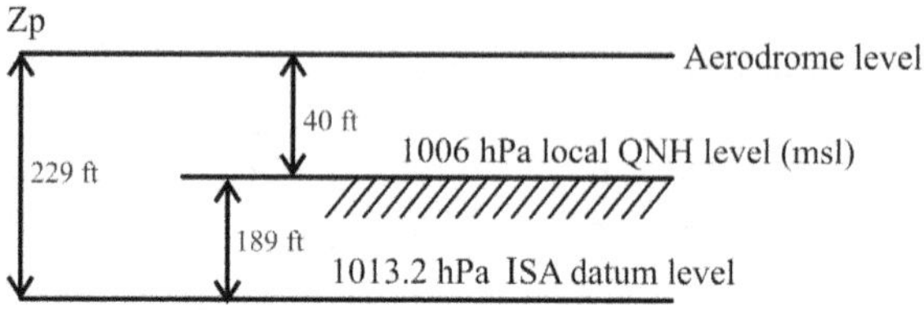

∴ The pressure altitude of the aircraft is 189 + 40 = 229 ft

Ex: Given QNH set is 1023 hPa and the aeroplane is on the ground whose elevation is 40 ft. Find the pressure altitude, of the aircraft.

Sol: The local QNH is given 1023 hPa, this means the ISA datum 1013.2 hPa level is above msl

Pressure deviation of QNH from ISA datum is about 10 hPa

(1023 – 1013.2 hPa)(ISA datum)

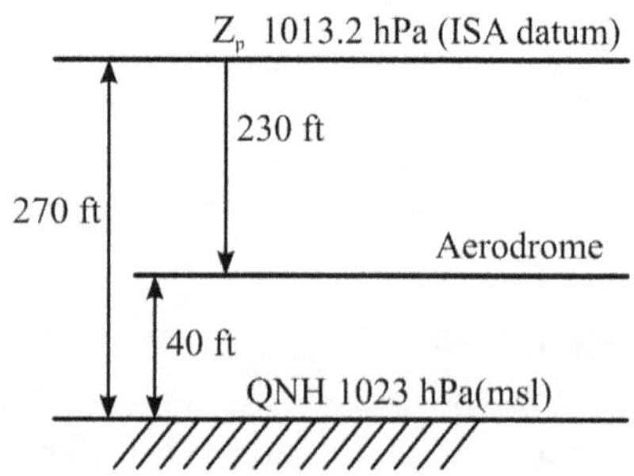

At msl 1 hPa pressure falls for every rise of 27 ft.

This means the msl is 10 × 27 = 270 ft below ISA datum.

∴ The pressure altitude of aircraft = – 270 + 40
= – 230 ft.

Ex: An aircraft is cruising at 4500 ft with QNH 1023 hPa set on the sub-scale. Find the pressure altitude of the aircraft.

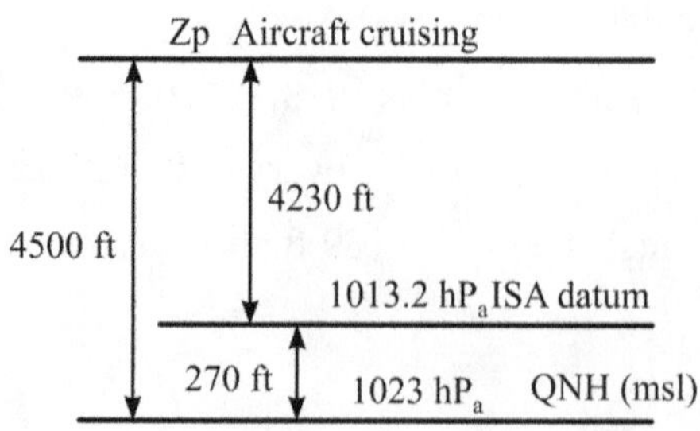

***Sol*:** The pressure deviation between QNH & ISA datum.

$$= 1023 - 1013.2 = 10 \text{ hPa.}$$

which is $10 \times 27 = 270$ ft at msl.

$\therefore$ the pressure altitude of the aircraft is $(4500 - 270) = 4230$ ft.

Ex: An aircraft is cruising at 5000 ft with QNH 1002 hPa set on the sub-scale. Find the pressure altitude of the aircraft.

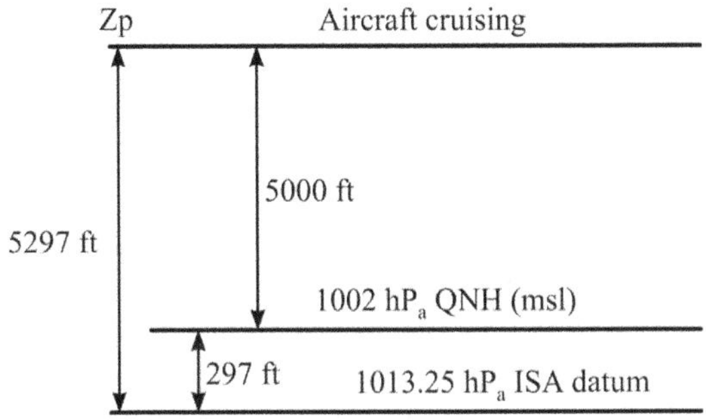

***Sol*:** The pressure deviation between QNH & ISA datum

$= 1013.25 - 1002 \simeq 11$ hPa, which is $11 \times 27 = 297$ ft.

$\therefore$ The pressure altitude of the aircraft $= 5000 + 297 = 5297$ ft.

Ex: QNH is 1006 hPa and the aeroplane is on the ground at an aerodrome whose elevation is 60 ft. Find the pressure altitude of the aircraft.

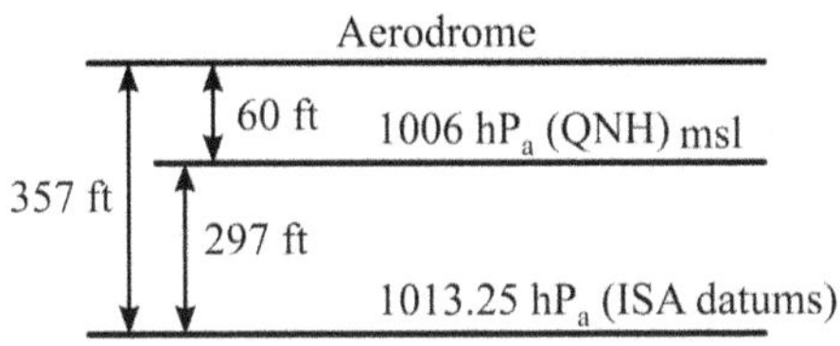

***Sol*:** The pressure deviation between QNH and ISA datum is

$1013.25 - 1002 \simeq 11$ hPa, that is theoretically $11 \times 27 = 297$ ft below msl, and Aerodrome elevation is 60 ft.

$\therefore$ pressure altitude of the aircraft $297 + 60 = 357$ ft.

Ex: What is the pressure altitude of (i) 990 hPa (ii) 900 hPa (iii) 850 hPa.

Sol: (i) Pressure difference of ISA datum (1013.2 hPa) and 990 hPa

$\simeq$ 23 hPa, which corresponds to height difference of $23 \times 27 = 621$ ft.

$\therefore$ The pressure altitude of 990 hPa pressure surface 621 ft.

(ii) Similarly

900 hPa surface is $113 \times 27 = 3051$ ft.

and 850 hPa is $163 \times 27 = 4401$ ft.

Ex: An aircraft is on the ground at an airfield whose elevation is 450 ft. The altimeter setting is 1003 hPa but the actual QNH is 993 hPa. What is the altimeter reading?

Sol:

QNH hPa	Altimeter setting hPa	True height ft Z	Zp Indicated height/ altimeter reading ft
993	1003	450 ft	?
Lo	Hi	Lo	Hi

Pressure deviation QNH, Altimeter setting is $1003 - 993 = 10$ hPa which is equivalent height difference of $10 \times 27 = 270$ ft.

From the thumb rule indicated height would be greater than true height by 270 ft.

$\therefore$ The indicated height/altimeter reading will be

$450 + 270 = 720$ ft

Ex: An aircraft is on an airfield whose elevation in 395 ft. Initial altimeter setting 1005 hPa and reading 125 ft, what is the actual QNH?

***Sol*:** Given

QNH	Altimeter setting	Z True height	Zp Altimeter reading
?	**1005 hPa**	**395 ft.**	**125 ft.**
Hi	Lo	Hi	Lo

Thumb rule

The height difference ($Z - Z_p$) is 395 – 125 = 270 ft, which is equivalent to 270/27 = 10 hPa pressure deviation

From the thumb rule QNH will be higher than altimeter setting by

10 hPa

∴ QNH = 1005 + 10 = 1015 hPa

Ex: What is altimeter reading if the setting is 982 hPa, the QNH = 990 hPa and the airfield elevation 560 ft?

Sol: Given

Thumb rule

QNH hPa	**altimeter setting hPa**	**True altitude (Z) ft**	**Altimeter reading/ indicating altitude(Z_p) ft**
990	982	560	?
Hi	Lo	Hi	Lo

Pressure difference of QNH & Altimeter setting

990 – 982 = 8 hPa,

which is equivalent to 8×27 = 216 ft.

From the thumb rule, altimeter reading (Z_p) is less than true altitude by 216 ft.

∴Altimeter reading 560 – 216 = 344 ft.

Ex: Given sub-scale pressure (altimeter setting), corresponding indicated altitude, find the true altitude of a place where QNH is given.

QNH (hPa)	**Altimeter setting (hPa)**	**True Altitude (Z) ft**	**Altitude / Altimeter reading (Z_p) ft**
(a) 1013	1007	-	3046
(b) 996	1013	-	6500

Sol: (a)

Pressure difference 1013 – 1007 = 6 hPa,

corresponding height difference 6 × 27 = 162 ft

Using thumb rule Hi Lo Hi Lo

The true altitude is higher than indicated altitude by 162 ft.

∴ The true altitude will be 3046 + 162 = 3208 ft.

(b)

Pressure difference 1013 – 996 = 17 hPa

corresponding height difference 17×27 =459 ft

Using the thumb rule in this case,

Lo Hi Lo Hi

The true altitude is lower than indicated altitude by 459 ft

∴ The true altitude will be 6500 – 459 = 6041 ft

Ex: The regional/local pressure setting is 1012 hPa, the altimeter setting is 1020 hPa and the indicated altitude is 4046 ft. Ahead there is a high mond of altitude 3500 ft. Will the aircraft clear the mond, and if so by what height?

Sol. Given

QNH hPa	Altimeter setting hPa	True altitude (Z) ft	Indicated altitude (Z_p) ft
1012	1020		4046

Pressure difference 1020 – 1012 = 8 hPa

corresponding height difference 8 × 27 = 216 ft

Applying thumb rule,

Lo Hi Lo Hi

True altitude is lower than indicated by 216 ft

∴ True altitude of the aircraft = 4046 – 216 = 3830 ft

The altitude of the high mond ahead is 3500 ft

∴ The aircraft will clear the obstacle by

3830 – 3500 = 330 ft

Ex: An aircraft is flying at 2500 ft indicated with the altimeter sub-scale set to 1020 hPa. Ahead there is a mountain range with elevation of 2000 ft. If during the flight the QNH in the area falls to 1010 hPa and the altimeter sub-scale is not reset, will the aircraft clear the mountain range, if so by what height?

Sol: Given

Local QNH hPa	Altimeter setting hPa	True altitude (Z) ft	Indicated altitude (Z_p) ft
1010	1020	?	2500
Lo	Hi	Lo	Hi

Pressure difference 1020 – 1010 = 10 hPa

corresponding height difference 10 × 27 = 270 ft

Applying the thumb rule

Lo Hi Lo Hi

True altitude is lower than indicated by 270 ft.

∴ The true altitude of the aircraft is 2500 – 270 = 2230 ft

The elevation of the mountain range is 2000 ft

∴ The aircraft will clear the mountain obstacle by

2230 – 2000 = 230 ft.

Ex: An aircraft at an airfield A with elevation 300 ft has the airfield QNH 1010 hPa correctly set. The altimeter setting is not changed. It landed at airfield B with elevation 500 ft, the aircraft altimeter reads 635 ft. What is the correct QNH at airfield B? (Assume 27 ft =1 hPa)

Sol: Given

QNH hPa	Altimeter setting hPa	True altitude (Z) ft	Indicated altitude (Z_p) ft
	1010	500	635

Height difference 635 – 500 = 135 ft

which corresponds to a pressure difference of $\frac{135}{27} = 5$ hPa

Using the Thumb rule

lo Hi lo Hi

The QNH at B is lower than altimeter set by 5 hPa

∴ The QNH at B = 1010 – 5 = 1005 hPa

Ex: The altimeter sub-scale is set to 1023 hPa and the altimeter reads 4000 ft. If local QNH is 1013 hPa, what is the altitude of the aircraft at that place?

Sol: Given data

local QNH hPa	**Altimeter Setting hPa**	**True altitude ft**	**Indicated altitude ft**
1013	1023	?	4000

Pressure difference is $1023 - 1013 = 10$ hPa, which corresponds to height difference of $10 \times 27 = 270$ ft.

Use the thumb rule

lo Hi Lo Hi

True altitude is lower than indicated by 270 ft

$\therefore$ The true altitude of the aircraft at that place is

$$4000 - 270 = 3730 \text{ ft}$$

Ex. QNH at an aerodrome is 1015 hPa, elevation is 240 m, what is the QFE ? (use 1hPa = 8 m)

Sol: Height difference between msl and aerodrome is 240m which corresponds to a pressure difference of $\frac{240}{8} = 30$ hPa.

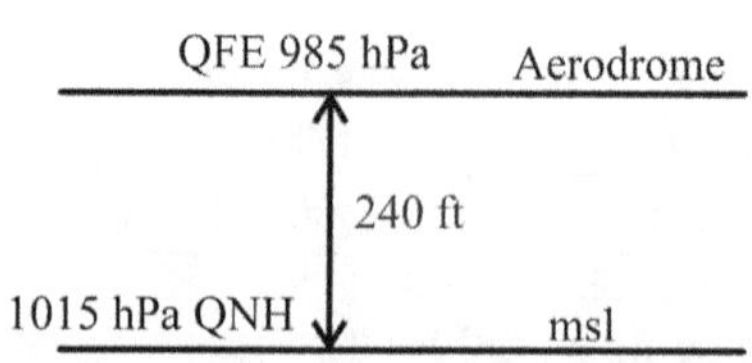

$\therefore$ The QFE = $1015 - 30 = 985$ hPa.

Ex: The QFE at an aerodrome is 998 hPa, whose elevation is 216 ft. What is the QNH?

Sol: Height difference between aerodrome and msl is 216 ft, which corresponds to pressure difference of

$$216/27 = 8 \text{ hPa}.$$

∴ QNH = 998+8 = 1006 hPa.

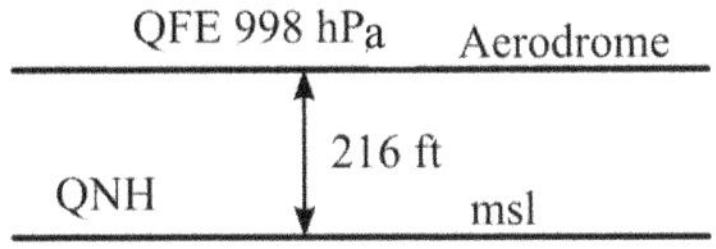

Ex: An airfield in Holland is 24 m below sea level, QFE is 1020 hPa. What is the value of QNH? (use 8 m= 1hPa)

Sol: Height difference between aerodrome and sea level is 24 m, which corresponds to a pressure difference of 24/8 = 3 hPa.

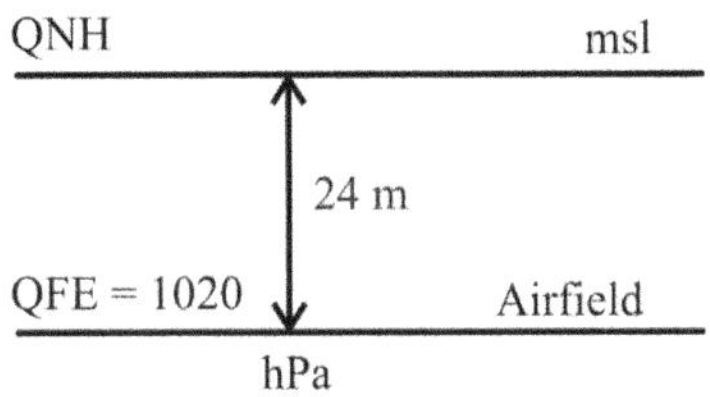

Since the aerodrome is 24 m below sea level the QNH will be less than QFE by 3 hPa.

∴ QNH is 1020 – 3 =1017 hPa

Ex: With a QNH of 1023 hPa, you set the SPS (standard pressure setting) at the transition altitude. What happens to the indicated altitude?

Sol: SPS is 1013 hPa

∴ Pressure change $1023 - 1013 \cong 10$ hPa, which corresponds to height change of $10 \times 27 = 270$ ft

∴ the indicated altitude drops by 270 ft (due to pressure drop).

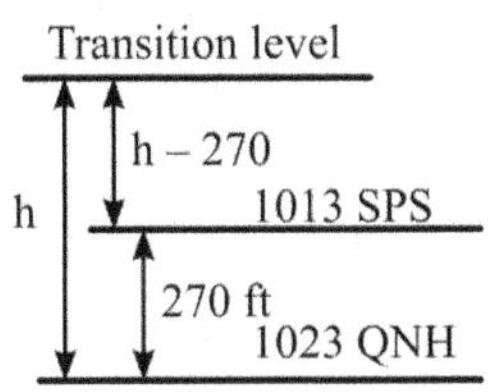

Ex: With a QNH of 1005 hPa, you set the SPS at transition altitude, what happens to the indicated altitude?

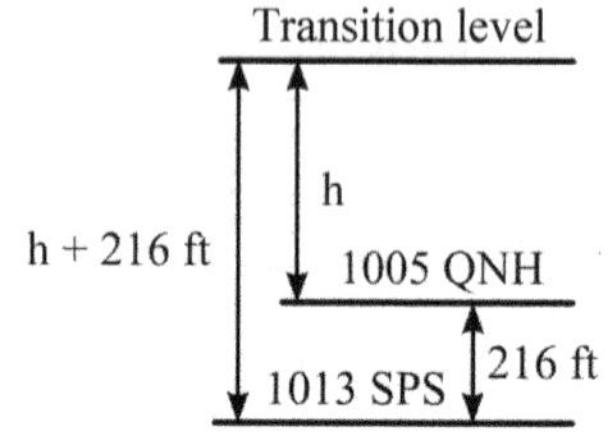

Sol: SPS is 1013.25 hPa

pressure change is 1013 – 1005 = 8 hPa, which corresponds to a height difference of $8 \times 27 = 216$ ft

∴ Indicated altitude rises by 216 ft.

Ex: The altimeter of an aircraft parked at an airfield indicated a height of 550 ft when the sub-scale of the instrument was set to 1013 hPa. If the QNH at the airfield was 998 hPa, find the elevation of the air field

(1 hPa pressure change equivalent to 27 ft)

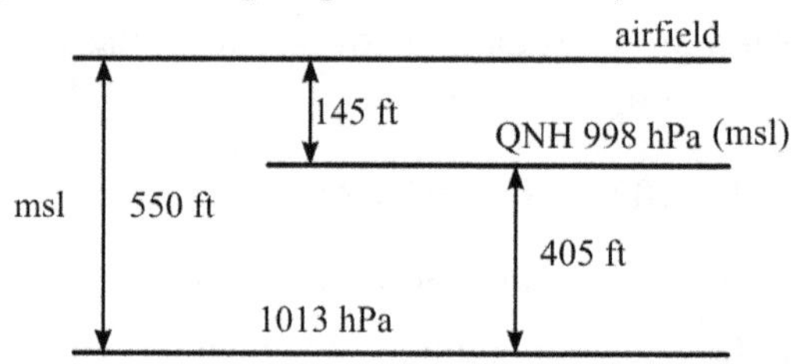

Sol.

Pressure difference 1013 – 998=15 hPa, which correspondent to a height change of 15× 27 = 405 ft i.e., the air field is 550 – 405 = 145ft amsl which is the elevation of the air field.

Ex: If an aerodrome is 2500 ft amsl and QNH 1033 hPa, what will be the actual height of an aircraft cruising at FL 80 at that aerodrome?

Sol. Pressure difference between 1SA datum and QNH = 1033 –1013 = 20 hPa, which correspondents to a height difference of $20 \times 27 = 540$ ft (= C)

$$D = FL\ 80 + C = B + C = 8000 + 540 = 8540 \text{ ft}$$

Height of aerodrome amsl = 2500 ft = A

∴ Cruising level of aircraft agl E= D – A = 8540 – 2500 = 6040 ft

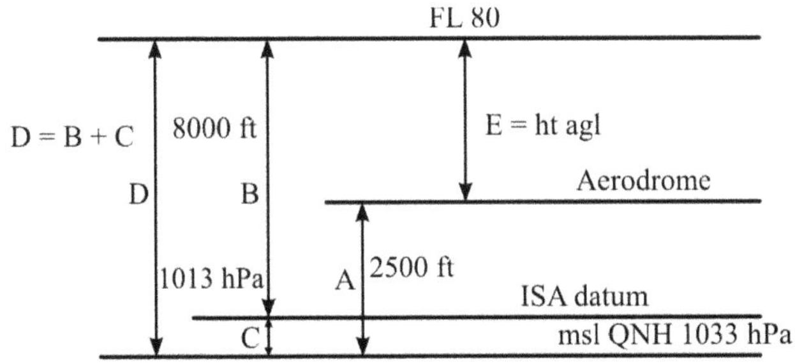

Questions

1. The basis of altimeter caliberation is:
   (a) Pressure-temperature relation
   (b) Temperature-height relation
   (c) Pressure-height relation
   (d) None above
2. To measure aircraft altitude or rate of change of altitude, altimeter uses:
   (a) Dynamic pressure (b) Pitot pressure
   (c) Static pressure (d) None above
3. Pitot pressure is equal to:
   (a) Dynamic pressure- static pressure
   (b) Dynamic pressure + Static pressure
   (c) Dynamic pressure
   (d) Static pressure
4. Pressure altimeter indicates the height of aircraft:
   (a) Above msl
   (b) Above 1000 hPa
   (c) Above aerodrome
   (d) Above a given datum pressure
5. In ISA atmospheric pressure decreases with increasing height at a constant lapse rate at msl and at 10000 ft of:
   height change for1 hPa
   (a) 20 ft, 40 ft (b) 27 ft, 34 ft
   (c) 34 ft, 41 ft (d) None above

6. An aircraft altimeter consists of a main scale caliberated in feet and sub-scale calliberated in hPa which is operated by knob to set any pressure value as datum. Corresponding to this datum the main scale indicates:

   (a) Freezing level (b) Tropopause level

   (c) Jet maxima level (d) Altitude

7. Barometric pressure reading at aerodrome elevation is called

   (a) QNH (b) QFF

   (c) QFE (d) QNT

8. When altimeter sub-scale set to QFE and the aircraft is on aerodrome, the main scale indicate the altitude:

   (a) Elevation of the aerodrome (b) Zero

   (c) Height above msl (d) None above

9. When the aircraft is over aerodrome and its altimeter sub-scale set to QFE, it will indicate (main scale):

   (a) Height of the aircraft a msl.

   (b) Height of the aircraft above aerodrome

   (c) Altitude of the aircraft

   (d) None above

10. When the aircraft is on the aerodrome and the sub-scale of altimeter is set to local QNH, then the main scale will indicate:

    (a) Elevation of the aerodrome (b) Indicate zero

    (c) Freezing level (d) None above

11. When the aircraft is over the aerodrome and the sub-scale of the altimeter is set to QNH then the main scale will indicate:

    (a) Height of the aircraft above aerodrome

    (b) It will indicate zero

    (c) Elevation of the aerodrome

    (d) Altitude of the aircraft

12. QFE, QNH are useful

    (a) Only for cross country flights

    (b) Only for enroute cruising

    (c) In the vicinity of an airfield only (for take-off, landing, terrain clearance)

    (d) For knowing freezing level

13. Standard Altimeter setting (SAS) is the msl pressure in ICAO ISA (1013.25hPa or 29.92 in Hg). Altitudes indicated in SAS setting is called:

(a) True altitude (b) Density altitude

(c) Pressure altitude (d) None above

14. Altitude correction or alticor required to be applied to the pressure altitude:

(a) To get indicated altitude

(b) To get elevation of the aircraft

(c) To get true altitude (to clear obstruction while flying)

(d) None above

15. Local QNH is 1006 hPa. Aircraft is on the aerodrome ground whose elevation is 40 ft. The pressure altitude of the aircraft is:

(a) 189 ft (b) 149 ft

(c) 229 ft (d) None above

(use 1 hPa. = 27 ft.)

16. An aircraft is on the ground at an aerodrome of elevation 40 ft. and QNH 1023 hPa. The pressure altitude of aircraft is:

(a) 270 ft. (b) – 230 ft.

(c) 310 ft. (d) None

(use 1 hPa. = 27 ft.)

17. An aircraft is cruising at 4500 ft with QNH 1023 hPa set on the sub-scale. The pressre altitude of the aircraft is

(a) 4770 ft. (b) 4500 ft.

(c) 4230 ft. (d) None

(use 1 hPa. = 27 ft.)

18. QNH is .... reduced to msl under ICAO ISA conditions

(a) QFF (b) QNT

(c) QFE (d) QNE

19. The vertical distance of an object/a point/a level measured above a specified datum (with known elevation)is called:

(a) Altitude (b) Height

(c) Elevation (d) None

20. The vertical distance of an object /a point / a level measured above mean sea level is called:
    (a) Altitude (b) Height
    (c) Elevation (d) None
21. The vertical distance of an object/a point/a level on or affixed to the surface of the earth, measured from mean sea level is called:
    (a) Altitude (b) Height
    (c) Elevation (d) None above
22. The QFE reduced to mean sea level using station temperature (which is isothermal lapse rate) between station and mean sea level is called:
    (a) QNH (b) QFF
    (c) QNT (d) QNE
23. When 1013 hPa set on the altimeter sub-scale, the main scale will read an altitude value on landing is known as:
    (a) QFE valene (b) QFF value
    (c) QNE value (d) None above
24. Altimeter setting is concerned with the phase of the aircraft flight namely:
    (a) Take off, landing
    (b) Cruising, and terrian clearance
    (c) Vertical separation in controlled space
    (d) All above (a), (b), (c)
25. The lowest flight level above which an aircraft will always fly on standard QNH of 1013.25 hPa is called: (a)
    (a) Transition level (b) Transition altitude
    (c) Transition layer (d) None above
26. The highest altitude below which an aircraft will always fly on local QNH is called:
    (a) Transition level (b) Transition altitude
    (c) Transition layer (d) None above
27. The lowest forecast mean sea level pressure in an area is
    (a) Regional QNH (b) Regional QFE
    (c) QNE (d) None above

28. The ICAO ISA standard QNH is:
    (a) 1000 hPa (b) 1013.25 hPa
    (c) 29.92 in Hg (d) Both (b) and (c)
29. To assess terrain clearance, latest and nearest.........has to be used:
    (a) QFE (b) QNH
    (c) QNT (d) None above
30. The pressure altitude of 900 hPa surface:
    (a) 3051 ft (b) 3000 ft
    (c) 2700 ft (d) None above
31. An aircraft over an airfield elevation 540 ft has an altimeter reading 783 ft with a setting of 1002 hPa. What is the local QNH?
    (a) 1011 hPa (b) 993 hPa
    (c) 1002 hPa (d) None above
32. What is the altimeter reading if the setting is 1005 hPa, QNH 1010 hPa, when the airfield elevation is 240 ft?
    (a) 240 ft (b) 375 ft
    (c) 105 ft (d) Cannot say
33. What is the altimeter reading if the setting is 982 hPa, the QNH 990 hPa and the airfield elevation 560 ft.
    (a) 560 ft (b) 756 ft
    (c) 344 ft (d) Cannot say
34. An aircraft is flying at 2000 ft. indicated with the altimeter sub-scal set to 1018 hPa. Ahead there is a high ground with elevation 1500 ft. If during the flight the QNH in the area falls to 1008 hPa and the altimeter sub-scale is not reset will the aircraft clear the high ground if so by what height? :
    (a) 270 ft (b) 230 ft
    (c) 500 ft (d) Will not clear the high ground
35. The altimeter sub-scale is set to 1020 hPa and the altimeter reads 3500 ft. If the local QNH is 1010 hPa, what is the altitude of the aircraft at that place?
    (a) 3770 ft (b) 3230 ft
    (c) 3500 ft (d) Cannot determine

36. QNH at an aerodrome is 1013 hPa, elevation is 240 m, what is the QFE? (use 1hPa = 8m)

(a) 983 hPa (b) 1043 hPa

(c) 1013 hPa (d) None above

37. The QFE at an aerodrome is 1000 hPa, whose elevation is 216 ft, the QNH is:

(a) 992 hPa (b) 1000 hPa

(c) 1008 hPa (d) None above

38. With a QNH of 1023 hPa, you set the SPS(standard pressure setting) at the transition altitude. Then the indicated altitude:

(a) Drops by 270 ft. (b) Rises by 270 ft.

(c) Does not change (d) Rising by 540 ft.

39. With a QNH of 1006 hPa, you set the SPS at transition altitude. The indicated altitude changes:

(a) Indicated altitude drops by 189 ft.

(b) Rises by 189 ft.

(c) Remains less changed

(d) Indicates 189 ft.

Answers

1. (c)	2. (c)	3. (b)	4. (d)
5. (b)	6. (d)	7. (c)	8. (b)
9. (b)	10. (a)	11. (d)	12. (c)
13. (c)	14. (c)	15. (c)	16. (b)
17. (c)	18. (c)	19. (b)	20. (a)
21. (c)	22. (b)	23. (c)	24. (d)
25. (a)	26. (b)	27. (a)	28. (d)
29. (b)	30. (a)	31. (b)	32. (c)
33. (c)	34. (b)	35. (b)	36. (a)
37. (c)	38. (a)	39. (b)	

CHAPTER 6

Density and Density Altitude

Density (ρ) of matter (solid, liquid or gas) is mass of a unit volume of the substance or the number of molecules in each unit volume.

$$\text{Density} = \rho = \frac{\text{mass}}{\text{volume}}$$

Example, the mass of 1 m^3 volume of water is 1000 kg. The density of water is 1000 kg/m^3

$$\rho_{\text{water}} = \frac{1000\text{kg}}{\text{m}^3}$$

$$\rho_{Hg} = \text{Density of mercury} = 13600 \text{ kg/m}^3$$

The atmospheric density is the mass of air contained in unit volume

The density of some common gases are given in Table 6.1.

Table 6.1

Gas	Density Kg/m^3	Gas	Density Kg/m^3
Air	1.29	Hydrogen	0.09
Nitrogen	1.25	Helium	0.18
Oxygen	1.43	Neon	0.90
Carbondioxide	1.98	Sulphurdioxide	2.93

The performance of an aircraft depends on a number of factors, of which ambian air density is important. It is not necessary for a pilot to have an equipment to measure density for he can use in its place two informations (upon which density depends), namely pressure altitude and temperature to derive density altitude. (Navigation Computers have the facility for calculating density altitude).

6.1 Density Altitude

Definition: Density altitude is the height in ICAO ISA conditions at which a given air density will occur and it is dependent on temperature.

If the local air temperature is same (coincident) as ISA temperature then the density altitude will be same as pressure altitude. If the local temperature is higher (or lower) than ISA, then the density altitude will be higher (or lower) than pressure altitude.

1°C temperature deviation from ISA is equal to 120 ft.

OR

Density altitude may be explained as follows:

The atmospheric air density (at various levels) expressed is terms of altitude in international stranded atmosphere (ISA) which corresponds to that (level) density.

The ideal gas law is pV = RT

or $$p = \rho RT$$

where P = pressure, V= Volume, $\rho = \frac{1}{V}$,

T = Temperature (°K), R = Gas constant.

Under ICAO ISA, at sea level

$P = 1013.25$ hPa, $T = (15 + 273) = 288^{o}$ K,

density of air $\rho = 1.225$ kg/m^3, which gives gas constant $R = 2.872$

The gas constant R for water vapour is 1.6 times that of dry air

ρ of water vapour is less than ρ dry air.

ρ of moist air is less than ρ of dry air.

Density of air decreases with height.

If ρ is taken as 100% at sea level, then

(i) at 6 km altitude (or 20,000 ft altitude) it drops to 50% (0.6125 kg/m^3) ;

(ii) at 12 km (40,000 ft) altitude it drops to 25% (0.30625 kg / m^3);

(iii) at 18 km (60,000 ft) altitude it drops to 10% (0.1225 kg/m^3).

In ICAO air density ρ = 100% at sea level, then densities at different altitudes are as follows:

(i) 56.4% of ρ at 5.5 km

(ii) 26.3% of ρ at 11.78 km

(iii) 15.3% of ρ at 15.24 km

(iv) 7.2% of ρ at 20.0 km and

(v) 1% of ρ at 32.0 km

From ideal gas law [$\rho = p/(RT)$] it is clear that (air density) ρ is directly proportional to atmospheric pressure (p), and inversely proportional to atmospheric temperature (T in degrees Kelvin).

At msl for air, $1.225 = \dfrac{1013.25}{R \times 288}$,

this gives $R = 2.872$

$$\rho = \frac{P}{RT} \qquad ...(6.1)$$

$$1.225 = \frac{1013.25}{R \times 288} \qquad ...(6.2)$$

From Eq. 1 & 2 we have [dividing eq. 6.1 by eq. 6.2]

$$\frac{\rho}{1.225} = \frac{P}{RT} \times \frac{R \times 288}{1013.25}$$

or

$$\rho = \frac{1.225 \times 288}{1013.25} \frac{P}{T}$$

$$\rho = 0.3482 \frac{P}{T} \text{ kg/m}^3$$

(i) water vapour obeys gas law ($P = \rho RT$), but the gas constant (R) for water vapour is 1.6 times that of dry air

(ii) 10 hPa pressure fall causes 1% decrease in air density

(iii) Increase of temperature by 3° C causes 1% decrease in density

(iv) Increase of height by 300 ft causes 1% decrease in air density

(v) On an overage, at sea level air density in equatorial region is 1.2 kg/m^3. This increases to 1.55 kg/m^3 pole wards over Saiberia in winter

(vi) In northern hemisphere summer, in association with low pressure and high temperatures the air density over southwest Asia, Africa falls to 1.15 kg/m^3

In general, at surface level air density increases with increase in latitude (minimum at equator and maximum at poles).

At 7.8 km (2,6000 ft) level, air density remains constant with increasing latitudes (that is air density at equatorial latitudes and poles are equal)

Above 7.8 km (26,000 ft) altitude level, air density decreases with increasing latitudes (that is, density is maximum at the equator and minimum at the poles, which is just reverse of density distribution at surface level), maximum deviation from the standard occurs at about 15 km (50,000 ft) altitude.

If the actual local temperature at any level (under consideration) is equal to the ICAO ISA temperature at that level, then the density altitude will be same as the pressure altitude.

If the actual local temperature is warmer (or cooler) than the ISA temperature at that level, then the density altitude will be higher (or lower) than the pressure altitude, which is to be calculated by the rule stated above (i.e., 120 ft for each 1° C deviation from ISA temperature at that level under consideration).

Buoyancy in air

We know an hydrogen filled balloon rises in the air because the density of hydrogen (0.09 kg/m^3) is approximately 1/14th of the air density (1.29 kg/m^3).

The total weight of the balloon $\alpha = V \times \rho_H\, g + m\, g$

where

V = volume of the balloon,

m = mass of balloon

ρ_H = density of hydrogen,

g = gravity;

while the weight of air displaced by balloon

$$\beta = V \times \rho_A\, g,$$

where ρ_A = density of air

$\beta > \alpha$, and by Archimedes principle the balloon rises.

The average volume of human being is about 0.06 m^3, density of air at msl is 1.225 kg/m^3.

The mass of volume of air displaced by human body

$$= 0.06 \times 1.225 \text{ Kg}$$

$$= 0.07350 \text{ Kg}$$

The average upthrust experienced by an human being

= weight of mass of air displaced by human being

$$= 0.0735 \times 9.8 \text{ N}$$

$$= 0.72 \text{ N}$$

Thus on an average, human being experiences an upthrust of about 0.72 N while standing on the earth due to atmospheric air.

We have learnt that in ICAO ISA, the msl temperature is +15 °C, pressure 1013.25 hPa, temperature lapse rate 1.98 °C/1000 ft or 2 °C/ 1000 ft approximately.

Under ISA conditions the temperature:

(i) at 5000 ft pressure altitude will be

$$15 - \frac{2}{1000} \times 5000 = +5\ ^\circ\text{C}$$

(ii) At 7000 ft pressure altitude will be

$$15 - \frac{2}{1000} \times 7000 = +1\ ^{\circ}\text{C}$$

(iii) At 10000 ft pressure altitude will be

$$15 - \frac{2}{1000} \times 10000 = -5\ ^{\circ}\text{C}$$

(iv) At 18000 ft pressure altitude will be

$$15 - \frac{2}{1000} \times 18000 = -21\ ^{\circ}\text{C}$$

Now, suppose the actual temperatures by radiosonde observation on a particular day at pressure altitude 5000 ft, 7000 ft, 10000 ft and 180000 ft were found to be + 7° C, – 2° C, – 9 °C and – 18° C respectively

The temperature deviation from ISA

(i) at pressure altitude 5000 ft is + 2 °C
(ii) at 7000 ft is – 3 °C
(iii) at 1000 ft is – 4 °C and
(iv) at 1800 ft is + 3 °C.

This means

(i) at 5000 ft pressure altitude the ISA temperature deviation is + 2 ° C.

(ii) at 7000 ft pressure altitude, the ISA temperature deviation is – 3 ° C

(iii) at 1000 ft pressure altitude the ISA temperature deviation is – 4 ° C, and

(iv) at 18000 ft pressure altitude, the ISA temperature deviation is + 3 °C.

The performance of aeroplane and engine depend on the air density. However pilot need not calculate the density. Pilot will have the information in his cockpit about pressure altitude and temperature. By this information he can assess density altitude (or density effect) indirectly, by using the approximate rule: 120 ft for each 1°C deviation from ISA temperature.

If the deviation is (x °C) positive (i.e., warmer than ISA) the density altitude will be higher than the pressure altitude by an amount $x \times 120$ ft.

If the deviation is (x °C) negative (i.e., cooler than ISA) then the density altitude will be lower than pressure altitude by an amount $x \times 120$ ft.

If the deviation is zero (i.e., actual temperature is equal to ISA temperature) then the density altitude will be equal to the pressure altitude.

In the above example the density altitude corresponding to pressure altitude of 5000 ft is

$$2 \times 120 + 5000 = 5240 \text{ ft (temperature deviation being } + 2\,^{\circ}\text{C)}$$

The density altitude corresponding to 7000 ft pressure altitude is

$$7000 - 3 \times 120 = 6640 \text{ ft (temperature deviation being } - 3\,^{\circ}\text{C)}$$

The density altitude corresponding to 10000 ft pressure attitude is

$$10000 - 4 \times 120 = 9520 \text{ ft (temperature deviation being } - 4^{\circ}\text{C)}$$

The density altitude corresponding to 18000 ft pressure altitude is

$$18000 + 3 \times 120 = 18360 \text{ ft (temperature deviation being } + 3^{\circ}\text{C).}$$

Example: on particular Sunday, Wednesday & Friday the radiosonde observation at Begumpet (India) recorded temperature at 10000 ft (FL 100) pressure altitude were – 5 °C, 0 °C and – 10 °C. What were the density altitude on those days?

Solution: the ICAO ISA temperature at 10000 ft pressure altitude is

$$15 - \frac{2}{1000} \times 10000 = -\ 5^{\circ}\text{C}$$

1. On Sunday the actual temperature at 10000 ft pressure altitude was – 5°C. That is the ISA temperature derivation is zero.

   ∴ The density altitude = pressure altitude = 10000 ft

2. On Wednesday the actual temperature was 0 °C. That is ISA temperature deviation was + 5°C

   ∴ The density altitude was 10000 + 5 × 120 = 10600 ft

3. on Friday the actual temperature was – 10 °C at 10,000 ft pressure altitude, while 1SA temperature was –5 °C. That is actual temperature was cooler than 1SA temperature by 5 °C (1SA temperature deviation was –5° C).

   ∴ The density altitude was 10000 – 5 × 120 = 9400 ft

Example: Calculate the density altitude at an aerodrome whose elevation is 900 ft if the current QNH is 996 hPa and the outside temperature (OAT) is +20 °C.

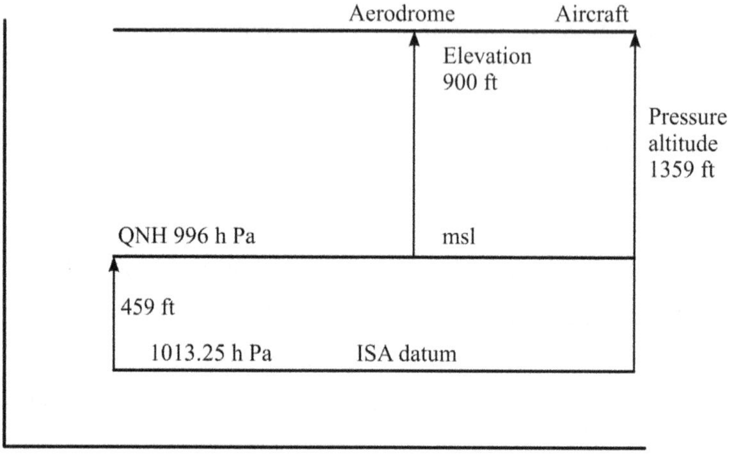

Solution: pressure deviation between QNH & 1SA datum

$$\simeq 1013 - 996 = 17 \text{ hPa}$$

This pressure deviation corresponds to height difference of

$$17 \times 27 = 459 \text{ ft}$$

∴ The pressure altitude of aircraft on aerodrome

$$= 900 + 459 = 1359 \text{ ft}$$

1SA temperature at this pressure altitude (1359 ft) is

$$= 15 - 1359 \times \frac{2}{1000}$$

$$\simeq 15 - 2.718$$

$$\simeq 12^{\circ}\text{ C}$$

Given actual temperature (OAT) at aerodrome elevation 20 ° C

∴ The 1SA temperature deviation (20 = ISA +8) is + 8 °C

1 °C ISA deviation corresponds to 120 ft

∴ The density altitude deviation from pressure altitude is

$$120 \times 8 = 960 \text{ ft}$$

Since actual temperature is warmer than 1SA by 8 ° C implies the density altitude is higher than pressure altitude by 960 ft.

∴ The density altitude of the aircraft is 1359 + 960 = 2319 ft

Note: Although the aircraft is at less than 1000 ft (900 ft) amsl, the engine will perform and the aircraft will fly as though it were at about 2300 ft (2319 ft) in the ISA.

That is, it will have very poorer performance.

Example: Calculate the density altitude of an aircraft on aerodrome whose elevation is 20 ft, current QNH 1033 hPa and outside air temperature (OAT) – 5 °C.

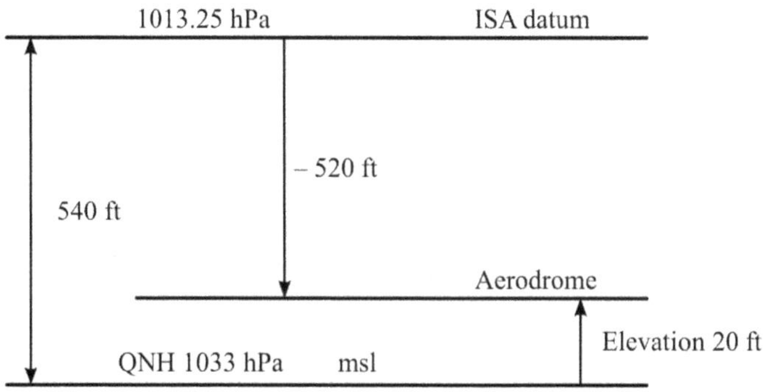

Solution: ISA (pressure) datum is clearly above msl.

Pressure deviation of QNH and ISA datum is 1033 – 1013 = 20 hPa.

This corresponds to a height difference of 20 × 27 = 540 ft.

The pressure altitude of aircraft on aerodrome is

= 540 – 20 = 520 ft below ISA datum i.e., – 520 ft

ISA temperature at – 520 ft would be

$$15 + 520 \times \frac{2}{1000} = 15 + 1.040$$

$$\simeq 16\ ^{\circ}\text{C}$$

(***Note:*** Temperature rises at the rate of 2 °C/1000 ft for levels below ISA datum)

Actual temperature (OAT) given – 5 °C

∴ ISA temperature deviation [–5 = 16 – 21] is – 21 °C

This temperature deviation corresponds to height (pressure altitude) deviation of 21 × 120 = 2520 ft

Since OAT is cooler than ISA, the density altitude would be less than pressure altitude by 2520 ft

The pressure altitude of aircraft on aerodrome is – 520 ft (calculated above)

$\therefore$ The density altitude will be $-520 - 2520$ ft $= -3040$ ft

i.e., 3040 ft below ISA pressure datum

Note: High QNH means high density, low temperature means high density. Temperature is a critical factor to watch out. In this example low aerodrome elevation and low temperature combined means a better performance of engine and aeroplane.

6.2 Aircraft Performance

Lift of an aircraft depends on:

(i) Wing shape of the aircraft
(ii) Angle of attack
(iii) Air density(ρ)
(iv) Wind velocity or Velocity of free stream,
(v) Wing surface area (s)

$$\text{Mathematically, lift} = C_L \frac{1}{2} \rho V^2 s$$

where C_L = Coefficient of lift.

C_L increases proportionally with angle of attack.

Lower air density (or increased density altitude) means longer take off distance. A high aerodrome elevation decreases aeroplane performance (since density decreases). Lower air pressure (associated with higher aerodrome elevation) means lower density. High-elevation aerodromes require (longer runway) longer take off distance.

In takeoff phase: About a 1000 ft increase in pressure altitude require an increased takeoff distance by 10 % (i.e., a factor of 1.1)

Also, 10° C increase in ambient temperature require an increased takeoff distance by about 10 % (i.e., a factor of 1.1)

In landing phase: An increase in pressure altitude of about 1000 ft or an increase in 10 °C require approximately an increase in landing distance by 5 %.

The thurst efficiency of an aircraft (particularly jet aircraft flight) is affected by density.

Cold air is denser than warm air. Flying from cold area into warm area altimeter under reads (the actual height).

Flying from warm area into cold area altimeter over reads (against the actual height).

Thumb Rule 6.1:Thus we have (i) C $\rightarrow$ W altimeter under reads

(ii) W $\rightarrow$ C altimeter over reads.

Dry air (consists of mainly nitrogen, oxygen molecules) is heavier than moist air because it includes water vapour molecules replacing some nitrogen, oxygen molecules. Moist air/saturated air is lighter than the dry air at the same temperature and pressure. Consequently on a dry/sunny day the atmosphere is denser as compared to foggy/misty day at the same temperature and pressure.

***Avogadro's Law*:** It states that equal volumes of ideal gases at the same temperature and pressure contain equal number of molecules [or the same number of molecules] i.e the number of molecules of any gas present in a volume is the same. In other words, moles of different ideal gases occupy the same volume at the same temperature and pressure.

***By virtue of density distribution* (aircraft performance):**

Low cruising (piston engine) aircraft performance operational efficiency, at any level upto FL 200-250, is greater in higher latitudes as compared to the equatorial/tropical region. In case of high altitude cruising aircraft (Jet aircrafts) performance operational efficiency at any level above FL 300 is greater in tropics /equatorial region as compared to the polar region/high latitudes.

In ICAO ISA the relation of pressure altitudes and corresponding to pressure levels, temperature and air densities are given in Table 6.2.

Table 6.2

Pressure altitude ft	Pressure level hPa	Temperature °C	Relative air density %
105,000	8.9	– 44.7	1.1
100,000	11.1	– 46.2	1.4
95,000	13.9	– 47.7	1.8
85,000	22.2	– 50.7	2.8
80,000	28.0	– 52.2	3.6

Table 6.2 Contd....

Pressure altitude ft	Pressure level hPa	Temperature °C	Relative air density %
75,000	35.4	– 53.7	4.5
70,000	44.9	– 55.2	5.8
65,000	56.4	– 56.5	7.4
60,000	72.3	– 56.5	9.5
55,000	91.8	– 56.5	12.0
50,000	116.6	– 56.5	15.3
45,000	148.2	– 56.5	19.5
40,000	188.2	– 56.5	24.7
35,000	239.1	– 54.2	31.0
30,000	301.5	– 44.3	37.5
25,000	376.5	– 34.5	44.8
20,000	466.0	– 24.6	53.3
15,000	572.0	– 14.7	62.9
10,000	696.9	– 4.8	73.8
5,000	843.1	+ 5.1	86.2
2,000	942.1	+ 11.0	94.3
0 (msl)	1013.25	+ 15.0	100.0
– 1000	1050.4	+ 17.0	103.0

The Table 6.3 gives the ICAO ISA pressure altitudes in kilometer and corresponding pressure levels, temperature and densities.

Table 6.3

Height/ pressure altitude in km	Pressure level hPa	Temperature °C	Relative air density %
32	8.9	– 44.7	1.1
30	11.9	– 46.6	1.5
27.5	17.4	– 49.1	2.2
25	25.5	– 51.6	3.3
22.5	37.5	– 54.1	4.9
20	55.3	– 56.5	7.2
17.5	81.8	– 56.5	10.7
15	121.1	– 56.5	15.8
12.5	179.3	– 56.5	23.5

Table 6.3 Contd....

Height/ pressure altitude in km	Pressure level hPa	Temperature °C	Relative air density %
10	264.9	– 49.9	33.7
7.5	382.9	– 37.7	44.5
5	540.4	– 17.5	60.1
2.5	746.9	– 1.2	78.1
1	898.7	+ 8.5	90.7
0.5	954.6	+ 11.7	95.3
0 (msl)	1013.25	+ 15.0	100.0
– 0.5	1074.8	+ 18.2	104.9

In aviation metrological offices the following contour charts are prepared for the cruising levels (flight levels)

Table 6.4

Pressure level hPa	FL	Approximate mean height ft
1013	0 (sea level)	0
850	050	5000
700	100	10000
500	180	18000
400	240	24000
300	300	30000
250	340	34000
200	380	38000
100	540	54000

Altimeters are calibrated for ICAO ISA conditions only. From the above table A, 5000 ft pressure altitude correspond to 843.1 hPa pressure level, with temperature + 5.1°C and density 86.2 $\rho/100$ (ρ= 1.225 kg/m^3); the thumb rule Flying into warmer than ISA, the real height (h_1) will be higher than altimeter indicted height (h), i.e., the altimeter under reads. On the contrary if an aircraft flying into the region colder than ISA, the real height (h_0) will be lower than the altimeter indicated height (h), i.e., the altimeter over-reads. referred is shown pictorially in Fig. 6.1.

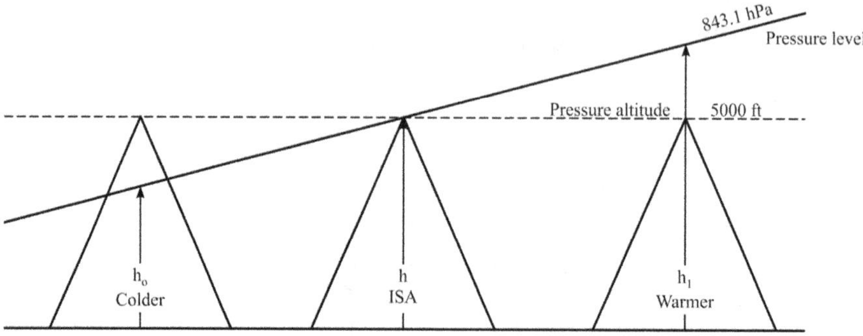

Fig. 6.1

When airfield elevation and temperature are related to ISA conditions have the same sign, QNH will be higher than QFF but when they are of different signs the QNH will be lower. These relations are shown in the thumb rule Table 6.5.

Table 6.5 Thumb rule.

Airfield elevation Stations	**Temperature of station compared to ISA**	**Signs**	**QNH Compared to QFF**
Above msl +	Above ISA +	Same signs (+, +)	QNH > QFF
Below msl –	Below ISA –	Same signs (–, –)	QNH > QFF
Above msl +	Below ISA –	Different signs (+, –)	QNH < QFF
Below msl –	Above ISA +	Different signs (–, +)	QNH < QFF

***Example* 1:** You are flying from Marseilles (QNH1010 hPa) to Palma de Mallorca (QNH 1010 hPa) at FL 100. You notice that the effective height above msl (Radio Altitude) increases constantly. Why?

Solution: Local QNH at both places being same 1010 hPa, but the effective height Marseilles to Palma increasing constantly indicates the air mass above Palma is warmer than over Marseilles.

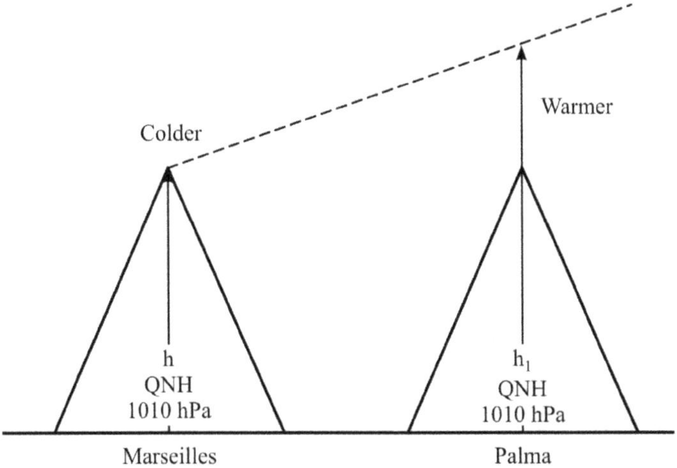

***Example* 2:** You are flying from Marseilles (QNH 1014 hPa to Palma de Mallorca (QNH 1014 hPa) at FL 100. You notice that the effective height above msl (Radio Altitude) decreases constantly. What might be the reason?

Solution: Local QNH at both places being same 1014 hPa but the effective height Marseilles to Palma de Mallorca decreasing constantly can happen when the air mass above Marceilles is warmer than that above Palma.

***Example* 3:** Flying at FL 140 above the sea, the Radio Altimeter indicates a true (height) altitude of 14000 ft. The local QNH is 1018 hPa. What does it indicate?

Solution: Flying at FL 140 above sea level the Radio Altimeter indicated altitude of 14000 ft. This implies the QNH 1013 hPa. Since the local QNH 1018 hPa and true altitude 14000 ft indicates that the crossed air mass on average colder than ISA.

***Example* 4:** What would cause your true altitude to decrease with constant indicated altitude?

Solution: Your true altitude decrease when fly into cold / low area with constant indicated altitude.

***Example* 5:** The QNH at an airfield 150 m above msl is 1010 hPa and air temperature is (ISA – 10)° C. what is the QFF?

Solution: Using the airfield elevation and temperature signs thumb rule we have + –, the product will be –, indicating QNH < QFF.

Hence QFF >1010 hPa.

***Example* 6:** An airfield in Holland is 30 m below sea level, QFF is 1018, hPa, temperature at the surfaces is + 28 °C. What is the value of QNH?

Solution: In ISA, temperature increase 0.65 °C/100 m below sea level.

∴ ISA temperature at aerodrome will be

$$15 + 0.65 \times \frac{30}{100} = 15 + 0.195 \simeq 15.2\ ^\circ\text{C}$$

Actual temperature given + 28 °C.

Using the aircraft elevation and temperature signs thumb rule –, +, product –ve indicate QNH < QFF

i.e. QNH < 1018 hPa

***Example* 7:** The QNH at an airfield 220 m above msl is 1010 hPa, air temperature is (ISA – 8 °)C. What is the QFF?

Solution: Using the airfield elevation and temperate signs thumb Rule, we have +, – , their product is – ve. Consequently QNH < QFF

∴ QFF is greater than 1010 hPa.

***Example* 8:** You are flying in atmosphere which is warmer than ISA; what might you expect?

Solution: Flying into warmer atmosphere, means low density. Consequently true altitude would be higher than the indicated altitude.

***Example* 9:** In Holland, an air field is 70 meter below msl, QFF 1025 hPa, temperature 8 °C below ISA (ISA – 8 °C). What is the QNH?

Solution: Using the thumb rule 6.2, elevation, temperature signs are

(–, –, product is +)

∴ QNH > QFF

i.e., QNH > 1025 hPa.

***Example* 10:** You are flying from Madrid (QNH 1009 hPa) to Paris (QNH 1013 hPa) at FL 100. What might be the reason if your true altitude and indicated altitude remain the same?

Solution: Since you are flying from low QNH to high QNH and still your true altitude and indicated altitude remain same means the air mass over Madrid is warmer than over Paris

Madrid	Pairs
QNH 1009 hPa	QNH 1013 hPa

***Example* 11:** An aerodrome located at msl has QNH 1018 hPa. What is QFF?

Solution: Since there is no change in pressure and temperature, the

QNH = QFF i.e., QFF = 1018 hPa.

***Example* 12:** If you did not reset your altimeter while flying from Rome (QNH 1012 hPa) to Glasgow (QNH 1020 hPa) at FL 210, why would your altitude remain constant (same) during the flight?

Solution: Though QNH increased from Rome to Glasgow but true altitude remained constant indicates warmer air mass over Rome than over Glasgow.

***Example* 13:** When flying from P (QNH 1010 hPa) to Q (QNH 1015 hPa), true altitude remains constant. Why?

Solution: As you are flying from P to Q, QNH is increases but true altitude remains constant, which implies the air mass over P is warmer than at Q.

***Example* 14:** Flying into an atmosphere warmer than ISA, what would you expect to occur?

Solution: Flying to an atmosphere warmer than ISA, the true altitude (h_2) to be higher than indicated altitude (h_1). (see fig below)

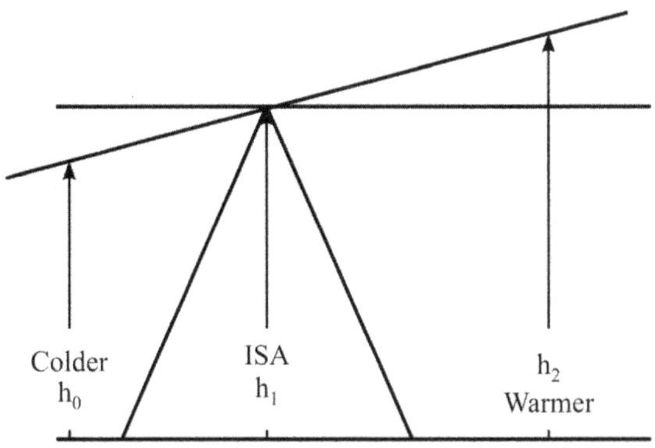

***Note*:** The density of air is important to a pilot for three reasons.

(i) The lift force generated by air molecules moving (flowing/circulating) around the wings supports the weight of the aeroplane.

(ii) Engine power is generated by burning of fuel and air molecules

(iii) For breathing air (as we go up oxygen content decreases with decrease of air density).

Questions

1. If the local air temperature is same as ISA temperature, then:
   (a) Density altitude is greater than pressure altitude
   (b) Density altitude will be same as pressure altitude
   (c) Density altitude is less than pressure altitude
   (d) None above
2. Density of water vapour is:
   (a) Greater than the density of dry air
   (b) Equal to the density of dry air
   (c) Less than the density of dry air
   (d) None above
3. In general, density of air drops to 50% as compared to the msl air density, at an altitude about:
   (a) 6 km (b) 12 km
   (c) 18 km (d) 30 km
4. In general, density of air drops to 25% as compared to the msl air density, at an altitude of about:
   (a) 6 km (b) 12 km
   (c) 18 km (d) 30 km
5. In general, density of air drops to 10% as compared to the msl air density, at an altitude of about:
   (a) 6 km (b) 12 km
   (c) 18 km (d) 30 km
6. Flying into warmer than ISA temperature, the real height will be:
   (a) Same as altimeter indicated height
   (b) Higher than altimeter indicated height
   (c) Lower than altimeter indicated height
   (d) None above

7. Flying into colder than ISA temperature, the real height will be:
   (a) Same as altimeter indicated height
   (b) Higher than altimeter indicated height
   (c) Lower than altimeter indicated height
   (d) We cannot say
8. Flying from warm area into cold area, the altimeter:
   (a) Does not alter (b) Over reads
   (c) Under reads (d) Cannot say
9. Flying from cold area into warm area, the altimeter:
   (a) Does not alter (b) Over reads
   (c) Under reads (d) Cannot say
10. Low cruising aircraft performance operational efficiency, below FL 250 is:
   (a) Lower in high latitudes compared to equatorial region
   (b) Greater in high latitudes compared to equatorial regions
   (c) Same in low & high latitudes
   (d) Can not say
11. High altitude cruising aircraft performance operational efficiency above FL 340 is:
   (a) Grater in tropics as compared to high latitudes
   (b) Lower in tropics as compared to high latitudes
   (c) Same in low and high latitudes
   (d) Cannot say
12. You are flying in winter from Bhopal (QNH 1010) to Srinagar (QNH 1010) at FL 140.You notice that the effective height amsl. (Radio altitude) decreases constantly indicating:
   (a) Air mass above Srinagar is cooler than over Bhopal
   (b) Air mass above Srinagar is warmer than over Bhopal
   (c) Same air mass over Bhopal and Srinagar
   (d) Faulty radio altimeter
13. An air field in Holland is 45 ft below sea level, QFF is 1016 hPa, temperature is (ISA+6) °C. The QNH is:
   (a) 1016 hPa (b) Less than 1016 hPa
   (c) More than 1016 hPa (d) Cannot say

14. An air field is 120 ft a msl, QNH is 1012 hPa and air temperature is (ISA+5) °C. Then QFF is:
    (a) 1012 hPa (b) More than 1012 hPa
    (c) Less than 1012 hPa (d) Cannot say
15. Air density is a function of temperature and pressure. A 10 hPa pressure fall causes (in vertical movement):
    (a) 1% decrease in density (b) 1% increase in density
    (c) 10% decrease in density (d) No change in density
16. Air density is a function of temperature and pressure. An increase of temperature by 3 °C causes:
    (a) Increase of density by 1%
    (b) Decrease of density by 1%
    (c) No change in density
    (d) None above
17. Air density is a function of height. An increase of height by 300 ft causes:
    (a) 1% increase in density
    (b) 3% increase in density
    (c) 1% decrease in density
    (d) 3% decrease in density
18. On an average, at sea level, air density in equatorial region (summer) and polar region (winter) varies:
    (a) 1.55 kg/m^3 to 1.2 kg/m^3
    (b) 1.2 kg/m^3 to 1.55 kg/m^3
    (c) 1.225 kg/m^3 to 1.225 kg/m^3
    (d) None above
19. In general, at surface level, air density changes from equator to pole:
    (a) Increases with increase in latitudes
    (b) Decreases with increase in latitudes
    (c) Increases up to lat 45^0 and then decreases
    (d) Decreases up to lat 45^0 and then increases

20. At about 7.8 km (26000 ft) altitude level air density varies from equator to pole:
    (a) Increases with increasing latitude
    (b) Remains constant
    (c) Decreases with increasing latitudes
    (d) None above
21. At about 15 km (50,000 ft) altitude level air density changes from equator to pole:
    (a) Increases with increasing latitude
    (b) Remains constant
    (c) Decreases with increasing latitudes
    (d) None above
22. If the actual local temperature at any level (under consideration) is equal to the ISA temperature at that level, then
    (a) Density altitude is greater than pressure altitude
    (b) Density altitude is smaller than pressure altitude
    (c) Density altitude is equal to pressure altitude
    (d) None above.
23. If the actual local temperature at any level is warmer (or cooler) than the ISA temperature at that level then:
    (a) The density altitude will be lower (or higher) than the pressure altitude
    (b) Density altitude will be higher (or lower) than the pressure altitude
    (c) Density altitude equal to pressure altitude
    (d) None above
24. Density altitude differs from pressure altitude for each 1 °C deviation from ISA temperature by:
    (a) 100 ft (b) 120 ft (c) 30 ft (d) 30 m
25. If at 5000 ft pressure altitude the actual temperature is (ISA + 2)°C, then the density altitude is:
    (a) 5000 ft (b) 5240 ft
    (c) 4760 ft (d) None above

26. If at 10000 ft pressure altitude the actual temperature is (ISA– 3) °C, then the density altitude is:
    (a) 10000 ft (b) 9640 ft
    (c) 10360 ft (d) None

27. If at 7000 ft pressure altitude, the actual temperature is ISA °C, then the density altitude is:
    (a) 7000 ft (b) 7120 ft
    (c) 6880 ft (d) None

28. If at 10000 ft pressure altitude the actual temperature is – 6 °C, then the density altitude is:
    (a) 10,000 ft (b) 9880 ft
    (c) 10120 ft (d) None

    (First find the ISA temperature deviation, calculating ISA temperature, which is – 5 °C)

29. In aviation meteorological contour charts, the flight level 100 corresponds to pressure level (hPa / height ft):
    (a) 850 hPa / 5000 ft (b) 700 hPa / 10000 ft
    (c) 100 hPa / 54000 ft (d) None above

30. In aviation meteorological contour charts, the flight levels 240, 300 correspond to pressure levels:
    (a) 24000 hPa / 30,000 hPa (b) 400 hPa / 300 hPa
    (c) 240 hPa / 300 hPa (d) None

31. When flying at FL 300 from Begumpet (QNH 1002 hPa) to New Delhi (QNH 1006 hPa) true altitude remaining constant indicate:
    (a) Air mass over Begumpet is cooler than over Delhi
    (b) Air mass over Begumpet is warmer than over Delhi
    (c) Same air mass over Begumpet and Delhi
    (d) Cannot say

Answers

1. (b)	2. (c)	3. (a)	4. (b)
5. (c)	6. (b)	7. (c)	8. (b)

9.	(c)	10.	(b)	11.	(a)	12.	(a)
13.	(b)	14.	(c)	15.	(a)	16.	(b)
17.	(c)	18.	(b)	19.	(a)	20.	(b)
21.	(c)	22.	(c)	23.	(b)	24.	(b)
25.	(b)	26.	(b)	27.	(a)	28.	(b)
29.	(b)	30.	(b)	31.	(b)		

CHAPTER 7

Water in the Atmosphere

Water is present in the atmosphere in varying degrees in all levels in troposphere. Beyond troposphere it is found in very minute quantities extending up to mesosphere. Water enters into the atmosphere by the processes of evaporation and transpiration. Water is generally found in the atmosphere in the form of water vapour. Water cycle over land and atmosphere is called hydrological cycle, which has three important stages – (i) evaporation (ii) condensation and (iii) precipitation. The annual average depth of water evaporation over the world is 1030 mm and the annual average depth of world over precipitation is also 1030 mm. It means the amount of water that enters into the atmosphere is wholly received back to the earth through precipitation. The average storage of water in the atmosphere is only about 2.5 mm but exchanges between oceans and land is large. The annual volume of evaporation over the world is about 525.1×10^{12} m^3 or 525.1×10^{12} metric tons of water. At any instant the amount of water vapour in the atmosphere (world over) is about 13000 km^3 and the residential time of water vapour in the atmosphere is about 9 days.

Near the surface of the earth water vapour can be found up to 4 % of the atmosphere by volume (about 3 % by weight), but at about 11 km altitude it is practically absent. As said earlier water vapour enters the atmosphere by evaporation and transpiration.

Vapour pressure: The pressure of a vapour (which is in equilibrium with liquid or solid) at any temperature is called the vapour pressure of the substance at that particular temperature. The vapour pressure is a function of temperature only and not of volume.

Atmospheric air is a mixture of gases (mainly nitrogen 78%, oxygen 21% and other gases together 1%). Each of which exerts its own pressure called partial pressure. The total pressure exerted by the atmospheric gases and particles is the sum of the pressures exerted by individual components (called partial pressures of these components), that is, each of the constituent gases of mixture behaves independently of others. The partial pressure of each of the component very nearly the same as if each component occupied the same volume as the mixture of gases. This is known as Dalton's law of partial pressures.

Dalton's empirical law of partial pressures: It states that the total pressure exerted by a mixture of gases is equal to the sum of their partial pressures, which would be exerted by each constituent if it alone filled the entire volume at that temperature of the mixture.

The partial pressure of water vapour in the atmosphere is of the order 0.005 to 0.05×10^5 Pa, or 5 to 50 hPa. The vapour pressure of water at different temperatures is given in Table 7.1.

Table 7.1 Saturation vapour pressure of water

Temperature ^{0}C	−60	−40	−30	−20	−10	0	5	10	15	20	30	40	60	100	200
Vapour pressure hPa	–	0.19	0.51	1.25	2.86	6.11	8.68	12.27	16.9	23.37	42.43	73.77	199.0	1013	15500

Air consisting of dry air and water vapour is called moist air.

Close to the surface of the earth large quantities of water vapour (vapour pressure about 30 hPa) is observed in tropics while small quantities observed in polar region, particularly over Antarctic region during winter. In vertical direction, water vapour in general decreases with increasing altitude (but on some occasions it may be seen increasing also in some pockets).

Water exists in nature in three states. Solid (ice), liquid (water) and gas (water vapour) and they may change one state to another either by absorption of heat or releasing heat (for change of state either heat is required or heat is liberated).

(i) The phase change (solid to liquid) ice to water is called thawing or melting. This process requires heat.

(ii) The phase change (liquid to gas) water to water vapour is called evaporation. This process requires heat.

(iii) The phase change (solid to gas) ice to vapour is called sublimation. This process requires heat.

(iv) The phase change (liquid to solid) water to ice is called freezing. In this process heat is released.

(v) The phase change (gas to liquid) vapour to water is called condensation. In the process heat is released.

(vi) The phase change (gas to solid) vapour to ice is called deposition or sublimation. In this process heat is released.

The energy involved in change of state from ice to liquid or liquid to ice is called latent heat of fusion. This involves 80 calories of heat per gram.

The change of state liquid to vapour or vapour to liquid involves 600 cal/gram at 0 °C. The latent heat of vaporization to evaporate 1 gm of water at 0 °C is 600 calories and at 100 °C it is 540 calories.

The change of state solid to vapour (sublimation) or vapour to solid (frost or deposition) involves 680 cal/ gm.

The process of solid to liquid and liquid to vapour formation, absorbs latent heat, whereas vapour to liquid and liquid to solid transformation the process releases heat. The process involved in water changes from one state to another is summarized in Table 7.2.

Table 7.2

Initial state or first state	At final state	Process
Ice	Water	Melting
Ice	Water vapour	Sublimation
Water	Water vapour	Evaporation
Water	Ice	Freezing
Water vapour	Water	Condensation
Water vapour	Ice	Deposition

Distinction between vapour and gas is shown in Table 7.3.

Table 7.3

	Vapour	Gas
1	The matter that exists below its critical temperature is called vapour	The matter that exists above its critical temperature is called a gas.
2	Saturated vapours do not obey Boyles law and Charle's law	Gases obey Boyle's law and Charles' law
3	A vapour can be liquefied by simply decreasing temperature alone or by increasing the pressure alone	A gas cannot liquefied by simply decreasing the temperature or simply increasing the pressure
4.	Vapours obey Dalton's law of partial pressures	Gases obey Dalton's law of partial pressures

Water Vapour pressure (e): When water (H_2O) is the only gas present, the vapour pressure is the pressure of this gas only.

When vapour (of water) is mixed with air (which is called moist air) the pressure exerted by the water molecules (H_2O) is called the vapour pressure (e). The maximum limit of vapour (i.e. amount of vapour) that can exist in gaseous state is called saturation vapour (e_s). e_s is a function of temperature only, because the water holding capacity of air depends on temperature.

Pressure of the atmosphere (p)

= pressure of atmosphere constituents
+ partial pressure of water vapour (e).

Partial pressure of dry air = p – e

Saturation vapour pressure is higher in tropical water bodies (lakes, rivers, seas/ oceans). Saturation vapour pressure increases with increase of temperature (since the water holding capacity increases with temperature), which is given in the Table 7.4

Table 7.4 Saturation vapour pressure over plane surfaces of water and ice

Saturation vapour pressure (hPa) at temperature (°C) of														
	– 60	**– 50**	**– 40**	**– 30**	**– 20**	**– 10**	**0**	**10**	**20**	**30**	**40**	**60**	**100**	**200**
Over ice	0.01	0.04	0.13	0.38	1.03	2.60	6.11	-	-	-	-	-	-	-
Over water	-	0.06	0.19	0.51	1.25	2.86	6.11	12.3	23.4	42.3	73.8	199.0	1013	15500

A second explanation of saturation vapour pressure is given below:

e_s (Saturation Vapour Pressure): Consider a plane surface of water at a given temperature. The water molecules of highest velocity escape from the free surface of water into the air above. A few of them return back to the water while others remain as gas molecules in the air above the water surface. A stage will come when the number of molecules escaping the water surface per second is equal to the number of molecules returning to the water. In this state the space above water is called saturated at that prevailing temperature. Under certain conditions more water vapour can be held in the atmosphere than saturation limit without condensation is called super saturation.

Note 1:Vapour pressure is partial pressure exerted by the water vapour in the atmosphere, usually expressed in Pascal or hecto-Pascal.

Note2: Saturated vapours do not obey Boyle's law and Charle's law, but obey Dalton's law of partial pressures.

Rate of evaporation: The process of vapour formation at the free surface of water is called evaporation. The rate of evaporation μ depends on the atmospheric pressure (P_o), velocities of the gaseous phase above free surface of water. It is given by the relation

$$\mu = \frac{KA(e_s - e)}{P_0}$$

where K = constant, A = Area of the surface of water, e = Pressure of vapour above free surface water, e_s = saturation vapour pressure.

Transpiration: Plants withdraw water from the soil through its roots, which moves upward to leaves. Through small openings in leaves some water escapes to the atmosphere. This loss of water from the leaves and stems of vegetation is called transpiration. The combined effect of moisture loss through evaporation and transpiration from a given surface area is called evapo-transpiration.

Density of moist air: We have studied that the density of dry air at msl under ICAO ISA conditions is 1.225 kg/m^3. The mean molecular weight of dry air is 28.96, while the molecular weight of water vapour is 18, which is 5/8th of mean molecular weight of dry air. This means water vapour molecule is lighter than dry air, consequently density of moist air is less than the density of dry air.

7.1 State Variables

The state of the atmosphere is determined by a number of parameters called state variables. The atmospheric state is said to steady or static or in equilibrium if it does not change with time. However, in general, atmosphere is dynamic (that is in motion). The principal state variables of the atmosphere are pressure (p), temperature (T), specific volume (α) or molar volume and moisture (or humidity). The most of the other atmospheric parameters are complex combination of these.

Ideal gas – Def: An ideal gas (perfect gas) is one in which there are no forces of molecular interaction.

In meteorology atmospheric gases are assumed to obey ideal gas law, namely

$$p\alpha = RT \qquad \text{or} \qquad p = \rho RT$$

Ideal gas obeys the following four laws:

(i) ***Boyle's law:*** Temperature remaining constant (isothermal process), the volume (V) of a given mass of gas is inversely proportional to the pressure (p),

$$V \propto \frac{1}{p}, \text{ when T is constant}$$

(ii) ***Charles's law:*** Pressure remaining constant (isobaric process), the volume of a mass of gas is directly proportional to the absolute (Kelvin) temperature (T),

$$V \propto T, \text{ when p is constant}$$

(iii) ***Gay-lussac's law:*** Volume remaining constant (isochoric process), the pressure of a given mass of gas is directly proportional to its absolute (Kelvin's) temperature,

$$P \propto T, \text{ when V is constant}$$

(iv) ***Avogadro's law:*** Equal volumes of different gases at the same temperature and pressure contain equal number of molecules irrespective of their physical properties and chemical nature. Moles of different gases occupy the same volume at the same temperature and pressure.

Moles of all ideal gases at STP (standard temperature and pressure) (T = 0° C or 273° K, p = one atmosphere or 760 mm of Hg) occupy 22.414 liters of volume or 1CC (cubic centimeter)of any ideal gas at STP contain 2.69×10^{19} molecules.

At STP, the mass of 22.414 liters of any ideal gas is equal to its molecular weight, expressed in grams.

Thus, the mass of 22.414 liters of oxygen (O_2), nitrogen (N_2) and hydrogen (H) at STP weighs 32 gm, 28 gm and 2 gm respectively.

7.2 Types of Atmospheric Physical Processes

(i) ***The Isobaric Process:*** A physical process which proceeds, where pressure remains constant, is called isobaric process.

(ii) ***The Adiabatic Process:*** A physical process in which the heat content remains constant (i.e. Heat is neither added nor removed), the process is called adiabatic.

If a moist air sample vertically moved up (or pushed down) adiabatically, it expands (or contracts) and cools (or warms). This results in cloud formation (dissipation) and condensation (evaporation).

(iii) ***The Isothermal Process:*** A physical process which proceeds, where temperature is kept constant (by adding heat or removing) is called isothermal process.

(iv) ***The Condensation Process:*** The conversion of water vapour into liquid form (very small water droplets) is condensation. If water vapour sent into a chamber containing saturated water vapour it condenses into water. Condensation depends on the presence of various sizes of nuclei (small, Atican, large, giant nuclei) and types (hygroscopic, non-hygroscopic). Condensation is the reverse process of evaporation.

Condensation linked parameters are: Air volume, temperature, pressure and humidity. Condensation occurs:

(i) when temperature decreases but volume remains constant,

(ii) when air is cooled to below dew point,

(iii) adiabatic expansion of volume which causes cooling (this is a dynamic process),

(iv) when both temperature and volume decreases.

In short condensation occurs in all such cases which causes temperature fall - like contact cooling, mixing of air masses of different temperatures, dynamic cooling of atmosphere.

As temperature decreases less water vapour is required for saturation. In atmosphere, as water vapour goes up (ascends) it cools, which causes saturation and further cooling by ascent results in condensation.

Note: On the surface of the earth, condensation results in the formation of dew, frost, mist and fog. In nature condensation on earth does not add much to the recorded precipitation.

The freezing Process: The conversion of liquid water into solid form (ice) is freezing. If pure water is kept undisturbed and cooled it remains in liquid state even below 0 °C (freezing point). This water is called super-cooled water. Super-cooled water is in unstable state and when disturbed it turns into ice. Introduction of an ice crystal or a freezing nuclei will cause freezing of super-cooled water. Laboratory experiments showed that super-cooled water can attain – 40 °C temperature, further cooling automatically freezes into ice crystals irrespective of the presence of nuclei present.

The Deposition Process: The process of conversion of water vapour directly into ice (without passing through liquid state) is called deposition or sublimation. However the word sublimation is used for the conversion of ice (solid) into vapour. The nuclei on which deposition takes place is called sublimation nuclei, which is different from condensation nuclei. In atmosphere sublimation nuclei is much less as compared to condensation nuclei.

Dew point (T_dT_d) : When a sample of moist air is cooled isobarically (keeping pressure constant) the temperature at which the sample of air becomes saturated is called dew point temperature (T_dT_d)or simply dew point. When moist is cooled below dew point temperature, condensation will occur.

Def. Dew point: The temperature to which moist air must be cooled to make it saturated with respect to water, pressure and mixing ratio keeping constant (unaltered i.e., without any addition of water vapour) is called dew point.

Frost point temperature: When a sample of moist air is cooled isobarically to saturate it with respect to a plane ice surface is called frost point temperature or simply frost point. If moist air is cooled below frost point, water vapour may be deposited as ice on some solid objects (including ice surface)

From table 7.4 it may be noted that moist air is still unsaturated with respect to water at some sub-zero temperature while it is saturated with respect to ice at that temperature. For example at – 30° C the saturation vapour pressure over ice is 0.380 while over water 0.509. In atmosphere cloud particles of ice generally do not form till saturation with respect water takes place. It appears condensation first takes place on condensation nuclei and if this nuclei also acts as freezing nuclei then the water which condensed on it may freeze. We shall study about different nuclei which are involved in cloud formation and precipitation processes.

7.3 Moisture Variables

Moisture content in the atmosphere is measured in a number of ways. Besides vapour pressure, absolute humidity, relative humidity, specific humidity and mass missing ratio are the important moisture variables or indicators, which are given below.

(i) ***Def*: *Absolute humidity* (ρ_v):** The mass of water vapour present in a given volume of air is called absolute humidity. It is expressed as grams / per cubic meter (g/m^3). OR, the mass of water vapour present per unit of volume is called absolute humidity. Absolute humidity, in fact is the density of water vapour.

(ii) ***Def* : *Relative humidity (RH)*** : The ratio of the mass of water vapour present in a unit volume of air to that of the mass of saturated water vapour in unit volume at the same temperature and generally expressed as percentage.

(Or)

The ratio of actual vapour pressure to that of saturation vapour pressure at the same air temperature.

$$\text{RH}(\%) = \frac{\begin{bmatrix}\text{Partial pressure water vapour} \\ \text{or Actual vapour pressure}\end{bmatrix}}{\begin{bmatrix}\text{Saturation vapour pressure} \\ \text{at the same air temperature}\end{bmatrix}} \times 100$$

$$= \frac{e}{e_s} \times 100 = \frac{\text{Amount of water vapour present}}{\text{Amount of water vapour it can hold at the same temperature}} \times 100$$

Suppose at certain temperature (T °C) the atmosphere had water 3 g/kg, where as at that temperature (T °C) it can hold a maximum of 12 g/kg.

then $$\text{RH} = \frac{3\ \text{g/kg}}{12\ \text{g/kg}} \times 100 = 25\%$$

The following Table 7.5 gives the saturation vapour pressure (e_s) of water at different temperatures.

Table 7.5 Saturation vapour Pressure (e_s) of water at different temperatures

*30	T °C	0	5	10	15	20	40	100
0.0424	e_s 10^s hPa	0.0061	0.00868	0.0119	0.0169	0.0233	0.0734	1.01

Example: If the partial vapour pressure of water (e) in the atmosphere is 0.0187×10^5 Pa (at air temperature 20 °C) and saturation vapour pressure at the same temperature is 0.0233×10^5 Pa.. Find RH

Solution: $$RH = \frac{e}{e_s} \times 100 = \frac{0.0187 \times 10^5 \text{ Pa}}{0.0233 \times 10^5 \text{ Pa}} \times 100$$

$$\simeq 80.3\%$$

Note: saturation vapour pressure (e_s) of water increases with temperature. At 100°C, e_s becomes one atmospheric pressure (10^5 P_a).

Man becomes uncomfortable during summer when RH is high because at that high humidity sweet does not evaporate, consequently body does not cool. Because of this man feels uppressive of heat of the environment.

Vapour pressure (e) varies with latitude and season. In January 'e' is about 0.2 hPa over northern Siberia and it is about 30 hPa in summer in tropics.

(iii) *Def. Mixing Ratio (ω)*: It is also called humidity mixing ratio. It is the mass of water vapour present in grams per kilogram of dry air.

$$\omega = \frac{\rho_v}{\rho_d} = \frac{\text{Absolute humidity}}{\text{Density of dry air}}$$

$$RH = \frac{\omega}{\omega_s} \times 100 = \frac{\text{Mass of water vapour in } V_{cc}}{\text{Mass of saturation vapour in } V_{cc}} \times 100$$

where ω_s = saturation mixing ratio.

When RH is 100%, air temperature = dew point temperature.

RH is a function of pressure, temperature and mixing ratio.

$$RH = \frac{e_s \text{ at dew point}}{e_s \text{ at air temperature}} \times 100$$

Example: Find the RH on a day when the temperature is 20^0C and dew point is 5^0C. What is the partial pressure of the water vapour?

Solution:

(i) $$\text{RH} = \frac{e_s \text{ at dew point}}{e_s \text{ at air temperature}} \times 100$$

$$= \frac{\text{saturated vapour pressure at } 5^0\text{C}}{\text{saturated vapour pressures at } 20^0\text{C}} \times 100$$

$$= \frac{0.00868 \times 10^5 \text{ Pa}}{0.0233 \times 10^5 \text{ Pa}} \times 100 \qquad \text{(using table 7.5)}$$

$$= 37.3\% \ (37.253\%)$$

(ii) To find partial pressure (e) water vapor we use the formula

$$\text{RH} = \frac{\text{partial pressure(e)}}{\text{saturated vapour pressure at the temperature}} \times 100$$

$$\text{A partial pressure of water vapour (e)} = \frac{\text{RH} \times e_s \text{at } 20^0\text{C}}{100}$$

$$e = \frac{37.253}{100} \times 0.0233 \times 10^5 = 8.679 \text{ Pa}$$

since we found above RH = 37.253 % at 20^0 C and e_s at that temperature is 0.0233×10^5 Pa (from Table 7.3).

Example: Find the dew point temperature on a day when air temperature was 20^0 C and RH 72.532 %. Also find the partial pressure of water vapour.

Solution:

(i) $$\text{RH} = \frac{e_s \text{ at dew point}}{e_s \text{ at air temperature}} \times 100$$

$$\therefore\ e_s \text{ at dew point} = \frac{\text{RH} \times e_s \text{at air temperature}}{100}$$

$$= \frac{72.532}{100} \times 0.0233 \times 10^5 \text{ Pa}$$

$$= 0.72532 \times 0.0233 \times 10^5 \text{ Pa}$$

e_s at dew point $\simeq 0.0169 \times 10^5$ Pa

from Table 7.5 the dew point is 15° C

(ii) e = partial pressure of water vapour

$$e = \frac{RH \times e_s \text{ at } 20°C}{100}$$

$$e = \frac{72.532}{100} \times 0.0233 \times 10^5 \text{ pa}$$

$$= 0.0169 \times 10^5 \text{ pa}$$

Note: At dew point $e = e_s$

Example : Find the RH on a day when air temperature is 21^{0}C, actual vapour pressure is 20 hPa and (e_s) saturated vapour pressure is 25 hPa.

Solution: $RH = \frac{e}{e_s} \times 100 = \frac{20}{25} \times 100 = 80\%$

Example : If air temperature is 15^0 C and partial pressure of water vapour (e) is 0.013×10^5 Pa , find RH

Solution: $RH = \frac{e}{e_s}$ at the same temperature × 100

$$= \frac{0.013 \times 10^5 \text{ Pa}}{0.0169 \times 10^5 \text{ Pa}} \times 100 = 77\%$$

[since $e_s = 0.0169 \times 10^5$ Pa at 15^0 C from table 7.5]

Example: Find the RH on day when dry bulb temperature (TT) and dew point temperature (T_d T_d) are as given below:

(i) TT = 30 °C, T_d T_d = 10 °C

(ii) TT = 20 °C, T_d T_d = 5 °C

(iii) TT = 40 °C, T_d T_d = 30 °C

(iv) TT = 10 °C, T_d T_d = 10 °C

Use the Table 7.5

Solution:

$$RH = \frac{e_s \text{ at dew point}}{e_s \text{ at air temperature}} \times 100$$

(i) $$RH = \frac{0.119 \times 10^5 \text{ Pa}}{0.0424 \times 10^5 \text{ Pa}} \times 100$$

$$\simeq 28\%$$

(ii) $$RH = \frac{0.00868 \times 10^5 \text{ Pa}}{0.0233 \times 10^5 \text{ Pa}} \times 100$$

$$\simeq 37\%$$

(iii) $$RH = \frac{0.0424 \times 10^5 \text{ Pa}}{0.0734 \times 10^5 \text{ Pa}} \times 100$$

$$\simeq 58\%$$

(iv) $$RH = \frac{0.0119 \times 10^5 \text{ Pa}}{0.0119 \times 10^5 \text{ Pa}} \times 100$$

$$\simeq 100\%$$

Note: When DB (TT) = Dew point (T_d T_d), then RH is 100%.

The see-saw diagram of relationship between temperature and RH is given below

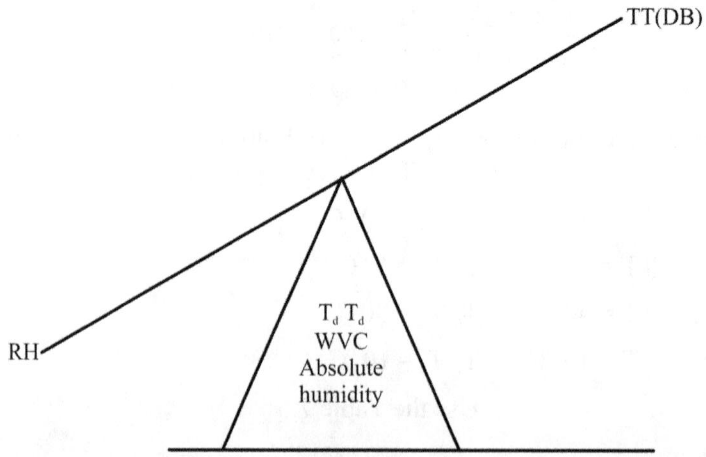

It is clear from the above see-saw diagram: As temperature goes down, relative humidity rises and vice versa, while there is no change in water vapour content (WVC), absolute humidity. Dew point stays the same in a constant air mass.

Diurnal variation of relative humidity: In general maximum temperature attains at about 1400 hrs local time, consequently RH can be lowest at that time. When temperature is lowest in the morning (is general it occurs in the morning half an hour after sunrise), RH can be maximum at that time. In either case there will be no change in WVC and dew point. Because of this reason fog will be thick just after dawn and dissipates (or melts) away after bright sunshine.

Humidity effects: Humidity has great effects on many organic substances. Completely degreased human hair is sensitive to variation of humidity. The length of human hair increases in length with increase in humidity and vice versa. This property is used in hair hygrograph.

The increase in length is not linear: The increase in length with RH from 0 – 100% is about $2\frac{1}{2}$% of its original length. Between 20 to 100% RH, the increase in hair length (Δl) is proportional to the logarithm of the change in RH [i.e. log (Δ RH)]

Let l_1 = initial length (when RH_1)

l_2 = final length (when RH_2)

$$\Delta l = l_2 - l_1$$

then $$\Delta l = \log [RH_2 - RH_1] \times \text{constant}$$

or $$\Delta l = \ln [RH_2 - RH_1] \times \text{constant}$$

Humidity may change (increase or decrease) with change in temperature and without adding or subtracting moisture from the sample air.

(iv) ***Def. Specific humidity (q)*** :

$$q = \frac{\text{Mass of water vapour}}{\text{Mass of moist air containing the vapour}}$$

$$= \frac{M_v}{M_v + Md} = \frac{\rho_v}{\rho}$$

where M_v = Mass of water vapour

M_d = Mass of dry air containing the vapour

$$RH = \frac{\text{Actual mixing ratio}}{\text{saturation mixing ratio at the same temperature}} \times 100$$

$$RH = \frac{\omega}{\omega_s} \times 100$$

$$\omega \simeq \in \frac{\rho}{P}$$

$$\omega_s = \in \frac{\rho_s}{P}$$

where ω = mixing ratio

ω_S = saturated mixing ratio

$$\in = \frac{m_v}{m_d} = \frac{\text{mole weight of vapour}}{\text{mole weight of dry air}}$$

e = vapour pressure

e_s = saturation vapour pressure and $e<<p$

(v) ***Def. wet bulb temperature (TwTw)***: The temperature to which a sample of air must be cooled by evaporating water into it to make the sample of air saturated, keeping pressure constant (in this case mixing ratio is not constant).

or

This is the lowest temperature attained by air when water is freely evaporated into it

or

The temperature to which air may be cooled by evaporating water into it isobarically, till it is saturated.

At lower humidity the difference between DB&WB (TT & TwTw) will be large, while the difference is zero when RH is 100%

i.e. at 100% RH, DB = WB = DP

TT = TwTw = TdTd.

Questions

1. Water enters into the atmosphere by the process of :
   (a) Condensation (b) Evaporation
   (c) Transpiration (d) Both (b) and (c)
2. Water cycle over land and atmosphere is called
   (a) Water cycle (b) Hydrological cycle
   (c) Nitrogen cycle (d) Carbon cycle

3. Evaporation causes the remaining water:
   (a) Rise in temperature by an amount proportional to the latent heat freezing
   (b) Decrease in temperature by an amount proportional to the latent heat of evaporation
   (c) Does not change the temperature
   (d) None above
4. Saturation vapour pressure of water:
   (a) Increases with temperature
   (b) Decreases with temperature
   (c) Does not Change with temperature
   (d) None above
5. Air consisting of dry air and water vapour is called:
   (a) Dew (b) Frost
   (c) Moist air (d) Saturated air
6. The phase change (solid to liquid) ice to water is called:
   (a) Evaporation (b) Melting or thawing
   (c) Sublimation (d) None above
7. The phase change (liquid to gas) ice to vapour is called:
   (a) Evaporation (b) Melting
   (c) Sublimation (d) Freezing
8. The phase change (solid to gas) ice to vapour is called:
   (a) Evaporation (b) Melting
   (c) Sublimation (d) Freezing
9. The phase change (liquid to solid) water to ice is called:
   (a) Evaporation (b) Melting
   (c) Sublimation (d) Freezing
10. The phase change (gas to liquid) vapour to water is called:
    (a) Evaporation (b) Condensation
    (c) Freezing (d) Sublimation
11. The phase change (gas to solid) vapour to ice is called
    (a) Evaporation (b) Condensation
    (c) Sublimation/ deposition (d) Freezing

12. The change of state (phase change) from ice to liquid or liquid to ice is called ....and latent heat required/liberated.... per gm.
    (a) Latent heat of vaporization and heat required 600 cal/gm at 0°C
    (b) Latent heat of fusion and heat repuired 80 cal/gm
    (c) Latent heat of sublimation and heat required 80 cal/gm
    (d) None above
13. The change of state liquid to vapour or vapour to liquid is called... and latent heat require/liberated is.......
    (a) Sublimation, latent heart released 680 cal/gm
    (b) Vaporization, latent heat required 600 cal/gm
    (c) Condensation, latent heat released 540 cal/gm
    (d) Freezing, latent heat required 540 cal/gm
14. Saturation vapour pressure of water is higher in:
    (a) Tropical water bodies
    (b) Mid latitude water bodies
    (c) Polar Circle
    (d) None above
15. Saturation vapour over plane surface of water changes
    (a) Increases with decreasing temperature
    (b) Increases with increasing temperature
    (c) Does not change with temperature
    (d) None above
16. Saturated vapours obey:
    (a) Boyle's law (b) Charle's law
    (c) Dalton's law of partial pressures (d) All above
17. Density of moist air is:
    (a) Greater than dry air (b) Less than dry air
    (c) Equal to dry air (d) None above
18. The principal atmospheric state variables are:
    (a) Pressure (b) Temperature
    (c) Specific volume and humidity (d) All above

19. Atmospheric gases are assume to obey:
    (a) Boyle's law (b) Charle's law
    (c) Gay- Lussacs law (d) All above
20. Equal volumes of different gases at the same temperature and pressure contain:
    (a) Different number of molecules
    (b) Equal number of molecules
    (c) More nitrogen molecules
    (d) None above
21. A physical process which proceeds where pressure remains constant is called:
    (a) Isothermal process (b) Isobaric process
    (c) Adiabatic process (d) Precipitation process
22. A physical process in which the heat content remains constant is called:
    (a) Isobaric process (b) Isothermal process
    (c) Adiabatic process (d) Convective process
23. A physical process which proceeds, where temperature remains constant is called:
    (a) Isothermal process (b) Isobaric process
    (c) Adiabatic process (d) Natural process
24. Condensation depends on:
    (a) Air volume and temperature (b) Pressure
    (c) Humidity (d) All above
25. On the surface of the earth condensation results in the formation of:
    (a) Dew (b) Frost
    (c) Fog and mist (d) All above
26. In nature condensation occurs mainly due to:
    (a) Contact cooling
    (b) Mixing of air masses of different temperatures
    (c) Dynamic cooling (adiabatic cooling)
    (d) All above

27. Supercooled water can attain temperature of:
    (a) 0°C (b) – 10°C
    (c) – 40°C (d) None above

28. If a sample of moist air is cooled to make it saturated keeping pressure and mixing ratio constant, the temperature attained is called:
    (a) Freezing temperature
    (b) Dew point temperature
    (c) Dry bulb temperature
    (d) None above

29. If a sample moist air is cooled isobarically to saturate it with respect to a plane ice surface is called:
    (a) Potential temperature
    (b) Dew point temperature
    (c) Frost point temperature
    (d) Dry bulb temperature

30. The mass of water vapour present in a given volume of air is called
    (a) Specific humidity (b) Relative humidity
    (c) Absolute humidity (d) Mixing ratio

31. The ratio $\frac{\text{Actual vapour pressure} \times 100}{\text{Saturation vapour pressure at the same temperature}}$ is called:
    (a) Specific humidity (b) Absolute humidity
    (c) Relative humidity (d) Mixing ratio

32. The mass of water vapour present, expressed in grams per kilogram of dry air is called:
    (a) Relative humidity (b) Absolute humidity
    (c) Mixing ratio (d) None above

33. In a constant air mass (no changes in the air mass) as temperature increases the relative humidity (RH):
    (a) Increases (b) Decreases
    (c) Remains constant (d) None above

34. Supposing in a day (24 hrs.) water vapour and dew point are remained constant, then RH attains maxima, minima at:
   (a) at about dawn; at 2 PM local time
   (b) at 2 PM; at dawn
   (c) at 10 AM; 10 PM
   (d) None above

35. The temperature to which a sample of air is cooled by evaporating water into it to make the sample of air saturated, keeping pressure constant is called:
   (a) Dry bulb temperature
   (b) Dew point temperature
   (c) Wet bulb temperature
   (d) None above

36. The difference between dry bulb and dew point (temperature spread) is maximum and minimum when:
   (a) RH is 0; 100%   (b) RH 100%; 0
   (c) RH 50%; 50%   (d) None above

37. When RH is 100%, then:
   (a) TT = TdTd   (b) TT = TwTw
   (c) TwTw = TdTd   (d) All above

38. The device that uses the property of matter (like expansion, contraction, electrical resistance) to measure .... is called thermometry.
   (a) Relative humidity   (b) Wind speed
   (c) Temperature   (d) Pressure

39. In caliberation of thermometers, internationally agreed two fixed points are:
   (a) Sublimation, evaporation of water
   (b) Freezing point, boiling point
   (c) Freezing point, condensation
   (d) Boiling point, evaporation

40. The phase change ice to water is called:
   (a) Sublimation   (b) Evaporation
   (c) Melting   (d) Condensation

41. The phase change ice to water vapour is called:
    (a) Sublimation (b) Evaporation
    (c) Melting (d) Deposition

42. The phase change water to water vapour is called:
    (a) Condensation (b) Evaporation
    (c) Melting (d) Freezing

43. The phase change water to ice is called:
    (a) Melting (b) Freezing
    (c) Evaporation (d) Deposition

44. The phase change water vapour to water is called:
    (a) Evaporation (b) Condensation
    (c) Deposition (d) Sublimation

45. The phase change water vapour to ice is called:
    (a) Condensation (b) Evaporation
    (c) Deposition (d) Sublimation

46. The latent heat of vaporization to evaporate 1 gm of water at 0°C and 100°C are:
    (a) 80 cal, 540 cal (b) 600 cal, 540 cal
    (c) 600 cal, 80 cal (d) 600 cal

47. The change of state (phase change) that requires or liberates heat energy without change in temperature is called:
    (a) Condensation (b) Evaporation
    (c) Latent heat (d) Freezing

48. When water vapour is mixed with dry air, the pressure exerted by the molecules of water is called:
    (a) Vapour pressure
    (b) saturated vapour pressure
    (c) Atmospheric pressure
    (d) Partial pressure of dry air

49. The maximum limit of vapour (water vapour) that can exist in a gaseous state is called:
    (a) Vapour
    (b) Saturation vapour
    (c) Atmospheric pressure
    (d) Partial pressure of dry air

50. If p = atmospheric pressure , e = partial pressure of water vapour, then the partial pressure of dry air is;
    (a) p (b) e (c) $p - e$ (d) $p + e$

51. The saturation vapour pressure is dependent on (or function of);
    (a) Water vapour content (b) Temperature only
    (c) Pressure only (d) Volume

52. Saturated vapours obey:
    (a) Boyle's law (b) Charle's law
    (c) Dalton's law of partial pressure (d) Gay-lusacks law

53. Plants that lose some water through small openings in the leaves and stems is called:
    (a) Evaporation (b) Transpiration
    (c) Evapotranspiration (d) Condensation

54. The combined effect of moisture loss through evaporation and transpiration from a given surface area is called:
    (a) Evaporation (b) Transpiration
    (c) Evapotranspiration (d) Condensation

55. The mean molecular weight of dry air is 28.8 and the molecular weight of water vapour is 18. Consiquently the density of moist air is:
    (a) More than dry air (b) Less than dry air
    (c) Equal to the density of dry air (d) None above

56. The principal state variables of atmosphere are:
    (a) Pressure (P), temperature (T)
    (b) Temperature (T), specific volume (α)
    (c) Specific volume, moisture
    (d) P, T, α, moisture

57. A Physical process which proceeds at the same pressure is called:
    (a) Adiabatic process (b) Isothermal process
    (c) Isobaric process (d) Precipitation process
58. A Physical process in which the heat content remains constant is called:
    (a) Adiabatic process (b) Isothermal process
    (c) Isobaric process (d) Precipitation process
59. A Physical process which moves at the same temperature is called:
    (a) Adiabatic proces (b) Isothermal process
    (c) Isobaric process (d) Condensation process
60. The conversion of water vapour into liquid form is called:
    (a) Evaporation (b) Condensation
    (c) Precipitation (d) Sublimation
61. On the surface of the earth, condensation results in the formation of :
    (a) dew (b) Frost
    (c) mist/fog (d) All above
62. If pure water is kept undisturbed and cooled it remains in liquid state even below 0°C. This water called:
    (a) Heavy water (b) Supercooled water
    (c) Super saturated water (d) Light water
63. Supecooled water can attain temperature of :
    (a) – 15 °C (b) – 25 °C
    (c) – 40 °C (d) – 60 °C
64. Supercooled water is:
    (a) in stable state (b) In unstable state
    (c) in neutral state (d) None above

Answers

1.	(d)	2.	(b)	3.	(b)	4.	(a)
5.	(c)	6.	(b)	7.	(c)	8.	(c)
9.	(d)	10.	(b)	11.	(c)	12.	(b)
13.	(b)	14.	(a)	15.	(b)	16.	(c)

17.	(b)	18.	(d)	19.	(d)	20.	(b)
21.	(b)	22.	(c)	23.	(a)	24.	(d)
25.	(d)	26.	(d)	27.	(c)	28.	(b)
29.	(c)	30.	(c)	31.	(c)	32.	(c)
33.	(b)	34.	(a)	35.	(c)	36.	(a)
37.	(d)	38.	(c)	39.	(b)	40.	(c)
41.	(a)	42.	(b)	43.	(b)	44.	(b)
45.	(c)	46.	(b)	47.	(c)	48.	(a)
49.	(b)	50.	(c)	51.	(b)	52.	(c)
53.	(b)	54.	(c)	55.	(b)	56.	(d)
57.	(c)	58.	(a)	59.	(b)	60.	(b)
61.	(d)	62.	(b)	63.	(c)	64.	(b)

CHAPTER 8

Temperatures, Laps Rates and Stability of Atmosphere

8.1 Temparature

We know heat is a form of energy and the main source of heat for the earth is sun's radiation energy. The degree of hotness or coldness of a object is called its temperature. Conversely temperature is a manifestation of heat.

Thermometry: The device that utilizes the property of a matter (expansion or contraction) or electrical resistance due to heat to measure temperature is called thermometry. In calibration of thermometers internationally agreed two fixed points are freezing point (or melting point) and boiling point (or steaming point) of water used.

The most commonly used liquids in thermometers are mercury and alcohol. Mercury freezes at – 39 °C and boils at 357 °C. Mercury thermometers are useless for measurement of temperature below – 39° C, and above 357° C. Alcohol freezes at – 115 °C and boils at 78 °C. Alcohol thermometers are used to measure temperature between – 30 °C and – 110 °C. High temperatures are measured with pyrometers.

Thermister: In this case change in electrical resistance of special ceramic chemical mixture with temperature property is used.

Temperature scale: In scientific measurements generally Celsius, Fahrenheit and Kelvin scales are used. In all these thermometric liquid taken is mercury.

In Celsius scale (also called centigrade), the lower fixed point (melting point of ice) is taken as 0 °C and upper fixed point (boiling point of water) is marked 100 °C. The interval between these two fixed points is divided into 100 equal parts and each degree is designated as 1 °C. In Fahrenheit scale, the lower fixed point is marked 32 °F and the upper fixed point as 212 °F. The interval between these two fixed points is divided into 180 equal parts. In Kelvin (also called absolute) scale, the lower limit is marked + 273 °K and the upper fixed point as 373 °K. The interval between these two fixed points is divided into 100 equal parts.

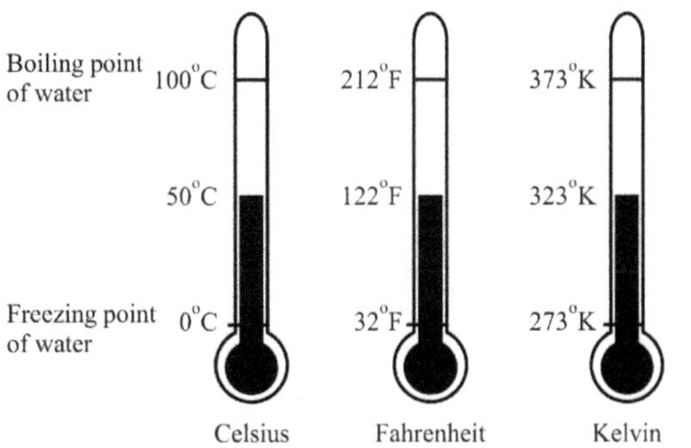

Conversion

$$C = \frac{5}{9}(F - 32) \quad \text{or} \quad t\,^{\circ}C = \frac{5}{9}(t\,^{\circ}F - 32)$$

$$K = C + 273 \quad \text{or} \quad T\,^{\circ}K = t\,^{\circ}C + 273.15$$

$$\frac{C}{100} = \frac{F-32}{180} = \frac{K-273}{100}$$

where

C is degrees in Celsius

F is degrees in Fahrenheit

K is degrees in Kelvin

Temperature and density of water is given below:

T °C	Density g/cm^3
0	0.99984
4	0.99997
10	0.99970
20	0.99820
50	0.98804
75	0.97486
100	0.95835

Example: The normal human body temperature is 98.6° F. Find the same in Celsius(C) and Kelvin.

Solution:

(i) $C = \frac{5}{9}(F - 32)$

$= \frac{5}{9}(98.6 - 32) = \frac{5}{9} \times 66.6$

$= 37\,^{\circ}C$

$\therefore$ The Normal human body temperature is 37 °C

(ii) $K = C + 273 = 37 + 273$

$= 310$

$\therefore$ The Normal human body temperature is 310 °K

Example: At what temperature reading both Celsius and Fahrenheit is same?

Solution:

Let X° reading both in Celsius and Fahrenheit

then $X = \frac{5}{9}(X - 32)$

$9X = 5X - 160$

or $4X = -160$

$X = -40$

i.e. $-40\,^{\circ}C = -40\,^{\circ}F$

***Bimetalic Thermometer*:** Two metal strips (Invar and brass or steel) or coils which have different coefficients of linear thermal expansion welded together along their lengths. When this bimetalic coil (or arcs) heated, it bends. Higher the rise of temperature greater is the bending of the strip. The amount of bending of the strip is used for measuring temperature. This is the basis of bimetalic thermometer, shown schematically in Fig 8.1. One end of the bimetalic spiral coil, is clamped and the free end attached to pointer, which moves over a graduated scale in degrees.

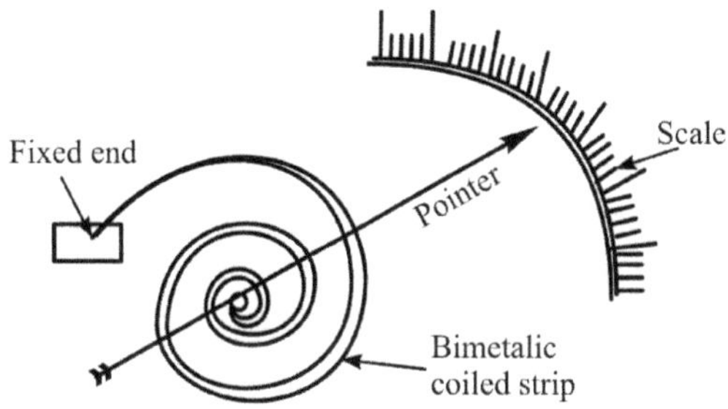

Fig. 8.1 Bimetallic thermometer.

***Surface air temperature*:** The free air temperature measured at a height of 1.25 m (4 ft) above ground level. Thermometers are housed in a wooden Stevenson screen (SS screen) box. Thermometers are always kept under shade. SS screen is designed to protect the thermometers to keep off from direct sunshine and exposed to free passage of air. The surface observations are taken for aviation purpose include:

(i) Dry bulb (DB) temperature (TT) or air temperature.

(ii) Wet bulb(WB) temperature (T_W T_W), which is used for calculation of dew point(DP) temperature (T_d T_d)

(iii) Maximum temperature (T_{max} T_{max})

(iv) Minimum temperature (T_{min} T_{min})

(v) In coastal observatories and ship observation, sea surface water temperature is included.

***Screen box*:** The elements measuring DB, WB, Max, Min temperatures and humidity are housed outdoors in a screen (4 ft/1.25 m height agl). The screen consists of slatted sides, a split base and a double top to allow free flow of air around the instruments. The box is painted white to minimize the absorption of solar radiation. The screen door will face north in the northern hemisphere and south in the southern hemisphere.

In aviation we use some special terms about temperature which are defined below.

***Static Air Temperature (SAT)*:** The correct temperature of ambient air (in flight) is called SAT. It is also called corrected outside air temperature (COAT). It is very difficult to obtain SAT directly from a temperature probe.

***Total Air Temperature (TAT)*:** If air is brought to rest completely at the temperature probe, there will be some compression of the air which will cause adiabatic heating. The resulting temperature is TAT.

The difference between TAT and SAT some times called stagnation rise. The proportion of the stagnation temperature which is sensed (felt) is called recovery coefficient (denoted by K). The device used by Rosemount probe has the value of K=1. The effect of compressibility heating is significant in case of modern fast moving aircrafts. The relation between SAT, TAT is given by the following formula.

$$T_A = T_R - \left(\frac{V}{100}\right)^2$$

where $T_A = \text{SAT in } ^\circ\text{C}$

$T_R = \text{TAT in } ^\circ\text{C}$

V = true air speed (TAS) in knots.

***Ram air temperature (RAT)*:** In case of some temperature probes the value of K is about 0.75 to 0.85. In this case the measured (probe) temperature is of air which partially brought to rest. This is called the Ram Air Temperature (RAT).

(RAT – SAT) is called ram rise.

It may be noted here that temperature has a direct bearing on aviation. Engine performance and cooling systems depend on the variation of temperature in the atmosphere. We have studied that density is inversely proportional to temperature. High temperature implies low density which adversely affects the engine performance, particularly during take-off/landing. In deciding run way length local temperature (T_{max}) has to be given due weightage.

The distribution of average temperature over the earth is directly related elevation of the sun. As a consequence, on an average temperature decreases equator to poles with maximum in equatorial region and minimum towards the poles. Temperature is also dependent on local soils, orography, land-sea proximity, vegetation and water bodies. The world record of highest day temperature is 58 °C, recorded at Alazizyah, Libya on 13 September 1922 and the lowest temperature recorded – 89.2 °C at Vostok, Antarctica on 21 July, 1983.

8.2 Diurnal Variation of Temperature

At the surface of the earth temperature depends on the nature of the surface. During the day ground temperature may be higher than the air temperature as much as 10 to 15 °C. Maximum temperature, in general, occurs at about 1400 hrs.

During night ground cools due to earth's longwave radiation emission. The minimum temperature at the ground is cooler than the air temperature, by about 5 °C. Minimum temperature, in general reaches at about sunrise. In contrast to this ground temperature variation, the sea surface water temperature hardly varies by about 1 °C.

In India, the range of temperature (Max-Min) is about 5 to 15 °C. The following climatological data (Table 8.1) explains the seasonal variation of temperature together with daily range.

Table 8.1 Variation of temperature together with daily range

Station	New Delhi Safdarjang		Portblair		Mumbai Santacruz		Kolkata Dum Dum	
Months	DM Max	DM Min	DM Max	DM Min	DM Max	DM Min	DM Max	DM Min
January	21.3	7.3	29.2	22.7	30.6	16.3	26.4	12.3
February	23.6	10.1	29.8	21.8	31.6	16.9	29.0	15.1
March	30.2	15.1	31.1	22.5	32.8	20.2	33.8	20.4
April	36.2	21.0	31.9	24.2	33.3	23.8	36.0	24.3
May	40.5	26.6	30.9	24.4	33.2	26.2	35.7	26.0
June	39.9	28.7	29.1	24.2	31.7	25.9	33.8	26.2
July	35.3	27.2	28.9	24.1	29.6	24.8	31.8	26.0
August	33.7	26.1	28.8	24.1	29.1	24.4	31.8	26.0
September	34.1	24.6	28.7	23.6	29.9	23.9	32.0	25.9
October	33.1	18.7	29.0	23.6	32.4	22.9	31.4	23.6
November	28.7	11.8	29.2	23.6	33.1	19.8	29.0	17.6
December	23.4	8.0	29.0	23.1	32.1	17.7	26.	13.0

Table 8.1 Contd....

Station	New Delhi Safdarjang		Portblair		Mumbai Santacruz		Kolkata Dum Dum	
Months	DM Max	DM Min	DM Max	DM Min	DM Max	DM Min	DM Max	DM Min
Annual Average	31.7	18.8	29.6	23.5	31.6	21.9	31.4	21.4
Average Range Max-Min	12.9		6.1		9.7		10.0	
Annual Range	19.2	21.4	3.2	2.6	4.2	9.9	9.6	13.9

D.M = Daily mean

Inland StationNew Delhi: In a year DM Max varies 21.3 to 40.5 °C, it's range is 19.2 °C. While DM Min varies 7.3 to 28.7 °C its range is 21.4 °C. Annual range of Max- Min is 40.5 to 7.3 = 33.2 °C

Island Station Port Blair: In a year DM Max varies 28.7 to 31.9 °C, its range is 3.2° C. While DM Min Varies 21.8 to 24.4° C its range is 2.6° C. Annual range of Max-Min is 31.9 to 21.8 = 10.1 °C

Coastal Station Mumbai: In a year DM Max varies 29.1 to 33.3 °C its range is 4.2 °C. While DM Min varies 16.3 to 26.2 °C, its range is 9.9 °C. Annual range of Max-Min is 33.3 °C –16.3 °C = 17.0 °C.

Above three stations data indicates the difference of inland, island and coastal station data. In addition to these **saliant** points we see the seasonal changes as well. We shall study more in climatology.

8.3 Adiabatic Lapse Rates

The decrease of temperature with height is lapse, and the rate of fall of temperature with height is called lapse rate. Lapse rate is normally calculated for 1000 ft or 1 km or 100 m. Rise of temperature with height is called inversion (in which lapse rate will be negative). Actual (real) temperature fall in the atmosphere with height is called environment lapse rate (ELR). Suppose on a particularly day the temperature height graph is as shown in the Fig. 8.2.

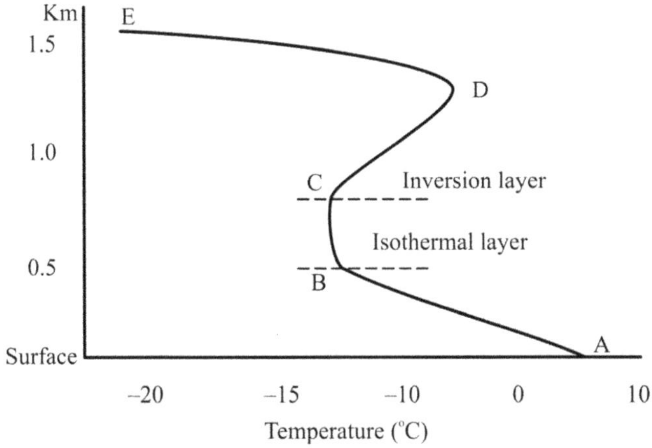

Fig. 8.2

In this fig the lapse rate is normal from A to B. Isothermal between B to C and inversion between C to D (C is base of inversion, D is top of inversion) and normal between D to E. That is, A to B there is temperature fall, between B to C the temperature is constant, between C to D there is temperature rise (inversion) and between D to E again there is temperature fall.

In the real atmosphere we can see following type of temperature curves (Fig. 8.3). Inversion may begin right at the surface or may be in some middle level. Isotherm also may begin at the surface or at any level. Isothermal and inversion layers are important in aviation, which affect the performance of aircraft. Isothermal and inversion layers in the atmosphere indicate some change in the weather and visibility.

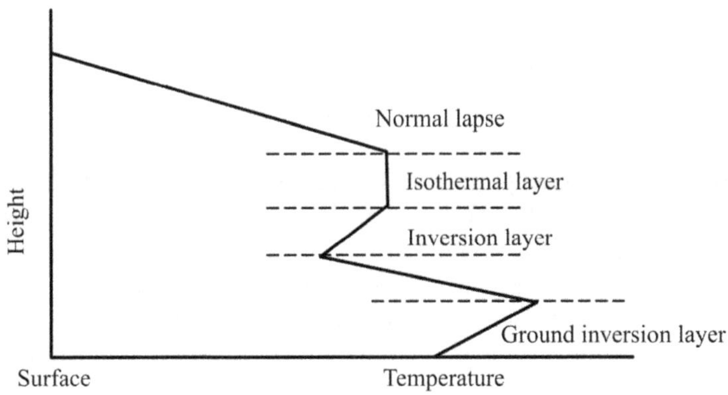

Fig. 8.3

The following are the main types of inversions:

(i) Tropopause inversion. (ii) Surface inversion. (iii) Subsidence inversion. (iv) Valley inversion. (v) Frontal inversion. (vi) Frictional inversion.

***Note*:** Ascent or descent of an air parcel in the atmosphere occur under adiabatic condition because:

(i) Air is a poor conductor of heat

(ii) Air parcel mixing with the environment is very slow and

(iii) The effective radiative process is very small in short periods of time.

Atmosphere is said to be stable (or unstable) at any point P (X_o, Y_o, Z_o) at any instant of time, when a parcel of air is displaced from its initial equilibrium position, acquires a relative motion in the environment is returning back to its initial position (or moving away). See Fig 8.4.

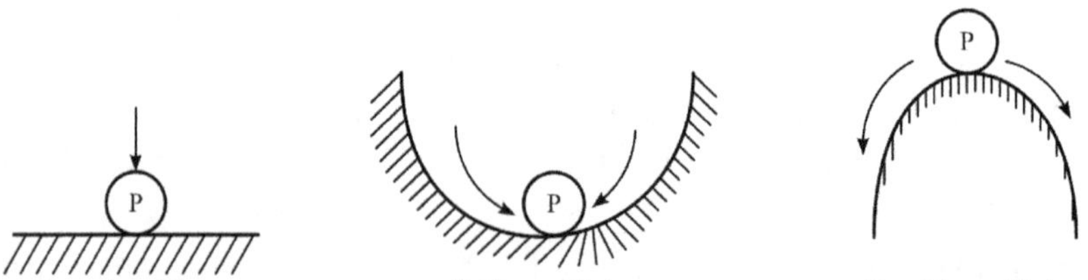

Fig. 8.4

In order to study the stability of a parcel, we have to consider two cases of lapse rate besides ELR.

(i) DALR (Dry Adiabatic Lapse Rate)

(ii) SALR (Saturated Adiabatic Lapse Rate).

Air parcel that continues to move up is unstable and as it moves up adiabatically expands and cools, which causes RH to rise. When it reaches 100% RH a heap cloud or convective cloud may develop. The air parcel that attains 100% RH and sinks back to its original level may produce a stratus cloud or layer cloud. One must remember that stable or unstable atmospheric conditions need not produce cloud. If the parcel doest not attain 100% RH, then a stable atmosphere may produce poor visibility and an unstable atmosphere may create excellent visibility.

***Adiabatic Lapse Rate*:** ELR is obtained from day to day radio sonde observations. Adiabatic lapse rate results in the parcel of air which was given an initial push. Air is forced to rise by several causes of which mountain barriers, frontal zone and advection are important.

Dry Adiabatic Lapse Rate (DALR): Any parcel of air with RH less than 100% is a dry parcel. If the parcel does not attain 100% RH by adiabatic cooling of parcel, the rate of fall of temperature will be DALR. For unsaturated air (RH < 100%) it will be calculated using gas equation that 3 °C /1000 ft or 9.8 °C/ km (or 1°C / 100 m). This is DALR denoted by Γd.

Saturated Adiabatic Lapse Rate (SALR): when a saturated parcel of air is given push it rises and cools which results into condensation. Condensation releases latent heat, which warms the parcel and thus adiabatic cooling is some what offset by this warming. The residual lapse rate is called SALR denoted by Γs, which is less than DALR. As the parcel goes up and attains low temperature, the water holding capacity goes on decreasing and when it reaches – 40 °C or less the whole water vapour would be condensed and the parcel attains dryness and DALR. Thus SALR is variable and at low temperatures (– 40 °C or less) it becomes DALR. At sea level SALR is about 5°C/ km and gradually approaches to 9.8 °C/km at temperature – 40 °C or below it . In lower levels the standard SALR is taken as 1.8 °C /1000 ft (0.6 °C/100 m). Table 8.2 gives the SALR at different temperatures.

Table 8.2 SALR at different temperatures

Temperature	**SALR °C/km**	
	Pressure 1000 hPa	**500 hPa**
30	3.6	-
20	4.5	-
10	5.6	4.2
0	6.9	5.4
– 10	8.1	6.8
– 20	8.8	8.4
– 40	9.8	9.8

Potential temperature (θ) : The temperature attained by a parcel (or sample) of dry (unsaturated) air when the parcel is compressed or expanded adiabatically from a given state to the pressure level 1000 hPa is called potential temperature (θ). θ is constant during an adiabatic process.

8.4 Stability and Instability of the Atmosphere

Parcel Method: Consider a parcel of air (for discussion, assume insulated) given an initial displacement which was in a state of equilibrium. After the initial displacement, if the parcel sinks back then the layer of atmosphere is said to be stable. If the parcel remains at the place where it is left, then the layer of atmosphere is said to be neutral. But if the parcel continues to move upward in the atmosphere then the layer of atmosphere is said to be unstable. In each of these three cases, vertical motion of the parcel of air depends on the environmental temperature. A comparison between ELR and DALR/SALR graphic lines establish the stability of the atmosphere.

Consider a plot of DALR and SALR graph. This gives the zones of stability. See Fig. 8.5.

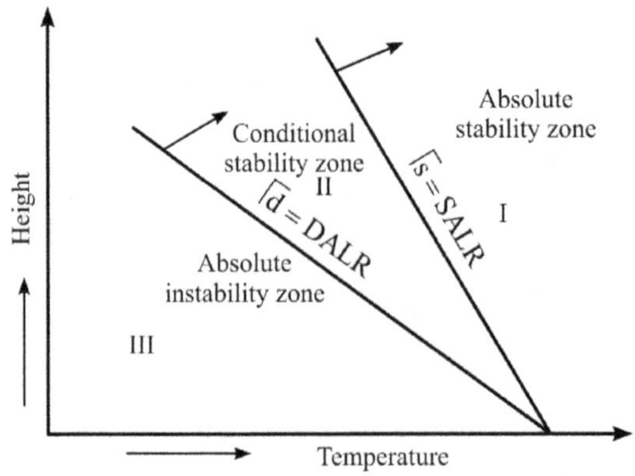

Fig 8.5

The plot shows Γs at 1.8 °C/1000 ft (0.6 °c/100 m), Γd at 3 °C/1000 ft (1. 0 °C/100 m), which divides the graph into three zones I, II and III.

Zone I, where temperature $t > \Gamma s$ is absolutely stable irrespective of RH.

Zone III, where temperature $t < \Gamma d$ is absolute instability irrespective of RH.

Zone II, where temperature (t) is $\Gamma_d < t < \Gamma_s$ is conditional instability. In zone II RH plays an important role. In zone II, temperature $t > \Gamma d$ but $t < \Gamma s$. In this zone RH 100% is unstable but RH < 100% is stable.

Principle of Buoyancy: According to principle of buoyancy, a parcel of air with higher density sinks till its density equals the surrounding air, while a parcel of air with lower density moves up till its density equals the surrounding air. We know from the gas law($p = \rho RT$) that density is inversely proportional to temperature. Consequently the above buoyancy law can be stated that, a parcel of air with lower temperature sinks till its temperature equal to that of surrounding air, while a parcel of air with higher temperature moves up till its temperature equal to that of its surrounding.

The basic graph to know stability we require graph having ELR in comparison with DALR / SALR graph, which is shown in Fig. 8.6.

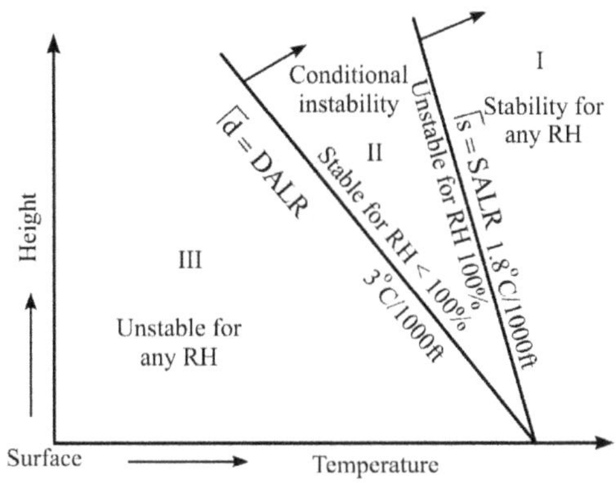

Fig. 8.6

Now we shall establish the criteria for:

(i) Absolute stability

(ii) For Absolute instability and then discuss conditional instability.

Absolute stability and Absolute instability means stability and instability irrespective of humidity.

8.4.1 Absolute Stability

This is best explained by an example.

Ex. Let surface temperature be +15 °C and 4000 ft temperature +11°C, RH < 100% say 80%. Discuss stability

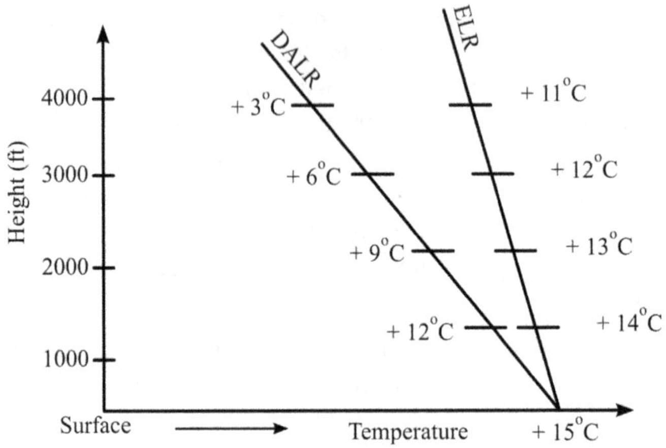

Fig. 8.7

$$\text{ELR} = \frac{\textit{Temperature Change}}{\textit{height difference}}$$

$$= \frac{15-11}{4000-0} = \frac{4^{\circ}C}{4000\,ft}$$

$$= 1^{\circ}\text{C} / 1000 \text{ ft}$$

$$\text{DALR} = 3^{\circ}\text{C} / 1000 \text{ ft.}$$

Fig. 8.7 gives the ELR, DALR graph.

Since ELR 1°C/1000 ft and surface temperature + 15 °C, the temperatures at 1000, 2000, 3000, 4000 ft are 14, 13, 12, 11 degrees Celsius while DALR temperature at these heights will be 12, 9, 6, 3 °C respectively.

It is given RH 80%, implies dry air. The parcel of air which is given an initial push follow DALR, that is the air parcel will have temperature of 12 °C at 1000 ft, 9 °C at 2000 ft, 6 °C at 3000 ft and 3 °C at 4000 ft. If we compare the temperature of the parcel with that of environment, temperature at these heights, the parcel is cooler at all heights as shown in Fig. 8.7. Consequently parcel will sink i.e. the environment is stable for the dry air.

Now consider the above example with RH 100% (saturated). In this case the parcel will follow SALR. The parcel temperatures at 1000 ft, 2000 ft, 3000 ft, 4000 ft will be 13.2 °C, 11.4 °C, 9.6 °C, 7.8 °C. The graph of SALR, ELR is shown in Fig. 8.8.

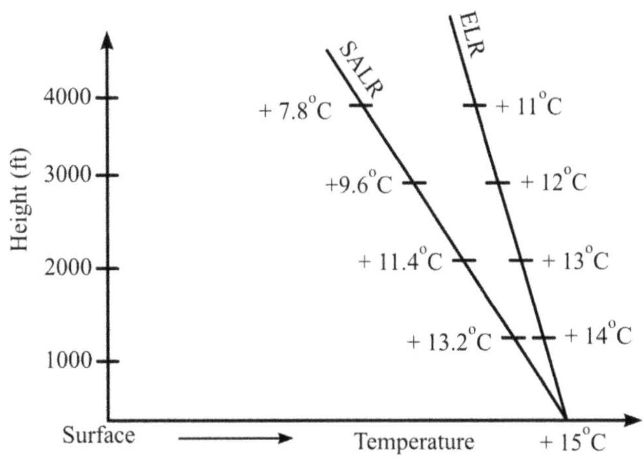

Fig. 8.8

It is seen that the parcel is cooler than the environment at all heights and sinks hence, the environment is stable for the saturated parcel.

In this example ELR < SALR & ELR < DALR

i.e. ELR < SALR < DALR

The environment is stable both for Dry air and saturated air, that is the environment is absolute by stable (in zone I)

Criteria for Absolute stability: ELR < SALR

Ex: An isothermal atmosphere is (absolutely) stable

since ELR (= 0) < SALR

8.4.2 Absolute Instability

Consider an example. Let surface temperature + 15 °C and a 3000 ft temperature 0 °C, RH unsaturated 80 % (RH $\angle$ 100%)

$$\text{ELR} = \frac{15-0}{3000-0} = \frac{15\ °\text{C}}{3000\ \text{ft}}$$

$$= 5°\ \text{C} / 1000\ \text{ft}$$

Since RH $\angle$ 100%, the parcel have dry air and follow DALR. The plot of DALR, ELR is shown in Fig. 8.9.

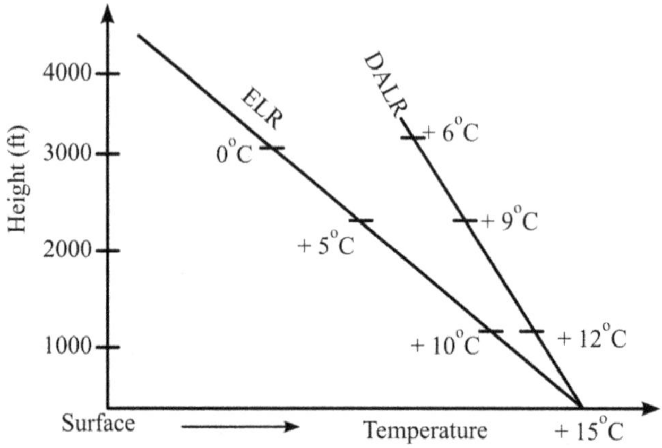

Fig. 8.9

It is obvious that the dry parcel is warmer than the environment at all heights indicating that the environment is unstable for dry air.

Now consider the above example with RH 100%, (saturated parcel). The parcel will follow SALR.

The plot of SALR, ELR is shown in Fig. 8.10.

It is obvious from this figure the parcel is warmer than the environment at all levels indicating the environment is unstable.

From these two examples it follows that the environment is unstable both for dry & saturated air i.e. the environment is absolutely unstable (Zone III).

Criteria for absolute instability is:

ELR > DALR (since ELR > DALR > SALR)

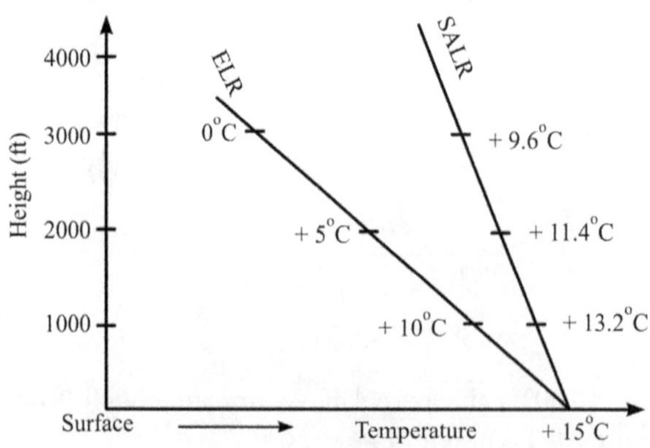

Fig. 8.10 Conditional Instability

8.4.3 Conditional Instability

Consider an example. Let surface temperature + 15 °C and at 3000 ft + 9 °C, RH 80% (RH < 100%, unsaturated parcel).

$$\text{ELR} = \frac{15-9}{3000-0} = \frac{6\ ^\circ\text{C}}{3000\ \text{ft}} = 2\ ^\circ\text{C}/\ 1000\ \text{ft}.$$

In this example DALR > ELR > SALR (zone II). Since RH < 100%, the parcel of air will follow DALR. The plot of DALR, ELR is shown in Fig. 8.11. From the figure, it follows that the dry parcel is cooler than the environment at all levels, indicating stability (zone II).

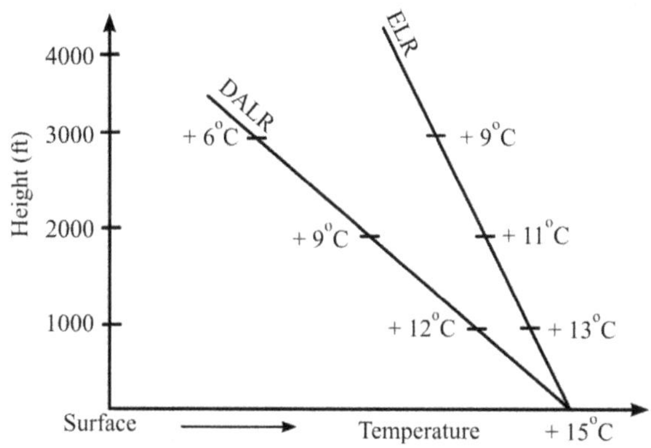

Fig. 8.11

For ELR < DALR, the environment is stable for dry air.

Now consider the above example with RH 100% (saturated air). Since the air is saturated, the parcel will follow SALR.The plot of ELR, SALR is shown in Fig. 8.12.

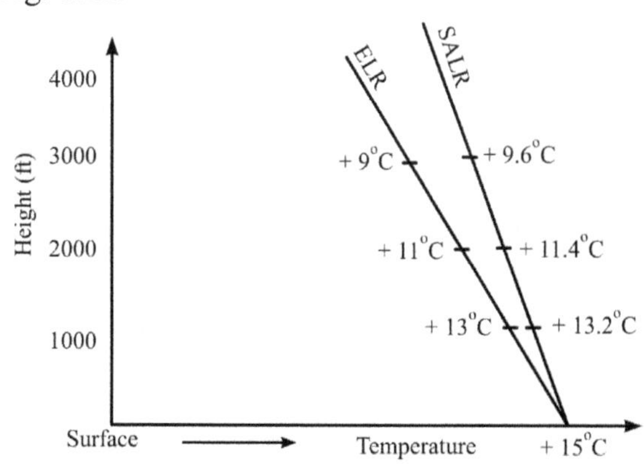

Fig. 8.12

From the figure it is seen that the parcel is warmer than the environment at all levels, indicating the environment is unstable for saturated air. ELR > SALR. (zone II)

From this example we can conclude that zone II is stable for dry air but unstable for saturated air. This is conditional instability criteria.

If *DALR > ELR > SALR* (zone II) *the environment is stable for dry parcel but unstable for a saturated parcel.*

Neutral stability: *when environment lapse rate and the adiabatic lapse rates are same* i.e. ELR = DALR in case of dry air

and

ELR = SALR in case saturated air.

then the environment is neutral. In this case the parcel remains there where it is left.

Summary

1. For dry air parcel

   If ELR < DALR, the atmosphere is stable

   If ELR = DALR, the atmosphere is neutral

   If ELR > DALR, the atmosphere is unstable

2. For saturated air parcel

   If ELR < SALR, the atmosphere is stable

   If ELR = SALR, the atmosphere is neutral

   If ELR > SALR, the atmosphere is unstable

3. If SALR < ELR < DALR, the atmosphere is conditionally unstable.

   (a) the atmosphere is stable for dry air (RH < 100%)

   but (b) the atmosphere is unstable for saturated air (RH 100%)

4. (For dry or saturated air)

   (a) Absolute stability

   If ELR < SALR, the atmosphere is absolutely stable.

   (b) Absolute instability

   If ELR > DALR, the atmosphere is absolutely unstable.

8.5 Flying Conditions in Stable Air

1. Smooth flying conditions
2. Low level poor visibility, fog may develop
3. Stratus cloud and steady precipitation
4. Steady winds

***Flying Conditions in unstable air*:**

1. Moderate to severe turbulence
2. Convective cloud developments
3. Showery precipitation
4. Good visibility except in precipitation
5. Gusty winds

***Latent Instability*:** A layer of atmosphere in which the lapse varies, being stable in the lower levels but unstable aloft (above it). In such case if the parcel is lifted to considerable height, the parcel enters into instability. This type of instability is called latent instability in the sense that instability is hidden aloft.

***Entrainment of mass*:** Incase of cumulus congests and cumulonimbus clouds, the vertical currents (updrafts) in the cloud suck in the surrounding air mass. This sucking of air into the cloud is called mass entrainment. The assumption adiabatic cooling in ascent (warming in descent) is not valid in case of mass entrainment into the cloud because moisture and heat content mixes up into the cell (parcel).

8.6 The Dew Point Lapse Rate

If we study the T-ϕ, we find the difference between dew point at 700 hPa level and 1000 hPa is 5 °C. Since 700 hPa level corresponds to 10,000 ft, the dew point lapse rate is

$$\frac{5^{\circ}\text{C}}{10{,}000 \text{ ft}} \quad \text{or} \quad 0.5\ ^{\circ}\text{C}/1000 \text{ ft}.$$

From the T-ϕ gram, we can calculate the height of cloud base (theoretically) using surface temperature, surface dew point, using DALR and dew point lapse rate using the empirical formula.

$$\text{Cloud base height} = (\text{TT} - \text{T}_d\ \text{T}_d) \times 400 \text{ ft}$$

where TT, T_d T_d are measured in °C.

To determine the stability and instability of atmosphere, the vertical distribution of temperature and humidity is required which is obtained on a day-to-day basis from radio-sonde observations. The values are plotted on thermodynamic diagram. The most commonly used thermodynamic diagram is T-ϕ gram. The stability, instability concept has wide application in atmospheric processes. Unstable layers, together with moisture are responsible for the development of adverse weather like Thunderstorms, hailstorms, Andhi, norwesters etc. This concept has wide application in day-to-day forecasting weather.

***Ex*:** Given surface temperature 18 °C and dew point 14 °C, find the height of the cloud base.

***Solution*:** (TT–T_d T_d) 400 ft.

Height of cloud base = (18 – 14) 400 ft. = 1600 ft

***Freezing level*:**

***Ex*:** Given surface temperature +18 °C, cloud base 4000 ft. find the freezing level inside the cloud

***Solution*:** surface to 4000 ft lapse rate will be (DALR) 3 °C / 1000 ft,

and inside the cloud lapse rate will be (SALR) 1.8 °C /1000 ft

Temperature at the base of the cloud

$$18 - \frac{3}{1000} \times 4000 = 18 - 12 = 6\ ^{\circ}\mathrm{C}$$

Inside the cloud lapse rate 1.8°C / 1000 ft

$$\text{Freezing level (0°C) inside the cloud} = \frac{6}{1.8} \times 1000 \text{ ft} + 4000 \text{ ft}$$

$$= 3333 + 4000 = 7333 \text{ ft}$$

$$\text{Freezing level outside the cloud would be} = \frac{6}{3} \times 1000 + 4000 = 6000 \text{ ft}$$

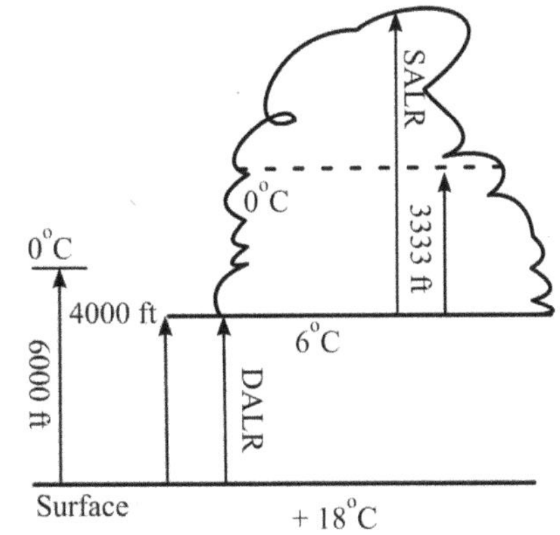

Note:

1. If moist stable air parcel is forced to rise over a mountain barrier stratiform clouds may develop.
2. The difference between TT & T_dT_d is maximum in dry air when RH $\simeq$ 0 % and minimum is zero in saturated air (RH $\simeq$ 100 %)
3. Stability depends on ELR.
4. ELR varies with moisture content at different levels or transport of heat by advection.

8.7 Thermodynamic Diagrams

Thermodynamic diagrams provide graphical display of atmospheric process, such as isobaric, isothermal, dry adiabatic, pseudo adiabatic processes that the atmosphere is subjected. These diagrams also contain constant values of saturation mixing lines. There are several thermodynamic diagrams (emagram, T-ϕ gram etc) but we shall learn about the T-ϕ gram which is used in IMD. To understand the thermodynamic diagram, a little bit of thermodynamic equations used in meteorology will be discussed presently.

The first law of thermodynamics states that the quantity of heat absorbed by a system is equal to the sum of external work performed by the system and increase in its internal energy

$$\delta Q = di + \delta W \quad \text{or} \quad dH = di + dW \qquad \text{....(8.1)}$$

where δQ = heat quantity absorbed by the system

di = increase in internal energy

$\delta W = dW$ = work done by the gas

dH = heat quantity added.

For ideal gas the thermodynamic fist law equation is

$$\delta Q = C_v\, dT + pd\,\alpha \qquad .....(8.2)$$

$C_v = \left(\frac{dh}{dT}\right)$ α –constant, where C_v = specific heat of a gas at constant volume.

For adiabatic process $dh = C_v\, dT + Pd\,\alpha$ .....(8.3)

or $$\frac{dQ}{dt} = C_p \frac{dT}{dt} - \alpha \frac{dp}{dt} \qquad .....(8.4)$$

where t is time, T = absolute temperature, C_p = Specific heat of a gas at constant pressure given by

$$C_p \left[\frac{dh}{dT}\right] P\,\text{constant}$$

equations (8.3) and (8.4) represent first law of thermodynamics used in meteorology.

Poisson equation is $\theta = T\left(\frac{P}{P_o}\right)^{-K}$ .....(8.5)

where θ = potential temperature,

P_o = 1000 hPa

P = pressure, hPa

$k = \frac{r-1}{r}$ (where $r = \frac{C_p}{C_v}$ adiabatic exponent)

Def. Potential Temperature **(θ):** The temperature attained by a sample of dry air when the sample is compressed or expanded adiabatically from a given state to the pressure level P_o = 1000 hPa. θ is constant in an adiabatic process.

Entropy (ϕ): The physical quantity which describes the ability of a system to do work is called entropy of the system.

Entropy may be taken as randomness of a process or disorder.

$$d\phi = \frac{dQ}{T} \qquad .....(8.6)$$

Change in entropy (b to a)

$$\Delta\phi = [\phi_b - \phi_a] = \frac{Q}{T} \qquad .....(8.7)$$

***Isentropic process*:** A thermodynamic process in which entropy of the system remains constant is called isentropic process.

The relation between entropy (ϕ) and potential temperature (θ) is given by the relation.

$$\phi = C_p \ln\theta + \text{constant} \qquad .....(8.8)$$

that is, entropy is proportional to the logarithmic of potential temperature.

***Geopotential* (ψ):** The potential energy imparted (given) to a unit mass, when it is lifted from mean sea level to a height (Z) is called geopotential. It is defined as

$$d\psi = gdZ \quad (\psi = 0 \quad \text{when} \quad Z = 0)$$

The value of 'g' at msl is 9.8 m/s^2,

A unit of geopotential meter (gpm) in defined as

$$\psi = \frac{1}{9.8}\int_0^Z gdZ \qquad ....(8.9)$$

where ψ (hight) is in geopotential meters.

8.7.1 T-ϕ gram

A thermodynamic diagram in which X-axis is T (temperature Celsius/Kelvin) and Y-axis is $C_p \ln\theta$ is called T-ϕ gram. The properties of T-ϕ gram used in IMD (see Fig. 8.1.3) are:

(i) The coordinate axes are: T and $\ln\theta$. One axis T is linear (isotherms are vertical straight lines), horizontal lines potential temperature or dry adiabats.

(ii) Isobars are logarithmic curves, which slope upward to the right. As temperature increases the slope of the isobar decreases.

(iii) The psuedo-adiabats are appreciably curved. Saturate adiabats curving left and gradually parallel to dryadiabats.

(iv) Saturation mixing ratio lines are nearly straight. Dashed lines are holygrics or lines of equal saturation (g/kg).

(v) The angle between isotherms and adiabats is 90°.

(vi) Area is proportional to energy.

Normand theorem: According to this theorem, if we produce the dry-adiabat from surface (level) dry bulb temperature (TT) point and the isohygric from the surface dew point (T_d T_d), and saturate adiabat from surface wet bulb temperature (T_w T_w) point, all three meet at a point. This point is LCL (lifting condensation level) or Normand point.

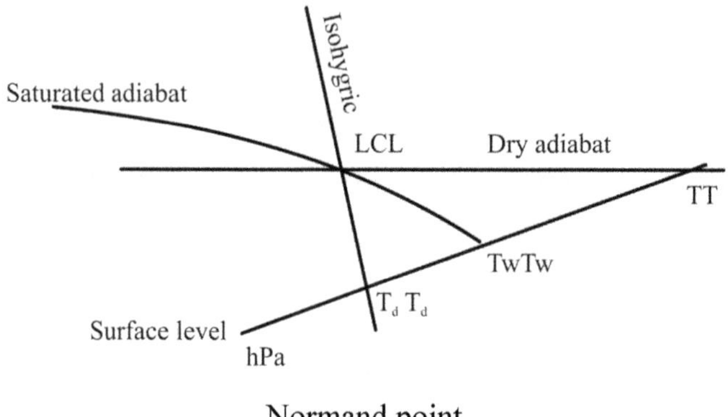

Fig. 8.13 T-ϕ gram sample.

Questions

1. An air parcel which is lifted in unstable atmosphere when reaches 100% RH may produce;
   (a) Stratiform clouds
   (b) Convective clouds or heap clouds
   (c) Fog/mist
   (d) None
2. An air parcel which is lifted in stable atmosphere when reaches 100% RH and sinks back may produce:
   (a) Stratiform clouds
   (b) Convective clouds or heap clouds
   (c) Fog/mist
   (d) None
3. stable or unstable atmosphere need not produce clouds. If the parcel does not attain 100% RH, then a stable atmosphere may produce:
   (a) Stratiform cloud (b) Poor visibility
   (c) Excellent visibility (d) Precipitation

4. Stable / unstable atmosphere need not produce clouds. If the parcel does not attain 100% RH, then unstable atmosphere may produce.:
   (a) Convective clouds (b) Poor visibility
   (c) Excellent visibility (d) Precipitation
5. If ELR < SALR but parcel lifted to RH 100% may produce:
   (a) Convective cloud (b) Stratiform cloud
   (c) Clear sky (d) None
6. If ELR > SALR and parcel lifted to RH 100% may produce:
   (a) Convective cloud (b) Stratiform cloud
   (c) No cloud (d) None above
7. If the surface air temperature is 25 °C, RH 60% , the temperature at 2 km height (assuming DALR prevailing up to 2 km from surface) is about:
   (a) 19 °C (b) 15 °C
   (c) 5 °C (d) None above
8. Warm and moist air moving over a cold ground gives rise to:
   (a) Thunder cloud (b) Fog and stratus
   (c) Frontal cloud (d) All above
9. An isothermal atmosphere is:
   (a) Stable (b) Unstable
   (c) Neutral (d) Can not say
10. A superadiabatic lapse rate atmosphere is:
   (a) Stable (b) Unstable
   (c) Conditionally unstable (d) Neutral
11. Radio-sonde is used to measure:
   (a) Cloud amount and its speed
   (b) Wind at various pressure levels
   (c) The temperature and humidity at various pressure levels
   (d) None above

12. Rawin is used to measure:
    (a) cloud amount and its speed
    (b) wind at various pressure levels
    (c) The temperature and humidity at various pressure level
    (d) All above

13. The rate of fall of pressure with height (pressure lapse rate) in warm air mass will be:
    (a) The same as in a cold air mass
    (b) More than in a cold air mass
    (c) Less than in a cold air mass
    (d) None

14. The rate of fall of pressure with height (pressure lapse rate) in a cold air mass will be:
    (a) The same as in warm air mass
    (b) More than in a warm air mass
    (c) Less than in a warm air mass
    (d) None

15. If an altimeter reads aerodrome elevation when the aircraft at ground, then the sub-scale must have been set to:
    (a) QEF (b) QNH
    (c) QNE (d) Can't say

16. If an altimeter reads zero when the aircraft is on aerodrome ground, then the sub-scale must have been set to:
    (a) QEF (b) QNH
    (c) QNE (d) None above

17. The density of moist air is:
    (a) Lower than dry air (b) Heavier than dry air
    (c) Equal to dry air (d) None

18. Under the same conditions of pressure and temperature, moist air is
    (a) Lighter than dry air
    (b) Heavier than dry air
    (c) Has the same density as dry air
    (d) None

19. Solar radiation received by the earth is called:
    (a) Convection (b) Insolation
    (c) Inversion (d) Long wave radiation
20. Earth radiates heat energy in the:
    (a) Short waves (b) Ultra short waves
    (c) Long waves (d) All above
21. DALR is about:
    (a) 5 °C /km (b) 10 °C/km
    (c) 3 °C/1000 ft (d) both (b) and (c)
22. SALR is:
    (a) Variable, at sea level 5 °C/km, gradually approaches to 9.8 °C/km at temperatures below – 40 °C
    (b) Variable, at sea level 10 ° C/km gradually fall to 5 °C/km at temperatures below – 40 °C
    (c) 1.8 °C/km and does not change
    (d) None above
23. In lower levels, the standard SALR is:
    (a) 1.8 °C/1000 ft (b) 0.6 °C/100 m
    (c) Both (a) and(b) (d) None
24. Any lapse rate less than SALR is:
    (a) Absolutely unstable (b) Conditionally unstable
    (c) Absolutely stable (d) Neutral
25. Any lapse rate more than DALR is described as:
    (a) Absolutely unstable (b) Conditionally unstable
    (c) Absolutely stable (d) Absolutely neutral
26. The density of air:
    (a) Varies directly as pressure and temperature
    (b) Varies directly as pressure and inversely as temperature
    (c) Varies inversely as pressure and directly as temperature
    (d) Varies inversely as pressure and temperature
27. The major constituent of atmospheric air by weight is:
    (a) Oxygen (b) Nitrogen
    (c) Hydrogen (d) Carbon dioxide

28. The major constituent of atmospheric air by volume is:
    (a) Oxygen (b) Nitrogen
    (c) Hydrogen (d) Carbon dioxide
29. In troposphere the density of air:
    (a) Increases with height (b) Decrease with height
    (c) Remains constant (d) None
30. The altimeter of an aircraft parked at airfield indicated 500 ft when the sub-scale of the instrument was set to 1013 hPa. If the QNH at the airfield was 1000 hPa and a change of pressure 1hPa represents a height change of 27 ft, what is the elevation of the air field?
    (a) 149 ft (b) 851 ft
    (c) 351 ft (d) None
31. An adiabatic process is one in which heat content of the system:
    (a) Increases (b) Decreases
    (c) Does not change (d) None above
32. The temperature to which moist air must be cooled at its existing pressure and humidity mixing ratio in order that it shall be saturated is called:
    (a) Virtual temperature
    (b) Dew point temperature
    (c) Wet bulb temperature
    (d) Dry bulb temperature
33. The atmosphere may be nearly taken to be dry if the difference DB and DP is:
    (a) 30 °C (b) 5 °C (c) 1 °C (d) 0 °C
34. When RH is 100%, then DB – DP is:
    (a) 10 °C (b) 5 °C (c) 2 °C (d) 0 °C
35. Dry air having a temperature of 30 °C on the surface when forced to rise adiabatically by 3000 ft would attain a temperature of:
    (a) 9 °C (b) 21 °C (c) 0 °C (d) 30 °C
36. Dry air parcel having temperature 21 °C at 3000 ft when forced to go down to surface adiabatically would attain a temperature:
    (a) 9 °C (b) 30 °C (c) 0 °C (d) 21 °C

37. A dry air parcel from the surface level with temperature 30 °C is lifted adiabatically to height of 4000 ft. Its temperature at this level will be:

(a) –10 °C (b) 18 °C (c) 22 °C (d) 0 °C

38. Assuming the water vapour content of the atmosphere to be constant throughout on a sunny day, the RH will very as follows:

(a) It is maximum in the morning, minimum in the afternoon and then rises

(b) It is minimum in the morning, maximum in the afternoon and then falls

(c) Since there is no change in the water vapour content of the atmosphere, RH remains constant during the day

(d) None

39. An air parcel is at a temperature of 25 °C and pressure 1000 hPa. It is cooled till it gets saturated without changing the water vapour content of the parcel and its pressure. The temperature at this stage is observed to 16 °C. This temperature of 16 °C is:

(a) Potential temperature of the air parcel

(b) The dew point temperature

(c) The wet bulb temperature

(d) The freezing temperature

40. The air temperature at the surface level is 25 °C and at (850 hPa) 1.5 km height agl is 27 °C recorded by radio sonde on particular day. Find the stability.

(a) The atmospheric layer surface to1.5 km level is unstable

(b) The layer is neutral

(c) The layer is stable

(d) The layer is conditionally unstable

41. Pressure decreases rapidly with height in:

(a) Warm air

(b) Homogeneous air of isothermal layer

(c) Cold air

(d) None

42. Pressure decreases less rapidly with height in:
    (a) Warm air
    (b) Homogeneous air of isothermal layer
    (c) Cold air
    (d) None above

43. The rate at which an ascending parcel of dry air cools is called:
    (a) Isothermal (b) Inversion
    (c) SALR (d) DALR

44. The rate at which an ascending parcel of saturated air cools is called:
    (a) Isothermal (b) Inversion
    (c) SALR (d) DALR

Note: (i) DALR means the rate which an ascending parcel of dry air cools (10 °C/km or 3 °C/1000 ft)

(ii) SALR means the rate at which an ascending parcel of saturated air cools (6 °C/km or 1.8 °C/1000 ft)

45. The dry air is said to be stable if:
    (a) ELR > DALR (b) ELR < DALR
    (c) ELR < SALR (d) Both (b)&(c)

46. Saturated air is said to stable if:
    (a) ELR < SALR (b) ELR = DALR
    (c) ELR = SALR (d) ELR > SALR

47. Dry air is said be unstable if:
    (a) DALR < ELR (b) DALR > SALR
    (c) DALR = ELR (d) DALR < SALR

48. Saturated air is said to unstable of:
    (a) ELR = SALR (b) ELR < SALR
    (c) ELR > SALR (d) DALR > SALR

49. If the value of (ELR > SALR, ELR < DALR) DALR > ELR > SALR the atmosphere is said be:
    (a) Stable (b) Unstable
    (c) Conditionally unstable (d) Neutral

50. If the air temperature near the surface is 15 °C, what will be the temperature of ascending air parcel of dry air at 5000 ft agl?
 (a) 0 °C (b) – 0.5 °C (c) + 0.5 °C (d) – 2 °C

51. If the air temperature near the surface is 15 °C, what will be the temperature of ascending air parcel of saturated air at 5000 ft?
 (a) 0 °C (b) 6 °C (c) – 6 °C (d) – 5 °C

52. The atmosphere is said to be conditionally stable if:
 (a) ELR > DALR (b) ELR < SALR
 (c) SALR < ELR < DALR (d) Does not exist

53. On a T-ϕ gram, from a surface level (i) DB, (ii) WB, (iii) DP temperature points respectively if we draw lines parallel to dry adiabat, saturated adiabat and isohygric all meet at a point, which is called:
 (a) Freezing point
 (b) Melting point
 (c) LCL (lifting condensation level) or Normands point
 (d) Stability critical point

Answers

1. (b)	2. (a)	3. (b)	4. (c)
5. (b)	6. (a)	7. (c)	8. (b)
9. (a)	10. (b)	11. (c)	12. (b)
13. (c)	14. (b)	15. (b)	16. (a)
17. (a)	18. (a)	19. (b)	20. (c)
21. (d)	22. (a)	23. (c)	24. (c)
25. (a)	26. (b)	27. (b)	28. (b)
29. (b)	30. (a)	31. (c)	32. (b)
33. (a)	34. (d)	35. (b)	36. (b)
37. (b)	38. (a)	39. (b)	40. (c)
41. (c)	42. (a)	43. (d)	44. (c)
45. (c)	46. (a)	47. (a)	48. (c)
49. (c)	50. (a)	51. (b)	52. (d)
53. (c)			

Chapter 9

Atmospheric Aerosols and Turbulence

An aerosol is suspension of solid or liquid particle matter in the atmosphere or gaseous medium.

The important constituents of atmospheric aerosols are: silicates, salts, acids, metal oxides, organic combustion products, Biological material, Volcanic material, Meteorites dusts. The radius of the aerosols vary from 6×10^{-4} μm to 150 μm (1 micron meter = 10^{-6} m).

Atmospheric ions (electrically charged particles) are classified into four groups according to their mobility B:

Small ions $1 > B \geq 10^{-2}$ cm^2 v^{-1} sec^{-1} ($6.6 < r < 78 \times 10^{-4}$ μm)

Large ions $10^{-2} > B \geq 10^{-3}$ cm^2 V^{-1} sec^{-1} ($78 < r < 250 \times 10^{-4}$ μm)

Langevion ions $10^{-3} > B \geq 2.5 \times 10^{-4}$cm^2 V^{-1}sec^{-1}($250 < r < 570 \times 10^{-4}$ μm)

Ultra large ions $B > 2.5 \times 10^{-4}$ cm^2 V^{-1} sec^{-1} ($r > 570 \times 10^{-4}$ μ m)

Electrically neutral particles are classified into three groups

(r = radius of the particle).

The Aitkin nuclei (r < 0.1 μm)

The large nuclei (0.1 < r < 1 μm)

The Giant nuclei (r > 1 μm)

The density (or the number concentration) of particles in the atmosphere is highly variable. The concentration of aerosol particles decrease with increasing size.

The life period of small ions in the atmosphere varies from few minutes to few seconds. The life period of Intermediate size particles in the lower troposphere varies 2 days to 2 weeks but 2 weeks to 4 weeks in the upper atmosphere.

Aerosols pay an important role, particularly large nuclei, in the formation of clouds and precipitation. The concentration of Giant nuclei in the atmosphere is very small and their role is limited. The concentration of Aitkin nuclei is high but they require super saturation for condensation process.

The average radii of cloud droplets, over land is 2 to 20 μm and over sea 3 to 22 μm. The average radii of rain drops exceed 100 μm.

***Formation of Aerosols*:** Aerosols are mainly formed as dispersion, combustion and photochemical action.

Dispersion is breaking of large particles or drops into smaller units which remain suspended in the atmosphere.

On combustion the volatile components of fuels (particularly carbon) and from mineral inorganic compound aerosols enter into atmosphere. On cooling they mix up with air and form very minute aerosols (diameter less than 0.1 μm) and found in great concentration.

The ultraviolet rays from the sun dissociate some of the constituents of the atmosphere into ions. Atmospheric lightning in the lower atmosphere produces nitrates and acidic salts. The photo-chemical action produces new chemical components but their concentration is small.

The diameter of some particles that found in the atmosphere are given is the Table 9.1.

Table 9.1

Nature of particle	Diameter size μm
Smog	0.01-1
Thick fog and cloud	3-80
Coal dust	1-100
Sulphuric acid mist	0.5-5
Bacteria	0.5-50
Soot	0.01-0.5
Cement dust	0.7-10
Drizzle drops	50-500
Raindrops	500-2500
Atmospheric dust	0.01-30

Note: (i) When RH exceeds 100%, the air is called supersaturated.

(ii) Aerosols which have affinity to water are called hygroscopic.

9.1 Formation and Dispersal of Clouds

As we have studied earlier that water is present in the lower atmosphere (troposphere) in varying degrees. Water vapour is not visible in the atmosphere but the formation of clouds provides us the visual evidence of presence of water in the atmosphere.

Clouds are formed due to the cooling of moist air in vertical ascent of air. The formation and dispersal of clouds is a very complex process but an elementary idea of the process is described here.

Condensation nuclei: condensation is a reverse process of evaporation. Condensation occurs in the atmosphere in the presence of minute particles (aerosols) which are called condensation nuclei or cloud condensation nuclei. Condensation of moisture in (dust free) clear air occurs with great difficulty or requires supersaturation (200-500% RH). Cloud drops remain supercooled upto – 25 to – 35 °C, but they freeze to ice crystals before reaching – 40 °C, even without the presence of nuclei.

Three types of nuclei

1. Particles which are insoluble in water and unwettable. These require supersaturation for condensation. Consequently they play small role as condensation nuclei.
2. Particles which are insoluble in water but wettable. These particles, in general irregular in shape and have flaky structure. Condensation

occurs on these particles even before saturation (due to capillary condensation in cavities or absorption of water in pores). These play important role in condensation process.

3. Particles soluble in water or droplets of solution. These particles require one third supersaturation as compared to above first and second type particles.

As a result of coagulation of wettable soluble and insoluble particles, there found a large number of particle types between type 2 and 3.

Generally condensation begins on nuclei when RH 100%. However on hygroscopic large nuclei condensation begins even before saturation. On sodium chloride aerosol (common salt) particles condensation begins at 78% RH. In contrast to this in purified air (free from aerosols) water vapour may not condense even at a supersaturation of 200% or more. Condensation on Aitkin nuclei requires supersaturation of 200-300% RH. Condensation on large nuclei begins at RH greater than 101% .

***Freezing nuclei*:** Water droplets in the atmosphere may not solidify when cooled to o °C. As said earlier, they may remain in liquid state even upto – 40 °C. Certain particles in the atmosphere help in freezing supercooled droplets about itself at much higher temperature (– 10 to – 25 °C) than – 40 °C. These nuclei are called freezing nuclei.

***Deposition or Sublimation nuclei*:** We have studied that change of water vapour directly into ice crystals without passing through liquid state is called deposition or sublimation. The particles upon which ice crystals form by deposition or sublimation are called sublimation nuclei. It has not been proved experimentally that sublimation nuclei is different from freezing nuclei. In meteorology the term freezing nuclei is used for ice forming nuclei, which is also called ice nuclei. The presence of ice nuclei is rare in the atmosphere. The majority of freezing nuclei are particles of certain clay minerals which were lifted high into atmosphere from the surface of the earth by turbulent mixing.

9.2 Atmospheric Turbulence

Anemograph trace indicate that the surface wind changes rapidly both in direction and speed. These fluctuations are the consequence of turbulent airflow with many eddies near the earth's surface. When wind blows over a rough surface it develops mechanical turbulence and eddy motions. Turbulence depends on a number of factors like wind speed, roughness of the surface, the environment lapse rate etc.

In meteorology turbulence is classified as (i) Thermal or Convection turbulence, (ii) Mechanical or frictional turbulence.

1. ***Thermal turbulence***: This develops due to convection currents set up by surface heating mainly caused by insolation. As the surface of the earth gets heated, temperature rises and environmental lapse rate increases. Air near the earth becomes unstable and convection currents become part of turbulent structure.

   In convective clouds (in thunderstorms) latent heat is released at cloud height, which creates updrafts and downdrafts or large eddy-type motions. In turn, these create a number of smaller turbulent eddies. Thermal turbulence or convection need not necessarily develop clouds. In Rajasthan Andhis (dust storms) are developed due to lack of moisture. In aviation the effect of turbulence on aircraft is called "bumpiness". If an inversion exists above the turbulent layer, which inhibits further vertical motion. Cloud layers become strato-cumulus.

2. ***Mechanical or frictional turbulence***: It develops due to the widespread roughness of the earth's surface. This turbulence is further enhanced by the airflow over buildings, trees, high ground, hills etc., and also generates by wind shear. In general, wind flow closer to the earth (that is in the planetary boundary layer) is strongly affected by the friction and is called frictional flow. Smooth or laminar flow occurs (frictionless flow) above planetary boundary layer where friction is absent. Mechanical turbulence is smaller over the open sea area.

   During stable atmosphere and light/calm winds the mechanical turbulence is greatly diminished.

   A severe form of turbulence occurs in clear air is called Clear Air Turbulence (CAT), which is hazardous for aircraft flying in those regions.

9.3 Temperature Inversions

In some cases temperature may increase with height in certain layers of atmosphere, which were called inversion. Inversion layer is stable and it acts as a lid for cloud development.

Trade wind inversion: The inversion of temperature in lower altitudes of trade winds is caused by the subsidence of air from the Subtropical High

Pressure Belt (STHB). This inversion persists in trade winds which move towards the equator and puts the lid on cloud development over ocean areas.

Trade wind inversion is prominent over western coasts of North Africa, South Africa and over North America, South America. In these areas the base of cloud comes down to about 500 m and the rise of temperature in inversion goes upto 10 °C. From the base of the temperature inversion RH decreases to the top of inversion by about 50%. Below the inversion layer air mass is moist and cool, and above the inversion layer air mass is dry and warm. The maximum intensity of inversion occurs in subtropical belt region-over northeast Atlantic and northeast Pacific and over southeast Atlantic and southeast Pacific. As it moves further closer to the equatorial trough the inversion dies down.

The temperature inversions are produced by a number of reasons, of which the following are important.

(i) Radiation or surface inversion, (ii) Turbulence, (iii) Subsidence, (iv) Frontal development, (v) Valley inversion and (vi) Tropopause inversion.

(i) ***Radiation Inversion*****:** A radiation inversion is produced as surface inversion. During night earth cools by nocturnal radiation (long wave radiation). Air close to the surface of the earth becomes cooler than the air above it. Radiation cooling occurs during night when the sky is clear, and wind is calm or light. If there is sufficient moisture mist/fog may develop in the early morning and in some cases ground frost may occur. During cloudy nights, radiation cooling also occurs at cloud tops, which produces inversion at cloud top layer. During cloudy nights at surface nocturnal cooling is restricted and also on occasions of strong surface wind inversions do not develop (due to quick mixing).

(ii) ***Turbulence inversion*****:** Sometimes, surface inversion is carried upward by mechanical turbulence. Radiational cooling produces surface inversion which is lifted by mechanical turbulence to spread. Turbulence, sometimes may produce an upper air inversion. In turbulence layer air is pushed up and down. Lifted air cools due to adiabatic expansion, while the pushed down air warms up due to adiabatic compression. This results in upper inversion.

(iii) ***Subsidence Inversion*:** In areas of surface level divergence and associated with upper air convergence over the same area, air sinks or subsides over a large area and hundreds of meters of thickness (see Fig. 9.1). In this process the subsided air warms up by adiabatic compression heating and develops inversion through a great depth near the surface of the earth. The subsiding air temperature is warmer than the original surface air temperature.

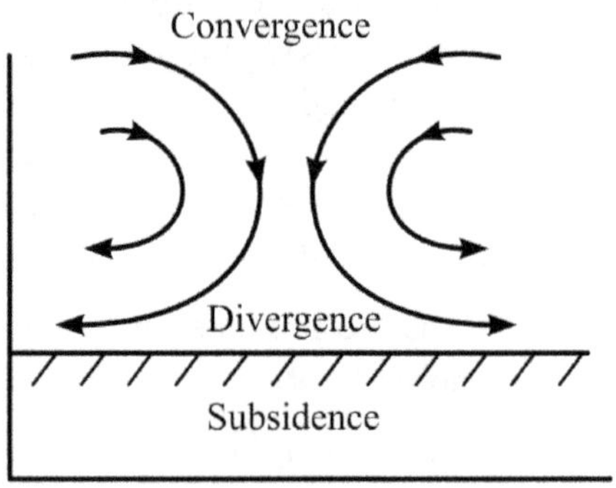

Fig. 9.1

(iv) ***Frontal inversion:*** when warm air mass is forced to move over cold air mass in a frontal zone, inversion develops. This is called frontal inversion. Depending upon the moisture convective clouds and precipitation results.

(v) ***Valley inversion:*** During clear nights, mountain slopes cool down and the air close to the slope becomes heavier and moves down into the valley. However during flow down it warms up due to adiabatic compression but the cold surface counteracts and keeps it cool. The cold air flows into valley replacing the warm air in it. Thus valley inversion develops. Mistral wind is of this kind.

(vi) ***Tropopause inversion:*** We have studied earlier that temperature falls with increasing height in troposphere. The top of the troposphere is called tropopause, where temperature begins to rise or isothermal. It is clear, troposphere is unstable and stratosphere is stable. Hence tropopause inversion.

The presence of inversion indicates the following facts: Stability of air, subsidence and (nearly) absence of turbulence.

During surface inversion fog, mist, smog may develop. Dust particles may be spread out below inversion level due to absence of turbulence, which will lead to poor visibility at surface. Inversion also effects radio wave transmission of wavelength (λ) less than 10 m.

In case of development of strong turbulence (after sunrise) the inversion dissipates into isothermal or completely dissipates depending upon temperature built up in the area. During winter, radiation inversion develops in the early morning due to nocturnal cooling under clear skies and calm / light winds. In such conditions if RH is 75% or more fog/mist may develop and on some occasions ground frost may develop.

Questions

1. Turbulence encountered in CB is to be classified as:
   (a) Frictional turbulence (b) Thermal turbulence
   (c) Clear air turbulence (d) None
2. Temperature decreases with increasing altitude in the:
   (a) Inversion layer (b) Troposphere
   (c) Isothermal layer (d) Stratosphere
3. Conditions most favourable for occurrence of Radiation fog during early morning hours at New Delhi in winter are:
   (a) RH about 90%, clear sky during previous night, strong surface winds about 20 kt.
   (b) RH about 90%, clear sky during previous night, light surface winds 2 to 3 kt.
   (c) RH about 90%, overcast sky during previous night and calm surface wind.
   (d) None above.
4. If the sky is overcast and it is heavily raining at the station, then:
   (a) Dew point will be much less than the DB temperature at the station.
   (b) Dew point will be higher than DB temperature at the station.
   (c) Dew point will be equal or nearly equal to DB temperature at the station.
   (d) None

5. Lapse rate is maximum on:
   (a) Cloudy day (b) Clear night
   (c) Sunny day (d) Rainy day

6. On a clear night when DB temperature and Dew point is same, you expect:
   (a) Rain (b) Fog
   (c) Haze (d) None

7. The rate at which an ascending parcel of dry air cools is called:
   (a) SALR (b) DALR
   (c) Isothermal (d) Super SALR

8. In a layer, increase of temperature with increasing altitude is called:
   (a) Inversion layer (b) Isothermal layer
   (c) DALR (d) SALR

9. From which measurement of the atmosphere can stability be determined?
   (a) Atmospheric pressure (b) The ambient lapse rate
   (c) DALR (d) SALR

10. Which conditions are favourable for the formation of surface based temperature inversion:
   (a) Clear, cool nights with calm or light wind
   (b) Area of unstable air, rapidly transferring heat from the surface
   (c) Broad areas of cumulus clouds with smooth, level bases at the same altitude
   (d) Cloudy, cool nights with calm or light wind.

11. Which aerosols play an important role in the formation of clouds and precipitation:
   (a) Aitkin nuclei (b) Large nuclei
   (c) Giant nuclei (d) Non-hygroscopic

12. The diameter of fog & cloud particles vary:
    (a) 1-10 μm (b) 3-80 μm
    (c) 50-500 μm (d) 1-100 μm
13. The diameter of drizzle drops vary:
    (a) 1-10 μm (b) 3-80 μm
    (c) 50-500 μm (d) 500-2500 μm
14. The diameter of rain drops vary:
    (a) 1-100 μm (b) 3-80 μm
    (c) 50-500 μm (d) 500-2500 μm
15. When RH exceeds 100% the air is called:
    (a) Saturated (b) Supersaturated
    (c) Dry (d) Moist
16. Visual evidence of presence of water in the atmosphere is:
    (a) Dry Bulb temperature (b) Fog; cloud
    (c) RH (d) Dew point
17. In the atmosphere, condensation occurs:
    (a) In the absence of dust particles
    (b) In the presence of minute particles, is called condensation nuclei
    (c) In the purified air
    (d) None
18. Condensation in atmosphere on Aitken nuclei occurs at:
    (a) RH 90% (b) RH 100%
    (c) RH 200-300% (d) None
19. In the atmosphere certain particles help in freezing supercooled droplets about itself are called:
    (a) Condensation nuclei
    (b) Freezing nuclei or ice nuclei
    (c) Sublimation nuclei
    (d) None

20. In the atmosphere certain particles upon which ice crystals form by deposition are called :
    (a) Condensation nuclei (b) Freezing nuclei
    (c) Sublimation nuclei (d) Ice nuclei
21. When wind blows over rough earth surface, it develops:
    (a) Thermal turbulence (b) Mechanical turbulence
    (c) Frictional turbulence (d) Both (b) and (c)
22. Thermal turbulence or convection currents:
    (a) Necessarily develop CB clouds
    (b) Not necessarily develop CB Cloud
    (c) Develop instability
    (d) Both (b) and (c)
23. Effect of turbulence on aircraft is:
    (a) Aircraft icng (b) Bumpiness
    (c) Engine icing (d) Both (a) and (c)
24. If an inversion exists above the turbulent layer:
    (a) Which favours vertical development of cloud
    (b) Which favours CB development
    (c) Which inhibits further vertical development of cloud
    (d) None
25. Mechanical turbulence depends on:
    (a) Widespread roughness of the earth
    (b) Wind speed
    (c) Instability
    (d) All above
26. Mechanical turbulence is:
    (a) More during stable atmosphere with light or calm winds
    (b) More during unstable atmosphere
    (c) More during strong winds
    (d) Both (b) and (c)

27. Inversion layer is :
    (a) Stable
    (b) Unstable
    (c) Acts as a lid for cloud development
    (d) Both (a) and (c)

28. Trade wind inversion is caused by:
    (a) Radiation inversion (b) Turbulence inversion
    (c) Subsidence inversion (d) Frontal inversion

29. Trade wind inversion in lower altitudes:
    (a) Aids vertical development of clouds
    (b) Develops instability
    (c) Inhibits vertical development of clouds and acts as a lid
    (d) Both (a) and (b)

30. Frictional turbulence is:
    (a) More over deserts
    (b) Less over sea areas
    (c) Less over rough hilly areas
    (d) None

31. Radiation inversion is produced:
    (a) During cloudy, calm nights
    (b) During clear, cool nights with calm / light winds
    (c) During clear nights with strong winds
    (d) None

32. During cloudy nights:
    (a) Radiational cooling and surface inversion develops
    (b) Radiational cooling produces inversion at cloud top layer
    (c) Radiational cooling occurs at surface
    (d) None above

33. In areas of surface level divergence associated with upper level ….. …………… produces subsidence inversion.
    (a) Divergence (b) Laminar flow
    (c) Convergence (d) None above

Answers

1.	(b)	2.	(b)	3.	(b)	4.	(c)
5.	(c)	6.	(b)	7.	(b)	8.	(a)
9.	(b)	10.	(a)	11.	(b)	12.	(b)
13.	(c)	14.	(d)	15.	(b)	16.	(b)
17.	(b)	18.	(c)	19.	(b)	20.	(c)
21.	(d)	22.	(d)	23.	(b)	24.	(c)
25.	(d)	26.	(d)	27.	(d)	28.	(c)
29.	(c)	30.	(b)	31.	(b)	32.	(b)
33.	(c)						

CHAPTER 10

Cloud Classification

Naturalist Lamark (1802) first classified clouds into five principal types, was published in France. However his classification was not useful for recognizing.

Luke Howard (in 1803) in England classified clouds into three classes – Cirrus, Cumulus and Stratus.

In 1840, the German meteorologist Kaemtz added stratocumulus to the Howard's form. In 1855, Renou (Director of observatories at Parc-Saint Maur and Montsouris) gave the present form.

Nomenclature: Cirrocumulus, Cirrostratus, Altocumulus and Altostratus.

In 1887, Hildebrandsson and Abercromby published a cloud classification which became the basis of international cloud classification as a result of the recommendation of International Meteorological Conference held at Munich in 1891 and the International Cloud Atlas came into being in 1896.

Clouds are continuously in the process of evolving in the atmosphere. Water vapour changes into water droplets and ice particles in the atmosphere and float. They become visible to our eyes as white clouds high in the sky, sometimes low gray are dark clouds. Meteorologists studied the clouds in great depth about its structure and development.

***Cumulifrom clouds*:** Which are generally separated from each other by clear spaces. They have vertical development and clearcut boundaries.

stratiform clouds : They form in sheets or layers covering large areas of the sky. Layer clouds have diffused boundaries.

A number of Latin words used to reveal the characteristics of clouds.

'*nimbus*' -(a shower) is added to the names of clouds which produce precipitation.

The prefix '*fracto*' (broken) appears in the beginning of the names of broken, wind-blown clouds.

The word '*cirrus*' (a hair) is used to indicate a cloud hair like appearance.

10.1 Clouds Genera

Clouds are grouped into ten genera.

Cirrus, Cirrocumulus, Cirrostratus, Altocumulus, Altostratus, Nimbostratus, Stratocumulus, Stratus, Cumulus and Cumulonimbus.

By convention clouds are divided into three etages or intervals – High, Middle or Medium and Low. The 'etages' overlap and their limits change with latitude. The approximate heights of the limits are given in the Table 10.1.

Tabel 10.1

Etages	Polar regions	Temperate regions	Tropical regions
High	3-8 km 10000-25 000 ft	5-13 km 16500-45000 ft	6-18 Km 20000-60000 ft
Middle	2-4 Km 6500-13000 ft	2-7 Km 6500- 23000 ft	2-8 Km 6500-25000 ft
Low	From earth's surface to 2 km/ 6500 ft	From the earths surface to 2 km /6500 ft	From the earth's surface to 2 km/ 6500 ft

The etages in which six of the genera are found as follows:

High level clouds: cirrus, cirrocumulus, cirrostratus,

Middle level clouds: Altocumulus

Low level clouds: Stratocumulus, Stratus.

The other four genera-grouped with following remarks.

(i) Altostratus found usually in middle etage but it extends to higher levels.

(ii) Nimbostratus found in middle etage but usually extends both downwards into low etage and upwards into the upper etage.

(iii) Cumulus and Cumulonimbus found in low etage, but vertically extend into middle and high etages. These clouds are also called clouds of vertical development.

***Definition of a cloud*:** According WMO (World Meteorological Organization) a cloud is a visible aggregate of minute particles of water, ice or both in the free atmosphere.

In simple terms, a cloud is a fog high in the air. From micro-physical point view, clouds are the result of the action of atmospheric nuclei. From macro-physical point view, clouds are the result of the cooling of moist air caused by adiabatic expansion and cooling during vertical motion, contact with colder air mass or by contact with colder land or water surfaces.

Clouds are formed in the atmosphere when the moist air from the earth's surface lifted and then condensed on nuclei due to cooling. The level at which moist air begins to condense is called lifting condensation level (LCL). Clouds are continuously in a process of formation or dissipation depending on the environmental conditions. They appear in an infinite variety of forms. There are, however, some characteristic forms which are frequently observed all over the world.

Clouds are classified based on their height, shape, reflection or refraction of light and colour. See Table 10.2. The WMO (International

Atlas) classification of clouds given below. It is based on Genara (10 main groups; Latin names), Species (sub divisional genara which is based on internal structure and shape), Varieties (special characteristics like arrangement and transparency), Supplementary features (attached to main clouds like udder like, hanging, protuberance etc), Mother clouds (a part of the cloud may develop into pronounced extension from the main cloud, which is called mother cloud).

10.2 Definition of a Cloud

According to WMO, a cloud is a visible aggregate of minute particles of water, ice or both in the free atmosphere. In simplest terms, a cloud is a fog high in the air. From micro-physical point of view, clouds are the result of the action of atmospheric nuclei, from a macro-physical point of view clouds are the result of the cooling of moist air caused by the expansion of air during vertical motion, contact with colder air masses or by contact with colder land or water surfaces.

Clouds are formed in the atmosphere when the moist air from the earth's surface lifted and then condensed on nuclei due to cooling. The level at which moist air begins to condense in called lifting condensation level (LCL). Clouds are continuously in a process of formation or dissipation depending on the environmental conditions and hence appear in an infinite variety of forms. There are, however, some characteristic forms frequently observed all over the world. Clouds are classified based on their height, shape, reflection or refraction of light and colour. The WMO (International Atlas) classification of clouds follows. It is based on Genera (10 main groups, Latin names), species (sub-divisional genera based on internal structure and shape), varieties (special characteristics such as arrangements and transparency), supplementary features (attached to the main clouds such as udder like, hanging, protuberance etc), Mother clouds (a part of the cloud may develop into pronounced extension from the main cloud, which is called mother cloud).

	Genera (Average height of cloud base)	Species	Varieties	Cloud temperature	Supplementary features	Mother cloud Genitus
Fig. 10.1 High Clouds (C_H)	Cirrus (Ci) (5000 – 13700 m)	Fibratus Uncinus Spissatus Castellanus Floccus	Intortus Radiatus Vertebratus Duplicatus	– 20 to – 60 °C	Mamma	Cc, Ac, Cb
	Cirrocumulus (Cc) 5000 – 13700 m	Stratiformis Lenticularis Castellanus Floccus	Undulatus Lacunosus	– 20 to – 60 °C	Virga mamma	– –
	Cirrostratus (Cs) 5000 – 13700 m	Fibratus nebulosus	Duplicatus Undulatus	– 20 to – 60 °C	–	Cc Cb
Fig. 10.2 Medium Clouds (C_M)	Altocumulus (Ac) 2000 – 7000 m	Stratiformis Lenticularis Castellanus Floccus	Translucidus perlucidus Opacus duplicatus radiatus lacunosus	+ 10 to – 30°C	Virga mamma	Cu, Cb
	Altostratus (As) 2000 – 7000 m	–	Translucidus Opacus duplicatus undulatus radiatus	+10 to – 30 °C	Virga praecipitatio pannus mamma	Ac, Cb
	Nimbostratus (Ns) 900 – 3000 m	–	–	+10 to – 15 °C	Praecipitatio Virga pannus	Cu, Cb
	Stratocumulus (Sc) 460 – 2000 m	Stratiformis lenticularis castellanus	Translucidus perlucidus opacus duplicatus Undulatus lacunosus	+15 to – 5	Mamma virga praecipitatio	As, Ns, Cu, Cb
Fig. 10.3 Low Clouds (C_L)	Stratus (St) surface to 460 m	Nebulosus fractus	Opacus translucidus undulates	+ 20 to – 5 °C	Praecipitatio	Ns, Cu, Cb
	Cumulus (Cu) 460 – 2000 m	Humilis, mediocris, congestus, fractus	Radiatus	+15 to – 5 °C	Pileus velum, virga, praecipitatio, arcus, pannus, tuba	Ac, Sc
	Cumulonimbus (Cb) 460 – 2000 m	Calvus capillatus	– –	+15 to – 5 °C	Praecipitatio virga, pannus, incus, mamma, pileus, velum, arcus, tuba	Ac, As, Ns, Sc, Cu

See Figs. 10.1, 10.2, and 10.3

Note: Cloud base height and cloud base temperatures are averages. They differ widely form equatorial regions to polar regions.

Table 10.2

Etage	Polar regions km	Temperate regions km	Tropical regions km
High (Ci, Cs, Cc)	3 -8	5 – 13	6 – 18
Middle (Ac, As, Ns)	2 – 4	2 – 7	2 – 8
Low (Sc, St, Cu Cb)	From the earth's surface to 2 km	From the earth's surface to 2 km	From the earth's surface to 2 km

(A) Definitions of Clouds

WMO (International Atlas) defined the ten cloud genera (10) is given below. Fig. 10.1 to 10.3

1. **Cirrus (Ci):** Detached clouds in the form of white, delicate filaments or white or mostly white patches or narrow bands. These clouds have fibrous (hair-like) appearance, or a silky sheen (lustre) or both.
2. **Cirrocumulus (Cc):** Thin, white patch, sheet or layer of cloud without shading, composed of very small elements in the form of grains, ripples etc., merged or separate, and more or less regularly arranged; most of the elements have an apparent width of less than one degree.
3. **Cirrostratus (Cs):** Transparent, whitish cloud veil of fibrous or smooth appearance, totally or partly covering the sky, and generally producing halo phenomena.
4. **Altocumulus (Ac):** White or grey or both white and grey, patch, sheet or layer of cloud, generally with shading, composed of laminae, rounded masses, rolls etc., which are sometimes partly fibrous or diffuse and which may or may not be merged; most of the regularly arranged small elements usually have an apparent width of between one and five degrees.
5. **Altostratus (As):** Greyish or bluish cloud sheet or layer of striated (small channeled), fibrous or uniform appearance, totally or partly covering the sky, and having parts thin enough to reveal the sun at least vaguely, as through ground glass. Altostratus does not show halo phenomena.
6. **Nimbostratus (Ns):** Grey cloud layer, often dark, the appearance of which is rendered diffuse by more or less continuously falling rain or snow, which in most cases reaches the ground. It is thick enough throughout to blot out the sun.

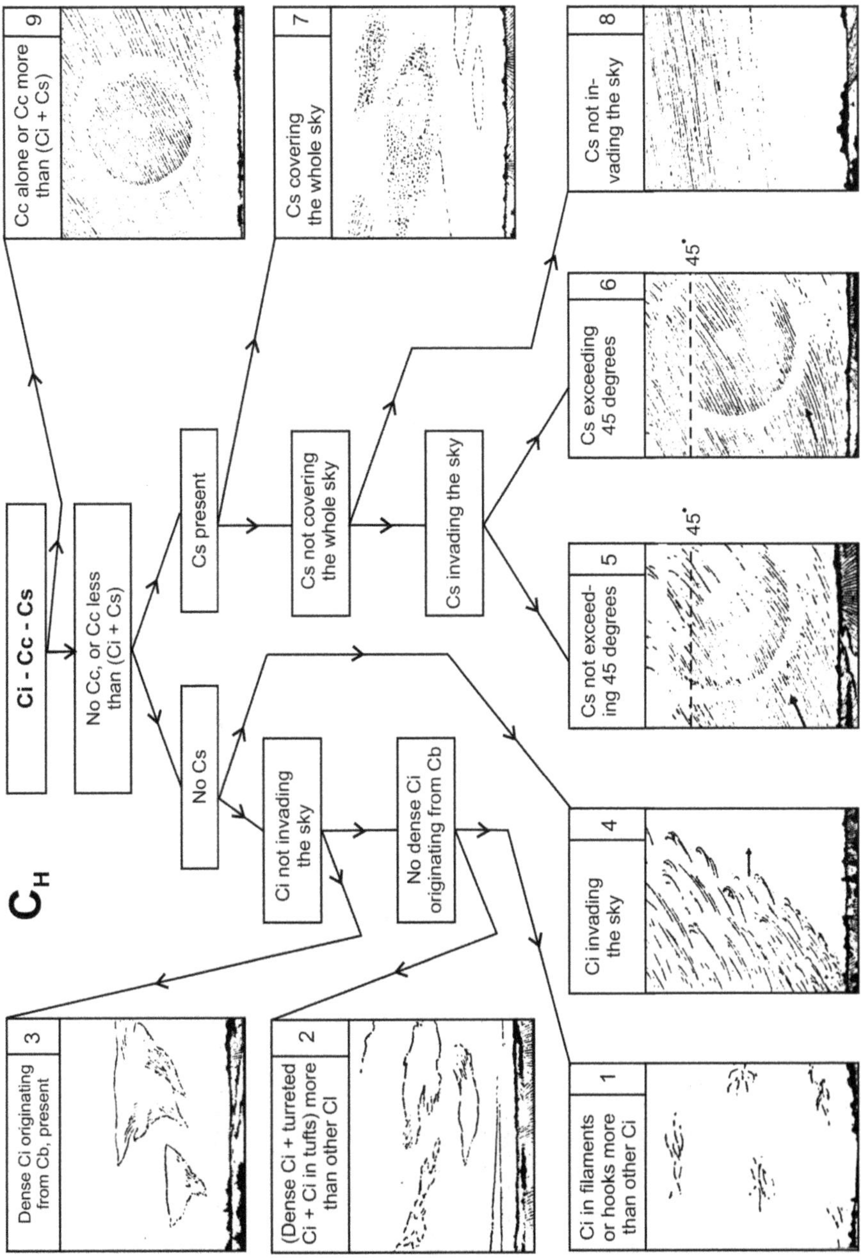

Fig 10.1 Pictorial guide for C_H- clouds.

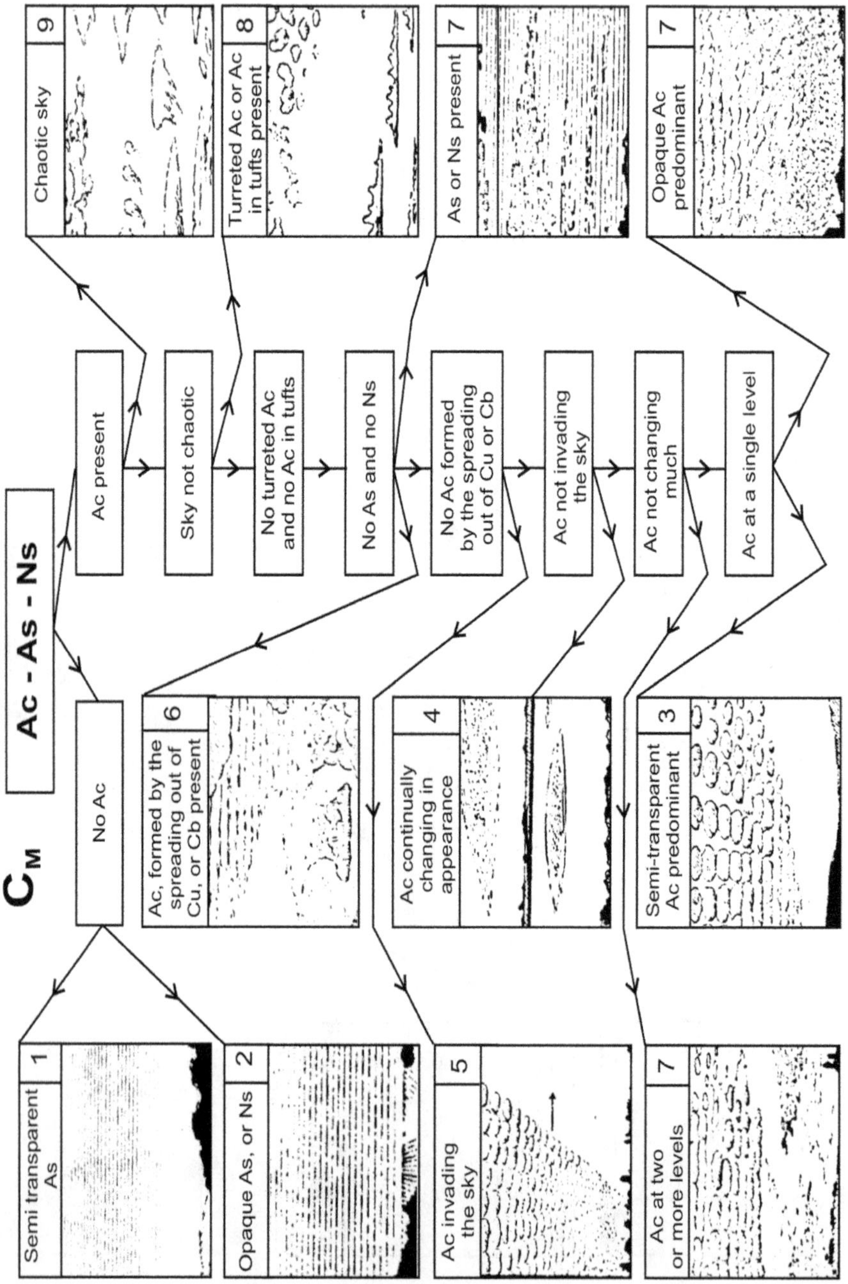

Fig 10.2 Pictorial guide for C_M - clouds.

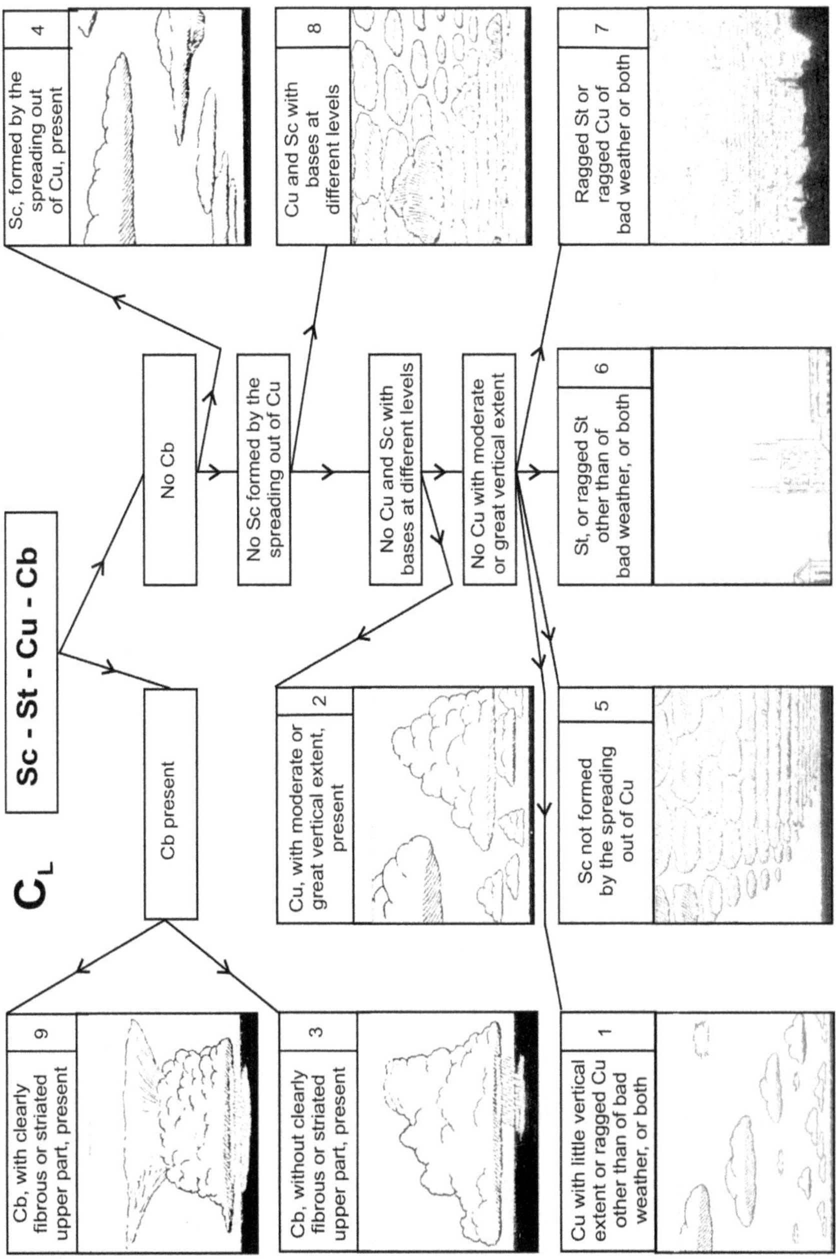

Fig 10.3 Pictorial guide for C_L - clouds.

Low ragged clouds frequently occur below the layer, with which they may or may not merge.

7. **Stratocumulus (Sc):** Grey or whitish or both gray and whitish, patch, sheet or layer of cloud which almost always has dark parts, composed of tessellations, (form into squares or checkers) rounded masses, rolls, etc., which are non-fibrous (except for virga) and which may or may not be merged; most of the regularly arranged small elements have an apparent width of more than five degrees.
8. **Stratus (St):** Generally grey cloud layer with a fairly uniform base, which may give drizzle, ice prisms or snow grains. When the sun is visible through the cloud, its outline is clearly discernible. Stratus does not produce halo phenomena except, possibly, at very low temperature.

   Sometimes stratus appears in the form of ragged patches.
9. **Cumulus (Cu):** Detached clouds, generally dense and with sharp outlines, developing vertically in the form of rising mounds, domes or towers, of which the bulging upper part often resembles a cauliflower. The sunlit parts of these clouds are mostly brilliant white; their base is relatively dark and nearly horizontal. Sometimes cumulus is ragged.
10. **Cumulonimbus (Cb):** Heavy and dense cloud, with a considerable vertical extent, in the form of a mountain or huge towers. At least part of its upper portion is usually smooth, or fibrous or striated, and nearly always flattened; this part often spreads out in the shape of an anvil or vast plume.

    Under the base of this cloud which is often very dark, there are frequently low ragged clouds either merged with it or not, and precipitation sometimes in the form of virga.

(B) *Definition*: Cloud Species (14)

1. **Fibratus:** Detached clouds on a thin cloud veil, consisting of nearly straight or more or less irregularly curved filaments which do not terminate in hooks or tufts.

   This term applies mainly to Cirrus and Cirrostratus.
2. **Uncinus:** Cirrus often shaped like a comma, terminating at the top in a hook, or in a tuft, the upper part of which is not in the form of a rounded protuberance.

3. **Spissatus:** _Cirrus of sufficient optical thickness to appear greyish when viewed towards the sun.
4. **Castellanus:** Clouds which present, in at least some portion of their upper part, cumuliform protuberances in the form of turrets which generally give the clouds a crenelated (loopholes) appearance. The turrets, some of which are taller than they are wide, are connected by a common base and seem to be arranged in lines. The castellanus character is especially evident when the clouds are seen from the side.

   This term applies to cirrus, cirrocumulus, Altocumulus and Stratocumulus.
5. **Floccus:** A species in which each cloud unit is a small tuft with a cumuliform appearance, the lower part of which is more or less ragged and often accompanied by virga.

   This term applies to Cirrus, Cirrocumulus and Altocumulus.
6. **Stratiformis:** Clouds spread out in an extensive horizontal sheet or layer.

   This term applies to Altocumulus, Stratocumulus and, occasionally, to cirrocumulus.
7. **Nebulosus:** A cloud like a nebulous (cluster or vague) veil or layer showing no distinct details.

   This term applies mainly to Cirrostratus and Stratus.
8. **Lenticularis**: Clouds having the shape of lenses or almonds, often very elongated and usually with well defined outlines; they occasionally show irisation . Such clouds appear most often in cloud formations of orographic origin, but may also occur in regions without marked orography.

   This term applies mainly to Cirrocumulus, Altocumulus and stratocumulus.
9. **Fractus:** Clouds in the form of irregular shreds, which have a clearly ragged appearance.

   This term applies only to Stratus and Cumulus.
10. **Humilis:** Cumulus clouds of only a slight vertical extent; they generally appear flattened.
11. **Mediocris:** Cumulus clouds of moderate vertical extent, the tops of which show fairly small protuberances

12. **Congestus:** Cumulus clouds which are markedly sprouting and are often of great vertical extent; their bulging upper part frequently resembles a cauliflower.
13. **Calvus**: Cumulonimbus in which at least some protuberances of the upper part are beginning to lose their cumuliform outlines, but in which no cirriform parts can be distinguished. Protuberances and sproutings tend to form a whitish mass, with more or less vertical striations.
14. **Capillatus:** Cumulonimbus characterised by the presence, mostly in its upper portion, of distinct cirriform parts of clearly fibrous or striated structure, frequently having the form of an anvil, a plume or a vast, more or less disorderly mass of hair. Cumulonimbus capillatus is usually accompanied by a shower or by a thunderstorm, often with squalls and sometimes with hail; it frequently produces very well -defined virga.

(C) Definition: Cloud Varities (9)

First six refer to arrangement of macroscopic elements and the last three the degree of transparency.

1. **Intortus:** Irregularly curved cirrus filaments.
2. **Vertebratus:** Cloud elements arranged like vertebrae, ribs or a fish skeleton. This applies to mainly cirrus.
3. **Undulatus:** Clouds in patches, sheets or layers showing undulations (wave like pattern).
4. **Radiatus:** clouds in parallel bands seem to converge towards a points on the horizon or towards two opposite points on the horizon called "radiation point(s)". This mainly applies to Ci, Ac, As, Sc and Cu.
5. **Lacunosus:** Cloud patches, sheets or layers (generally thin) arranged in a manner that appears as a net or a honey-comb. This mainly applies to Cc, Ac and on rare occasions to Sc.
6. **Duplicatus:** Superposed cloud patches, sheets or layers, at slightly different levels or partly merged. This mainly applies to Ci, Cs, Ac, As and Sc.
7. **Translucidus:** Clouds in an extensive patch, sheet or layer, the greater part of which is translucent to reveal the position of the sun or moon. This mainly applies to Ac, As, Sc and St.

8. **Perlucidus:** An extensive cloud patch, sheet or layer, with small spaces between the elements. The spaces allow the sun, moon etc., or overlying clouds to be seen. This mainly applies to Ac and Sc.
9. **Opacus**: An extensive cloud patch, sheet or layer, the major part of which is opaque to mask the sun or moon. This mainly applies to Ac, As, Sc and St.

(D) Definition: Supplementary Features of Clouds (6)

1. **Incus:** The upper part of Cb cloud spreading into anvil with a smooth, fibrous or striated appearance.
2. **Mamma:** Hanging protuberances, like udders, on the under surface of a cloud. This mainly applies to Ci, Cc, Ac, As, Sc and Cb.
3. **Virga:** Vertical or inclined trails of precipitation (fall-streaks) attached to the under surface of a cloud, which do not reach the earth's surface. This occurs mostly with Ce, Ac, As, Ns, Sc, Cu & Cb.
4. **Praecipitatio:** Precipitation (rain, drizzle, snow, ice pellets, hail etc.,) falling from a cloud and reaching the earth's surface. This feature is accompanied with As, Ns, Sc, St, Cu and Cb.
5. **Arcus:** A dense, horizontal roll with more or less tattered edges (lace like edging) situated on the lower front part of certain clouds with the appearance of a dark, menacing arch. This is associated with Cb and sometimes with Cu.
6. **Tuba:** Cloud column or inverted cloud cone, protruding from a cloud base; showing more or less intense vortex. This is associated with Cb and sometimes with Cu.

(E) Definition: Accessory Clouds (3)

1. **Pileus:** An accessory cloud of small horizontal extent, in the form of a cap or hood above the top or attached to the upper part of a cumuliform clouds which often penetrates it. Several pileus clouds may fairly often be observed in superposition. Pileus principally occurs with Cu and Cb.
2. **Velum:** An accessory cloud veil of great horizontal extent, close above or attached to the upper part of one or several cumuliform clouds which often pierce it. Velum occurs principally with Cu and Cb.

3. **Pannus:** Ragged shreds, sometimes constituting a continuous layer, situated below another cloud and sometimes attached to it. Pannus mostly occurs with As, Ns, Cu and Cb.

Definition: Mother Clouds

Clouds may develop in either of two ways: (a) They may form in clear air (b) They may form from other clouds, called mother clouds.

10.3 Special Clouds

1. Nacreous clouds:

Definition: Clouds resembling Ci or Ac lenticularis with marked irisation similar to that of mother-of-pearl; the most brilliant colours are observed when the sun is several degrees below the horizon.

The physical constitution of nacreous clouds is still uncertain. They are rarely observed in Norway, Scotland, Scandinavia and sometimes in France. Height 21-30 km. During evening they appear as pale cirrus. Just after sunset they show brilliant colours with irisation. Irisation reaches maximum brilliance when the sun is several degrees below the horizon. Afterwards it shows general colouration which changes from orange to pink, then grayish and subsequently fadeout. Before dawn the above sequence of colours and brilliance repeated in the reverse order.

Nacreous clouds are stationary or move very slowly (this does not mean the wind speed at that level is low).

2. Noctilucent clouds:

Definition: Clouds resembling thin cirrus, but generally with bluish or silvery or sometimes orange to red colour; they stand out against the dark night sky.

The physical constitution of noctilucent clouds is still uncertain, may be composed of very fine cosmic dust. They are rarely seen during summer in higher latitudes of northern hemisphere, when the sun is 5 to 13 degrees below the horizon. Height varies 75 to 90 km.

First appear as grayish (with increasing brilliance with time), then bluish-white (like tarnished silver), sometimes reddish in the vicinity of horizon. The sequence of changes is then repeated in reverse order. These clouds appear brilliant after midnight. They move at about 50 mps (100 kt) to more than 250 mps (500 kt), usually from northeast (NE) to east (E).

3. Condensation trails (contrails):

Definition: Condensation trails are visible streaks of condensed or sublimated water vapour formed in the wake of a moving aircraft.

Contrails are clouds formed due to natural process in the wake of aircraft when the atmosphere at flying level is sufficiently cold and humid. First they appear as brilliant streaks, then pendant swelling (like inverted mush room). Generally they are short lived but may persist several hours when Ci, Cs, is present. Contrails may produce halo phenomena with exceptional pure colours. The cause of formation is cooling of the exhaust gases which have high water vapour content together with condensation nuclei.

Importance of contrails to aviation: When an aircraft leaves a visible trail behind it, which reveals its position and heading to an observer on the ground. This is disadvantageous to military aviation. In mountainous areas where radar surveillance is very difficult, but the contrails become the main tracking of aircraft by enemy observers.

The two end products of aircraft – exhaust water vapour and heat – play part in contrail formation. Water vapour aids for improving humidity and while heat acts in the reverse. At high altitudes where temperature is low, water vapour out-weighs the reduction due to heat in the formation of saturation. Contrails form only when the ambient air temperature is low (in subzero scale). The critical temperature at which contracts form on theoretical basis is called MINTRA (Minimum trail) temperature and the level at which this temperature is likely is called Mintra level. It is not necessary at Mintra level trails form but trails does not form below this level. The flight level at which temperature – 45 °C occurs is a good estimate of Mintra level for jet aircraft. When ambient air is absolutely dry (RH 0 %) but temperature is very low, contrails form. In such cases the exhaust moisture itself is sufficient for saturation and sublimation. This level is called DRYTRA level, and above this level contrails form. Drytra level is about 2 km above Mintra level.

In lower tropical stratosphere, temperature rises with height. At a certain level ambient air temperature becomes greater than the critical temperature for contrail formation. Above this level no contrails can form. This level is called MAXTRA level (maximum trail level).

In India, contrails are very rare below 7 km. Contrails generally form between 11-14 km height. Mintra level is lower in winter as compared to summer. In extreme north India the low Mintra level in winter comes to about 9 km.

***Wing Tip Trials*:** Vortices are formed in the form of eddies off the tips of airplane wing in flight. If the vertices are strong and humidity is high, this cause cooling condensation. White cloud like trails form off the wing tips called wing tip trails.

4. Clouds from water falls:

High water falls produce water spray, which saturate the air in the vicinity. The downdrafts developed by the falling water in general compensated by the up currents in the vicinity of the waterfalls. This lifts the saturated air and forms a cloud which reassembles cumulus. The spray itself causes brilliant rainbow when the sun rays fall from the opposite of water fall.

5. Clouds from Fire:

In big fires, like forest fire, fire in petrol stores etc., the combustion products assume as dense, dark, sprouting clouds. These grow vertically to great heights (like TCu, Cb)

Fire clouds often diffuse into ambient atmosphere as smoke or haze. Combustion products are carried by wind to great distances and then they appear as startiform veils. When observe the sun or moon through this veil they appear as blue. In case of forest fires, sufficient water vapour is released which develop into Cu, Cb and lead to thunderstorms. In Prairie fires there is little smoke but convective clouds are visible.

6. Clouds from volcanic eruptions:

Volcanic eruptions create clouds which appear as developed Cu with protuberances. When these clouds spread to both horizontally and vertically the sky takes a peculiar tint which may persist for several weeks. The volcanic clouds may produce strong electrical manifestations. These clouds contain watervapour, dust, ash, solid particles of different sizes. Some volcanoes produce pasty siliceous lava, emit vitreous filaments, the fall of which may appear as snow shower from a distance. The following historical examples illustrate the real situations.

The eruptions of Tambora (1815, which took a human toll of 12000 lives) spewed out about 80 cubic kilometers of ash and rock pieces into the atmosphere. Some particles entered into stratosphere and remained there. This caused cooling. As a result there was no summer in 1816 in northern parts of U.S. and it caused extensive crop failure. The volcanic dust, ash, watervapour reached high altitudes and were carried by winds around the globe. The precipitation of this dust back to earth took years. The ejected particles (into the atmosphere) dispersed sunlight and temporarily increased the earth's albedo.

The Krakatova (in East Indies) eruption on August 27, 1883, was the most powerful volcanic eruption in the history of man kind till that date. The eruption of clouds reached the heights about 30 km and turned day into night in Batavia (about 150 km away). A Tsunami wave, caused by the fall of broken away part of Krakatova island travelled around the world. Rare optical phenomena were observed in Europe in the end of November 1883. The sky remained purple for several hours during sunsets. This was caused by the dispersion of sunlight by a layer of dust particles (size about 2 μm) injected by the volcano into the stratosphere. The dust /ash particles did not settle down on earth for several years Successive years of weather over the entire earth was cooler than normal. This was due to the increase of earth's albedo and partly due to the mixing of ocean waters by Tsunami waves.

The violent eruption of volcano Gunung Agung in Bali (1963) injected particulate matter into the upper troposphere and stratosphere which gave rise to the spectacular sunsets throughout the world. The eruption caused the rise in stratospheric temperature by about 5 ^{o}C in the height range of 18-20 km. The volcanic particles consisted of lava and crystalline material (Calcium and Ammonium sulphate).

7. Clouds resulting from industry:

These clouds have diverse origins.

(1) Clouds of smoke and steam in industrial areas.

(2) Smoke clouds created for frost protection purposes

(3) Clouds of insecticide gas or powders in agricultural areas.

8. Clouds resulting from explosions:

When an explosion is very large, it is usually accompanied by a cloud of smoke and dust. Above this cloud velum or pileus are often seen. Sometimes the propagation of shock waves is manifested by dark rings or bands moving with extreme rapidity.

Questions

1. Which type of clouds are produced in stable atmosphere?
   (a) Stratiform (b) Cumuliform
   (c) Medium clouds Ac / As (d) None
2. Which type of clouds are produced in unstable atmosphere?
   (a) Stratiform (b) Cumuliform
   (c) High clouds Ci / Cs (d) None
3. Cloud formed by turbulence (Turbulence cloud) is:
   (a) Unstable, Cu / Cb (b) Stable, St / Sc
   (c) Neutral, Ci / Cs (d) None
4. Turbulence cloud (i.e., cloud formed by turbulence) after mixing:
   (a) Increase the surface temperature
   (b) Decrease the temperature at the top of mixing layer
   (c) The midpoint temperature remains the same
   (d) All above
5. Convective cloud produces:
   (a) Rain (b) Drizzle
   (c) Shower (d) Stability
6. Nimbostratus in winter produces:
   (a) Rain (b) Snow
   (c) Drizzle (d) Shower
7. If moist stable air parcel is forced to rise over a mountain, which is most likely to form:
   (a) Thunderstorm (b) Stratiform clouds
   (c) Inversion (d) Cirri form clouds
8. If moist unstable air parcel is forced to rise over a mountain, which is most likely form:
   (a) Thundustorm / Convective cloud
   (b) Stratiform cloud
   (c) Inversion
   (d) Cirri form clouds

9. What type of cloud is associated with thundustorm?
   (a) Stratus (b) Cumulonimbus
   (c) Altostratus (d) Cirrostratus

10. What type of cloud is associated with drizzle?
   (a) Stratus (b) Cb
   (c) As (d) Cs

11. "Nimbus" is added to the names of clouds that produce:
   (a) Drizzle (b) Precipitation
   (c) Thunderstorm (d) Fog

12. By convention clouds are divided into:
   (a) Convective (b) Stratiform
   (c) High, Medium & low (d) None

13. The high level clouds are:
   (a) Ci, As, Cb (b) Ci, Cs, Cc
   (c) St, Sc, Ns (d) Cu, Cb

14. The usual middle etage clouds are:
   (a) Ci, Cs, Cc (b) Ac, As
   (c) Cu, Cb (d) Cc, Ns

15. A cloud is a visible aggregate of minute particles of ....in the free atmosphere
   (a) Water (b) Ice
   (c) Dust (d) Both (a) & (b)

16. Clouds are the result of:
   (a) Cooling of dry air during vertical motion
   (b) Cooling of moist air during vertical motion
   (c) Moist air contact with colder air mass or land / water surfaces
   (d) Both (b) & (c)

17. The level at which moist air begins to condense is called:
   (a) Freezing level
   (b) Lifting condensation level
   (c) Cloud dissipation level
   (d) None

18. Clouds are classified based on:
    (a) Their height and their shape
    (b) Their reflection and refraction of light,
    (c) Their Colour
    (d) All above

19. WMO classified cloud genera consists of:
    (a) Ci, Cs, Cc (b) As, Ac, Ns
    (c) St, Cu, Sc, Cb (d) All above

20. Clouds having the shape of lenses or almonds, often found during cloud formation of orographic origin are called:
    (a) Lenticularis (b) Fibratus
    (c) Stratiformis (d) Calves

21. The precipitation descending from the base of a cloud but not reaching the surface is called:
    (a) Calves (b) Lenticularis
    (c) Virga (d) Fibratus

22. Detached (or cellular) clouds with sharp outlines, developing vertically like domes or towers, base dark and horizontal is:
    (a) Ac (b) Cu (c) Cb (d) Ns

23. Heavy & dense cloud, with considerable vertical extent, upper portion nearly flattened in the form of anvil is:
    (a) Ac (b) Cu (c) Cb (d) Ns

24. The cloud which produces halo phenomena is:
    (a) Stratus (b) Cs (c) As (d) Ns

25. Grey cloud with dark appearance, diffused boundaries with continuously falling rain/snow, thick enough to blot the sun is:
    (a) Cu (b) Cb (c) Ns (d) Cs

26. Cloud column or inverted cloud cone, protruding from a cloud base (Cu or Cb) showing intense vortex is called:
    (a) Virga (b) Mamma (c) Tuba (d) Incus

27. Clouds resembling Ci or Ac lenticularis with brilliant colours observed when the sun is several degrees below the horizon, height 21- 30 km observed in Norway, Scotland, Scandinavia are called:
    (a) Noctilucent clouds (b) Nacreous clouds
    (c) Condensation trails (d) Volcanic clouds

28. Clouds resembling thin cirrus with bluish or silvery or orange to red clour, they stand out against the dark night sky, height 75 to 90 km, rarely observed in northern latitudes when the sun is 5 to 13 degrees below the horizon are called:
    (a) Nacreous clouds (b) Noctilucent clouds
    (c) Contrails (d) St. elmo'sfire

29. Clouds visible streaks that formed in the wake of high level (jet) aircraft movement due to condensed or sublimated water vapour are called:
    (a) Nacreous clouds (b) Noctilucent clouds
    (c) Contrails or condensation trails (d) St. elmosfire

30. Volcanic eruption clouds contain:
    (a) Strong electrical manifestation
    (b) Water vapour, dust, ash & solid particles
    (c) Extend up to 30 km height and exhibit cooling
    (d) All above

31. Clouds of vertical development are:
    (a) Cu / TCu (b) Cb
    (c) Ns (d) Both (a) & (b)

32. Nimbostratus falls under the etage of:
    (a) Low clouds (b) Medium clouds
    (c) Both (a) & (b) (d) High clouds

33. Stratiform clouds appear in:
    (a) Heaps / lumps
    (b) Layers with diffused boundaries
    (c) Cellular form with sharp boundaries
    (d) Layer with sharp boundaries

34. Cumuliform clouds appear in:

(a) Heaps / lumps with vertical development

(b) Layers with sharp boundaries

(c) Deck on the surface

(d) Lumps with diffused boundaries

35. Cloud on deck is:

(a) Cumulus (b) Cirrus

(c) Nimbostratus (d) Fog

36. The base of stratus lies between:

(a) 6500–1300 ft (b) Surface–1500 ft

(c) Surface–6500 ft (d) 1000–6000 ft

37. The base of cumulus lies between:

(a) 1000–6500 ft (b) 1500–6500 ft

(c) Surface–1500 ft (d) Surface–6500 ft

38. The base of cumulonimbus lies between:

(a) 1000–6500 ft (b) 1500–6500 ft

(c) Surface–6500 ft (d) Surface–1500 ft

39. Mother of pearl cloud sighted:

(a) In Stratosphere, height 20–30 km, just before surise or just after sunset

(b) In Mesosphere, height 80–85 km, midnight summer, in bluish – silver to yellow orange Colour

(c) In Troposphere, height 10–20 km, during midnight, with rainbow colours

(d) In ozone layer, midnight summer, illuminated by the direct sun light

40. Noctilucent clouds observed:

(a) During midnight summer, height 80–85 km, illuminated by direct sunlight in the northern horizon of Europe, with bluish silver to yellow / orange Colour.

(b) In stratosphere, height 20–30 km, just before sun rise or just after sunset, with rainbow colours

(c) In troposphere height 10–20 km, in southern hemisphere towards South Pole

(d) In ozone layer, in winter midnight, in southern hemisphere.

Answers

1.	(a)	2.	(b)	3.	(b)	4.	(d)
5.	(c)	6.	(b)	7.	(b)	8.	(a)
9.	(b)	10.	(a)	11.	(b)	12.	(c)
13.	(b)	14.	(b)	15.	(d)	16.	(d)
17.	(b)	18.	(d)	19.	(d)	20.	(a)
21.	(c)	22.	(b)	23.	(c)	24.	(b)
25.	(c)	26.	(c)	27.	(b)	28.	(b)
29.	(c)	30.	(d)	31.	(d)	32.	(c)
33.	(b)	34.	(a)	35.	(d)	36.	(b)
37.	(b)	38.	(a)	39.	(a)	40.	(a)

CHAPTER 11

Meteors, Cloud Formation and Dissipation

11.1 Meteors

Definition

According WMO, a meteor is a phenomenon, other than cloud, observed in the atmosphere or on the surface of the earth, which consists of a precipitation, a suspension or a deposit of aqueous or non-aqueous liquid or solid particles, or a phenomenon of the nature of an optical or electrical manifestation.

Note: Astronomical meteors that enter the atmosphere such as shooting stars are different from meteorological meteors.

Meteors are classified into four categories, They are:

(i) hydrometeors,
(ii) litho-meteors,
(iii) photometeors and
(iv) electrometeors

(i) ***Hydrometeors***: *Definition:* A hydrometeor is a meteor consisting of an ensemble of liquid or solid aqueous particles, falling through or suspended in the atmosphere, blown by the wind from the earth's surface or deposited on objects on the ground or in the free air (hydro means water).

The most common hydrometeors that are frequently referred in meteorology are given below.

Precipitation: This refers to falling of hydrometeors that finally reach the ground. This will not include a virga, which is seen streaming from the bottom of a cloud, but does not reach the earth's surface.

Rain: Precipitation of liquid water particles, either in the form of drops of more than 0.5 mm diameter or of smaller widely scattered drops.

Shower: liquid precipitation of big size usually marked with sudden onset. The falling drops are more widely scattered than rain. The diameter of the drop size is more than 0.5 mm. Showers generally fall from convective clouds (Cu or Cb).

Freezing rain: Rain, the drops of which freeze on impact with the ground or with objects on the earth's surface or with aircraft in flight.

Drizzle: Fairly uniform precipitation composed of exclusively of fine drops of water, whose diameter is less than 0.5 mm and that these drops fall very close to one another. The impact of the drops on water surface is imperceptible.

Freezing drizzle: Drizzle, the drops of which freeze on impact with the objects on the earth's surface or ground or with aircraft in flight.

Snow: Precipitation of ice crystals, most of which are branched (or star shaped).

Sleet: Rain and snow falling together or snow melting as it falls.

Snow pellets: Precipitation of white and opaque grains of ice. The grains are spherical or conical with diameter ranging 2 to 5 mm.

Snow grains: Precipitation of very small white and opaque grains of ice. These grains are fairly flat or elongated and have diameters less than 1 mm.

Ice pellets: Precipitation of transparent or translucent pellets of ice, whose shapes are spherical or irregular, but rarely conical, diameter less than or equal to 5 mm. These ice pellets are of two types:

(a) Frozen rain drops or largely melted and re-frozen snowflakes,

(b) Pellets of snow encased in a thin layer of ice.

Hail: Precipitation of small balls or pieces of ice (called hailstones) diameter ranging 5 to 50 mm (or sometimes more), falling either separately or agglomerated into irregular lumps.

Ice prisms: A fall of un-branched ice crystals in the form of needles, columns or plates, often so small that they appear to be suspended in the air. These crystals may fall from a cloud or from a cloudless sky.

Fog: A suspension of very small water droplets in the air, mostly reducing the horizontal visibility at the earth's surface to below one kilometer (1000 m). Fog normally occurs when the surface wind is calm or very light and relative humidity 75% or more.

Ice Fog: A suspension of a large number of minute ice crystals in the air that reduces the horizontal visibility at the earth's surface to less than one kilometer. Ice fog creates luminous pillars, small haloes etc.

Mist: A suspension of microscopic water droplets or wet hygroscopic particles that reduces the horizontal visibility at the earth's surface to (1-2 km) less than 2 km but more than one kilometer. Relative humidity is generally less than 100% but at least 75%.

Drifting snow and blowing snow: An ensemble of snow particles raised from the ground by a sufficiently strong and turbulent wind, that is snow blown off the ground into air after it has already fallen.

(a) *Drifting snow*: An ensemble of snow particles raised by strong wind to a height less than 2 meters above ground, but the horizontal visibility is not deteriorated to mist or fog level.

(b) *Blowing snow*: An ensemble of snow particles raised by strong wind to moderate, to great heights above ground level. Horizontal visibility below 2 meter height sufficiently deteriorated to below 2 km.

Spary: An ensemble of water droplets torn by the wind from the surface of an extensive water body, generally from the crests of waves and carried up to a short distance into the air.

Dew: A deposit of water drops on objects at or near the ground, produced by the condensation of water vapour from the surrounding clear air.

or

Moisture condensed from atmosphere on exposed surfaces. It is a common phenomenon of night or early morning with calm and clear sky.

White dew: A deposit of white frozen dew drops.

Hoar frost: Water vapour of the clear air directly transformed into ice crystals of the form scales, needles, feathers or fans.

Frost: Crystalline ice deposit formed on objects near the ground similar to dew when the temperature is below freezing point.

Rime: A deposit of ice composed of grains more or less separated by trapped air, sometimes adorned with crystalline branches.

Glaze (clear ice): Generally a homogeneous and transparent deposit of ice formed by the freezing of supercooled drizzle droplets or raindrops on objects, the surface temperature of which is below or slightly above 0 °C.

Spout: A phenomenon consisting of an often violent whirl-wind, revealed by the presence of a cloud column or inverted cloud cone (funnel cloud), protruding from the base of a Cb and of a 'bush' composed of water droplets raised from the surface of the sea or of dust, sand or litter, raised from the ground.

(ii) *Lithometeors*

A lithometeor is a meteor consisting of an ensemble of particles most of which are solid and nonaqueous. The particles are more or less suspended in the air or lifted by the wind from the ground. The common lithometeors are described below.

Haze: A suspension in the air of extremely small, dry particles (smoke, dust etc but not of hydrometeors) invisible to the naked eye but sufficiently large number and cause the air an opalescent appearance. Visibility deteriorates 2 km to 10 km or less but more than 2 km.

Dust haze: A suspension in the air of dust or minute sand particles, raised from the ground prior to the time of observation by a dust storm or sand storm, reducing the horizontal visibility 2 to 10 km and relative humidity less than 75%.

Smoke: A suspension in the air of small particles emanated by combustion.

Smog: A combination of smoke and haze (or dust haze) that reduces the horizontal visibility 2 to 5 km.

Note: Combination of smoke and fog is also called smog, generally observed around industries and cities in the morning.

Drifting and Blowing Dust or Sand: An ensemble of particles of dust or sand raised from the ground, at or near the station to a small or moderate heights (20 to 30 m) by strong and turbulent wind. In drifting dust / drifting sand, the visibility is not sensibly deteriorated at the eye level (or 1.8 m height). In blowing dust / blowing sand the horizontal visibility at eye level is appreciably reduced (2 to 5 km).

Duststorm or Sandstorm: An ensemble of particles of dust or sand energetically lifted to great heights (about 100 m or more) by a strong and turbulent wind causing horizontal visibility to less than 1000 m.

The further classification of dust/sand storm are:

Slight (wind speed 20-49 kmph, visibility less than 1000 m)

Moderate (wind speed 39-74 kmph, visibility 500 m or less)

Severe (wind speed 75 kmph or more, visibility 200 m or less)

This is defined on the basis of wind speed and visibility.

Dust Whirl or Sand Whirl (Dust Devil): An ensemble of particles of\ dust or sand, sometimes accompanied by small litter, lifted from the ground in the form of a whirl column of varying height with a small diameter and an approximately vertical axis.

or

A narrow column of whirling dust or sand going up in a spiral form.

(iii) *Photometeors*

A photometeor is a luminous phenomenon produced by reflection, refraction, diffraction or interference of light from the sun or the moon.

The common photometeors are:

(i) Halo phenomena,

(ii) Corona,

(iii) Irisation,

(iv) Glory,

(v) Rainbow,

(vi) Bishops ring,

(vii) Mirage,

(viii) Shimmer

(ix) Scintillation,

(x) Green flash,

(xi) Twilight colours

These are briefly described below:

Halo phenomena: A group of optical phenomena in the form of rings, arcs, pillars or bright spots produced by the refraction or reflection of light by ice crystals suspended in the atmosphere (cirriform clouds, ice fog etc).

Corona: One or more sequence of colored rings of relatively small diameter, centered on the sun or moon.

Solar Halo: A ring of light round the sun by thin veil of cirrus cloud. It is often white but sometimes red near the sun, then orange, then yellow. In most cases the ring has a radius of 22° surrounding the sun.

Lunar Halo: A circle round the moon similar to solar halo.

Solar Corona: A ring of light round the sun, much smaller than halo. Its inner edge is brownish red, while the sky between the ring and the sun has a distinct bluish white color. The radius of the ring varies from 5° to 8°.

Lunar Corona: A ring round the moon similar to the solar corona.

Irisation: Colors appearing on clouds, sometimes mingled, sometimes in the form of bands nearly parallel to the margin of the clouds. Green and pink predominate, often with pastel (delicately tinted or colored crayon) shades.

Glory: One or more sequence of colored rings, seen by an observer around his own shadow:

(i) on a cloud consisting mainly of numerous small water droplets,

(ii) on fog or,

(iii) very rarely on dew.

Rainbow: A group of concentric arcs with colors ranging from violet to red (VIBGYOR – Violet, Indigo, Blue, Green, Yellow, Orange and Red) produced on a screen of water drops (raindrops, droplets of drizzle or fog) in the atmosphere by light from the sun or the moon.

The primary rainbow shows violet on the inside (radius 40°) and red on the outside (radius 42°).

Secondary rainbow is much less bright, shows red on the inside (radius 50°) and violet on the outside (radius 54°).

Note: If the size of a drop exceeds a certain size, its shape in no longer spherical and hence rainbow is not observed (as dispersion does not take place). Similarly if the droplets are too small, even though droplets are spherical, rainbow may not occur. The fog bow is a primary rainbow consisting of a white band which appears on a screen of fog or mist; it is usually fringed with red on the outside and blue on the inside.

Bishop's Ring: A whitish ring, centered on the sun or moon with a slightly bluish tinge on the inside and reddish brown on the outside.

Mirage: An optical phenomenon consisting mainly of steady or wavering, single or multiple, upright or inverted, vertically enlarged or reduced images of distant objects. This occurs due to total internal reflection of light.

Shimmer: The apparent fluttering of objects at the earth's surface when viewed in the horizontal direction.

Scintillation: Rapid variations, often in the form of pulsation of the light from stars or terrestrial light sources.

Green flash: A predominantly green coloration of short duration, often in the form of a flash, seen at the extreme upper edge of a luminary (Sun, moon or sometimes even a planet) when disappearing below or appearing above the horizon.

Twilight colors: various colorations of the sky and of the peaks of mountains at sunset and sunrise. A number of these color effects have been given special names.

Note:

1. Astronomical twilight ends in the evening (or begins in the morning) when the sun's centre is 18^{0} below the horizon.
2. Civil twilight ends in the evening (or begins in the morning) when the Sun is 6^{0} below the horizon. Twilight begins (ends) when the sun just recedes (rises) below (above) the horizon in the evening (morning).
3. The duration of twilight varies with the latitude.
4. Twilight means half light after sunset or before dawn (sunrise)

(iv) ***Electrometeors***

An electrometeor is a visible or audible manifestation of atmospheric electricity.

Some important electrometeors are described below.

Thunderstorm: One or more sudden electrical discharges manifested by a flash of light (lightning) and a sharp or rumbling sound (thunder).

Lightning: A luminous manifestation accompanying a sudden electrical discharge which takes place from or inside a cloud or less often from high structures on the ground or from mountains.

Thunder bolt: or ground discharge which takes place between the thunder cloud (Cb) and ground.

Sheet lightning: or cloud discharge which takes place within the thunder cloud (Cb).

Air discharge: which pass from a thunder cloud (Cb) to the air but do not strike the ground.

Thunder: A sharp or rumbling sound which accompanies lightning.

Saint Elmo's fire: A more or less continuous luminous electrical discharge of weak or moderate intensity in the atmosphere, emanating from elevated objects at the earth's surface (lightning conductor, wind wanes, masts of ships) or from aircraft in flight (wing tips, propellers etc).

Polar Aurora: A luminous phenomenon which appears in the high atmosphere, in the form of arcs, bands, draperies or curtains. More will be discussed in optical phenomena of atmosphere.

Note: Fine weather: Sky cloudless or with isolated cirrus floating in the blue sky and showing signs of dissolving or with a small amount of pure stratiform cloud at a fixed level, but with no clouds having vertical development.

Fair weather: Thin cirrus covering considerable part of the sky but not increasing or forming a continuous layer or sky with "Fair weather cumulus" or Altocumulus with characteristic changes in the courses of the day.

Fair weather cumulus appears in patches in the afternoon and disappears in the evening. It never develops into cumulonimbus.

Precipitation Associated with Cloud Genera: Hydrometeors are closely related with certain cloud genera. The Table 11.1 gives various precipitations associated with cloud genera.

Table 11.1

Hydrometeors	**Cloud genera**					
	As	**Ns**	**Sc**	**St**	**Cu**	**Cb**
Rain	+	+	+		+	+
Drizzle				+		
Snow	+	+	+			+
Snow pellets			+			+
Snow grains				+		
Ice pellets	+	+				+
Ice prisms				+		
Hail						+

An aircraft observer may encounter ice prisms under Ci, Cc, Cs Ac clouds.

The Rate of fall of Water Droplets

Any objects falling through the atmosphere is subjected to three forces.

(i) The force of gravity,

(ii) The up thrust due to buoyancy and

(iii) The air resistance due to its motion.

A freely falling object in the atmosphere, first it accelerates due to force of gravity, overcoming the upward effect of thrust and air resistance. However, after a while the forces are balanced and it then comes down with a constant velocity. This is called terminal *velocity*. A parachutist at first accelerates downwards, but after a while when his parachute opens he descends at a constant speed. Similarly falling rain drops initially accelerate downwards and after a while achieving certain maximum speed they come down with a constant velocity called terminal velocity. The terminal velocity of rain drops depends on its size, which are given in the Table 11.2.

Table 11.2 Terminal velocity of rain drops (water droplets)

Diameter (μm = micron meters)	Terminal velocity (mps)
2	0.12×10^{-3}
8	0.192×10^{-2}
100	0.27
200	0.72
500	2.06
1000	4.03
2000	6.49
5000	9.09
Drizzle r < 250mm	1 mps
Rain drop 0.1 – 2.2 mm	9 mps
Hail (r) 0.5 – 5 mm	20 – 50 mps

where r = radius

It may be noted here that the shape of the rain drops are deformed as they fall through the air. If the drops are larger in size and if they are spherical initially they loose their spherical shape as they fall down and they may break down into smaller drops. This property accounts in non-formation of rainbow.

11.2 Methods of Cloud Formation

Clouds are formed by the vertical lifting of moist air, then cooling on condensation nuclei on saturation. The cooling of moist air may take place due to:

(i) long wave radiation of air mass itself from upper surfaces of clouds,

(ii) contact with cold surface and

(iii) mixing of two saturated air masses at different temperatures.

The sequence of adiabatic cooling already discussed. However the various processes that are involved in vertical lifting will be discussed here.

The vertical motion and cloud formation results mainly by four ways, namely:

(i) Mechanical turbulence

(ii) Convection

(iii) Orographic ascent

(iv) Slow widespread ascent.

1. **Mechanical turbulence clouds:** we have already discussed about mechanical or frictional turbulence. The air in the friction layer is thoroughly mixed by frictional forces (see Fig. 11.1).

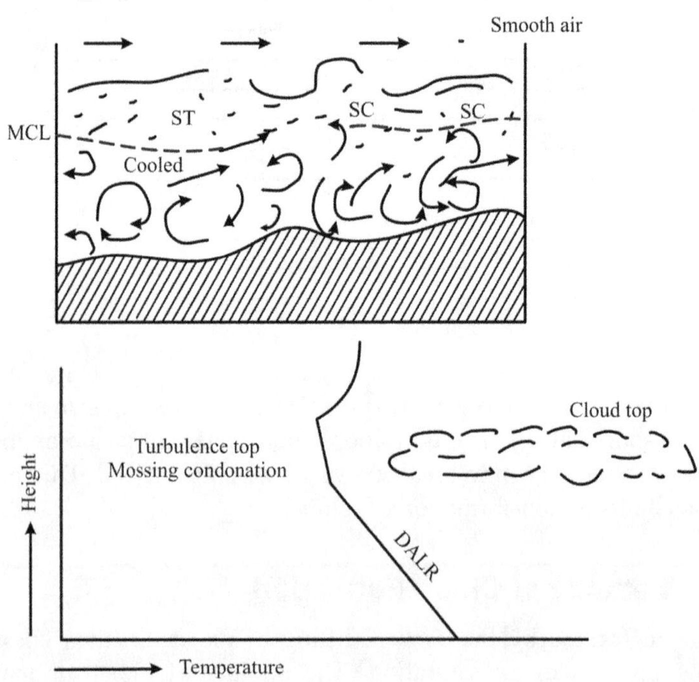

Fig. 11.1

Initially if the layer is stable, the upper part of the parcel cooled and the lower portion warmed. This establishes DALR in the layer if air remains unsaturated but it mixes up moisture evenly. If there is sufficient moisture in the lower level, it may become saturated at some height below the top of friction layer and condensation begins. This height is called mixing condensation level (MCL). The temperature of the air at condensation level is not the same as the dew point temperature at the surface level. MCL represents the base of the cloud. When cloud formed in this manner, air will obey DALR up to MCL and there after SALR. In this process stratus cloud is formed initially and subsequently it may develop like wave form and transform into stratocumulus. Turbulence clouds may develop below precipitation clouds such as Ns, As, Cb. Turbulence clouds that form may be Fractostratus or Fractocumulus of bad weather. In a moist layer, above friction layer, if wind changes with height Sc or Ac may form. Initially if the layer is unstable, the turbulent layer is greater than that of stable air. This produces cumuliform clouds with vertical development (see Fig. 11.2). Turbulence cloud at higher level (say Ac) may become Cu because of developing instability.

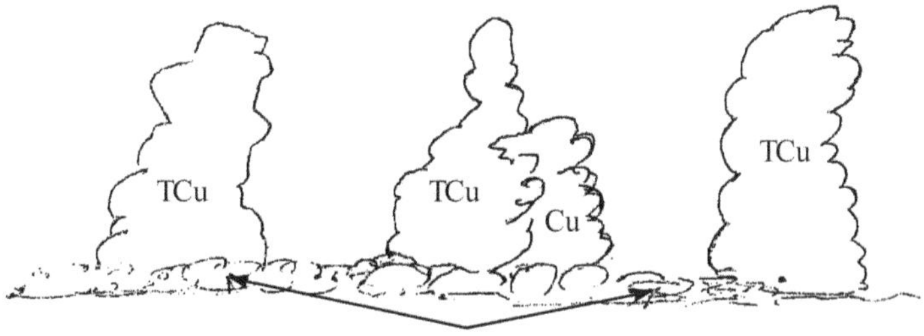

Fig. 11.2

2. Convection Clouds

The transfer of heat by means of actual motion of material is called convection. If a material is forced to move (by a blower or pump), the process is called forced convection. If the material flows due to density difference (caused by thermal expansion), the process is termed free convection (or natural convection). In meteorology all the three processes of convection are involved. We have already studied about thermal turbulence; we shall now discuss about convection/ thermal turbulence clouds.

***Fair weather cumulus*:** In the morning, as the day advances, the surface of the earth gets heated up first by insolation, then the air close to it. This develops convection currents in the atmosphere. This process is called thermal convection or thermal turbulence. Initially, the convection currents start. Suppose the environment is stable, the currents do not reach the condensation level. As the day advances surface temperature increases, convection currents strengthen, and instability develops. As long as air remains unsaturated ELR tends to be DALR and this lapse rate (ELR = DALR) may extend up to condensation level, which is called in this process, convection condensation level (CCL) clouds begin to form at this level (CCL). Upward development of cloud depends on ELR above CCL. If ELR > SALR, the atmosphere is unstable for saturated air parcel, which continue to move up. This leads to the formation of cumulus. Isolated convection cells of moderate height, called fair weather cumulus develops, which will not give rain. (see Fig. 11.3). Initial atmosphere is stable, the vertical development of cloud is restricted, rising currents form Cu and spread out side ways and transform into stratocumulus.

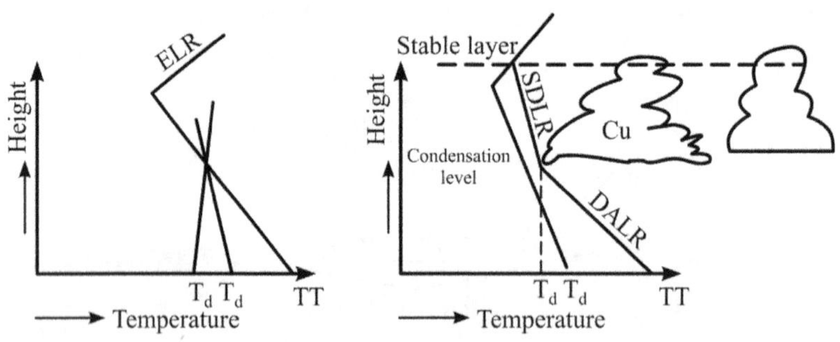

Fig. 11.3

Towering Cu/Cb: suppose the environment is unstable through a depth more than 3 km (10,000 ft). As day advances the surface heating increases with isolation or due to advection of cool air over warm surface, the saturated air parcel from the surface continue to move up while the cooling due to SALR remains warmer than the environment (see Fig.11.4). The rising currents produce condensation which continue to great heights. If the tops of cloud extend to the height where ice crystals form, in such case cumulonimbus (Cb) develops. The veil of floating ice crystals at the top gives fibrous (cirrus like) appearance, called false cirrus, which distinguishes Cb from Cu. Depending on the degree of instability and cloud height, slight to heavy precipitation occur from the cloud. During showers ragged turbulence cloud may develop below the cloud base which may even reach the surface. The cloud top reaches a stable inversion layer and then spreads horizontally, which is called anvil of Cb. The tops of Cb may reach to tropopause but only in some intense cases it may extend beyond tropopause.

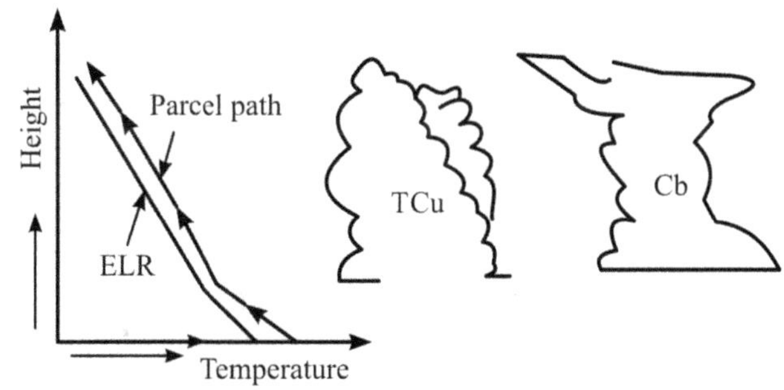

Fig 11.4

3. ***Orographic clouds*:**

Orographic ascent is mainly concerned with mountains and hilly barriers. When the wind flow reaches mountain / hilly barrier, air is forced to rise at all levels right from the foot to the top of barrier. This causes vertical ascent of air and associated adiabatic cooling and cloud formation. Depending on moisture content, orographic ascent and other factors St, Cu may develop. Orographic clouds form without concern to the time of day or season of the year. If St is formed in this manner it will have a flat base and little thickness of cloud. In general clouds form on the windward side of the barrier and dissolve on lee ward side. On the lee ward side at top of the barrier lenticular clouds are formed which appear to be stationary and associated with rain. On the lee ward side foot of the range standing waves form which affect the flight performance of aircraft. We shall consider two important cases.

(a) *Stable air*

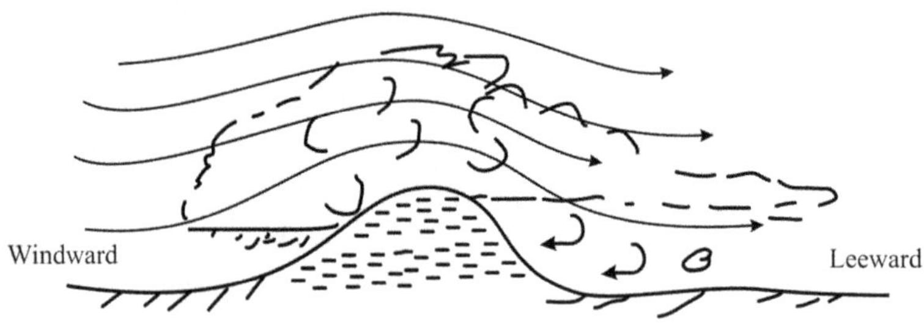

Fig. 11.5

If an airstream is moist through a deep layer but stable will produce a thick layer of stratiform cloud (see Fig. 11.5). The high ground will be covered by stratiform clouds or hill fog and lee side cloud base at higher level. In the wind ward side of the high ground there may be drizzle or snow and rain shadow on the lee side.

Hazards associated with orographic clouds

(i) obscuring high ground with stratiform cloud or hill fog,

(ii) air frame icing,

(iii) aircraft bumping or turbulence.

The orographic stable air may develop lenticular clouds which may be low, medium or high cloud as shown in the Fig. 11.6. Lenticular clouds have sharp boundaries and appears stationary. Clouds form in ascending currents, but evaporate further downstream where the air descends.

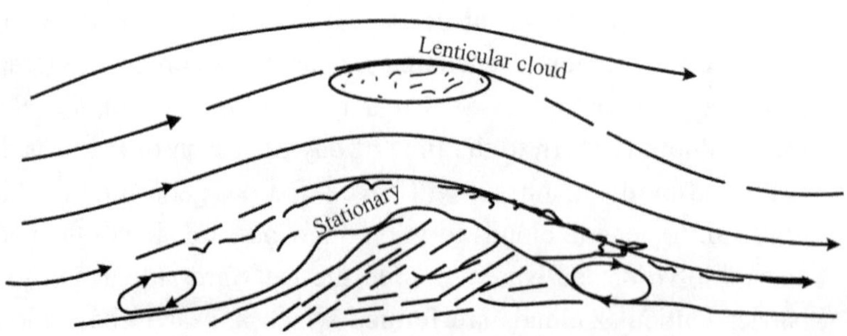

Fig. 11.6

Presence of orographic cloud indicate area of ascending air and turbulence, however the absence of orographic cloud does not indicate smooth undisturbed flow over mountains / high grounds.

(b) **Unstable air:** If the airstream is moist and unstable (introducing potential instability) cumuliform or convective clouds develop on mountain ridge and in leeward crest waves. Depending on moisture and instability cumuliform cells or Cb with showers may develop (Fig.11.7)

Fig. 11.7 (a) and (b)

11.3 Vertical Current Developed by Ground Contours

The turbulent wind over level surface may develop eddies, which include upward and downward motion and horizontal variations in wind speed and direction. When the size of the eddies is comparable with an aircraft they cause short period bumpiness in aircraft movement.

The effects of a terrain on wind flow is complex. In case of isolated symmetric hill or long ridge, change in the undisturbed flow pattern can be described to some extent.

A schematic representation of flow in three cases in shown in Fig. 11.8 (a, b c)

(a)

(b)

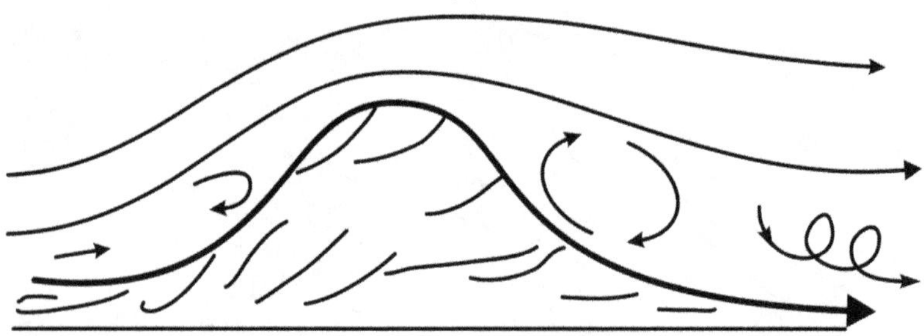

(c)

Fig. 11.8 (a), (b) and (c)

The magnitude of the vertical velocity is about $U_o \tan \alpha$, where U_o = horizontal wind speed, α = slope of the ground and valid for $\alpha > 45°$

For slopes of ground less than 10° (=α) or less than one in five, the flow remains unattached to the topography on both sides [Fig 11.8 (a)].

For slopes 1 in 3 ($\alpha \simeq$ 18° or greater) the resulting eddies are in upstream and downstream of the hill [Fig. 11.8 (b)].

The region of eddies are highly turbulent. With steep topography, the turbulent region extended downwind with vortex circulations these are shown in Fig. 11.8 (c).

In case of large topography, extensive waves develop to the lee side of mountain / barrier. In stable atmospheric conditions standing waves develop with air moving through them. Over the mountain the flow takes the shape (profile) of the ground. Beyond the mountain a lee wave flow develops which will have regular undulations (which have no bearing to the ground below). See Fig.11.9.

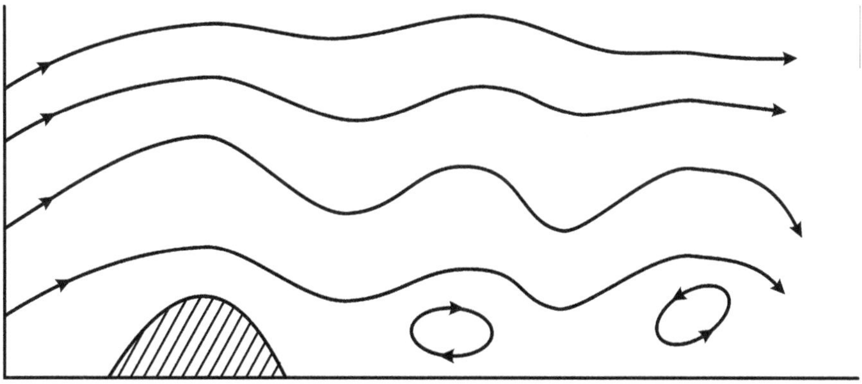

Fig. 11.9

If the air stream consists of a deep current with little or no change in direction in relation to height but with increase of speed with increasing height (extending to tropopause) develops strong lee waves. See Fig. 11.10.

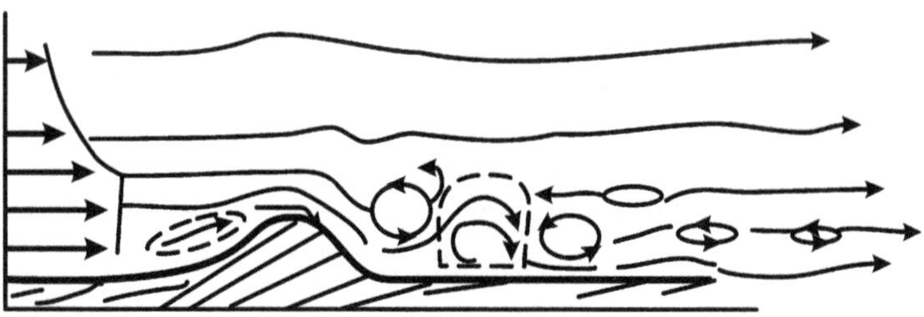

Fig. 11.10

In case of small mountains, wind speed 15 kt or more at the ridge (crest) and nearly perpendicular to the orientation of the ridge (within 30° to the perpendicular) or in case of big mountain, wind speed of 30 KT or more at the ridge and flow is perpendicular to the ridge, will create developed waves with the rotors in the lee of the barrier. Wavelength is a function of wind speed and atmospheric stability.

Light winds with stable atmosphere create short wavelengths, while strong wind and slight stability create long wavelengths. Wave motions are dampedout in an unstable air stream. Higher the mountain the greater the amplitude of the lee waves.

11.3.1 Standing Lee Waves

A series of stationary waves may form in the lee of a range mountain / hills. If the air mass forced to rise is stable, the up motion increases with height. On the lee the downstream air descends through a greater distance as compared to the ascending air. Subsequently further downstream, the air rises once again. This process is repeated many times with gradually decreasing as moved from the mountain / hilly barrier. The air mass particles follow an undulation path and create standing waves. Conditions for the generation of standing waves is the presence of very stable layer at the surface, then an unstable layer and then again another very stable layer above it. It may be noted that the undulatory flow may be present even in the absence any cloud formation due to lack of moisture in the air. Standing waves generally effect the performance of aircraft.

4. **Cloud formation under slow widespread ascent**

   In the above three cases we have discussed cloud formation under small scale turbulence, local convection currents and orographic barries of a few kilometer extent. We have studied earlier subsidence inversion and associated cloud formation. Now we shall discuss cloud formation under widespread vertical motion developed by large scale pressure / wind systems such as depressions, cyclones and anticyclone subsidence. In case of depressions a low level convergence and higher level divergence over the same area over a widespread region causes slow vertical rise of moist air. The air ascent persists for many days over a large area which causes lapse rate increase leading instability. This causes condensation and widespread cloud formation. Cloud masses may develop several kilometer thickness, which under upper air divergence produce widespread cloud precipitation. This is exactly happens in low pressure systems in tropics and frontal systems of subtropical and polar frontal depressions. We will study about these systems in detail in subsequent chapters.

Note: The vertical development of the cloud and turbulence in the cloud depend on the stability or instability of the environment in which the air parcel forced to ascend. Also the flight conditions depend on the turbulence that developed in the cloud.

5. ***Cloud dispersal***

Cloud dispersal is a reverse process of cloud formation. When any process of cloud formation ceases to operate then the cloud dissipation begins. In general the warming of cloud airmass, precipitation, and mixing of drier airmass (or entrainment) from the surrounding, the cloud dispersal takes place. Stable conditions at the surface level (ELR < SALR) inhibits cloud formation and thus aids dispersal of cloud. Low level divergence with associated upper air convergence over the same area, surface High or Anticyclone or upper air subsidence favours dispersal of cloud. Fair weather cumulus which forms by insolation in the morning, reaches maximum height in the afternoon and then dissipates towards evening. Entrainment of dry air (mixing of dry air) from the surrounding leads to the dissipation of fair weather cumulus.

Questions

1. A meteor consisting of an ensemble of liquid or solid aqueous particles, falling through or suspended in the atmosphere blown by wind from the earth's surface or deposited on objects on the ground or in the free air is called:

   (a) Photometeor (b) Lithometeor
   (c) Hydrometeor (d) Electrometeor

2. Falling of hydrometeors other than virga, that finally reach the ground is called:

   (a) Rain (b) Snow
   (c) Fog (d) Precipitation

3. Fairly uniform precipitation, composed of fine drops of water with diameter less than 0.5 mm, that does not impact water surface is:
   (a) Rain (b) Drizzle
   (c) Snow (d) Hail

4. Precipitation of small balls or pieces of ice, diameter ranging 5 to 50 mm, falling either separately or agglomerated into lumps is:
   (a) Snow (b) Ice pellets
   (c) Hail (d) Sleet

5. A phenomena which occurs at the surface if the surface wind is calm or very light, sky is clear during the night and humidity is more than 75% is:
   (a) Haze (b) Fog
   (c) Dust storm (d) Sand storm

6. A luminous phenomenon produced by reflection, refraction, diffraction or interference of light from the sun or the moon is called:
   (a) Hydrometeor (b) Lithometeor
   (c) Photometeor (d) Electrometer

7. A group of optical phenomena in the form of rings, arcs pillars or bright spots, produced by the light from the sun or the moon, by the refraction, or reflection of light through ice crystals suspended in the atmosphere (cirriform clouds, ice fog etc.) is called:
   (a) Carona (b) Glory
   (c) Rainbow (d) Halo phenomena

8. A group of concentric arcs with colours ranging from violet to red (VIGBYOR) produced on a screen of water drops in the atmosphere by light from the sun or the moon is called:
   (a) Carona (b) Glory
   (c) Rainbow (d) Halo phenomena

9. A more or less continuous luminous electrical discharge of weak or moderate intensity in the atmosphere, emanating from elevated objects at the earth's surface or from aircraft wing tips or propellers in flight aircraft is called:
   (a) Thunder (b) Air discharge
   (c) Saint Elmo's fire (d) Thunder bolt

10. The cooling of moist air may take place by:
    - (a) Long wave radiation of airmass itself from the upper surface of clouds
    - (b) Contact with cold surface
    - (c) Mixing two saturated air masses at different temperatures
    - (d) All above (a), (b) and (c)

11. If there is sufficient moisture, the air parcel if lifted it may become saturated at some height. This height is called:
    - (a) Dew point level
    - (b) Mixing condensation level
    - (c) Ground friction level
    - (d) Both (a) & (b)

12. The mixed condensation level represents the base of the cloud. Air will obey up to the top of cloud:
    - (a) DALR throughout
    - (b) SALR throughout
    - (c) DALR upto base of the cloud then SALR
    - (d) None

13. Frictional turbulence cloud in stable atmosphere is:
    - (a) Cumulus (b) Cirrus
    - (c) Stratus (d) Altocumulus

14. If the environment is stable and convection currents develop as the day advances, this may produce:
    - (a) Stratus (b) Fair weather cumulus
    - (c) Cb (d) Cs

15. If the environment is unstable and the saturated air parcel lifted by the convection currents may produce in the afternoon:
    - (a) Stratus (b) Cb and Precipitation
    - (c) Ns (d) Ac

16. Orographic clouds form:
    - (a) In the morning
    - (b) In the afternoon
    - (c) In particular season
    - (d) Without concern to the time of day or season of the year

17. In orographic form of clouds that develop on the barrier top is:
    (a) Heap shaped
    (b) Lense shaped called leutrcularis
    (c) Fair weather cumulus
    (d) Stratus

18. Lanticular clouds are:
    (a) Stationary
    (b) Lense shaped with clear boundaries
    (c) Indicate aircraft performance bumpiness
    (d) All above

19. If the airstream is moist, unstable lifted by orography may produce:
    (a) St (b) Sc (c) Cu / Cb (d) Ac

20. If the slope of mountain / barrier is 1 in 3 or more, air stream flow may produce turbulence and:
    (a) Eddies in upstream (b) Eddies in down stream
    (c) Smooth flow (d) Both (a) & (b)

21. On the lee side of mountains you would normally expect to find:
    (a) Down drafts (b) Updrafts
    (c) Eddies and down drafts (d) Smooth flow

22. In general, the presence of orographic clouds indicate:
    (a) Down drafts
    (b) Ascending air and possible turbulence
    (c) Smooth undisturbed flow
    (d) Both (a) & (c)

23. Ascending wind flow in mountainous region over high ground causes:
    (a) O°C isotherm (freezing level) increases
    (b) Freezing level decreases
    (c) Increases severity of ice accretion on aircraft in cloud
    (d) Both (b) & (c)

24. The safety height for an aircraft overflying mountain waves over high ground should be:
    (a) Increased to allow possible downdrafts
    (b) Decreased to allow possible updrafts
    (c) Safety height need not be changed
    (d) Both (b) and (c)

25. Over the high mountain the wind flow is deflected by the high ground. Beyond the mountain a lee wave establishes with:
    (a) Smooth flow (b) Regular undulations
    (c) Develop rotors (d) Both (b) and (c)

26. Mountain standing lee waves develop under the conditions:
    (a) Wind blows nearly at right angles (within 30°) to the barrier, with 15 kt or more and little or no change in direction, stable thick layer of air.
    (b) Wind blows parallel to the barrier, with frequent change in wind direction & speed, unstable layer of environment
    (c) Wind blows nearly perpendicular to the barrier with light and variable wind, and unstable environment
    (d) None above

27. The orographic cloud formed on the mountain is called cap cloud and above it and on the lee there will be:
    (a) Stratus cloud (b) Lenticular cloud
    (c) Cb cloud (d) Cs

28. Rain occurs from the clouds:
    (a) St (b) As, Ns
    (c) Cu, Cb (d) Both (b) & (c)

29. Drizzle occurs from the clouds:
    (a) St (b) As, Ns (c) Cu, Cb (d) Sc

30. Snow occurs from the clouds:
    (a) St (b) Cu (c) As, Ns, Sc (d) None

31. Snow grains & ice prisms occur from the clouds:
    (a) As, Ns (b) Cu, Cb (c) St (d) Sc

32. Hail occurs in the cloud:
    (a) Cu (b) St (c) As (d) Cb

33. A whitish ring, centred on the sun or moon with a bluish tinge on the inside and reddish on the outside is called:
    (a) Halo (b) Carona
    (c) Bishop's ring (d) Glory

34. One or more sequence of coloured rings of relatively small diameter, centred on the sun or moon is called:
    (a) Halo (b) Carona
    (c) Bishop's ring (d) Glory

35. One or more sequence of coloured rings seen by an observer around his own shadow on a cloud consisting mainly numerous water droplets, on fog, or rarely on dew is called:
    (a) Halo (b) Carona
    (c) Bishop's ring (d) Glory

Answers

1. (c)	2. (d)	3. (b)	4. (c)
5. (b)	6. (c)	7. (d)	8. (c)
9. (c)	10. (d)	11. (d)	12. (c)
13. (c)	14. (b)	15. (b)	16. (d)
17. (b)	18. (d)	19. (c)	20. (d)
21. (c)	22. (b)	23. (d)	24. (a)
25. (d)	26. (a)	27. (b)	28. (d)
29. (a)	30. (c)	31. (c)	32. (d)
33. (c)	34. (b)	35. (d)	

CHAPTER 12

Precipitation Processes

Precipitation is defined as the falling hydrometeors, which reach the ground finally. It does not include virga. In other words precipitation is drizzle, rain and shower, sleet and snow (that water drops or ice crystals falling from clouds).

Precipitation process refers to the return of hydrometeors to the ground.

Precipitation process has four steps; condensation nuclei, initiation of condensation, growth of condensation products and precipitation. The first two steps were dealt in aerosols of the atmosphere and condensation nuclei. Now we shall study the other two steps.

In cloud condensation the cloud droplets grow to the size of about 15 μm (0.015 mm) while rain drops diameter ranges 500-1000 μm or more. In condensation, initially droplets grow from nucleus size to visible size (cloud) rapidly but there after the growth of droplets slow down. Condensation alone is unlikely to produce droplets of average size larger than 30 μm. This hints that precipitation process differs form cloud condensation process. In cloud condensation process, minute nuclei (Aitken nuclei) transforms to small particles of higher concentration. In precipitation process large number of small particles convert into a small number of much larger particles which come down to earth under gravity as precipitation. Depending on the mass and size of water droplets, there will be a falling velocity called terminal velocity of the water droplets. Heaviest the drop greater is the terminal velocity. We have studied in meteors that the sizes of drizzle drops, rain drops, hail size vary from 10-250 μm, 100-2200 μm, and 500-5000 μm and they acquire a fall speed of 1 mps, 9 mps and 20-50 mps respectively. Radius of drizzle drops 0.05-0.5 mm have terminal velocity 0.7 to 2.0 mps, radius of rain drops 0.5-2.5 mm have terminal velocity 3.9 to 9.1 mps.

The condensation nuclei concentration over land is found to be much larger than over sea. On an average the cloud droplets are found to be of smaller radii ranging 2 to 10 μm over continents, while in maritime cloud droplets size vary from 3 to 22 μm. The concentration of giant salt nuclei (hygroscopic nuclei) is very small, one per liter both on continents and over oceans which produce cloud droplets of size radius 20-30 μm or more. Due to ascending currents in the clouds, droplets may acquire a size of more than 100 μm radius and fall as rain drops with terminal velocity of about 1 mps under gravity.

Clouds whose tops lie entirely below freezing level (0° C isotherm) are called warm clouds. They fully consists of water droplets.

Clouds whose base lie entirely above freezing level are called cold clouds. These clouds contain supercooled water droplets and ice crystals. Clouds whose base lie below freezing level and tops lie above freezing level are called mixed clouds. Mixed clouds contain water droplets, supercooled water droplets and ice crystals. Generally clouds appear in mixed form.

The growth of cloud droplets depend mainly on moisture (RH), nature of nuclei, effects of surface tension and the amount of heat transferred to the surroundings (by way latent heat of release). Studies indicate that more than 80% of clouds containing liquid water were warmer than –10 °C, half of them were mixed with liquid and ice crystals. By – 20 °C, only 10% contain liquid drops and 30% contained both supercooled droplets and ice crystals.

In warm clouds (cloud) droplets grow to precipitation size by collision and coalescence process, while in cold and mixed clouds droplets grow to precipitation size by Bergeron (ice crystals) process. These processes are described below.

12.1 Growth of Droplets by Collision

We have studied earlier that different size droplets have different rates of fall and attain different terminal velocities. Largest droplets fall fastest and collide with smaller ones in their path. Droplets having opposite charges are attracted. Droplets of similar sizes are attracted towards each other by the reduction of pressure between them when they move in the same direction side by side.

12.2 Growth of Droplets by Coalescence

Coalescence means coming together and form one (that is fusing together). In this process cloud droplets grow by direct collision and coalescence. Droplets lifted by updraughts (ascending currents) fall relative to the rising air and by this larger (heavier) droplets fall faster relative to smaller droplets. When falling droplets size exceeds 18 μm, they overtake and collect smaller droplets in their path, thereby they become larger, fall faster and sweep up small droplets more rapidly. In warm clouds collision and coalescence process alone occurs for the growth of droplets to precipitation.

***Note*:** Steam occupies a volume which is about 1600 times (1600V) greater than the liquid water (V). In gaseous state the molecules of water, on an average, about 12 times further apart than they were in liquid state.

Calculations show that it requires 50 minutes for a droplet size of 30 μm to grow by coalescence to the size of 300 μm in a typical cloud of average droplet size 8 μm and liquid content 1 g/m^3. The Table 12.1 gives the growth of initial droplet size of 30 μm falling through a homogeneous clouds mass of average drops of 20 μm containing 1 g/m^3 liquid water.

Table 12.1 Droplet Growth

Diameter μm	Time (cumulative minutes)	Falling distance (cumulative distance) meters
30 initial	0	0
40	45	65
60	75	165
100	90	320
200	105	650
500	116	1475
1000	123	2675

12.3 The Bergeron (Ice Crystal) Process

This process deals with the initial growth of ice crystals in mixed clouds.

In 1933, a Swedish meteorologist propounded the basic principles of growth of ice crystals in mixed clouds. It is assumed that ice crystals develop initially by freezing of supercooled water over ice nuclei. When ice crystals exists in the presence a supercooled water droplets, the system becomes unstable because the saturation vapour pressure over ice is less than over water at sub-zero temperatures. As a result the water droplets evaporate and sublimate (transform directly into ice particles) and is deposited on the ice nuclei or ice crystals. This is called accretion. Thus ice crystal grow at the expense of water droplets. This transformation is not evident at slightly below 0 °C, but it is conspicuous at –10 °C and below, and at temperature –15 °C the transformation of vapour to ice crystals is most efficient. However the growth rate decreases as crystals become larger. Thus Bergeron process essentially deals with the initial growth of ice crystal.

Once the ice crystals grew appreciably (become heavier than water droplets) they begin to fall relative to the water droplets and colloid with them. Collision of ice crystals with supercooled water droplets lead to freezing of water droplets on ice crystals. The accretion of supercooled water droplets on ice crystals results into formation of rime, glaze and snowflakes. Finally, when water droplets or ice crystals grow to a size which the air currents (updrafts /downdrafts) may not be able to hold, they fall under gravity colliding with different size particles and descend below the cloud base as precipitation. In some cases the precipitation products may evaporate below the cloud base as virga or ice crystals melt or sublimate. After the survival of this, the hydrometeors reach the surface of the earth as precipitation viz as drizzle, rain, shower, snow fall or hail.

***Note*:**

1. Fog, dew and frost are also involved in the transfer of moisture from atmosphere to the surface of the earth but they are not considered as precipitation.
2. Water droplets in clouds grow by coalescence, while ice particles grow by accretion of supercooled droplets on ice particles.

Summary of Precipitation Process

Clouds form and develop in the atmosphere by adiabatic expansion and cooling of ascending moist air. When the moist air is lifted above condensation level, active condensation nuclei grow into droplets. Further cooling below 0 °C may or may nor freeze. In the absence of ice forming nuclei supercooled droplets form (up to about – 35 °C). Ice nuclei is different from condensation nuclei.

Initially water vapour condenses and water droplets grow at faster rate and become larger and subsequently the growth rate decreases. In a similar fashion the growth of ice crystals by Bergeron process is rapid initially but the growth rate decreases as the size increases.

Further water droplets growth to precipitation size occurs by collision and coalescence. Similarly further growth of ice crystals to precipitation size occur by collision and accretion of supercooled liquid droplets on ice crystals.

Finally when water droplets or ice crystals grow to a critical size which the ascending air currents (pdrafts) cannot hold, they come out of the cloud base under gravity. If these condensation/sublimation products survive the evaporation in the unsaturated air below the cloud base and reach the ground then it is called precipitation, but if it evaporates it is called virga.

12.4 Effects of Precipitation in Aviation

Precipitation effects avition in three ways:

(i) visibility,

(ii) aircraft icing and

(iii) runway contamination.

The visibility in rain depends on the size of rain drops and intensity.

Light rain has little effect, moderate rain reduces visibility to 2 to 4 km while heavy rain or showers reduce the visibility to less than 1000 meters. In drizzle visibility may be as low as 500 meters but generally it is 2.4 km. Drizzle is hazardous in association with low stratus clouds. Similarly the visibility in snow fall may be less than 1000 meters and at times it reduces to less than 500 meters in heavy snowfall.

Aircraft icing will be discussed in a separate chapter.

Runway surface conditions are greatly changed due to precipitation. A wet runway will require a longer landing distance. A low friction runway surface (like wet, slick, icy) will not allow effective braking, consequently a longer landing distance required.

Aquaplaning on a wet-surface required because this gives extraordinary increases in the stopping distances. Aquaplaning is the phenomenon of a tyre skating along on a thin film of water and not rotating even though it is free to do so.

If the runway is wet/icy the available runway length may not be sufficient to the problem created by runway surface conditions. Accelerate-stop distance will increase by 15% on wet runway and 50% on snow and 75% on water flooding more than 3 mm and 100% for ice cover on runway.

Aquaplaning and procedures are given in the aircrafts flight manual and company's operational manual which must be followed. The contamination of runways may create problems of additional drag, power loss due to ingestion, reduced braking or aquaplaning, directional control reduction and likely structural damage.

12.5 Diurnal and Seasonal Variation of Rainfall

In tropics convective activity is more common and which is vigorous in the afternoon and as a consequence precipitation is common in the after-noon. In coastal areas and valley (NE India) rainfall is maximum during night or early morning. Summer season is the main rainy season as compared to the others.

In temperate latitudes maximum rainfall is due to snowfall that occurs in winter in association with frontal activity (extra-tropical depressions). Extreme northwestern part of India gets rainfall of this type during winter.

12.6 Precipitation Measurement

Precipitation is expressed as the depth to which it would cover a horizontal projection of the earth's surface, in the absence of no losses by evaporation, run off, infiltration and the part of precipitation, that fallen snow or ice melted. The average annual depth of precipitation over the land surface of the earth is 760 mm, overseas/oceans 1140 mm and over the entire globe average 1030 mm. The factors that determine average amount of precipitation at any point of the earths surface are: latitude, altitude, distance from the source of moisture, prevailing winds, relation to mountain ranges (windward or leeward) and relative temperature of land and neighboring oceans.

Note:

(a) A stable atmosphere is condusive for development of stratiform clouds, while unstable atmosphere for cumuliform/convective clouds.

(b) Turbulence cloud is stable St or Sc.

(c) Convective clouds give showers.

(d) Ns in winter gives snow

(e) Generally the base of low cloud is below 6500 ft

Definition. Rainy Day: A rainy day is defined as when 24 hours rain is 2.5 mm or more.

Intensity of rainfall is given in the Table12.2 and precipitation in Table 12.3.

Table 12.2 Rainfall intensity

Rainfall amount in 24 hrs	Description of intensity
0.1-2.4 mm	Very light rain
2.5- 7.5 mm	Light rain
7.6-34.9 mm	Moderate rain
35.0-64.9 mm	Moderate rain
65.0-124.9 mm	Heavy rain
125.0 mm or more	Very heavy rain

Table 12.3 Intensity of precipitation

Intensity	Rainfall Rate	mm/hr	Snow Accumulation Cm/hr
	Rain	Rain/hail Showers	
Slight	Less than 0.5	Less than 2	Less than 0.5
Moderate	0.5 to 4	2 to 10	0.5 to 4
Heavy	More than 4	10 to 50	More than 4
Violent		More than 50	

Note: The water equivalent of snow is usually one-tenth,

i.e., 1 cm snow = 1 mm of water.

12.7 Record Rainfall in India

Mumbai recorded 944 mm of rain (in 21 hrs) at 0830 hrs IST on Wednesday the 27th July 2005, which surpassed the earlier record of 838 mm recorded in 24 hr at Cherapunji in 1910. The earlier record rainfall of Mumbai was 575.6 mm (24 hrs) in 1974.

Questions

1. Precipitation includes:
   (a) Rain, Drizzle (b) Shower, Sleet, Snow
   (c) Dust storm, Sand storm (d) Both (a) and (b)
2. Precipitation process includes:
   (a) Condensation Nuclei, initiation of condensation
   (b) Growth of condensation products, Precipitation
   (c) Dust storm, Sand storm
   (d) Both (a) and (b)
3. Condensation allows droplet growth:
   (a) From nucleus size to visible size
   (b) Visible size to Rain drop size
   (c) Visible size to precipitation size
   (d) None
4. The diameter of drizzle drops size and terminal velocity are
   (a) 10-250 μm, 1 mps (b) 100-2200 μm, 9 mps
   (c) 500-5000 μm, 25-50 mps (d) Both (b) and (c)
5. The diameter of rain drops size and terminal velocity are:
   (a) 10-250 μm, 1 mps (b) 100-2200 μm, 9 mps
   (c) 500-5000 μm, 25-50 mps (d) None above
6. The diameter of hail size and terminal velocity are;
   (a) 10-250 μm, 1 mps. (b) 100-2200 μm, 9 mps
   (c) 500-5000 μm, 25-50 mps (d) None above
7. Clouds whose tops lie entirely below 0o C isotherm are called:
   (a) Low clouds (b) Warm clouds
   (c) Cold clouds (d) High clouds

8. Clouds whose tops lie entirely above freezing level are called
   (a) Low clouds (b) Warm clouds
   (c) Cold clouds (d) Mixed clouds
9. Clouds whose base lie below freezing level and tops above freezing level are called:
   (a) Low clouds (b) Warm clouds
   (c) Cold clouds (d) Mixed clouds
10. Mixed clouds contain:
    (a) Water droplets
    (b) Super cooled water droplets
    (c) Ice crystals
    (d) All above
11. Warm clouds contain:
    (a) Water droplets
    (b) Supercooled water droplets
    (c) Ice crystals
    (d) All above
12. Cold clouds contain:
    (a) Water droplets
    (b) Supercooled water droplets
    (c) Ice crystals
    (d) both (b) and (c)
13. In warm clouds droplets grow to precipitation size by the process of:
    (a) Collision (b) Coalescence
    (c) Accretion (d) Both (a) and (b)
14. In cold clouds, could droplets grow to precipitation size by the process of:
    (a) Collision
    (b) Accretion (ice crystal Process)
    (c) Coalescence
    (d) Both (a) and (b)

15. In Bergeron process ice crystal grow by accretion:
    (a) Between 0 to 10° C
    (b) At the expense of supercooled droplets between – 10 to – 35 °C
    (c) Between + 10 to – 10° C
    (d) Above 0 °C

16. Precipitation includes:
    (a) Fog at surface (b) dew at surface
    (c) Frost surface (d) Rain

17. In clouds ice particles grow by:
    (a) Condensation
    (b) By accretion of super cooled droplets on ice particles
    (c) Condensation of water droplets on ice
    (d) Both (a) & (c)

18. Effects of precipitation in aviation are:
    (a) Visibility deterioration (b) Aircraft icing
    (c) Run way contamination (d) all above

19. Runway surface conditions are greatly changed by precipitation. A wet runway will require:
    (a) Shorter landing distance
    (b) Longer landing distance
    (c) Same as dry run way
    (d) None above

20. On wet runway surface, accelerate–stop distance:
    (a) Will increase by 50% (b) Increase by 15%
    (c) Does not change (d) None above

21. Aquaplanning is a phenomenon of:
    (a) Tyre staking along a thin film of water
    (b) Friction increasing on tyres
    (c) Not rotating even though it is free to do so
    (d) Both (a) and (c)

22. Precipitation is measured in:
    (a) hPa (b) % of RH
    (c) depth in cm (d) in °C

Answers

1.	(d)	2.	(d)	3.	(a)	4.	(a)
5.	(b)	6.	(c)	7.	(b)	8.	(c)
9.	(d)	10.	(d)	11.	(a)	12.	(d)
13.	(d)	14.	(d)	15.	(b)	16.	(d)
17.	(b)	18.	(d)	19.	(b)	20.	(b)
21.	(d)	22.	(c)				

CHAPTER 13

Thunderstorms

In meteorology the scales of weather phenomenon ranges from small eddies to thunderstorms, tornadoes, depressions, cydenes and to very large features of hemispherical or global scales. The typical scales of atmospheric motions are given in Table 13.1

Table 13.1 Different scales of atmospheric motions

Types of motion	Order of horizontal scales (in meters etc.,)
Molecular mean free path in atmosphere	10^{-9} m $= 10^{-3}$ µm
Minute turbulent eddies	(10^{-2}-10^{-1} m) = $10^4 - 10^5$ µm
Small eddies	(10^{-1}–1 m) = 10^5–10^6 µm
Dust devils	1-10 m
Gusts of wind	10-10^2 m
Tornadoes	10^2-10^3 m
Thunderstorm (Cb cell)	10^3 m = 1 km
Fronts and squall lines	10-10^2 km

Table 13.1 Contd.....

Cydenes /hurricanes	10^2-10^3 km
Planetary waves/ Rossly waves	10^4 km
Zonal wind / Jet streams	10^4 km
Planetary scale or Macro scale } Horizontal length	5000 km
Synoptic scale Horizontal length	1000 km
Meso scale Horizontal length	100 km
Micro scale Horizontal length	$\leq$ 1 km

13.1 Time Scale Order

Macro / planetary-scale	$\geq$ 5 days
Synoptic-scale	1 day
Meso-scale	5 hours
Micro-scale	$\leq$ 10 minutes

The average diameter of a molecule (aerosol) in air is about 3.7×10^{-10} m (= 3.7×10^{-4} μm)

Number of molecules of air in one cubic centimeter = 3×10^{19}

Number of molecules of air in one cubic meter = 3×10^{25}

Mean free path of an air molecule 6×10^{-8} m (=6×10^{-12} μm)

***Micro-meteorology*:** In Greek micro meaning small. It deals with the weather processes like turbulence, evaporation, heat fluxes or transfers. These processes occur in a layer close to the ground.

Meso-Meteorology: In Greek meso meaning middle. It deals with the weather phenomena like thunderstorm, Land/sea breezes etc. These whether phenomena are very small to be identified or located on synoptic charts but they have profound effect particularly on aviation. Identification depends on local observations.

***Synoptic-Meteorology*:** In Greek synoptic meaning having the same view point. It deals with large weather phenomena like depressions, anticyclones, tropical cyclones. Network of observation stations vary 100 to 1000 km apart.

Macro-meteorology: In Greek macro meaning large. It deals with very large weather features that can be detected on hemispherical weather maps. It includes the weather phenomena of major atmospheric circulation, Rossly waves, jet streams, mean zonal winds etc.

Cosmic-soale: This includes weather phenomena which involves inter-planetary environment, like solar wind, radio black-out.

13.2 Thunderstorm

A thunderstorm is an electrometer, which is defined as a visible or audible manifestation of atmospheric electricity.

Definition: According WMO, a thunderstorm is defined as, one or more sudden electrical discharges, manifested by flash of light (called lightning) and a sharp or rumbling sound (called thunder).

In a thunderstorm cloud is invariably cumulonimbus and generally associated with precipitation. The sound of thunder is due to electrical discharge. In electrical discharge enormous heat develops which heats the air channel between the cloud and environment or earth. The conducting air channel is heated up to 30000 ^{o}K, which is about five times the temperature of the surface of the sun's photosphere (6000 ^{o}K). Because of sudden heating of the conducting air channel in a very short period, the diameter of the conducting channel expands abruptly from a few millimeters to a few centimeters (ten folds). This sudden expansion of conducting channel generates shock waves, which spread out as sound of thunder. [A shock wave is a wave surface that spreads in solid or fluid medium, at which there is abrupt pressure increase accompanied by variations in density, temperature and velocity of the medium). Thunder is generally heard upto 10-15 km and rarely beyond 25 km. At any instant there are about 2000 to 3000 active thunderstorms around the globe, and most of these occur in tropics. Simultaneously at any moment there are about 100 lightning flashes per second.

A thunderstorm day is defined as a local calendar day on which thunder is heard, irrespective of the actual number of thunderstorms on that day. The highest annual average of 242 thunderstorm days was recorded at Kampla, Uganda (0 ^{o}20' N, 32° 30' E) in Africa.

Thunderstorms occur almost in all parts of the world. They are frequent in tropical regions and rare in polar regions and infrequent at latitudes higher than 50° N and 50° S. From the aviation point of view a thunderstorm is a severe hazard. It is accompanied by thunder and lightning, strong vertical currents (up draughts and down draughts) severe gusts and squalls, heavy rain/showers, poor visibility in association with heavy rain and low celing of cloud base and sometimes hail.

13.2.1 Types of Thunderstorms in Classical Approach

It was classified as: (a) *Airmass*: orographic, heat thunderstorm, (b) *Frontal*: particularly cold front storms.

Orographic thunderstorms can occur at any time in 24 hours.

Airmass Heat thunderstorms depend on surface heating which triggars convection cloud. Generally develop in the middle-afternoon, continue till night. Frontal thunderstorms are associated with cold fronts. Thunderstorms develop along trough line and move with trough line in a direction perpendicular to the trough line. They may develop at any time.

In general thunderstorms move in the direction of local wind at 700 hPa (10,000 ft) level, however in Col regions and slack pressure gradient regions, they move in any direction (that is movement is erratic).

From the modern view point, thunderstorms depend on local meteorological environment in which the storms develop. The United States Weather Service defined as a severe thunderstorm is any storm that develops a tornado [wind speed 26 mps or more] or hail (size : diameter greater than 1.9 cm or 0.75 inches).

A Thunderstorm consists of several convective cells each having a distinctive convective circulation. The dynamical building block of a thunderstorm is the cell. A cell is a large compact region having updrafts (about 10 mps). Most of thunderstorm cells have short period life. At any instant of a thunderstorm, it consist of cells which are at different stages of evolution and each interacts with neighboring cells or outside environment. Thunderstorms are local affairs and rarely exceed diameter of 10 km(100 Km^2 area) and duration of one to two hours, height may reach to about 10 km (33,000 ft) and usually do not create violent weather at the ground surface.

13.3 Conditions Favorable for the Development of Cb

1. ELR>>SALR in a considerable depth of layer (5-6 Km depth) extending to several thousand feet above freezing level ($0^{o}C$).
2. Adequate supply of moisture from below.
3. A triggering mechanism (or a process) which produces saturation in the region of steep lapse rate or suitable synoptic situation to cause low level convergence and upper level divergence in that area.
4. Suitable upper air flow that advects warm moist air in the lower troposphere and cold dry air in the upper troposphere. This increases the instability.

5. The triggering mechanism which sets off thunderstorm are:
   (i) Local convection (isolation)
   (ii) Orographic lifting
   (iii) Convergence
   (iv) Frontal lifting
   (v) Radiational or Katabatic cooling

13.4 The Life Cycle of a Thunderstorm Cell

A thunderstorm cell has three stages of life cycle. (i) Cumulus or growing stage, (ii) Mature stage, and (iii) Dissipating stage (see Fig.13.1)

Fig. 13.1

Cumulus Stage

This stage is characterized by up drafts throughout the cloud, height grows and takes the shape of cauliflower type domes on the top and the cloud top temperature may be about –20 to –30° C. In all stages of the cloud, unsaturated air mass entrainment takes place at all levels of the cloud. Marked temperature difference ($\Delta T = T_u - T_e$) is observed between updraft temperature T_u and the ambient environment temperature (Te) of the cloud. ELR > SALR, hence T_e being comparatively warmer in updrafts as it acquires ELR while the descending air which acquires SALR is cooler. The temperature difference ΔT is the measure of the strengths of updrafts (V_u). The cloud base may be about 10 km, height of the top of the cloud may be 8-10 km, updraft strength 100 ft/second (30's mps) common when $\Delta T = 4°C$. Below the cloud base horizontal convergence of wind may be of the order $5\text{-}10 \times 10^{-4}\ S^{-1}$ and wind speed 3-4 mps observed. The life period of cumulus stage may be 15-20 minutes.

Mature Stage

This stage is characterized by both updrafts and downdrafts. Down-draughts are prominent below freezing level. The lowest temperature in the cloud is observed in downdrafts. Downdrafts generally begin in the cloud at a level where could temperature is equal to the ambient air temperature. Severe turbulence occurs in this stage and downdrafts extend below the cloud base. The temperature of the air in downdrafts (T_D) attains minimum value compared to the environment (T_e) [i.e; $T_D < T_e$]. Squall wind speed is a measure of downdrafts. This stage lasts for about 30 minutes, when the cell attains maximum height (12 to 18 km). In this stage severe lightening, precipitation and squalls occur. The strength of the updrafts may be 10-30 mps (20-60 kt) and downdrafts 5-15 mps (10-30 kt). Downdrafts below the cloud base produces divergence of cold air (gust-front), which can spread to 32 km and up to a height of 6000 ft. In the wake of this cold divergence, new cells develop in the neighbourhood and thus trigger chain action of thunderstorm. The life period of this gust-front is about 0-75 minutes.

Note: When the upper part developing above 0 $^{\circ}$C level, the precipitation is producing down draughts in the cloud, it is warming at about SALR.

Dissipating stage

This stage is characterized by only downdrafts (speed 10 mps) throughout the cell except the top, where feeble up currents may be present. The downdrafts below the cloud base may spread as gusts or squalls. The speed of the squall at ground may be 10-15 mps. In this stage the entire cloud mass temperature becomes cooler than the ambient air temperature. No more condensation occurs. The height the cloud dissipates with dissipation of anvil which consist of the ice crystals called false cirrus. In the first two stages energy is generated for about 45 minutes and the energy is spent in Cb (about 2 hours). Clod air divergence commonly spreads below the cloud base and downdrafts cease to exist. Average life of the cell in this stage is about 10-30 minutes.

13.5 Regeneration of Thunderstorms or Chain Action or Self Propagating Storms

The cellular nature of thunderstorm helps in regenerating thunderstorms in the neighbourhood as we find a group of clouds arranged. We have studied that when a well developed thunderstorm cell dissipates, its cold gust-front moves away at the low level close to the ground. This clod gust-front can activate a nearby cumulus cell by pushing its mass from the surface level by creating instability at the base. By this cumulus cell rapidly develops into cumulonimbus. The Cb in turn in its dissipation help generate another cell into Cb. In this way in a quick succession thunderstorm belt can extend to hunders of kilometers. The violent local thunderstorm in India called Norwester or Kalbaisaki propagate the thunderstorm activity in summer which affect Bengal, Bihar and Orissa in this fashion.

Multiple Cell Thunderstorms

In nature, cells of clouds develop simultaneously but they acquire different stages of life cycle. If the cells develop very close by, these cells organize into a multiple-cell storm, squall lines or a supercell. All these cells develop due to local mesoscale disturbances. Multiple cell storms have several updrafts and downdrafts zones in proximity to one another. Within this cluster one cell may influence others before it weakens, and then the another cell in the cluster or compact. In squall-lines, thunderstorms regenerate continuously with gust-front.

In association with low pressure systems or with frontal wave depressions, Mesoscale convective systems develop which transform into multi-cell thunderstorms or into supercell storms. These storms create havoc by their torrential rain, serve thunderstorm activity with poor visibility and squall-lines. All these phenomena associated with mesoscale thunderstorms are very hazardous for aviation both at take off / landing and in cruise of flight.

13.6 Lightning

In 1600 AD, William Gilbert, the physician of Queen Elizabeth, studied about static electricity. He showed that a glass / amber rod when rubbed with silk / fur attract small pieces of paper. He named this phenomena electric. In 1744 Benjamin Franklin found out positive and negative charges. In 1752 he showed, with kite experiment, that thunderstorms have electricity. In 1753 he found that thunder clouds have negative charges, but later in one experiment he also found positive charge in the thunderstorm cloud. From his experiments Franklin propounded that a majority of thunder cloud base possess negative charge but rarely positive charge may be seen. In 1756 be developed the lightning conductor for protection of buildings, which is still in use.

The Electrical Field of the Atmosphere

Thunderstorm is the most spectacular electrical phenomena that naturally occurs in the atmosphere. The atmosphere carries a net positive charge (even during fine weather) while the ground surface carries a negative charge. In fair weather electrical potential near the ground is about 150 volt meter. It is assumed that the electrical potential of the ground is zero. The rate of change of potential with height is called potential gradient . In middle troposphere, potential is of the order hundred of thousands of v/m. The gradient decrease with altitude in free air, rapidly in the first one kilometer and slowly there after. The gradient at the surface varies during the day, attaining maxima at 1900 UTC and minima at 0400 UTC. The diurnal range is about 35% of the daily mean. During haze / fog the potential gradient near the surface of the earth is about 2000 v/m. In such weather conditions electric sparks (or static electric discharges) may take place from the extremities of metallic conductors connected to the earth.

Critical Potential Gradient

An electric discharge (spark) takes place in the atmosphere when the electrical potential between two points reaches a certain value, which depends on the conducting air and the distance between the two points. In clear air with normal density the critical potential gradient is about 3×10^6 V/m. Dry air is a very poor conductor of electricity and requires a large potential gradient for electrical discharge. However the presence of water droplets in the atmosphere increases the conductivity of the atmosphere. In Cb cloud the lightning discharge occurs at a potential gradient of 1.0×10^6 V/m. In thunderstorm zone the field strength is of average field strength within a thunderstorm cell is about 10^5 V/m.

13.7 The Lightning Discharge

In a thunderstorm the lightning stroke starts from the cloud in the form of stepped leader, with average step length of 50 m. The time period of each successive step is about 50 μ seconds. The average velocity of the individual step is about 5×10^4 km/sec and the velocity of total step is about 1.5×10^2 km/sec. The total time required for stepped leader to reach the earth is about 0.01 sec. After leader reaches the earth, a reverse stroke starts from the earth to the cloud through the channel made by the leader with an average velocity of 15×10^4 km / sec. After this first discharge there may be other discharges which start from the cloud but little different from that of stepped leader. It will be a continuous leader, also called dart leader, with average velocity of about 2×10^3 km / sec.

When a lightning flash strikes near an observer, about 100 m of radius, the sound consists of (i) first a click, (ii) then a crack (like that of a whip) and (iii) finally characteristic continuous rumble of thunder. The click is caused by a discharge streamer directed upward from the ground (towards the leader stroke before commencement of the return stroke). The crack is caused by the intense return stroke in the lightning channel nearest to the observer. The rumble comes from the multiplicity of sound sources distributed along the lightning channel. When the lightning strikes away from the observer (several hundred meters) the first sound heard like the tearing of cloth.

13.7.1 Charge Separation in Clouds

In convective clouds charges (positive and negative) are observed from the formative stage. The positive and negative charges formed in the cloud are separated by the following mechanisms. (i) By the breaking of big rain drops, (ii) by capture of ions by rain drops and ice particles in the atmosphere, (iii) by collision of ice particles at different temperatures, (iv) by freezing the super-cooled water droplets.

In the cloud the charges are separated by the combination of the above mechanisms.

In a well developed convective cloud (matured stage) updrafts transport positive charge to the top the cloud, while downdrafts and falling rain drops transport negative charge to the base. Small ice particles or ice nuclei carry positive charge to the top of the cloud while big ice particles, soft hail carry negative charge to the base of the cloud. Generally positive charge is accumulated at the top of the Cb and negative charge at the base. In contrast to this classical dipole model, modern theory assumes that there can be much lateral displacement between two charges (see Fig.13.2). Most lightning flashes either transfer the lower negative charge to the ground or achieve effective partial neaturalization of the cloud electrification by internal cloud discharge.

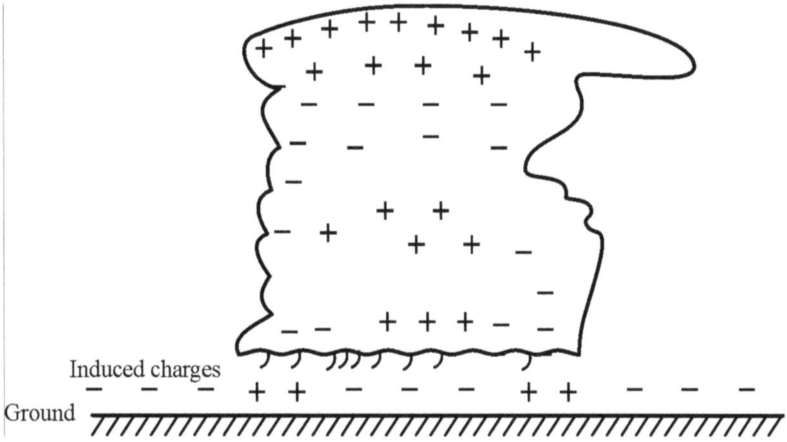

Fig. 13.2 Charge distribution.

13.7.2 Lightning Discharge within between Clouds and Earth

Lighting discharge occurs, more than 60% of the occasions within the cloud cell or cloud-to-cloud between positive and negative charges. In another form discharge takes place between cloud and the earth, which is commonly called thunderbolt.

An ionized air column that is formed by the accelerated electrons in a thunderstorm electric field is a leader stroke. The stroke constitutes air channel in the form of progressive steps of 50-100 m from the base of cumulonimbus cloud to the earth.

Generally the first leaderstorke is not visible, but the first one that steps on the earth is visible. This looks like an illuminated river with tributaries. The speed of the leader stroke may range 160 km/sec to 1600 km/sec. The return stroke which goes (in the reverse direction) from the earth to the cloud base appears as a flash. This we call lightning. The speed of the return stroke may be 14×10^4 km/sec, which is approximately half the speed of light (speed of light 3×10^8 m/s or 186000 miles/s). The period of lightning discharge may vary from a few micro-seconds to a full seconds. If T denotes the duration of time between the flash and thunder heard by an observer, then the distance of thunderstorm from the observer is approximately T/3 km (or T/5 miles), that is every three seconds is equal to one kilometer (or every five second is equal to one mile)

An electrically active thunderstorm cell has a life period of 30 minutes to one hour. During this period flashes may occur 1 to 10 per minute with a maximum of 20 flashes per minute after first flash. During the life period of cell, the mean flash rate is about 3 per minute.

13.8 Different Types of Electrical Discharges in the Atmosphere

The following are the main types of electrical discharges in the atmosphere.

1. ***Ribbon lightning***: This occurs between the cloud and the earth when the wind is very strong (speed exceeding 30 kmph) which displaces the conducting air channel into a ribbon form.
2. ***Sheet Lightning***: Sheet lightning is observed due to distant thunderstorms at the horizon, which illuminate the sky by its lightning flashes.
3. ***Air Discharge***: This occurs between the cloud base and the air below, but it will not reach the ground. Air discharge generally takes place in deserts.
4. ***Bolt from the Blue***: The electrical discharge which begins as an air discharge but after traveling a distance away from the cloud area, the leader stroke reaches the ground. The discharge may reach the ground in some cases about 15 km away from the cloud where the sky is completely clear. That is why it is termed Bolt from the Blue.
5. ***St. Elmo's Fire***: This form of discharge generally takes place at the top of high masts or elevated objects. When the thunderstorm cell passes over the elevated masts a positive glow will be seen at the top of the mast, while the negative charge flows down. This type of discharge is observed over ship masts, on aerial masts of flying aircrafts etc.
6. ***Ball lightning***: There is no scientific explanation about ball lightning. Ball lightning occurs in the form of electric balls of diameter ranging 1 cm to about 2 meters. These lightning balls may fall from thunderstorm clouds and explode or the balls may roll down a hillock and strike against some objects and then explode. The electric balls appear like soap bubles. These balls some times may roll down into rooms through doors, windows or electrical discharges.

13.9 Static Electricity and its Effects in Aviation

A spark discharge may occur when the potential difference between the aircraft and the ambient air exceeds critical (certain) value. The static charge may accumulate on the aircraft from cloud particles (ions), from precipitation or from contact with dust/sand charged particles. Radio noise on high and medium frequencies is the first sign of electrical discharge in the atmosphere, however this will have little effect on VHF reception. The increase in radio noise is indicative of severity of static electricity. In extreme cases of severity a visible discharge takes place on parts of aircraft which is called St. Elmo's fire/brush discharge/corona. In general, static electricity is not dangerous to aircraft, but on rare occasions the discharge my damage (break) wind screens and plastic panes. It may be noted here that every thunderstorm causes electrical effects including emissions of electromagnetic waves. Static effects Amplitude modulated signals in lower frequency bands mostly, and in some cases it may mask transmitted signals at long ranges. (Static interference is caused by electrical disturbances in the atmosphere). Background static (from electrical disturbances in the atmosphere) can be received at very long ranges. This causes attenuation in the Non Directional Beacons (NDB) between 190 KHz & 1750 KHz. ICAO recommends that the signal/noise ratio of NDBs should be minimum of 3:1 within their rated coverage. As pointed out earlier lightning from thunderstorm close to the aircraft may produce signals which are stronger than the wave from the NDB. This may confuse the Automatic Direction Finder (ADF), causing totally wrong indications. Similarly the omega navigation system is affected by static interference from thunderstorms.

13.9.1 Precipitation Static

The static electricity generated on aircraft when the aircraft flight passes through precipitation is called precipitation static. The electricity carried by precipitation particles hits the aircraft charge is accumulated on the aircraft wings, aerial mast/propellers. Charge may be generated by breaking of snow crystal/precipitation particles when they hit the aircraft or from the friction with solid particles. Charges accumulated on the aerial contribute to the radio noise.

13.9.2 Cloud Burst

In general, thunderstorms move with the prevailing wind in the middle layer of the cell. If the height of the cell is 8-10 km, the prevailing wind direction between 4 to 5 km layer may be taken as a first approximation as the direction of movement. This is only an empirical movement direction. The speed of movement of cell is less than the empirical wind speed.

In case of super cell, when the winds aloft are weak or variable, the movement of thunderstorm cell will be very little or no movement. If other conditions are fulfilled, particularly steep lapse rate in the lower and middle levels associated with sufficient moisture in these levels, there would be very heavy showers confined to a small area below the cell and that too in short time. This is popularly called cloud burst. The very heavy precipitation of 241.5 mm of rain recorded on 24 August 2000 at Begumpet (Hyderabad) mostly between 2100 hours to 400 hours IST is an example of cloud burst.

13.10 Diurnal and Seasonal Variation of Thunderstorms over India

Thunderstorms over India mostly triggered by insolation and their frequency of occurrence is maximum in the afternoon of summer. However a secondary maxima occurs close to the mountains/hills in the early morning hours. In well defined valleys the most of thunderstorms occur at night and minimum during day time. The annual distribution of thunderstorms over India is shown in Fig. 30.1. The figure shows highest frequency of thunderstorm belt from northeast India to Chattisgarh, east Madhyapradesh and a secondary maxima over southwest India, Kerala and neighborhood. The lowest frequency less then 10 days over Saurastra, Kutch.

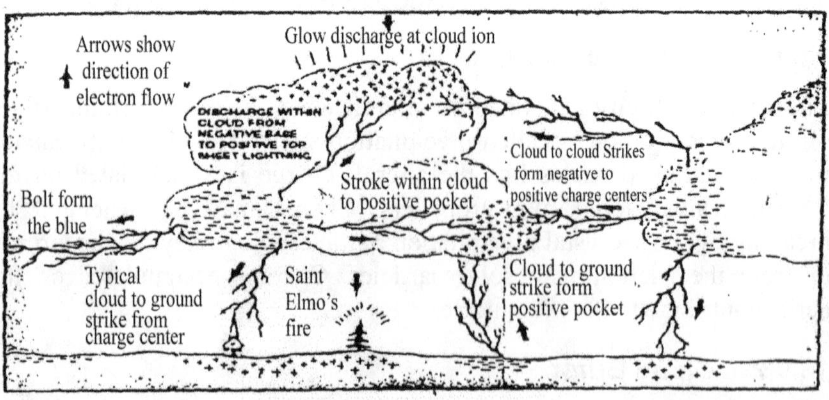

Fig. 13.3 Electrical activity in a thunderstorm.

In case of frontal thunderstorms, the frequency is maximum in winter and the storms move in the direction of associated front.

Summary

In General,

(a) Over the plain areas, thunderstorms occur during afternoon and dissipate towards evening/night.

(b) Over the valleys and foothills thunderstorms occur during night and/or early morning.

(c) Over sea areas, thunderstorms are frequent during night. In tropics and middle latitudes thunderstorms are frequent during summer while frontal type thunderstorms are frequent in winter in association with cold fronts.

Over India

(a) Maximum thunderstorms activity occurs during 1500-2100 hours IST, in all seasons. This is followed by 2100-0300 hours IST.

(b) The lowest frequency of thunderstorm activity occurs during 0300-0900 hours IST, except in premansoon (March–May) when it occurs during 0900-1500 hours IST.

Questions

1. The weather processes that occur in a layer close to the ground like turbulence, evaporation, heat fluxes or transfers is called

   (a) Micro-Meteorology (b) Meso-Meteorology

   (c) Synoptic-Meterology (d) Macro-Meteorology

2. The weather phenomena like thunderstorms, land/sea breezes etc which cannot be identified or located on synoptic charts, which have profound effect on aviation is classified as:

   (a) Micro-Meteorology (b) Meso-Meteorology

   (c) Synoptic-Meteorology (d) Macro-Meteorology

3. The weather phenomena like depressions, anticyclones, tropical storms etc., which can be identified with the help of weather charts is classified as:

   (a) Micro-Meteorology (b) Meso-Meteorology

   (c) Synoptic-Meteorology (d) Macro-Meteorology

4. The weather phenomena like major atmospheric circulation, Rossby waves, jetstreams etc., which can be identified on hemispherical weather maps is classified as:
   (a) Micro-Meteorology (b) Meso-Meteorology
   (c) Synoptic-Meteorology (d) Macro-Meteorology

5. According WMO, one or more sudden electrical discharges, manifested by flash of light and sharp or rumbling sound is called:
   (a) Hailstorm (b) Dust storm
   (c) Thunderstorm (d) Snowstorm

6. Thunderstorm is associated with the cloud:
   (a) Ns, (b) Cu (c) Cb (d) Ac

7. The electrical discharge which suddenly heats the conducting air channel expands abruptly from a few mm to a few centimeters which generates shock waves that spread out as:
   (a) Seismic wave (b) Sound of thunder
   (c) Rossby wave (d) Planetary wave

8. In a thunderstorm:
   (a) First thunder is heard, then lightning is observed
   (b) First lightning is observed, then the sound of thunder heard
   (c) Both lightning and thunder are visible and heard simultaneously
   (d) None above

9. Thunder normally heard upto:
   (a) 10-15 km, rarely beyond 25 km
   (b) 25-30 km, rarely beyond 50 km
   (c) 0-10 km
   (d) None above

10. The highest annual average of thunderstorm over globe occur at:
   (a) Beijing, in china
   (b) Veraval in India
   (c) Bogor, in Indonesia
   (d) Kampala, Uganda in Africa

11. Thunderstorms occur almost in all parts of the world but they are frequent in:
    (a) Polar regions (b) Sub tropics
    (c) Tropics (d) None above
12. Thunderstorms classified as:
    (a) Air mass: orographic / heat thunderstorms
    (b) Frontal
    (c) Convective/Stable
    (d) Both (a) & (b)
13. In general thunderstorms move in the direction of local wind flow at:
    (a) 500 hPa, (b) 700 hPa,
    (c) 850 hPa (d) 1000 hPa
14. Thunderstorms are:
    (a) Regional affairs, exceed diameter 500 km, duration 1 to 2 days
    (b) Local affairs, rarely exceed diameter of 10 km, duration 1 to 2 hrs
    (c) Micro-scale, diameter rarely exceed a few meters. Duration 1 to 5 minutes
    (d) None above
15. The triggering mechanism which sets off a thunderstorm:
    (a) Local convection, orgraphic lifting,
    (b) Surface convergence, frontal lifting
    (c) Radiational or katabatic cooling
    (d) All above
16. In cumulus stage of thunderstorms cell:
    (a) Both updrafts & downdrafts occur
    (b) Only updrafts occur
    (c) Only downdrafts occur
    (d) ELR < SALR

17. In mature stage of thunderstorm cell
    (a) Both updrafts & downdraft occur
    (b) Only updrafts occur
    (c) Only downdrafts occur
    (d) None above

18. In dissipating stage of thunderstorm cell:
    (a) Both updrafts & downdrafts occur
    (b) Only updrafts occur
    (c) Only downdrafts occur
    (d) Neither updrafts nor downdrafts occur

19. In mature stage of thunderstorm downdraughts below the cloud base produces cold air divergence or gust-front. This gust front can spread:
    (a) To 10 km horizontally and 3 km (10,000 ft) vertically
    (b) To 20 km horizontally and 5 km (16,000 ft) vertically
    (c) To 32 km horizontally and 1.8 km (6000 ft) vertically
    (d) To 50 km horizontally and 1 km (3300 ft) vertically

20. If a group of cloud cells arranged along a trough line, a well developed thunderstorm dissipates and forms a cold gust-front at the surface, this will:
    (a) Dissipate the other cells
    (b) Generates another cell into thunderstorm and dissipates other cells
    (c) Generates another thunderstorm which on dissipation regenerates another cell and thus continues
    (d) No change take place in the neighboring cells

21. In multi-cell thunderstorm, squall lines generate within the cluster of cells:
    (a) Thunderstorms regenerate continuously
    (b) Thunderstorms do not regenerate
    (c) The cluster of cells die down
    (d) None above

22. Meso-scale thunderstorms are:
    (a) Not hazardous for aviation
    (b) Very hazardous for aviation
    (c) Meso-scale convective systems do not develop into thunderstorms
    (d) None above

23. In a fair weather, electrical potential near the ground is about 150 v/m. During haze/fog the potential gradient near the ground is about:
    (a) 100 v/m (b) 2000 v/m (c) 0 v/m (d) none

24. During fog conditions, electrical discharges (sparks) may take place:
    (a) On the ground
    (b) On the extremities of metallic conductors, connected to the earth
    (c) From tall trees
    (d) None above

25. Thunderstorms most likely over Europe during:
    (a) Just after down (b) Midday
    (c) Late afternoon (d) Midnight

26. The fastest moving type of thunderstorms are:
    (a) Thermal (b) Air mass
    (c) Frontal (d) Orographic

27. In summer cumulonimbus cloud contains:
    (a) Water droplets
    (b) Ice crystals
    (c) Water droplets and ice crystals
    (d) Water droplets, ice crystals and supercooled water droplets

28. From what cloud does hail fall?
    (a) Ns (b) St (c) Cb (d) Cs

29. From what cloud does drizzle fall?
    (a) Ns (b) St (c) Cb (d) Cs

30. In which stage would you expect a rotor cloud ahead of Cb?
    (a) Developing stage (b) Mature stage
    (c) Dissipated stage (d) None of above

31. Normally in which direction an air mass thunderstorm move?
    (a) In the direction of the movement of front in which it is associated
    (b) 10,000 ft (3 kt) prevailing wind
    (c) 700 hpa prevailing wind
    (d) Both (b) and (c)

32. In which direction a frontal thunderstorm move?
    (a) In the direction of the movement of front in which it is associated
    (b) 10,000 ft (700 hPa)
    (c) Earratic
    (d) None above

33. In which temperature range zone severe icing expected in Cb?
    (a) + 10 °C to – 10 °C (b) 0 °C to – 10 °C
    (c) 0 °C to – 30 °C (d) 0 °C to – 45 °C

34. In which temperature zone frequent lightning expected in Cb?
    (a) + 10 °C to – 10 °C (b) 0 °C to – 10 °C
    (c) 0 °C to – 30 °C (c) 0 °C to – 45 °C

35. Normally over plain areas, when do thunderstorms occur and dissipates?
    (a) Occur in the morning and dissipate in the evening
    (b) Occur during afternoon and dissipate towards evening/night
    (c) Occur at any time and dissipate
    (d) Occur during night and dissipate in the morning

36. Over the valleys and foot hills, when do thunderstorms occur and dissipate?
    (a) Occur in the morning and dissipate the evening
    (b) Occur during afternoon and dissipate in the night
    (c) Occur during night/early morning and dissipate by noon
    (d) Occur at anytime and dissipate

37. Over sea/oceanic areas thunderstorms are frequent:
    (a) During morning (b) During afternoon
    (c) During night (d) Both (a) and (b)

38. Normally frontal type thunderstorms in which season you except and in which front?
    (a) Summer, warm front (b) Winter, cold front
    (c) Winter, warm front (d) Summer cold front

39. Orographic thunderstorms occur during:
    (a) Morning (b) Afternoon
    (c) Night (d) Any time (24 hrs)

40. In a thunderstorm electrical discharge begins from:
    (a) Positive charge to negative charge
    (b) Negative charge to positive charge
    (c) In positive charge only
    (d) In negative charge only

41. Lightning discharge normally occurs:
    (a) Within cloud (b) Cloud to cloud
    (c) Cloud to ground (d) All above

42. In cumulus stage of thunderstorm the maximum speed of ascending currents (updraughts) in individual cell is:
    (a) 100 ft/sec (b) 500 ft/sec
    (c) 1000 ft/sec (d) None above

43. In mature stage of thunderstorm cell the speed of updraughts and downdraughts are of the order:
    (a) 20-60 kt/ 10-30 kt (b) 10-30 kt/ 5-15 kt
    (c) 0-10 kt/ 0-5 kt (d) None above

44. In dissipating stage of thunderstorm cell, the speed of up currents and down currents in the cell are of the order
    (a) 0/20 mps (b) 0/10 mps
    (c) 10 mps/ 10 mps (d) 10/0 mps

45. In dissipating stage of thunderstorm, the speed of the squalls below the cloud base at ground is of the order:
    (a) 20 kt or more (b) 30 kt or more
    (c) 40 kt or more (d) 10 kt or more

46. In mature stage of thunderstorm at the base of the cloud, cold (divergence) horizontal gust-front spreads up to:
    (a) 10 km (b) 20 km (c) 32 km (d) 50 km

47. In a cluster of cells, a dissipating thunderstorm cold gust-front on the ground moves which triggers:
    (a) Dissipation of other cells
    (b) Activates other cells to thunderstorms
    (c) Activates next cell to thunderstorm which on dissipation activates the next and continues
    (d) Dissipates next cell and continues

Answers

1.	(a)	2.	(b)	3.	(c)	4.	(d)
5.	(c)	6.	(c)	7.	(b)	8.	(b)
9.	(a)	10.	(d)	11.	(c)	12.	(d)
13.	(b)	14.	(b)	15.	(d)	16.	(b)
17.	(a)	18.	(c)	19.	(c)	20.	(c)
21.	(a)	22.	(b)	23.	(b)	24.	(b)
25.	(c)	26.	(c)	27.	(d)	28.	(c)
29.	(b)	30.	(b)	31.	(d)	32.	(a)
33.	(c)	34.	(a)	35.	(b)	36.	(c)
37.	(c)	38.	(b)	39.	(d)	40.	(b)
41.	(d)	42.	(a)	43.	(a)	44.	(b)
45.	(b)	46.	(c)	47.	(c)		

CHAPTER 14

Aviation Hazards Associated with Thunderstorms

The main hazards associated with thunderstorms to aircraft are:

(i) Squall (ii) Heavy showers (iii) Ice accretion. (iii) Hail (iv) lightning. These will be discussed briefly.

Squall: A squall is a sudden burst of strong wind that lasts for a few minutes.

Definition: A sudden increase of wind speed by at least three stages on Beaufort scale (16 kt) and the speed reaching BF6 (24 kt) or more and lasting at least one minute.

For example, the sudden increase of wind speed from 16 kmph (BF3) to 44 kmph (BF6) (or 9 kt to 24 kt) and remains at 44 kmph for at least one minute, is a squall.

Squall line: A violent squall associated with the passing of a long line or arch of dark cloud and accompanied by thunder and lightning, rain or hail and a sudden cooling with a shift of wind direction. A squall line may be of short duration, may blow off trees and houses.

In association with thunderstorm, the initial downdraught from the Cb cloud spreads out horizontally on reaching the ground and the leading edge appears as squall with velocities ranging 40–100 KT. The strongest wind speeds move ahead of the cloud with respect to the direction of movement of cloud. Pressure fluctuations are common with severe thunderstorms. The first gust/squall produces sharp pressure rise which will be conspicuous on barographic chart.

Heavy showers: In association with thunderstorm heavy showers surface visibility temporarily deteriorate to less than 1000 meters.

Fractostratus clouds: In association with temporary thunderstorm fractostratus cloud base may lower to less than 1000 ft above ground.

Draughts: In matured Cb cloud, strong updraughts of order 20-60 kt and down draughts of order 10-30 kt may toss the aircraft up and down by 3000 to 4000ft at the flying altitude. By this aircraft may suddenly gain or loose the height above ground level and in some cases of hilly areas may lead to aircraft crash.

Gusts: A gust is a sudden increase in wind speed relative to its mean value over a period of time which is shorter than squall duration.

In a developed Cb cloud, in the vicinity of the boundary of up-draughts and down-draughts, vigorous eddies are formed. These eddies travel within the cloud and disturb the horizontal flow and also vertical currents. They give rise to gusts like the ones on the surface wind. These gusts cause severe bumpiness if the aircraft is inside the cloud. In extreme case aircraft may be out of control and there may be structural damage. Many aircraft accidents are attributed to this phenomenon (see Fig. 13.3). Gusts increase the load on the wings of aircrafts hence the speed of aircraft should be decreased in gusty air. However in landing phase a little higher speed may be maintained to keep the aircraft under positive control. In case of tail wheel airplane landing in gusty air, wheel landing should be made.

Ice Accretion: The freezing of supercooled water droplets on ice crystals or sublimation of water droplets on ice nuclei or ice crystals is ice accretion. In mature Cb cloud the accretion process is common but when this accretion leads icing on the aircraft which may harm the structural properties of aircraft. This shall be dealt in detail in ice accretion in flight in a separate chapter.

***Hail*:** Hail forms in a matured well developed Cb cloud. These may strike the aircraft and damage it. This also will be dealt in a separate chapter.

***Lightning*:** In a thunderstorm cell lightning may occur at any level but in test flights it is observed most frequently between isotherms of + 10 °C to – 10 °C (that is a band of 3 km or 10000 ft, about 5000 ft above and 5000 ft below 0 °C isotherm). Lightning distracts the attention of the pilot with temporary dazzling his eyes. Lighting interferes in the radio communication and may affect the performance of magnetic compass seriously. Listening on R/T or on inter-com is seriously hampered by the thunder sound.

14.1 Avoidance and Flying Techniques in Thunderstorm

In the area of active thunderstorm it is always safer to avoid flying. However in case of any uncertainty following safety measures to be followed.

1. Take-off of an aircraft should be postponed.
2. Landing / or approach phase it may be delayed or diversion carried out on advice of ATC (Air Traffic Control). The worst phase of thunderstorm activity normally lasts about half an hour. In case of transport aircraft generally diversion not required. However in case of fighter aircraft, the endurance at overhead position will be the main consideration.
3. In case of cross-country flight areas of possible thunderstorm may be clarified during meteorological briefing. Flight should avoid thunderstorm route and the flight level likely to be affected by thunderstorm.
4. Flying through a thunderstorm cell should be avoided. If there is sufficient terrain clearance, fly around the storm keeping thunderstorm cell to your left.

In unavoidable circumstances of penetration of thunderstorm, the following evasive steps be observed to minimize the risk.

(a) Select a level below freezing level or above 25000 ft (7.5 km / 400 hpa), because this height zone is worst for bumpiness and hail formation, that is, avoid flying in the layer between FL 100 to FL 250.

(b) Securely fasten safety belts and loose articles, lest they will be tossed about.

(c) Select a safe speed for penetration of turbulent zone, so that stalling may not occur due to gusts. Most of the aircrafts have been specified speeds of penetration, if not specified then select safest speed of 1.6 times the stalling speed with flaps and undercarriage retracted.

(d) Earth a trailing aerial and reel it in, lest it enhances the chance of lightning strike.

(e) In a thunderstorm cell positive, negative electrical charges are separated. When an aircraft flies through it, some positive charges may be downloaded on it. This causes continuous luminous electrical discharge (called St. Elmo's fire) along the wing tips and around the propeller tip. This Fire is awesome but generally harmless.

(f) Disengage the autopilot and check all instruments before entering the storm.

(g) If you have a second pilot, wear goggle and ask the second pilot to keep an eye on the instruments. Turn ON full cockpit lighting.

(h) Inside thunderstorm cell, the aircraft instruments (like Altimeter, Air speed Indicator) may not function accurately and they may show spurious reading. The pitot-tube may be partially blocked due to heavy rain and give lower readings of the airspeed indicator (ASI). Consequently do not depend fully on aircraft instrumental readings.

(i) While within the storm area, maintain constant heading and avoid coarse movement of control. It is safer to let the aircraft "ride the storms". If control is lost even temporarily, the subsequent recovery action may lead to structural failure due to combined loads of gusts and recovery action.

(j) Do not try to control altitude or execute turns, which may increase strain on control surfaces and may lead to dangerous attitude of the aircraft,

In view of the very hazardous weather phenomena of thunderstorms, part of UK Aeronautical information circular AIC 72/ 2001 (pink 22) regarding (pink 22) regarding (3) procedures and flying Techniques reproduced below.

3.1 Thunderstorms should be avoided, either visually or by use of radar but if this is not possible the following procedures and techniques evolved from research and operational experience, are recommended.

(a) If it is found, necessary to penetrate an area of cloud which may contain cumulonimbus cloud.

(i) Ensure that crew members safety belts or harnesses are firmly fastened and secure any loose articles.

Switch on the seat belt notices and make sure that all passengers are securely strapped in and that loose equipment (eg. Cabin trolleys and galley containers) are firmly secured. Pilots should remember that the effect of turbulence is normally worse in the rear of the aircraft than on the flight deck.

(ii) One pilot should control the aircraft and the other monitor the flight instruments continuously.

(iii) Select an altitude for penetration bearing in mind the importance of ensuring adequate terrain clearance.

Investigations have shown that although in some thunderstorms there is little turbulence at lower levels, in others there is a great deal : altitude is not necessarily a guide to the degree of turbulence.

(iv) Set the power to give the recommended speed for flight in turbulence, adjust the trim and note its position so that any excessive changes due to auto-pilot or Mach trim can be quickly assessed. Turbulence penetration speeds quoted in flight or operations manuals provide a single or a speed bracket; increasing height will decrease the buffet margin and up-currents may force the aircraft into buffet owing to an increased angle of attack.

(v) Check all flight instruments and electrical supplies.

(vi) Ensure that the pitot heaters are switched on.

(vii) Check the operation of all anti-icing and de-icing equipment and operate all these systems in accordance with manufacturers or operations instructions.

The operation of leading-edge, expanding boot type de-icers should be delayed until some ice has formed, otherwise their effectiveness will be greatly reduced. In the absence of specific instructions, ensure that all anti-icing systems, including windscreen heaters, are on.

(viii) Disregard any radio navigation indications subject to interference from Static, EG ADF (see AIC limitations of Non-Directional Beacons and Automatic Direction Finding Equipment for the effects of thunderstorms on ADF / NDBs).

(ix) Turn the cockpit lighting fully on and lower the Crew seats and Sun visors to minimize the blinding effect of lightning flashes.

(x) Follow the manufacturers or operator's recommendations on the use of the flight director, auto-pilot and manometric locks. If these are not stated, height, Mach, rate of climb or descent and airspeed locks should be disengaged but the yaw damper(s), if fitted, should be operative. On many aircrafts the auto-pilot, when engaged in a suitable mode (turbulence or basic attitude modes), is likely to produce lower structural loads than would result from manual flight; however, if major trim movements occur due to the auto-pilots automatic trim, the auto-pilot should be disengaged. Note that Mach trim operation may also occur on some aircraft but the Mach trim should remain engaged.

(xi) Continue monitoring the weather radar in order to select the safest track for penetration, to minimize the time of exposure whilst avoiding areas of intense activity.

(xii) Be prepared for turbulence, rain, hail, snow, icing, lightning, static discharge and windshear. In turbulence powered aircraft switch on the continuous ignition system (to reduce the possibility of flame-out due to water ingestion) ensuring that limitation, if any, on its use are not exceeded.

(xiii) Avoid flying over the of a thunderstorm whenever possible. Overflying small convective cells close to large storms should also be avoided, particularly if they are on the upwind side of the large storm, because they may grow very quickly. Similarly do not contemplate flying beneath the cumulonimbus cloud. As well as the dangers associated with turbulence, rain, hail, snow or lightning, there may well below cloud base; poor visibility, and possibly low level wind shear.

(b) Within the Strom Area

(i) Control the Aircraft regardless of all else.

(ii) Concentrate on maintaining a constant pitch attitude appropriate to climb, cruise or descent, by reference to the attitude indicators, carefully avoiding harsh or excessive control movements. Do not be misled by conflicting indications on other instruments. Do not allow large attitude excursions in the rolling plane to persist because these may result in Nosedown pitch changes.

(iii) Maintain the original heading. It is usually the quickest way out. Do not attempt any turns.

(iv) Do not correct for attitude gained or lost through up and down draughts unless absolutely necessary;

(v) Maintain the trim settings and avoid changing the power setting except when necessary to restore margins from stall warning or high speed buffet. The target pitch attitude should not be changed unless the mean IAS differs significantly from the recommended penetration speed.

(vi) If trim variation due to the auto-pilot (auto-trim) are large the anto-pilot should be disengaged. Movement of the Mach trim, where it occurs, is however necessary and desirable. Check that the yaw-damper remains engaged.

(vii) If negative 'G' is experienced, temporary warnings (e.g. low oil pressure) may occur. These should ignored.

(viii) On no account climb in an attempt to get over the top of the storm,

(ix) Light aircraft operators should ensure that their aircraft are adequately secured on the ground when severe thunderstorm activity is forecast.

4. Concluding Remarks

(i) Do not take-off if a thunderstorm is overhead or approaching.

(ii) At destination hold clear if a thunderstorm is overhead or approaching. Divert if necessary.

(iii) Avoid severe thunderstorms even at the cost of diversion or an intermediate landing if avoidance is impossible. The procedures recommended in this circular should be followed.

(iv) Pilots of turbo-jet swept-wing transport aircraft are advised to ensure that they are fully conversant with the control problems which may be met in turbulence with the type of aircraft they fly.

(v) If it is suspected that an aircraft may have been damaged by severe weather conditions, record the fact as a defect in the technical log, so that a proper investigation occurs.

Questions

1. A squall is a sudden increase of wind speed at least by three stages of Beaufort Scale (BFS) (16 kt) and the speed reaching or exceeding…… and lasting at least…….

   (a) B.FS3; 5 minutes (b) B.FS6; 1 minute

   (c) B.FS9; 30 seconds (d) B.FS5; 2 minutes

2. The initial downdraft from Cb, spreads horizontally on reaching the ground. The leading edge appears as a squall which is associated with:

   (a) Sharp fall of pressure (b) Sharp rise of pressure

   (c) Without change of pressure (d) None above

3. A matured cumulonimbus cloud may be associated with:

   (a) Updrafts of order 20-60kt

   (b) Downdraft of order 10-30kt

   (c) Updrafts & downdrafts may toss the aircraft up & down by 3000 to 4000ft

   (d) All above

4. A gust is a sudden increase in wind speed relative to its …….. value, and lasts over a period…….

   (a) Lowest value; few minutes

   (b) Mean value; less than one minutes

   (c) Highest value; two minutes

   (d) BFS 3; few minutes

5. The vigorus eddies formed in a thunderstorm cell may cause bumpiness in aircraft flight and lead to:

   (a) Aircraft out of control

   (b) Cause structural damage

   (c) Aircraft crash

   (d) All above

6. Lightning distracts the attention of the pilot with temporary dazzling his eyes and:
   (a) Improves radio communication and magnetic compass performance
   (b) Interferes in radio communication but improves in magnetic compass performance
   (c) Interferes in radio communication and affects magnetic compass performance
   (d) Does not interfere in radio communication, magnetic performance
7. In thunderstorm cell lightning may occur at any level, but most frequently occurs between:
   (a) 0° C to – 25°C level
   (b) + 10°C to – 10°C level
   (c) + 25° C to – 10°C level
   (d) + 20° C to – 20°C level
8. In active thunderstorm area:
   (a) Climb are fly over the cell
   (b) Descend and fly beneath cell
   (c) Fly around the storm keeping thunderstorm cell to your left
   (d) Don't fly
9. In unavoidable circumstances of penetration of thunderstorm, select a flight level:
   (a) Below freezing level or above FL 250 to avoid worst bumpiness
   (b) FL 150
   (c) FL 350
   (d) Freezing level
10. In case of unavoidable circumstances of penetration of thunderstorm to overcome turbulent zone, select safe speed of penetration specified for the aircraft or:
    (a) Stalling speed
    (b) 1.6 times stalling speed with flaps and under carriage retracted
    (c) less than stalling speed
    (d) no change in airspeed

11. In case of unavoidable circumstances of penetration of thunderstorm, inside thunderstorm cell:
    (a) Depend fully on aircraft instrument reading
    (b) Do not depend on aircraft instrument, which may show faulty readings
    (c) Maintain constant heading and avoid coarse movement of control. Let the aircraft ride the storm
    (d) Both (b) and (c)
12. If a thunderstorm is overhead or approaching at destination:
    (a) Hold till clear or divert in consultation of ATC
    (b) Don't wait, land immediately
    (c) Avoid severe thunderstorm even at the cost of diversion or intermediate landing
    (d) Both (a) and (c)

Answers

1.	(b)	2.	(b)	3.	(d)	4.	(b)
5.	(d)	6.	(c)	7.	(b)	8.	(c)
9.	(a)	10.	(b)	11.	(d)	12.	(d)

CHAPTER 15

Radar and its Application in Weather Detection

The word Radar is derived from "Radio Detection and Ranging". Radar was developed during second world war to detect enemy planes, warships in all weather conditions and round the clock. Now weather radar is extensively used for the detection of thunderstorms, tornadoes, waterspouts, tropical cyclones, hurricanes and for the detection of clouds and precipitation. In aviation radars are mainly used for:

(i) Air traffic control, (ii) Inflight fixing position of aircraft, (iii) Guidance to aircraft in approach and landing at an aerodrome, and (iv) Detection of thunderstorm activity.

Explanation of some terms used in Radars given below.

Pulses: Extremely short bursts of Super High Frequency (SHF) radio energy [SHF: frequency f : 3-30 GHz, wavelength λ : 10-1 cm]

Radar principle: The principle of radar involves the use of radio waves [by reflection or scattering of radio waves] by airborne objects (or targets) to determine the presence of objects / targets, its distance and direction and its velocity.

Pulse energy transmitted provides range; directional aerial rotation gives bearing of the object / target.

Pulse width: The length of time that a radar pulse transmitted. Usually measured in microseconds.

Pulse length: The length occupied by transmitted pulse in space. This is alternative to pulse width.

Pulse repetition interval or period: The time between begining of two consecutive pulses. Measured in microseconds.

Pulse repetition frequency: The number of pulses per second.

Attenuation: The weakening of the outgoing radar energy by the intervening clouds, raindrops, suspended particulate matter (aerosols) is called attenuation.

Beam width: Angular width of beam transmitted. Beam width depends on aerial design.

15.1 Principle of Radar

A radar station consists of an ultra-short-wave radio transmitter and a receiver with a single (common) transmitting and receiving antenna which produces a pencil of beam (a narrow directional beam) (see Block diagram Fig. 15.1).

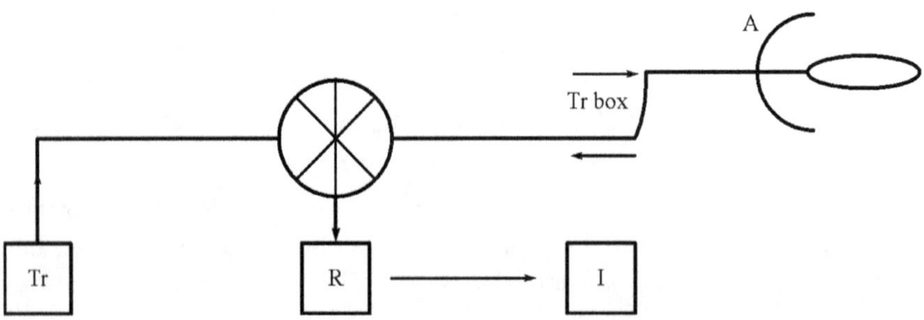

Fig. 15.1 Block diagram of a radar.

Where

Tr = Transmitter

R = Receiver

A = Antenna

TR = Transmitter-Receiver

I = Indicator / Display

Transmission is in the form of short pulses of period of 10^{-16} seconds (micro seconds). During the interval between successive (or consecutive) pluses the antenna is automatically switched over to reception of the echo-signal reflected from the target (also called object or obstacle). The distance of the target (radial distance from the antenna) is determined from the time interval between each transmitted pulse and reception of the echo-signal. Radar uses ultra short waves (in the bands of decimeter, centimeter and millimeter). The most successful operation (or detection) depends on the nature and size of the target (or object) being detected. The object (target) size should be much larger than the wavelength (of radar transmission wavelength λ). At present radars are used most efficiently and profitably in air navigation, maritime navigation and in detection of hazardous weather (like thunderstorm or convective clouds, clouds, micro bursts, cyclones, tornadoes, precipitation etc).

The functioning of a radar depends on various factors, like wavelength (λ), transmitting power (T_r), size and shape of antenna and number of scopes (displays). If a radar antenna radiates pulse energy 500 kilowatts power, its receiver can receive as small energy as 10^{-14} watts. The weather radar frequency (f) of radio waves normally used in the range of 1500 Mega Hertz (cycles) per second to more than 3000 mega Hertzs per second. The energy at these frequencies is called microwave energy. It is generally referred to wavelength (λ) instead of frequency. The common microwave spectrum is given in the Table 15.1.

Table 15.1 Microwave bands/Radar frequency bands

Radar Band name	λ - wavelength	f-frequency
mm-band	1-7.5 mm	40-300 GHz
K-band	(0.75 – 2.5 cm) 1 cm	12 – 40 GHz
X-band	(2.5 – 4.0 cm) 3 cm	8 – 12 GHz
C-band	(4.0 – 8.0 cm) 5 cm	4 – 8 GHz
S-band	(8.0 – 15.0 cm) 10 cm	2 – 4 GHz
L-band	15 – 30 cm	1 – 2 GHz

K or X-band radars are used for the detection of clouds, smaller size particles. X-band radars are also called Storm Detection radars in meteorological departments. S-band radars are used for the detection of rain, precipitation, cyclones, hurricanes, depressions. These radars are also called cyclone detection radars (used for the detection and tracking of cyclones).

Standard Radar Frequency Letter-band Nomenclature

Band Designation	Frequency Range	Specific Radiolocation (radar) bands based on ITU assignments for region 2
HF	3-30 MHz	
VHF	30-300 MHz	138-144 MHz 216-225
UHF	300-1000 MHz	420-450 MHz 890-942
SHF	3000-30000 MHz 3-30 GHz	
L	1000-2000 MHz	1215-1400 MHz
S	2000-4000 MHz	2300-2500 MHz 2700-3700 MHz
C	4000-8000 MHz	8500-10680 MHz
X	8000-12000 MHz	8500-10680 MHz
Ku	12.0 – 18 GHz	13.4-14.0 GHz 15.7 – 17.7
K	18-27 GHz	24.05-24.25 GHz
Ka	27-40 GHz	33.4-36.0 GHz
mm	40-300 GHz	

The reflectivity of an airborne object/target depends on the size (relative to the wavelength of signal) and shape (signal beam like a mirror reflection). A narrow directional beam transmitted through an aerial (which are reflected by the airborne object/target contained in the beam). For this purpose (an aerial is used) a parabolic antenna is used in the transmitter. The same antenna is used for the receiver.

The reflected radar signal (by the airborne object / target) is called echo, which is displayed on a screen with phosphorescent (glowing) coating. The width of the screen represents the maximum distance that can be tracked. The screen has fixed range markers from the transmitter (which is the centre of the screen). The reflected signal from the target (echo) is seen on the screen gives the range (which can be interpolated with reference to fixed range markers) from the transmitter. With the polar diagram of the display screen, the bearing of the target from the receiver can be identified (by Azimuth, Elevation and Range).

15.2 The Plan Position Indicator (PPI)

The antenna of the radar can be rotated horizontally 0 – 360° (in a circular sweep) and vertically 0 – 90°. As described earlier the target (airborne object) is illuminated by the radar beam and its reflection received on the screen as echo. The display unit of a radar is a cathode ray tube (CRT) (Fig. 15.2) which is calibrated for Azimuth (0 – 360°) and range details. On the horizontal scanning of antenna (keeping elevation angle fixed, generally 1^{O} or 2^{O}, as long as the reflection in received, the echo of the object/target or painting is seen on the screen (CRT). A moving target will leave a trail of fading points. To know more about the moving target, a separate antenna or the same can be moved back and forth in a sector scan. If the scan is vertical it will give the height of the target (which is Range Height Indicator). Two targets will be identified separately (in the range) if the distance between them is more than 0.5 pulse width.

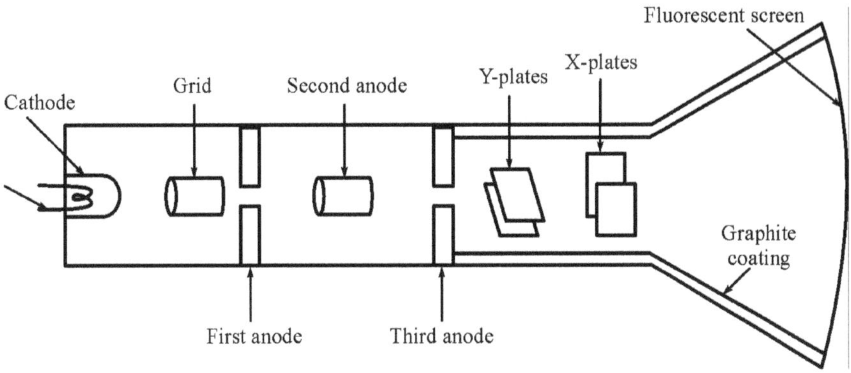

Fig. 15.2 Cathode Ray Tube components.

In modern radars the antennas do not always require to rotate physically to sweep the beam. A number of transmitting / receiving antennas in a large array electronically switched by a computer such that the resultant beam direction changes as though the aerial was actually moving. The actual received signals are not displayed. A computer converts the received signals to a digital input, which is processed and displayed on the computer monitor.

Radar Equation

The simple radar equation is

$$P_r = \frac{P_t G^2 \lambda^2 \sigma}{(4\pi)^3 R^4} \qquad .....(15.1)$$

where Pt = Power of the radar transmitter

Power density from isotropic antenna $= \dfrac{P_t}{4\pi R^2}$

R = distance from the radar (Power density = watts per unit area)

G = Antenna gain (direction antennas channel the radiated power P_t into som particular direction). G is a measure of the increased power radiated in the direction of the target as compared with the power that would have been radiated from an isotropic antenna.

Power density from a directive antenna $= \dfrac{P_t G}{4\pi R^2}$

The measure of the amount of incident power intercepted by the target and reradiated back in the direction of the radar is denoted by σ (called radar cross section), and is defined by the relation.

Power density of echo signal at radar $= \dfrac{P_t G}{4\pi R^2} \dfrac{\sigma}{4\pi R^2}$

Radar cross section σ has units of area

The radar antenna captures a portion of the echo power

The power (P_r) received by the radar is:

$$P_r = \frac{P_t G}{4\pi R^2} \frac{\sigma}{4\pi R^2} A_e = \frac{P_t G A_e \sigma}{(4\pi)^2 R^4}$$

$$\therefore \quad P_r = \frac{P_t G}{(4\pi R^2)^2} \cdot \frac{G\lambda^2}{4\pi} \sigma = \frac{P_t G^2 \lambda^2 \sigma}{(4\pi)^3 R^4}$$

Antenna theory gives the relationship between the transmitting gain and the receiving effective area of an antenna as

$$G = \frac{4\pi A_e}{\lambda^2} \text{ or } A_e = \frac{G\lambda^2}{4\pi}$$

Let σ_i denote cross section of meteorological targets of rain, snow, hail etc located within radar resolution cell. Then $\Sigma\sigma_i$ denote the average total back scatter cross section of the particles in unit volume, (summation of σ_i is carried out over the unit volume).

∴ σ = radar cross section $= V_m \Sigma\sigma_i$ where V_m = volume of the radar resolution cell.

The volume V_m occupied by a radar beam of vertical beam width ϕ_B, horizontal beam width θ_B, and a pulse of duration τ is approximately

$$V_m = \frac{\pi}{4}(R\theta_B)(R\phi_B)\left(\frac{C\tau}{2}\right)$$

Where $C = 3 \times 10^8$ m/s (speed of light) or velocity of propagation.

In general the radar pulse extent h (in units of length) is used instead of $C\tau$ and the factor $\frac{\pi}{4}$ accounts for the elliptical shape of the beam area.

Assuming a Gaussian-shape antenna pattern, the volume V_m given above will be reduced by a factor 2 *l*n 2 to describe the equivalent volume that accounts for the echo power received by the two-way antenna pattern distributed clutter under theses assumptions the radar equation (15.1) reduces to

$$\left[\bar{P}_r = \frac{P_t G^2 \lambda^2 \theta_B \phi_B C\tau}{1024(ln2)\pi^2 R^2}\sum_i \sigma_i\right]$$

$$\bar{P}_r = \frac{P_t G^2 \lambda^2}{(4\pi)^3 R^4}\frac{\pi R}{4}\theta_B . R\phi_B \frac{C\tau}{2}.\frac{1}{(2ln2)}\sum_i \sigma_i$$

$$= \frac{P_t G^2 \lambda^2 \theta_B \phi_B C\tau}{1024(ln2)R^2}\sum_i \sigma_i$$

$$\bar{P}_r = \frac{P_t G \lambda^2 C\tau}{1024(ln2)R^2}\sum \sigma_i \qquad .....(15.2)$$

Where ($\bar{P}$ average power over many radar sweeps)

where fn Gaussian beam shape $G = \frac{\pi^2}{\theta_B \phi_B}$

Note: Volume of antenna resolution cell is assumed to be completely filled with uniform precipitation) when the radar wavelength (λ) is large compared with circumference of a scattering particle diameter D (Rayleigh scattering region) the radar cross section is

$$\sigma_i = \frac{\pi^5 D^6}{\lambda^4}|K|^2 \qquad .....(a)$$

where $|K|^2 = \frac{\in -1}{\in +2}$ ε = dielectric constant of the scattering particles.

The volume of $|K|^2$ for water varies with temperature and wavelength at 10 °C and 10 cm wavelength it is approximately 0.93.

Its value for ice at all temperatures is about 0.197 and is independent of frequency in the centimeter wavelength region.

Substituting (a) in eq. (15.2) we get

$$\overline{P}_r = \frac{P_t G \lambda^2 C \tau}{1024(ln 2) R^2} |K|^2 \frac{\pi^5}{\lambda^4} \sum_i D^6$$

$$\overline{P}_r = \frac{P_t \pi^5 G C \tau}{1024(ln 2) R^2 \lambda^2} |K|^2 \sum_i D^6 \qquad \text{.....(15.3)}$$

Since D^6 appears in the above equation it implies in any distribution of precipitation particles the small number of large drops will contribute most to the echo power.

The above eq. (15.3) does not include the attenuation of the radar energy by precipitation, which can be considerable at the higher micro wave frequencies.

Eq. (15.3) applies to Rayleigh scattering and hence may be used for measuring with radar the (Σ D^6) sum of the sixth power of the raindrop diameter in a unit volume. The Rayleigh approximation is normally for heaviest rains, is a good approximation at X band (3 cm) Rayleigh scattering does not apply above X-band. Above X-band attenuation due to precipitation prevents quantitative measurements.

Σ D^6 is denoted by Z and called radar reflectivity factor $Z = \sum_i D^6$

However experimental results indicate $Z = ar^b$ .....(15.4)

where r = rainfall rate and a, b are empirically determined constants.

Thus the above eq. (4) helps, received echo power relating to rainfall rate.

The widely used constants a, b (by virtue of measure for stratiform rain) are a = 200, b = 1.6 i.e., $z = 200\ r^{1.6}$.

Where Z is in mm^6/m^3 and r is in mm/hr

For orographic rain $Z = 31\ r^{1.71}$

For thunderstorm rain $Z = 486\ r^{1.37}$

Single expression need not be used but Z-r relation can be made on the basis of type of rain.

For stratiform rain ($Z = 200\ r^{1.6}$), $|K| = 0.93$ the radar eqn. (3) becomes

$$\bar{P}_r = \frac{2.4 P_t G \tau r^{1.6}}{R^2 \lambda^2} \times 10^{-8} \qquad .....(15.5)$$

Where r is in mm/hr, R and λ in meters,

τ in seconds and P_t in watts

The above eqn. (5) is for the Rayleigh scattering when the scattering is not Rayleigh, a quantity similar to Z is defined, which is called equivalent radar reflectivity factor denoted by Z_e and is given by

$$Z_e = \frac{\lambda^4 \eta}{\left(\pi^5 |K|^2\right)} \qquad .....(15.6)$$

Where η = actual radar reflectivity or back scatter cross section per unit volume and $|K|^2 = 0.93$

Instead of the rainfall rate r, the intensity of precipitation is stated in terms of the dB reflectivity facto $Z = 200\ r^{1.6}$ or dBz =10 log z.

A rainfall rate of 1 mm/hr = 23 dBz

A rainfall rate of 4 mm/hr = 33 dBz

A rainfall rate 16 mm/hr = 42 dBz

***Scattering from snow*:** The following two expressions are used

$$Z = 2000\ r^2 \qquad .....(15.7a)$$

$$Z = 1780\ r^{2.21} \qquad .....(15.7b)$$

By measurement it is suggested

$$Z = 1050\ r^2 \text{ for dry snow (temp} < 0\,^\circ\text{C)} \qquad .....(15.8a)$$

$$Z = 1600\ r^2 \text{ for wet snow (temp} > 0\,^\circ\text{C)} \qquad .....(15.8b)$$

or $Z = 1000\ r^{1.6}$ for lower surface temperature .....(15.8c)

Brightband reflectivity is 12 to 15 dB greater than the reflectivity from snow above it, and about 6 to 12 dB greater than the rain below.

15.3 Types of Ground Radars used in Aviation

(i) *Primary surveillance or search Radars*: An enroute surveillance radar has a range of 200 miles (320 km) or more. It has good range and azimuth resolution, short pulse width and narrow beam width. The frequencies used around 400MHz (1MHz = 10^6 Hertz). PPI displays, Terminal Area Surveillance Radars (TARs) are used for guidance close to the aerodromes, have range about 25 nm (40 Km). These radars have sector scanning facility. In modern airports these are replaced by Secondary Surveillance Radars.

(ii) *Precision Approach Radars*: Two sweeping radars are used to give accurate aircrafts position (in three dimensions) in approaching the runway. One height finder is used for glide path guidance and the second one sweeps horizontally for Centerline guidance. These radars have short range about 10 miles (16 km). The wavelength is about 3 cm (X – band radars),

(iii) *Surface Movement Radars* (SMRs): At large airports, areas are hidden from Air Traffic Control (ATC) particularly during poor visibility (fog). In such airports Ground Controllers use SMRs to help the aircrafts in directing movement on the aprons, taxiways and runways. The range of these radars are very low, about 3 miles (5 km).

(iv) *Weather Radars*: Both Meteorological Services and Air Traffic Control use mostly X–band radars to identify thunderstorm areas or hazardous weather phenomena and to track Radiosonde balloons,

(v) *Surveillance Radar Approach* (SRA): These are used in operation of Aircraft Approach at aerodrome. The Controller Vectors the aircraft towards the Approach Path, giving instructions to descend (when the aircraft is 1500 ft or 450 m) above aerodrome before it arrives on the extended runway Centerline. Once the aircraft established on the Centerline, the Controller will Vector the aircraft towards the final approach fix and instructs the pilot to commence descent towards the threshold.

15.4 Some Characteristics of Weather Radar Echoes

Weather radar echoes may be classified into three groups. (1) Feature type, (2) Synoptic type and (3) Seasonal type.

1. ***Feature type eachoes*** *on radar scopes sub-classified as:* (i) Convective (ii) Stratiform (iii) Bright band (iv) Dry holes (v) Wind shear (vi) Protuberance (vii) Spiral band.

(i) *Convective type*: On PPI scope they look circular or oval shaped. Relatively small in size with clear cut boundaries (sharply divided edges). Each echo is a single cell or a group of closely packed cells. On RHI scope, Convective Cells appear as tall, narrow columns. Intensity is quite bright and can be detected even on long range. They do not exhibit Bright Band except when the convective cell is in decay stage. They show iso-echo (contours). The brightness of each echo is due to the high water content, updrafts and downdrafts in the cell. Bright portion of echo denotes severe turbulence. Extra-ordinary brightness of echo indicates the presence of hail in the cloud. Blurring edges of convective cell echo is the first symptom of decay of the cell. On reduction of receiver gain, the reduction in size of convective echo is comparatively less than that of stratiform cloud echo.

(ii) *Stratiform*: On PPI scope the stratiform cloud echoes appear large in size but without clear cut boundaries. The echoes are less bright, have uniform intensity with diffused boundaries. Height-width ratio is small and does not show any vertical development. Distant echoes may not be detected. Intervening precipitation may mask the stratiform echoes beyond precipitation. Stratiform echoes exhibit bright band and do not display iso-echo contour. At low gain the stratiform echoes may disappear. In decay stage of convective echo degenerate into stratiform echo.

(iii) *Bright Band*: A Bright Band or a bright line is associated with stratiform echoes, which can be seen on RHI scope on reduction of receiver gain. This bright band is found in the vicinity of 0 ^{o}C isotherm and differentiates the cloud particles above and below the freezing level. It helps to decide the height of freezing level and the stability of the atmosphere or turbulence free cloud. Bright Band appears in the decay stage of Cb but not in the developing stage of Cb.

(iv) *Dry holes*: In a bright large convective echo, sometimes a small area embedded in it is found without any echo. This is called dry hole. This is different from iso-echo contour. The cause of this is not fully understood. Dry holes are associated with severe thunderstorms with heavy precipitation, strong surface wind. It is also an indication of a tornado or hurricane. In association with cyclonic storm dry hole is observed besides the eye of the cyclone.

(v) *Wind shear*: A shower associated with a thunderstorm can be seen on RHI scope as separate columns of falling rain. These vertical columns often slant out, which represents vertical wind shear.

(vi) *Protuberance and Hook's*: On PPI scope, the protuberance or 'comma' or six ',' figure shaped echoes indicate very intense convective cloud which is associated with hail or tornado. These can be seen even with higher antenna elevations and low receiver gain.

(vii) *Spiral Bands*: On PPI scope, spiral band shape echoes indicate tropical cyclone. These bands persist for a long period.

2. **Synoptic Type Cloud Echoes:** These include thunderstorms, hailstorms, showers, squall lines, cyclones. On PPI scope squall line echoes appear with length-breadth ratio of (10:1) ten to one or more. Air mass front echoes appear like squall line but they are thick and more uniform.

3. **Seasonal Type Cloud Echoes:** These include rainy weather echoes, which appear diffused with ill-defined edges. Due to attenuation, echoes beyond precipitation may not be seen. Others are globular, hurricane, tornadoes, water spouts and tropical cyclones.

   In a tropical cyclone, five echo regions can be identified:

   (i) At the periphery convective cellular echoes are observed. These are associated with squally weather.

   (ii) Inward spiraling echoes with slight movement is observed. The outer bands towards the radar side will be bright, while the banding portion farther from the radar will be weak. Inward spiraling echoes will be 500 to 600 km away from the storm centre. The gap between the banding will be about 50 to 80 km, which is free from echoes.

   (iii) After banding a large area of intense echo will be observed.

   (iv) Next to this intense echo, a cellular structure near the Eye which has thick band of echoes arranged in a definite spiral band.

   (v) The Eye of the storm is a cloud free zone, which is either circular or oval shaped. With the help of these patterns the centre of the storm can be located exactly. In a decay stage of a storm, the spiral bands appear diffused. With the continuous monitoring of a cyclone on a radar, a running commentary of cyclone intensification, decay, speed of movement can be given accurately as long as it is on the sea. For tracking a cyclone an S-band radar (or 10 cm radar) is most suitable with less attenuation. Generally the range of the radar is 400 km. Radar is a best tool to track a cyclone over the sea when it is in the range of the radar.

A tornado is specified by a strong hook echo. On RHI the vertical extent will be about 16 to 18 km higher. Some small echoes move at right angles to the general motion of the wind towards the right side. A waterspout gives a thunderstorm type echo.

***Anomalous Propagation*:** Non-precipitation echoes and ground clutter. This is due to the permanent objects such as hills, tall buildings, lofty trees etc.

***Normal Super Refraction Echoes*:** By bending of the radar beam we get some echoes. This occurs mainly in severe winter due to inversion of temperature.

Abnormal Refraction Echoes: These are multiple refraction echoes, also called Radio duct.

15.5 Iso-Echo Contour Display

In a normal radar screen the display of the returns (echoes) will be black and white. White returns indicate strong reflecting rain drops or hailstones. However this display does not distinguish or identify with certainty. In modern radars black and white displays (echo-returns) can be reversed to present the return as black on white, which under specified lighting conditions make the shape of the echo to identify easily. A device is made to present levels of intensity by processing returns as they are received. Any part of a return above certain specified strength (datum) is phase changed. A return that enters the antenna will be more than the specified strength, which will be processed. This means the part above the specified (datum) level is at negative phase, which is added to the positive phase of the part below reduces the displayed return. A return twice the datum (specified) strength indicates heavy concentration of water droplets, infact appears to have no strength and makes no paint on the screen. Thus a very strong return produces more negative phase (black paint) than positive signal (white paint) and the CRT screen does not display this negative phase signal (appears black). However it is surrounded by the weaker return (white). As a result very intense signal returns appear as a black centre with surrounded area (of weak return) as white [or vice versa if the display has been reversed]. This display is called iso-echo or contour display.

Colour Display of Echoes: Modern displays use colours for different return intensities. The colours-Green indicate low intensity return (no hazard), Yellow for moderate intensity return (warning signal) and Red (or Magenta) for a very strong return (danger signal).

15.6 The Main uses of Weather Radars

1. Instantaneous location of clouds, precipitation and their position, movement (speed).
2. To find the heights of base and top of clouds.
3. To find the intensity of precipitation and estimation of quantity of precipitation and rate.
4. Identification of thunderstorms, duststorms, hail storms.
5. Identification of areas of turbulence.
6. Tracking of cyclones and estimating their intensity.
7. Finding upper winds and temperatures (tracking of radiosonde balloons).

Questions

1. In aviation, radars are used:
   (a) Air Traffic Control and inflight fixing position of aircraft
   (b) Guidance to aircraft in approach and landing
   (c) Detection of thunderstorm activity
   (d) All above
2. The principle of radar involves the use of radio waves:
   (a) By reflection or scattering by airborne objects or target
   (b) Pulse energy transmitted provides range of the object
   (c) Directional aerial rotation gives bearing of the target
   (d) All above
3. The functioning of a radar depends on:
   (a) Wavelength
   (b) Transmitting power
   (c) Size, shape of antenna and number of scopes
   (d) All above

4. For the detection of clouds, small size particles, storm detection:
   (a) X-band radar used
   (b) S-band radar used
   (c) C-band radar used
   (d) Both b & c band radars used
5. For the detection of rain, precipitation, cyclones, depressions:
   (a) X-band radar used
   (b) C-band radar used
   (c) S-band radar used
   (d) Both X and C band radars used
6. The horizontal scanning of radar antenna provides:
   (a) azimuth of the target
   (b) range of the target
   (c) vertical height of the target
   (d) both (a) & (b)
7. The vertical scanning of radar antenna provides:
   (a) Azimuth of the target
   (b) Range of the target
   (c) Vertical height of the target
   (d) All above
8. Convective clouds appear on radar screen as:
   (a) Circular or oval shaped with clear boundaries
   (b) On RHI scope they appear as all narrow columns but do not show bright band except in decay stage.
   (c) Intensity is bright and can be detected at long range
   (d) All above
9. Stratiform clouds appear on radar screen as:
   (a) Large in size with diffused boundaries
   (b) Height-width ratio is small and bright band seen
   (c) Distant echoes may not be seen, intervening precipitation masks stratiform echoes
   (d) All above

10. A Bright Band or a bright line is seen on RHI scope in association with stratiform echoes. It indicates:
    (a) Tropopause level (b) Maximum wind level
    (c) Freezing level (d) Jet stream level
11. A Bright Band echo appears in:
    (a) Stratiform clouds (b) Mature Cb clouds
    (c) Decay of Cb clouds (d) Both (a) & (c)
12. In a bright large convective radar echo, a small area of without echo embedded in it is found, which is called dry hole. The presence of dry hole indicates:
    (a) Cb
    (b) Severe thunderstorm, heavy precipitation & strong winds or tornado
    (c) It indicates eye of cyclonic storm or hurricane
    (d) Both (b) & (c)
13. On PPI scope the protuberance or 'comma' or '6' figure shape echo indicates:
    (a) Convective cloud
    (b) Intense convective cloud with hail storm
    (c) A tornado
    (d) Both (b) & (c)
14. On PPI scope bands of echoes persisting, spiraling in shape indicate:
    (a) Tropical cyclone (b) Tornado
    (c) Thunderstorm (d) Dust storm
15. A device is made to identify the intensity of echo return. Any part of a return above a certain specified strength (datum) is phase changed. A return that enters the antenna will be more than the specified strength. A very strong return produces negative phase black paint on the screen with a positive signal (white) weaker strength around it. This display is called:
    (a) Iso-echo or contour display
    (b) Bright band
    (c) Ground clutter
    (d) Anomalous propagation

Answers

1.	(d)	2.	(d)	3.	(d)	4.	(a)
5.	(c)	6.	(d)	7.	(c)	8.	(d)
9.	(d)	10.	(c)	11.	(d)	12.	(d)
13.	(d)	14.	(a)	15.	(a)		

CHAPTER 16

Hailstorms

Hail is one of the byproducts of a severe thunderstorm. A thunderstorm that is associated with hail is called hailstorm. Hail is formed in a well developed cumulonimbus cloud (Cb), which is characterized by severe updrafts. A typical hailstone when cut shows layer structure like that of an onion. Hail is a climatic element varies with time and space. The frequency occurrence of hailstorm is more in middle latitudes and it abruptly reduces to zero towards poles. Over the world, the western coasts of continents have maximum frequency during spring season. Hailstorms are not confined to any part of the world. Most hailstorms occur around lat 10° N in April to September period and around lat 8 $^{\circ}$S in October to March. However the whole of the northern hemisphere between lat 20 $^{\circ}$N and 50 $^{\circ}$N is prone to the incidence of hailstorm in summer. Generally the size of the hail is small (pea size, diameter less than 10 mm) and occurs over an area less than one square kilometer. But they cause heavy damage to fruit trees. It is estimated that on an average about 1% of the world's crop is damaged by hail. The mass of the hail in general does not exceed 10% of the mass of that thunderstorm.

Hail has several shapes and is classified by its size compared with the familiar objects. If the diameter is less than 5 mm it is called shot, 5-10 mm is called pea size; 10-20 mm is called grape size, 20-30 mm is called walnut size, 30-50 mm is called golf ball size, 50-60 mm is called hen egg size and if it exceed 60 mm it is called tennis ball size (See fig. 16.2). However, in general, the solid precipitation from cloud is classified into four types.

(i) *Grauple or soft hail*: They are white, opaque and conical in shape, diameter is less than 5 mm, density 800 kg / m^3.

(ii) *Small hail*: They are partly transparent, round or conical in shape, diameter up to 5 mm and density 800-900 kg / m^3.

(iii) *Ice pellets*: They are transparent, spherical or irregular in shape. Diameter less than 5 mm, density same as the of ice (920 kg/m^3 or 0.92 gm/cm^3). They form by freezing of rain drops.

(iv) *Hail*: They are lumps of ice and water with air inclusion. Diameter more than 5 mm (up to 5 cm). They are partly transparent or opaque. They have alternating layers of ice and air bubbles. Generally they are spherical or conical in shape. See Fig. 16.1.

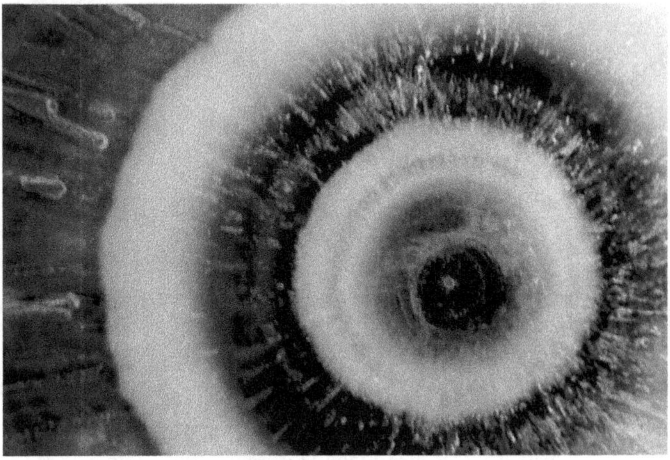

Fig. 16.1 Reproduce section of large hail.

Largest Hail: On 30 April 1888, Moradabad in Bihar (India) reported large hailstones of cricket ball size, which took the life of 246 human beings and 1600 goat & sheep. In Germany in 1925, Talman measured an ellipsoidal hail 26 × 14 × 12 cm in size and weighed 2.04 kg. On 11 March 1957, Begumpet Airport (Hyderabad) India reported average hailstone of size 5-8 cm in diameter. On 27 May, 1959, in Delhi (India) hail size of 200 mm diameter caused holes in aircraft frames and there were reports of hail size 250-375 mm diameter. On 3 September 1970, at Coffey Ville, Kansas, USA, a hailstone was weighted 758 gm, diameter 190 mm and circumference 444 mm. The substantiated reports of hail of 1.9 kg fell in Kazakisthan, 970 gm hail full near Strasbourg France, in August 1958. The other reported hail weights are 4.5 kg recorded in China, 4 kg in Hungry and 3.4 kg in India. During 11-13 March 1981, there were widespread hailstorm over Telangana (AP) India, which caused damage to 33000 acres of standing crop, damaged 85000 houses and taken a toll of 18 human lives and 13000 livestock. The size of hail range 100 to 1200 gm.

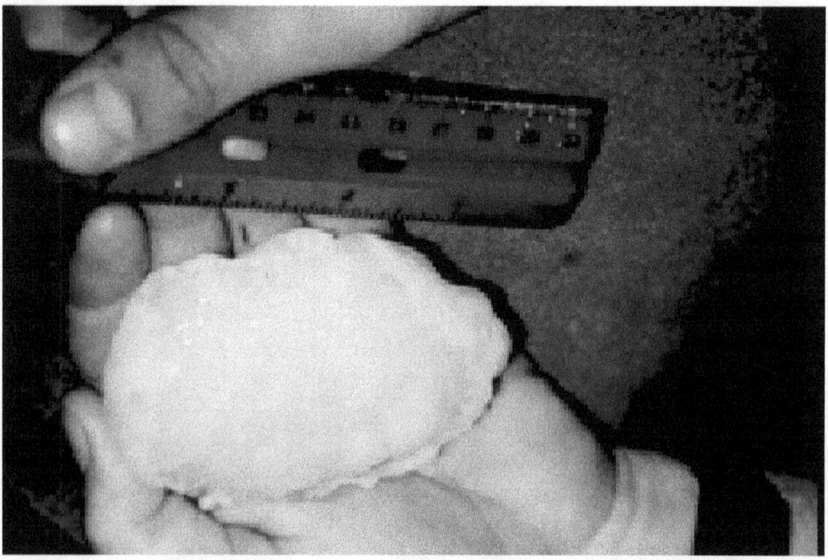

Fig. 16.2 Reproduced Hail tennis ball size.

During February Hailstorm damages (1976 – 2003) in India given below

Date	Places affected	Loss
28 – 2 – 1982	Kanpur, Mathura Unnoa Districts of UP	8 persons killed, 25 injured, 60% Standing Rabicrop damaged
9 – 2 – 1986	Around Nagpur	125 cattle killed, 16 persons injured, Rs 10 croers worth standing crop damaged
11 – 2 – 1986	Yawatmal (Maharastra)	4 persons killed, 5 injured, Rs 2 crores property damaged
19 – 2 – 1987	Faridabad, near Delhi	Crops worth Rs 20 crore damaged
16 – 2 – 1988	Sambalpur, Puri. Size 2 to 3 kg	Several people injured 2000 acres crop damaged
11 – 2 – 1991	Delhi, Faridabad district	3 persons killed, 4 injured 5 persons killed 4 injured
28 – 2 – 1991	Karimnagar & Hyderabad (AP)	Hundereds of sheep and goats perished Crop worth Rs 10 crore damaged
21 – 2 – 2001	Bashirhat (West Bengal)	200 persons injured, 400 cows and many cattle injured

16.1 Hail Hazard in Aviation

In flight hail is encountered rarely and even if it occurs only in a small region of the cloud. Small hail may cause superficial damage, but large hailstones may cause severe damage. It may cause holes in the aircraft, splinter the windscreens, break canopies into piece, ripp-off de-icer boots and bend radiator fins. Damage by hail increases with the speed of the aircraft.

In July 1957, a jet aircraft flying over the English Channel, at flight level 295, encountered severe turbulence, rime icing and hail as it flew into the anvil cirrus. The aircraft suffered structural damage by hail on its nose. The ambient air temperature at the onset of icing was –35° C. All the leading edges were battered and caused depressions of the order 6 × 8 inches. Protective sheathing on some electrical leads in the engine intakes was frayed away. As said earlier hailstorm damage is proportional to the aircraft speed. Damaging hail can be encountered from any part of thunderstorm and can be encountered up to 45000 ft (FL 450). In a thunderstorm there is a danger of air frame and engine icing in 0° to – 45 ° C temperature range and severe icing between 0° C to – 30° C. This will be further dealt in subsequent chapter.

16.2 Hail Formation in the Cloud

It is believed that some kind of ice nuclei (clay particle) when falls in the supercooled water, hail embryos are formed. It then grows by collision and accretion process. Hailstones have concentric of clear and opaque ice. When embryo is carried aloft by updraft it is frozen. By repeated circular trips within updrafts of thunderstorm cell [that carry hailstone above and below the freezing level] hailstone is formed. Depending on the time spent above freezing level milky hail layer forms by rapid freezing, while the clear ice layer is formed below the freezing level. The number of trips above and below freezing level gives the number of layers within the hailstone. In one hailstone twenty five separate layers have been counted. Calculations show that an updraft of 95 kmph (51 kt) is required to support the hail of diameter 25 mm. A hailstone of 80 mm diameter requires an updraft of 200 kmph (108 kt) to support, and a 130 mm diameter hailstone requires an updraft of 377 kmph (200 kt) to support.

16.3 Favourable Synoptic Conditions

Various studies indicate the following favourable synoptic conditions for the occurrence of hailstorm.

(i) Surface high humidity and also in the lower levels of the atmosphere.

(ii) Windshear in middle troposphere (850 to 200 hPa)

(iii) Low level convergence of moist air masses of contrasting characteristics (such as continental and maritime air masses)

(iv) In tropics, movement of upper air trough in zonal westerlies jetstream (200 hPa), preferable location of hailstorm is along the axis of trough.

(v) North-south trough on sea level chart with associated wind discontinuity/cyclonic circulation in the lower troposphere.

(vi) Steep lapserate of temperature in the lower levels (surface to 500 hPa)

A number of factors must be considered simultaneously, and it is possible that a non-linear contribution of parameters will be required. For example, wind shear plays an important role in the organization of the most severe hailstorm, but it is likely that if shear is too strong, storm development will be inhibited. A major source of non-linearity in forecasting problem is the existence of several distinct type of hailstorms. Some principal types are: (i) an airmass or pulsating multiple bubble storm, (ii) a multi-cell storm (iii) large severe super cell storm.

The first type requires great thermal instability and occurs in the absence of wind shear / strong winds. The second type requires moderate to strong instability and a proper vertical wind profile. The third type may occur without great instability, requires strong winds with a properly organized vertical wind profile and is associated with a strong organized downdraft fed by rain cooled air from the middle troposphere. It can produce very large stones, great crop damage and severe wind damage. It seems to be a type of storm associated with tornadoes.

Divergence exists from a trough to downstream crest in upper tropospheric (westerly) wave. In a sinusoidal wave, maximum upper level divergence is located about midway between trough and ridge. Its intensity is greatest for waves with short length and large amplitude and large wind speed in the jet stream. It seems that jet stream is a major factor in the development of severe thunderstorm.

A study revealed that wind speed from surface to 500 hPa was very different from that compared to the wind speed from 500 hPa to 250 hPa. The difference between the two was related to the size of the hailstones produced. A wind shear of 65 kmph (35 kt) between the two layers correspond to heavy hail (3-5 cm diameter), 61 kmph (33 kt) to moderate hail (1-2 cm diameter) and 48 kmph (26 kt) to light hail (diameter less than one centimeter) and less than 44 kmph (24 kt) indicate no hail formation.

The terminal velocity of hail is given by an approximate formula

$$V = \sqrt{2W / C_d \, \rho_a \, A}$$

where W = Weight of the hail

A = Cross section area

C_d = drag coefficient

ρ_a = density of air, which varies with height.

The fall speed of large hail is of the order 40-100 mps.

Low level windshear: Low level windshear is associated with the following meteorological situations.

(i) Ground layer inversion during winter months, which result in strong jet like winds aloft.

(ii) At the boundaries of air masses.

(iii) Sea-breeze circulation

(iv) Sharp changes due to topographical features over aerodromes.

(v) In the vicinity of thunderstorm.

16.4 Microburst

Downdrafts in thunderstorms encircling horizontal gusts of large magnitude is called downburst or microburst. A downburst may be defined as a localised intense downdraft with vertical currents exceeding a downward speed of 3.6 mps (12 ft/sec) at an height of 91 m (300 ft) above ground. This speed is comparable to the normal descent rate of jet aircraft during landing. This threshold speed of downburst tends to double the sinking speed of such an aircraft to adjust balance (or trimmed) for normal approach near touch down at usual 3° glide slope below 150 m (500 ft). A microburst is smaller but rather more intense localized version of the downburst. The horizontal extent of downburst is 4-10 km, life period 5-20 minutes, and micro burst is 1-4 km, life period 4 minutes. When down burst/microburst reach the surface, air spreads out radially on the ground forming a ring of diverging air, about 0.3 to 1 km deep. The burst causes strong gusty surface wind with average maximum gust of 50 mps (97 kt) to 60 mps (117 kt). In such gusts encounter, aircraft loses balance and may crash. Microbrusts are very hazardous to aviation when they are overhead. Micro busts cause many accidents, a few noted mishaps area given below.

(i) 24 June 1956, BOAC at Kano, Nigeria. 32 Killed, 11 injured

(ii) 30 January 1974, PAA 759 at Pago. 96 killed

(iii) 24 June 1975, Eastern Airlines B 727 at JKF New York. 112 killed, 12 injured

(iv) 07 August 1975, Flight 426 at Denver. 15 injured

(v) 14 May 1976, Royal Jordanian. Flt 600 at Qatar. 45 killed, 15 injured

(vi) 23 June 1976. Flight 121 at Philadelphia. 86 injured.

(vii) 7 August 1977 at Stapleton International Airport (Denver Colorado).

(viii) 2 August 1985 at Dallas-Fort worth International Airport in few minutes apart.

(ix) Two accidents 1983 Eastern Flight 66 and Pan American flight 769 due to multiple vortex rings.

Microbursts are found to be asymmetrical a few patterns are shown in Fig. 16.3.

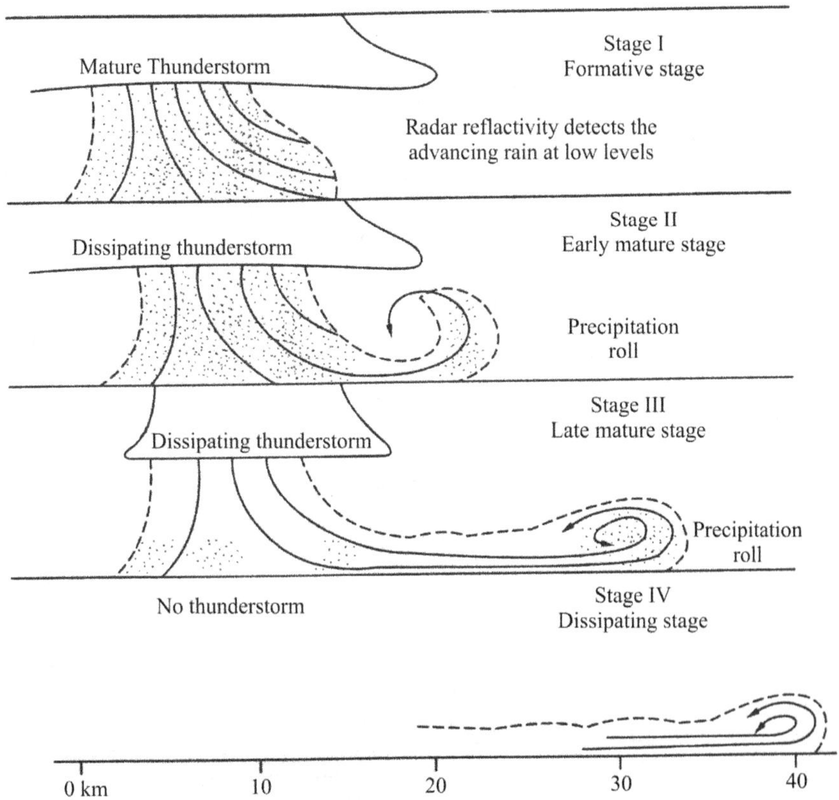

Fig. 16.3(a) Four stages of cold downdraft air spreading on ground.

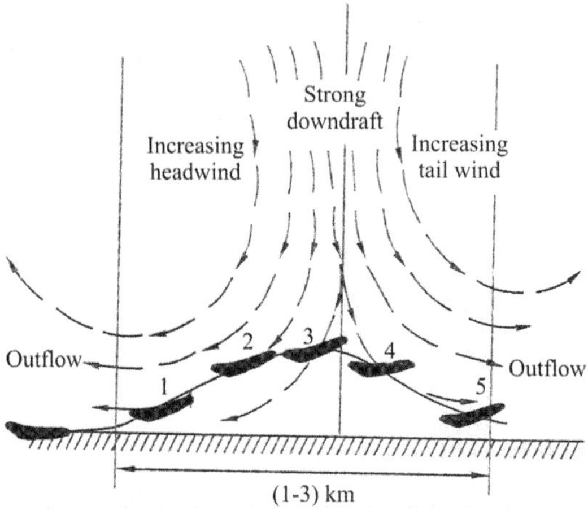

Fig. 16.3(b) Microburst.

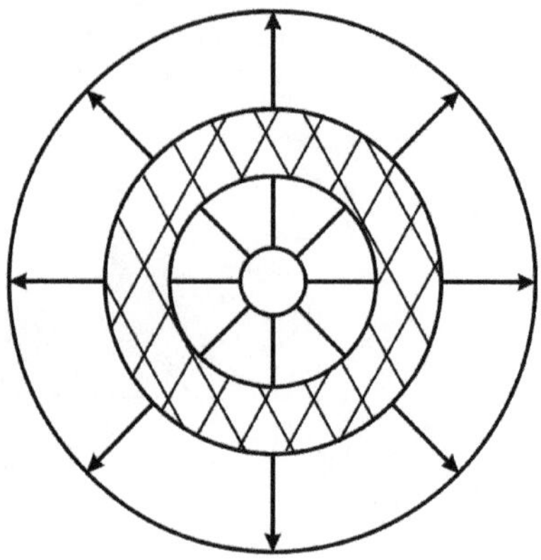

Statecravy

Fig. 16.3(c) Stationary.

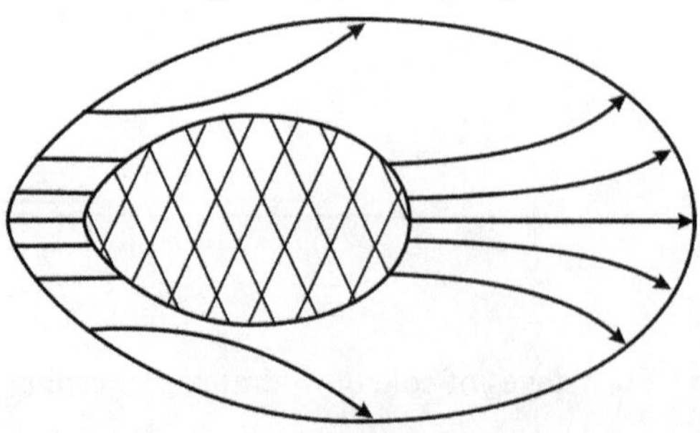

Moving

Fig. 16.3(d)

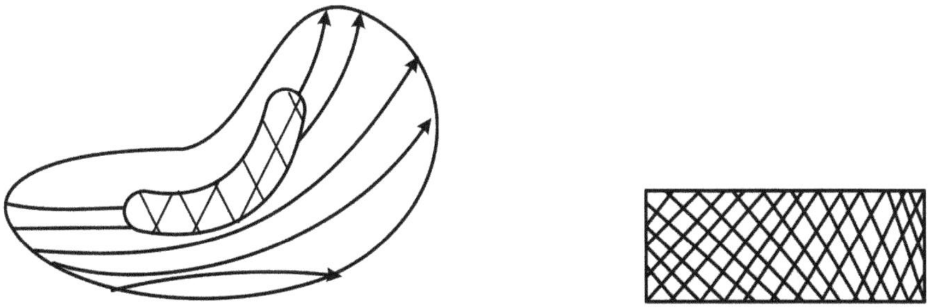

Twisting Strong winds

Fig. 16.3(e)

16.5 Microburst Models

The salient features of study models of downburst/microburst developed by National Oceanic and Atmospheric Administrations, US Department of Commerce are given below.

Microburst – small scale severe thunderstorm downdrafts are great dangers to aircraft during take-off or landing. The best microburst detectors are Doppler Radars and Lidars. All international airports have installed or installing Terminal Doppler Weather Radars (TDWR) for the detections of downbursts, microbursts. TDWR can provide information of potential threat of microbursts and thus alert ATC and flight crew.

(a) Gust front: The gust front is formed along the leading edges of large domes of rain-cooled air that result from downdrafts from individual thunderstorm cells.

Surface High pressure results due to cool denser air that piles up over the surface in a dome. At the leading edge of this current (the gust front) produces wind shift, temperature drop and gusty wind that precede a thunderstorm.

(b) Downburst and Microburst: Downburst refers to a concentrated severe downdraft of horizontal dimension exceeding 4 km and poses a serious hazard to aircraft. A small scale downburst with maximum horizontal extent of 4 km. In some cases downburst may begin as microburst which subsequently spread as a larger scale.

The model characteristics include:

(i) Precipitation (or dust) curl that is carried by the wind back up toward cloud base.

(ii) Horizontal bulging near the surface in a precipitation shaft creating a foot shaped appearance

(iii) Increase of wind speed as microburst expands out over the ground.

(iv) Abrupt surface wind gusts followed by rapid dissipation in some cases.

The Thunderstorm Project report (BYERS and BRAHAM 1949) relates pressure areas smaller than thunderstorm meso highs (called pressure noses). They were less than 8 km in diameter, life period few minutes. These pressure noses occur at about the commencement of mature stage of a thunderstorm cell, when rain first begin to reach the ground. Tighter and stronger pressure noses indicate a microburst. In a microburst pressure nose of 2.4 hPa, lasting only 3 minutes was observed in the Florida Area cumulus experiment (FACE) on 1 July 1975.

In a plane crash at Stapleton International Airport (Denver, Colorado) on 7 August 1977, upward curl of microburst was observed with rainshaft which ascended several kilometers towards the base of cumulonimbus. Dust and smoke were seen blowing opposite to the motion of the upward curl.

On 26 June 1978, a microburst at Dulles International Airport a 5 hPa pressure nose observed. A microburst that passed over Andrews AFB was accompanied by winds more than 120 kt (62 mps), just after Air Force One landed on 1 August 1983. Peak winds were less than 10 kt (5 mps) only 4 km to the left of the track of microburst.

(c) Vortex-ring Circulation in a microburst: A dynamical feature that cause a ring of low pressure on the edge of a downdraft in some kind of circulation as a vortex ring surrounding the downdraft core (see Fig. 16.4)

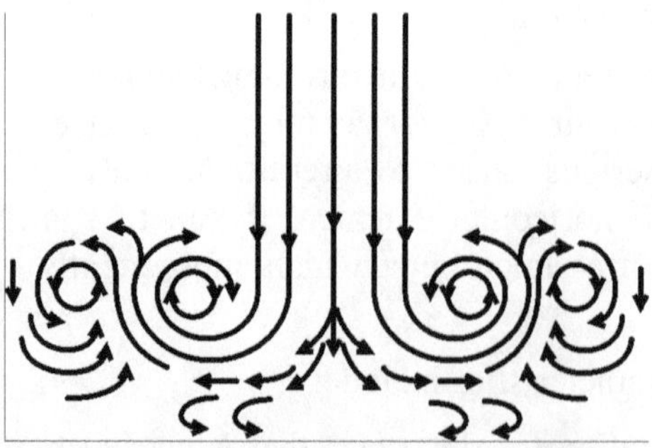

Fig. 16.4 Cross section of a conceptual Vortex ring Model of a microburst.

Airflow in a microburst may be structured as a vortex ring with a downward circulation at the centre that reinforces the downdraft and a return updraft around the periphery.

The vortex ring model of a microbursts circulation was corroborated by observations taken during the field experiments.

There is evidence that not only one, but several vortex rings may organize at the base of a severe downdraft, giving embedded elements of smaller scale, extreme windshear and turbulence within the dangerous large scale microburst. This might be the case in two accidents – Eastern Flight 66 and Pan American Flight 769 (Caracena et al. 1983, Fujita 1983) in which periodic microbursts were observed several minutes apart.

Dry, Wet Microbursts

Different environments can develop different microburst but two extreme types – extremely dry environment and extremely wet environment are important.

(a) **Dry Microburst:** Dry microburst producing storms have high base, little or no rain at the surface and even no lightning. This type of microbursts are common in Rajasthan, which raises dust leading to dust storms (See Fig.16.5). Cloud tops may extend 7-15 km. In this case dust ring under virga shaft descend from high-base cumulonimbus. In dry microburst clouds of dust is raised in the area.

(b) **Wet Microburst:** Low based clouds also produce heavy rain with foot shaped appearance, of an embedded microburst. In this case precipitation blows well under the rainfree cloud base. A descending wet microburst first appears as a darkeness mass of rain descending through light rain. A storm of moderate to heavy rain is a potential source of wet microburst. This type of wet microbursts occur in northeast India.

In general dry microbursts occur in arid or semi-arid regions of the country (like Rajasthan) and wet-microbursts in humid areas (like Assam).

Microbursts normally occur in mid afternoon in midsummer but there are exceptions. Regardless of type and time a microburst is very hazardous in aviation and cause aeroplane crash, should be taken very seriously.

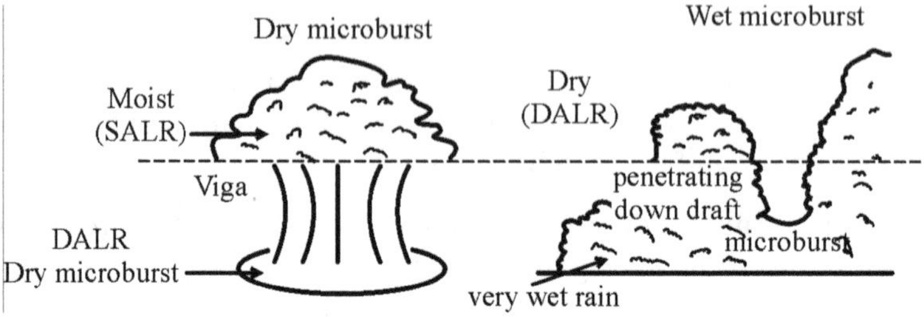

Fig. 16.5

Safety Measures: The best safety measure against Microburst is to avoid it completely. In almost all International Airports windshear warnings are issued. In case of turbulence, the pilots of an aircraft in flight must follow the prescribed speed as given in the flight manual. If the aircraft speed in flights is more than the prescribed speed in the turbulence, there is a risk of structural failure. If the speed is less than the prescribed speed it may lead to the aircraft stalling. All precautious to be taken by the crew as described in the chapter of thunderstorms.

Questions

1. Precipitation in the form of crystals of white ice, opaque, and star like structure, diameter 4 or 5 mm is called:

   (a) Snow (b) Ice pellets

   (c) Sleet (d) Hail

2. Snow and rain falling together or snow melting is called

   (a) Snow (b) Ice pellets

   (c) Sleet (d) Diamond dust

3. Very small crystals of ice appear suspended in the air is called:

   (a) Snow (b) Diamond dust

   (c) Snow grains (d) Ice pellets

4. White opaque ice particles, flat or elongated, diameter less than 1 mm are called:

   (a) Sleet (b) Snow

   (c) Snow grains (d) Diamond dust

5. Transparent ice particles, spherical or irregular or rarely conical in shape, diameter less than 5 mm are called:
   (a) Hail (b) Snow grains (c) Ice pellets (d) Sleet

6. Hail is formed in a well developed:
   (a) Cumulus (b) Cumulonimbus
   (c) Nimbostratus (d) Cirrus

7. The frequency occurrence of hailstorm is more in:
   (a) Equatorial latitude (b) Middle latitudes
   (c) Polar latitudes (d) Near poles

8. Generally the size of hail is small (diameter less than 10 mm) and occurs over an area:
   (a) Less than one square kilometer (b) Less than 10 km^2
   (c) Less than 25 km^2 (d) Over large area

9. The mass of hail in general does not exceed:
   (a) 1% of the mass of that thunderstorm cell
   (b) 10% of the mass of that thunderstorm cell
   (c) 25% of the mass of that thunderstorm cell
   (d) None above

10. Hail has several shapes and classified by its size and compared with familiar objects. If the diameter is 20–30 mm, it is called:
    (a) Grape size (b) Walnut size
    (c) Golf ball (d) Hen egg size

11. If the diameter of the hail is 30-50 mm, it is called:
    (a) Walnut size, (b) Hen egg size
    (c) Golf ball size (d) Tennis ball size

12. If the diameter of hail is 50-60 mm, it is called:
    (a) Golf ball size (b) Tennis ball size
    (c) Hen egg size (d) Walnut size

13. If the size of hail diameter exceeds 60 mm it is called:
    (a) Hen egg size (b) Tennis ball size
    (c) Golf ball size (d) Grape size

14. In the Cb cell an updraft of .... or more required to support the hail of diameter 25 mm

(a) 51 kt
(b) 100 kt
(c) 200 kt
(d) None above

15. In the Cb cell, to support the hail of diameter 80 mm an updraft required:

(a) 51 kt or more
(b) 100 kt or more
(c) 200 kt or more
(d) None above

16. In the Cb cell, to support the hail of diameter 130 mm an updraft required:

(a) 51 kt or more
(b) 100 kt or more
(c) 200 kt or more
(d) None above

17. A study revealed that wind speed difference between 500 hPa and 250 hPa related to the size of hail stones. A windshear of 35 kt, 33 kt, between these levels correspond to hail sizes of diameter:

(a) 1-2 cm; 2-3 cm
(b) 3-5 cm; 1-2 cm
(c) 11-20 mm; 1-10 mm
(d) 0-10 mm; 1-10 mm

18. Wind shear between 500 hPa and 250 hPa of the order ... required for the formation of light hail diameter less than 1 cm:

(a) 26 kt
(b) 33 kt
(c) 35 kt
(d) Less than 24 kt

19. Low level wind shear is associated with the meteorological situation:

(a) Ground layer inversion
(b) At the boundaries of air masses
(c) In the vicinity of thunderstorm
(d) All above

20. Downdrafts in thunderstorms encircling horizontal gusts of large magnitude is called:

(a) Clout burst
(b) Squall
(c) Downburst / microburst
(d) Gustiness

CHAPTER 17

Tornadoes, Waterspouts, Dust Devils

17.1 Tornadoes

Tornadoes are the most violent storms on earth, but they miss detection on synoptic chart. A Tornado is a violent rotating column (spiral motion) of air. It appears as pendent cloud extending from cumulonimbus cloud base to the ground. The column (or funnel) does not always extend to the ground and may be masked by the dust. A funnel cloud is similar to a tornado, except that the funnel does not reach the ground. The diameter of the column is about 100 m. In general tornadoes are more frequent in extra-tropics than in tropics.

The word Tornado is derived from the Spanish word Tronada, meaning thunderstorm. In northern and western parts of Africa tornado still refers to a thunderstorm. In Latin Tornare means "to turn", thus tornadoes are also called twisters. Tornadoes have different names in different parts of the world. In France and Germany they are called

CHAPTER 17

Tornadoes, Waterspouts, Dust Devils

17.1 Tornadoes

Tornadoes are the most violent storms on earth, but they miss detection on synoptic chart. A Tornado is a violent rotating column (spiral motion) of air. It appears as pendent cloud extending from cumulonimbus cloud base to the ground. The column (or funnel) does not always extend to the ground and may be masked by the dust. A funnel cloud is similar to a tornado, except that the funnel does not reach the ground. The diameter of the column is about 100 m. In general tornadoes are more frequent in extra-tropics than in tropics.

The word Tornado is derived from the Spanish word Tronada, meaning thunderstorm. In northern and western parts of Africa tornado still refers to a thunderstorm. In Latin Tornare means "to turn", thus tornadoes are also called twisters. Tornadoes have different names in different parts of the world. In France and Germany they are called

Trombe, in Spain and Italy Tromba, in Russia Symerch, in JapanTatsumaki, in India Hatishnura (meaning Elephants trunk).

Life Cycle of a Tornado

Tornado funnel may be seen as thin rope, conical shape, cylindrical shape or thick dense cloud mass touching the ground. Generally tornadoes rotate in an anti clockwise direction but there may be some that rotate in clockwise direction. During the life span of a tornado the funnel undergoes many changes. The life cycle of a tornado may be divided into five stages.

1. ***Dust Whirl Stage*:** In this stage whirling of dust is observed below cumulonimbus cloud but it does not touch the ground.
2. ***Organising Stage*:** In this stage the funnel from the base of Cb cloud touches the ground and intensity increases.
3. ***Mature Stage*:** In this stage funnel attains maximum diameter (100-250 m). Central pressure in the eye drops (25-200 h Pa) and attains maximum wind speed. Circulation usually stays in continuous contact with the ground through stages 3, 4 and 5. The funnel creates havoc, destroys buildings, poles, trees and sucking debris and dust raises to great heights into the air. Sometimes motor cars, animal and heavy objects sucks in, lifts aloft and thrown away at considerable distance. The destruction of buildings is also caused by explosive effect due to sudden fall of pressure (in less than a minute) by over 50 h Pa or more. The large pressure defference between inside a closed building and the outside atmosphere leads to an explosion, which bursts the walls and the ceiling outwards. According to some empirical theory, a pressure drop of 100 h Pa causes wind speed 600 kmph around the vortex.
4. ***Shrinking Stage*:** In this stage the width of the funnel decreases and tilt of vertical axis increases, fury drastically decreases.
5. ***Decay Stage*:** In this stage the shape of the funnel spreads like a spiral of rope, decreases in height and ultimately disappears.

Major tornadoes pass through all five stages life cycle, while minor tornadoes abort in stages one – two – five.

Size of a Tornado

The diameter of the funnel generally varies 100-250 m, rarely reaches to 1000 m. The vertical depth of tornado circulation extends to the middle of Cb cloud (about 10 km). The average path length is 5-10 km but it may range 30 m to 500 km. Life span varies 2-3 minutes to 3 hours. Short lived tornadoes have wind speed 50 mps, path width 100 m and travel length 2 km, whereas long life period tornadoes have wind speed 100 mps, path width 200-600 m and travel length 200 km. Life period varies 15 seconds to 8 minutes at a point. An extreme duration of 7 hours along the ground observed in Illinois on 26 May 1917. They generally move in a straight line path over a flat country for a long distance. Movement becomes zig-zag when it passes through hills, tall buildings and limits speed, on some occasions dies down. The average energy of tornado is 10^{-11} times the solar energy received by the earth (3.67×10^7 cal/min).

Central Pressure Drop in a Tornado

The central pressure drop (in the eye) may be 100 to 200 h Pa. On 20 Aug 1904, at Minnesota, aneroid barometer recorded a pressure fall of 200 h Pa. The eye of a tornado behaves like a vacuum, consequently strange things happen near the vortex such as, trees are stripped of their barks, sheep lose their wool, chickens lose their plumes, corks of bottles fly off, chests explode and splinters fall in all directions and bursting of closed buildings etc. People who had the experience narrated that they had the bursting experience in their chest and ears.

Synoptic Situations Favourable

The first signs of a Tornado development is a large active cumulonimbus cloud with mammatus. The cloud may acquire green and yellow colour. Green colour lightning or ball lightning also be observed. On some occasions a Tornado may occur without any thunderstorm activity but there will be heavy rain, showers and hail. Cb is associated with every tornado. It is probable a supercell type convective cloud is the genesis of a tornado. The synoptic conditions favourable for the development of a supercell type convective cloud are:

(i) Low level convergence with associated upper air divergence,

(ii) lot of moisture in low levels,

(iii) super adiabatic lapse rate in the lower-mid levels,

(iv) strong vertical shear in the horizontal wind (upto 300-250 hPa or more).

Radar Echoes

As mentioned earlier in thunderstorms hook (6) type echo is associated with a tornado, which is an indication of meso-cyclone. It may be noted here that every hook type echo is not a tornado, it only indicates severe thunderstorm. The echo reflectivity is to the order of 4 or more.

Tornado Vortex Signature (TVS)

A pulsed Doppler Radar when directed to a stationary cyclonic vortex, it will show particles moving away from the radar to the right (positive velocities) of the line joining the radar and the centre of vortex and the left of this line particles will appear to be moving towards the radar (negative velocities). Doppler radar will indicate positive velocities to the right of the centre of the vortex and negative velocities to the left of the vortex centre and zero along the line of the vortex centre. The velocity distribution configuration detected by Doppler Radar is called Tornado Vortex Signature (TVS). TVS is detected in the middle region of Cb. A tornado takes about 15 to 20 minutes to touch the ground after TVS is detected.

Electrical Phenomena

Increased lightning activity will be observed about half an hour before touch down of a tornado.

Acoustic Phenomena

A peculiar whining sound like buzzing of an army of bees is heard when the funnel cloud is high up in the air. When the funnel touches the ground a terrific roar, like the sound of a cannon fire, is heard for a few minutes. Snake hissing sounds heard before tornado touches the ground. This is attributed to the vibrations of the air masses rotating around the funnel.

Precipitation

Heavy rain sometimes associated with hail precede and follow a tornado. The eye of the tornado, like the cyclone eye, is generally free from precipitation.

Tornado Hazards

Prediction of a tornado is still in infant stage as its mechanism of formation is not clearly understood. However with the help of radar nowcasting is done (forecasting its movement and severity in a very short period) to alert people to save their lives. Property damage due to tornadoes seems at present inevitable (see Fig. 17.1). The average death toll in United States of America is about 150-250 per annum, but the injury seems many folds of this. This is caused by the flying objects hitting the people and the subsequent fire. The greatest loss of life due to a single day tornado was 689 people recorded on 18 March 1925, while the greatest loss on a single day was about 1200 people recorded on 19 February 1884.

Fig. 17.1 The tornado (upper photo) which struck Edmonton, Canada on 31 July 1987 left 26 dead and cause damage amounting to US$350 million (*lower photo*)

(***Photos :*** *Atmospheric Environment Service Canada*).

Some Recorded Tornado Incidence

On 30 May 1879 an iron bridge weighing 108 tons was lifted over the River Big Blue in Irving (USA). This incident was recorded in great detail in a book by Finley (1881). A tornado in Minnesota lifted a passenger car with 117 passengers, total weight about 72 tons. In 1879, in Missouri, a horse was lifted and carried a distance of few hundred meters and deposited without any injury. In New Delhi (India) on 17 March, 1978 a tornado in about three minutes took a toll of 28 people, caused injuries to about 700 people and damaged property worth more than one crore rupees. The same tornado lifted a passenger bus with 70 passengers and deposited twenty meters away in a canal. A tempo vehicle was also lifted from a petrol pump and thrown 100 meters away. In a Germany city market a man who was standing under the vortex was lifted high in the air, pelted with hailstones and thrown. When he came back to senses he found himself under two men, one woman and a horse. Another man who was standing close to a post in front of an emergency ward of a hospital in Lefortova was lifted five meters high in the air, carried a distance of 100 m in the garden and deposited on the lawn.

Development of Tornadoes Associated with Fire and Volcanic Eruptions

Under favourable circumstances widespread fire may develop a tornado. On 1 September 1923, a fire broke out in Tokyo after an earthquake. Within 24 hours about 120 tornadoes and smoke devils developed in Tokyo after the great earthquake. Some tornado acquired intensity of wind 50-70 m/s, picked up cars, people outside the fire line. About one lakh forty two thousand (142000) people died in Tokyo, more by fire than by earthquake itself.

Tornadoes/waterspouts may develop after volcanic eruptions over land/water surface. In 1963, there was volcanic eruption of Surtsey in the middle of the sea, off the coast of Iceland. Each intense volcanic eruption gave rise to an enormous dense cloud. From these clouds tornadoes of different sizes developed on the leeward side.

Safety Measures

People can adopt safety measures to protect themselves by listening to tornado advisory bulletins broadcast on national TV channel. The following measures to be adopted as in case of lightning strokes and cyclones. Shut off immediately, after alert, electric power, gas supplies and extinguish all fires. Seek shelter quickly in a tornado cell or in sturdy reinforced pucca concrete building. Do not drive a car which may be hit by missile debris or it may even be lifted and thrown away by strong winds. If caught in an open place, run to a nearby culvert or ditch, lie down and hold on to a fixed object to protect from being blown off. Protect your head from the injuries of flying objects.

Fujita-Pearson (FPP) Tornado Intensity Wind Scale

This was proposed by Fujita in 1971. This wind scale designed to link the Beaufort scale of Force 12 with Mach No.1 in twelve steps. The scale is derived from the formula

$$V = 6.30\ (F + 2)^{3/2}$$

where V = wind speed in mps. The scale specifications are given in Table 17.1.

Table 17.1

Scale	Wind	Damage description
F0	18-32 mps	***Light damage***: such as breaking of tree branches, falling of sign boards
F1	33-49 mps	***Moderate damage***: Beginning of hurricane wind speed. Moving autos pushed off the roads. Peel surface off roofs. Mobile homes overturned
F2	50-69 mps	***Considerable damage***: Box cars pushed over; Mobile homes demolished. Large trees uprooted or snapped. Light object missiles generated.
F3	70-92 mps	***Severe damage***: Roofs and walls tornoff well constructed homes. Trains overturned, Heavy cars lifted off ground and thrown. Most forest trees uprooted.
F4	93-116 mps	***Devastating damage***: Well constructed houses levelled; cars thrown and large missiles enerated, structures with weak foundation blown off some distance.
F5	117-142 mps	***Incredible damage***: Automobile sized missiles fly through the air beyond 100 meters; strong frame houses lifted off foundations and disintegrated at considerable distance. Trees debarked; incredible phenomena will occur.
F6-F12	142 mps to Mach-1	The maximum wind speeds of tornadoes (= speed of sound) not likely to reach F6 wind speeds.

Tornado Statistics

Tornadoes can occur in any latitudes, but they are more frequent in extra-tropics than in tropics, and rare in equatorial belt. The annual frequency of occurrence of tornadoes in USA is about 200. The Mississippi valley has maximum occurrence in the world. Generally tornadoes move from south westerly direction (in northern hemisphere) at the average speed of 15 mps. The maximum wind speed does not exceed 142 mps. The average travel length is about 2 km (range 100 m to 600 km), in USA 7 km. The average width of track is 50 m to 250 m. There may be erratic movement as well. Average life span varies from 2 minutes-2 hours.

17.2 Waterspouts

Waterspouts are of two types. Type one that develops downward from a cumulonimbus cloud and it is also called fair-weather waterspout. Type two is simply a tornado over the water. Type two develops upwards from the surface of water and is not directly associated with a cloud. In both of these types water gets sucked up into air along with aquatic creatures such as fish, frog etc. These aquatic creatures may be found in rain associated with waterspout. In the world over Florida keys and Palm beach Florida have the highest frequency of incidence of waterspouts about five per annum per 100 sq km.

Waterspouts undergo a life cycle similar to that of tornado on land. It has five stages of life cycle which are briefly described below.

1. ***Initial or Dark Spot Development Stage***: First visible sign of a vortex is a dark spot on the sea surface. In this initial stage a short funnel pendant may develop from the super cell (Cb cloud). The dark spots may occur in bunches of two or more, of which one may be dominating the others. The dominant one becomes waterspout and the others decay. The life span in this stage may be about 1-20 minutes. Many may dissipate after this stage. From the droppings of tracers it is gathered that the dark spots are developed by the rotation imposed from above.
2. ***Spiral Stage***: In the second stage, formation of spiral occurs on the sea. Only one dark band (150-1000 m) comes out from a nearby shower band. Spiral bands indicate the lines of confluence or difluence on sea surface while the regions of flow away from the vortex seen along the surface.
3. ***Ring Spray Stage***: In this stage wind strength increases beyond a critical value of about 22 mps and throws up a ring of spray from the surface. In this stage funnel increases in size and the axis of funnel tilts. It begins to move rapidly along the surface as it comes under the influence of the wind shift line. The wind shift line is believed to be associated with the cool breeze from close by shower.
4. ***Mature Stage***: Mature stage lasts about 2 to 18 minutes. This stage is characterized by the strongest wind, funnel tilt and associated with forward wind speed 4-8 mps. In this stage the funnel may acquire double eye wall like that of cyclone and generally moves along a gentle curved path. The spray ring becomes spray vortex.

Waterspout may not be visible all through from cloud to water surface. On radarscope hook echo may be seen.

5. ***Decay Stage***: The last stage is decay stage. This stage generally lasts about 2 to 3 minutes. Spiral and funnel disappears, and rain cooled air takes over waterspout. However in decay stage spiral rain may still be seen which lasts about 5 minutes.

Waterspout Statistics

The maximum tangential wind speed estimated by photogrammetric technique is 85 mps in the lowest levels of 10-15 m (amsl). Waterspouts carry objects weighing 5 tones to a height of about 40 m over coastal waters out at sea. The helical funnel of waterspout may go upto about 300 m in air. Waterspout vortex may generate waves of moderate to high amplitude over the sea surface. According to Schroder (1977) statistics Hawaiian island has maximum incidence of water spouts about 7 per annum. Photogrammetrically estimated wind speed at an altitude of 38 m is about 55 mps and vertical velocity 25 mps.

According to aircraft penetration of waterspouts observation the principal funnel features are:

(i) A warm central core region.
(ii) Funnel core deficit 1 to 10 hPa depending on waterspout intensity.
(iii) In the core upward vertical velocities 5-10 mps.
(iv) Tangential velocities at penetration altitude of 400 m is about 30 mps.
(v) Both cyclonic and anti-cyclonic vortices encountered.

In general waterspouts are more common than tornadoes and the average frequency is about 400 per annum per 10^4 sq km in favorable locations. Compared to tornadoes, central pressure of waterspouts have higher pressure and lower pressure defect and relatively less ferocious/violent.

17.3 Dust-Devils

Dust-devils are a frequent phenomena in tropics and sub-tropics, however they can be seen at any latitude under suitable environmental conditions. Generally dust-devils are observed during scorching hot sun in deserts and semi deserts. They are generally much smaller in size and less violent than tornadoes. The funnel core is warm with temperature anomaly of 1 to 10 °C. The core has low pressure with pressure defect 1 to 4 h Pa.

Dust and sand whirls in a funnel shape but no pendant cloud is involved. The rotation of wind may be cyclonic or anticyclonic. The horizontal rotation and upward flow of wind in dust-devils exceed 30 kmph with average height of column being 200 m. Duration of life varies from few minutes to 7 hours with horizontal path distance of about 50 km.

Favourable condition for formation of Dust-Devils

(i) Hot or intense insolation (ii) Dry dusty ground (iii) Steep lapse rate of temperature at the ground (iv) calm or light surface wind.

Questions

1. A violent rotating column (spiral motion) of air, most violent storm on earth that cannot be detected on synoptic chart is a:
   (a) Super cydone (b) hurricane
   (c) Depression (d) tornado
2. A tornado has five stages of life cycle. Dust whirl stage, organizing stage, mature stage, shrinking stage and decay stage. In organizing stage the funnel:
   (a) The funnel is observed below cumulonimbus
   (b) The funnel touches the ground from the base Cb
   (c) The funnel attains maximum diameter
   (d) The funnel spreads like a spiral rope
3. In mature stage of a tornado the funnel attains maximum diameter, central pressure drops (26-200 hpa). The fury of a tornado mainly depends on:
   (a) The diameter of tornado (b) Central pressure drop
   (c) Length of travel (d) All above
4. The average path length of a tornado is 5-10 km, life span about 3 hours. Short lived tornadoes wind speed, path width and travel length are:
   (a) 25 mps, 50 m, 1 km
   (b) 50 mps, 100 m, 2 km
   (c) 50 mps, 200 m, 100 km
   (d) 100 mps, 400 m, 200 km

5. The long period tornadoes wind speed, width and travel length are:
   (a) 25 mps, 50m, 1 km
   (b) 50 mps, 100m, 2 km
   (c) 100 mps, 200-600 m, 200 km
   (d) 25 mps, 1000 m, 20 km
6. The velocity distribution configuration detected by Doppler Radar is called Tornado Vortex Signature (TVS). After detection of TVS, a tornado takes about:
   (a) 10 minutes to touch the ground
   (b) 15 to 20 minutes to touch the ground
   (c) 20-30 minutes to touch the ground
   (d) None above
7. Fujita-pearson proposed tornado intensity wind scale to link the Beaufort scale force 12 with Mach No. 1 in 12 steps. F5 stands for wind speed:
   (a) 70-92 mps (b) 93-116 mps
   (c) 117-142 mps (d) 142 mps to Mach-1
8. Fujita-Pearson wind intensity scale F6-12 stands for:
   (a) 70-92 mps (b) 93-116 mps
   (c) 117-142 mps (d) 142 mps-Mach-1
9. The helical funnel of waterspout may go up into air:
   (a) 10 m (b) 100 m
   (c) 200 m (d) 300 m
10. As compared tornadoes, the frequency of water spouts, its central pressure and pressure defects are:
   (a) more; more; less (b) Less; more; more
   (b) less; less; less (d) More; more; more
11. Dust devils are frequent in tropics and sub tropics. They are observed (i) during (ii) with core temperature anomaly and (iii) pressure defect of:
   (a) Morning; cooler; 1-5 °C; rise 1-4 hPa
   (b) Afternoon; warmer 1-10 °C; fall 1-4 hPa
   (c) Night; cooler; fall
   (d) Afternoon; cooler; rise

Answers

1.	(d)	2.	(b)	3.	(b)	4.	(b)
5.	(c)	6.	(b)	7.	(b)	8.	(d)
9.	(d)	10.	(a)	11.	(b)		

CHAPTER 18

Ice Accretion on Aircraft

Freezing level or 0° C isotherm is one of the important element in aviation route forecasts. The hazards associated with ice accretion in aviation are:

1. Airframe ice accretion,
2. Engine icing.

Generally airframe icing will be under control, but the engine icing significantly effects the aircraft performance. Aircraft icing may occur (i) in cloud, (ii) in freezing rain and (iii) in freezing drizzle.

Clouds which are composed of ice crystals (like cirrus) are not hazardous, but the clouds which have supercooled water droplets are potentially hazardous in respect icing. Airframe ice accretion takes place when aircraft passes through cloud or rain above freezing level, while engine icing occurs even in clear air and temperature above 0°C.

Supercooled water droplets are in unstable state, which transform into ice and in this process latent heat of fusion of ice (80 calories per gram) is released. In route forecasts, the severity of icing is denoted by light ψ, moderate $\psi\!\psi$ and severe $\psi\!\psi\!\psi$ icing. Airframe icing develops below 0 °C increases upto – 7 °C (servere icing). Moderate icing devlopes in the temperature range – 8 °C to – 12 °C, light icing develops in the temperature range – 13 °C to – 20 °C and very light icing occurs in the temperature range – 21 to – 40 °C.

Light icing is not significant and does not create problems, and its effects can be easily nullified by de-icing or anti-icing equipments. Moderate icing is hazardous even over a short period if neglected. Moderate icing can be countered by using de-icing or anti-icing equipment or diversion. In case of severe icing de-icing or anti-icing equipment working is not sufficient and diversion is necessary. It is to be noted that airframe icing will not be blown off (removed) by increasing speed in level flight or during take-off. Modern aircrafts are fitted with anti-icers or de-icers to prevent ice formation or to remove ice after it has formed. This is achieved through (i) fluids (ii) rubber boots or (iii) heating devices.

(i) A fluid is an anti-icing-device, because it makes it difficult in ice formation. These are fluids which are released through slinger rings or porous leading edge members to flow over the blades of the propellers and on the surfaces of the wings

(ii) Membranes of rubbers are attached to the leading edges. The rubber membranes can be made to pulsate which cracks the ice and broken off after it has formed. This is de-icing device.

(iii) Heating of vulnerable areas to prevent ice forming by hot air from engine or through special heaters ducted to the leading edges of wings, empennages etc. Electrically heated coils protect pilot, pressure sources, propellers etc. This is called heating device.

Ice accretion on aircraft wings, on the surface near leading edge results in decrease lifting ability, disturbs the airflow and may prevent an aircraft from becoming air borne.

18.1 Types of Airframe Icing

In the atmosphere ice may deposit on an object mainly in two ways. (i) By sublimation process (direct transformation of water vapour into ice) or (ii) by freezing process (conversion of liquid water into ice). We have learnt that on ground sublimation produces Hoar frost and freezing process produces Rime. However an aircraft in flight faces a number of icing conditions, of which hoar frost, rime ice and clear ice (glaze ice) are important.

Hoar frost

It occurs in clear air on a surface whose temperature is below frost point of the air which is in contact. In a clear (i.e., cloudless) atmosphere, if moist air is cooled by aircraft frame which flew from colder air layer and if the aircraft surface temperature is below 0 °C and to below its frost point then hoarfrost forms. Such situations can occur by rapid descent of aircraft or take-off in marked surface inversion (temperature being negative i.e., sub-zero at the surface). These situations are shown by typical examples in Fig. 18.1 (a), 18.1 (b).

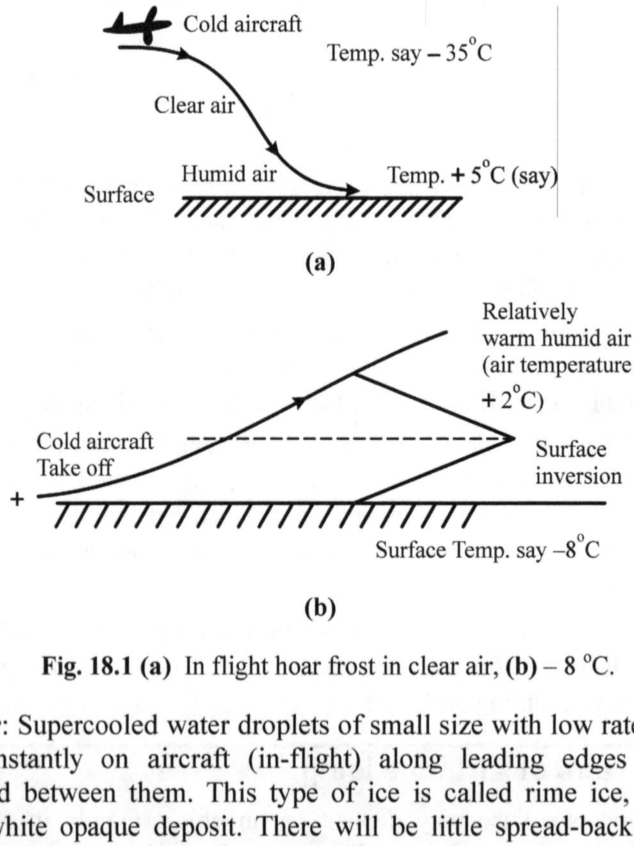

Fig. 18.1 (a) In flight hoar frost in clear air, **(b)** – 8 °C.

Rime ice: Supercooled water droplets of small size with low rate of catch freeze instantly on aircraft (in-flight) along leading edges with air entrapped between them. This type of ice is called rime ice, which is rough, white opaque deposit. There will be little spread-back over the main planes.

(a) Clouds of limited vertical height in the temperature range 0° C to – 40° C (like Stratus, Stratocumulus, Altocumulus and fair weather Cumulus) give rise to Rime ice of slight intensity (ψ).

(b) Upper parts of dense clouds in the temperature range – 10 °C to – 40 °C (like Nimbostratus) also give rise to light intensity rime ice.

(c) But in the clouds of vertical development, like towering Cumulus or Cumulonimbus in the temperature range – 20° C to - 40° C give rise to moderate intensity (Ψ̸) rime ice in the lower parts of the cloud and light intensity rime in the upper part of the cloud. These are shown in Fig. 18.2 (a), (b) & (c)

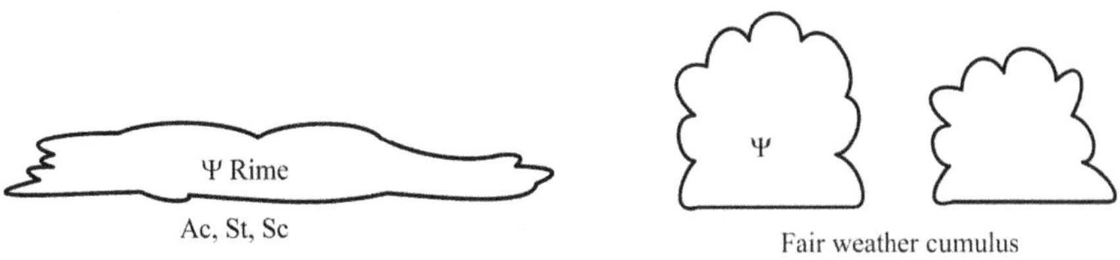

(a)

Fig. 18.2 (a)

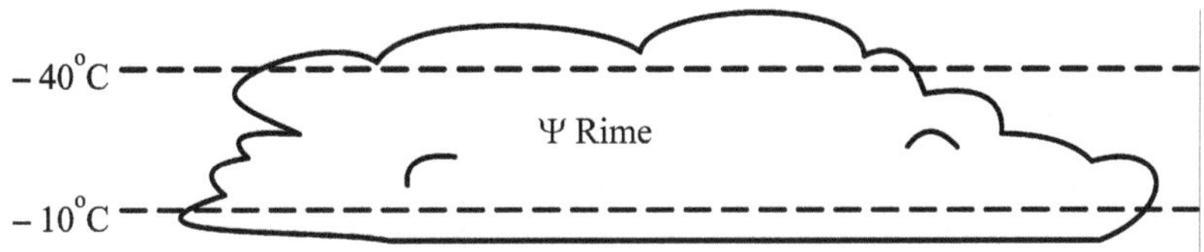

(b)

Fig. 18.2 (b)

Fig. 18.2 (c)

***Clear ice (or glaze ice)*:** clear ice forms on aircraft in flight when the aircraft encounters large suppercooled water droplets with high rate of catch (see fig. 18.3). This can occur in dense clouds like T Cu or Cb of vertical development or in rain. The temperature of the cloud + 0° C to – 45 ° C. Clear ice occur in the cloud in temperature range 0 °C to –10 °C and also 0° C to – 20° C (see fig. 18.3) . Clear ice forms near 0° C and below it, but not at very low temperature.

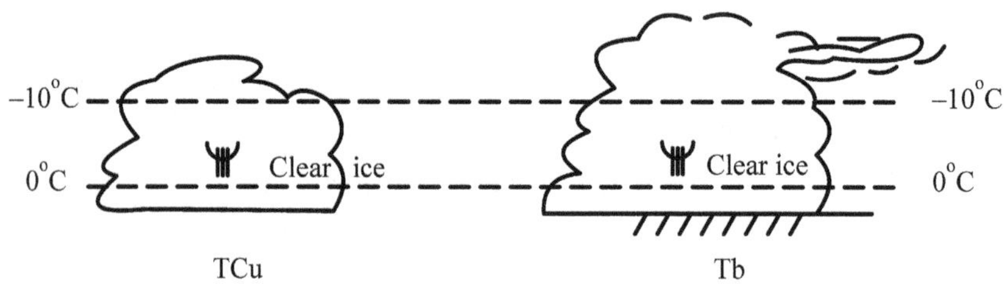

Fig. 18.3 Clear ice formation.

***Cloudy or Mixed Ice*:** Rime ice and glaze ice are the main forms of ice accretion on aircraft inflight in cloud or rain. However supercooled water droplets exist upto – 40 °C, hence a wide variety of icing exists between 0 °C to – 40 °C. These varieties usually called as cloudy or mixed ice. Smaller supercooled drops with lower temperatures produce rough ice or large drops with temperature near 0 °C produce smooth or glassy ice with spread back over the airframe.

18.2 Effect of Airframe Icing on Performance

(a) *Aerodynamic effects*: Ice formation on the leading edges of the airframe changes the airflow pattern round the affected part. It causes increase in drag, a decrease in lift, an increase in stalling speed. Ice accretion on leading edges of the fin, rudder, movable parts interfere with the airflow and cause serious effect on control and effects aerodynamic properties. Thin objects like aerials, struts, leading edges of propellers etc., are more prone to icing than airframe, nose of the fuselage. Ice formation under the wing is dangerous (normally which is out of sight), which hazard may be inferred from a change in aircraft performance.

(b) *Effect of Weight*: In general, increase in weight due to ice accretion is not very important. But an unequal distribution of ice seriously effects, particularly on the propeller. It creates unbalance when part of ice breaks away. This leads to serious vibration. A similar hazard may occur with aerial masts, exposed balanced weights and the arms actuating control surfaces.

(c) *Effects on instruments*: The pitot-static system aerial masts are susceptible to gather ice. It may cause vibration and fracture, may reduce the efficiency (of the part affected), may cause serious errors in airspeed measurement and loss to communication facilities.

(d) *Effect on control surfaces*: In some positions of control, ice may accumulate in the gaps and also on the leading edge of the movable surface. This may cause jam in the control. This risk is more on small aircraft as compared to large aircraft.

(e) *Other effects*: Formation of ice coating on windscreens and canopies may dim (blur) the vision. The ice coating may occur due to flight in icing cloud or due to hoar-frost formation in clear air.

18.3 Engine Icing

We shall study engine icing separately as (a) piston engines (b) turbine and jet engines and (c) helicopter icing.

A. Piston Engines

Ice formation in the air intake and induction system (of a reciprocating engine) leads to loss of engine power as it (ice) obstructs the air passages. It creates disturbance in the fuel metering and makes mobile parts defunct. Engine icing may occur at low temperatures as well above 0 °C, in clear air, in cloud and precipitation.

There are three types of engine icing. (1) Impact ice, (2) fuel icing and (3) carburettor icing.

Impact ice: It is similar to airframe icing. Ice forms and accumulates at air intakes and filters, when either the ambient temperature or temperature of aircraft is below 0 °C. For turbo charged engines this is the only icing hazard.

Fuel Icing: When the water is held in suspension in the fuel, it precipitates and freezes at elbows in the induction piping. This type of icing occurrence is in general not common.

Fuel ice can form downstream of the jet where the fuel is introduced into airstream which is fed to carburetor. It vapourises causes reduction in temperature (due to latent heat of absorption).

When the temperature of fuel/air mixture drops and lies between 0 °C and – 8 °C water precipitates from the incoming moist air and freezes on any surface which it faces. This will seriously affect airflow and causes engine's output drop.

Fuel ice can occur even at ambient air temperature of + 20 °C to + 30 °C when the RH is more than 50%.

Carburettor Icing: In piston engine the throttle effect combined with fuel evaporation icing is called carburettor icing. Carburettor icing is very hazardous and most common in occurrence.

In moist atmosphere icing is possible in the induction system between temperatures – 5 °C to + 30 °C. At the air induction process the pressure is reduced, which causes expansion and adiabatic cooling (as much as 25 to 30 °C), which is sufficient for condensation and supercooling. This results in icing on adjoining surfaces (see Fig 18.4). Ice gradually builts up on ventury walls, throttle value and blocks the ventury. This upsets the fuel/air ratio and thus causes slow loss of power.

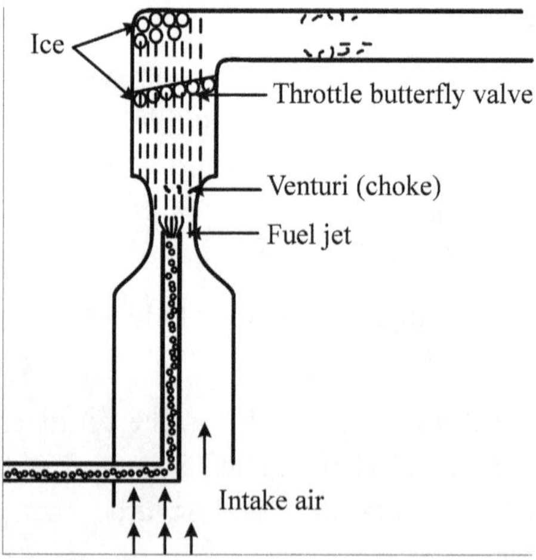

Fig. 18.4 Carburettor ice.

In aviation accidents, carburettor icing accounts for a great many accidents. These accidents can be prevented by greater pilot awareness and vigilance.

Carburettor icing will occur when the ambient temperature is well above 0 °C. Carburettor ice is most likely to from when the temperature is between – 10 °C and + 20 °C and RH is high. The icing hazard will be much more on humid summer day than on cold winter day (because the water vapour content of the air will be less in winter day). Carburettor icing can occur in cloud or in clear air as it depends on humidity. The minimum RH necessary for icing is 50%. The hazard potential increases with the increase of RH. It must be noted that carburettor icing hazard is less at temperatures much below – 5 °C, because liquid water content steadily decreases on descent with low power and just before landing in humid air it is normal to apply full carburettor heat to avoid carburettor icing. However avoid use of carburettor heat on the ground (except during critical pre-take-off check). The pre-take-off check of carburettor heat should be done on a hard surface. It involves, applying full carburettor heat and observe rpm drop in about 5 seconds. Subsequently return the control to Full Cold or OFF and note that the rpm returns to the original value (that indicates no ice is formed).

***Carburettor Heat*:** Carburettor heat system is to combat icing. This is done by passing the induction air past the hot engine exhaust manifold. The carburettor heat control normally located near the throttle in the cockpit. By selecting it fully ON, heated air is passed into the carburettor.

***A note of caution*:** Partial use of carburettor heat may increase the temperature of the induction air in the range which is favorable to the formation of carburettor icing.

B. Turbine and Jet Engines

The intakes of turbine and jet engines are subject to icing in a similar way as airframe icing when the flight is through supercooled droplet cloud. The susceptible parts are: The rim of the intake, any struts across the intake, the vanes in the early stages of the compressor.

Engine icing is directly proportional to the rate of airflow through the engine or the number of engine revolutions per minute. The rate of icing decreases with decrease in engine revolutions. During jet engine operation the intake pressure is much less than the pressure outside. Due to expansion and adiabatic cooling the temperature fall is of the order 5 °C is common. If the indrawn air is moist and temperature is near 0 °C, prolonged operation results in condensation and ice formation (but this will not occur in airframe icing).

When the flight is in icing cloud or when the aircraft is landing or take-off under freezing fog, the effect of icing will be markedly more.

In flight icing in jet engines are expected under the same conditions of airframe icing.

Airframe icing and engine icing intensities are different, because airframe icing depended on the air speed, where as the engine icing is depended on both air speed and engine speed (r.p.m).

At high air speeds the engine will face ram effect, while at low speeds (less than 250 kt) air is sucked in. Icing rate increases markedly with increase of air speed above 250 kt, while engine icing tends to be constant with decreasing airspeed but airframe icing decreases with decreasing airspeed.

C. Helicopter Icing

In comparison to fixed-wing aircraft, helicopters are sensitive to the effects of ice accretion. The rotor blades are finely balanced to avoid vibration and rotor head is subjected to large centrifugal stresses. Even small ice deposite on blades have serious effect on performance, which destroys the aerodynamic lift by causing vibrations and by increasing the centrifugal stress on the rotor head. Consequently dangerous situations may arise within minutes of the entry of copter into icing conditions.

Ice accretion on the head alone interfere with the control of blades. During running up on the ground, any vibrations set up by ice formation on the blades will be transmitted to the springs of the under carriage, in turn becomes large by sympathetic vibrations and may lead to overturning of the helicopter.

The other problems are ingestion of snow by engines and the icing of windows and aerials.

Helicopters are normally operated in cloud free atmosphere, hence airframe icing are limited to flights at temperatures below 0 °C in rain, drizzle, wet fog or snow.

18.4 Icing on Parked Aircraft

Icing on parked aircraft can occur in three ways.

(i) By sublimation process (conversion of water vapour directly to ice),

(ii) By freezing of water, and

(iii) Depositing on (snow flakes) ice already formed.

Hoar frost on parked aircraft must be removed before take off, because it causes dimming vision through transparencies, disturbs boundary layer airflow by reducing total lift and increasing drag. Rime forms on the parked aircraft on windward side and builds out into the wind.

Glazed ice poses serious problems for all forms of transport because of its weight. It brings down overhead wires, power cables etc. In short all types of icing on the parked aircraft must be removed before take off.

In general, there are eight types of frost, which are briefly given below:

(i) *Ground Frost*: It forms when the grass thermometer measures sub-zero temperature.

(ii) *Air frost*: It forms when the air temperature at Stevension screen level is below 0 °C (at 4 ft level).

(iii) *Radiation frost*: It is a particular case of air frost, caused by radiation cooling of the ground to below 0 °C.

(iv) *Wind frost*: It is also a particular case of air frost caused by sub-zero temperature developed elsewhere but advected by wind to the place.

(v) *Black frost*: It is an air frost but not hoar frost on the ground.

(vi) *White frost*: It is a kind of hoar frost.

(vii) *Glazed frost*: It is a frost formed on the ground.

(viii) *Silver frost*: A surface deposit of frozen dew.

18.5 Aircraft Icing in Flight

In flight aircraft icing will be of the same forms as it occurs on the ground but the contributory factors change in many ways. They are: (1) temperature and supercooling water and its size, (2) kinetic heating, (3) concentration of liquid water, (4) air speed, (5) cloud base temperature, (6) chord and high ground.

Temperature

In flight temperature changes as the aircraft moves and super cooled water droplets may be encountered up to – 40 °C, but the size decreases. Supercooled water droplets solidify on impinging aircraft and releases latent heat (of fusion of ice 80 cal/gm). One-eightieth ($1/80^{th}$) of the supercooled drop freezes on impinging (the aircraft) for each degree of Celsius of supercooling.

Example

At – 25 °C, $\frac{25}{80}$ part of the supercooled water droplet size freezes on the aircraft frame and the remaining $\frac{55}{80}$ part of droplet remains in the liquid state.

At –10 °C, only $\frac{10}{80}$ part of the supercooled water droplet freezes on the aircraft frame and the remaining $\frac{70}{80}$ part remains in liquid state.

At – 3 °C, only $\frac{3}{80}$ part of the supercooled water droplet freezes on the airframe and the remaining $\frac{77}{80}$ part of droplet remains in liquid state.

These are shown in the Fig. 18.5 (a), (b), (c), (d) .

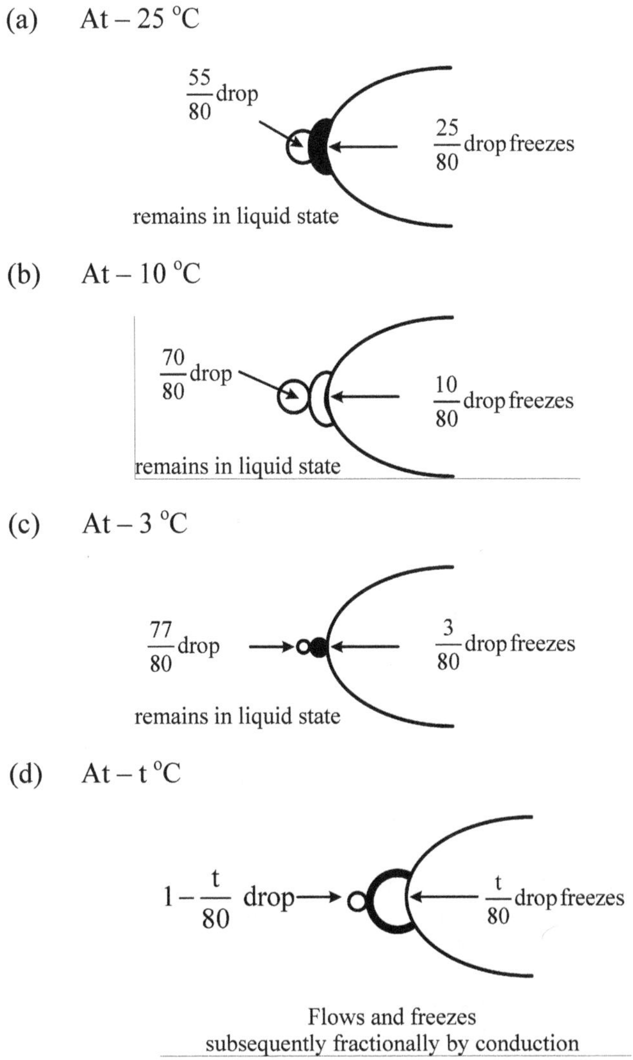

Fig. 18.5 supercooled droplet freezing.

Supercooled water droplets freeze partially on impact. In Fig. 18.5 (a) $\frac{25}{80}$ drop size freeze on impact and the remaining $\frac{55}{80}$ drop flows over the aircraft structure which freezes fractionally later on conduction.

Similarly in Fig. 18.5 b, c, d, $\frac{10}{80}$ drop size, $\frac{3}{80}$ drops size, and $\frac{t}{80}$ drop size freeze on impact and the remaining

$\left(1-\frac{10}{80}=\right)\frac{70}{80}, \left(1-\frac{3}{80}=\right)\frac{77}{80}, \left(1-\frac{t}{80}\right)$ drop flows as liquid over the aircraft structure, which freeze fractionally later on conduction.

The liquid part of the drop flows over the aircraft structure which freezes fractionally later on conduction. Closer to (– 0 °C) sub-zero temperature water droplet flows-back will be more and resulting ice subsequently worse.

Kinetic Heating: In the above Fig. 18.5 (a), (b), (c), (d) it is shown that the aircraft acquires/attains the ambient air temperature at the flight level. This is roughly true only for aircraft with speed below 250 kt. However in case of fast speed aircraft, kinetic heating occurs due to airflow around the aircraft by compressibility and frictional effect. The kinetic heating temperature rise is given by the formula:

$$\text{Temperature rise } (^{\circ}\text{C}) = \left[\frac{\text{True Air speed kt}}{100}\right]$$

For example for speed of 600 kt, 400 kt the temperature rise would be $\left(\frac{600}{100}\right)^2 = 36^{\circ}\text{C}, \left(\frac{400}{100}\right)^2 = 16\ ^{\circ}\text{C}$ respectively.

Thus kinetic heating may offset (compensate) to some degree. If the aircraft is flying through the cloud then the wet surface evaporates the moisture by kinetic heating and releases latent heat of vaporization (at 600 cal/gm). In practice, only indicated air temperature is used for assessing icing risk (because the temperature probe itself will be subjected to kinetic heating and cooling due to latent heat of vaporization).

Concentration of liquid water and airspeed: Ice accumulation on aircraft in flight is proportional to the water content that it encounters and the airspeed. As stated earlier, the latent heat released causes the spreading of water, which subsequently solidifies as clear ice. In this water spreads over the aircraft surface and merging with adjoining drops produces more clear ice.

Cloud base temperature: If the cloud base temperature is more or warmer then there would be greater concentration of liquid water, which effects the icing process. The base height of cumuliform cloud is more or less same whether it is in tropics or middle latitudes. However the cloud base temperature in tropical latitudes is more than in higher latitudes, implying there is more condensation in lower parts of the clouds in tropics as compared to higher latitudes. Because of warmer cloud base temperatures in tropics towering cumulus, cumulonimbus in tropics/summer will have more liquid concentration than those in higher latitudes/winter.

Effect of Highground

Over flat terrain cloud height development is less as compared to high ground. In general clouds over flat terrain contain small droplets, with low density create slight to moderate icing. On the contrary clouds over high terrain have vertical development, large droplets with relatively high density create severe icing. These are shown in Fig. 18.6 (a), (b).

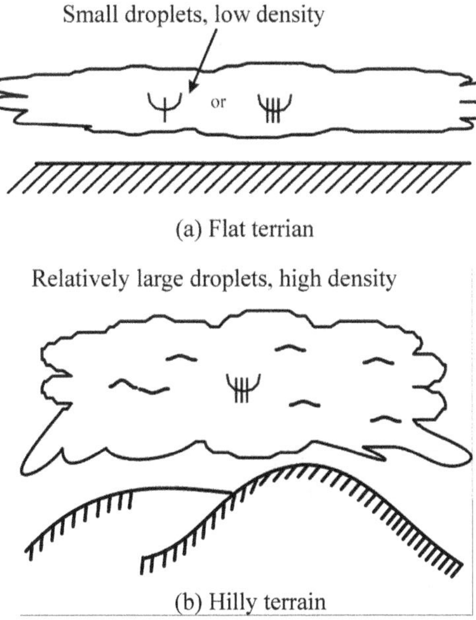

Fig. 18.6 (a) Flat terrain, (b) Hilly Terrain.

Effect of Supercooled Water Droplet size

Smaller droplets form rime ice over smaller area of the aircraft, while larger droplets spread over large area and form clear ice See Fig. 18.7 (a), (b)

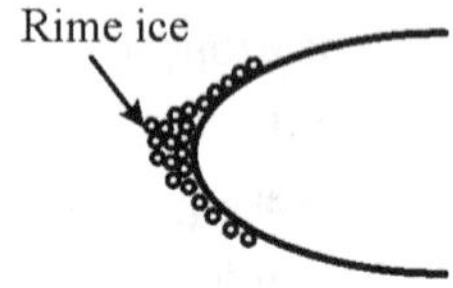

Fig. 18.7 (a) Small droplets form rime ice over a small area

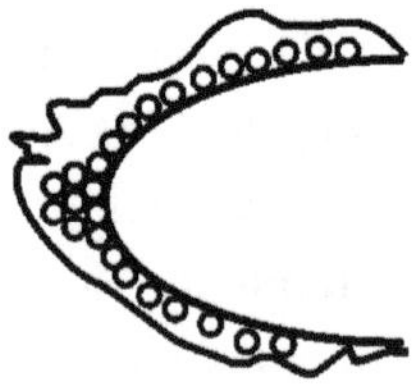

Fig. 18.7 (b) Large drops form clear ice over large area

Chord

For a given airspeed a thin chord wing accumulates more ice (being relatively cooler) as compared to the thick chord wing. On thin chord the droplets impinge on the leading edge, while on the thick chord wing droplets flow over the wing due to advance compression.

18.6 Icing in Clouds and in Precipitation

In discussion of clouds and precipitation it was said that Ci, Cc, Cs, Ac and St clouds contain ice prisms, As contains snow, ice pellets and rain, Ns contains snow, ice pellets and rain; Sc contain snow, snow pellets and rain; St contains drizzle, snow grains and ice prism; cumulus contains rain and T Cu, Cb contains rain, snow, snow pellets, ice pellets and hail.

Ac contains supercooled water droplets up to – 30 °C. In Ac and As ice accretion will be low. Only in prolonged flights in these clouds appreciable ice accretion is possible and there will not be difficulty in getting out of these cloud by change of flight level because they are generally of shallow depth.

Sc is a layer cloud in the lower part of the atmosphere with vertical extent 3000 ft and rarely exceeds this and pilot can get out of this cloud by change of flight level. Light to moderate icing is common and rarely severe icing occurs in this cloud. However if Sc has formed orographically or convectively then there is possibility of embedded Cu and high rate of icing.

(a) ***Convective Clouds*****:** Ice accretion is likely to be severe in towering Cu and Cb. By increasing flight level one cannot expect decrease in ice accretion unless the aircraft reached to temperature below – 25 °C. The risk of icing is negligible only if the aircraft reaches level where temperature is –40 °C or less. When icing occurs during flights in (TCu or Cb) large convective clouds, flight level should be decreased subject to terrain clearance, where temperature is above 0 °C or flight should be continued until emergence from cloud. It is always safer to avoid convective clouds altogether.

(b) ***Cold fronts*****:** In cold front formation of convective clouds are likely to be cellular and fairly widespread and moderate to severe icing is common over a long distance. As said earlier it is safest to avoid convective clouds. In unavoidable case the belt of convective clouds can be crossed by flying at right angles to the belt. In continuous cloud or precipitation pilot has to adopt one of the following three alternatives to avoid icing See Fig. 18.8.

(i) Descend to a layer where temperature is definitely above 0 °C, subject to terrain clearance,

(ii) Climb to a level where icing is likely be negligible (an inversion where temperature is above 0 °C), this is possible at altitude where there is change in wetted area of the aircrafts surface (– 40 °C or blow). Finally after de-icing, aircraft should be able to continue flight carrying the ice already formed.

(iii) Turn back if the aircraft still has fully manoeuvrability. This is the safest option.

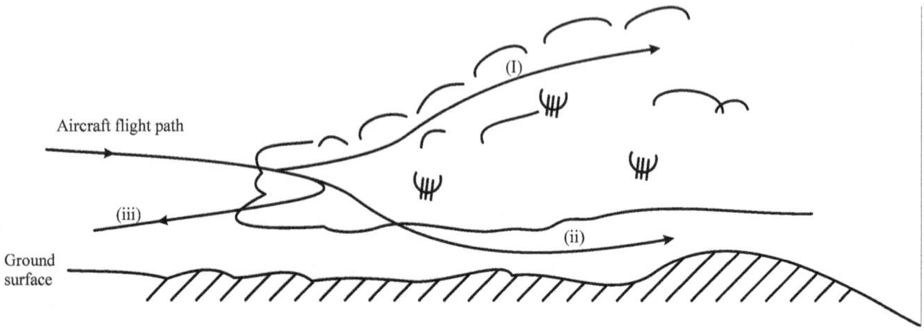

Path (i) climb, if sufficient power reserve

Path (ii) descend, subject to terrain clearance

Path (iii) turnback if has full manoeuvrability

Fig. 18.8 Avoiding icing in continuous cloud/precipitation.

(c) ***Cold front clouds*****:** As said earlier cold front clouds are cellular and convective, and there would be gap between them. It is safe to avoid them.

(d) ***Warm front clouds*****:** In a warm front extensive horizontal layer clouds are observed. Near the surface front (Nimbostratus) cloud base is low, steady rain, snow or mixture of both. The main cloud sheet near the front, vertically extends from few hundred feet to 5000 ft to 10,000 ft. Warm frontal cloud contain supercooled drops. If there is large area of uniform frontal precipitation, severe icing is likely within the cloud in temperature range 0 °C to – 15 °C (a layer of about 10,000 ft). Supercooled rain may be present beneath the cloud. In the cloud, at temperatures below – 15 °C, mostly ice crystal may be present. Depending on the practicability, fly below freezing level (subject to terrain clearance) or at a level where temperature is below – 15 °C. Occasionally embedded Cb clouds are found in warm front, which should be avoided.

18.7 Preflight Action

Parked aircraft on cold nights should be protected from hoar frost, rime, snow or rain ice. Ideally keep it in heated hanger. Any frozen deposit should be completely removed before take off. Route forecasts contain (0 °C height) freezing level and expected icing in cloud like (ψ, Ψ, Ψ) light, moderate or severe. Even if the forecast does not indicate any icing, pilot should be prepared for icing when the clouds expected to encounter with temperature near or below 0 °C.

De-icing fluid of correct strength required to clear off any frost and ice. Fit all orifice gaurds, covers and engine blanks before de-icing begins.

Remove all covers, gaurds, blanks etc., and check that all orifices are clear of ice, snow or slush. Make double check that all inlets and outlets, vents, hinges, exposed operating mechanisms, seals and gaps between surfaces are completely clear of contamination. Make sure that de-icing fluid has not cleared away grease from vital areas.

If there is likely to be hold, do not taxi, lest all de-icing procedures will be undone.

When taxing on contaminated runways, use engine anti-icing, carburettor heat and propeller de-icing as necessary.

Never put your aircraft in the wake of hot engine exhaust of another aircraft for the sake of de-icing. Just before take-off, re-check all the visible surfaces are clear and that all anti-icing devices (available) have been set.

Once airborne, be ready to operate the de-icing equipment as necessary.

Questions

1. In respect of ice accretion on aircraft, clouds ________ are potentially hazardous:
   (a) Which have ice crystals,
   (b) Which have supercooled water droplets at about freezing level
   (c) Which have both ice crystals and super cooled droplets at – 20 °C
   (d) Which have ice crystals at – 40 °C
2. Airframe ice accretion may take place when aircraft passes through:
   (a) Clear air and temperature below 0 °C level
   (b) Cloud and/or rain above freezing level
   (c) Cloud below freezing level
   (d) Cloud at temperature – 40 °C

3. Engine ice accretion may occur in:
   (a) Cloud temperature – 25 °C
   (b) In clear moist air at about freezing level (both above and below)
   (c) Cloud, below freezing level
   (d) In clear dry air at – 25 °C temperature

4. Aircraft in flight likely to be affected seriously when passes through cloud at level of:
   (a) Light intensity Ψ icing
   (b) Moderate intensity Ψ icing
   (c) Severe intensity Ψ icing
   (d) Extreme low intensity

5. Diversion in flight is necessary, in case of icing of:
   (a) Light intensity (b) Moderate intensity
   (c) Severe intensity (d) Clear air

6. To counter the ice accretion effect on aircraft in flight, modern aircrafts are fitted with devices of:
   (a) Anti-icers (b) De- icers
   (c) Aerials (d) Both (a) & (b)

7. An anti-icing and de-icing devices are:
   (a) A fluid, membranes of rubber
   (b) A fluid, fluid
   (c) Rubber membrane, rubber membrane
   (d) Both (b) & (c)

8. Ice accretion on aircraft wings, on surface near leading edge results in:
   (a) Decreases in lift, increases air flow, and adds ice weight
   (b) Decreases in lift, disturbs air flow and increases weight
   (c) Increases lift, disturbs airflow, decreases weight
   (d) Increases lift, increases air flow and adds ice weight

9. Hoar frost may form on an aircraft in flight while:
   (a) Descending through Stratus
   (b) Descending through Nimbostratus
   (c) Descending through cloudless air into moist air which is cooled (by aircraft) to below 0 °C
   (d) In an area of supercooled water

10. Hoar frost may form on an aircraft in flight while:
   (a) Climbing through Stratus
   (b) Climbing through Nimbostratus
   (c) Climbing through an inversion below 0 °C to above 0 °C into moist air
   (d) In an area of supercooled water

11. Worst airframe icing on aircraft in flight is expected in:
   (a) Snow (b) Freezing rain
   (c) Freezing fog (d) Hail

12. Worst icing is expected on aircraft in flight in the range of temperature:
   (a) – 10 °C to – 25 °C (b) 0 °C to – 15 °C
   (c) – 25 °C to – 40 °C (d) Near freezing level

13. Icing and turbulence would be most servere in:
   (a) Stratocumulus (b) Nimbostratus
   (c) Cumulonimbus (d) Altocumulus

14. Snow is likely fall more from:
   (a) Stratocumulus (b) Nimbostratus
   (c) Cumulonimbus (d) Stratus

15. Supercooled water droplets are found in:
   (a) Clouds only
   (b) Clouds, fog and precipitation
   (c) Fog and precipitation
   (d) Precipitation

16. Worst aircraft icing depends on:
   (a) Shape of airfoil and speed
   (b) Droplet size and temperature
   (c) Freezing level
   (d) RH and temperature

16. (a) What safest action to be taken in case flight encounter of freezing rain?
   (a) Climb into air above
   (b) Descend
   (c) Accelerate
   (d) Turn round before loss of manoeuvrability

17. In flight clear ice is a problem because:
   (a) It is translucent and forms along leading edges
   (b) It not translucent and forms along leading edges
   (c) It is considerably heavy and can affect aircraft control and surfaces
   (d) Forms in clear air

18. Non-orographic cloud instensification gives rise to moderate icing in:
   (a) Cumulonimbus and Nimbostratus
   (b) Altostratus and Altocumulus
   (c) Nimbostratus and cirrostratus
   (d) Nimbostratus and cirrocumulus

19. With ground temperature – 10 °C, freezing level at FL 35, Nimbostratus at FL 40 with continuous rain, icing is expected at:
   (a) Above FL 40
   (b) From surface to FL 35
   (c) Icing is not expected
   (d) Only in Ns

20. With the same true air speed (TAS) through a cloud of supercooled water drops, we expect:
   (a) Same amount of icing likely on thick wing and thin wing aircraft
   (b) More icing on thick wing than the thin wing
   (c) More icing on thin wing than the thick wing
   (d) None on either wing type

21. Freezing rain results when:
    (a) Rain hitting the ground and freezing on impact
    (b) Rain falling into warmer air
    (c) Rain falling into colder air and freezing into pellets
    (d) Rain falling from an inversion into an area below 0 °C

22. Over Central Europe freezing rain occurs in:
    (a) Warm front, (b) Cold front
    (c) Warm occlusion (d) Cold occlusion

23. For the same TAS a thin chord wing accumulate ------ ice than a thick chord wing:
    (a) Same (b) More
    (c) Less (d) None above

24. If an aircraft climbs or descends from sub-zero temperature into warm moist clear air _______ forms:
    (a) Clear ice (b) Hoar frost
    (c) Rime ice (d) None above

25. Airframe icing ________ over high ground
    (a) Decreases (b) Worsens,
    (c) Does not change (d) None above

26. Serious carburettor icing likely in descent, in temperature over _________ and if RH greater than ________
    (a) + 10 °C; 10% (b) + 30 °C; 30%
    (c) – 20 °C; 20% (d) None above

27. Larger the droplet size, the rate of catch will be more and result in:.................
    (a) Rime ice will be formed
    (b) Heavy clear or translucent ice with glassy surface will be formed
    (c) No ice will be formed
    (d) None above.

28. The warmer clouds contain more free water than the clouds with lower base temperature. The warmer the cloud base temperature, the........
    (a) Lower the icing severity
    (b) Greater the severity of icing
    (c) Same severity of icing
    (d) None above

29. Clear ice is the most dangerous form of icing; the closer to 0 °C......
    (a) The lesser clear ice
    (b) The more clear ice
    (c) There is no change clear ice
    (d) None above

30. One-eightieth per degree below zero will freeze as rime ice and the remainder ...
    (a) Also freeze into rime ice
    (b) Will flow back and freeze as clear ice
    (c) Will evaporate
    (d) None above

31. Kinetic heating may warm the airframe out of the icing range or ....
    (a) Cool it but out of icing range
    (b) Warm it into the icing range
    (c) Cool it into icing range
    (d) None above

32. The smaller the drop size, lower the rate of catch and the lower the temperature _________
    (a) Smooth clear ice will be formed
    (b) More rough rime like cloudy ice is formed
    (c) Rough clear ice will be formed
    (d) None above

33. The larger the drop size, the nearer the temperature to 0 °C, the rate of catch .... and ...... ice forms
    (a) Will be less; rough rime
    (b) Will be more; smoother glassier
    (c) Becomes zero; no ice forms
    (d) None above

34. Smaller supercooled water droplets, the rate of catch .....and ...... ice forms:
    (a) Will be more, and clear
    (b) Will be less, and rough opaque rime
    (c) Will be less, and clear
    (d) None above

Answers

1.	(b)	2.	(b)	3.	(b)	4.	(c)
5.	(c)	6.	(d)	7.	(a)	8.	(b)
9.	(c)	10.	(c)	11.	(b)	12.	(b)
13.	(c)	14.	(b)	15.	(b)	16.	(b)
16. (a)	(d)	17.	(c)	18.	(b)	19.	(b)
20.	(c)	21.	(d)	22.	(a)	23.	(b)
24.	(b)	25.	(b)	26.	(b)	27.	(b)
28.	(b)	29.	(b)	30.	(b)	31.	(b)
32.	(b)	33.	(b)	34.	(b)		

CHAPTER 19

Visibility and Fog

Visibility is very important in all transportation services, let it be on land (buses, trains, trucks etc) water (ships, travelers, boats etc) or in air (aeroplanes, fighters, helicopters, ballooning etc). Many transportation accidents occur because of poor visibility and lack of its understanding. It is common feature in India, as seen a number of accidents occur in winter in the wake of morning thick fog. Visibility is one of the very important parameter in aviation particularly during aircraft take-off and landing under Visual Flight Rules (VFR). In aviation, flight visibility means air-to-air visibility or visibility forward, from the cockpit of an aircraft in flight.

Visibility denotes the transparency of the atmosphere, with reference to normal human vision, and it is expressed as certain distance. Meteorological visibility refers to the greatest distance at which a black object of suitable size can be seen and recognized against the horizon sky. The object should subtend at least an angle of 0.5 degrees, both in horizontal and vertical directions at the observers eye. However the horizontal dimension of the object should not subtend an angle more than 5 degrees. Visibility depends on the state of atmosphere, size, shape, colour of the object, background illumination. A dark building or object on the skyline (horizon) is easily recognized as compared to a sheep in a less distant bare paddock (small field).

The criterion used for day light visibility cannot be used for night measurements. Night visibility is the greatest horizontal distance at which suitable objects can be recognized for what they are in day light, or at which lights of special candlepower can be seen at night. For practical purposes the following equivalent may be expressed.

The greatest distance at which a light of luminous intensity equal to 100 Candela can be seen or the luminous intensity of light that is just visible a specified distance.

In meteorological routine weather report (METAR) contains two different visibility measurements. One, which is commonly called Met visibility or Met vis and the other one is RVR (Runway visible range)

According WMO (World Meteorological Organization), "Met vis' refers to Meteorological Optical Range (MOR) defined above.

RVR is defined as follows:

RVR is the maximum distance in the direction of take-off or landing at which the runway or the specified lights delineating the runway can be seen from a position on the centerline at a height corresponding to the average eye-level of the pilot at touch down.

Visibility is estimated by choosing some prominent objects (called visibility land marks) situated at standard distance as laid down in the visibility code or as near to them as possible (within 10% margin). A list of credibility land marks showing their respective distance and direction from the observatory with corresponding code figures kept ready. Visibility observations consists in seeing which is the farthest object that is visible and the nearest object not visible. The Table 19.1 provides a sample of visibility observation code.

Table 19.1

Landmark	Distance from the place of observation	Direction	Observation as to while land mark visible and which is not	Code figure
A	50 m	( )	A is not visible	90
B	200 m	( )	A is visible but B is not visible	91

Table 19.1 Contd...

Landmark	Distance from the place of observation	Direction	Observation as to while land mark visible and which is not	Code figure
C	500 m	( )	B is visible but C is not visible	92
D	1 km	( )	C is visible but D is not visible	93
E	2 km	( )	D is visible but E is not visible	94
F	4 km	( )	E is visible but F is not visible	95
G	10 km	( )	F is visible but G is not visible	96
H	20 km	( )	G is visible but H is not visible	97
I	50 km	( )	H is visible but I is not visible	98
	Object visible at 50 km or more	( )	I or farther objects visible	99

Visibility at might may be determined in the same manner as during the day by night land marks.

When visibility is different in different direction, the lowest (code) figure is reported with making entries of visibility in different directions. For example 95 E, 96 NW. This means that visibility in the east corresponding to code 95 and visibility in the northwest corresponds to code 96, 95 being the lowest it is reported. In airports visibility meters (skopograph, Transmission meter) are used for measuring RVR.

RVR is reported only when the visibility is 1500 m or below. RVR measured by instrument (visibility meters) is called instrumented runway visual range (IRVR).

19.1 Factors that affect the Visibility

Visibility is affected by the concentration of (aerosols) particulate matter or pollutants. They may be (i) falling materials to the ground, (ii) rising materials from the ground, or (iii) suspended materials in the air (Hydrometers or lithometeors).

The main particulate metter (aerosols) are:

(a) Precipitation, (b) fog and mist (hydrometeors–water droplets or ice particles) (c) wind blown spry from sea (water droplets) (d) Oils (oil drops), (e) smoke (solid impurities), (f) dust and sand (solid impurities or lithometeors), (g) salt (solid particles).

In precipitation visibility deteriorates due to the presence of water droplets or snow ice particles. Reduction in visibility depends on the intensity of rain and size of raindrops. Light rain may cause little change in visibility. Moderate rain may reduce the visibility 10 to 3 km. Heavy rain may reduces the visibility 500 m to 50 m or less. Similarly light snow may reduce the visibility to less than 5 km, moderate snow may reduce to less than 1 km and heavy snow reduce the visibility to 200 m to 50 m or less.

In drizzle visibility may range 500 m to 3 km depending on the intensity. If fog develops the visibility may reduce to 500 m or less.

In high latitudes, when temperatures are low, the snow (dry and powdery) is blown off the ground. This is called blowing snow which affects high latitude aerodromes. In this wind plays an important role and observed in high latitude continental winters. Then there is a whiteout with no discernible horizon in any blizzard.

Fog: Fog is defined as a suspension of very small water droplets (in some case ice particles (ice fog), radius 1-70 μm or more) that remain suspended in air in such concentration that the horizontal visibility near the earth's surface reduces to less than 1000 m, with RH about 100%.

Mist: Suspension of very small water droplets that reduces visibility 1 to 2 km, with RH 95% to less than 100%. In fact fog is cloud (stratus) on the surface of the earth. Lifting of fog due to insolation and turbulence results into stratiform cloud.

Haze: Deterioration of visibility by suspended particulate matter or lithometeors (dust smoke etc) is called haze, in which visibility is more than 2 km and ranges to 10 km. RH less than 95%.

Ice fog: Suspension of ice crystals in high latitudes when temperature is below – 20° C with light wind, ice fog develops. Visibility reduces to less than 1000 m with RH about 100%.

Effect of Wind blown spray form sea: Visibility in coastal areas deteriorates on account of wind-blown spray. With the increase in wind speed to gale strength (41-47 kt) the crests of waves begin to break and spray forms. Strong winds, streak of foam are blown through the air.

When wind speed reaches hurricane force (64 kt or more) visibility may be seriously affected.

***Effects of Oils*:** In cities the visibility deteriorates due to the presence of oil particles which enter the atmosphere from motor vehicle or from at petrol bunks. Air pollution is also caused by industrial sources and from motor vehicles.

***Effect of smoke*:** Smoke from industries and domestic use, contribute carbon of smoke which produces haze, black haze and deteriorates the visibility. Larger carbon particles settle down but the fine particles remain suspended in the air and thus cause pollution. Black haze is the byproduct of soot and various tarry substances that evolve due to burning of coal and wood.

***Effect of smog*:** The combination of smoke and fog is called smog. Smog generally forms when surface inversion exists and winds are light, humidity is more than 95%. Smog reduce the visibility and also affects the health (health hazard) due to toxic combustion products. Both fog and smog are hindrances in road, rail transportation and also in air traffic at aerodromes.

***Effect of Dust and Sand*:** During strong surface winds or intense surface convection currents dust and sand particles are raised into the atmosphere which remain suspended and cause severe deterioration in visibility. When visibility deteriorates to less than 1 km, it is called dust storm or sand storm. Sand particles are lifted to a height of 20 to 30 m in the air but they settle down quickly at a short distance from the source. However the dust particles are carried to a great heights and remain suspended for days or even weeks. This phenomena is common in Rajasthan where it is called Andhi, in which visibility may deteriorate to hundreds of meters or even to few meters.

***Effect of salt particles*:** When sea spray evaporates after blown into atmosphere it leaves salt particles which act as condensation nuclei. These salt particles are hygroscopic and condensation begins on them at about 70% of RH. For the same concentration of salt particles and smoke, deterioration of visibility is considerably low in case of salt particles. The haze developed by sea salt particles is whitish in appearance. The salt particles are corrosive with respect to iron and some other metals which effect is commonly observed in coastal towns/villages.

19.2 Types of Fogs

We know fog forms when the air temperature is cooled to below its dew point. Fog is formed in the presence of sufficient number of condensation nuclei and moisture saturation. Saturation of moisture occur in two ways: (a) by cooling of air, (b) evaporation of water into it. Based on these processes fogs are classified as: (i) Cooling fog, (ii) Evaporation fog ***Cooling fogs are***: (a) Radiation fog, (b) Advection fog, (c) Upslope fog and (d) Inversion fog.

(a) ***Radiation fog***: It is also called ground fog. It is formed by the cooling of the ground (heat loss from the ground) by nocturnal radiation (long wave radiation).

Conditions favorable

(i) High relative humidity (more than 75%),

(ii) Clear sky during night or little high clouding (one or two octas),

(iii) A wind of 2 to 8 kt near the surface considered favorable.

Under these conditions nocturnal cooling becomes prominent which results in fall of surface temperature and development of surface inversion of temperature. This causes turbulent free (stable) layer near the ground and condensation of moisture in the early morning. On sunrise minute stirring (turbulent mixing) takes place near the ground which causes increase of fog (spreading vertically) in the morning. When insolation effect becomes sufficiently strong (turbulence increases) fog slowly lifts from the surface of the earth and slowly develops into stratus cloud. The effects of surface cooling, inversion and formation of heavy dew, fog and stratus is shown in Fig. 19.1.

Fog dispersal begins as the day advances due to increased insolation and breakdown of the inversion or increasing turbulent missing. Maximum frequency of fog occurs at about dawn (about one hour after sunrise, when the minimum temperature occurs). In summer radiation fog is seldom and unlikely to persist much after sunrise.

Radiation fog does not form over the sea because the diurnal variation of temperature is insignificant. On the contrary in polar regions when surface temperature is below – 30° C, fog contains minute ice crystals which are called ice fogs or diamond dust because of glitter in the light.

(b) ***Advection fog***: Advection fog occurs when air moves over a cooler surface or sea. The surface temperature should below dew point temperature of the moving air. Over land advection fog is likely in winter when RH is high.

Over the sea advection fog occurs in three ways. One is transportation of relatively warm air from land to colder sea area (which occur in spring and summer). The second is the flow of air from warmer part of sea to over much colder sea current. This type is seen on Newfoundland Grand Banks caused by cold Labrador current. Here air is blown from Gulfstream to Labrador current. The third type is extensive air mass over sea as it moves over to higher latitudes is cooled. The third type is extensive air mass over sea as it moves over to higher latitudes is cooled. There is hardly any diurnal variation in advection fog over the sea. It may be noted here that both radiation fogs and advection fogs may be described as air mass fogs.

Advection fog is not confined to any particular pressure distribution, it follows the only criteria air should move over to a colder surface.

(c) ***Upslope fog***: When moist air is lifted up (to lifting condensation level) due to orography along a sloping plain or hill, air cools down a adiabatically. On saturation upslope fog develops. In this case conditional stability of air mass take place. This types of fog frequently observed on east of Rocky mountains. In Wyoming it is locally called "Cheyenne fog". Before formation of this fog, light precipitation is observed over the sea.

Upslope fog metamorphosises into stratus cloud on further heating by insolation (convection) and turbulence, generally by midday.

(d) ***Inversion fog***: Fog that forms at the top of a moist layer under subsidence is called inversion fog. In an anticyclone region, when a cold air mass under subsidence comes in contact with warm water bodies inversion fog occurs. In this region ground inversion of temperature is observed besides inversion stratus and light precipitation.

(e) ***Frontal Fog***: The sloping boundary surface between the contrasting air masses is called front. Frontal fog is the result of interaction of two adjacent air masses. Frontal fog occurs in two ways.

(i) Cloud coming down to the surface with the passage of a front. This type is common over hills and rare over low ground.

(ii) It may develop due to continuous rain preceding the warm front or warm occlusion of a depression. In this case air becomes saturated by continuous rain (dew point reaches without cooling of lower air). These conditions generally develop in the cold air ahead of a warm front. Fogs associated with warm fronts (or prefrontal fogs) usually very widespread.

(f) ***Steam fog*****:** When cold air passes over a warm water surface, water vapour evaporates into the air from the water. In this case cold air advection is responsible. The effect is akin to the fog formation when we take hot water bath in cold winter.

Favourable conditions for development of steam fog are: (a) Air should be much cooler than the water. (b) Marked surface inversion in the air before it moves to the sea (or inland water body). (c) A low temperature about O °C or below. This assures super saturation. When water vapour enters cold air, it condenses and saturation leads to fog formation. Steam fog confined to water surfaces near the source of cold air. Steam fog is unstable and observed during autumn in middle and high latitudes, in and around water bodies. This mixing of Cold Labrador current with Warm Gulfstream results in thick fog.

Note: The droplets of fog diameter ranges 25 – 75 μm with average of 45 μm. Mean liquid water content in air is 0.13 gm/m^3 with a range of 0.01 to 0.30 gm/m^3.

The other examples are: Ice Landic and Norwegian fjords, Arctic Sea smoke, and frost smoke.

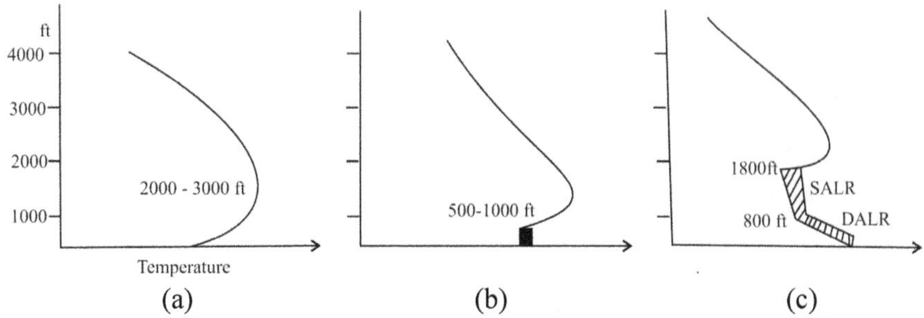

Fig. 19.1 Effects of Surface cooling w.r.t ELR.
(a) Heavy dew formation, (b) fog formation, (c) Stratus development

Ice fog

In extremely cold, calm conditions ice fog forms in moist air by sublimation process. Ice fog is formed when additional water vapour added to the air through fuel combustion. Ice fog may suddenly appear when an aircraft engine is started.

VMC: Visual Meteorological Conditions indicate that visibility, distance from cloud and ceiling are equal to or better than the specified aerodrome minima under VFR (Visual Flight Rules).

IMC: Instrument Meterological conditions indicate that visibility, distance from cloud and ceiling are below specified aerodrome minima and flight can be conducted only under IFR (instrument flight rules).

19.3 Visibility Measurements in Airport

In the day time, visibility is estimated with the help of objects placed along side the runway as visibility markers. At night runway lights are used which are at fixed distance.

Transmissometer

It measures the transmission of light through a portion of the atmosphere (ranway). It consists of an intense source foccused light at one end and a photocell at the other end of runway. The instrument is used at both day and night with modulated light and tuned telephotometer. This gives a continuous record of runway visibility. The output of photocell depends on the turbidity (water particles and dust) of the atmosphere (on runway) between the source and the receiver and gives a measure of transparency of air on the runway. The display units are located in Met. Office and ATC tower. AVRA (Automatic Visual Range Assessor) is also available in some airports.

19.4 Fog Dispersal

Cold fog over the runways may be dispersed by seeding dry ice or silver iodide smoke or by releasing propane gas through expansion nozzle which produces freezing and precipitation. Alternately, strategically placed jet engines placed alongside the runway and spraying with propane. However the dispersal of warm fog (having droplet temperature above freezing point) is comparatively difficult. Warm fog dispersal was attempted by seeding hygroscopic particles, which grow by condensation and then settledown by gravity or by precipitation. It was calculated that hydroscopic particles of size 30 μm (diameter) dispersed at the rate of 5 gm/m^3 may improve the visibility from about 100 m to 1000 m in about ten minutes. However, in practice, it was observed that visibility again deteriorated after a while with the reformation of fog. Consequently this method abandoned. In western countries, particularly in America and Europe, warm fog dispersal is made by ground based heating. During second world war this brute force of dissipating fog, called FIDO, was used in 15 airfields in England successfully. This system is however expensive. In FIDO, the direct application of heat is made by burning of oil in long lines of burners on either side of the runway. Fog is also dispersed by dropping electrified sand (electrically charged particles) through fog.

19.5 Horizontal, Slant and Vertical Visibility

When an airfield is covered with a shallow fog or haze layer, the horizontal visibility on ground (airfield) is poor. The pilot of an aircraft flying over the air filed above could see the airfield clearly, in position A, (see fig 19.2) but on approach for landing in position B, he would encounter poor visibility or he may not be able to see the runway. In position A, the vertical visibility is obscured by a shallow layer of fog/ mist, but in position B the slant visibility is obscured by a thick layer and hence the pilot would not be able to see the runway.

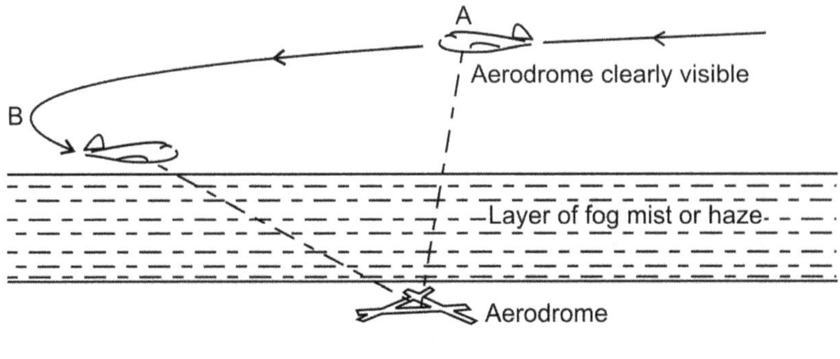

Fig. 19.2 Variation of Visibility during descent position B.

Questions

1. Visibility denotes the transparency of the atmosphere with reference to normal human vision and it expressed as:
   (a) hPa (b) $^{\circ}C/Km$
   (c) Distance (km, meters) (d) Volume (m^3)
2. Meteorological visibility refers to the greatest distance at which a black object of suitable size can be seen and recognized against the horizon sky. The object, should subtend at least angle of ….. at the observers eye.
   (a) 0.5 degrees (b) 5 degrees
   (c) 10 degrees (d) None above
3. Visibility depends on:
   (a) The state of the atmosphere
   (b) Size, shape, colour of the object
   (c) Background illumination
   (d) All of the above
4. When visibility is different in different directions, the reported visibility indicates:
   (a) Lowest visibility in any direction
   (b) Longest visibility in any direction
   (c) Average visibility of all directions
   (d) None above
5. In routine weather report, RVR is reported when the visibility is:
   (a) Less than 5000 m (b) Less than 3000 m
   (c) 1500m or less (d) Less than 10,000 m
6. Visibility deteriorates due to the presence and concentration of:
   (a) Hydrometeors (b) Lithometeors
   (c) Electrometeors (d) Both (a) and (b)
7. In moderate rain visibility may reduce to:
   (a) 10 to 3 km (b) 500 m to 50 m
   (c) 50 m to nil (d) None above

8. In heavy rain visibility may reduce to:
   (a) 10 to 3 km (b) 500 m to 50 m
   (c) 0 – 50 m (d) None above
9. In light snow fall, visibility may reduce to:
   (a) 200 m to 50 m or less (b) 1 Km or less
   (c) 5 km or less (d) 50 m to 0 m
10. In moderate snow fall visibility may reduce to:
   (a) 200 m to snow fall (b) 1 km or less
   (c) 5 km or less (d) None above
11. In heavy snow fall visibility may reduce to:
   (a) 200 m to 50 m or less (b) 1 km or less
   (c) 5 Km or less (d) None above
12. In drizzle, depending on the intensity visibility may range:
   (a) 1 to 3 km (b) 3 to 5 km
   (c) 500 m to 3 km (d) 50 m to 500 m
13. Fog forms when RH is ……… and visibility reduces to…..
   (a) Close to or at 100%, 1 to 2 km
   (b) 95% or less, 1 km
   (c) Close to or at 100%, less than 1000 m
   (d) More than 95%, 100 to 22 m
14. Mist forms when RH is ………. and visibility ranges….
   (a) About 100% or more
   (b) 95% or more, 1000 to 2000 m
   (c) 95% or less, 5000 m or more
   (d) None above
15. When visibility ranges 2 to 10 km and humidity is less than 95% it is called:
   (a) Fog (b) Mist
   (c) Haze (d) Snow fall
16. Radiation fog occurs only:
   (a) In air (b) On land
   (c) Over sea (d) None above

17. Radiation fog occur when sky is clear or 1 to 2 oktas cirrus and:
    (a) Temperature and dew point are far apart
    (b) Temperature and dew point are more or less same
    (c) Temperature 0 °C less and dew point 5° C or more
    (d) None above
18. Advection fog may form when:
    (a) Cold dry air moves over to warm surface
    (b) Warm moist air moves over to cold surface
    (c) Warm dry air move over to cold surface
    (d) Cold moist air moves over to warm surface
19. Radiation fog mostly occurs at about:
    (a) Dawn (b) Midnight
    (c) Midday (d) 1 hr after sunrise
20. Steaming fog likely to form with the advection of:
    (a) Cold air over land
    (b) Warm air over land
    (c) Warm air over cold sea
    (d) Cold air over a warm sea
21. In a tropical downpour visibility may reduce to:
    (a) 2000 to 1000m
    (b) 1000m to 500m
    (c) 200 to 100m
    (d) Less than 100m
22. Smoke from industries and domestic use contribute carbon of smoke, which produces:
    (a) Fog (b) Mist
    (c) Haze / black haze (d) Smog
23. Smog is the combination of smoke and fog. It generally forms when:
    (a) Surface inversion exists, winds are light, RH is 95% or more
    (b) Steep DALR, strong wind RH less than 75%
    (c) SALR, light winds, RH more than 75%
    (d) Isothermal at surface, light winds, RH about 75%

24. During intense surface convection currents, dust/sand particles are raised into the atmosphere which remain suspended and visibility goes down below 1 km, it is called:
    (a) Thunderstorm (b) Hailstorm
    (c) Sand / duststorm (d) Tornado

25. When sea spray evaporates after blown into atmosphere it leaves salt particles which act as condensation nuclei. Condensation may begin on them:
    (a) At about 95% RH (b) At about 80% RH
    (c) At about 70% RH (d) At about 50% RH

26. Transmissometer provides the continuous record of:
    (a) Atmosphere pressure (b) RH
    (c) RVR (d) Temperature

27. Cold fog over runway can be dispersed by:
    (a) Seeding of dry ice
    (b) By spreading of silver iodide smoke
    (c) Releasing propane gas
    (d) All above (a), (b), (c)

28. Warm fog over runways dispersed during second world war by:
    (a) Direct application of brute force of heat, called FIDO
    (b) By seeding dry ice
    (c) By spreading silver iodide smoke
    (d) Spreading propane gas

29. A pilot flying in aircraft over an air field saw the runway clearly but on immediate approach for landing he encountered poor visibility because of:
    (a) Thick fog advected over the airfield
    (b) The fog / mist existing slant thickness was much more than the vertical thickness
    (c) Accumulation of snow fog on wind screen
    (d) All above

Answers

1. (c)	2. (a)	3. (d)	4. (a)
5. (c)	6. (d)	7. (a)	8. (b)
9. (c)	10. (b)	11. (a)	12. (c)
13. (c)	14. (b)	15. (c)	16. (b)
17. (b)	18. (b)	19. (b)	20. (c)
21. (d)	22. (c)	23. (a)	24. (c)
25. (c)	26. (c)	27. (d)	28. (a)
29. (b)			

CHAPTER 20

The Surface Wind

Air in motion is called wind and is expressed in terms of direction and speed. In meteorology wind refers to the flow of air either at surface of earth or in the free atmosphere. Wind velocity is a vector quantity, has both magnitude and direction. The magnitude of the wind is called speed which is a scalar quantity.

Wind direction is regarded as the direction from which wind blows and speed is the rate of movement of air in its instantaneous direction. Wind direction is determined with reference to true north and expressed in degrees and measured clockwise from geographical north in terms of points of the compass 0 – 360°, where East (E) denotes 0 90°, South (S) denotes 180°, West (W) denotes 270°, North (N) denotes 0° or 360° the table 20.1 gives the wind direction in 32 points of compass with code figures used in meteorology.

Table 20.1 Surface wind direction

Direction of Compass point	Degrees	Code
Calm	-	00
N by E	11.25	01
NNE	22.5	02
NE by N	33.75	03
NE	45	05
NE by E	56.25	06
ENE	67.5	07
E by N	78.75	08
E	90	09
E by S	101.25	10
ESE	112.5	11
SE by E	123.75	12
SE	135	14
SE by S	146.25	15
SSE	157.5	16
S by E	168.75	17
S	180	18
S by W	191.25	19
SSW	202.5	20
SW by S	213.75	21
SW	225	23
SW by W	236.25	24
WSW	247.5	25
W by S	258.75	26
W	270	27
W by N	281.25	28
WNW	292.5	29
NW by W	303.75	30
NW	315	32
NW by N	326.25	33
NNW	337.5	34
N by W	348.75	35
N	360	36
Variable	-	99

***Note*:** Read N by E: North by East, NNE: North North East etc.

Note: In meteorological reports or forecasts (in METAR, TRFOR, ROFOR etc) wind direction is given in degrees with reference to true north (TN), while ATC (Air Traffic Control) and ATIS (Automatic Terminal Information Service) reports it is given in degrees magnetic (M) with reference to magnetic north.

Units: Wind speeds are given in knots (nautical miles per hour), (KT), meters per second (MPS) or kilometers per hour (or KMH). The conversion of these measurements are:

1 KT = 1.852 kmph, 1 KT = 0.5144 MPS

The other aviation measurements include statute mile, feet.

1 statute mile (sm) = 5280 ft

1 Nautical mile (nm) = 6080 ft = average length of 1 minute of latitude

= length of 1 minute arc along any great circle.

1 km = 1000 m (meters)

1 knot = 1 nm per hour, and $\frac{76}{66}$ sm = 1 nm

1 KT = 1.15 mph

1 kmph = 0.54 KT

Table 20.2 Gives the conversion of kt to kmph (or KMH)

Table 20.2

I		II		III		IV		V	
Kt	**Kmph**	**Kt**	**Kmph**	**Kt**	**Kmph**	**Kt**	**Kmph**	**Kt**	**Kmph**
1	1.9	11	20	21	39	31	57	45	83
2	3.7	12	22	22	41	32	59	50	93
3	5.6	13	24	23	43	33	61	55	102
4	7.4	14	26	24	44	34	63	60	111
5	9.3	15	28	25	46	35	65	65	120
6	11.1	16	30	26	48	36	67	70	130
7	13.0	17	31	27	50	37	69	75	139
8	14.8	18	33	28	52	38	70	80	148
9	16.7	19	35	29	54	39	72	85	157
10	18.5	20	37	30	56	40	74	90	167
								95	176
								100	185

Wind Speed Estimation

Wind speed estimation without instruments was initially developed by Admiral Sir Francis Beaufort (1905) for ship navigation. It is still being used where wind instruments are not available or cannot withstand in gales. The Table 20.3 and 20.4 gives the Beaufort scale and equivalent wind speed with specification.

Table 20.3

Beaufort Number	Description	Kt	mps	Kmph	Miles per Hour
0	Calm	<1 (0)	0-0.2 (0.1)	<1(0) <1	(0)
1	Light air	1-3 (2)	0.3-1.5(0.8)	1-5(3)	1-3(2)
2	Light Breeze	4-6 (5)	1.6-3.3(2.5)	6-11(9)	4-7 (5)
3	Gentle breeze	7-10 (9)	3.4-5.4(4.4)	12-19(16)	8-12 (10)
4	Moderate breeze	11-16 (13)	5.5-7.9(6.7)	20-28(24)	13-18 (15)
5	Fresh breeze	17-21 (18)	8.0-10.7(9.3)	29-38(34)	19-24 (21)
6	Strong breeze	22-27(24)	10.8-13.8(11.8)	39-49(44)	25-31 (28)
7	Near gale or moderate gale	28-33 (30)	13.9-17.1	50-61(55)	32-38 (35)
8	Gale or fresh gale	34-40 (37)	17.2-20.7	62-74(68)	39-46(43)
9	Strong gale	41-47(44)	20.8-24.4	75-88(82)	47-54(51)
10	Storm	48-55(52)	24.5-28.4	89-102(96)	55-63(59)
11	Violent storm	56-63 (60)	28.5-32.5	103 -117 (110)	64-72(68
12	Harricane force	$\geq$ 64	32.7 and more	118 and more	73 and more

Table 20.4

B. No	Description	Speed estimation over land	Over sea
0	Calm	Calm, smoke rises vertically	Sea like a mirror.
1	Light air	Direction of wind shown by smoke-drift but not by wind vanes.	Ripples with the appearance of scales are formed, but without foam crests.
2	Light breeze	Leaves rustle, ordinary vanes moved by wind	Small wavelets, still short but more crests have a glassy appearance and do not break.
3	Gentle breeze	Leaves and small twigs in constant motion, wind extends light flag.	Large wavelets, crests begin to break. Foam of glass appearance perhaps scattered

Table 20.4 Contd...

B. No	Description	Speed estimation over land	Over sea
4	Moderate breeze	Raises dust and loose papers, small branches are moved	Small waves becoming longer, fairly frequent white horses.
5	Fresh breeze	Small trees in leaf begin to sway, orested wavelets form on inland waters	Moderate waves taking a more pronounced long form, many white horses are formed (chances of some spray).
6.	Strong breeze	Large branches in motion; whistling sound heard in telegraph wires, umbrellas used with difficulty	Large waves begin to form, the white foam crests are more extensive everywhere (probably some spray)
7	Near gale or Moderate gale	Whole trees in motion, inconvenience felt when walking against the wind.	Sea heaps up and white foam; breaking waves begin to be blown in streaks along the direction of the wind (spindrift begins to be seen)
8	Gale or Fresh gale	Breaks twigs off trees, generally impedes progress	Moderate high waves of greater length; edges of crests break into spindrifts, the foam is blown in well-marked streaks along the direction of the wind.
9	Strong gale	Slight structural damage occurs (chimney pots and slates removed)	High waves; dense streaks of foam along the direction of the wind. Crests of waves begin to topple, tumble and roll over; spray may affect visibility.
10	Strom	Seldom experienced inland; trees uprooted; considerable structured damage occurs.	Very high waves with long over hanging crests. The resulting foam in great patches is blown in dense white streaks along the direction of the wind. On the whole, the surface the sea takes white appearance. The rolling of the sea becomes heavy and shock like. Visibility is affected.

Table20.4 Contd…

B. No	Description	Speed estimation over land	Over sea
11	Violent storm or severe storm	Very rarely experienced; accompanied by widespread damage.	Exceptionally high waves (small and medium sized ships might for a time be lost behind waves). The sea is completely covered with long white patches of foam lying along the direction of wind. Every-where the edges of the wave crests are blown into froth. Visibility badly affected
12	Hurricane or super cyclone	Severe and extensive damage	The air is filled with foam and spray. Sea is completely white with driving spray. Visibility is very seriously affected

Measurement of Wind Direction

The wind direction is measured by wind vane. It is a balanced lever which freely turns about a vertical axis. In a common type of wind vane one end of the lever exposes a broad surface to the wind, while the other end, which is narrow, points to the direction from which the wind blows. Under this movable system there is a fixed rigid cross, whose arms are set to the four cardinal directions – North, East, South and West. Some wind vanes are provided with eight directions N, NE, E, SE, S, SW, W and NW.

Exposure

The wind vane must be freely exposed and be high enough above buildings and trees so that it is not affected by the eddies created by them. In a very open site the vane can be installed on steel or wooden lattice tower or mast, 6 meters high and well guyed within the observatory enclosure. Where the site is obstructed by trees, buildings etc, it may be erected on a building or a high mast so that it is at least higher than the highest obstacle (in the vicinity) by 3 meters. Alternately a separate site for the mast may be selected outside observatory, such as the roof of a building. When both anemometer and the wind vane are fixed on the same platform, they should be at least 2 meters apart.

Anemometer or Cup Anemometer

Wind speed measurement instrument is called Anemometer. IMD uses (i) Cup Anemometers (ii) Pressure Tube anemometers. The cup Anemometer consists of three large semi-conical cups (diameter 127 mm) with beaded edges fixed at the ends of three rods. The cups are mounted symmetrically about a vertical axis so that the diametral plane of each cup is vertical. As the force on the concave side of any cup, due to wind is greater than that on convex side in a similar position, the cup wheel rotates. For any given anemometer, the dimensions and design are such that the speed at which the cup wheel rotates depends solely on the wind speed provided the wind speed is steady and is greater than the minimum required to set the cups in motion; the lower limit is due to the effect of friction on the bearings of the cup wheel. The rate of rotation does not depend on the direction of the wind nor to any appreciable extent on the density of the air.

The cups are attached to a central spider which is mounted on a spindle carrying a worm. The worm engages with a gearwheel and drives a revolution counter mounted in a water proof aluminium housing. The gear ratio between the cup and counter spindle is so chosen in relation to the factor of the anemometer, that the counter indicates directly the run of wind in kilometers and tenths, when the instrument is suitably exposed.

To obtain the run of the wind in km and tenths over a given period (T) the counter is read at the beginning (R_1) and at the end of the period (R_2). The mean wind speed during this period is given by

$$\left(\frac{R_2 - R_1}{T}\right) \text{ where T is in minutes.}$$

Exposure

The wind velocity near the surface of the earth varies rapidly with height and is also largely affected by the presence of irregularities in the ground or nearby obstacles such as trees, buildings. For synoptic reports of the surface wind and for general climatological records, it is therefore necessary to define the height and conditions under which the measurement should be made. The standard exposure of wind instruments over level open terrain is 10 meters above ground. Open terrain is defined as an area where the distance between the anemometer and any obstruction is at least 10 times the height of the obstruction. This ideal condition is difficult to obtain but care should be taken to select the best possible.

Wind Speed at the Hour of Observation

Surface wind speed for synoptic purposes is reported in knots. To determine the wind speed at the time of observation, take two successive readings (R_1, R_2) of the anemometer at an interval of 3 minutes. Then 20 ($R_2 - R_1$) gives the wind speed in kmph. To convert into knots multiply by 0.54 i.e., $20 \times 0.54 \times (R_2 - R_1)$ gives wind speed in knots.

Example 1

Let the counter reading at the beginning of the observation $R_1 = 2063.4$ and let the reading after 3 minutes be $R_2 = 2065.1$

$$\text{Wind speed in kmph} = (2065.1 - 2063.4) \times 20$$
$$= 1.7 \times 20 = 34 \text{ kmph.}$$
$$\text{Wind speed in knots} = 34 \times 0.54 = 18.36 \simeq 18$$

Example 2

Average wind speed during the day. Let the counter reading at 0830 hrs IST on 10.11.2000 be 9985.5 and at 0830 hrs IST on 11.11.2000 be 0097.6 (which is 10097.6).

$$\text{The average wind speed in kmph} = \frac{10097.6 - 9985.5}{24}$$
$$= \frac{112.1}{24} = 4.67 \approx 4.7$$

$$\therefore \text{ Average wind speed on 10.11.2000 in kt} = 4.67 \times 0.54 = 2.5218$$

Accuracy

The anemometer is expected to have frictional and other errors less than 10%. The anemometer normally begins to rotate at wind speeds of the order of 1 ½ – 2 kt and between 2 – 5 kt. Its indications depend to a large extent on the bearing friction and the state in which the instrument is maintained.

Pressure Tube Anemometer

In a pressure tube type anemometer, a wind vane at the top of the mast keeps the opening end of the tube facing the wind. As the wind blows into open and it produces pressure rise inside, which is proportional to the wind speed. This pressure rise is transmitted to the indicating device. An outer tube just below the wind vane contains small holes. The wind blowing against the holes reduces the pressure inside which again proportional to the wind speed. This effect is also transmitted to the indicator by a suction tube. The combination of these two effects forms a system which is independent of the slight pressure difference if any inside and outside the building in which the recorder is kept.

Theory

In streamline flow, the Bernoulli's equation $\frac{1}{2}\rho\upsilon^2 + p = \text{constant}$, gives that total head (velocity head + static pressure) is constant. When an open end tube is facing wind, the total pressure acting at its mouth is constant (= C). If the difference between the total head and static pressure is measured, the result is proportional to v^2 (v = velocity). This is the theoretical basis of Dines pressure tube anemometer or its modification.

Anemograph is an instrument which gives continuous recorded data of wind speed and direction. A density correction has been applied by W. Ferrel when the pressure tube anemometer located at heights which are sufficiently different from the place of standardization. This is given by

$$p = \frac{0.002698}{1+0.004t} V^2 \frac{P}{P_o}$$

where p = pressure in pounds per sq. ft

P_o = standard barometric pressure (760 mm of Hg)

P = Barometric pressure at observation station

t = temperature in ^oC, V = wind speed in feet/sec

At a temperature 15 ^oC, P = 760 mm of Hg, the above formula reduces to $p = 0.00255\ V^2$

Dines Pressure Tube Anemograph

This is a continuous recording wind speed instrument. In this device the pressure and suction are transmitted to a float chamber, which controls levels of floating cylinder in water. The float is attached to the indicator which marks on a graph chart wound on a cylinder driven by a clock device.

Wind velocity and pressure relationship at various exposed heights.

Height (meters)	Horizontal wind velocity (Kmph)	Horizontal pressure (Kg/m^2)
0 (surface)	80	40
3	96	58
6	108	73

***Squall*:** A squall is a sudden burst of strong wind that lasts for a few minutes. More specifically, a sudden increase of wind speed by at least three stages on Beaufort scale (16 Kt) and the speed reaching B.F.6 or more and lasting at least one minute.

For Example: The sudden increase of wind speed from 16 kmph (BF 3) to 44 kmph (BF 6) [or 9 kt to 24 kt] and remains at 44 kmph for at least one minute, is a squall.

Squall line: A violent squall associated with the passing of a long line or arch of dark cloud and accompanied by thunder and lightning, rain or hail and a sudden cooling with a shift in wind direction. A line squall although of short duration may blow off trees and houses.

***Gale*:** Wind speed of B.F. 8 or more, blowing continuously and doing damage to trees, houses etc. is called gale.

Gust: A gust is a sudden increase in wind speed relative to its mean value over a period of time which is shorter than squall duration.

Diurnal Variation of Surface Wind Speed

Clockwise change of direction of wind is called veering and anticlockwise changes of direction of wind is called backing. It is observed that wind changes during day and night both in direction and speed particularly in mountain or hill slopes, valleys, coastal areas. This change is diurnal (or daily) variation. Prominent local winds and their cause of change discussed separately. The relation between pressure gradient and wind speed discussed in Geostrophic wind. Over land, wind speed generally strengthens during midday and weakens towards evening.

The maximum wind speed is observed during afternoon (at about 1400 hrs local time) and minimum at dawn. During the day time ground surface is heated up by insolation as the day advances and causes convection in the wind layer closer to the ground.

At night the air close to the ground is cooled (on earth being cooled) and tend to remain or sink at lower levels. Thus day heating strengthen the wind speed while nocturnal cooling be calms it.

It will be discussed later that in northern hemisphere Coriolis force deflects wind to the right of its direction and in southern hemisphere it deflects towards its left. Due to surface friction surface wind rarely blows parallel to the isobars. In general it flows cross isobars from High to Low pressure area at an angle varying 10 degrees over sea to 30^{0} over land. Surface friction effect has been discussed in planetary boundary layer. In 1857 Buys Ballot, a Dutch meteorologist, gave a relationship between wind direction and isobars which is popularly called Buys Ballots law, which states as follows:

If an observer stands with his back to the wind the low pressure is on his left in the northern hemisphere, and on his right in the southern hemisphere.

Wind Veering: clockwise change of wind direction with time is called veering.

Wind Backing: Anticlockwise change of wind direction with time is called backing.

Cyclonic Circulation: In Northern Hemisphere (S.H) the anticlockwise (clockwise) spiralling in of wind (at any level) is called cyclonic circulation.

Anti Cyclonic Circulation: In Northern Hemisphere the clockwise spiralling out of wind (at any level) is called anti-cyclonic circulation.

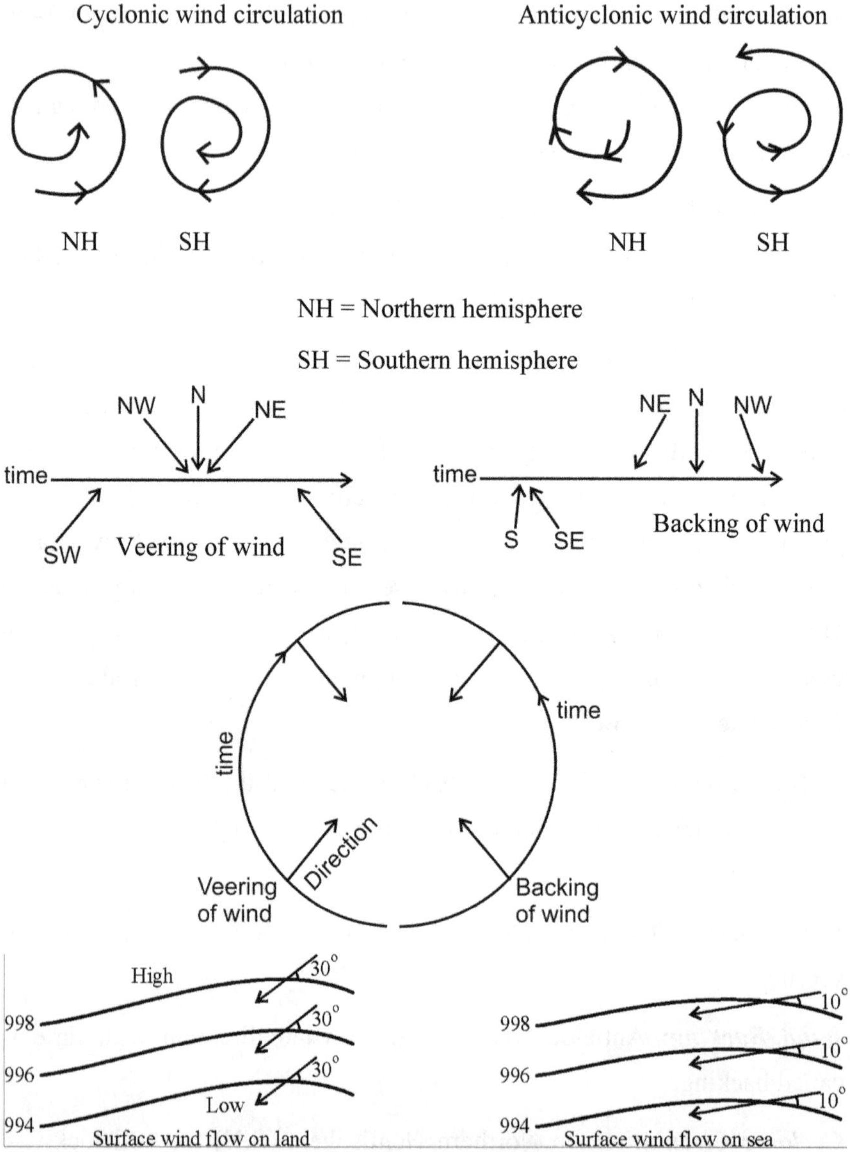

Fig. 20.1 Wind circulation.

Variation of surface wind speed: The lowest layer of atmosphere surrounding the earth extending up to one kilometer is called planetary boundary layer (PBL) or simply friction layer. The rate of flow of wind is greately affected by the roughness of the surface like trees, buildings, hillocks etc., creates turbulence. Wind speed in friction layer depends on lapse rate of temperature and roughness of the surface. In PBL wind speed increases from surface upwards. Over land wind speed generally strengthens during day till midday and weakens towards evening. The maximum wind speed observed during afternoon (about 1400 hrs local time) and minimum at dawn. During the day ground surface is heated up by insolation as the day advances and convection in the wind closer to the ground. At night the air close to the ground is cooled (as earth cools) and tend to remain or sunk to lower levels. Thus day heating strengthens the wind speed while nocturnal cooling calms (weakens the wind) it.

20.1 Winds below 2000 ft (agl)

At the surface there will be maximum friction effect and decreases with height agl (above ground level). For this purpose we consider 2000 ft height agl as non-frictional layer, where the frictional effect will be least. Diurnal variation of temperature also plays role. This change is considerable over land, but not over sea, where the variation of temperature is about 1 °C. As a result surface wind over sea are will be same during day and by night. The table 20.5 gives the surface winds for day and night over land and sea as compared to the wind at 2000 ft. In general winds will be backed and reduced in strength.

Table 20.5 Variation of wind direction and speed as compared to the wind at 2000 ft (agl)

	Day	**%**	**Night**	**%**
	Degrees backed	**Speed reduced**	**Degrees backed**	**Speed reduced**
Over land	30°	50%	40°	70%
Over sea	15°	30%	15°	30%

***Example*:** Let the wind at 2000 ft be 270°/30 kt. Then during the day over land it will be 240° /15 kt (direction backed by 30° and speed reduced by 50%) and at night compared to same wind at 2000 ft, it will be 230° / 9 kt (direction backed by 40° and speed reduced by 70%). Over sea, (as compared to the wind at 2000 ft 270°/30 kt), both during day and night it will be 255° 21 kt (direction backed by 15° and speed reduced by 30%)

20.2 Diurnal Variation of Wind within Friction Layer

For all practical purposes 1500 ft agl is considered as friction layer. Here we will consider variation of surface wind by day with surface wind by night and also the wind at 1500 ft by day and by night.

During day time insolation causes turbulence mixing effect more. Consequently during day the surface wind will be stronger than at night, while at 1500 ft the wind will be lower in speed as compared to night. The table 20.6 gives the horizontal comparison, day winds at the surface with night winds at the surface or day winds at 1500 ft (friction layer) with night winds at 1500 ft, (all winds are backed and slowed closer to the ground surface).

Table 20.6 Diurnal variation of winds in friction layer

	Day as compared to night	**Night as compared to day**
At Surface	Veered + faster	Backed + slowed
At 1500 ft	Backed + slowed	Veered + faster

Example

(i) Let during night surface wind be 270°/25 kt then during day it would be 290°/35 kt at surface

(ii) Let during day surface wind be 230°/25 kt then during night it would be (at surface) 200°/15 kt.

(iii) Let during night wind at 1500 ft be 260°/30 kt then during day it would be (at 1500 ft level) 230°/20 kt.

(iv) Let during day wind at 1500 ft be 220°/20 kt then during night it would be (at 1500 ft level) 250°/30 kt.

20.3 Local Winds

Local conditions such as earth's elevation, rocky areas, forests, deserts, hills, mountain slopes, valley, lakes, reservoirs, sea coasts, towns and cities etc., have great influence on surface wind. Diurnal variations of surface wind and temperature are decisively affected by these conditions. We shall consider a few important cases.

Land and Sea Breezes

(A) ***Sea Breeze*****:** A few hours after sunrise sea breeze develops on calm summer days. It has cooling effect in tropics and continues throughout the day (light hours) and dies down around sunset. After this seaward blowing land breeze develops which is confined to coastal area. The direction of sea breeze does not remain constant day long. It sets in with gustiness. Diurnal variation of wind in a coastal area is a striking example. Near the sea coast on shore wind generally sets in at about 9–10 AM and reaches peak at about 2 PM (1400 hrs local time) and then slowly weakens and dies away at about 6–7 PM. The strength of the sea breeze is stronger on warm days and weaker on cloudy occasions. Maximum sea breeze velocity varies (4–7 mps) 14 to 25 kmph. The theoretical aspects of sea breeze is considered in solenoids. The specific heat of sea water is 40% more than that of soils. During day time land gets heated up (forms local low) and sea remains comparatively cooler (forms local high). This tilts the isobaric surfaces downwards near the coast. Thus wind blows at surface level from sea to land, and land to sea in lower levels of atmosphere, forming wind circulation as shown in Fig. 20.2 (a). The depth of the sea breeze is about (150-1000 m) 1 km, but it thins at advancing edge. Generally sea breeze penetrates over land to about 50 km (by 2100 hrs). In tropical regions sea breeze reaches 50 to 60 km, sometimes even 125 to 145 km interior.

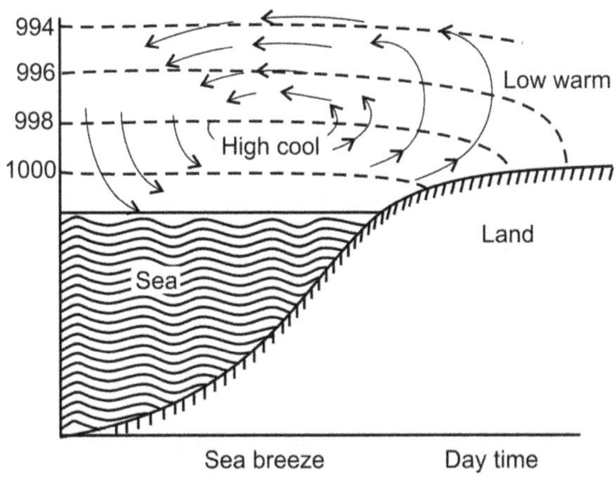

Fig. 20.2(a) Sea breeze.

A similar situation is observed (at smaller scale) near large lakes where we find lake breeze. Monsoon circulation is also caused by differential heating of land and sea, resembles to sea breeze but over a very large scale (This will be discussed in great deal subsequently). In this case the land area is a continent and sea area is ocean [The diurnal variation of temperature over land is about five times that of the diurnal variation of temperature over sea]

(B) ***Land Breeze*:** During night the reverse of sea breeze occurs. Coastal land area cools down (local high forms comparatively) while the sea area remains warm (local low forms). As a consequence wind blows from land to sea at surface level and from sea to land in the lower levels of the atmosphere as shown in the Fig. 20.2(b). In general land breeze is not so prominent as the sea breeze. However in tropics land breeze is also marked, and in some areas thunderstorms may occur off the coast at about dawn. In this case the isobaric surfaces lifted up near the coast and wind blows from land to over sea. The maximum speed of land breeze is about (2 mps) 7–8 kmph and depth is about ½ km and seaward range of land breeze is very small about 5–10 km only (in tropical regions extends up to 25–30 km).

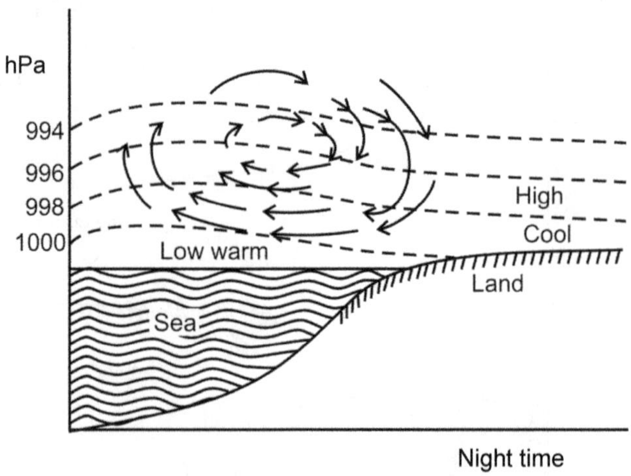

Fig. 20.2(b) Land breeze.

***Note*:** Land and sea breeze are prominent in tropical regions with great regularity where its occurrence is 80%, of days, elsewhere its occurrence is less than 50% of days and with lesser regularity. In polar regions land and sea breeze may be hardly seen during summer sphere. In tropical monsoon regions, during summer monsoon period, land and (150 m to 1000 m) sea breeze are masked and may be seen at a small scale and strength on less than 20% of the days of the season.

Katabatic Winds (Mountain Winds)

The phenomena of daily wind change along the axis of large valley is well known in mountainous regions. During clear nights, on mountain slopes wind flows down the slopes. At night mountain ridge slope cools down and the air close to it becomes heavier (denser) and moves down into (river) valleys. These down slope winds are called Katabatic winds (also called mountain winds) Fig. 20.3(a). It may be noted that as the air flows down the slope it warms up due to adiabatic warming, but the cold surface counteracts and keeps it cool (compared to the open air at that height). Thus the wind continues to flow down the slope. Katabatic winds are in general light but they acquire considerable speed if the slope is quite steep and is covered by snow or ice. Down slope wind when reaches the bottom (open valley) it replaces the warm wind and attains maximum speed (5-6 mps) at about sun rise (0700 hrs), when the diurnal cooling is maximum. Thickness is about 100–150 m. Example, Mistral wind belongs to this category.

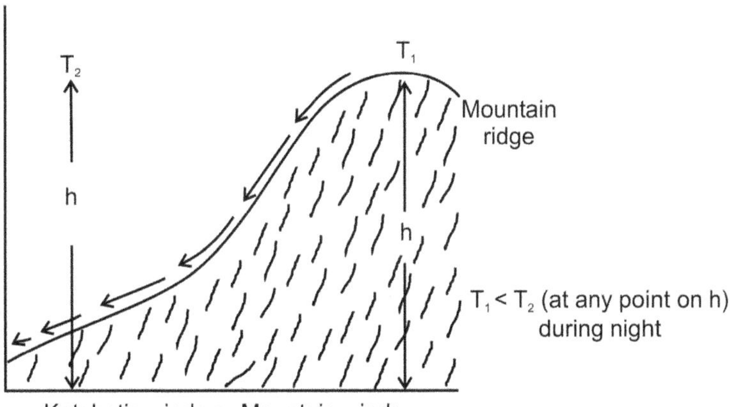

Fig. 20.3(a) Katabatic winds or mountain winds.

Anabatic Winds (Valley Winds)

On warm clear day upslope winds, a reverse process of katabatic winds occur on mountain slopes. These are called anabatic or valley winds. In Greek Katabatic means 'go down' and Anabatic means 'go up'.

On a clear warm day, the mountain slope (hill slope) gets heated up and the air close to it becomes warmer (lighter) than the free air at that level. As a result air close to the mountain slope, being lighter than the free air, ascends Fig. 20.3(b). As the wind moves upward adiabatic cooling takes place, but the cooling is countered by the warm air close to the slope. Thus, the air continues to move up slope of mountain during afternoon, at about 1400 hrs LMT, thickness is 100–200 m and attain maximum wind speed of 2–4 mps. In general anabatic winds are lighter than katabatic winds due to friction and gravity. Mountain and valley winds (Katabatic and Anabatic winds) are well developed in the wide and deep valleys of the Alps. They are conspicuous during summer fair weather when high pressure area develops there.

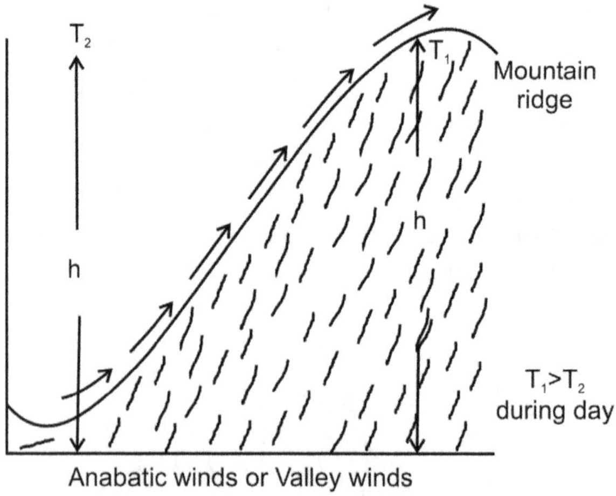

Fig. 20.3(b) Anabatic winds or valley winds.

Fohn Winds

When moist air is forced to rise a mountain barrier (due to pressure gradient), it cools adiabatically and attaining saturation it condenses into precipitation. After crossing the barrier on leeward side, the dry air (mostly) moves down the mountain slope and warms up adiabatically. At the foot of the mountain, on leeward side, it blows as pleasant warm wind. This phenomenon occurs on the southern side slopes of the (mountain) Alps. These are called Fohn winds. Fohn are well developed in Alps (Europe), Switzerland, Tirol. They are seen in Greenland (from the inland ice over the mountain ranges down to the coast). In north America it is known as the Chinook. In Argentina it is called Zonda and blows down the Andees. The clouds associated with Fohn are lens-shaped (lenticularis). Any such physical process which warms the wind on leeward side of the mountain is now called Fohn winds. Fohn also occurs on lee of the mountains of the Caucsaus and central Asia. At Tashkhant (in central Asia) in Russia, where average temperature in winter is about 0 ^{O}C, which rises to about 20 ^{O}C due to Fohn.

In Fig. 20.4 the rising wind up to B (A to B) moves dry adiabatically (DALR = 10 ^{O}C/km), from B to C moves at saturated adiabatic lapse rate (SALR = 6 ^{O}C/km) and from C to D moves again at DALR. In between B and C precipitation (pptn) occurs. If the ascending air is completely dry or ascending air does not precipitate to release latent heat (between B and C), then the descending air at D regains to the original temperature (T_1= 18 ^{O}C) without warning to T_3 = 22 ^{O}C.

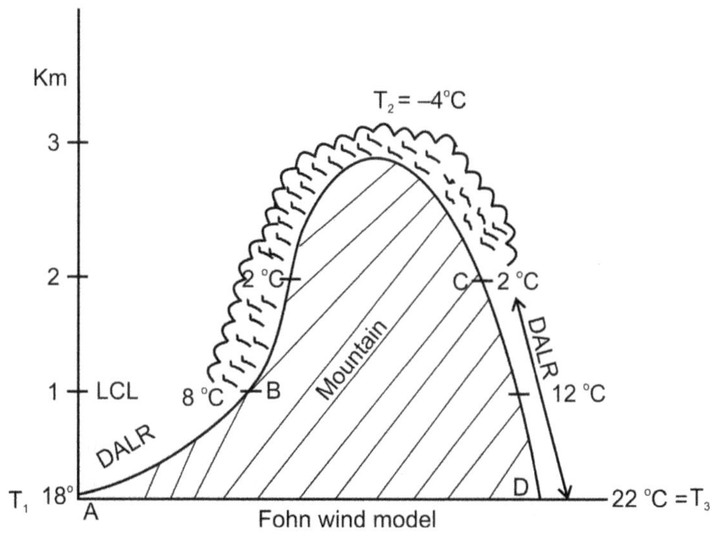

Fig. 20.4 Fohn wind model.

The Mistral

The Mistral is a Katabatic (or gravity) wind which is prominent along Mediterranean coast of France. In winter high pressure ridge frequently extends from the Azores to France. It drains from the higher lands and snow-capped mountains of the north. After draining down it is channelled by Rhone valley. It abruptly changes the climate of Riviera with coolness and dryness which otherwise is warm and pleasant during winter. Mistral may last for about a week. The annual frequency of Mistral speed exceeding 21 kt is about 105 days and 33 kt is 30 days.

Bora

Bora is another example of Katabatic wind which flows down from the plateau of Yugoslavia on to the northern coast of Adriatic sea and northern Aegean sea. In winter very cold continental wind from Russia, travelling over Dalmatia mountain arrives to the coast of Adriatic sea as cold wind and replaces the normally existing warm wind. This cold gusty wind causes damage. There are lulls between strong gusts of about 50 mps, called Reffoli, which are associated with pressure variation of about 4 mm of Hg. Bora does not extend to a long distance over the sea, but it causes a cloud of mist over the sea which is locally called Fumarea. The Bora of the Adriatic sea has marked diurnal variation of strength and frequency. It attains maximum intensity at about 0700 (LMT) and minima at about midnight (2400 hrs). It has maximum frequency of occurrence at about 0600 hrs in the morning and minimum frequency in the afternoon at about 1400 hrs.

Harmattan

In certain regions winds blow with some regularity and characteristics of the region in association with pressure systems. Harmattan is one such wind which is cool, dry and dusty. In winter it comes from Sahara and flows on to Guinea coast of north Africa and replaces humid tropical air. Due to its cooling effect locally it is called Doctor.

Scirocco

Hot, dusty and dry winds of continental tropical origin blows in north Africa in summer. These winds locally called by various names. In Algeria and the Levant they are called by Scirocco, in south-east Spain Leveche, in Egypt Khamsin. During summer depressions in this region move west to eastward and these local winds move ahead of the depressions in a northerly direction.

Chinook

Like Fohn, Chinook is a warm dry air which blows on the leeward side of Rocky mountains. These winds have the origin over Pacific ocean. While crossing over the Rockies they loose their moisture in precipitation over the mountain. In Canada Chinook's effect is found over a large distance.

It has been recorded a variation of about 22 ^{O}C temperature during the passage of Chinook in just five minutes. Winter maximum temperature records during the passing of Chinook. In local Red Indian language Chinook means snow eater. In winter months at Medicine Hat the lowest minimum temperature goes to about – 45 ^{o}C while the highest maximum is associated with Chinook goes to about + 20 ^{o}C. Similarly western suburbs of Calgary records about 0 ^{o}C where Chinook effects while the eastern suburb records about – 15 ^{o}C (where there is no Chinook effect).

***Note*:** It is impossible to give precise rules for variation of wind in friction layer. Winds that were discussed in 20.1, 20.2 are empirical, because wind variations depend on several meteorological factors particularly on local conditions.

20.4 Pressure and Wind

The state of the atmosphere is determined by a number of parameters. The atmospheric state is said to steady or static if it does not change with time or in equilibrium state.

However, in general, atmosphere is dynamic, that is in motion. The principal state variables are pressure (p), temperature (T), specific volume (or molar volume) and moisture (Relative humidity). The most of the other parameters of atmosphere are the complex combination of these and a few of them considered in thermodynamic variables. An ideal gas is one in which there are no force of molecular interaction. Atmospheric gases are treated as ideal gases for practical purposes and they obey gas laws, namely Boyle's law, Charle's law, Gay-lusac's law and Avogadro's law. In addition, the gases behave like fluids. We know fluids flow from higher level to a lower level. Similarly air flows from high-pressure region to low pressure region. This movement is the primary cause of wind. The fluid flow speed is function of shape or gradient, similarly the speed of movement of air is a function of pressure gradient or speed is proportional to the pressure gradient. Pressure gradient is change in pressure divided by horizontal distance and its direction of measurement is from High to Low.

An ideal fluid is a continuous fluid substance that cannot exert any shearing stress however small. Atmosphere is regarded as a continuous fluid medium. The physical quantities pressure, temperature, density, velocity, mass, composition are assumed to have unique values at any point in the atmosphere. These field variables and their derivatives are assumed to be continuous functions of space and time.

In the above discussion, we note that pressure gradient force acts at right angles to the isobars and that wind or parcel of wind moves from High to Low implying that wind also moves at right angles to the isobars.

This statement is true only when the pressure gradient is the only force when present. However, there are other forces whose presence and absence also matter, particularly rotation of the earth, friction and gravitational force. Depending on their presence we have different wind models (Geostropic wind, Gradient wind, Cycloslrophic wind etc), which will be discussed.

Effect of the Earth's Rotation: The effect of the earth's rotation on moving objects over the surface of the earth was explained by Gaspard Coriolis, hence it is called Coriolis force by meteorologists. An object in motion on the surface of the earth is deflected at every stage in its motion relative to the moving earth. This was demonstrated by Coriolis with an experiment of rotating turntable by drawing a line (see Fig. 20.5). The deflection caused by apparent force (a fictitious force, not real) is called Coriolis force or Geostrophic force. Coriolis force deflects the wind to the right of its motion in the northern hemisphere and to the left in the southern hemisphere.

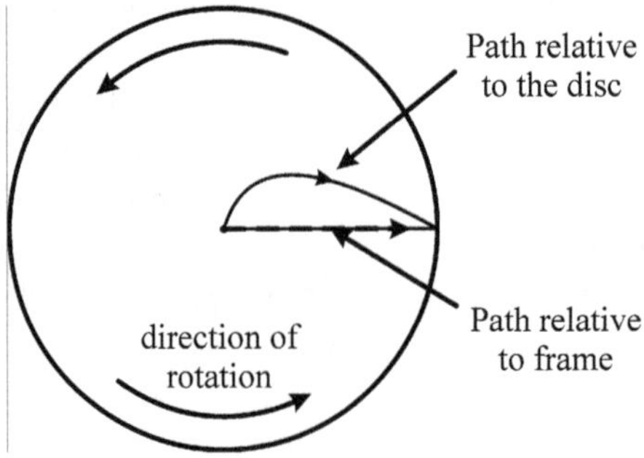

Fig. 20.5 Coriolis deflecting force operating on body moving outward from the centre of the rotating turntable.

The effect of Coriolis force is given by the formula

$$2\ \Omega\ \rho V \sin \phi$$

Where Ω = angular velocity of the earth

$= 7.29 \times 10^{-5}$ radians/sec

V = horizontal wind speed

ϕ = Latitude of the place

ρ = density of air

From the above formula it is obvious that there is no deflection at the equator ($\phi = 0$, $\sin \phi = 0$) but slowly increases and attains maximum value at the poles ($\phi = 90$, $\sin \phi = 1$).

Because of Coriolis force effect, near the surface of the earth the North east and South east Trade winds have developed.

Buys Ballot's law: The law states that if you stand with your back to the wind, the low pressure will be on your left in the northern hemisphere and to your right in the southern hemisphere.

We shall now consider some important wind models which lend support to numerical weather forecasting.

20.5 Geostrophic Wind

If we consider frictionless flow (the flow of air in the free atmosphere), and if we assume the flow is free from acceleration and constant and overlook gravity and friction, then the wind flow is termed geostrophic wind. The equation motion for geostrophic wind is given by

$$\alpha \, \nabla p = -2 \, \vec{\Omega} \times \vec{v_g}$$

where α = specific volume = $\frac{1}{\rho}$ (ρ = density of air)

∇p = Pressure gradient

$\vec{\Omega}$ = Earth's angular velocity

$\vec{V_g}$ = Geostrophic wind vector

Definition of Geostrophic Wind: The horizontal wind occurring in a frictionless flow, when the pressure gradient force is in balance with the coriolis force or geostrophic force is called geostrophic wind. The Fig. 20.6 indicates the geostrophic balance.

Geostrophic wind as a finite difference is given by

$$V_g = \frac{1}{\rho f}\frac{\Delta p}{\Delta n}, \text{ where } \rho f = 2\,\Omega\,\rho \sin\phi$$

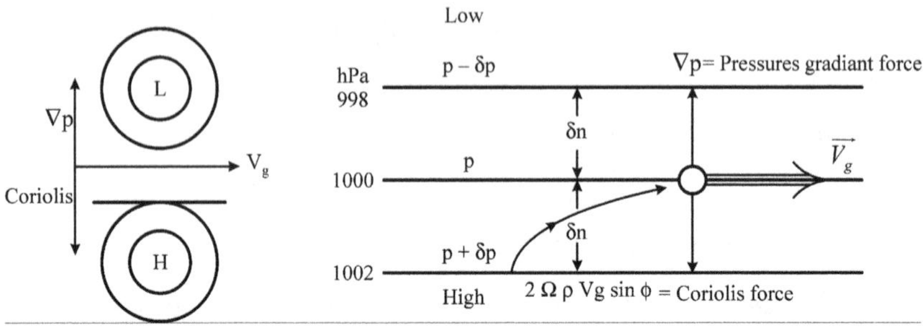

Fig 20.6 Geostrophic balance.

Properties of Geostrophic Wind

1. Pressure gradient force is normal or perpendicular to the isobars and directed from High to Low
2. Geostrophic wind obeys Buys Ballot's law
3. Geostrophic wind is proportional to the pressure gradient (closer the isobars stronger the wind or vice versa)
4. Coriolis force or geostrophic force is perpendicular to the isobars and is directed from Low to High and in balance with the pressure gradient force
5. Geostrophic approximation applies only when the isobars are straight and parallel and pressure distribution is steady not varying with time
6. Geostrophic approximation is not valid in case of local winds (like sea and land breeze), when isobars are curved.
7. Geostrophic approximation is a good estimate of actual wind above friction layer, where it is parallel to the isobars
8. Geostrophic approximation rarely occurs between latitudes 15°N & 15°S (In these latitudes Coriolis force is generally weak)
9. Geostreophic wind blows along the isobars with low pressure on its left in the northern hemisphere (which is in agreement with the Buys Ballot's law), on its right in southern hemisphere
10. Geostrophic speed is inversely proportional to the sin ϕ (sine of latitude). The geostrophic approximation fails at the equator. Thus the formula is applicable in extra tropical latitudes.

The table 20.7 gives the geostrophic wind speeds in various latitudes for the same pressure gradient

Table 20.7 Geostrophic wind speeds at various latitudes for the same pressure gradient ∇p.

Latitudes	15°	30°	45°	60°	75°	90°
Speed (KT)	65	33	24	20	17	16
	82	42	30	25	22	21

Note: For different pressure gradients wind speeds will be in the same ratio.

The Fig 20.7 shows the direction of winds about a Low and a High in northern hemisphere. Broken arrows indicate the direction of upper winds (frictionless flow) and solid arrows indicate direction of surface wind (frictional flow) (a) on land area (b) on sea area.

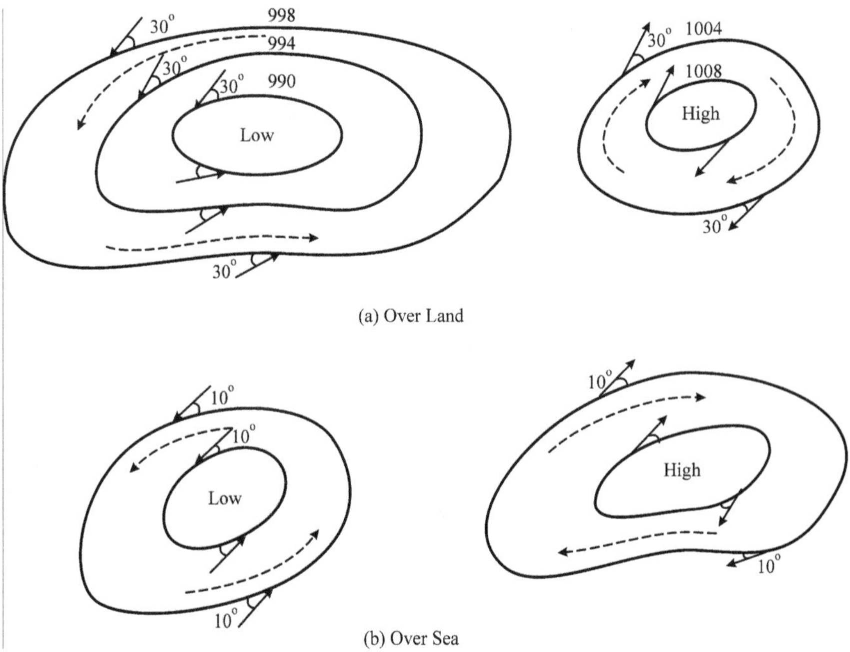

Fig. 20.7 Wind flow around Low and High.

20.6 Gradient Wind

Definitions

Streamline: A streamline is a curve drawn in a fluid such that at any instant of time, the direction of the tangent at any point of the curve coincides with the direction of the velocity of the fluid particle at that point.

Path line: The path line or trajectory is a curve (drawn in a fluid), which a particular fluid particle describes during its motion. It should be noted here that streamlines show how each particle is moving at a given instant, while the path line or trajectory shows how a given particle is moving at each instant.

Gradient flow: A frictionless, horizontal motion at a constant speed in a curve is called *gradient flow*

or

The wind occurring in a frictionless flow when there is a balance between pressure gradient force, coriolis force and centripetal force (or negative centrifugal force) is called Gradient wind.

[*Geostrophic wind is a particular case of Gradient wind in which the centripetal force is negligible*]

In this case the magnitude of pressure gradient force and the coriolis force are not equal. Because of this, the parcel of air moves horizontally at constant speed but not parallel to the straight isobars. It is curved to the left in case of flow around a Low in circular or curved path and curved right around a High in a circular or curved path [see in Fig. 20.8 (a), (b)]

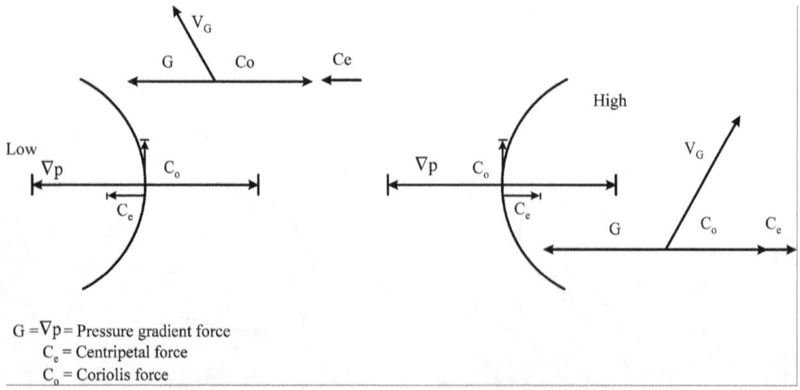

(a) Cyclonic motion (b) Anticyclonic motion

Fig. 20.8 Gradient wind (balance) in northern hemisphere.

$G = \nabla P$ = Pressure gradient force

Ce = Centripetal force

Co = Coriolisforce

Mathematically

$$\alpha \nabla p = \vec{\Omega} \times \vec{R} - 2\,\vec{\Omega} \times \vec{V_g}$$

where R is radius of curvature

(i) In case of cyclonic motion we have from Fig. 20.8 (a)

$$\nabla p = \rho V^2 / R + 2\Omega \rho\, V \sin \phi$$

Dividing by $2\Omega\ \rho \sin \phi$, we have

$$V = G/(2\Omega\rho \sin \phi) - V^2/(2R\Omega \sin \phi)$$

or $V = Vg - V^2/2R\ \Omega \sin \phi$ .....(1)

where

$G = \nabla p$

V = gradient wind speed or V_G

V_g = geostrophic wind speed

From the above equation it follows gradient wind speed (for the same pressure gradient) is subgeostrophic (smaller than geostrophic wind).

(ii) Incase of anticyclonic motion we have from Fig. 20.8 (b)

$$\nabla p = \rho\, V^2/R - 2\ \Omega\ \rho V \sin \phi$$

Dividing by $2\ \Omega\ \rho \sin \phi$, we have

$$V_G \text{ or } V = V_g + V_G^2/(2R\ \Omega \sin \phi) \qquad \text{..... (2)}$$

From the above equation, it follows that in anticyclonic flow (for the same pressure gradient) the gradient wind is supergeostrophic (greater than geostrophic wind).

Example: For radius of curvature R = 800 km and $\phi = 45^\circ$ and V = 40 KT, then in case of cyclonic motion gradient wind speed will be 34 KT sub-geostrophic and for anticyclonic motion gradient wind speed will be 58 KT super-geostrophic.

Example: For radius of curvature R = 1600 km, $\phi = 45^o$, V = 40 KT, then for cyclonic motion gradient wind speed will be 36 KT (sub-geostrophic) and for anticyclonic motion gradient wind speed will be 46 KT (super-geostrophic).

Note: At a single point the flow may be nearly gradient or else over an extended path. In case of a single point, the direction is tangential to the isobar, but for extended flow the parcel motion will be along the isobars (curved path) throughout the trafectory. This will be true only when isobars are not changing with time.

Example: In a cyclonic flow, find the difference between geostrophic wind (V_g) and gradient wind (V_G) at latitude 45 °N, when R = 1000 km, R = 100 km, assume V_g = 10 m/s.

Solution: $\Omega = 7.3 \times 10^{-5}$ / s, R = 1000 km

$$V_G \simeq 9.16 \text{ m/s}, \qquad V_g - V_G = 10 - 9.16 = 0.84 \text{ mps}$$

When R = 100 km, $V_G = 6.17$ mps

$$V_g - V_G = 3.83 \text{ mps}$$

Example: In an anti-cyclonic flow, find the difference between gradient wind (V_G) and geostrophic wind (V_g) at latitude 45°, V_g = 10 mps, and radius of curvature is R = 400 km, R = 1000 km

Solution: When R = 400 km, V_G = 20 mps

$$V_G - V_g = 20 - 10 = 10 \text{ mps}$$

When R = 1000 km, V_G = 11.27 mps

$\therefore$ $V_G - V_g = 1.27$ mps

Note:

1. $V_G < V_g$, cyclonic gradient wind is sub-geostrophic and VG $\rightarrow$ Vg when R $\rightarrow \infty$.

   [When V_g = 10 mps, R = 200 km at lat 45°, we get V_G = 7.3

   But when R $\rightarrow \infty$, $V_G = V_g$]

2. $V_G > V_g$, anticyclonic gradient wind is supergeostrophic and $V_G = 2\,V_g$ when R $\rightarrow \infty$.

Maximum anticyclonic gradient wind is twice the geostrophic wind.

Summery: For the same pressure gradient or same isobar spacing we have:

1. A Low pressure system has relatively lighter wind speeds than geostrophic wind or High (anticyclone)
2. A High pressure system has relatively strong wind speeds than the geostrophic wind or the low pressure system
3. Movement of any system (L or H) towards or near to the poles the wind speeds decrease.
4. Movements of any system (L or H) towards or near to the equator the wind speeds increase.

The above properties are shown in the Fig. 20.9

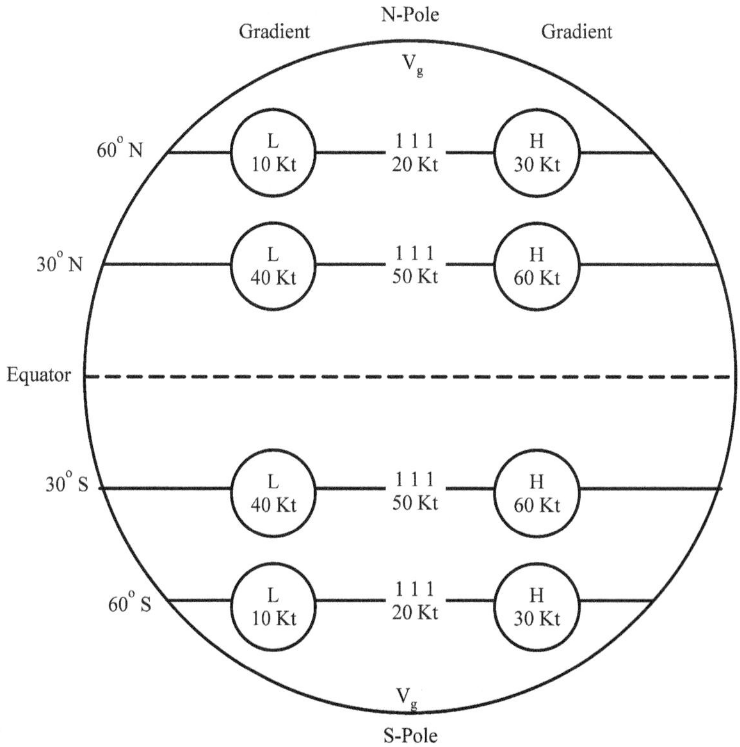

Fig. 20.9 For the same pressure gradients relationship between V_G (gradient wind) at various latitudes around Low and High and V_g geostrophic wind.

Fig. 20.9 shows that V_G (gradient wind) around High is faster than V_G around Low at the same latitude. V_g (geotropic wind) at High and at Low have the same isobar spacing or pressure gradient and they are between the two systems, consequently have same speed. For the same pressure gradient, winds are slow in Low and fast in High. Also winds are faster at lower latides as compared to higher latitudes.

Irrespective of the pressure system wind speed is directly proportional to pressure gradient (i.e closer the isobars stronger the wind)

20.7 Cyclostrophic Wind

***Definition*:** The wind occurring in a frictionless flow or in free atmosphere when the pressure gradient force is in balance with the Centripetal force, in comparison to these forces Coriolis force is negligible is called Cyclostrophic wind.

Suppose the wind flow in the free atmosphere is curved and that compared to the pressure gradient force and centripetal force (which is necessary for the air parcel to move in circular or curved path) the coriolis force is negligible. If there is a balance between pressure gradient force and centripetal force (which is negative of centrifugal force) which is called cyclostrophic wind.

Mathematically $\rho V^2/R = \nabla p$ or G

The above formula is valid incase of windflow around tropical cyclones or in a circular tornado, which give good approximation of actual wind around the tropical cyclone or tornado.

Cyclonic and Anticyclonic Flow

It was stated earlier that, in northern hemisphere the flow of wind in anti-clockwise direction is called cyclonic flow and the flow of wind in clockwise direction is called anticyclonic flow. In southern hemisphere the clockwise wind flow is called cyclonic flow and anti-clockwise wind flow is called anticyclonic flow.

Advection of Temperature

The change in temperature in horizontal adiabatic flow is regarded as geostrophic advection or simply advection.

Veering of Wind with Height

Clockwise change of wind direction in two levels L(lower level) and U (upper level) is called veering of wind with height.

Backing of wind with Height

Anticlockwise change of wind direction in two levels L and U is called backing of wind with height.

Veering of geostrophic wind with height indicates warm air advection.

Backing of geostrophic wind with height indicates cold advection.

20.8 Thermal Wind

Definition: Wind aloft at any level is considered to have two components, (i) Geostrophic wind above friction layer and (ii) a wind blowing along the mean isotherms with low temperature to its left in the northern hemisphere (and to its right in the southern hemisphere). The second component is called the thermal wind.

Consider two levels - L(lower) and U(Upper).

Let $\overrightarrow{V_{gL}}$ = Geostrophic wind vector in lower level

$\overrightarrow{V_{gU}}$ = Geostrophic wind vector in upper level

The vector difference $\vec{V}_T$ (thermal wind) is given by the relation

$$\vec{V}_T = \vec{V}_{gU} - \vec{V}_{gL}$$

$\vec{V}_T$ is called thermal wind in the layers L and U

$\vec{V}_T$ is parallel to the isotherms, see Fig.20.10.

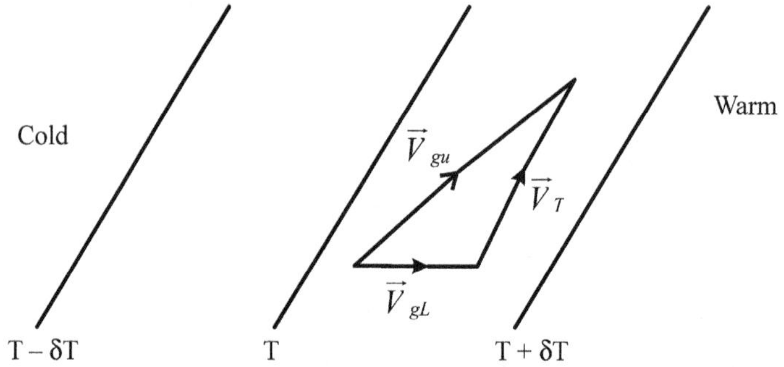

Fig. 20.10 Thermal wind.

Thermal wind is a fictitious wind and it has no real existence. However, the concept is of great importance in practical weather prognosis.

The empirical formula of thermal wind is given by

$$V_T = \frac{g}{f\bar{T}}\frac{\partial \bar{T}}{\partial n}\delta z,$$

where $\frac{\partial \bar{T}}{\partial n}$ = temperature gradient

$\bar{T}$ = mean temperature of the layer

δz = thickness between the layers

Thermal wind is caused by a temperature difference between two adjacent air masses.

Consider two air mass columns A and B which are adjacent. Suppose at the surface both have the same pressure P_0 and A is warmer than B. Let the surface temperature of the column of A be T_1 and that of B be T_0. $T_1 > T_0$. Since A is warmer than B, the number of molecules at the surface of these columns m_0, m_1 where $m_1 > m_0$. This is true by virtue of A is warmer than B. Now consider the situation at a height h agl. The number of molecules in column A at h be mh_1 and that B at the same height be mh_0. Since both columns exert the same pressure at the surface we have $mh_1 > mh_0$, while $m_1 > m_0$ at the surface. This implies that in areas where temperature high (warm) the pressure in the upper air also tends to be high and where the temperature is low (cold) the upper air pressure also tends to be low. This shows wind at the surface being calm (pressure being same) but at high levels tends orient itself from high to low with lower temperature on the left in the northern hemisphere. In the above assumption there is no pressure gradient between column A and column B at the surface and hence wind is calm while the geostrophic wind at an upper level is determined by the horizontal distribution of mean temperature. Thus we can conclude that the wind vector difference between two levels is related to horizontal distribution of mean temperature of the intervening layer of air. Thus thermal wind $\left(\vec{V}_T\right)$ is defined in layer as the wind which should vectorially be added to the geostrophic wind ($\vec{V}_g$) at the lower level to get geostrophic wind at the upper level $\left(\vec{V}_g\right)$.

i.e., $\vec{V}_T + \vec{V}_{gL} = \vec{V}_{gU}$

or $\vec{V}_T = \vec{V}_{gu} - \vec{V}_{gl}$

The direction of thermal wind ($\vec{V}_T$) is parallel to the mean isotherm with lower temperature on the left in northern hemisphere and on the right in the southern hemisphere, which is akin to the Buys Ballots law. The speed of the thermal wind is proportional to the temperature gradient or, a modified Buys Ballot's law may be stated as: if you stand with your back to the thermal wind, low temperature will be on your left in the northern hemisphere (and on your right in the southern hemisphere).

If we consider high altitude jetstreams as thermal winds, then we may state that "If you stand with your back to a westerly jet, cold air will be to your left (north) in the northern hemisphere and to your right (south) in the southern hemisphere" .

Contour Charts or Thickness Charts

The thickness between two constant pressure levels such as 1000 hPa and 500 hPa or 200 hPa and 500 hPa are prepared for a range of temperatures. The thickness is shown in geopotential meters (or geopotential decameters). At any point the thickness (difference in height) depends only on the mean temperature of the intervening air column and isopleths of thickness coincide with the mean isotherm temperature. The contour lines are lines joining points of equal height or thickness (also called isohypses). Lower heights indicate cold pools (being denser). The upper wind strength can be calculated by the spacing of contours, which shows the rate of change of temperature. Using thermal wind in two layer concept the upper wind will be inferred. At present this has become obsolete in tropics.

Questions

1. In metrological aviation reports wind direction is given in ________ while ATC and ATIS reports it is given in ________
   (a) Degree (TN), Degree (M)
   (b) Degree (M), degree (M)
   (c) Degree (TN), degree (TN)
   (d) Celsius degrees, fahrentiat
2. 1 Nautical mile (1KT) is equal to:
   (a) 6080 ft
   (b) Average length of 1 minute latitude
   (c) Length of one minute arc along great circle
   (d) All above
3. Squall is defined as sudden increase of wind speed by ________ stages on Beaufort scale and the speed reaching or exceeding Beaufort scale __________ and lasting at least_________
   (a) 5 stages; BF 6; few seconds
   (b) 3 stages; BF 6; one minute
   (c) 6 stages; BF3; few minutes
   (d) 2 stages; BF8; one second
4. Katabatic winds are mountain slope__________during _________
   (a) Up slope winds; warms clear day.
   (b) Down slope winds; clear nights
   (c) Down slope winds; clear day
   (d) Upslope winds; clear night
5. Anabatic winds are mountain slope_____________ during _______
   (a) Upslope winds; warm clear day
   (b) Upslope winds; clear night
   (c) Down slope winds; clear day
   (d) Down slope winds; clear night

6. When moist air forced to rise a mountain barrier it cools adiabatically and precipitates. After crossing the barrier on lee ward side_______ is called Fohn wind.

   (a) Moves down, and cools adiabatically. This cool wind

   (b) Moves down and warms up adiabatically. This warm wind.

   (c) Moves up and forms renticular clouds which are

   (d) Moves up and dissipates in rotors, which

7. Thermal wind is caused by temperature difference between two adjacent air masses. The wind vector difference between two levels is related to horizontal distribution of mean temperature.

   If $\vec{V}_T$ = Thermal wind vector

   $\vec{V}_{gl}$ = Geostrophic wind vector at lower level

   $\vec{V}_{gu}$ = Geostrophic wind vector at upper level

   Then thermal wind vector is given by the relation:

   (a) $\vec{V}_T = \vec{V}_{gl} + \vec{V}_{gu}$ (b) $\vec{V}_T + \vec{V}_{gl} = \vec{V}_{gu}$

   (c) $\vec{V}_T + \vec{V}_{gu} = \vec{V}_{gl}$ (d) $\vec{V}_T - \vec{V}_{gL} = \vec{V}_{gu}$

8. In northern hemisphere, the direction of thermal wind is parallel to the mean isotherm with:

   (a) Low temperature to the right

   (b) Low temperature to the left

   (c) Same temperature on both sides

   (d) None above

9. A modified Buys Ballot's law for thermal wind may be stated as: if you stand with your back to the thermal wind, in northern hemisphere,............

   (a) Low temperature will be to your right

   (b) Low temperature will be to your left

   (c) Same temperature on your both sides

   (d) None above

10. If we consider high altitude jetstreams as thermal winds, then it may be stated that, if you stand with your back to the westerly jetstream, in northern hemisphere...............
    (a) Cold air will be to your left (north)
    (b) Cold air will be to your right (south)
    (c) Isothermal temperature on your both sides
    (d) None above
11. Heading south in the southern hemisphere pilot experience port drift when:
    (a) Flying towards high pressure
    (b) Flying towards low pressure
    (c) Flying towards low temperature
    (d) Flying towards warm temperature
12. Contours on weather charts indicate:
    (a) Heights of pressure levels
    (b) Contour height of ground
    (c) Thickness between two pressure levels
    (d) Distance between pressure levels
13. At (FL 180) 18000 ft in the northern hemisphere, with a wind from the left and heading of 360°, the temperature:
    (a) decreasing from sourth to north
    (b) decreasing from north to south
    (c) cant say without pressure
    (d) None above
14. If an Isohypse on constant pressure chart of 500 hPa shows 580, this shows:
    (a) Topography of 580 m amsl
    (b) Topography of 580 decameters amsl
    (c) Pressure of 580 hPa
    (d) A high pressure surface

15. Flying towards a depression/cyclone in northern hemisphere at FL 15, pilot experiences:
    (a) Left drift
    (b) Right drift
    (c) light tail wind and starboard drift
    (d) light head wind and port drift
16. Flying towards a High (anticyclone) in northern hemisphere at 1500 ft (FL 15), pilot experiences.
    (a) Wind from left
    (b) Wind from right
    (c) Wind from left and behind
    (d) Wind from right and on the nose (head wind)
17. In the northern hemisphere with a jet stream south to north, cold air would be from:
    (a) East and below (b) West and below
    (c) East and above (d) South and above
18. The geostrophic wind around a Low is stronger than the gradient wind because:
    (a) Centripetal force adds to the gradient force
    (b) Centripetal force opposes the gradient force
    (c) Coriolis force adds to the gradient force
    (d) Coriolis force opposes the gradient force
19. Every physical process of weather is the result of:
    (a) Air movement
    (b) Pressure variation
    (c) Heat exchange/heating of the earth
    (d) None above
20. Wind is the result of:
    (a) Temperature (b) Advection
    (c) Pressure difference (d) Moisture difference
21. The wind at 2000 ft (agl) above ground level in the free atmosphere is westerly, then wind at the surface would be:
    (a) Northwesterly (b) Southwesterly
    (c) Westerly (d) None above

22. In Northern hemisphere, the wind in its motion is deflected:
    (a) towards left due to coriolis force
    (b) towards right due to coriolis force
    (c) towards right due to frictional force
    (d) None above
23. In northern hemisphere wind blows around a Low:
    (a) In a clockwise direction
    (b) In an anti-clockwise direction
    (c) Wind remains calm
    (d) None above
24. In Southern hemisphere wind blows around a High:
    (a) In a clockwise direction
    (b) In an anticlockwise direction
    (c) Wind remains calm
    (d) None above
25. For a given pressure gradient, geostrophic wind speed is:
    (a) Maximum at the poles and minimum in equatorial region
    (b) Maximum in equatorial region and minimum at poles
    (c) Same at all latitudes
    (d) None above
26. In southern hemisphere wind flows around low .... and around high..........
    (a) Anticlockwise; clockwise
    (b) Clockwise; anticlockwise
    (c) In both places parallel to the isobars
    (d) In both places perpendicular to the isobars
27. In northern hemisphere, near the surface friction causes wind to change:
    (a) By 30° to the isobars towards low
    (b) By 30° to the isobars towards high
    (c) By 10–15° to the isobars towards low
    (d) By 10–15° to the isobars towards high

28. In southern hemisphere near (over) the sea surface, friction causes the wind to change:
    (a) By 30° to the isobars towards low
    (b) By 30° to the isobars towards high
    (c) By 10-15° to the isobars towards low
    (d) By 10-15° to the isobars towards high
29. In northern hemisphere the direction of wind flow around a closed isobar low (1) at surface and (2) above (2000 ft) friction levels:
    (a) Clock wise into low by 15°; and parallel to the isobars
    (b) Anti clock wise into low by 30°; and parallel to the isobars
    (c) Anti clock wise into high by 15°; and parallel to the isobars
    (d) Clockwise into high by 30°; and parallel to the isobars
30. Over a span of time clockwise change of wind direction is called:
    (a) Veering (b) Backing
    (c) Stalling (d) Gusting
31. Over a span of time anticlockwise change of wind direction is called:
    (a) Veering (b) Backing
    (c) Stalling (d) Gusting
32. Duration of (i) gusts and lulls …… and (ii) squalls ….:
    (a) Minutes; seconds (b) Seconds; minutes
    (c) Hours; minutes (d) Minutes; hours
33. Surface wind is measured (i) above surface at a height of.... agl…(ii) by (instrument)…….
    (a) 10 ft; barometer (b) 10 m; anemometer
    (c) 6 ft; thermometer (d) 30 ft; altimeter
34. Gale or fresh gale wind speed is:
    (a) 8 or more Beaufort scale, (b) 34 kt or more
    (c) 30 mps or more (d) Both (a) and (b)
35. The cause of wind is:
    (a) Temperature differential (variation)
    (b) Pressure differential (variation)
    (c) Air masses difference
    (d) Rotation of the earth

36. In a land breeze, surface wind blows:
    (a) From the sea by night
    (b) Form the sea by day
    (c) From the land by night
    (d) From the land by day
37. In a sea breeze, surface wind blows:
    (a) From the sea by night
    (b) From the sea by day
    (c) From the land by night
    (d) From the land by day

Answers

1.	(a)	2.	(d)	3.	(b)	4.	(b)
5.	(a)	6.	(b)	7.	(b)	8.	(b)
9.	(b)	10.	(a)	11.	(b)	12.	(c)
13.	(a)	14.	(b)	15.	(c)	16.	(d)
17.	(b)	18.	(a)	19.	(c)	20.	(c)
21.	(b)	22.	(b)	23.	(b)	24.	(b)
25.	(b)	26.	(b)	27.	(a)	28.	(c)
29.	(b)	30.	(a)	31.	(b)	32.	(b)
33.	(b)	34.	(d)	35.	(b)	36.	(c)
37.	(b)						

CHAPTER 21

Divergence, Vortical Motion and Vorticity

Streamlines are lines of instantaneous air motion, which may be straight or curved. When streamline converge (come together) it is called confluence, when they go apart it is called difluence or divergence.

In case of convergence of horizontal flow, mass convergence (accumulation) takes place, while in divergence of horizontal flow there will be mass divergence (or dispersal) takes place. Convergence is associated with vertical motion and divergence is associated with subsidence (see Fig. 21.1).

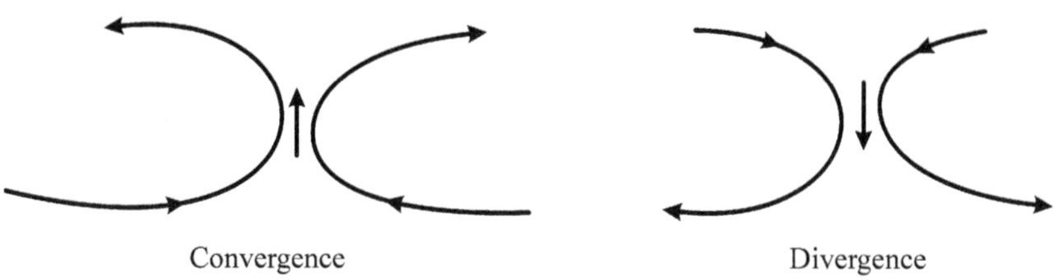

Fig. 21.1 Convergence and Divergences at the surface.

At a surface level, in an area of convergence loss of mass and fall of atmospheric pressure are associated, while in an area of surface divergence gain of mass and increase in pressure are associated.

The velocity of rotation (angular velocity) of an infinitesimal fluid element (or air parcel) is called vorticity.

Analysis of weather charts (surface and upper air) enables the meteorologist to understand the weather at the time of observation and to infer the future weather.

The three terms divergence, vertical motion and vorticity are required to understand to unravel the weather on synoptic scale. Let us study briefly how to infer divergence, vertical motion and vorticity from the synoptic charts.

21.1 Fields of Divergence

The wind field is developed by the variation of wind speed along the streamlines (as said earlier) coming together (confluence) or going apart (difluence) of streamlines. Based on the wind speed the divergence of wind is inferred as below.

1. When the wind speed increases downstream in a field of parallel streamlines divergence occurs (see Fig. 21.2 (a)).

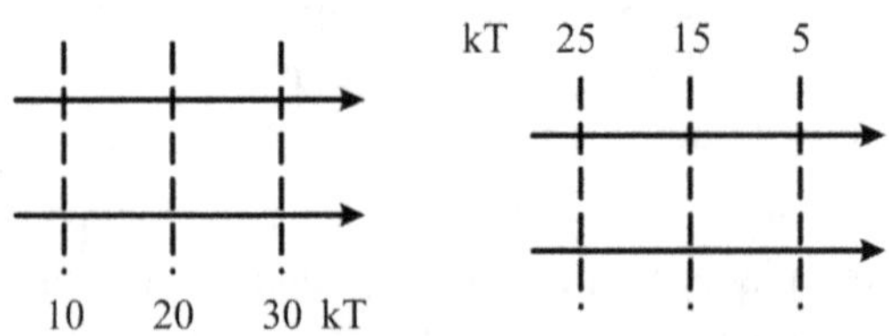

(a) Divergence (b) Convergence

Unbroken line: Stream line .... isotach line is doted line or equal wind speed line

Fig. 21.2 (a)

2. When the wind speed decreases downstream in a field of parallel streamlines convergence occurs (see Fig. 21.2 (b)).
3. Fanning out of streamlines is called difluence and converging (coming together) of streamlines is called confluence (see Fig. 21.3).

Difluence Confluence

Fig. 21.3

4. If streamlines fanout in an area of uniform wind speed along streamlines divergence occurs (see Fig. 21.4).

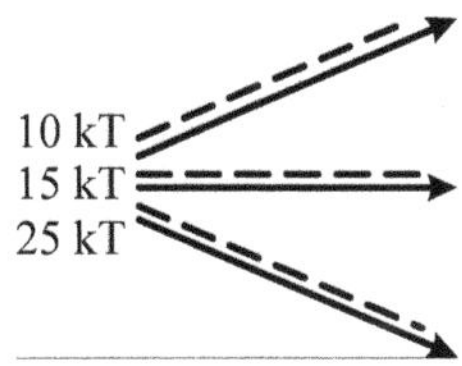

Divergence

Fig. 21.4

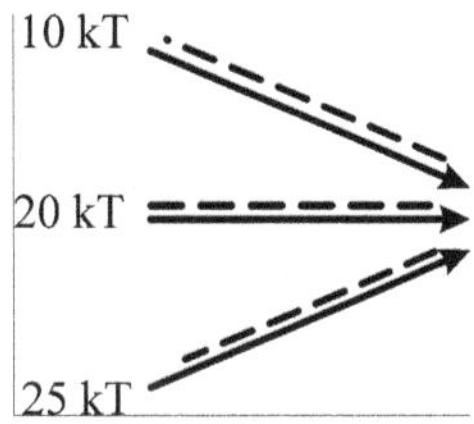

Convergence

Fig. 21.5

5. If streamlines come together in an area of uniform wind speed along streamlines convergence occurs (see Fig. 21.5).

6. If wind speed increases downstream and simultaneously streamlines fanout indicates strong divergence (see Fig. 21.6).

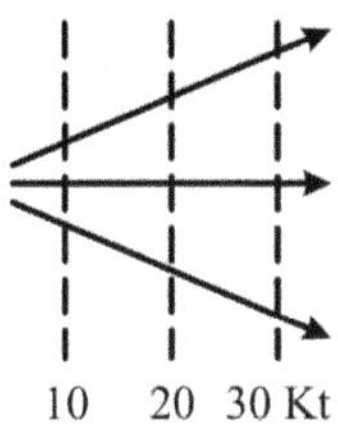

Strong divergence

Fig. 21.6

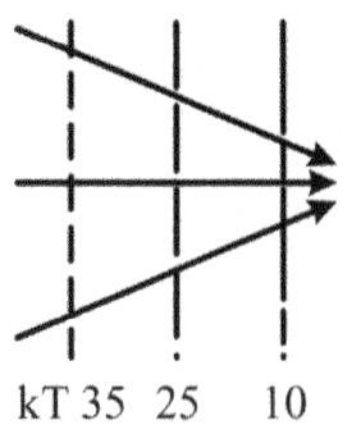

Strong convergence

Fig. 21.7

7. If wind speed decreases downstream and simultaneously streamlines converge (come together) indicates strong convergence (See Fig. 21.7).

8. If wind speed decreases downstream while streamlines fan out OR if wind speed increases downstream while streamlines

converge (come together) in such area convergence or divergence cannot made out. Such situations arise in middle and high latitudes (see Fig. 21.8).

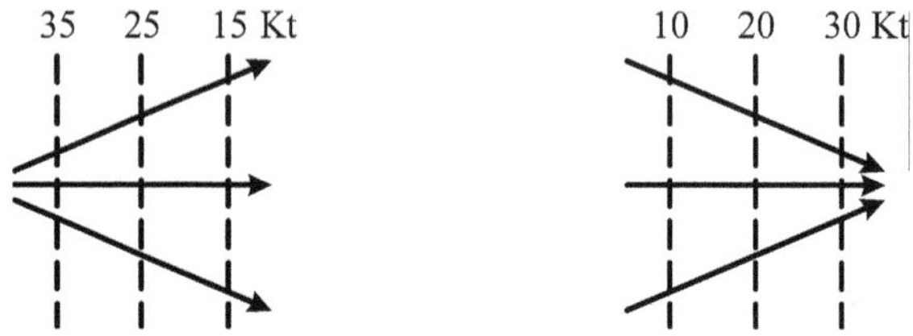

Fig. 21.8 Neither convergence nor divergence can be made out.

9. Areas of divergence indicate generation of mechanical kinetic energy, while convergence indicate dissipation of kinetic energy.

10. Horizontal convergence (or divergence) of wind near the surface of the earth causes vertical motion ascent (or descent) of air.

11. In case of low level convergence and simultaneous upper level divergence over the same place vertical motion results. Such situations are very conducive for triggering convective activity (in the presence of low level moisture), precipitation and other adverse weather phenomena. Contrary to this, low level divergence and with simultaneous upper level convergence in the same area leads to subsidence, generally fair-weather or smog/fog (that is poor visibity) may occur and inversion of temperature near the surface of the earth. A level at which neither convergence nor divergence effectively observed is called a level of non-divergence (LND). Below LND either convergence (or divergence) begins and above LND either divergence (or convergence) begins (see Fig. 21.9).

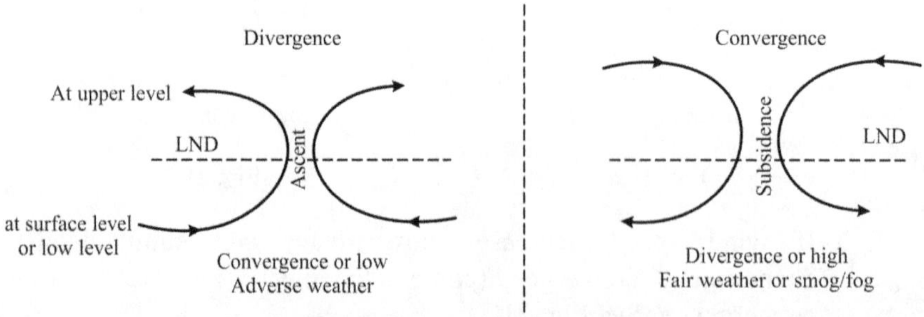

Fig. 21.9 Level of Non-Divergence.

Starboard (right drift) and port (left drift).

21.2 Flight in the Fields of low level Convergence (Low) and Divergence (High)

Flight in friction layer, below 2000 ft/FL FL020 or into low or low level convergence will give an aircraft slight tail wind with drift towards right in the NH (northern hemisphere). Right drift is also called starboard drift (see Fig. 21.10).

Similarly, in NH, flight into High or low level divergence in friction layer (below 2000 ft/FL 020) an aircraft will experience slight head-wind with left drift, also called Port drift (see Fig. 21.11).

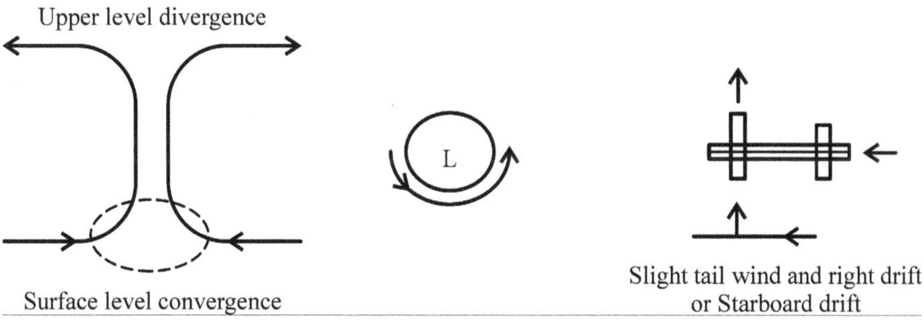

Fig. 21.10 In NH, starband drift or starboard drift or right drift.

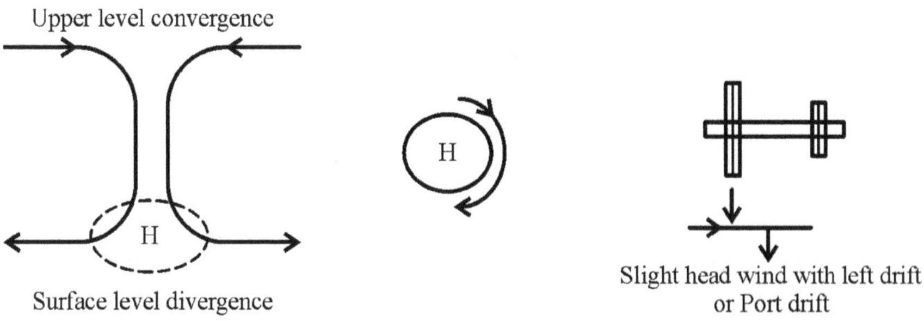

Fig. 21.11 In NH, Port drift or left drift.

21.3 Fields of Vorticity

The vertical component of vorticity (ζ) is given by the formula

$$\zeta = \frac{V}{R_s} - \frac{\partial V}{\partial n}$$

It has two components

(i) curvature $\left(\frac{V}{R_s}\right)$ and

(ii) another shear $\frac{\partial V}{\partial n}$

where V = velocity

R_s = radius curvature

∂n = thickness of layer

There are six combinations of these two components which are illustrated for the northern hemisphere.

1. If the flow is of uniform speed or isotachs (equal wind speed lines; indicated in figure by broken lines) are normal (perpendicular) to the sreamlines then curvature determines the sign of vorticity (positive for cyclonic and negative for anticyclonic) (see Fig. 21.12).

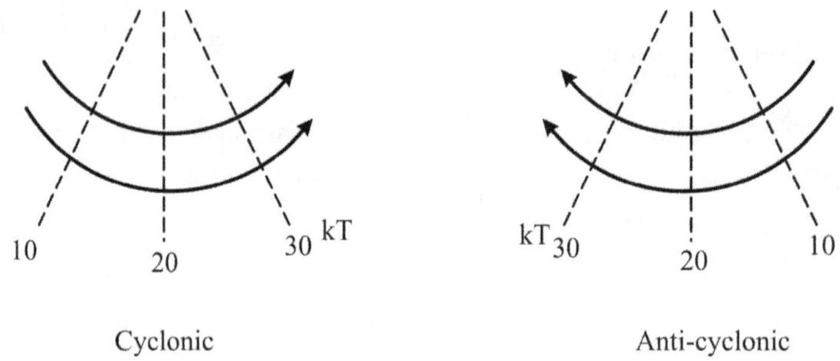

Fig. 21.12

2. In a straight flow, looking down stream, decreasing of wind speed to the left of the current indicate cyconic vorticity (see Fig. 21.13).

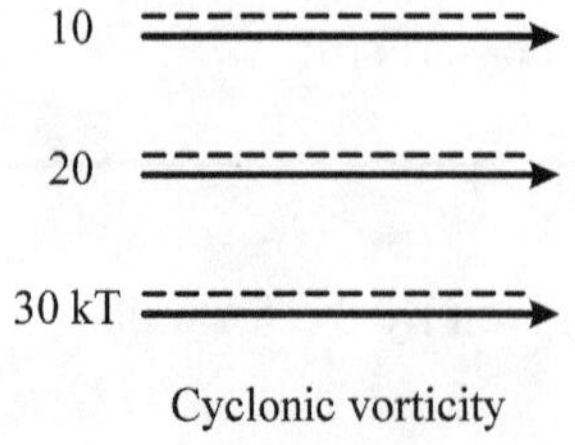

Fig. 21.13

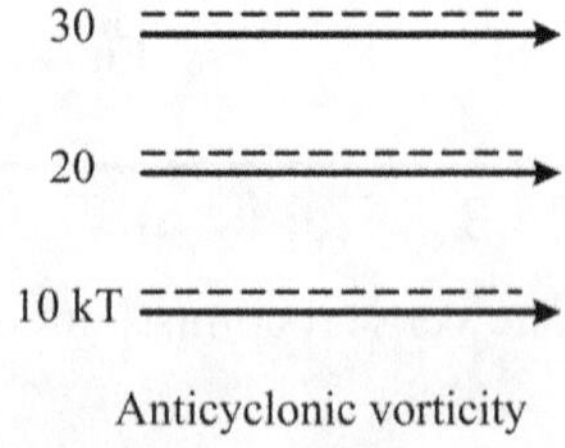

Fig. 21.14

3. In a straight flow, looking down steam, increasing of wind speed to the left of the current indicate anticyclonic vorticity (see Fig. 21.14).

4. In streamlines of cyclonic curvature, decreasing of wind speed to the left of the current (looking downstream) indicate strong cyclonic vorticity (see Fig. 21.15).

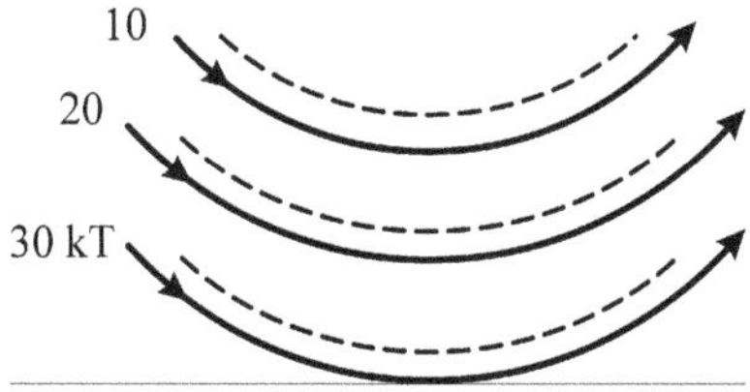

In N.H strong cyclonic vorticity

Fig. 21.15

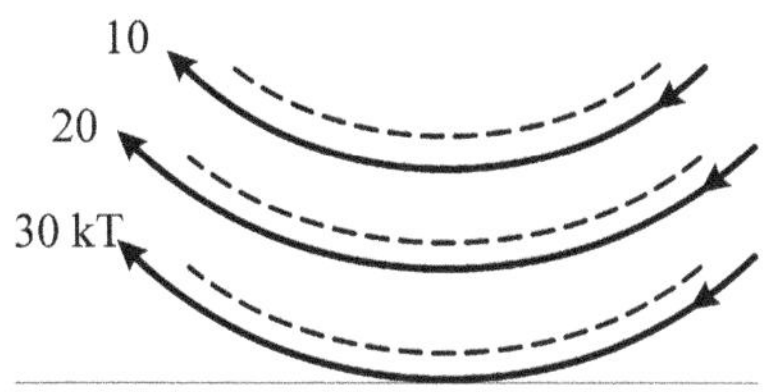

In N.H Strong Anticyclonic vorticity

Fig. 21.16

5. In streamlines of anticyclonic curvature, increasing of wind speed to the left of the current (looking downstream) indicate strong anticyclonic vorticity (see Fig. 21.16).

6. If the curvature of the streamlines is cyclonic, while the wind speed increases to the left of the current (looking downstream) (see Fig. 21.17) or if the curvature of the streamlines is anticyclonic while the wind speed decreases to the left of the current (looking down steam, see Fig. 21.18), in these cases it is difficult to determine the vorticity (indeterminate). In these cases relative vorticity cannot be determined by the inspection.

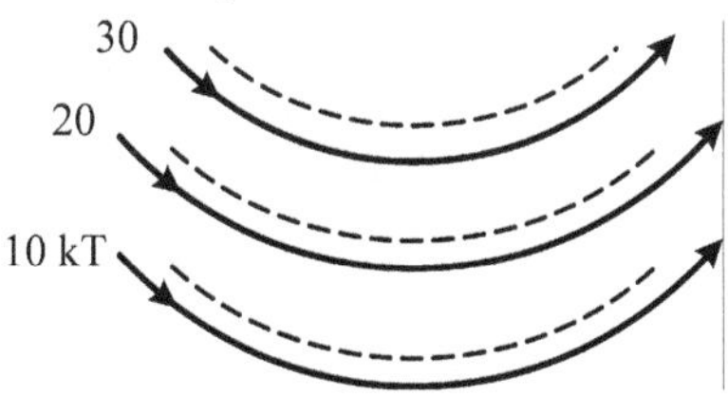

Fig. 21.17

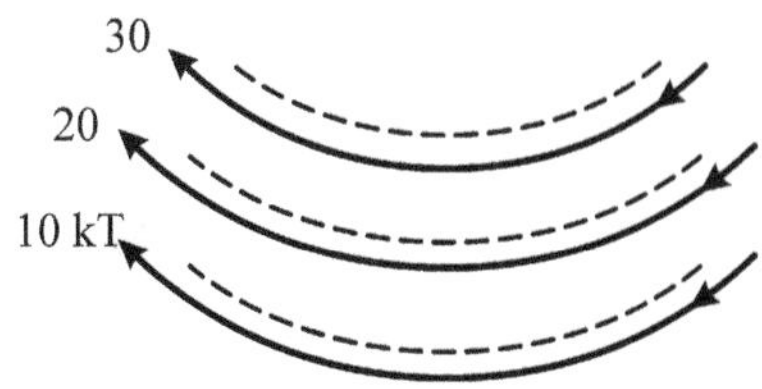

Fig. 21.18

Indeterminate vorticity

The vorticity of a current that crosses the equator with uniform curvature, changes the vorticity sign, but it is an obstacle in understanding the equatorial flow.

In jetstream zones, a velocity maxima is present. Inspection shows signs of vorticity in some parts of the map only. The following streamline, isotach analysis presents two such models.

In Fig 21.19, the velocity maxima is located in the trough. North of the axis of the jet core, wind has cyclonic vorticity (NH). However south of the jet core it is indeterminate because the curvature is cyclonic while the windshear is anticylonic.

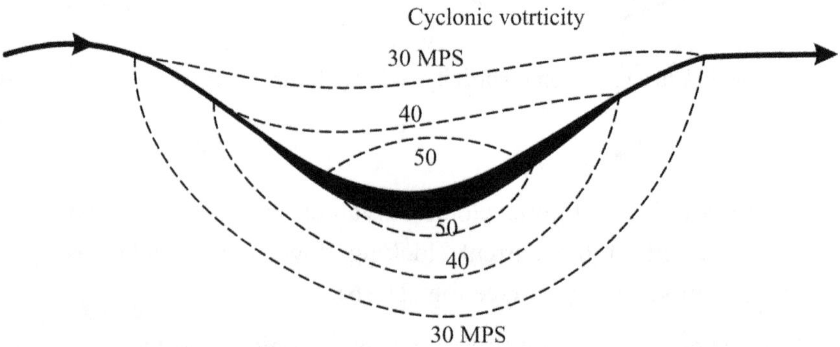

Fig. 21.19

In Fig. 21.20, the velocity maxima is centred in the ridge. To the north of the axis of jet core, wind has anticyclonic vorticity (in NH). However south of the jet core it is indeterminate, since the curvature is anticyclonic while the wind shear is cyclonic.

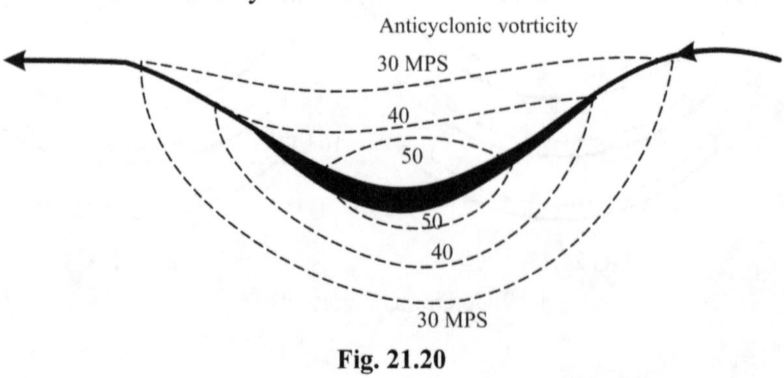

Fig. 21.20

Air within a depression may be regarded as having infinite number of small air parcels, each rotating cyclonically about an axis perpendicular to earths surface. Vorticity is related to the air motion about a cyclone or anticyclone and also related to the location of the pressure system on the rotating earth.

21.4 Convergence and Divergence due to Surface Friction in Northern Hemisphere

Surface friction reduces the airflow near the surface to lower than the geostrophic value and it causes the airflow across the isobars (which is called ageostrophic) towards low. When we consider the total flow around a closed isobar Low (low pressure with close isobars), the wind flow observed to converge towards the centre of low (see Fig, 21.21).

Similarly the total flow around a closed isobar High, winds appear to diverge from the centre of High with cross isobar wind flow (ageostrophic) (see Fig 21.21). This type of convergence of wind which is confined to friction layer may cause at best some clouding and light rain because in this case vertical motion of wind is poor.

Fig 21.21 In NH convergence and Divergence in surface friction.

21.5 Vertical Motion

Though the scale of vertical motion of air is much smaller as compared to horizontal motion (of wind), yet it plays very important role in cloud development and weather phenomena. Vertical motion of air in atmosphere begins from local irregular gusts and lulls (which last for a few seconds) to widespread slow movements (in low pressure systems) which last for days.

Vertical motions may be classified into two categories:

1. Localized vertical motion (extending over a few square kilometers).
2. Large scale vertical motion (extending over hundreds of square kilometers).

The causes of loclised vertical motions are:

(a) Ground frictional effects (like trees, buildings, boulders etc).

(b) Ground contour heights (like hills, mountain ranges)

(c) Thermal heating of ground.

Large scale vertical motions are caused by:

(a) Pressure systems (like Depressions, Cyclone, anticyclones).

(b) Frontal systems (like warm front, cold front).

Friction Eddies: Surface friction depth varies from 500 ft (160 m) to 1500–3000 ft (450 m-1 km). If we carefully observe the recording of Dines pressuretube anemograph or electrical anemograph, the direction and wind speed records show varying width caused by oscillations in quick successions without any regularity. However from these records it is definitely possible to have an idea of mean position of wind (both direction and speed) at any intant of time. The width of the trace gives the degrees of gustiness. The gustiness factor is defined as:

$$\text{Gustiness factor} = \frac{\text{Range of fluctuation in gusts and lulls}}{\text{mean wind speed}}$$

Gustiness factor is generally expressed as percentage

Example: Let the mean wind speed 25 KT, gusting up to 35 KT and lulls down to 10 KT. Then the range is 35 – 10 = 25 KT

$$\text{Gustiness factor} = \frac{25}{25} \times 100 = 100\%$$

There is another measure of gustiness, called gust ratio, which is defined as:

$$\text{Gust ratio} = \frac{\text{Maximum gust}}{\text{mean wind speed}}$$

In the above example, gust ratio $= \frac{35}{25} = 1.4$

Generally gust ratio varies between 1.3 over open sea, 1.6 over flat open country and 2.1 near centre of large city. In case of storm, the gust ratio may vary 2.5 to 2.9. Anemograph records show that the surface wind is very rarely steady. The roughness of the ground is responsible for gustiness. The roughness does not allow smooth laminar flow of air near the ground. Near the ground surface air stream breaks up into small irregular cells, which are called frictional eddies. The frictional eddies travel with the wind and have small vertical motion, because of which the horizontal wind speed shows fluctuations (gusts and lulls) on either side of the mean value across the stream. An aircraft flight in friction layer encounters these frictional eddies, which results the flight bumpy. This type of frictional effect is called mechanical turbulence. The mechanical turbulence as said earlier extends up to 1 km agl. The degree of mechanical turbulence and the height to which it persist depends on environmental temperature lapse rate. If ELR is close to DALR it favours strengthening of frictional eddies but an inversion tends to dampen the eddies.

Mechanical / frictional turbulence is important in fog formation and its lifting to form stratus clouds.

***Vertical motion induced by ground contours*:** This was discussed earlier in chapter 11, in methods of cloud formation. However a summary of it is given below.

When an obstacle is placed in the laminar flow depending on its size, shape and orientation the wind flow is affected in three ways and all these involve vertical motion

(i) the airflow (air stream) is deflected around as well as over the obstacle

(ii) Smooth laminar flow is disturbed which generates turbulence (if it is already tubulent flow, turbulence will be intensified)

(iii) On the leeward side of the obstacle (after descending) the airstream forms wavy motion for some distance before laminar flow is restored.

In case of an isolated hill (which makes some angle to the direction of flow, other than parallel), the flow is partly around (sidewise) and partly over the top of the hill. If the air is stable (or if there is an inversion), vertical motion is dampened and air tends to flow around the hill (and there will be or no surmount of the top). If the air is unstable or with steep lapse rate the air stream will surmount the hill top.

In case of continuous line of hills stretching across the wind, then the air stream passes over the hill tops. In the absence of inversion but lapse rate is stable, then airflow is affected up to about four times the height of the hills above surrounding terrain, after which it will tend to be laminar.

***Thermal Eddies*:** This was also discussed in chapter 11, under convection clouds, a brief revision is made here.

During day, ground is heated by insolation. The heating of the ground is uneven because of absorbing power of different soils, colour, vegetation cover etc. Black soil absorbs more heat than white sandy soil. Air close to the hotter ground becomes warmer than the environment and the flow breaks off the environment and forms thermal eddies which drifts with the prevailing wind. Depending on the lapse rate, thermal eddies rise to great heights. In case of steep lapse rate, thermal eddies form a shallow layer or column which is called thermal turbulence. Thermal turbulence or column has no well defined boundary. In the presence of moisture thermal turbulence generates convective clouds also called hot towers or columns which is known as convection. Thermal eddies are much larger in dimensions as compared to frictional eddies. The aircraft flight in updrafts and downdrafts of convective clouds is very hazardous, which we have already studied.

Glider pilots make use of these thermals for gaining height.

Vertical motion in convergence and divergence we have discussed in the beginning.

Vertical Motion in a Frontal Zone

When two airstreams of markedly different temperatures meet, they do not mix up immediately. A sloping boundary surface forms in which warmer lighter airstream ascends, while the relatively colder, denser air stream under cuts the warmer airstream. The speed of vertical motion depends on the speed of airstreams and the slope of the boundary. These boundaries are called fronts in middle latitudes and they are called discontinuities in tropics.

Questions

1. In case of convergence of horizontal flow, mass convergence and (i) …… takes place, while in divergence of horizontal flow mass (ii) ….. and (iii) …… takes place

	(i)	(ii)	(iii)
(a)	Vertical Motion;	Depletion or Dispersal;	Subsidence
(b)	Subsidence;	Mass Accumulation;	Vertical Motion
(c)	Vertical Motion;	Mass Accumulation;	Subsidence
(d)	Subsidence;	Depletion;	Vertical motion

2. When wind speed increases downstream in a field of parallel streamlines, it results in:
   (a) Convergence (b) Divergence
   (c) Cyclonic vorticity (d) Anticyclonic vorticity
3. If wind speed decreases downstream and simultaneously streamlines come together this results in:
   (a) Convergence (b) Divergence
   (c) Strong convergence (d) Strong divergence
4. If wind speed decreases downstream while streamlines fan out, the result is:
   (a) Convergence
   (b) Divergence
   (c) Neither convergence nor divergence can be made out
   (d) Vorticity
5. Areas of divergence indicate generation of ….. while convergence indicate……..
   (a) Mechanical kinetic energy; no loss of energy
   (b) Mechanical kinetic energy; dissipation of kinetic energy
   (c) Potential energy; dissipation of kinetic energy
   (d) Potential energy; dissipation of kinetic energy
6. Horizontal convergence of wind near the earth's surface causes:
   (a) Generation of kinetic energy
   (b) Vertical motion
   (c) Subsidence motion
   (d) Dissipation of potential energy

7. In a straight flow, looking downstream, decreasing of wind speed to the left of the current indicate:
   (a) Convergence (b) Divergence
   (c) Cyclonic vorticity (d) Anticyclonic vorticity

8. In a straight flow, looking downstream, increasing wind speed to the left of the current indicate:
   (a) Convergence (b) Divergence
   (c) Cyclonic vorticity (d) Anticyclonic vorticity

9. In northern hemisphere, in streamlines of cyclonic curvature, decreasing wind speed to the left of the current looking downstream indicate:
   (a) Convergence
   (b) Divergence
   (c) Strong cyclonic vorticity
   (d) Strong anticyclonic vorticity

10. In northern hemisphere, if the curvature of the streamlines is anticyclonic while wind speed decreasing to the left of the current (looking downstream) indicate:
    (a) Cyclonic vorticity,
    (b) Anticyclonic vorticity
    (c) Relative vorticity cannot be determined
    (d) None above

11. In the following figure, the velocity maxima is located in the trough of the jet core.

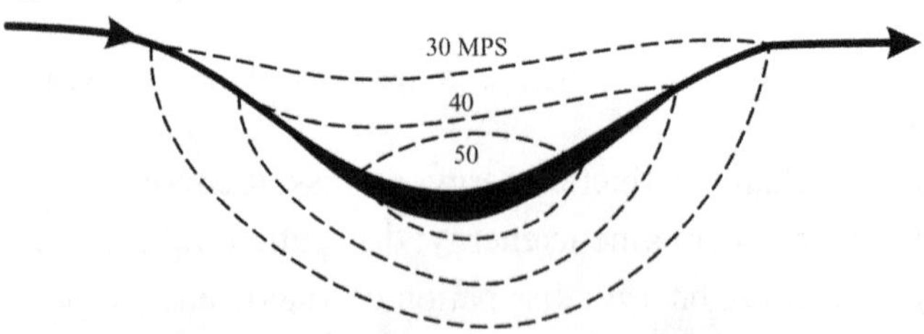

North of jet axis has .... vorticity and south of the jet axis has ..... vorticity
    (a) Anticyclonic, cyclonic
    (b) Cyclonic, indeterminate
    (c) Anticyclonic, indeterminate
    (d) Cyclonic, anticyclonic

12. Surface friction reduces the air flow near the surface to …… and it causes air flow ………. the isobars.
    (a) More than the geostrophic value, parallel to
    (b) Less than the geostrophic value, across
    (c) Less than the geostrophic value, parallel to
    (d) More than the geostrophic value, across

13. Given the mean wind speed 25 KT, gusting upto 40 KT and lulls down to 10 KT. The gustiness factor is:
    (a) 100% (b) 75% (c) 120% (d) 90%

14. Given the mean wind speed 25 KT, gusting upto 40 KT and lulls down to 10 KT. The gust ratio is:
    (a) 1.2 (b) 1.6 (c) 1.0 (d) 0.75

15. Depending on the lapse rate thermal eddies rise to great heights. Incase of steep lapse rate these eddies form a column or shallow layer which is called:
    (a) Turbulence (b) Thermal turbulence
    (c) Mechanical turbulence (d) Frictional eddies

16. Glider pilots make use of ……..for gaining height
    (a) Propellers (b) Downdrafts
    (c) Thermals (d) Updrafts

Answers

1.	(a)	2.	(b)	3.	(c)	4.	(c)
5.	(b)	6.	(b)	7.	(c)	8.	(d)
9.	(c)	10.	(c)	11.	(b)	12.	(b)
13.	(c)	14.	(b)	15.	(b)	16.	(c)

CHAPTER 22

Wind Shear, Clear Air-Turbulence and Jet Streams

22.1 Wind Shear

A common definition of wind shear is a change in wind direction and / or speed in space. Wind shear has profound effects on aircrafts in landing, takeoff and cruising phases. It is very hazardous to overlook wind shear phenomena. It is certain that every aircraft and every pilot will be affected by wind shear. Light wind shears a daily phenomena in flying needs little action. Moderate wind shears requires positive action from the pilot, while a severe wind shear that occurs rarely makes the aircraft out of control. A number of accidents and looses caused by severe wind shears that forced the ICAO to introduce wind shear warnings.

***Definition*: wind shear:** The rate of change of wind (vector) in space is called wind shear.

Thus wind shear arises from changes in wind speed or wind direction or both.

For aviation purposes wind shear and turbulence defined as below.

Wind Shear: Variations in the wind along the aircraft flight path of a pattern, intensity and duration that displace an aircraft abruptly from its intended path such that substantial control input and action is required to correct it.

Turbulence: Variations in the wind along the aircraft flight path of a pattern, intensity and duration that disturb the aircraft's attitude about its major axis but do not significantly alter its flight path.

Low Altitude Wind Shear: Wind shear along the final approach path or along runway and along take-off and initial climb-out flight paths.

Vertical Wind Shear: Change in wind velocity with height. Mathematically,

$$\text{Vertical wind shear} = \frac{\overrightarrow{V_2} - \overrightarrow{V_1}}{\overrightarrow{Z_2} - \overrightarrow{Z_1}}$$

where $\overrightarrow{V_1}$, $\overrightarrow{V_2}$ are wind velocities at heights Z_1 , Z_2 respectively, $Z_2 > Z_1$

Horizontal Wind Shear: Change in wind velocity in horizontal direction. Mathematically

$$\text{Horizontal wind shear} = \frac{\overrightarrow{V_2} - \overrightarrow{V_1}}{\overrightarrow{d_2} - \overrightarrow{d_1}}$$

where $\overrightarrow{V_1}$, $\overrightarrow{V_2}$ are wind velocities at distances d_1, d_2 respectively from some origin.

Updraught or downdraught Shears: Change in the vertical component of wind vector with horizontal distance. Mathematically

$$\text{Updraught or downdraught shear} = \frac{\overrightarrow{W_2} - \overrightarrow{W_1}}{\overrightarrow{Z_2} - \overrightarrow{Z_1}}$$

where $\overrightarrow{W_1}$ = Vertical component of wind vector at Z_1

$\overrightarrow{W_2}$ = vertical component of wind vector at Z_2

Wind shears are associated with the following phenomena:

(i) Thunderstorms, microbursts, Tornodoes / waterspouts (funnel cloud) and gust fronts.

(ii) Frontal surfaces (Temperature difference between air masses 5 °C or more across the front or movement of front 30 KT or more)

(iii) Strong surface wind coupled with local topography (that is, in turbulent planetary boundary level).

(iv) Sea breeze fronts

(v) Mountain waves

(vi) Low level temperature inversions

(vii) Low level jet.

22.1.1 Wind Shear Warnings

Wind shear warnings are prepared by aerodrome meteorological offices. It contains brief information of the observed or expected wind shear, which could adversely affect aircraft on approach path or take-off path or during circling approach between runway level and 500 m (1600 ft) above that level and aircraft on the runway during landing roll or take-off run.

Existence of wind shear can be inferred from:

(a) ground based wind shear detection remote sensing equipment (Ex: Terminal Doppler weather Radars),

(b) ground based wind shear detection equipment (Ex: A system of surface wind and /or pressure sensors located in an arry monitoring a specific runway and associated approach and departure paths),

(c) aircraft observations during climb-out or approach phases of flight,

(d) other meteorological information (Ex: appropriate sensors located on existing masts or towers in vicinity or the aerodrome or nearby areas of high ground)

Note: Significant wind shear likely:

(a) when there is temperature difference of 5 °c or more between the air masses across a front or wind discontinuity in tropics

(b) when active front is moving at a speed 30 KT or more.

Pilots in flight can infer the pressure of wind shear by the changes in:

(i) vertical speed,

(ii) pitch attitude,

(iii) deviation from glide slope,

(iv) aircraft attitude,

(v) abrupt or large drift and

(vi) situations that require to change power settings

Examples of Wind Shear Warnings

1. " WS WRNG MBST APCH RWY 27"
2. " WS WRNG B747 REPORTED MOD WS IN APCH RWY 09 AT 1510"
3. " WS LDG RWY 24 L"
4. " WS WRNG 25 KT 3 NM FINAL"- meaning expect encounter wind shear between 3 nm and touch down with a maximum intensity of 25 kt.

22.1.2 Effects of Wind Shear on Take-off and Landing Performance

Wind shear may cause airspeed changes more than 15 K T or vertical speed changes 500 ft / min or more.

Turbulence associated with shear is a major safety problem during take-off or landing approach phase of flight. High altitude wind shears in frontal zone, mountain induced waves, CAT have considerable effect on passenger safety.

Examples

A. Wind Shear Report

1. SURFACE WIND 320 / 10 KT WIND AT 60 M 360 / 25 KT IN APCH

   or

   SURFACE WIND 320 / 20 KMH WIND AT 60M 360 / 50 KMH IN APCH
2. "MOD TURB AND ICE INC IN CLIMB-OUT"

   "REFZRA", "WS RWY 12" or 'WS ALL RWY"

B. Wind Shear Warning

1. "WS WRNG SURFACE WIND 320/10 KT WIND AT 60M 360/25 KT IN APCH"

   or

   "WS WRNG SURFACE WIND 320/20 KMH WIND AT 60M 360/50 KMH IN APCH".
2. "WS WRNG 30 KT AIRSPEED LOSS 2 NM FINAL RWY 09"

or

"WS WRNG 60 KMH AIRSPEED LOSS 4 KM FINAL RWY 09"

A Head wind reduces the gliding distance over the ground, while a Tail wind increases the glide distance over the ground, however wind does not effect the rate of descent. This means a head wind reduces the ground run and landing distance.

Taking off into head wind implies climb into an increasing head wind (due to decreasing frictional effect) and increase the angle of climb-out (or gradient of climb-out). Flying into up draft will increase the rate of climb and also increase the angle of climb relative to the ground. Thus climb performance will increase when flying into an increasing head wind, a decreasing tail wind or into updraft. This will cause an aircraft a tendency to overshoot or go above the original flight path and hence termed overshoot effect.

Taking off into tail wind or down wind implies climb into increasing tail wind, which lowers the angle of climb-out. Flying into a downdraft will decrease the rate of climb and also decrease the angle of climb. Thus the climb performance will decrease when taking off downwind and flying into increasing tailwind or a decreasing head wind or into downdraft. This will cause a tendency to undershoot or go below the original flight path and hence called undershoot effect.

Turbulence associated with wind shear is a major safety problem during take-off or landing/approach phase of flight. High altitude wind shear in frontal zone, mountain induced waves, clear Air turbulence have considerable effect on passenger safety.

Summarily we have:

1. Descending on an airfield where wind speeds are decreasing downstream, an aircraft may undershoot.
2. Descending on an airfield where wind speeds are increasing downstream an aircraft may overshoot.
3. Take-off (or climb) an aircraft over an airfield where wind speeds are decreasing with increasing height, the angle of climb decreases with height.
4. Take-off (or climb) of an aircraft over an air field where wind speeds are increasing with increasing height, the angle of climb increases with height.

22.1.3 Wind Shear Alerts and Warnings

The enhanced ground proximity warning system (EGPWS) of Honeywell has wind shear detection system, It compares the air speed information from the air data computer (ADC) with ground speed information from navigation computer to find the changes in wind affecting the aircraft and likely wind shear turbulence.

If an increasing headwind (or decreasing tailwind) and / or a severe updraft exceeding a defined threshold observed, the warning computer gives wind shear cautions (Alerts). These conditions are characteristics of an encounter with a microburst. The computer alert (caution) light appears and words "caution wind shear" or "wind shear ahead" are spoken.

If a decreasing headwind (or increasing tailwind) and / or a severe downdraft exceeding a defined threshold observed, the warning computer gives warning. These conditions are characteristics found within a microburst or just afterwards. The computer warning Red lights appear and the words "Wind shear, Wind shear, Wind shear" are spoken.

The Alert and Warning thresholds depend on available climb performance, flight path angle, air speed changes and fluctuations in static air temperature. Wind shear Warning takes priority over all others, however wind shear Alerts are low priority.

22.2 Clear Air Turbulance (CAT)

Turbulence in the atmosphere is caused in four ways:

1. Mechanical or frictional turbulence
2. Thermal turbulence
3. Orographic ascent or mountain waves
4. Clear air turbulence

We have studied the first three processes in methods of formation of clouds.

The air flow near the surface of ground is affected by friction, which is associated with the roughness of ground surface, tall buildings, trees, big rocks, hillocks, steep coast etc. Thermal turbulence caused by (ground heating by isolation) temperature or thermal currents. When strong wind blows roughly perpendicular to a orographic barrier (within 30°) (like the Western Ghats in monsoon) it is forced to move up and creates turbulence on wind ward side and mountain standing waves on leeward side. Clear air turbulence is associated with stable atmosphere and shears in wind (both vertical and horizontal) and temperature. CAT is a sub-synoptic micro-scale feature associated with Kelvin-Helmholtz instability waves which are caused by vertical wind shear in a statically stable layer. These waves amplify break and tumble over into chaotic motion like that of a breaking ocean waves.

CAT is predominantly a patchy phenomena. Its intensity is less than the turbulence associated with thunderstorms. The typical dimensions of CAT are: 30 km by 5 to 15 km across between heights 4 to 5 km. Thickness about 300 m to 500 m, life period 30 minutes to about 3 hours. Individual turbulence or patches generally occur in close proximity to one another both vertically and horizontally. High level turbulence is associated with strong horizontal gusts and the value of Ri (Richardson number) less than one (about 0.65). The trump frequency is maximum at higher levels as compared to lower levels.

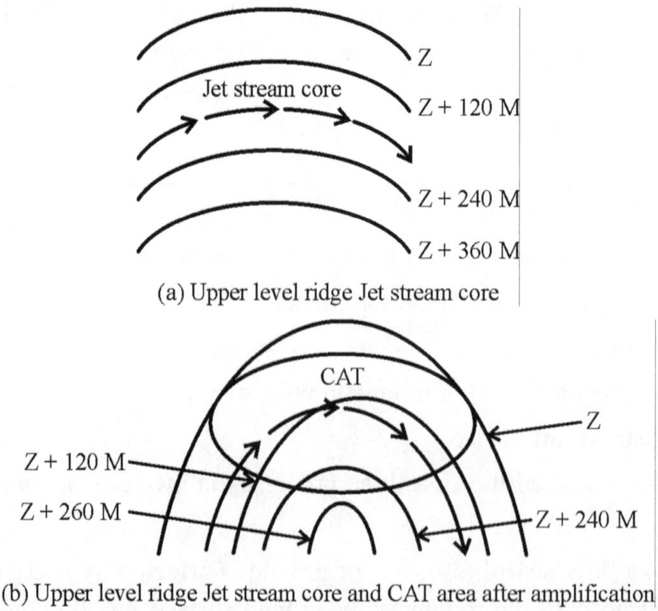

Fig. 22.1 CAT area associated with an amplifying upper level ridge.

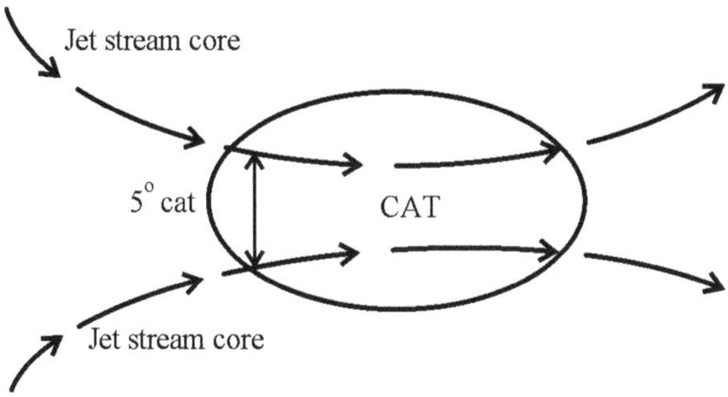

Fig. 22.2 CAT associated with confluent jet stream cores.

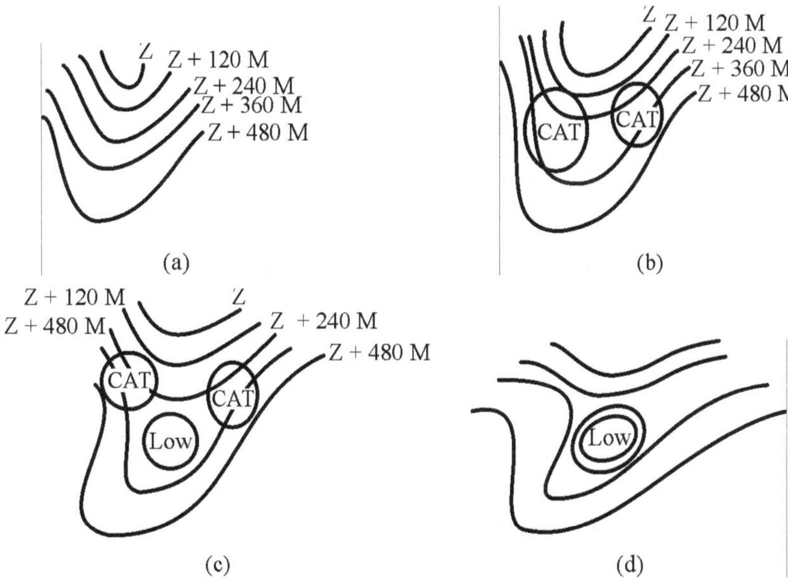

Fig. 22.3 CAT associated with the formation of upper level Low.

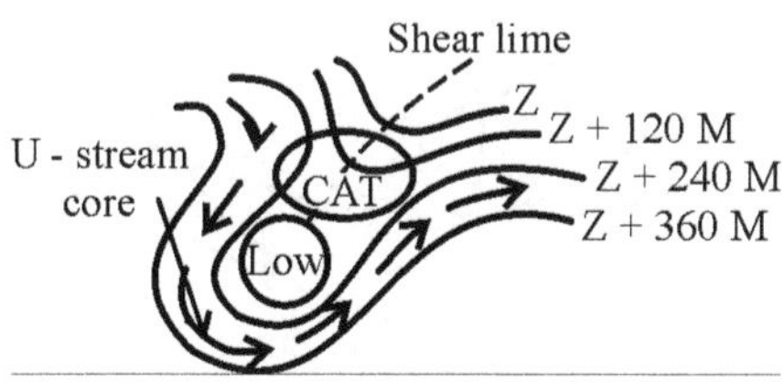

Fig. 22.4 CAT associated with shear line in the throat of an upper level cut-off low.

$$\text{Ri} \simeq \frac{\left[\begin{array}{l}\text{Rate of destruction of turbulent energy by}\\ \text{the negative buoyancy of the stable layer}\end{array}\right]}{\left[\begin{array}{l}\text{Rate of production of turbulent energy by}\\ \text{the Vertical wind Shear}\end{array}\right]}$$

$\text{Ri} \leq 0.25$ is a necessary but not sufficient condition for the occurrence of shear induced CAT.

CAT is a sub-synoptic, micro-scale phenomena.

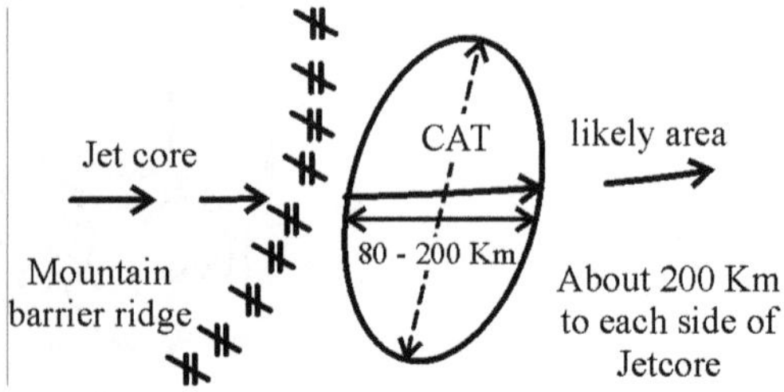

Fig. 22.5 CAT associated with mountain waves.

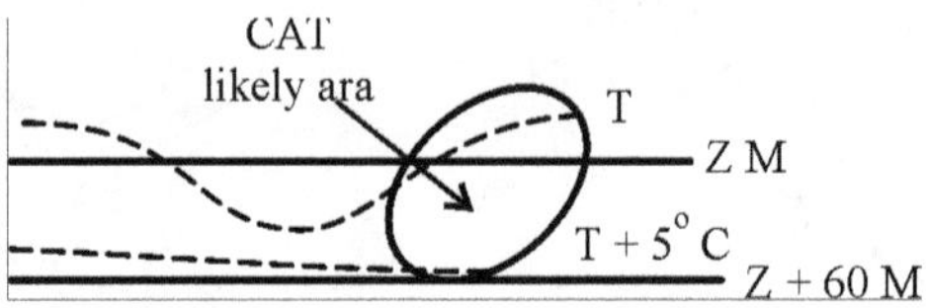

(a) CAT area associated with minor temperature trough

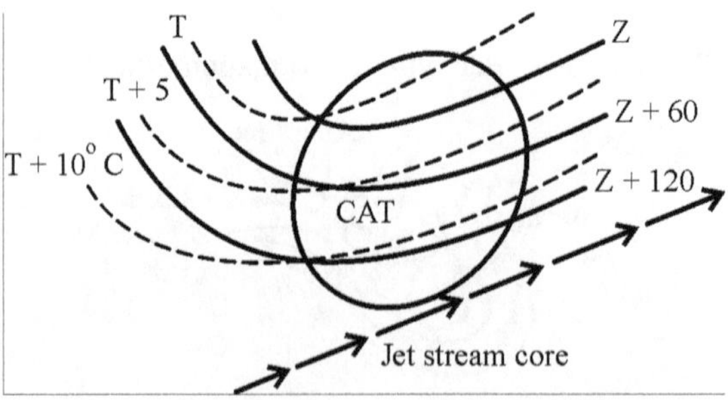

(b) CAT area associated with close isotherm packing by cold advection.

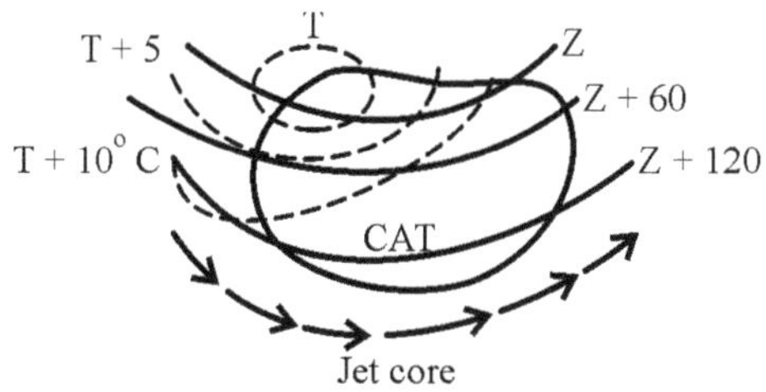

(c) CAT area associated with cold core advancing down the rear side of a trough.

Fig. 22.6 CAT associated with cold advection in upper air trough near the Jet-stream core.

***Note*:** CAT area extends vertically from 600 meters above the tropopause to about 2000 meters below the tropopause.

A strong association exists between jet streams and CAT, and also standing mountain waves and CAT. CAT is frequent at tropopause. The favorable locations of CAT are upper level sharp troughs or ridges, zones with vertical wind shear greater than 4 KT / 300 m, temperature shear 1°C / 30 KM. Moderate turbulence associated with wind shear of 25 KT / 150 km or more, and severe turbulence with shear 50 KT / 150 km or more. Over land CAT occurs mostly on cyclonic side of the jet stream, while over water (sea / ocean areas) mostly on anticyclonic side on the jet stream, and in the zones of convergence of two jet streams. On ridge and leeward side of mountain ridges with conspicuous temperature inversion. The association of CAT with jetstreams, mountain waves, upper level troughs, ridges, lows etc., are shown in Fig. 22.1 to 22.6.

Hazards of CAT in Aviation

CAT may cause temporarily aircraft stalling, loss of control and damage to airframe. Depending on the changes in altitude and/or attitude of an aircraft, CAT is defined light, moderate and severe.

Light Turbulence: Instant turbulence that cause slight changes in aircraft altitude and/or attitude, without fluctuations in IAS (indicator of air speed). IAS may fluctuate 5-15 KT. Changes in accelerometer readings less than 0.5 g at the aircraft centre of gravity. Passengers (or occupants) may feel slight strain against seat belts, but no difficulty in walking. Loose objects may be slightly displaced.

Moderate Turbulence: Instant turbulence that cause moderate changes in aircraft altitude and / or attitude but the aircraft remains in positive control at all times. IAS may fluctuate slightly 15-25 KT. Changes in accelerometer reading of 0.5 to 1.0 g at the centre of gravity. It may cause jolts or rapid bumps in aircraft with moderate changes in altitude and / or attitude. Passengers experience strain against seat-belts and slight difficultly in walking. Loose objects are displaced considerably.

Severe Turbulence: Large and sudden changes in aircraft altitude and / or attitude and momentarily aircraft may be out of control. IAS may fluctuate in excess of 25 KT. Changes in accelerometer readings greater than 1.0 g at the aircraft's centre of gravity. Passengers are violently forced against seat belts, great strain in walking. Loose objects are tossed about.

Detection of CAT: An air borne radar can detect CAT at a distance of 30 km with a 20 dB improvement in sensitivity. Vertical wind shear (which is the best indicator for the presence of CAT) can be detected with Doppler beat frequencies between spaced laser impulses. For this purpose 100 fold increase in sensitivity and coherence of laser beam are required. IR and microwave radiometers measure temperature fluctuations, temperature anomalies associated with CAT along the flight path.

It may be noted here that there is as yet no clear cut way of predicting CAT. As such it is essential that all pilots who encounter CAT should report to ATS (Air Traffic Services) unit with which they have radio contact, giving details of: time of occurrence, location, flight level, type of turbulence, type of aircraft.

Ex:

(i) Over Westernghats, at 0900 UTC intermittent moderate CAT. FL 330 B 747

(ii) From 80 km east of Jaipur, 1005 UTC, intermittent moderate/ severe CAT FL 330 B 737.

22.2.1 Avoidance of Turbulence

Avoid the region if:

(i) wind speed is more than 15 KT and speed increasing with height,

(ii) the angle of attack is within 30° to the perpendicular to a mountain barrier,

(iii) stable region at mountain top with either isothermal or inversion of temperature,

(iv) Lenticular (or lense shaped) clouds, rotors on mountain top and leeward side indicate orographic turbulence. Avoid flight near lenticulars or rotors. Avoid low altitude flight on lee ward side of mountain to about 200 Km downstream. Vertical wind shear generally reported in Kt / 100 ft, while horizontal wind shear reported in kt / 1000 ft. In the vicinity of jet stream marked wind shears both vertical and horizontal observed. By changing flight level, avoid flight through jet core and its vicinity, CAT is prominent between 1300 m above and below the troppopause in the proximity of high level jet stream core just downstream of mountain ridge line.

Avoidance of CAT

When CAT is encountered in flight, the following evasive action is recommended.

1. Turn towards north or south to get our of CAT zone (in relation to westerly jet stream).
2. Climb or descend to a level where the vertical wind shear is known to be less.
3. Climb or descend to level where the temperature lapse rate is known to be less. A layer next above cirrus or cirrostratus is likely to be free from CAT.
4. Avoid flying below the tropopause.
5. Flight above one kilometer of tropopause may be free from CAT.
6. Fly on wind ward side of the high ground

22.2.2 Empirical rules favorable to the occurrence of CAT

CAT occurs in the vicinity of:

(a) Vertical wind shear greater than 4 KT / 300 m

(b) Horizontal temperature shear 5°C/150 Km or more

(c) Moderate turbulence if horizontal wind shear is greater than 25 KT / 150 Km and severe turbulence if wind shear is 50 kt / 150 km or more.

(d) Polar side of the jet stream and next below tropical tropopause. These is strong association between jet streams and CAT

(e) In the convergence zone of two jet streams.

(f) In inversion layers next to leeward side of mountain ranges.

(g) Above upper air sharp troughs and ridges.

(h) High terrain has positive influence on CAT

(i) CAT is associated with surface cyclogenesis.

(j) CAT is associated with the formation zone of a upper level low.

(k) CAT is associated with the shear-line in the throat of an upper level Cut-off Low.

(l) CAT is associated with the formation of difluent upper flow pattern.

(m) CAT is associated with close isotherm packing at 300 hPa level in the tropics.

22.2.3 Incidence of CAT over India

1. Over central and north India the CAT frequency is highest during October to May, while over southern India the CAT frequency is highest during July-August.
2. In association with the peak activity of sub-tropical westerly jet steam, the maximum frequency of CAT is from December to February.
3. CAT zones are usually patchy in nature. Average dimensions of CAT zones are 150 km in north south direction and 300 km in east west direction. Vertical thickness may be 300 to 600 m.
4. Severe CAT occurs during December-February while moderate CAT during remaining months of the year.
5. The high chances of CAT are in a zone of 300 km south of the axis of sub-tropical jet stream.
6. In Western Himalayas (Ladakh area) CAT is most frequent in October when the sub-tropical jet stream appears over the area.
7. In Eastern Himalayas (Assam and NEFA area) CAT is most frequent in mid-winter months.
8. In peninsular India, CAT is conspicuous with easterly jet stream during southwest monsoon.

22.3 Jet Streams

Definition: According to WMO, "A jet stream is a strong narrow current, concentrated along a quasi horizontal axis, in the upper troposphere or in stratosphere, characterized by strong vertical and lateral wind shears and featuring one or more velocity maxima. The speed of the wind must be greater then 30 mps or 60 KT".

Rossby gave the name "Jet Stream" on the analogy of "Gulf Stream".

Generally a Jet stream is thousands of kilometers (km) in length, hundreds of kilometers in width and a few kilometers in thickness. Vertical wind shear is of the order of 5-10 mps/Km, lateral wind shear is of the order 5 mps/100 km. The path of the maximum speed is called axis and the tubular volume around it is called core.

Jet streams are embedded in the atmospheric wind flow in extremely shallow layers with strong wind speed. Horizontal dimension is measured normal to the jet axis. Wind speed reduces to (V/2) half its value at jet core (V) at about 200 Km left of the jet core and 300 Km right of the jet core (that is cyclonic shear is more than anti-cyclonic shear about the jet core). In vertical, the half value (V/2) speed occurs at about 5 Km above or below the jet core (Maximum wind) (see Fig.22.7).

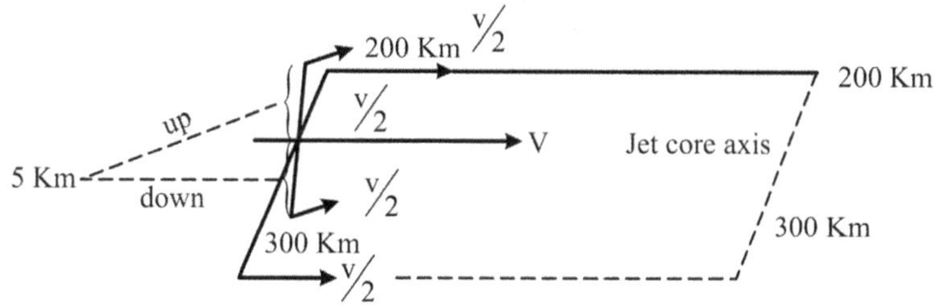

Fig. 22.7 Wind field around jet core.

Jet streams form/develop mainly from the existence of strong thermal gradient. The wind velocity in jet stream is due to large latitudinal temperature gradients, which may extend to great depth of atmosphere and contribute a large thermal wind component.

Normal velocity of jet streams is about 100 KT, but it may reach 200 KT, occasionally it may reach to about 300 KT, over western Pacific. Maximum jet stream speeds found immediately below the level of trapopause. Jet streams have mainly westerly flow but marked changes of directions can be seen. In northern hemisphere jet streams mainly observed in two areas, one between latitudes 30 t0 35 °N and another about latitudes 40 to 45 °N.

In aviation Met. (Meteorological) forecasts jet streams are important weather phenomena. It was discovered during Second World War. Pilots flying military aircrafts first flew into jet streams. They used them in navigation for bombing operations. In northern hemisphere, fly west they avoided them, but fly east they flew along the jetstreams.

The jet streams play important role in the steering and formation of large scale weather phenomena, like cyclones and monsoons and abnormal weather conditions (like hailstorms). Besides windshears, there is thermal gradient near jet streams, which may be about 3 °C per degree latitude. This high temperature gradient may be due to transition from polar air to tropical air in the vicinity of the jet stream axis.

22.3.1 Types of Jet Streams

Jet streams essentially are of two types:

1. Polar front jetsteam (PFJ) and
2. Sub-tropical Jetstream (STJ).

   In addition to these, there are jet streams called:

3. Tropical Easterly Jet stream (TEJ),
4. Arctic Jet stream (AJ),
5. Low level Jet stream (LLJ),
6. Stratospheric Polar night Jetstream (SPNJ),
7. QBO jet in lower equatorial stratosphere

1. ***Polar Front Jetstream* (PFJ)**

   The sub-tropical westerlies and polar easterlies (which have contrasting air mass characteristics) meet at sub-polar lows. These are called Polar Fronts. Polar Front Jet stream generally located near polar front (lat 40°–45 °N) at an altitude of 9.0 km (300 hPa) attains maximum intensity 100 KT in winter. PFJ migrates in winter to south as far as lat 30 °N and in summer to as far as lat 65–70 °N. During summer its strength is less than 80 KT. PFJ is close to middle latitude tropopouse. In the vicinity of PFJ core, cyclonic shear (left of the core) is more than anticyclonic shear (right side of the core). Jet core is observed to be right of tropopouse break.

2. **Sub Tropical Jet stream (STJ)**

   STJ is similar in structure to PFJ, but occurs at higher potential temperature than PFJ. It is generally located above sea level subtropical ridgeline (latitude 25-30°N) at an altitude of 12 km (200 hPa) and attains maximum speed about 250-300KT (over Japan and south Indian Ocean). It migrates to about latitude 25°N in winter and in summer it moves to north latitude 35°N. STJ is more intense in winter than in summer and it is persistent and in direction and location. STJ also occur between tropopouse breaks. It lies close to the middle-latitude tropopouse seen only to the north. Towards the south of the core the tropical tropopouse lies aloft (about 4 to 5 km above the Jet stream core). STJ is dependent on temperature gradient of upper troposphere, while PFJ is dependent on steep temperature gradient caused by the interaction of polar and tropical air masses. The strongest cyclonic shear (left side of Jet core) is observed below the Jet core and anticyclonic shear (right side of jet core) observed slightly above the core.

3. **Tropical Easterly Jet stream (TEJ)**

TEJ is found over India, North Africa during summer at an altitude of 14 to 16 km (150-100 hPa) with speed 60-100 KT between latitudes equator to 15 °N and longitudes 50-80 °E. TEJ is not observed over tropical Atlantic and Pacific. The average easterly wind speed at 200 hPa is about 50 KT. During summer easterlies are found in tropics around the hemisphere but they do not reach the jet speed. During winter the easterlies weaken and found about 20 KT (maximum) over Indonesia and over Central Africa. The most important characteristic of TEJ is related to intense summer monsoon rainfall in its presence. Monsoon rainfall in general, increases with the intensification of TEJ and decreases with its weakening.

4. **Arctic Jet Stream (AJ)**

AJ is found in the high level westerlies of the Artic region, at a mean height of about 7-8 km close to the polar tropopause

5. **Low level Jet stream (LLJ)**

There are strong winds that occur between 1 to 3 km asl but they do not mostly satisfy WMO definition of jet stream. These are strong winds as compared to monthly and seasonal average winds. These strong winds are referred to as Low Level Jet streams. LLJ are effectively influenced by orography, friction and diurnal heating. They are limited in horizontal and vertical extents and attain average speed of about 50 KT. LLJ are prominent over East African Coast and over peninsular India during summer. The other locations are West Central America (USA), Nambian coast of south Afica. Favourable locations are mountain slopes, continental coasts and narrow mountain gaps (obeying Buys Ballots law) like the Palghat gap in South India.

East African coast LLJ appears in summer months. This originates near Mauritius, moves in a southeasterly direction, crosses the equator and moves northwards along the East Coast of Africa

(Kenya, Somali). It then moves in a southwesterly direction in Arabian sea to west coast of India. It acquires maximum speed in the late night or early morning. Based on observational evidence, LLJ along Kenya and Somali acquires the speed of 60 KT or more (satisfies to WMO definition of jet stream) on many occasions. The strongest current is found in July and August at an altitude of 1 to 1.5 km. Mean maximum wind speed is about 30 KT near north Malagasy and 35 KT near Somali. It has steady southerly direction (190°).

The LLJ over West Central USA

This occurs in summer. It runs parallel to the Rocky mountains (direction SSW) over the eastern slopes at an altitude of about one kilometer, with mean maximum speed about 25 mps (50 KT). It also acquires maximum speed during late night or early morning.

The LLJ along Peru coast runs parallel to the western slopes of Andes from southerly direction at a very low altitude of 0.3 km asl. It acquires a maximum speed about 12 mps (25 kt) for most part of the year. Because of its very low altitude it touches the sea water, cools it and aids upwelling along the Peru coast. This subsidence of air is presumed to be the cause of desertification along the coast. The LLJ along the Nambian coast is similar to the Peru coast LLJ. It moves along the west coast of South Africa in south-east Atlantic, from southerly direction. It also causes upwelling (off the coast) in the cool Benguela Sea current.

Marsabit (Africa) and Palghat gap (south India) winds are examples of wind channeling effect. In both cases wind acquires jet stream speed when wind flows perpendicular to the barriers and passes through the mountain gaps. Such orographic wind channeling effect may be found in many places. These gaps are very useful places for extracting wind energy.

6. **Stratospheric Polar Night Jet Stream (SPNJ)**

   SPNJ is located in the stratosphere around Arctic and Antarctic circles.

   These are westerlies in winter and easterlies in summer located above 20 km (50 hPa) at Lat 70 N/S and speed 100-200 KT.

   These are associated with stratospheric sudden explosive warming.

7. **QBO Jet in lower Equatorial Stratosphere**

   During July-August easterlies are strong in upper troposphere in equatorial region. These penetrate into lower stratosphere. If the

QBO in July-August coincides in phase with easterly regime, then a jet appears in lower stratosphere. Similarly, a westerly jet appears in lower stratosphere during winter season.

22.3.2 Jet Streams over India

The principal jet streams over India are:

(i) STJ and

(ii) TEJ

However during monsoon period on some occasions LLJ is observed over peninsular India, which is the extension of LLJ found off Somalia Coast in Africa.

(i) ***STJ over India*:** It is a westerly jet stream, generally seen during non-monsoon months (October to May) over north India. It is strongest in winter months. Its mean position is about latitude 27 °N, altitude 12 km (200 hPa) and mean wind speed at core level 100-120 KT. The southernmost position of STJ over India is latitude 22 °N, occurs during February with wind speed 150-200 KT, direction mainly westerly. During monsoon months STJ migrates to north of India, weakens and disappears. The mean speeds at jet core are lowest 60-75 KT in October and May and altitude about 12-14 km.

STJ over India is confined to troposphere only. It appears there are two maxima cores. One is located over heat low area of west India, Jodhpur (27 °N, 72 °E), at an altitude of 12 km (200 hPa) and mean speed about 100 KT. Another core is located over northeast India, Gauhati (26 °N, 92 °E) at an altitude of 10.5 km, mean wind speed about 100 KT. In both cases direction is westerly. Mean wind speed of 60 KT or more is found, north of lat 20 °N at an altitude of 12 km. North of latitude 23 °N this speed begins at 300 hPa level over west and in Assam (over east) at 400 hPa. In these latitudes wind speed of 60 KT or more extends upto 150 hPa level. Horizontal windshear is about 3 KT/100 km and vertical wind shear is about 10 KT/100 km at 3 km below Jet core maxima. Horizontal wind shear is more to the north of the core 30 KT/deg latitude, while south of the core 10 KT/deg latitude. Temperature increases from north to south below the core and reverses (viz) south to north above the core. Temperature gradient below the core is about

1 °C/deg latitude. At the jet core gradient is slack. Reversal of meridional temperature gradient exists near the jet core level.

(ii) ***TEJ over India*:** TEJ is prominent during monsoon period (June-August) over India. It extends from South China Sea to Africa through Malaysia, Indian Penensula. It covers more or less global monsoonal area. TEJ occurs between 14 to 16 km (150-100 hPa) level, close to the tropical tropopause, along lat 15 °N, with mean speed of about 70 KT. TEJ is not found over tropical Atlantic and Pacific oceans. It weakens to less than 50 KT over India in September (at an altitude of 14-16 km). When TEJ appears over Indian area, the STJ over north India disappears. This occurs more or lass simultaneously. During break monsoon conditions TEJ moves northwards up to lat 20 °N (see Fig. 22.8).

In July winds are easterly 10-20 KT at 300 hPa level south of 27°N and westerly north of 32°N with transition in between. Over the Peninsular India easterlies strengthen with height up to 150 hPa/100 hPa level, where it attains maximum wind speed .At Chennai, mean wind at 150 hPa is 085° / 60 KT, at 100 hPa 085° / 80 KT and at 070 hPa 080° /25KT. The easterly jet core occurs over India at latitude 13°N and height 100 hPa. Easterly regime at 150 hPa level extends southwards up to 5 °S and northward up to latitude 25-30 °N. At Delhi (28 ° 30'N) it is easterly 10 Kt, while at Gan Island (1 °S) 55 Kt. Easterlies do not extend north of 30 °N. Ambala (30 ° 30'N) shows westerlies between 150 hPa - 070 hPa.

The cause of easterly jet stream over Peninsular India is the effect of south ward decrease of temperature over the region in the entire troposphere. In fact it is the position of the sun 20°N at this time of the year. Horizontal wind shear north of the jet core (100 hPa level) is about 7 Kmph/ 100 km, while south of the core 9 Kmph/ 100 Km, vertical wind shear between 300 hPa and 100 hPa varies between 15 -20 Kmph/ 100 Km and the maximum shear occurs 5 Km below the core. Above the core the shear between 100 hPa and 070 hPa is about 22 Kmph / 100 Km. In September strong easteries are found south of lat 20°N. At Chennai a maximum speed of 50 Kt is found at 150-100 hPa.

The temperature distribution is characterized by:

1. Large temperature gradient exist below jet level
2. A little or no gradient of temperature at the jet level, and

3. A reversal of temperature gradient above jet level. Temperature anomaly charts indicate that jet is located in the col region.

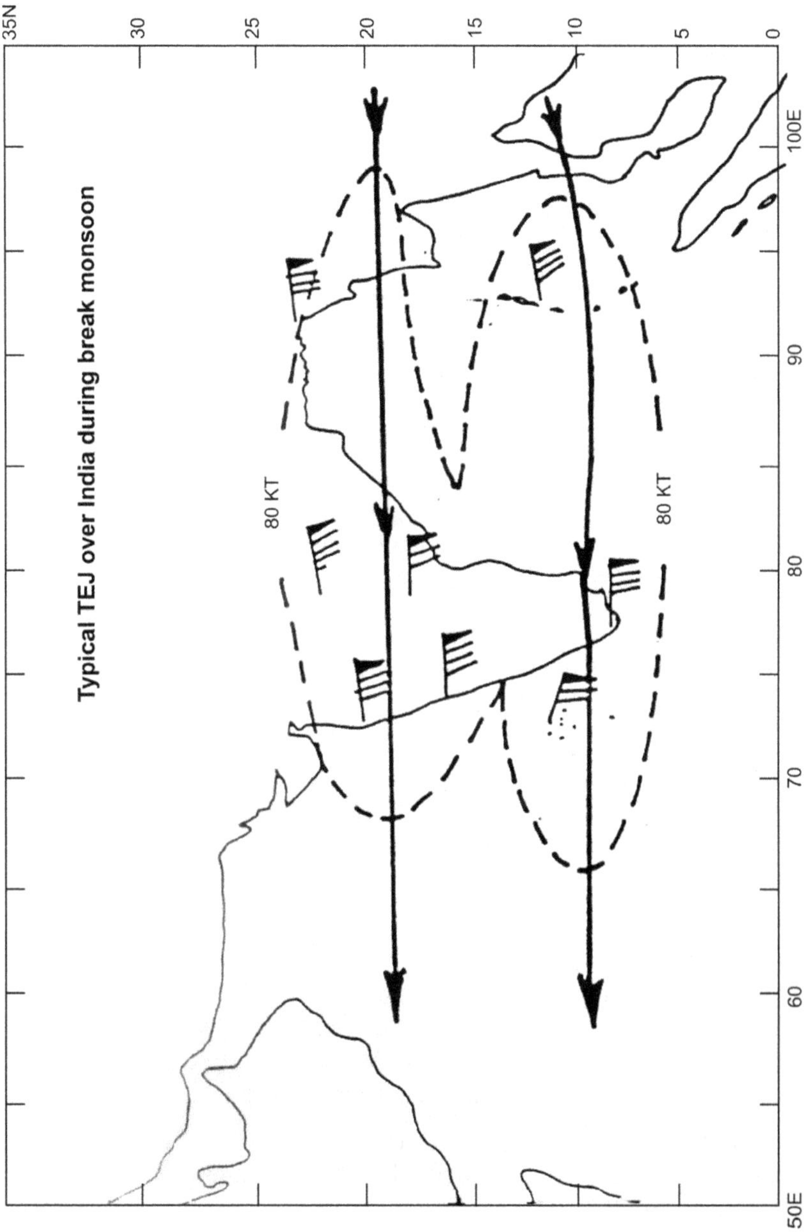

Fig. 22.8 Maximum wind 0000 UTC 0f 148.88 a break monsoon condition.

22.4 Westerly Waves

In the northern hemisphere, the barriers of the Rocky mountains and the Himalayas set up huge atmospheric waves called Rossby (or long) waves, which move from west to east in middle and higher latitudes. These waves are sinusoidal, 4 – 5 waves caver a hemisphere and move slower than average wind speed in middle and upper troposphere. Westerly waves are superimposed on these long waves and have shorter wavelength and amplitude. Westerly waves move faster and are completely guided by the movement of Rossby waves. Westerly waves are not seen on surface charts but observed in upper levels as waves. On surface chart it may be seen as trough and sometimes we may get a closed isobar if drawn at 1 hPa interval when the waves is well marked. Westerly waves cause polar front to deviate south over land mass and north over oceans.

Weather disturbances originate in these westerly waves or these waves act as breeding places for extra-tropical depressions/cyclones. The area of convergence lies ahead of the trough line, where convective clouds, broken to overcast sky with heavy precipitation is observed. In the rear of the trough line, generally fair weather cumulus is observed.

22.5 Easterly Waves

In tropical zonal easterlies sinusoidal oscillations often develop, particularly during monsoon period. Troughs in mid-tropospheric easterlies may form closed systems, however they are not seen as a closed low on sea level pressure chart. The main characteristics of these easterly wave troughs are: they form in ITCZ, have wave length 200-2500 km, period 3-4 days, speed of propagation 8 mps (or 6-7 longitudes per day).

These waves generally originate between lat 15^{0} and equator, move east to west between altitudes 600 h Pa–200 h Pa. These waves have a tilt of northeast to southwest in the northern hemisphere and northwest to southeast in southern hemisphere. They are located south of the easterly jet core. These waves are steered by tropical easterly current. The waves become weak, stagnate or even dissipate over land area.

Easterly waves are found in tropics on both sides of the equator throughout the year. They are marked during monsoon months June-September. Tropical disturbances originate in these waves. Easterly waves seems to originate in west central Pacific as weak low pressure areas and propagate. These disturbances develop into depressions/cyclones in the vicinity of ITCZ. The remnants of west Pacific systems enter Bay of Bengal after crossing Myanmar. They rejuvenate and under favorable conditions they transform into monsoon depressions over Head Bay of Bengal. Easterly waves may cross Peninsular India, enter Arabian sea as weak systems. From there they appear over north Africa where they gain intensity. They cross Africa enter into central Atlantic, move to west Atlantic (Caribbean). Due to long travel over Atlantic, these easterly waves intensify and develop into Hurricanes. Some cross central America and enter into east central Pacific as weak systems. From there they move to west Pacific and develop into typhoons in south China sea. During its global journey these easterly waves undergo a lot of changes in strength, size, horizontal and vertical tilt, cloud structure etc. In general southern hemispheric easterly waves are similar to northern hemispheric easterly waves but they are comparatively less marked except in its summer (Dec-March).

It is generally accepted that the energy for easterly waves is provided by both barotropic and baroclinic instability. A model of easterly wave is shown in Fig. 22.9.

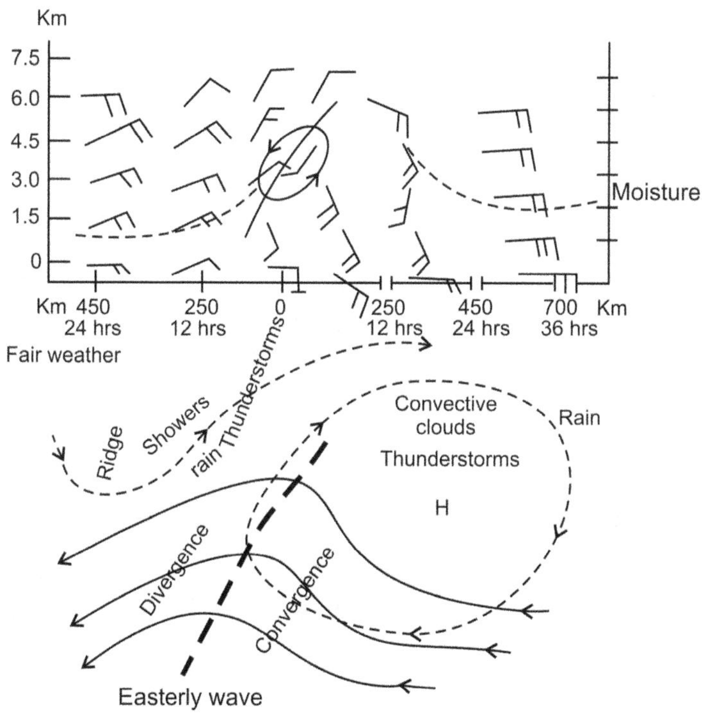

Fig. 22.9

Cloud and weather sequence over sea during the passage of typical easterly wave, are given below.

1. **In Ridge:** Trade Cu average height, no precipitation.
2. **Ahead of Trough:** Cumulus humilis, strong haze, no pptn.
3. **Close to the Trough Line:** Large cumulus, Ci, Ac, scattered showers, improvement of visibility
4. **At Trough Line:** Cumulus congestus, broken to overcast Ci, Ac, frequent rain or showers
5. **Rear of the Trough Line:** Winds veer, decrease of temperature, Cumulus congestus, Cb with layers of Sc, As, Ac and Ci. Frequent moderate to heavy showers with light rain between showers. Thundershowers.
6. **Eastern Outskirts:** large Cu, Cu-congesturs, isolated Cb cells, SCT, Sc, Ac, Ci. Moderate showers-decreasing.

   (SCT = Scattered)

Questions

1. The rate of change of wind vector in space is called
   (a) Thermal wind (b) Geostrophic wind
   (c) Wind shear (d) Wind veering
2. Variations in wind speed along the aircraft flight path of a pattern, intensity and duration that displace an aircraft abruptly from its intended path, such that substantial control input and action required is called:
   (a) Frictional turbulence (b) Thermal turbulence
   (c) Wind shear (d) None above
3. The shear in wind along the final approach path of an aircraft or along runway and along take-off and initial climb-out flight paths is called:
   (a) Vertical wind shear (b) Horizontal wind shear
   (c) Low latitude wind shear (d) Turbulence

4. Variations in the wind along the aircraft flight path of a pattern, intensity and duration that disturb the aircrafts attitude about its major axis but not significantly alter its flight path is called:
   (a) Wind shear (b) Turbulence
   (c) Up draught (d) All above

5. Change in wind velocity with height is called:
   (a) Wind veering
   (b) Horizontal wind shear
   (c) Vertical wind shear
   (d) Up draught

6. Change is wind velocity in horizontal direction is called:
   (a) Wind convergence (b) Wind divergence
   (c) Horizontal wind shear (d) Vertical wind shear

7. Change in the vertical component of wind vector with horizontal distance is called:
   (a) Horizontal wind shear (b) Vertical wind shear
   (c) Up drafts / downdrafts shear (d) Wind backing

8. Wind shears are associated with:
   (a) Frontal surfaces
   (b) Mountain waves
   (c) Low level jet and low level temperature inversion
   (d) All above

8. (a) Low level wind shear is likely to be maximum:
   (a) At the top of surface inversion during clear nights
   (b) At the top of friction layer during strong / bright sunshine
   (c) At freezing level
   (d) At tropopause

9. Wind shear warnings are issued by:
   (a) ATC
   (b) ATS
   (c) Aerodrome Meteorological Office
   (d) All above

10. Significant wind shear likely when there is temperature difference of ...... between the air masses across a front or wind disconuity in tropics

(a) 0 °C (b) 5 °C (c) 10 °C (d) 2 °C

11. Significant wind shear likely when active front is moving at speed of:

(a) 10 KT or more (b) 20 KT or more
(c) 30 KT or more (d) 40 KT or more

12. Pilots inflight can infer the presence of wind shear by the changes in:

(a) Vertical speed, deviation from glide path
(b) Pitch attitude, aircraft altitude
(c) Abrupt or large drift
(d) All above

13. Wind shear may cause airspeed changes more than ....or vertical speed changes .... or more

(a) 5 KT; 200 ft/min (b) 10 KT; 350 ft/min
(c) 15 KT; 500 ft/min (d) 0 KT; 0 ft/min

14. A head wind reduces the gliding distance over the ground, while a tail wind .... over the ground

(a) Increases the gliding distance
(b) Decreases the gliding distance
(c) Does not change the gliding distance
(d) None above

15. A head wind (in a aircraft take-off) .... the ground run and also...... landing distance

(a) Increases; decreases (b) Decreases; decreases
(c) Decreases; increases (d) Increases; increases

16. Flying into updraft ...... the rate of climb and ...... the angle of climb relative to the ground

(a) Will decrease; increase
(b) Will increase; also increase
(c) Will increase; decrease
(d) Will decrease; also decrease

17. The climb performance ...... when flying into increasing head wind/decreasing tail wind or into ....
    (a) Will decrease; updrafts
    (b) Will increase; updrafts
    (c) Will increase; downdrafts
    (d) Will decrease; downdrafts

18. Flying into an increasing head wind or decreasing tail wind the tendency of aircraft is.......
    (a) Under shoot (b) Over shoot
    (c) No change (d) None above

19. Flying into a downdraft will ...... the rate of climb and ...... the angle of climb
    (a) Decrease; increase
    (b) Decrease; also decrease
    (c) Increase; decrease
    (d) Increase; also increase

20. The climb performance will .... when taking off downwind and flying into increasing tail wind or into ....
    (a) Increase; updrafts (b) Decrease; updrafts
    (c) Decrease; downdrafts (d) Increase; down drafts

21. When taking off downwind and flying into increasing tail wind or into ...., the tendency of aircrafts to ....
    (a) Downdraft; overshoot
    (b) Downdraft; undershoot or go below the original flight path
    (c) Updraft; undershoot
    d) Updraft; overshoot or go above the original flight path

22. Descending on an airfield, where wind speeds are decreasing downstream, an aircraft may ....
    (a) Overshoot (b) Undershoot
    (c) Retain the original flight path (d) None above

23. Descending on an airfield, where wind speeds are increasing downstream, an aircraft may ....
    (a) Over shoot
    (b) Undershoot
    (c) Stall
    (d) Retain the original flight path

24. Take-off, of an aircraft over an airfield where wind speeds are deceasing with increasing height, the angle of climb .... with height
    (a) Increases (b) Decreases
    (c) Does not change (d) None-above

25. Take-off of an aircraft over an airfield where wind speeds are increasing with increasing height, the angle of climb.......with height
    (a) Increases (b) Decreases
    (c) Does not change (d) None-above

26. Turbulence in the atmosphere is caused by:
    (a) Mechanical (turbulence), thermal
    (b) Orographic, mountain waves
    (c) Clear air
    (d) All above

27. Clear air turbulence is associated with:
    (a) Stable atmosphere (b) Wind shears
    (c) Temperature shears (d) All above

28. The typical dimensions of CAT are ...... across, thickness....... and life period about......
    (a) 30 Km by 5-15 Km; 300 m to 500 m; 30 mts to 3 hours
    (b) 30 Km by 30 Km; 500 m; 6 hrs
    (c) 50 Km by 25 Km; 1000 m; 24 hrs
    (d) 20 Km by 10 Km; 200 m; 10 to 20 mts.

29. A strong association exists between CAT and:
    (a) Standing mountain waves (b) Jet streams
    (c) Tropopause (d) Both (a) and (b)

30. Favourable locations of CAT are:
    (a) Upper level sharp troughs or ridges
    (b) Zones of vertical shears greater than 4 KT/300 m
    (c) Temperature shear 1°C/30 Km or more
    (d) All above

31. The favourable locations of CAT are:
    (a) Over land on cyclone side of jet stream
    (b) Over water (ocean area) on anticyclonic side of jet stream
    (c) On ridge and lee ward side of mountain ridge with conspicuous temperature inversion
    (d) All above

32. CAT can be detected with:
    (a) Airborne radar at a distance of 30 Km with 20 dB improvement in sensitivity
    (b) CAT associated vertical wind shear best be detected with Doppler radar with Doppler beat frequencies between spaced laser impulses
    (c) IR and microwave radiometers measure temperature fluctuations associated with CAT along flight path
    (d) All above

33. Orographic turbulence is associated with:
    (a) Wind speed 15 Kt or more increasing with height and angle of attack within 30° to the perpendicular to the barrier.
    (b) Stable region at mountain top with isothermal or inversion of temperature
    (c) Lenticular clouds, rotors on mountain top and lee-ward side
    (d) Above all

34. When CAT encountered inflight, the recommended evasive action:
    (a) Turn towards north and south to get out of CAT zone; climb or descent to a level where vertical wind shear is less
    (b) Climb or descend to a level where temperature lapse rate is less or a layer above Ci/Cs
    (c) Flight above 1 Km of tropopasue
    (d) Above all

35. In case of moderate CAT/severe CAT the horizontal wind shear is:
    (a) Greater than 15 Kt/150 Km and 25Kt/150 Km
    (b) Greater than 25 Kt/150 Km and 50 Kt/150 Km
    (c) Greater than 35 Kt/150 Km and 70 Kt/150 Km
    (d) Greater than 10 Kt/150 Km and 20 Kt/150 Km

36. Over central and north India the CAT frequency is maximum during:
    (a) June-Sept (b) Oct-May
    (c) July-Aug (d) Jan-Feb

37. Over southern India, the CAT frequency is maximum during:
    (a) Jan-Feb
    (b) March-June
    (c) July-Aug associated with easterly jet
    (d) Sept-Dec

38. In western Himalayas (Ladakh) CAT is most frequent during
    (a) Jan-Feb (b) March-May
    (c) June-Sept (d) October

39. In eastern Himalayas (Assam) CAT is more frequent during:
    (a) Summer (b) Rainy season
    (c) Mid winter months (d) Oct-Dec

40. The speed of wind in jet stream must be greater than:
    (a) 30 KT (b) 60 KT
    (c) 75 KT (d) 100 KT

41. The vertical and lateral wind shear in a jet stream is of the order:
    (a) 5-10 Kt/Km, 5KT/100 Km
    (b) 5-10 mps/Km, 5mps/100 Km
    (c) 5-10 ft/Km, 5ft/100 Km
    (d) None above

42. Jet streams form mainly from the existence of latitudinal strong:
    (a) Pressure gradient (b) Thermal gradient
    (c) Wind gradient (d) All above

43. Maximum jet stream speeds found:
    (a) Immediately above the level of tropopause
    (b) Immediately below the level of tropopause
    (c) At the level of tropopause
    (d) None above

44. Besides wind shears, there is thermal gradient near jet sreams, which may be about:
    (a) 1 °C per degree latitude
    (b) 2 °C per degree latitude
    (c) 3 °C per degree latitude
    (d) 5 °C per degree latitude

45. Polar front jet stream generally located near polar front at an altitude of ……, attains maximum intensity during………:
    (a) 16 Km, summer and 80 Kt
    (b) 8 Km, winter 100 Kt
    (c) 14 Km, extropical depressions
    (d) 5 Km, tropical depressions

46. Polar front jet Stream in NH migreates in winter to … and in summer to ………..
    (a) Lat 70°N, Lat 30°N (b) Lat 30°N, Lat 70°N
    (c) Equator, N-pole (d) None above

47. Sub-tropical Jet is generally located at Latitude….., Altitude……… and attains maximum speed in winter to……..
    (a) 40-50°N; 8 Km; 100 Kt
    (b) 25-30°N; 12 Km; 250-300 Kt
    (c) 20°N; 16 Km; 150 Kt
    (d) 60°N; 11 Km; 200 Kt

48. Sub tropical jet migrates in winter to about latitude……. and in summer to about lat…… and lies close to………
    (a) 25°N; 35°N; middle latitude tropopause
    (b) 45°N; 70°N; polar tropopause
    (c) 35°Nm N pole; 8 Km
    (d) None above

49. STJ is dependent on temperature of ...... while PFJ is dependent on ....
    (a) Lower tropopause; upper tropopause
    (b) Middle tropopause; steep temperature gradient caused by interaction air masses
    (c) Upper troposphere; steep temperature gradient caused by the interaction of polar and tropical air masses
    (d) Tropical tropopause; polar tropopasue
50. Tropical easterly jet found over India during...and attain maximum speed.... at height about.....
    (a) Winter; 250 Kt; 12 Km (200 hPa)
    (b) Summer 300 KT; 6 Km (500 hPa)
    (c) Monsoon; 100 KT; 14-16 Km (150-100 hPa)
    (d) None above
51. Stratospheric Polar Night Jet stream associated with sudden explosive warming located at,.... .... at lat ... and speed........
    (a) 100 hPa; 25°N; 100 Kt
    (b) 070 hPa; 50 N/S; 150 Kt
    (c) 050 hPa; 70 N/S; 100-200 Kt
    (d) None above
52. STJ over India strongest in ...... Located at.... and altitude.........
    (a) Monsoon; Lat 15°N; 100 hPa
    (b) Winter; Lat 27 °N; 200 hPa
    (c) Premonsoon; lat 20°N; 300 hPa
    (d) Cyclones; cyclone center; 400 hPa
53. TEJ over India strongest in ...... located at latitude....... and altitude...........
    (a) During monsoon 70 Kt; lat 15°N; 100 hPa
    (b) During winter 150 Kt; lat 25°N; 200 hPa
    (c) During western disturbances; lat 30°N; 500 hPa
    (d) During heat waves; lat 5°N; 850 hPa3

CHAPTER 23

Air Masses and Fronts

Introduction

In earlier chapters we have considered various meteorological elements that are significant for aviation (for operating aircraft). Now we shall consider a synoptic view of weather conditions in a local flying area or a particular route. This is called synoptic meteorology. Synoptic means having the same view point. Based on the analysis of synoptic charts (both surface and upper air) weather inference and synoptic forecasting is made. For this purpose air mass analysis and pressure analysis are important. In this chapter we shall learn about air masses and fronts.

23.1 Definition of Air Mass

A large body of air, which has uniform characteristics horizontally level by level in respect of temperature, moisture, lapse rate is called an air mass.

In an air mass isobaric and isothermal surfaces are parallel. Isobaric, Isosteric (constant density) surfaces do not intersect. The region of an air mass is stable and non-turbulent.

CHAPTER 23

Air Masses and Fronts

Introduction

In earlier chapters we have considered various meteorological elements that are significant for aviation (for operating aircraft). Now we shall consider a synoptic view of weather conditions in a local flying area or a particular route. This is called synoptic meteorology. Synoptic means having the same view point. Based on the analysis of synoptic charts (both surface and upper air) weather inference and synoptic forecasting is made. For this purpose air mass analysis and pressure analysis are important. In this chapter we shall learn about air masses and fronts.

23.1 Definition of Air Mass

A large body of air, which has uniform characteristics horizontally level by level in respect of temperature, moisture, lapse rate is called an air mass.

In an air mass isobaric and isothermal surfaces are parallel. Isobaric, Isosteric (constant density) surfaces do not intersect. The region of an air mass is stable and non-turbulent.

Formation of an Air Mass

When air is calm or under subsidence it remains over the same area for a long time (about a week or more). It acquires a state of balance (or equilibrium) with the underlying surface below in respect of temperature and moisture. This balance extents vertically a few kilometers over a period of time. An air mass retains these characteristics for some time more when it is displaced from the source region.

***Definition Front*:** When two air masses of different characteristics meet, the interface region of these two air masses is called a front or frontal region. It is similar to that of an army front of two enemy battalions.

23.2 Classification of Air Masses

Considering the general circulation of the atmosphere, a single front called the polar front may be defined in each hemisphere. Polar front circles in each hemisphere (in middle latitudes) oscillates equator ward and pole wards. In northern (/ southern) hemisphere south (/ north) of the front there is warm or tropical air mass and north (/ south) of the front there is cold polar air mass. Tropical air mass includes air from polar and sub-polar regions. This simple classification may be true to some extent in middle and upper troposphere. However in the lower troposphere it is much complicated because of semi-permanent Lows and Highs and due to Continental and Maritime air.

Based on air mass source region the following classification is preferred by some meteorologists.

1. Equatorial air mass (E)
2. Tropical air mass (T)
3. Polar air mass (P) and
4. Arctic (or Antarctic) air mass (A).

Air Mass Symbols

Based on source region, temperature and moisture the following simplest classification is described below. Tropical (T), Polar (P) air masses are considered as warm and cold air masses. Maritime (m) and continental (C) air masses considered as moist and dry respectively. There is little difference between Equatorial (E) and Tropical (T) air masses. Similarly, there is little difference between Arctic (or Antarctic) (A) air masses and Polar (P) air masses. The following notations are used to distinguish them.

A - Arctic air mass

T_m - Tropical maritime

E_m - Equatorial maritime

T_C - Tropical continental

P_m - Polar maritime

P_C - Polar continental

The table 23.1 gives the principal characteristics of these air masses.

Table 23.1

Air masses (Symbol)	**Place of Origin**	**Characteristics**
Arctic (A)	Polar regions	Low temperature. Low specific RH but high summer RH. Of all winter air masses it is coldest
Polar continental (P_c)	Sub-polar continental areas	Low temperature. In NH temperature increases With southward movement. low humidity
Polar Maritime (P_m)	Sub-polar and Arctic oceanic areas	Low temperature. Temperature increases with movement. High humidity
Tropical Continental (T_c)	Sub-tropical High pressure land areas	High temperature Low moisture content
Tropical Maritime (T_m)	In NH, southern border of oceanic Sub-tropical Highs	Moderately high temperature, High specific humidity High relatively humidity
Equatorial Maritime (E_m)	Equatorial/Tropical seas	High temperature High humidity

Cold Air Mass

In northern hemisphere the cold air mass regions are located over Anticyclones of Siberia, northern Canada and over High of Arctic region. These cold sources have profound effect on the air in the lower levels due to low moisture content (0.1–0.5 gm/kg) near the surface. Because of subsidence (of air) over these areas the air is stable with surface inversion extending to about 1.5 km (850 hPa) is winter. During summer this inversion disappears due to ground heating and convection particularly over Canada and Siberia, while inversion depth becomes shallow over Artic region.

Warm Air Mass

In both hemispheres warm air mass regions are located over Sub-tropical High pressure (STH) regions, particularly in summer. In northern hemisphere (NH) it is located over Central Asia, while in southern hemisphere (SH) over north Africa. The warm regions have profound effect on the air in lower levels (due to ground heating) with steep lapse rate of temperature. However convective clouds are not prominent in these areas because of low humidity. During winter, stratiform clouds form over north Africa under the influence of maritime air mass, while over central Asia no such effect is observed due to lack of moisture (only clear skies prevail).

General Characteristics of Air masses movement

	Cold air mass	**Warm air mass**
Stability	unstable	satable
Turbulence	turbulent air	smooth air
Visibility	good	poor
Cloud	cumuliform clouds	sratiform clouds/fog
Precipitation	Showers, thunderstorms, hail	drizzle

23.3 Modifications of Air Masses

Consider first the modifications of temperature (of air mass) starting from source region to a new place.

Consider cold air mass moving over to warmer surfaces / latitudes. By the time it moves over to a new place the original air mass slowly becomes unstable in the lower levels. If the air mass remains moist at new place, it generates towering cumulus or cumulonimbus clouds and showers (see Fig. 23.1).

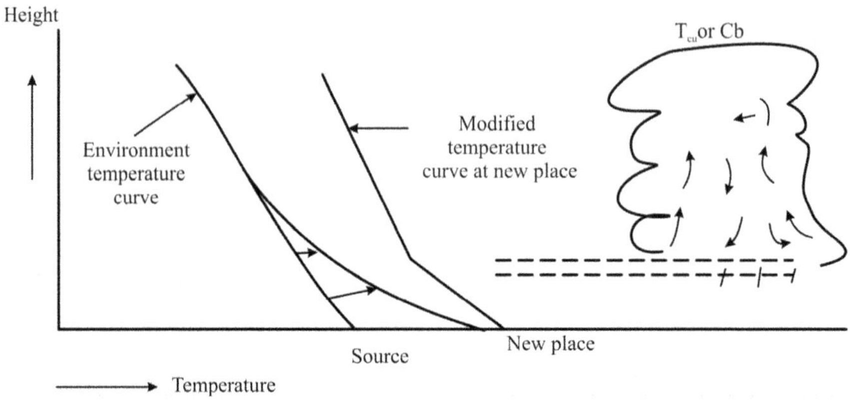

Fig. 23.1 Cold air mass moving over to warmer area becoming unstable.

Now consider a warm moist air mass slowly moving over to colder surface area. In this movement it will cool from below and the air mass becomes stable in the new area, with temperature inversion. This may generate stratus and drizzle. In case of warm dry air mass moving over to colder area / latitudes at new place, it stay stable with poor visibility due to formation of fog / mist at new place. See Fig.23.2.

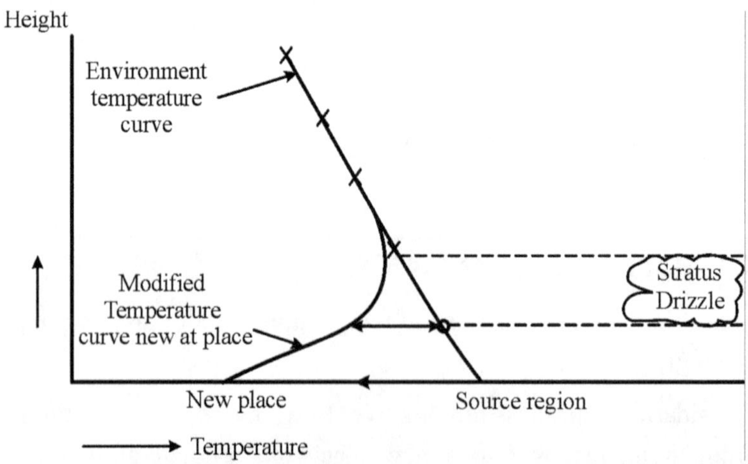

Fig. 23.2 Warm air mass moving over to colder area become stable.

As the air masses change/modify their characteristics once they move away from their source regions, marked effects may occur on moving over to new area which may be warmer or colder surface.

Based on the discussions (Fig. 23.1 and 23.2) we have sub-classification as:

(i) Air colder than the surface over which it moves (K) (Fig. 23.1).

(ii) Air warmer than the surface over which it moves (W) (Fig. 23.2.)

Combining these symbols with source region symbols we have the following.

KT_m – Tropical maritime air mass moves over to warmer surface

KT_c – Tropical continental air mass moved are to warmer surface

KP_m – Polar maritime air mass moved over warmer surface

KP_c – Polar continental air mass moves over to warmer surface

In all these above cases thermal instability may develop in lower levels and generate Cu, Cb and showers.

WT_m - Tropical maritime air mass moved over to colder surface

WT_c - Tropical continental air mass moved over to colder surface

WP_m - Polar maritime air mass moved over to colder surface

WP_c - Polar continental air mass moved over to colder surface.

In all these above cases thermal stability may develop in lower levels with surface inversion of temperature. Fog/mist may form.

23.4 Air Masses over the Indian Region

Over the Indian area usually two types - tropical and equatorial air masses prevail in all seasons, however in winter on rare occasions incursion of polar air also takes place. Sometimes two or three air masses may reach and undergo modification in situ.

(a) **Tropical maritime air (T_m):** The source of T_m air is sub-tropical highs over North Pacific, which comes over to India during monsoon season through the countries of south east Asia (Malaysia, Thailand, Mayanmar as residual systems of Pacific). The air mass characteristics over source region are high temperature, high relative humidity and high dew point. All these characteristics are retained as it moved over to summer low over northern India. The air mass is characterized by excellent visibility, large diurnal temperature variation, development of convective clouds with thunderstorms and showers. Tm air may be seen occasionally in other seasons in association with depressions and cyclones, particularly over south peninsula.

(b) **Tropical continental air (T_c):** The source of origin is the Siberian High. Tc air prevails over India during winter season as cold dry current which is associated with northeast monsoon. The visibility is moderate due to some dust particles. There is little moisture and hence little clouding. During monsoon period (June-September) it is confined to northwest India. During premonsoon period (March– May) it undergoes surface heating and little sea travel it becomes unstable in the lower levels, consequently gives rise to duststorms and thunderstorms.

(c) **Polar Continental air (P_c):** P_c air is cold and dry and makes intrusion into northern India in winter. P_c current enters India in the wake of strong sub-tropical depressions or western depressions which move over Kashmir and Punjab. This air mass gives rise to severe cold waves which affect northern India.

(d) **Equatorial Maritime air (E_m):** E_m, which is tropical maritime air mass, actually originates in the sub-tropical high of the South Indian Ocean (also called Mascarene High). It crosses the equator and enters India as equatorial maritime air (E_m).

During southwest monsoon period it prevails over India south of latitude 25 °N. On isolated occasions it enters extreme south peninsula during winter. At its source region (E_m) it is cool and humid (but convectively stable or neutral). As it crosses the equator, it undergoes surface heating, lapse rate becomes steeper. It gives rise to convective clouds. Visibility is good except in showers. Diurnal variation of temperature is small. During strong monsoon surges (the indraught of the air mass is vigorous) it causes widespread showers, and squally weather. Thunderstorms are rare except during first advance or during fresh revivals after prolonged breaks.

23.4.1 Air Mass Identification

Air mass identification is made by weather phenomena and with T-ϕ grams. The principal identifications are: Temperature and dew point and their changes. The source of origin indicated by stream lines on upper air charts. State of visibility and structure of clouds also throw light on air mass identification.

Fig. 23.3 Principal air masses affecting British Isles.

23.5 Air Mass Fronts

In 1883, using weather charts, the Ralph Abercromby first delinated a subtropical depression on msl synoptic chart by drawing isobars. In 1918, at the end of world war I, the Norwegian meteorologists introduced the term Front, and the concept of frontal surface, and in 1928 the concept of air masses was introduced.

It was earlier defined that when two air masses of different characteristics meet the interface region of these two air masses is called an air mass front or frontal region. A sharp interface or boundary between two air masses is next to impossible, however, on synoptic charts a line is drawn to demarcate them and is called a Front. In a vertical direction a front is not a vertical plane but a curved surface (see Fig. 23.4). The angle of slope of the surface is variable and averages about 1 in 100.

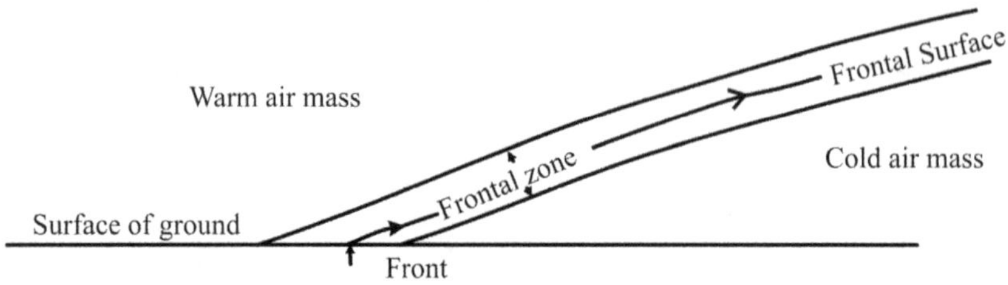

Fig. 23.4 Frontal zone.

The global atmospheric circulation develops a number of regions/areas around the world, which experience air masses discontinuity. Of this Arctic front, Polar front, and ITCZ (Inter Tropical Convergence Zone) are important. The interaction of two air masses at a front or frontal surface in the sub-tropics is the main breeding ground for the formation of clouds and precipitation.

(a) **Arctic Front:** This is a boundary between Arctic and Polar air masses. It is located mainly near the Arctic Circle. It displaces southwards into temperate latitudes in winter.

(b) **Polar Front:** This is a boundary between Polar and Tropical air masses. It is located in temperate latitudes in both north and south hemispheres. Its position varies with season and from day to day. In North Pacific its position is south of the Aleution Islands and in Atlantic it extends from Florida to southwest of England in winter and from Newfoundland to Scotland and thence to Scandinavia in summer. Often it extends eastwards into Europe and at times it displaces southwards into the Mediterranean.

(c) **Inter-tropical Convergence Zone (ITCZ):** This lies within tropics and it is a broad zone of separation of northeast (NE) and southeast (SE) trade winds of the two hemispheres. It is also called Equatorial Trough. (see Fig. 23.5) It is greatly modified by monsoonal winds in Asia. ITCZ extends from western Africa to Southeast Asia across Indian Ocean, and another branch across the Atlantic and the Pacific oceans. The wind discontinuity occurs between NE trades and SE trades. In general, no temperature difference exists on two sides of ITCZ. The region of ITCZ is a breeding ground of tropical cyclones and disturbances. It produces bad weather condition over an extended wide area.

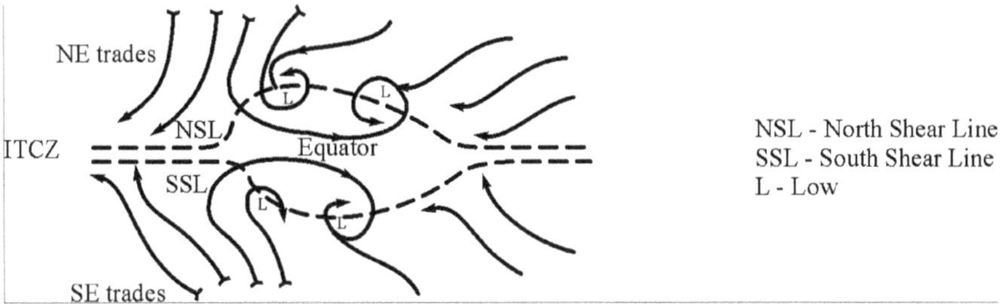

Fig. 23.5

In brief the front where warmer air mass displaces the colder air mass is called a Warm front, and where the colder air mass displaces warm air mass is called Cold front. At other times there is little change in position and the front is called quasi-stationary. An active front lies along the axis of trough of low pressure. A weak front develops a shallow trough and the isobars are U-shaped, while a front with large difference between the air masses develops V-shaped isobars (see Fig. 23.6).

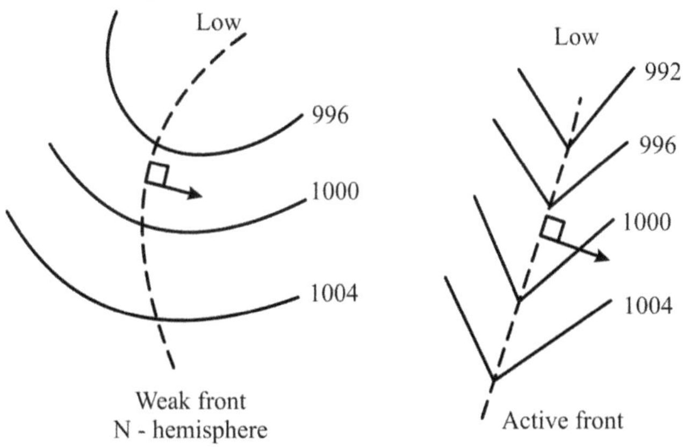

Fig. 23.6 Isobars in active and weak fronts.

In general the slope of front is 1 in 100 (one in hundred) and may range 1 in 50 to 1 in 200. An average frontal surface rises 1000 ft in 30 km (100, 000 ft) and always slopes upward from warm to cold side. A front moves along the winds and the geostrophic winds are not convergent at the front. In equilibrium at a front (at frontal equilibrium) the winds are geostrophic, and there will be no tendency of warm air to rise over the cold (because it is in equilibrium) and no cooling by ascend of air and no cloud formation. That is, the presence of a front does not imply clouds and rain. The necessary property of a front is the difference in temperature between the two sides of a front, which implies disturbed weather is associated with disturbed conditions at the front.

The frictional convergence of surface winds at a front may cause a net inflow of air near the ground but this may give light (slight) rain.

The convergence in a front, which sharpens / activates the front is called frontogenesis. The divergence in a front, which dissipates/weakens the front is called frontolysis.

23.5.1 Warm and Cold Fronts

We know density is inversely proportional to temperature. When light (warm) and heavy (cold) air masses are placed side by side a sloping surface of separation is formed (frontal surface), with the heavier air mass lying below the lighter one. This simply means

" Warm air upglides over the colder air over a sloping surface".

or

" Colder air undercuts warmer air beneath a sloping surface".

Classification of Fronts: Air mass fronts are classified as: (a) Warm front, (b) Cold front and (c) Occluded front.

(a) *Warm Front*: If a warm (light) air mass advances into a cold air mass, it will glide over the cold air mass. In this case the frontal surface formed is called warm front. On synoptic chart it is denoted as :

The warm frontal surface has an average slope of 1 in 100 to 1 in 150. Due to upgliding of warm air, clouds are formed on the frontal surface. About 800 km, ahead of the surface front Ci clouds formed at a height of about 9 km. As one moves closer to the surface front, the height of cloud progressively decreases with Ci, Cs, As and Ns. About 300 km from the surface front rain occurs from As clouds. The rainfall is continuous till the surface front where it stops and rain bearing clouds will be cleared. On crossing the front, the wind veers and air temperature and dew point increase. Pressure which had been falling becomes more or less steady.

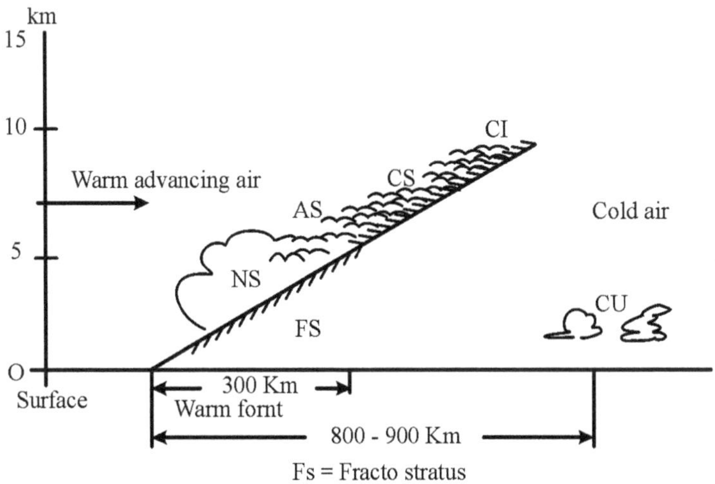

Fig. 23.7 Vertical cross section of a warm front.

(b) Cold front: If cold air mass advances into warm air mass, it undercuts warmer air beneath the sloping surface. The frontal surface formed is called cold front. On synoptic chart it is represented as ▲▲▲. The cold frontal surface is much steeper than the warm front. Its average slope is 1 in 50. Dense clouding of Cu/Cb and thunderstorm, turbulent wind, heavy precipitation are common features at surface cold front area. Weather is spread over a shorter area in the rear. Due to friction cold front is its retardation near the surface. It gets a bent in the friction layer (up to 1 km). This is called nose of the cold front. This is unstable layer formed by friction and destroyed by turbulent mixing. This unstable shallow layer develops roll cloud characteristics of severe cold front, namely squally weather during the passage. In the rear of the cold front clouding, precipitation stops rapidly.

Pressure rises steeply, wind veers, temperature and dew points fall.

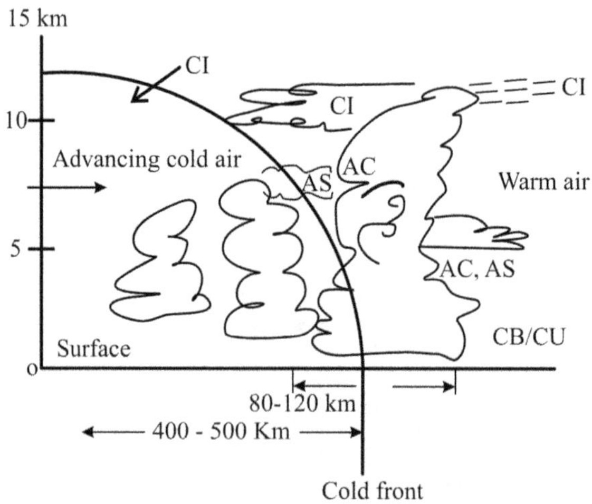

Fig. 23.8 Vertical cross section of a cold front.

(c) **Occluded front:** The cold front generally moves faster than the warm front and if it overtakes it, the warm air sector of the two fronts mixes and forms an occluded front or composite front. On synoptic chart it is represented by

Occlusions are of two types:

(i) Cold front occlusion (ii) Warm front occlusion.

(i) ***Cold front occlusion*:** In this case the composite front is of the cold front, in which very cold air in the rear replaces the cold air ahead. See Fig. 23.9.

Let T_{cc} be the temperature of the cold air sector of the cold front and T_{cw} be the temperature of the cold air sector of the warmfront. If $T_{cc} < T_{cw}$ (cold air of the cold front is cooler than the cold air of the warm front), the cold front undercuts the warm front. This is called cold front occlusion.

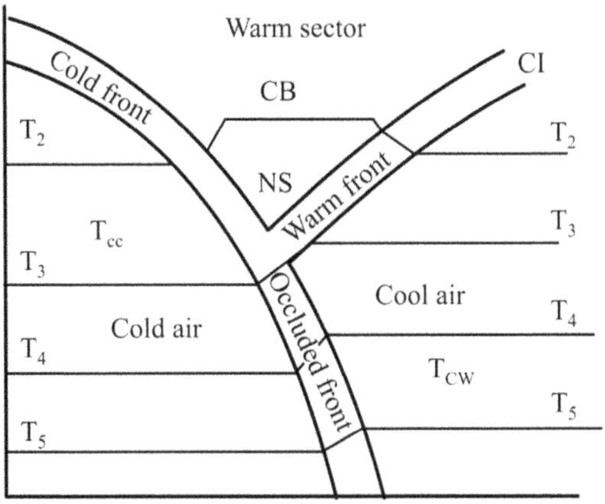

Fig. 23.9 Cold front occlusion where temperature T_5, T_4, T_3, T_2.

(ii) ***Warm front occlusion***: In this case the composite front is of the warm front, in which less cold air in the rear replaces colder air ahead (see Fig. 23.10).

If $T_{cw} < T_{cc}$ (cold air under warm front sector is cooler than the cold air of the cold front) the warm front undercuts the cold front. This is called warm front occlusion.

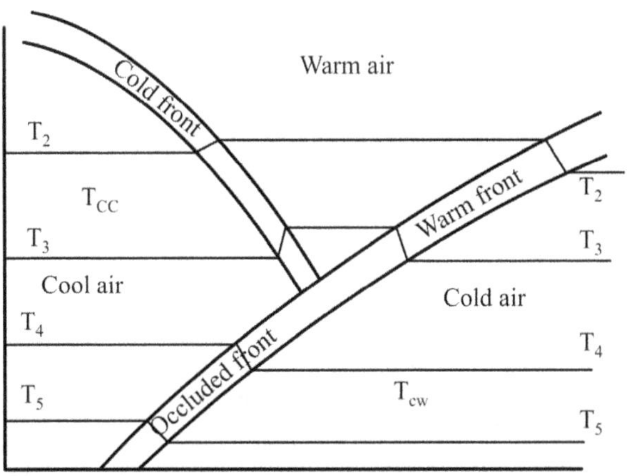

Fig. 23.10 Warm front occlusion.

In both types of occlusions cold air lies at the surface and warm air above it. In case of extra-tropical cyclones the system dissipates.

23.5.2 Summary of Frontal Characteristics

The tables 23.2 and 23.3 gives the summary of warm front and cold front characteristics as the front passes over a station.

Table 23.2 Warm front

Met-element	In advance	During Passage	In the rear
Pressure	Falls	Fall ceases	Little change (or steady)
Wind	Backs, increasing	Veers, decreasing	Steady direction
Temperature	Steady or slow-rise	Rises	Steady or little change
Dew point	Rises in rain	Rises	Steady
Relative humidity	Rises in precipitation	Rises further	Little change
Clouds	Ci, Cs As and Ns in succession, F_s below	Low Ns and Fs	Rapid clearances, some St or Sc
Weather	Continuous rain or snow	Sudden cessation of rain or snow	Fair, at times drizzle or fog
Visibility	Good except in precipitation	Poor, often mist/fog	Moderate to poor mist/fog may persist

Table 23.3 Cold front

(1) Met. Element	(2) In advance	(3) During passage	(4) In the rear
Pressure	Steady and then falls	Sudden rise	Slow rise progressively
Wind	Backs and becomes squally	Sudden veering squally	After squall veers
Temperature	Steady	Sudden fall	Steady, variable in showers
Dew point	Steady	Sudden fall	Steady
Relative humidity	Rise in precipitation	Rise	Fall
Clouds	St or Sc, then Ac, As and Cb in succession	Cb with Fs or Fc or low Ns	Rapid clearance, later Cu, Cb for a short spell
Weather	Continuous rain or snow, thunderstorm,	Sudden cessation of rain or snow Heavy rain or snow	Fair, at times drizzle or fog
Visibility	Moderate or poor in fog	Good	Very good except in later snow showers

From the Tables 23.2 and 23.3 a front can be located from synoptic charts based on station reports of:

1. Wind direction (will veer)
2. Pressure tendency (rise or fall or steady)
3. Temperature (rise, steady or fall)
4. Dewpoint (rise, steady or fall)
5. Cloud sequence (for cold anafront : Cb, Cu. Warm front; St)

Note:

(i) Cold fronts move faster than warm fronts.

(ii) In cold front occlusion, colder air behind than in front. Colder air mass under cuts cold air ahead.

(iii) In warm front occlusion colder air ahead than behind. Cold air mass over rides the colder airmass.

(iv) Cold occlusions occur in summer, while warm occlusion occur in winter.

23.5.3 Representation of Fronts on msl Synoptic Chart

The WMO accepted notation of fronts on msl synoptic chart are given in the table 23.4.

In (v) alternate barbs (▲) semicircles (●) pointing in the direction of motion.

Table 23.4

<table>
<tr><td>(i) Cold front</td><td>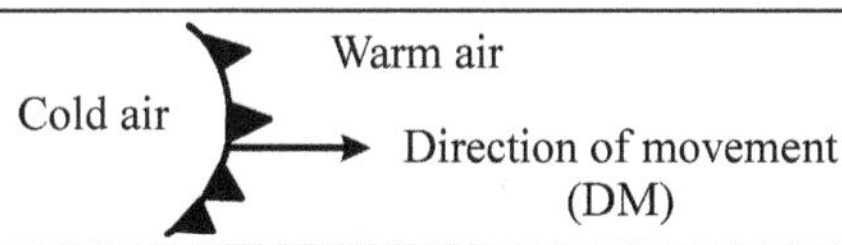
</td></tr>
<tr><td>(ii) Warm front</td><td>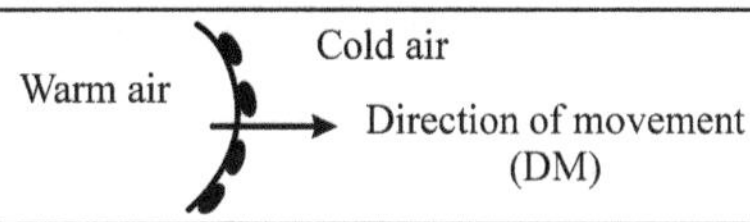
</td></tr>
<tr><td>(iii) Stationary front</td><td>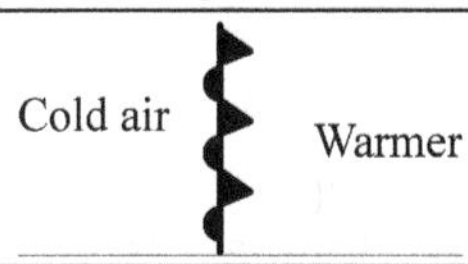
</td></tr>
<tr><td>(iv) Frontal wave</td><td>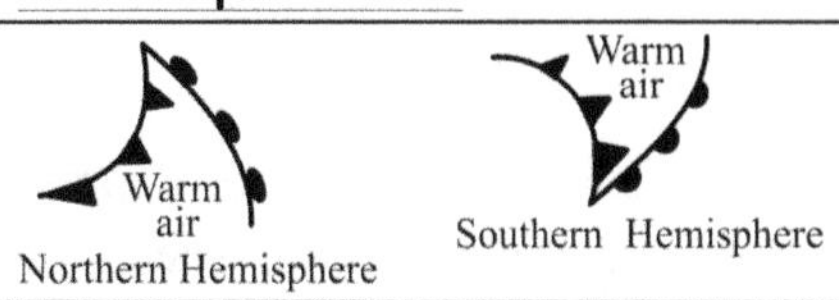
</td></tr>
<tr><td>(v) Occluded front AB</td><td>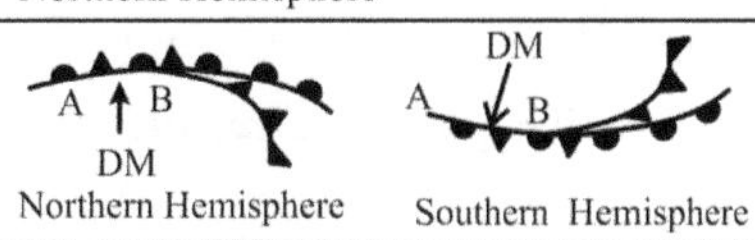
</td></tr>
</table>

23.5.4 Frontal Depressions

The sequence of development of frontal depressions was first described by J. Bjerknes (of the Norwegion group of meteorologist). These are also called Polar front Depressions or Extra-tropical depressions.

According to this Polar front theory, extra-tropical depressions / cyclones develop on the polar front. However it has been found that extra-tropical depressions may develop from a wave like twist on any front. These frontal waves are: (i) Stable or, (ii) Unstable (dynamically). Stable waves do not undergo any change (that the atmosphere is barotropic, in which isobaric surfaces are also surfaces of constant density). These waves finally fade out. Unstable waves undergo rapid changes with atmosphere becoming baroclinic (i.e., which is not barotropic). The various stages of development of frontal depressions in northern hemisphere (NH) described below.

Consider a stationary or quasi-stationary polar front (NH) with warmer air to the south and colder air to the north (see Fig. 23.11 (a)]. Due to surface discontinuity between two air masses, initially a small wave of warm air forces (intrudes) into the cold air. This is called frontal wave. It is similar to the sea wave formation. In light winds ocean waves are stable without change of form, but when winds are strong, the waves are unstable. Both stable and unstable waves occur in a similar way at the sloping frontal surface between the two air masses. When the wave motion is unstable, the amplitude increases and a depression is formed. Studies indicate frontal waves are unstable only when the wavelength is of the order 1200-3000 Km. Wih the unstable waves, pressures falls in the place where warm air intruded into the cold air, and a cyclonic circulation develops. The front passes through the stages of Fig. 23.11 (b) and (c) to (d). The region between warm and cold fronts on the warmer side is called Warm sector while the remaining portion is called Cold sector. Continually falling pressure shows convergence into the depression.

Polar front/Extra-tropical depressions/cyclones move eastwards or east-north-east wards with speed of about 15 kt. Their movement is steered by the westerly currents. The speed reduces in the state of occlusion or dissipation. They are virtually stationary before they fill up.

23.5.5 Weather Associated with a Mature Extra-tropical Wave Depression

We have studied earlier the sequence of weather that follows, warm front, cold front as passes over a station. Similarly we shall consider the weather when depression approaches and passes over a station if it is at a stage (d) of frontal depression (in Fig. 23.11). This implies the weather associated first with warm front, then warm sector and finally the cold front passes over the station.

With the approach of warm front, pressure falls and more rapidly as the front comes closer. Simultaneously cloudiness and precipitation increases and also humidity. Temperature remains more or less constant or slowly increases till the surface front reaches the station, then there is a sharp rise in temperature. With the passage of front, there is decrease in cloudiness or clearing. Now the warm sector is at the station.

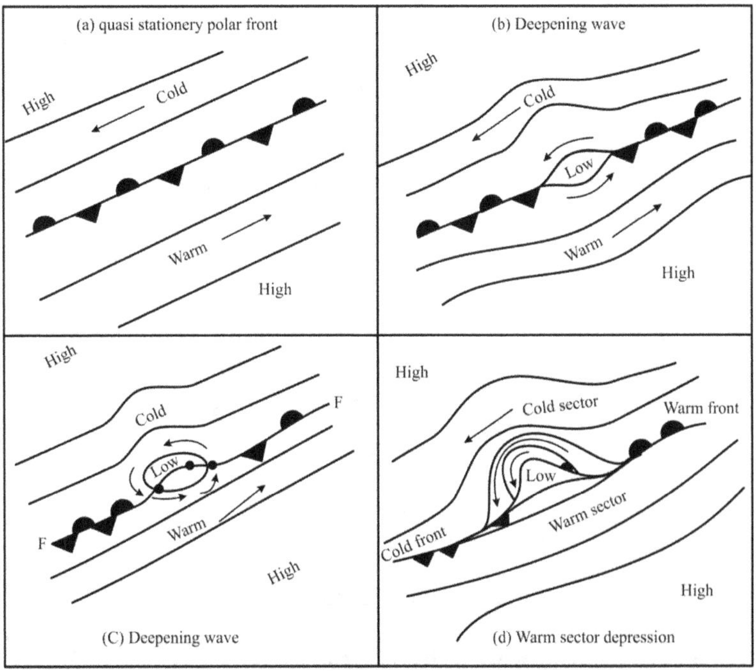

Fig. 23.11 Formation of warm sector, depression on a stationary front.

In the warm sector temperature remains high, pressure is steady or slight fall but marked fall occurs as cold front approaches. Cloudiness depends on moisture and lapse rate.

At the end, when cold front (well developed) comes over the station cumuliform clouds approach. As the cold front comes closer, cumulonimbus cloud, precipitation starts. In the wake of this, squally weather and sharp change in wind direction takes place. After the arrival of cold front, marked rise in pressure occurs and thereafter rapid clearance takes place.

It must be noted that the above sequence of changes in weather elements is associated with an idealzed case and the same may not be followed by every wave depression. Cloud precipitation activity in any wave depression depends on the temperature, moisture and lapse rate conditions in the air mass.

23.5.6 Families of Frontal Depressions

When one wave depression forms, moves along the front and ultimately occludes, the cold front trails back from the point of occlusion. Conditions may become favaourable for the formation of a fresh depression at a lower latitude and the process may be repeated. In this way there forms a series or family of 3 to 4 depressions, each at a lower latitude in the wake of first, second and so on. The family may move eastwords almost together as a train (See Fig. 23.12)

Fig. 23.12 Families of frontal depressions.

23.5.7 Secondary Depressions or Lows

From Fig. 23.13 we find a weak low within the area of circulation of strong low / depression. This type of low moves along the strong low. The stronger low is called primary and the weaker low within the whole circulation is called secondary.

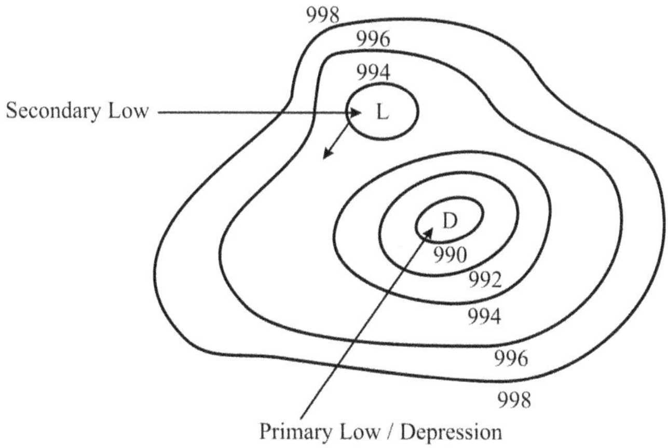

Fig. 23.13 Primary and secondary lows.

Secondary lows may form (i) along with a quick moving polar front depression or (ii) at the top of the warm sector of a partly occulded depression or (iii) on the cold front well within the circulation of a frontal depression.

Secondary depressions form on the trailing edge of the primary system of cold front. In northern hemisphere secondary depressions rotate in anticlockwise direction around the primary system.

23.5.8 Flying Hazards Associated with Polar Front Depressions

The main flying hazards associated with frontal depressions are:

- (i) Ice accretion in As, Ns clouds associated with warm fronts
- (ii) Poor visibility in fog/mist/drizzle in warm sector
- (iii) Turbulence and hazards of thunderstorms passing through cold front. Hail and severe icing
- (iv) Sudden changes in wind direction and speed
- (v) Sudden/gradual changes in pressure
- (vi) Squally weather in cold fronts

23.5.9 Other Depressions

We shall briefly learn about low pressure areas formed other than frontal characters. These include:

- (a) Orographic depressions which form in the leeward side of mountain ranges

(b) Thermal lows, associated with surface heating
(c) Cold air lows, associated with instability
(d) Tropical cyclones and
(e) Tornadoes and waterspouts

(a) ***Orographic depression***: When a current of air meets a large mountain barrier at large angle, a shallow depression is formed to the lee of the mountains and a ridge of high pressure on the windward side (see Fig. 23.14). Sometimes it may produce only trough on lee side. Weather in the low is influenced by the fohn effect which is fine dry.

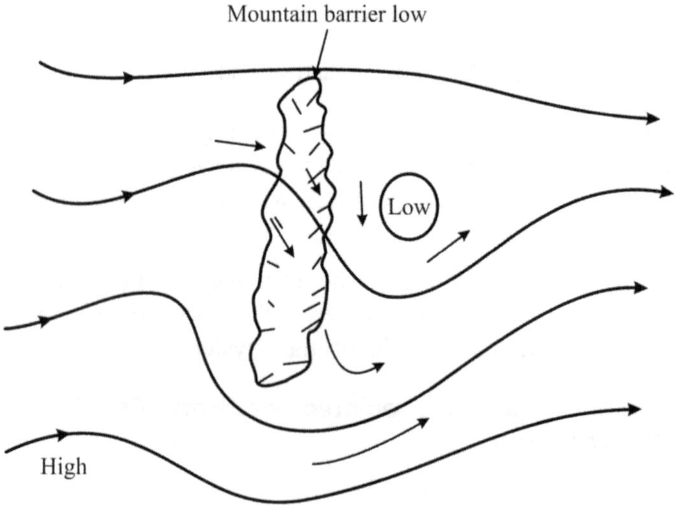

Fig. 23.14 Orographic depression.

When a cold front reaches a mountain range after being roughly parallel and if there is warm stagnant air on the lee side. If part of the front is able to skirt around the end of the mountain range, while the rest of the front is held up by the high ground, then a low may intensify into depression which is associated with bad weather.

(b) ***Thermal Lows / Depressions***: Insolation heats the ground surface, which depends on the inclination of the sun and soil (black soil, sandy soil), vegetation cover etc. Soil heats up the air close to it, which expands, becomes less dense and lifts the air by thermal currents. This sets up low level convergence of winds and a low at the surface and upper air divergence or anticyclone at higher levels. Thus it develops a cyclonic circulation at lower levels and anticyclonic circulation in the upper levels. This is called heat low or thermal Low / depressions. Also called as instability depressions, because the surface heating creates unstable lapse rate which is responsible for the development of low / depression. Monsoon lows and Equatorial trough and ITCZ develops due to thermals. Summer lows on land and inland water lows also belongs to this type.

(c) ***Cold Air Lows***: In temperate latitudes depressions develop in homogeneous polar air by virtue of vertical instability. They are circular in shape and move in the direction of existing air current. Precipitation is confined to local showers. These polar lows form during winter months and give snowfall to windward coasts and high ground of Scotland.

(d) ***Tropical Cyclones***: This will be dealt separately.

(e) ***Tornadoes and Waterspouts***: These have been dealt in chapter 17.

Questions

1. Polar continental air mass (P_c) originates ...., has (temperature) .... and (humidity) ....:
   (a) Polar region; low temperature; high humidity
   (b) Sub-polar continental areas; low temperature; low humidity
   (c) Sub-tropical high; relatively warm temperature; high humidity
   (d) None above
2. Polar maritime air mass (P_m) originates ...., has (temperature) .... and (humidity).............:
   (a) Sub-polar and Arctic ocean areas; low temperature; high humidity
   (b) Polar circle; warm temperature; low humidity
   (c) Sub-polar continental area; low temperature; low humidity
   (d) None above

3. Tropical continental air mass (T_c) originates ...., has .... (temperature) and .... (humidity).
   (a) Equatorial region; relatively high temperature high humidity
   (b) Sub-tropical high continental area; high temperature; low humidity
   (c) Sub-polar ocean areas; low temperature; high humidity
   (d) Non above

4. Cold air mass originates over .... and has .... air
   (a) Anticyclone, stable
   (b) Sub-polar low, unstable
   (c) Equatorial low, unstable
   (d) None above

5. The source of tropical maritime air mass over India is .... and has characteristics (temperature, humidity, dewpoint) .... and visibility ....:
   (a) Sub-tropical high over Pacific; high temperature/humidity/ dew point; excellent except in showers.
   (b) Indian ocean; low humidity / temperature / dew point; poor visibility
   (c) Subtropical high over Siberia; low temperature / humidity / dew point; good
   (d) None above

6. The source of Equatorial maritime air mass over India is .... and has characteristics.........
   (a) Africa, low humidity; high temperature, unstable air
   (b) Sub-tropical High over Indian ocean; relatively cool, humid but convectively stable prevails during monsoon
   (c) Sub-tropical high over Siberia; cool, dry, stable
   (d) None above

7. The tropical continental air mass (T_c) source over India is..... and has characteristics ....
   (a) Sub-tropical high over marcarene; warm dry current, stable, good visibility
   (b) Sub-tropical high over Siberia; cold dry current, convectively unstable with moderate visibility
   (c) Sub-tropical high over pacific; warm, humid current, stable, good visibility
   (d) None above

8. The principal air mass identifications are :

   1. .... 2. ....

   3. .... 4. ....

   (a) Rainfall; humidity; visibility; clouds
   (b) Temperature; dew point and their changes; cloud structure and state of visibility; source identification by streamlines
   (c) Convergence; jet stream; stability; visibility
   (d) None above

9. A boundary between polar and tropical air masses is called .... It is located in both hemispheres in ....
   (a) Arctic front; around Arctic/Antarctic circles
   (b) Polar front; in temperate latitudes
   (c) Monsoon front; equatorial regions
   (d) None above

10. The broad zone of separation of NE-trades and SE-trades of the two hemispheres is called................
    (a) Monsoon front (b) Arcitc front
    (c) ITCZ (d) Dol drums

11. The front where warmer air mass displaces the colder air mass is called ....
    (a) Arctic front (b) Warm front
    (c) Cold front (d) Occluded front

12. The front when the colder air mass displaces warmer air mass is called ....
    (a) Warm front (b) Cold front
    (c) Tropical front (d) Polar front

13. The slope of a front may range .... An average frontal surface rises .... and always slopes upward ....
    (a) 1 in 50 to 1 in 200; 1000 ft in 30 km; from warm to cold
    (b) 1 in 100; 1 km in 30 km; from cold to warm side
    (c) 1 in 50 to 1 in 100; 1000 m in 30 km; does not slope
    (d) None above

14. The convergence in the front which activates the front is called .... The divergence in the front which .... the front and is called ....
    (a) Frontolysis; activates the front; frontogenesis
    (b) Frontogenesis; weakens; frontolysis
    (c) Frontogenesis; activates the front; frontogenesis
    (d) None above

15. In cold front occlusion, colder air behind than in front. Colder air mass...... cold air ahead. Cold occlusion occurs in ......
    (a) Upglides; winter
    (b) Undercuts; summer
    (c) Mixes at the same height; rainy season
    (d) None above

16. When warm front passes over a station, how temperature and wind behave (i) ahead, (ii) during passage, (iii) in the rear ?
    (a) Temperature-steady, rise, little change; wind-backs, veers, steady
    (b) Falls, steady, rise; veers, backs, steady
    (c) Steady, falls, rises; steady, backs, veers
    (d) None above

17. When a cold front passes over a station how pressure and clouds display, ahead, during passage, in the rear?

(a)	Fall,	Sudden fall,	Steady,
	St, Ns	Ns,	St or Sc
(b)	Steady,	Sudden rise,	Slow rise
	St, then, Ac, As and Cb in succession	Cb with Fs or Fc or low Ns	Rapid clearance
(c)	Rise,	Fall	Steady
	Ci, As, Ns	Ns, Fs	Rapid clearance
(d)	None above		

18. According to polar front theory, extra-tropical depressions develop from frontal waves. Dynamically stable waves...... and unstable waves ....
    (a) Develop into depressions; fadeout
    (b) Develop into low pressure areas; develop into low pressure areas
    (c) Fade out; undergo rapid changes and form depressions
    (d) None above

19. Frontal waves are........., pressure ...... in the place where warm air intruded into cold air and a ...... develops
   (a) Stable; rises; cyclonic circulation develops
   (b) Unstable; falls; cyclonic circulation develops and becomes depressions
   (c) Unstable; rises; circulation dissipates
   (d) Stable; falls; circulation dissipates
20. The sequence of weather that follows when a mature extra-tropical wave depression passes over a station: ahead, during passage and in the rear are:
   (a) Warm front weather; warm sector weather; finally cold front weather
   (b) Cold front weather; cold sector weather; finally warm front weather
   (c) Cold front weather; warm sector weather; warm front weather
   (d) Warm front weather; cold sector weather; cold front weather
21. Cloud, precipitation activity in any wave depression depends on:
   (a) Temperature, moisture and lapse rate conditions in air mass
   (b) Pressure, wind, wave stability
   (c) Pressure, temperature, divergence
   (d) None above
22. The main flying condition associated with polar front depressions are:
   (a) Pressure fall, steady wind, precipitation
   (b) Turbulence, icing, squally weather, poor visibility
   (c) Smooth flying, CAT, good vision
   (d) All above
23. Frontal waves which from in.... (waves) become sub-tropical depression and move............
   (a) Stable waves; west wards
   (b) Stable waves; east wards
   (c) Unstable waves; eastwards
   (d) Unstable waves; westwards

24. Secondary depressions from on the .... of the primary system of ... front. In NH, secondary depressions rotate in .....direction around the primary system.

    (a) Leading edge; warm; clockwise

    (b) Trailing edge; cold; anticlockwise

    (c) Trailing edge; cold; clockwise

    (d) Leading edge; warm; anticlockwise

25. Orographic depression form in the .... mountain ranges and are associated with.........weather

    (a) Lee of; good weather

    (b) Wind ward side; bad weather

    (c) Lee ward side; bad weather

    (d) Wind ward side; good weather

26. The cloud sequence of warm anafront is:

    (a) As, Ns, Cu, Cb (b) Ci, Ac, Cb, Ns

    (c) Ci, Cs, As, Ns (d) Cu, Cb

27. The cloud sequence of cold anafront is:

    (a) As, Ns, Cu, Cb (b) Ci, Ac, Cb, Ns

    (c) Ci, Cs, As, Ns (d) Cu, Cb

28. In a frontal depression, warm sector clouds .... and weather .... are:

    (a) St, Sc; drizzle and fog

    (b) Cu, Cb; showers, hail, continuous rain

    (c) St, Ns; rain

    (d) Ac, As; rain

29. In a frontal depression cold sector clouds .... and weather.......... are:

    (a) Cl, Cs, As, Ns, Cu, Cb; continuous rain, showers, hail

    (b) Cl, CS, AS; drizzle fog

    (c) Cu, Cb, NS; showers

    (d) AC, As; rain

30. Warm front slope is .... while cold front slope ....

(a) 1 : 100; 1: 150 (b) 1 : 50; 1 : 150

(c) 1: 150; 1 : 50 (d) 1 : 100; 1 : 200

31. In cold occlusions have .... air behind than in front and colder air ....cold air ahead and generally occur in ....

(a) Cold; upglides; winter

(b) Colder; undercuts; summer

(c) Colder; upglides; summer

(d) Cold; undercuts; winter

32. In warm occlusions have .... air ahead than behind and cold air, ...the colder air and generally occur in ....

(a) Cold; undercuts; summer

(b) Cold; upglides; winter

(c) Colder; upglides / orrerides; winter

(d) Colder; under cuts; summer

33. The type of low generally associated with frontal activity is:

(a) Heat low (b) Mountain low

(c) Polar front low (d) ITCZ

34. A person at Visakhapatnam observed a low pressure (area) passing him to the north. What wind will he experience?

(a) Veering then backing (b) Constantly veering

(c) Constantly backing (d) Backing then steady

35. A pilot flying cross-country at Srinager at 5000 ft first observed Ns, As, Cs then Ci, what could he expect?

(a) Decreasing temperature

(b) Veering of wind

(c) Pressure rise

(d) Increasing of temperature

36. If a pilot observes at Kabul while flying at 5000 ft Ns, As, Cs and then Ci, What could he except?
    (a) Rise in temperature
    (b) Decrease in temperature
    (c) Encounter of downdrafts
    (d) Wind shear

37. What type air movement expected in trough?
    (a) Diverging, ascending
    (b) Converging, ascending
    (c) Diverging, descending
    (d) Converging, descending

37(a). Where does tropical continental air in winter in India originates?
    (a) Siberian high (b) Mascarene high
    (c) Arabia (d) Australia

38. Where does equatorial maritime air during monsoon in India originate?
    (a) Siberian high
    (b) Mascarene high
    (c) Heat low over Rajasthan
    (d) Equatorial trough over Australia

39. An air mass that can effect Indian sub-continent during winter could be:
    (a) Tc (b) Tm (c) Em (d) Pc

40. An air mass that can effect Indian sub-continent during monsoon could be
    (a) Tc (b) Pc (c) Em (d) Pm

Answers

1. (b)	2. (a)	3. (b)	4. (a)
5. (a)	6. (b)	7. (b)	8. (b)

CHAPTER 24

Tropical Meteorology

The region between Tropic of Cancer (lat 23½ ^{O}N) to Tropic of Capricorn (lat 23½ ^{O}S) is called Tropics. The regions between Tropic of Cancer to Arctic circle (lat 66½ ^{O}N) and Tropic Capricorn to Antarctic circle (lat 66½ ^{O}S) are called sub-tropics or extra-tropics. However the weather systems are not restricted to geometrical limits of these latitudes. Weather mainly depends on the migration of the sun from Tropic of Cancer (summer solstice) to Tropic of Capricorn (winter solstice). It may be noted here that the Hadley cell which is located in lower latitudes of each hemisphere with descending air reaching the earth's surface at about lat 30°. For all purpose we consider the region between the 30th parallels of latitude which are the centres of sub-tropical high pressure belts. In many respects the weather systems of tropics and sub-tropics differ greatly, which are given below.

1. In the lower and middle troposphere the tropical atmosphere is convectively unstable while extra-tropical atmosphere is convectively stable.

CHAPTER 24

Tropical Meteorology

The region between Tropic of Cancer (lat 23½ ^{o}N) to Tropic of Capricorn (lat 23½ ^{o}S) is called Tropics. The regions between Tropic of Cancer to Arctic circle (lat 66½ ^{o}N) and Tropic Capricorn to Antarctic circle (lat 66½ ^{o}S) are called sub-tropics or extra-tropics. However the weather systems are not restricted to geometrical limits of these latitudes. Weather mainly depends on the migration of the sun from Tropic of Cancer (summer solstice) to Tropic of Capricorn (winter solstice). It may be noted here that the Hadley cell which is located in lower latitudes of each hemisphere with descending air reaching the earth's surface at about lat 30°. For all purpose we consider the region between the 30th parallels of latitude which are the centres of sub-tropical high pressure belts. In many respects the weather systems of tropics and sub-tropics differ greatly, which are given below.

1. In the lower and middle troposphere the tropical atmosphere is convectively unstable while extra-tropical atmosphere is convectively stable.

2. In tropics, in lower atmosphere, weather systems move from east to west while in extra-tropics they move from west to east.
3. The tropical regions are affected by both westerly and easterly waves, while extra-tropics are affected by only westerly waves.
4. In tropics the mean meridional (north-south) circulation (Hadley cell) is direct while in extra–tropics the mean meridional circulation (Ferrel cell) is indirect.
5. In tropics atmosphere gains westerly momentum (that it exports momentum to extra–topical regions) while in extra-tropics westerly momentum is lost.
6. In tropics cyclones are of warm core, while in extra-tropics the cyclones are of cold core.
7. In tropical stratosphere QBO (Quasi Biennial Oscillations) are prominent while in extra tropics QBO is not marked.
8. In tropics the relative vorticity is equivalent to coriolice parameter (f), while in extra-tropics the relative vorticity is smaller than coriolice parameter.

We shall now consider some important tropical systems.

Equatorial Trough

The trade winds of each hemisphere tend to converge from sub-tropical High pressure belt towards relatively low pressure area near equator. This area is called equatorial trough, commonly known as the doldrums. [The equatorial trough is centred at about lat 5°S in January and lat 12 to 15°N in July. It migrates through 20° latitudes between seasons. This migration of trough influences the seasonal march of cloudiness and rainfall and also the position of formation of tropical cyclones. The annual mean position of the equatorial trough lies at about lat 5°N rather than at the geographical equator. This mean position of lat 5°N is called "meteorological equator".]

The idealized average mean sea level pressure and wind circulation in January and July are given in the Fig. 24.1 and Fig. 24.2 respectively.

The average pressure distribution as given in Fig. 24.1 and Fig. 24.2 indicates that: (a) In sub-tropics each hemisphere has a pressure maxima of about 1015 h Pa and 1020 h Pa in summer and winter respectively. (b) The axis of equatorial trough is along lat 5°S in January and 12 to 15°N in July. The annual mean position is called the meteorological equator which lies along lat 5 °N. (c) The equatorial trough migrates about 20 latitudes (5°S to 15°N) annually. This influences the seasonal march cloudiness, rainfall and formation of tropical cyclonic storms. (d) The sub-tropical high pressure belts are asymmetric with respect to the geographical equator. (e) In January the ridges are centred near lat 30 °N and 35 °S while in July they are centred near 35 °N and lat 30 °S. (f) The area between equatorial trough and subtropical ridge is broader in winter hemisphere as compared to summer hemisphere. (g) In each hemisphere the mass of air in the tropical belt decreases from winter to summer. As a consequence of this mass does not escape fully to the polar regions, which implies that a net export of mass takes place from summer hemisphere to winter hemisphere across the equator.

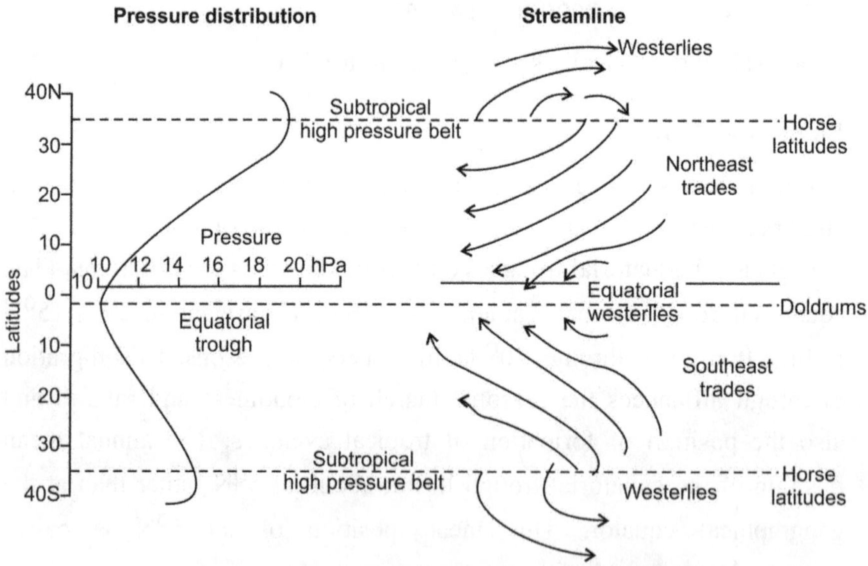

Fig. 24.1 January mean sea level pressure and streamlines.

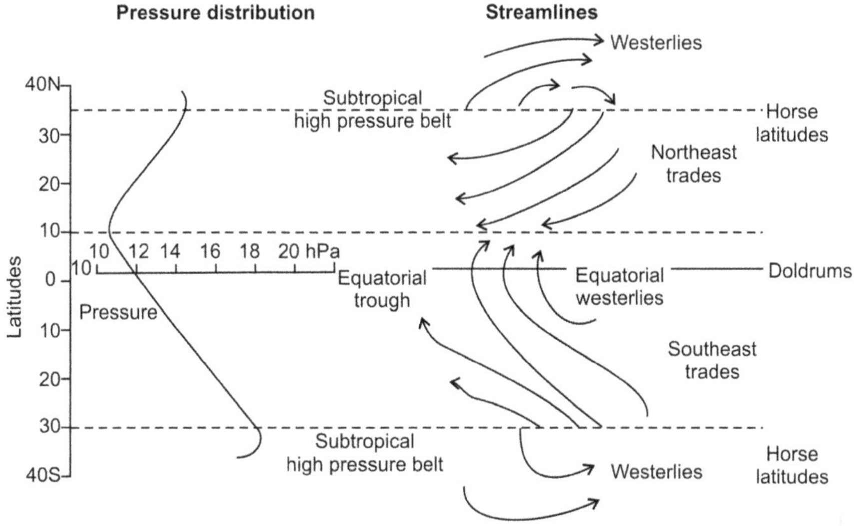

Fig. 24.2 July mean sea level pressure and streamlines.

Some Special Features of Tropics

Weather systems in tropics are different in many ways from extra–tropics. The quasi-geostrophic approximation model works satisfactorily in extra–tropics but fails in tropics, where primitive equation models work well. The theory of geostrophic adjustment of extra–tropical region, namely in a relatively small period of time, adjusting of wind field to pressure field results in quasi geostrophic balance, but this fails in tropics, where pressure field adjusts itself to the wind field over a longer period of time and thereby acquires quasi-geostrophic balance. Geostrophic approximation which suites extra–tropics is not valid in tropics. Similarly Frontal models which suites extra–tropics are not applicable to tropics. In tropics meridional temperature gradient is of the order 1 ^{o}C/100km, while in extra–tropics it is of the order 10 ^{o}C/100 km. Similarly 24–hour temperature changes in tropics and extra–tropics are of the order 1°C and 5 ^{o}C respectively. Pressure gradients (24 hour pressure changes) in tropics are weak (1 hPa) while in extra-tropics they are large (10 hPa). In tropics weather is seasonal in extra-tropics it is continuous. The tropical atmosphere consists of persistent wind systems, with their associated composition and thermodynamic properties, the trades (Trade winds) and monsoon (seasonal winds). Diurnal variations of meteorological elements in tropics are approximately governed by the perturbations of these systems attributable to thermal and orographic disturbances of known origin.

Note: Persistency is defined as the ratio of the mean vectorial wind speed to the mean speed without regard to direction.

Trade Wind Inversion

Trade winds are generally persistent and steady. Over ocean areas cumulus clouds may develop with base about 1km and tops about 12 km. However along coastlines facing trade winds the cloud development is higher and associated with showers. On the lee side of the mountains the region is free from clouds. The restricted cloud height development and fair weather in trade wind zone is attributed to trade wind inversion. The inversion of temperature in the lower altitudes of trade winds is caused by the subsidence of air from the subtropical high pressure belt (STHPB). This inversion persists in trade winds which move towards equator and puts the lid on cloud development over ocean areas. However as the air moves towards equatorial trough, the inversion weakens and cloud height development increases due to instability. This results in heavy, frequent precipitation in equatorial trough zone.

Trade wind inversion is prominent over western coasts of North Africa, South Africa, and over North America and South America. In these areas the base of cloud comes down to about 500 m and the rise of temperature in inversion goes up to 10 ^{o}C. From the base of temperature inversion RH decreases to the top of inversion by about 50%. Below the inversion layer air mass is moist and cool and above the inversion layer the air mass is dry and warm. The maximum intensity of inversion occurs in the subtropical belt region; over northeast Atlantic, northeast Pacific and over southeast Atlantic and southeast Pacific and decrease as it moves closer to equatorial trough.

Inter Tropical Convergence Zone (ITCZ)

The northeast and southeast trade winds of the two hemispheres converge (meet) in a narrow zone on a large scale in the lower levels. This is called ITCZ. Generally these trade winds are separated by a wide doldrum area (or equatorial trough), but in some places they meet, where surface wind discontinuity with horizontal convergence in an east-west direction takes place. ITCZ area is associated with cloudiness and thunderstorms, squalls, overcast skies and heavy rains. ITCZ meanders with the movement of the sun. Wind discontinuity exists between westerlies (SE trades deflected after crossing the equator) in the equatorial region and the easterlies (NE Trades) on the other side. ITCZ extends from western Africa to southeast Asia across Indian ocean and another across Atlantic and the Pacific oceans, the wind discontinuity occurs between NE-Trades and SE-Trades. This was earlier termed as Inter-tropical front and Inter-tropical Discontinuity.

Some Features of ITCZ

ITCZ is the equator ward limit of upward limb of Hadley cell. On sea level, averaged over the globe, ITCZ lies north of the equator in the northern hemisphere summer and south of the equator in the southern hemisphere summer. The extreme positions of ITCZ during July passes over south Asian continent and in January penetrates deep into the African continent, in Australia it lies on land, in south America it lies south of the equator, in ocean region it lies north of equator. On land surface ITCZ coincides with highest temperature zone, on sea it coincides with the region of highest sea surfacewater temperature (SST). For this reason ITCZ is regarded as thermal equator. The position of ITCZ on surface differs from the lower levels at 850, 700 and 500 h Pa levels. The north south migration of ITCZ is maximum over south Asia, it goes up to 28 ^{O}N over India, in July, and over east Africa it goes up to 18°S in January. In such extreme positions, the air masses on either side of the ITCZ may not belong to different hemispheres. There is cyclonic wind shear across ITCZ with westerly component on the equatorial side and easterly component on the pole ward of ITCZ. Generally no temperature difference exists on two sides of ITCZ. Relatively cold moist maritime air lies on equator side of the ITCZ and dry (less moist) air on pole ward side of the ITCZ. For this reason heat transport takes place from ITCZ to middle latitudes. ITCZ produces bad weather conditions over an extended wide area, with cloud tops penetrating tropopause altitudes (up to 18 km), cloud base may be less than 1 km or even may come down to the earth's surface.

24.1 Tropical Discontinuities

The Common boundaries between air masses in tropics are called discontinuities or convergence zones. These are of two types. (i) Stable discontinuity, (ii) Active discontinuity

(i) **Stable Discontinuity:** In tropics, when air masses on either side of the discontinuity have parallel streamline are called stable discontinuity (see Fig. 24.3).

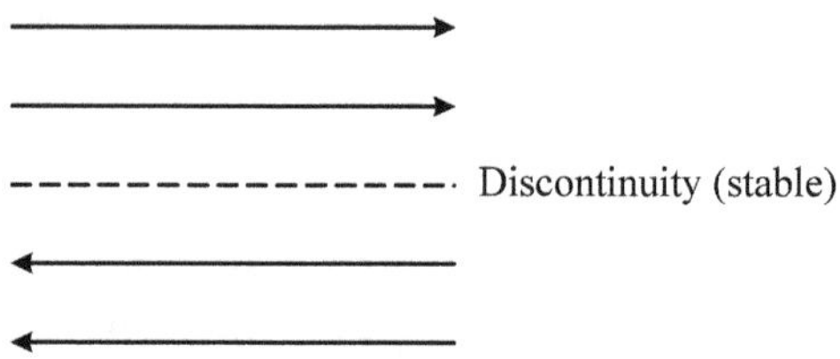

Fig. 24.3 Stable Discontinuity.

(ii) **Active Discontinuity:** In tropics, when two air masses converge (asymptotically) on to the discontinuity, it is called active discontinuity (see Fig. 24.4).

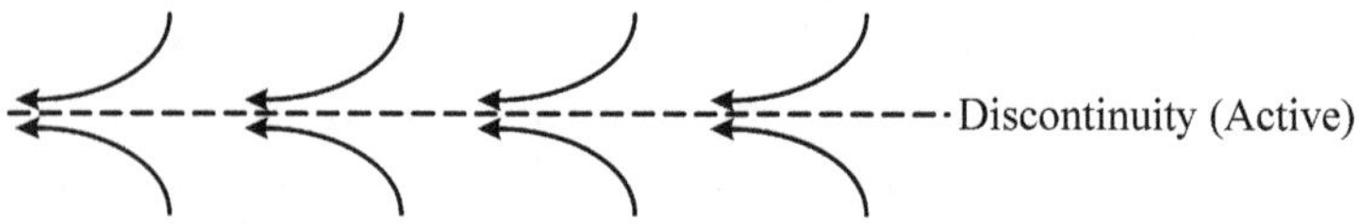

Fig. 24.4 Active Discontinuity.

Cloudy with convective clouds form in a zone of 5–10 km wide along the discontinuity and decreases up to 50–75 km wide.

In tropics the temperature differences between two air masses across the discontinuity are small. Ground heating and cooling greately affects the stability of air in tropics. The most important discontinuity in tropics is the inter tropical convergence zone (ITCZ).

Over Peninsular India we get discontinuity south of latitude 23 °N in hot summer (April, May). This discontinuity is generally formed by Tc (tropical continental) air mass from north and Tm (tropical maritime) air mass from Bay of Bengal or Arabian Sea. This discontinuity gives rise to afternoon thunderstorms and at times hail.

Weather in tropics is mostly concentrated along the discontinuities (where convergence takes place).

In middle latitudes weather disturbance originate in frontal zones and depend on the air mass differences / contrasts. In tropics the disturbances form in three ways.

(i) Disturbances in the single air mass

(ii) Disturbances which form and then move on to the ITCZ

(iii) Disturbances which form on ITCZ, develop intense circulation and then move away from ITCZ.

Surge: A surge is a tropical disturbance in a single air mass, developed by abrupt strengthening of air current (see Fig 24.5). Surges are common feature in monsoon current.

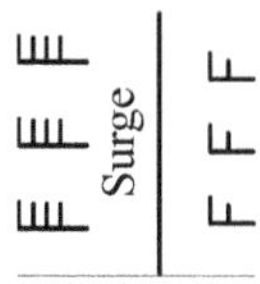

Fig. 24.5 Surge.

In surge, velocity convergence takes place that results in vertical ascent and development of Cu. Towing Cu gives rise to frequent showers accompanied by squalls or gusts.

The surge line moves along the wind. On reaching the coast, it looses its convergence characteristics. It then merges in ITCZ. On west coast of India, the surge plays an important role in precipitation. On arrival of surge, sky becomes overcast, having a number of large cumulus clouds, produces frequent sharp showers. Winds become strong, gusty or squally. Surge line plays role in the formation of low pressure systems in ITCZ.

Shear Line (Horizontal)**:** A typical horizontal shear line is shown in Fig. 24.6 during monsoon.

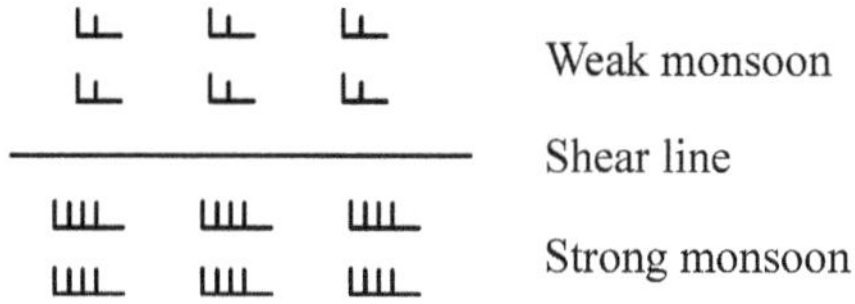

Fig. 24.6 Shear line effect on monsoon.

In the Fig. 24.6, winds are westerlies 15 kt north of the shear line, while they are 40 kt in the south.

The combination of surge and shear line produces thick cloud layers of Cu, Cb and heavy precipitation along western coasts of India and Myanmar during monsoon.

24.2 Waves and Depressions on ITCZ

We have learnt earlier that ITCZ is a breeding place for disturbances, whose strength varies from weak (trough) to severe (cyclone). The disturbances may be classified into three categories: (i) Wave on ITCZ, (ii) Depression, (iii) Cyclones.

Waves on ITCZ do not move away from ITCZ during their life period and retain their air mass contrasts to a large extent. While depressions and tropical cyclones form on ITCZ but move away from it and loose the air mass contrast quickly.

***Formation of Waves on ITCZ*:** Consider a part of ITCZ over sea area where Tm air lies to the north of it and weak E_m (equatorial maritime) air to the south. This state may be called stationary or undisturbed ITCZ. If surge now appears in E_m and penetrates ITCZ a bent or hump forms (see fig. 24.7). The streamlines over the hump become cyclonic with a shallow low pressure are develops. This weak low pressure area moves along the ITCZ from east to west as a wave and dissipates.

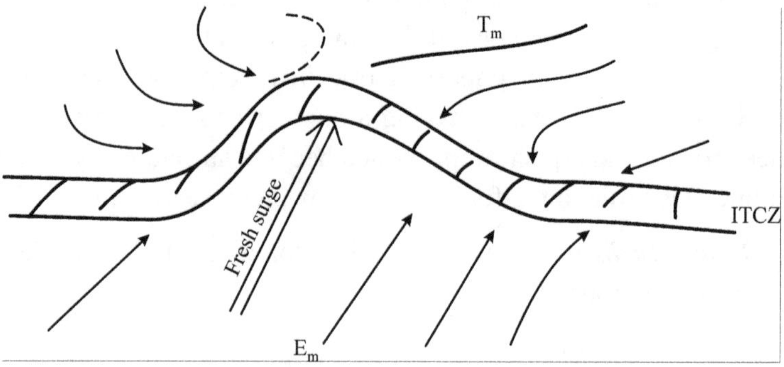

Fig. 24.7 Wave formation on ITCZ.

A wave on ITCZ can be identified on surface chart as a trough or feeble low and in upper air chart as well marked trough. The wave can also be identified on vertical time sections with strong changes in the upper winds as it passes over the station. The sequence of weather changes associated with wave passage is given below.

1. Ahead of ridge	Fair weather Cu
2. Ahead of trough	large Cu
3. Close to the trough line	Large Cu, few Ac and Ci, scattered showers
4. At trough line	Increase in large Cu. Broken to overcast As and Ci. Frequent showers
5. Rear of trough line	TCu, Cb with layers of Sc, As, Ac and Ci.

	Frequent moderate to heavy showers, light rain in between thunderstorms
6. At eastern periphery	Large Cu, occasional Cb, decrease in cloud coverage of Sc, Ac and Ci. Moderate showers decreasing
7. Further east of periphery	Fair weather Cu

In flight very hazardous weather encountered immediately to the east of trough line.

When the waves move over land the wave form undergoes radical change due to orography and also weather.

The wave motion is steered by tropical easterly current (at high altitude) of low latitudes. Consequently these waves are also called easterly waves. The speed of wave is about 15 kt over sea area and reduces as it comes over land where waves dissipate.

***Effects of easterly waves over the Indian area*:** During winter, when the ITCZ is located south of the equator the easterly waves do not affect India. However on rare occasions a surge may push the ITCZ to the extreme south Bay of Bengal. On such occasions disturbed weather will prevail for 3-4 days. The bad weather moves from Bay Islands to Tamilnadu coast, interior Tamilnadu to Kerala coast, thence into Arabian Sea Islands. This weather sequence can be located on successive synoptic charts.

During summer, ITCZ shifts to north of the equator, with this migration of ITCZ the incidence of easterly waves over India increases. The arrival of easterly waves over Southeast Asia, north Bay of Bengal can also be located on charts. When conditions are favourable, depression are formed in these waves.

Development of Tropical Depressions: On certain favourable conditions the waves on ITCZ develop tropical depressions. The favourable conditions are: (i) Adequate moisture, (ii) Tropical continental (Tc) air incursion into the circulation and (iii) a fresh surge of E_m air into the ITCZ.

When a wave on the ITCZ moves over to a location, where these conditions are fulfilled the wave intensifies, cyclonic circulation strengthens, pressure falls in the central region leading to the formation of depression. It may take one to two days transform wave form into a well developed depression. This period (one to two days) is called unsettled conditions.

Ex: "Conditions are unsettled in the northwest angle of the Bay of Bengal where a depression may form".

Questions

1. In the lower and middle troposphere the tropical atmosphere is convectively.........., while extra-tropical atmosphere is convectively..........

   (a) Stable; unstable (b) Unstable; stable

   (c) Neutral; unstable (d) Stable; neutral

2. In general, lower troposphere weather systems in tropics move .... while in extra-tropics move.........

   (a) East to west; west to east

   (b) West to east; east to west

   (c) West to east; west to east

   (d) East to west; east to west

3. The tropical regions are affected by ..... (waves) while extra-tropicas are affected by....(Waves)

   (a) Wasterly waves; westerly waves

   (b) Both easterly and westerly waves; westerly waves

   (c) Westerly waves; both easterly and westerly waves

   (d) Both easterly and westerly waves; both easterly and westerly waves

4. In tropics cyclones are... (core), while in extra-tropics cyclones are....(core)

   (a) Cold core; warm core (b) Warm core; cold core

   (c) Cold front; warm front (d) Warm front; cold front

5. In January equatorial trough is centred about... and in July it is centred about..................

   (a) Latitude 5 °N; latitude 12-15 °N

   (b) Latitude 5 °S; latitude 12-15 °N

   (c) Latitude 12-15 °S; latitude 5 °N

   (d) Latitude 12-15 °S; latitude 5 °S

6. In a year equatorial trough migrates....and its annual mean position is ....which is called.............
   (a) 15 ° latitudes, 5 °S; magnetic equator
   (b) 20 ° latitudes, 5 °N; thermal or meteorological equator
   (c) 10 ° latitudes, 0 °; pressure equator
   (d) 30 ° latitudes; 10 °N, equator
7. In sub-tropics, each hemisphere has pressure maxima in summer .... and winter.............. respectively
   (a) 1020 hPa, 1015 hPa (b) 1015 hPa, 1020 hPa
   (c) 1020 hPa, 1000 hPa (d) 1000 hPa, 1020 hPa
8. The sub-tropical high pressure belts are asymmetric with respect geographical equator. In January the ridge positions are:...., while in July ridge positions are:....
   (a) 30 °N, 35 °S; 35 °N, 30 °S
   (b) 35 °N, 30 °S; 30 °N, 35 °S
   (c) 25 °N, 25 °S; 30 °N, 30 °S
   (d) 15 °N 35 °S; 35 °N, 15 °S
9. In tropics, the meridional temperature gradient is.... while in sub-tropics is..............
   (a) 1 °C / 100 km; 10 °C / 100 km
   (b) 10 °C / 100 km; 1 °C / 100 km
   (c) 5 °C / 100 km; 15 °C / 100 km
   (d) 15 °C / 100 km; 5 °C / 100 km
10. In tropics, diurnal temperature variation is of the order.... while in sub-tropics..........
   (a) 2 °C, 4 °C (b) 0 °C, 5 °C
   (c) 1 °C, 5 °C (d) 5 °C, 1 °C
11. In trade wind inversion areas, the cloud base comes down to......, temperature rise in inversion about......and RH decreases lower to top of inversion by....................
   (a) 100 m; 5 °C; 30% (b) 500 m; 10 °C; 50%
   (c) 800 m; 12 °C; 70% (d) 1000 m; 15 °C; 80%
12. The common boundaries between air masses in tropics are called.
   (a) Discontinuities (b) Convergence zones
   (c) Divergence zones (d) Both (a) and (b)

13. In tropics, when the air masses on either side of the discontinuity have parallel streamlines is called.........
   (a) Stable discontinuity (b) Unstable discontinuity
   (c) Active discontinuity (d) Neutral discontinuity
14. In tropics, when two air masses converge on to the discontinuity it is called...........
   (a) Stable discontinuity (b) Active discontinuity
   (c) Frontal discontinuity (d) None above
15. Over Peninsular India, in summer (April, May) tropical discontinuity lies about lat 23 ^{O}N, which is formed by tropical continental air mass (T_c) from...... and tropical maritime air mass (T_m) from.......
   (a) North, Bay of Bengal and / or Arabian Sea
   (b) Bay of Bengal, Afghanistan
   (c) Africa, Pacific
   (d) None above
16. A surge is a tropical disturbance in.......,developed by......
   (a) Two different air masses; smooth air current
   (b) Single air mass; abrupt strengthening of air current
   (c) Single air mass; smooth air current
   (d) Two different air masses; abrupt strengthening of air current
17. In a surge,.......takes places, gives rise to.......and the surge moves.......
   (a) Velocity convergence; frequent showers along with squalls or Gusts; along the wind
   (b) Velocity divergence; rain or fog; perpendicular to the wind
   (c) Velocity convergence; drizzle or fog; against the wind
   (d) Non above
18. During monsoon, along western coast of India, the combination of surge and shear line produces......
   (a) Fair weather,
   (b) Fog and upwelling
   (c) Thick layers of Cu, Cb and heavy precipitation
   (d) None above

19. ITCZ is a breeding place for……
    (a) Extra-tropical disturbances
    (b) Tropical disturbances / cyclones
    (c) Frontal depression
    (d) None above
20. If surge appears in E_m which penetrates ITCZ a hump forms which becomes the genesis of……
    (a) Extra-tropical cyclone,
    (b) Polar wave depression,
    (c) Tropical cyclone
    (d) None above
21. An observer at Minicoy notes the sequences of changes in weather: first Cu, Ac, Ci, scattered showers, next large Cu, broken to overcast As, Ci, frequent showers, then TCu, Cb, layers of Sc, As, Ci, with frequent moderate to heavy showers or thundershowers. What might be the system that passed over the station?
    (a) Extra-tropical cyclone (b) Typhoon
    (c) Wave in ITCZ (d) Polar front depression
22. The favourable conditions that waves on ITCZ develop into tropical depressions are:
    (a) Steep lapse rate, divergence, T_c air mass incursion
    (b) Adequate moisture, incursion of T_c air mass and a fresh surge of E_m air
    (c) Divergence, incursion of T_c and Tm
    (d) Divergence, shear line and incursion of warm front air mass
23. When a wave on ITCZ moves to a location, cyclonic circulation strengthens, pressure falls in the central region but depression not declared. This period is called…………
    (a) Conditions are favourable for weakening of a system
    (b) Unsettle conditions
    (c) Tropical cyclone likely to form
    (d) A low pressure area may form

Answers

1.	(b)	2.	(a)	3.	(b)	4.	(b)
5.	(b)	6.	(b)	7.	(b)	8.	(a)
9.	(a)	10.	(c)	11.	(b)	12.	(d)
13.	(a)	14.	(b)	15.	(a)	16.	(b)
17.	(a)	18.	(c)	19.	(b)	20.	(c)
21.	(c)	22.	(b)	23.	(b)		

CHAPTER 25

Tropical Cyclone

Tropical cyclone are the most destructive phenomena of atmospheric nature. They are foremost in their violence, destruction and duration. They strike the coastal areas extending thousands of square kilometers with their ferocity of wind, torrential rain and inundate the area with storm tides. Compare with the other natural disasters - an earthquake causes much more damage to life and property but it is short lived, only one or two minutes. Tornadoes are the most violent storms on earth but they are small and cause destruction in their narrow path for a short period. The extra-tropical cyclone of winter are the largest phenomena of atmosphere but they are much milder as compared to their tropical cousins. It is estimated that a normal cyclone precipitates water about 2 giga (10^9) tons in an hour and dissipates about 36×10^{10} kilowatt-hours of kinetic energy.

Tropical cyclone are associated with large pressure defect and well developed cyclonic wind circulation. The origin of tropical storms is mostly confined to western portions of the oceans. Tropical storms are known by different names over the globe which are given below.

Geographical Area	Name
Indian ocean and south Pacific	Cyclone
Western Pacific (when wind speed 64 kt or more)	Typhoon
Atlantic and eastern Pacific	Hurricane
Pacific – (Phillipines)	Baguios
Pacific – (Australia)	Willy Willy
Mexico (Atlantic)	Cordonazo

Our knowledge of tropical cyclones is gradually built up as and when more and more sophisticated instruments are invented. Initially our information of cyclones was based on ship reports and coastal observations when they approached the coast. Subsequently aircraft reconnaisance reports, radar observations were added up. Since 1960 weather satellite observations immensely contributed to enhance our knowledge and now with these observations none of the cyclones escape our detection.

25.1 What are Tropical Cyclones?

Tropical cyclones are intense low pressure systems that form mostly over sea areas. They move from east to west as gaigantic whirls of violent winds and severe weather. They form over warm sea waters where sea surface water temperature is 26 °C or more. They die (dissipate) if they enter cold waters. Equatorial regions between latitudes 5 °N and 5 °S are not favourable for the formation of cyclones due to the weak or absence of coriolis force. The average life span of a cyclone is about six days until they land or recurve into temperate latitudes. Some storms last for only few hours while others last as long as two weeks.

The classification of tropical disturbances by IMD using wind speed over sea areas is given in Table 25.1.

Table 25.1 Classification of Tropical Disturbances

Name of pressure system	Beaufort scale	Wind Speed		Satellite picture intensity	Central Pressure defect hPa	Gradient / 250 km hPa
		Knots	Kmph			
Low pressure	4	Less than 17	19-28	-	-	5
Depression	5, 6	17-27	29-49	1.5	4.0	5-7.3

Table 25.1 Contd,,

Name of pressure system	Beaufort scale	Wind speed		Satellite picture intensity	Central Pressure defect hPa	Gradient / 250 km hPa
		Knots	Kmph			
Deep Depression	7	28-33	50-61	2.0	4.5	7.3-12
Cyclonic Storm	8,9	34-47	62-88	2.5-3.0	6.1-10	13-18
Severe Cyclonic Storm	10 , 11	48-63	89-117	3.5-4.0	15-21	18
Very severe cyclonic storm	12 or more	64-119	118-220	4.5-6.5	28-80	18 or more
Super cyclone	More than 12	More than 120	More than 220	More than 6.5	More than 80	More than 18

Tropical cyclone resembles a giant heat engine that derives its energy mainly from the transfer of sensible and latent heat from sea to air. The main energy input is water vapour (a form of latent heat). A suitable wind arrangement acts as the starting mechanism. The kinetic energy of these winds provide the initial independent energy. It acquires spin from the coriolis effect. Condensation of water vapour during vertical ascent transforms latent heat into sensible heat. Upper level outflow (divergence) together with low level inflow (convergence) develop radial motion. This circulation in the vertical plane must stand the influences that tend to destroy it. High tropospheric currents provide the cooling system which carries the excess heat (from equatorial regions) to other regions of the globe. The thermal structure of tropical disturbances have cold core. When they fully develop they are transformed into warm core systems. The core temperature in a fully developed cyclone may be about 10 °C warmer than its surroundings up to about 15 km, but above this it is colder than the surroundings. It seems the release of latent heat in condensation and precipitation transforms the cold core system into warm core. Huge Cb clouds extending to great heights act as hot towers in transporting heat to the upper levels.

25.2 Life Cycle of a Tropical Cyclone

Tropical cyclones have four stages of life cycle. (i) Formative, (ii) Immature, (iii) Mature and (iv) Decay stage. There is no set duration that a storm may be in one stage. The storm may skip any one stage or go through in such a short period that it is not possible to distinguish with the available data. On some occasions it is difficult to say in any one stage of development.

(i) *Formative Stage*: In this stage winds are variable, squally with thunderstorms and unsettled conditions. This stage spans from the genesis of cyclonic circulation as a low pressure area to the stage when the system intensifies into cyclonic storm. Pressure falls, winds strengthen and are confined to a small area and a closed vertex may emerge. Central pressure may be 10 hPa below normal. Formative stage is a very slow process and may spread over a few days.

(ii) *Immature or Developing Stage*: In this stage central pressure falls rapidly, winds strengthen and these two reach their maximum limits. This may occur in 24 hours or sometimes develop from deep depression to severe cyclonic storm in a few hours. Strong winds will be confined to a small area, hurricane wind speed form a tight ring band around the centre. Clouds and rainfall bands organized in to a bands spiralling inward.

(iii) *Mature Stage*: In this stage circulation expands outward but central pressure and winds remain steady (constant). Symmetry is lost as the area of squally weather and gales extend further. The area of hurricane winds (strength more than or equal to BF 12) increases from the centre. The size of mature storm varies greately but exhibits four distinct parts (see Fig. 25.1).

(a) *Eye*: A calm (or light winds) central area about 20 km in diameter with clear skies or with partly cirrus, lowest pressure in the centre. This is called eye of the cyclone.

(b) *Wall cloud region or inner ring of hurricanes*: A ring shaped inner region surrounding the eye with towering Cb cells (or wall clouds) with heavy rain and maximum wind is called wall cloud region. The width of the ring varies 10 to 20 km. According to some eye witness reports that the area looks like a burning inferno due to sheet lightning.

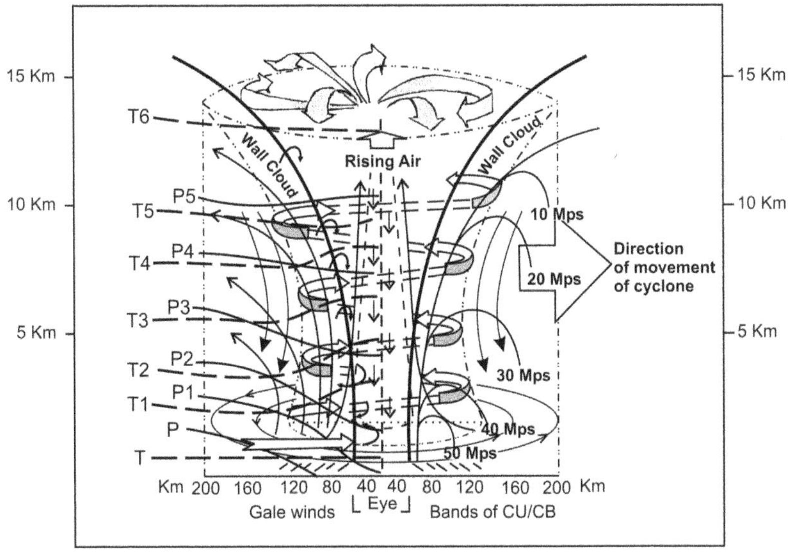

Fig. 25.1 Schematic diagram of a mature cyclone, with cloud, temperature, pressure and wind distribution.

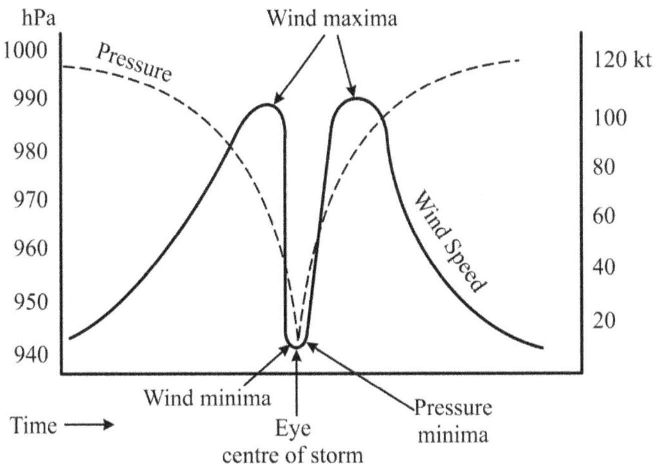

Fig. 25.2 Eye of a cyclone-wind and pressure distribution.

(c) *A Belt of Hurricane Winds*: This area extends surrounding the wall cloud region up to 80 km from the centre. This region contains hot towers (Cb cells) about 100 to 200 which produce squall lines, torrential rain, thunder and lightning. It is a belt of hurricane winds with squalls exceeding 300 kmph (162 kt).

(d) *Outer Region*: In the outer region (extending 200-500 km from the centre) of the cyclone winds decrease radially outwards. This region is also characterized by convection but to a limited extent. The diameter of a tropical cyclone is of the order 100 km to 1000 km or more, some have diameter as small as 30 km.

(iv) *Decay Stage*: In this stage system weakens, which may be due to storm crossing the coast or entering cold water or recurve and move over to extra-tropical latitudes. Decay stage may also be rapid. Between the belt of strong winds and outer region, there is relatively clear or Moat area with little convection.

The greatest damage caused by a cyclone is due to the hurricane winds and tides. The area of hurricane wind is always small. No one knows precisely how high wind speeds can rise in tropical storm. At sea, storms produce a distinctive heavy swell that effects ocean shipping. As the storms approaches a coast, the piling up of water by strong winds produce a disastrous storm surge (tidal wave). A hurricane wind speed of 100 kt creates a roughly a wave height of about 30 m and the KE of about 10^{19} J. The liberation of this much energy and its conversion into heat however does not warm the ocean surface water but cools it. It mixes the ocean water to a depth of about 200 to 400 meters in an area of about 50-100 km of radius.

A Belt of Hurricane Winds

A belt of hurricane winds (speed 118 Kmph or more) surround the wall cloud region, radius extending to 80 Km from the centre. This region contains hot towers (Cb clouds with heavy convection) about 100 to 200, which produce squall lines, torrential rain, thunder and lightining. Wind in squalls exceed 300 Kmph

Outer Region

In the outer region (extending 200-500 km from the centre) of the cyclone winds decrease radially outwards. This region is also characterized by convection but to a limited extent. The diameter of a tropical cyclone is of the order 100 to 1000 km or more. Some have diameter as small as 30 km.

Between the belt of strong winds and outer region, there is relatively clear or Moat area with little convection.

Tropical Storm Statistics

In northern hemisphere maximum tropical storms occur during August-September months when the sea surface temperatures are high, while in southern hemisphere the maximum occurs during February-March. However in Indian seas the maximum frequency is during April-May and October-November.

The global annual average of Tropical cyclones (based on about 100 years data) is 80, and its annual variation is about 8 (i.e., 10%) see Fig. 25.3. Year to year variation of cyclonic disturbances and cyclonic storms are large. The average life period of global Tropical cyclones is 6 days. In Indian seas the annual average of cyclonic disturbances is about 16 (Bay of Bengal 13) and standard deviation is 3.1. Its variation is 7 in 1984 and 23 in 1927. The annual average of Tropical cyclones is about 6 (or 7% of the global total), standard deviation is 1.85. Its variation is 1 in 1949 and 10 in each of the year 1893, 1926, 1930 and 1976. The average life period of cyclones in Indian seas is 1.5 days. In northern Indian ocean 65% of the cyclonic disturbances do not recurve, while the remaining 35% recurve.

The number of Tropical cyclones (cyclonic storms and severe cyclonic storms) that approach or strike the east coasts of Srilanka, Tamilnadu, Andhra Pradesh and Bangladesh is highest during October to December, while that affects the Orissa coast, west Bengal coast and Arakan coast is highest during June to September and March to May (Post monsoon and monsoon period).

The worlds longest record life period of TC is 31 days, in case of Hurricane Ginger, formed in Atlantic ocean in 1972, while in Indian seas 14 days (2-15 Nov 1886, 16-29 Nov 1964). The radius of global Tropical cyclone (TC) vary 50-100 to 2000 km. A large number of the storms in north Indian ocean had diameter less than 100 km.

Maximum TCs form in western parts of north Pacific (about 33% of the global TC may form through the year). Southern parts of eastern Pacific and south Atlantic are free from TC.

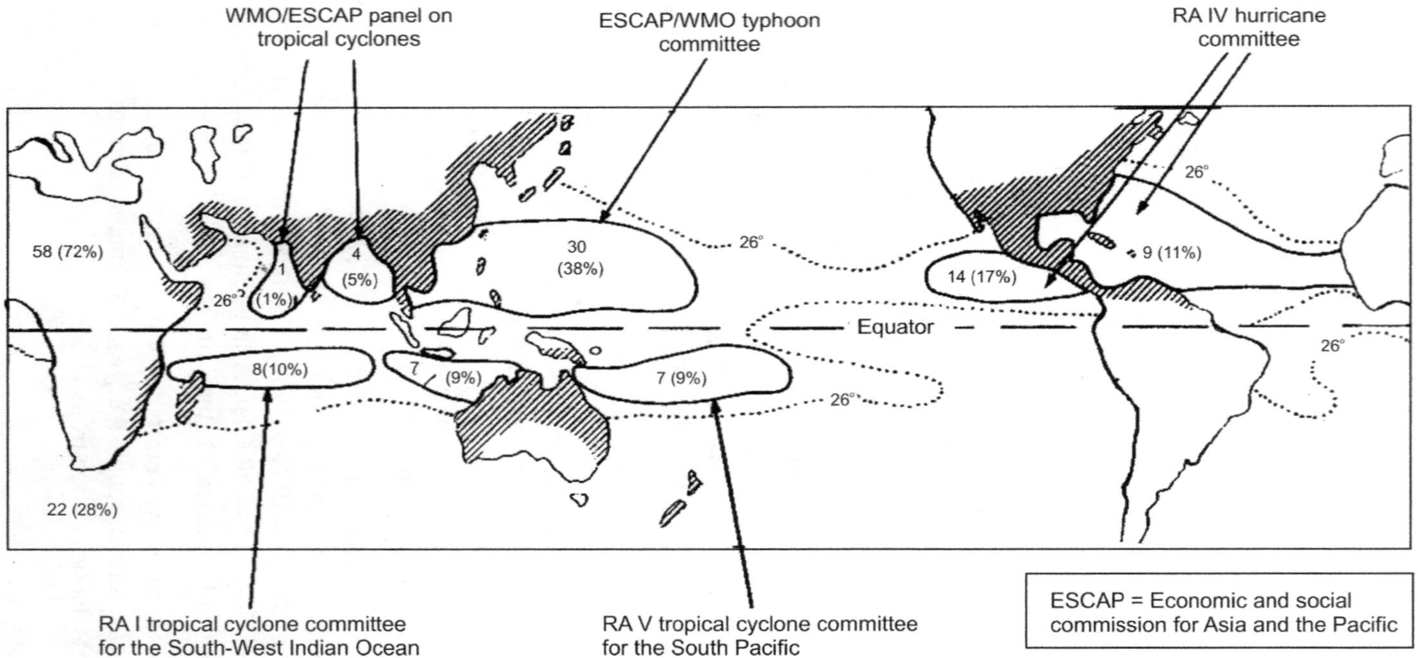

Fig. 25.3 The global occurrence (numbers and percentages) of tropical cylones. Cyclones form only where sea temperature exceed 26 ^{O}C (dotted line); the land areas affected are shown hatched (after W.M. Gray. 1975). The names of tropical cyclone bodies and basins covered by their programmes are also indicated.

In Indian seas most of the cyclonic storms form in the months of April, May and in October, November and December. The cyclonic disturbances over Bay of Bengal are more frequent than over the Arabian sea. Tropical disturbances of land origin are confined to southwest monsoon period (June to September). During January, February and March the tropical storms do not form in the Arabian sea. In the Bay of Bengal they are a few and far between. They usually originate between latitudes 5 °N and 8 °N, move in a westerly or north–westerly direction and strike the North Tamilnadu coast or the east coast of Srilanka. During April, May they form between latitudes 8 °N and 15 °N, move in a north–westerly or northerly direction initially and then recurve. In Bay of Bengal the whole of east coast of India. Coastal areas of Bangladesh and Arakan coast of Myanmar are likely to be hit by the storms, particularly in May. In the Arabian sea these storms move towards the coast of Arabia. A few move in a northerly direction towards Maharashtra, Gujarat coast. In June there is a striking change from May. Storms in Bay of Bengal originate between latitudes 16 °N and 21 °N and west of longitude 92 °E. They move in a north–westerly direction and weaken after crossing the coast. In Arabian sea most of the stroms confined to the area north of lattitude 15 °N and east of longitude 65 °E. During July and August stroms in the Bay of Bengal form between lattitudes 16 °N and 21 °N and west of longitude 92 °E. They move in a north–westerly or west–north–westerly direction between latitudinal belt of 20 °N and 25 °N. There is an abrupt fall in the frequency of storms in the Arabian sea from June to almost nil in July, August and September. In September Bay storms originate north of latitude 15 °N and west of longitude 90 °E, move initially in a westerly or North westerly direction and later recurve towards north-north-east. During October, November the frequency of storms increases rapidly. Bay storms originate between latitudes 8 °N and 14 °N, move initially in a northwesterly or west-north-westerly direction. Most of them later recurve to north-east wards. During these months north coastal Tamilnadu, Andhra coasts and Bangladesh coast are relatively vulenerable to the incidence of storms. In Arabian sea initially they move in westerly or north-westerly direction up to lat 15 °N, then recurve north-east wards and strike the Maharashtra – Gujarat coasts. During December the frequency of disturbances fall. Most of the Bay storms originate between latitude 5 °N and 10 °N, move initially in a northwesterly direction and strike Tamilnadu coast or northeast coast of Srilanka. A few of these cross over peninsula and enter into Arabian sea. It may be noted here that the position formation of storms migrates with the movement of the sun both north wards and south wards.

Flying Conditions in Cyclones

Flying conditions within the storm area are extremely hazardous. Cloud formation in the storm area tends to be in bands spiraling inwards into the centre (eye). Cloud bases may go down to the surface and tops may reach 16-20 km. Severe turbulence in TCu and Cb. Due to strong winds frictional turbulence is also common in lower levels. All hazards of thunderstorms are also a common feature. Visibility becomes very poor in torrential rain. Pressure changes are also very high en-route.

Cyclone Warnings in India: IMD issues cyclone warnings to various interests like railways, telephones, telegraphs, public works. Irrigation and the public on radio / DD. Port warnings are issued to hoist various visual signals.

A Name Scheme of US Weather Bureau: Since 1953 the US weather Bureau used names of girls to identity tropical storms to avoid confusion when two or more storms existed simultaneously. The name scheme which is in use since 1973 with semi-permanent list for 4 years with alphabetic order omitting (Q, U, X, Y, Z) and beginning from A in each year, given below. After 4 years, the sets are used again.

"Anna, Blanche, Carol, Debbie, Eve, Francalia, Gerda, Holly, Jenny, Kara, Laurie, Martha, Netty, Orva, Peggy, Rhoda, Sardie, Tanya, Virgy and Wenda"

The frequencies of global tropical cyclones are given in the Table 25.2.

Table 25.2 Global Annual Frequencies of tropical cyclones (Wind speed exceeding 50 KT)

Location	Annual frequency
Western North Atlantic	10
Western North Pacific	27
Eastern North Pacific	14
North Indian Ocean	5
North Hemisphere	56
Western South Pacific	8
Southwest Indian Ocean	7
Southeast Indian Ocean	14
Southern Hemisphere	29
Global Total	85

Questions

1. The most destructive phenomena of atmospheric nature is:
   (a) Tornado (b) Kalbaisakhi
   (c) Extra-tropical cyclone (d) Tropical cyclone
2. The most violent storms on earth is
   (a) Tornado, (b) Kalbaisakhi
   (c) Extra-tropical cyclone (d) Tropical cyclone
3. The largest phenomena of atmosphere is:
   (a) Tornado (b) Dust storm
   (c) Extra-tropical cyclone (d) Tropical cyclone
4. The tropical cyclones in western Pacific (wind speed 64 kt or more) are called:
   (a) Cyclones (b) Hurricances
   (c) Willy-willies (d) Typhoons
5. The tropical cyclones in Atlantic and eastern Pacific are called:
   (a) Cyclones (b) Hurricanes
   (c) Willy-willies (d) Typhoons
6. The tropical cyclones adjoining Australia in Pacific are called:
   (a) Cyclones (b) Hurricances
   (c) Willy-willies (d) Typhoons
7. The tropical cyclones in Indian Oceans and in south Pacific are called:
   (a) Cyclones (b) Hurricances
   (c) Willy-willies (d) Typhoons
8. Tropical cyclones form where.......and they do not form in latitude......
   (a) Sea surface water temperature is more than 26 °C; belt of 5 °N to 5 °S
   (b) Surface temperature is more than 26 °C; belt of 23 °N to 66° ½ N
   (c) Lapse rate is 10 °C / km or more; 10 °S to 23 °S
   (d) None above

9. According to IMD convention, when sea surface wind ranges 48 to 63 kt, satellite picture intensity 3.5 to 4.0, pressure defect at centre 15-21 hPa, the system is called:

   (a) Deep depression (b) Cyclonic storm

   (c) Severe cyclonic storm (d) Super cyclone

10. According to IMD convention, when sea surface wind ranges 64-119 kt, satellite picture intensity 4.5-6.5, pressure defect at centre 28-80 hPa, the system is called:

    (a) Cyclonic storm (b) Severe cyclonic storm

    (c) Very severe cyclonic storm (d) Super cyclone

11. According to IMD convention, when sea surface wind strength exceeds 120 kt, satellite picture intensity 6.5 or more and central pressure defect 80 h Pa or more, the system is called

    (a) Hurricane, (b) Severe cyclonic storm

    (c) Super cyclone (d) Deep depression

12. According to IMD convention, when sea surface wind strength ranges 34-47 kt, satellite picture intensity 2.5 to 3.0 and central pressure defect 6-10 h Pa, the system is called

    (a) Cyclone (b) Hurricance

    (c) Typhoon (d) Super cyclone

13. In a fully developed cyclone, the core temperature may be……than its surroundings up to about 15 km height

    (a) 5 °C warmer (b) 10 °C warmer

    (c) 5 °C colder (d) 10 °C colder

14. In a formative stage of a cyclone, pressure……,winds……and a closed vertex emerges

    (a) Falls to 10 h Pa below normal; strengthen

    (b) Pressure remains steady; winds strengthen

    (c) Pressure rises; remain steady

    (d) Pressure rises; strengthen

15. In developing stage of a cyclone, pressure……winds……to maximum limit but confined to a small area
    (a) Falls readily, strengthen
    (b) Falls rapidly, strengthen
    (c) Remains steady, remain steady
    (d) Rises, remains steady
16. In a mature stage of a cyclone, the central pressure……………but……………….wind strength exceeds ……….on BFS
    (a) Remains steady; wind circulation expands; BF 10
    (b) Falls; wind circulation reduces; 6 BF
    (c) Rises; wind circulation remains steady; 10 BF
    (d) Rises; wind circulation reduces; 8 BF
17. In mature stages of a cyclone, horizontal distribution exhibits four parts: (i) …(ii)…(iii)…(iv)
    (a) Dense cloud at the centre; lowest pressure at the centre; squally weather; lightning
    (b) Clear eye; wall cloud region; belt of hurricane winds; outer region
    (c) Diffused eye; inner thick cloud region; squall lines belt; region
    (d) Dense cloud at centre; which extends to the outer region; outer region
18. In northern hemisphere, the maximum number of cyclones develop during……while in southern hemisphere it occurs during……
    (a) August-September, December
    (b) August-September, February-March
    (c) January-February , June-July
    (d) November-December, April-May
19. In Indian seas, maximum frequency of cyclones occur during…...
    (a) January-February and October-November
    (b) April-May and October-November
    (c) June-September and January-February
    (d) None above

Answers

1. (d)	2. (a)	3. (c)	4. (d)
5. (b)	6. (c)	7. (a)	8. (a)
9. (c)	10. (c)	11. (c)	12. (a)
13. (b)	14. (a)	15. (b)	16. (a)
17. (b)	18. (b)	19. (b)	

CHAPTER 26

Monsoon

The word monsoon is derived from the Arabic word mousam, meaning season. It cannotes seasonal reversal of wind, relief from sweltering heat and plenty of rainfall (75% or more of annual rainfall). The characteristic features of monsoon air mass is that is moist maritime and extends vertically to a depth of 3 to 9.5 km. Consider globally the areas of monsoon in the northern summer (June-September – summer monsoon) prevailed over large parts of Indian Ocean, Pacific Ocean and Atlantic Ocean. During this period southern hemispheric air mass crosses the equator and enters into the northern hemisphere in India, Southeast Asia and parts of Africa. During southern summer (or northern winter) (December-March) northern hemispheric air mass crosses the equator and enters into South America (Brazil), eastern Africa and Northwest Australia. These are main winter monsoon regions where the air mass of the colder hemisphere penetrates into the warmer hemisphere. The areas influenced by monsoons are:

Indian sub-continent, South east Asia, Japan, southeast china, west Africa, eastern Africa, northwest Australia and southeast America (Brazil).

Asian Monsoon: The Asian monsoon region extends from latitude 10 °S to 50 °N, with three distinct circulations over this region (i) Indian sub-continent monsoon, (ii) Malaysian monsoon, (iii) Japanese monsoon. Like the active cold front, the seasonal reversal of winter continental winds penetrate into summer maritime winds causing changes in temperature and precipitation patterns.

The countries affected by monsoons are: India, Pakistan, Bangladesh, Myanmar, Thailand, Combodia, Laos, Vietnam, Malaysia, Singapore, China and Japan.

In these regions summer monsoon is reflected by tropical disturbances (typhoons, cyclones). The frequency of disturbances is highest in Philippines. Summer monsoon commences withdrawal from northern most region in the last week of August and completes withdrawal by November end. Philippines experiences both summer and winter monsoons - summer monsoons extend from May to September. Easterly waves travel close to ITCZ during June-December. Winter monsoons extend from November to February.

African Monsoon: Over west coast of Africa the ITCZ remains north of the equator throughout the year. In July it occupies the northern most position and in January the southernmost position. ITCZ oscillates with seasonal march northwards or seasonal retreat southwards. Rainfall follows the ITCZ. Westerlies prevail in the lower troposphere and easterlies aloft. Westerlies decrease in strength while easterlies increase in strength with height in the lower troposphere. Westerlies are cool and moist.

Winter Monsoon: During winter (October-March) the Sibarian anticyclone (High) intensifies by the passages of migratory extra-tropical cyclone waves. As a result at the eastern end of the High, north-easterlies blow over east China and give winter precipitation. The Siberian High periodically propagates the cold surges with north-easterly bursts. These northeasterlies move over South-China and then to near equatorial trough (NET). Easterly waves move in the neighbourhood of NET and germinate Depressions / Cyclones. These systems give copious rain in the equatorial region and Malaysia, north Indonesia and South China. This is winter monsoon over China.

Australian Monsoon: North Australia gets 90 % of annual rainfall during summer (November to April) and experiences dry season during winter (May to October). During onset of monsoon rainfall amount suddenly increases while in withdrawal phase it drops suddenly. Like the

Indian summer monsoon (June-September), Australia experiences tropical disturbances (lows, depressions, cyclones - called willy-willies).

Australian monsoon wind circulation shows westerlies in the lower troposphere and easterlies in the upper troposphere. Monsoon triggering mechanism is believed to be cold surges of South China sea and low level south easterlies along the western coast of Australia.

26.1 Indian Summer Monsoon

The characteristic features of Indian summer monsoon air mass is moist maritime and extends vertically to a depth of 3 to 9.5 km (300 h Pa). More than 80% of the annual rainfall occurs in monsoon period (June-September). Indian monsoon is an annual oscillation of atmosphere with great regularity. The description of monsoon strength over land and sea areas differ. According to IMD convention, the strength of monsoon over sea and land areas is given in table 26.1.

Table 26.1

Over sea area

Description	**Wind speed reported or estimated to be existing**	
Weak monsoon	upto 12 kt (23 kmph)	
Moderate monsoon	13 to 22 kt (24-42 kmph)	 to 
Strong monsoon	23 to 32 kt (43-60 kmph)	 to 
Vigorous monsoon	33 kt or more (61 kmph or more)	≥ 

Over land Area

Description	**Rainfall compared with long period average called normal at a place or sub–division**
Weak monsoon	RF less than half the normal
Normal monsoon	RF half to less than 1½ times the normal
Active monsoon	(i) RF 1½ to 4 times the normal (ii) RF in at least two stations should be 5 cm if the sub-division is along west coast and 3 cm if it is elsewhere in India. (iii) RF in that sub-division should be FWD or WD.
Vigorous monsoon	(i) RF more than 4 times the normal (ii) RF in at least two stations should be 8 cm if the sub-division is along west coast and 5 cm if it is elsewhere in India. (iii) RF in that sub-division should be FWD or WD.

Normal RF: It is the average RF of long period (morethan 30 years) of a station.

Isolated (one or two places) means RF over a sub-division is less than 25% of the stations of the sub-division reporting rainfall.

Scattered RF (at a few places) means 25 to 50% of the station of the sub-division reporting rainfall.

Fairly wide spread (FWD or at many places) stands for 50 to 75% of the stations of the sub-division reporting RF.

Widespread (WD or at most places) stands for 75 to 100% of the stations of the sub-division reporting RF.

A rainy day is defined when 24 hrs (0830 to 0830 hrs IST) is 2.5 mm or more.

Rainfall intensity in 24 hrs

Rain Fall	Description
0.1-2.4 mm	Very light rain
2.5-7.5 mm	Light rain
7.6-64.9 mm	Moderate rain
65.0-124.9 mm	Heavy rain
125.0 mm or more	Very heavy rain

The time and space variation of RF during monsoon is an inherent phenomena. Monsoon dynamics is viewed (like land and sea breeze circulation) as the cause of differential heating of vast land (Asian continent) on the north and the sea (Indian ocean) in the south. There is a great rhythm or cyclicity of onset and withdrawal of mansoon phenomena. The normal dates of onset of southwest mansoon and withdrawal are shown in Fig 26.1(a) & (b). The mansoon rainfall over India and its variation are shown in Figs. 26.2(a) & (b).

Monsoon rainfall over India varies from more than 800 cm to less thaan 40 cm. The coefficient of variation is 20% in Northeast India and Konkan to 60% (maximum 80%) in Rajasthan and south Tamilnadu.

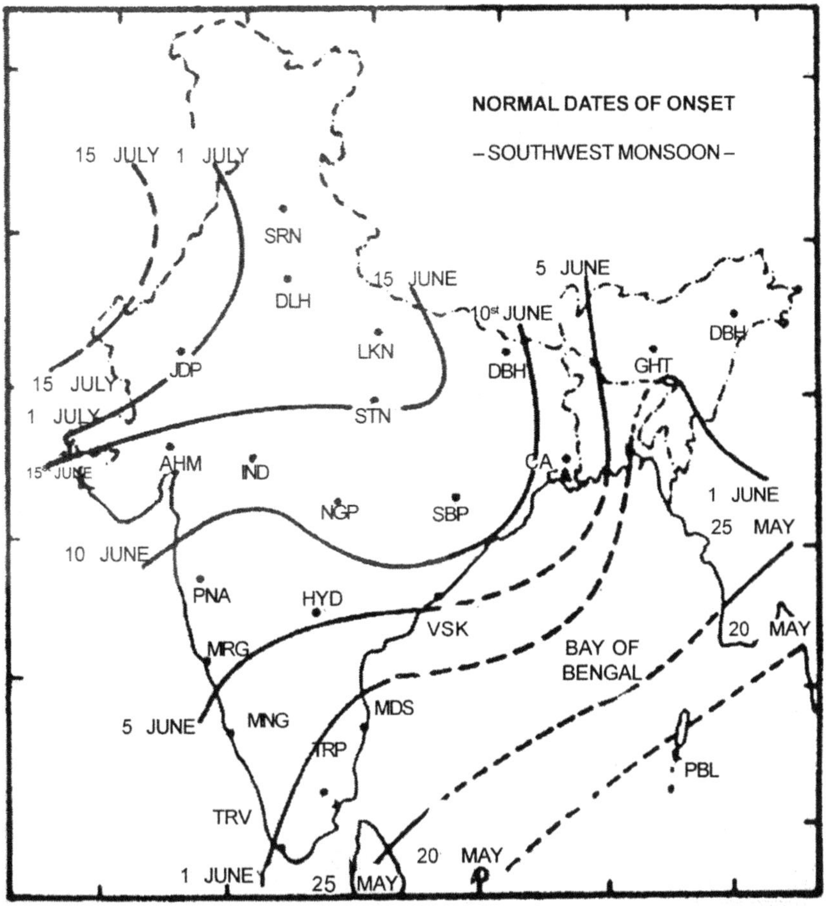

Fig. 26.1(a) Normal dates of onset southwest monsoon over India (Rao Y.P 1976).

Inter-Monsoon Rainfall Variability

Inspite of regular annual recurrence of summer monsoon there are large variation of rainfall within the season. These variations are attributed to synoptic scale disturbances, oscillations of monsoon trough, mid-latitude effects. Synoptic scale disturbances such as Lows, Depressions, Cyclonic storms, mid-tropospheric cyclones, off shore vortices cause rainfall variations in periods of 5-7 days. The intensity, frequency of disturbances have dominant role in floods. The oscillation or meandering of monsoon trough (on sea level chart) from its normal position causes rainfall variability in periods of 10-15 days. Shifting of the eastern end of monsoon trough to the foot hills of Himalayas generally cause floods in Brahmaputra, while most part of the country receives scanty rainfall. On the contrary if the monsoon trough shifts south of its normal position widespread rainfall occurs over most part of the country.

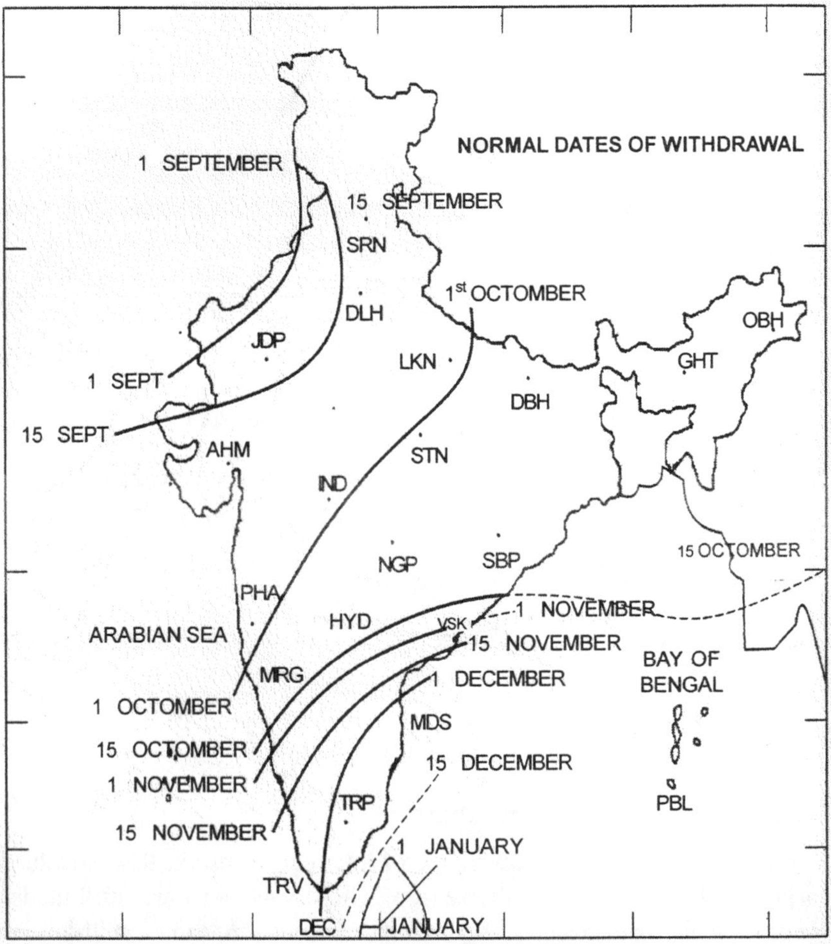

Fig. 26.1(b) Normal dates of withdrawal of southwest monsoon over India (Rao Y.P. 1976).

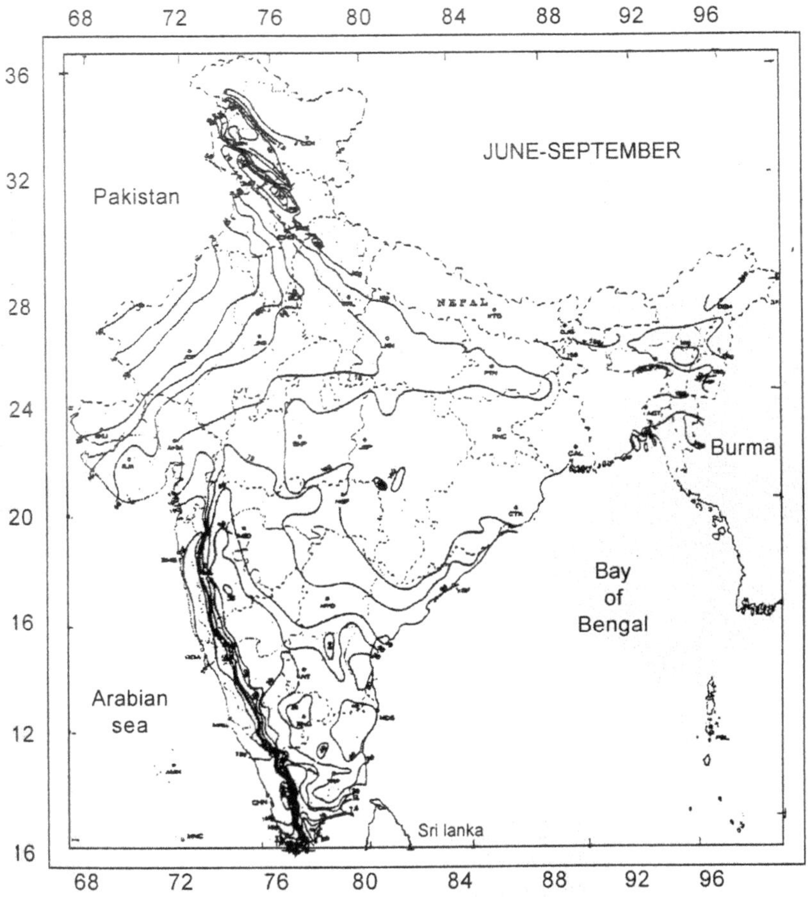

Fig. 26.2(a) Monsoon Rainfall over India.

There are five semi permanent features of monsoon which are described below.

(i) ***Heat Low:*** The formation of heat low over Pakistan and adjoining Rajasthan is a semi-permanent feature of Indian monsoon. Heat low is very shallow extends upto 1.5 km asl. Above the heat low there exists a marked ridge (part of Sub-Tropical High). Rainfall associated with this low is very small in that area and is the cause of formation of Thar desert in Rajasthan. The surface air temperatures in the area are high which evaporates the moisture. However the intense heat low acts on a suction devise of moist air along the monsoon trough to some extent related to good monsoon over India. During weak heat low monsoon rainfall over most part of India is deficient or scanty.

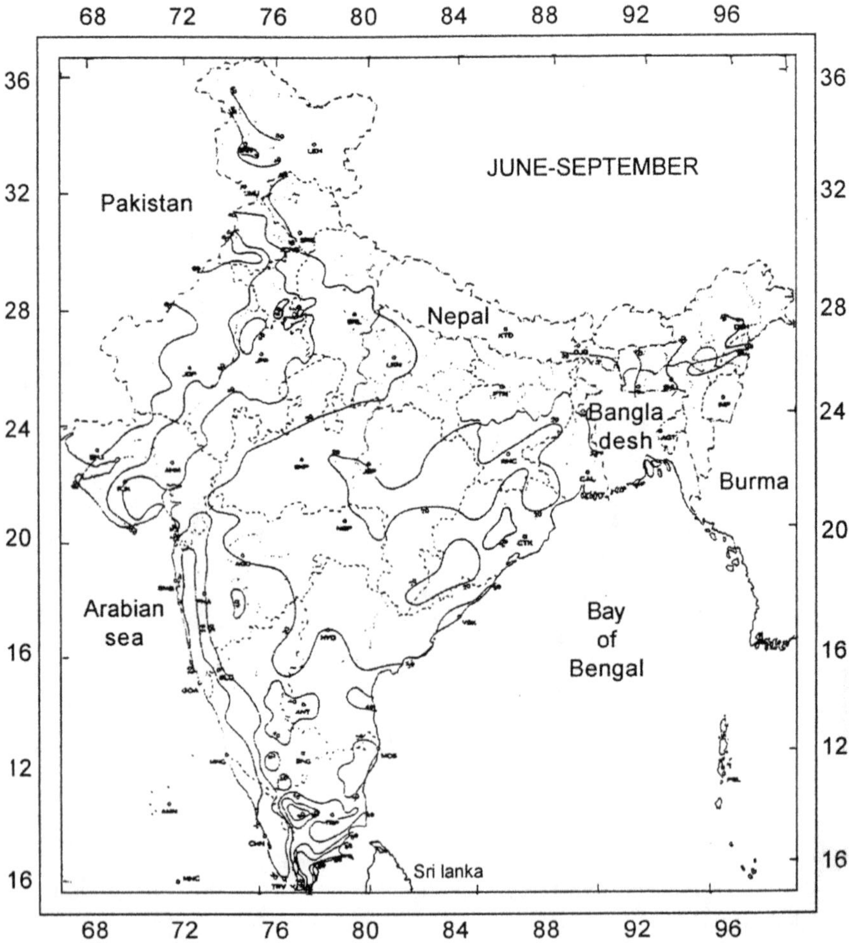

Fig. 26.2(b) Coefficient of variation of rainfall.

(ii) ***Monsoon Trough:*** During monsoon period an elongated low pressure area extends from heat low over Pakistan to Head Bay of Bengal. This is called monsoon trough. This is a semi-permanent feature of monsoon circulation. It migrates sometimes southwards and sometimes northwards. Southward migration results in active/vigorous monsoon conditions over major part of the country, while its migration northwards leads to break monsoon conditions over major part of the country, however heavy rains occur along the foot hills of Himalayas.

(iii) ***Mascarene High:*** During monsoon period a High pressure area is located around Mascarene Islands (in South Indian Ocean). This is responsible for cross-equatorial flow and acts as a southern hemispheric linkage. The variation in the intensity of high pressure causes monsoon surges across equatorial flow which is responsible for heavy rains along west coast of India.

(iv) ***Tibetan High:*** It is a large warm anticyclone situated over Tibetan Plateau (Centre at lat 28^0 N, long 98^0 E) in middle/upper troposphere during June-September. Shifting its position east or west causes variation in monsoon activity over India. West movement causes monsoon transients from Bay to northwards.

(v) ***Somali Jet:*** Low-level (upto 1.5 km asl) inter-hemispheric cross equatorial flow of air attains jet speed at the west end of monsoon regime along the east coast of Africa. The jet originates near Mauritius and northern part of Madagascar in the southern hemisphere. The jet reaches Kenya Coast (about 3^oS) covers the plains of Kenya, Ethiopia and to Somali coast at about 9^0 N. During May it moves further north into eastern Africa, then into Arabian sea and reaches west coast of India in June. It attains maximum strength in July. The Low Level Jet gives rise to strong monsoon over peninsular India.

Monsoon Disturbances

During monsoon (June-September) a series of low pressure systems (called monsoon disturbances) form over Head Bay of Bengal and generally move in a west-north-westerly direction along the monsoon trough and weaken over west MP or Rajasthan. On their way these system give copious rain. Some of these pressure systems intensify into Depressions or Deep Depressions and rarely become cyclones (Def. of the systems refer to Table 2.1). The average occurrence of monsoon disturbances is 2-3 per month or 10 per season. The frequency of these disturbances in the Bay of Bengal is much more compared to Arabian sea and over land. Monsoon Depressions generally move in west-north-westerly direction and weakens over Northwest India. Along the path of these disturbances monsoon strengthens to active/vigorous conditions and cause floods.

26.2 Breaks in the Monsoon

Monsoon fully establishes over India by the end of June and thereafter sustained monsoon conditions prevail over whole of India during July and August. With-drawal begins in September middle. However during July, August months there were occasions of short period, say about a week, when rainfall over most part of the country is absent. This low rainfall activity is termed "Breaks in monsoon'' which is associated with change in established monsoon circulation. A convention is followed that if the monsoon trough is absent on sea level chart and at 850 h Pa level for more than two days then it is called break in monsoon. According to this convention the longest break of 20 days occurred in 1906 (29 July to 18 Aug). A statistical study of 80 years data (1887-1967) shows that the average period of breaks in July 5.8 days and August 6.5 days and most frequent duration in these months 4 and 3 days respectively and the longest period in these months being 17 and 20 days respectively. Standard deviation is about 3 days. There was no break in 12 years out of 80 years indicating the probability of occurrence of break is surface 0.85. A typical surface chart of break monsoon is shown in Fig. 26.3 and weak/break monsoon condition trough axis are shown in Fig. 26.4.

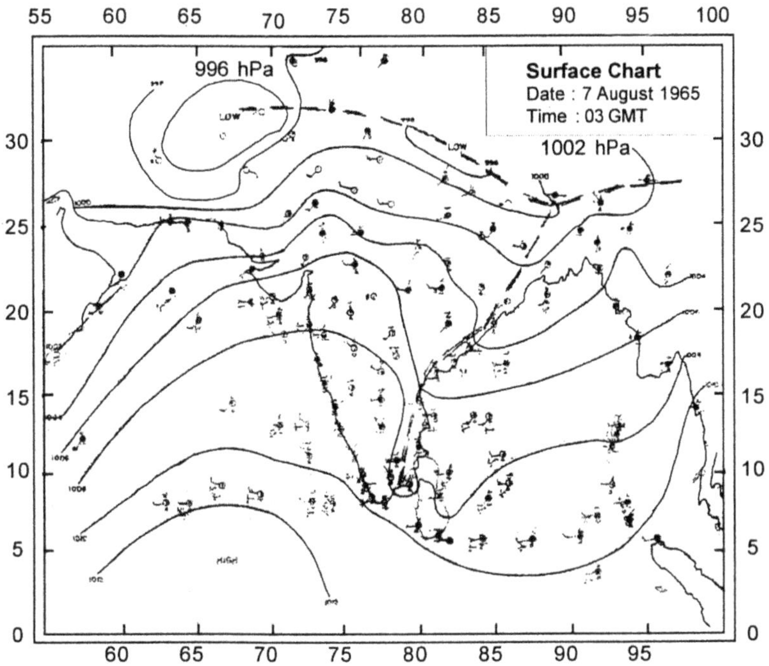

Fig. 26.3 Break monsoon conditions typical surface chart.

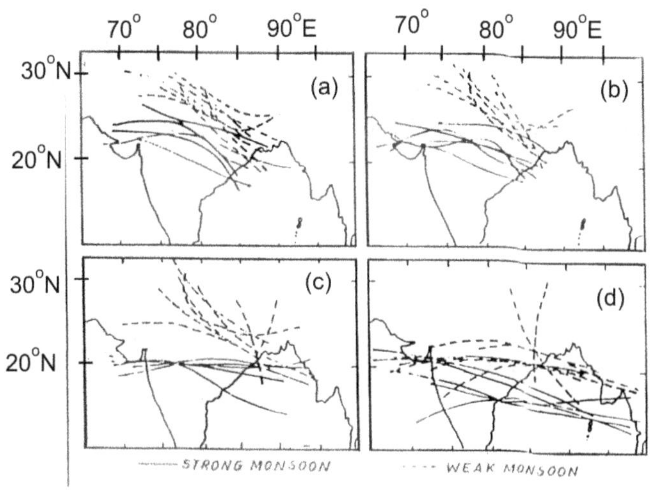

Fig. 26.4 Mean positin of the monsoon trough during strong (normal) and weak (break) monsoon; (a) at 900 hPa, (b) at 850 hPa (c) at 700 hPa (d) 500 hPa.

Synoptic Situations During Breaks

Shifting of Monsoon trough to the foot hills of the Himalayas causes break. This may be caused by :

(i) The eastward movement of active western disturbance north of the country,

(ii) Pronounced low index westerly circulation in middle latitudes,

(iii) Eastward movement of large amplitude trough in westerlies (between 500-300 h Pa) over north of India across Tibetan Plateau,

(iv) Movement of Bay depression northwards to foot hills of the Himalayas,

(v) Movement of a western Pacific Typhoon to north of 30°N over China or its recurvature northeastwards.

Break in monsoon may be caused by the formation of a low in lower latitudes which moves westwards along lat 10 ^{O}N over south peninsula or by the formation of High pressure ridge at 500 h Pa level over Central India.

During break monsoon conditions mid-latitude upper tropospheric ridge (at 300 h Pa) shifts southwards to about 26 ^{O}N, mean sea level pressure falls to below normal over south peninsula (south of 13 ^{O}N) where surface temperature becomes above normal. This warming is observed in the entire troposphere. Over central parts of India sea level pressure rises to the order of 4 h Pa.

During break monsoon, heavy rains occur along foot hills of the Himalayas and over southern parts of India (Tamilnadu, south Andhra Pradesh), while rainfall activity drastically reduces over western and central parts of India. In upper troposphere easterly Jet shifts northwards as far as 23 ^{O}N (normal position is 13 ^{O}N). Two Jet core maxima are located, one at about 12 ^{O}N (south of its normal position) and another at about 22 ^{O}N. Two westerly jet cores are also observed, one at about 32 ^{O}N and another at about 40 ^{O}N.

26.3 Northeast Monsoon

In India the period October to December is called northeast monsoon. In this period south coastal Andhra Pradesh, Rayalaseema, Tamilnadu and neighbour- hood gets bulk of the annual rain. For Tamilnadu (TN) and neighborhood it is main rainy season. Coastal districts of TN get 60% of annual rainfall while interior districts get 40-50% of annual rain. Rayalaseema gets about 36% of annual rain during September, October months, which are the rainiest months. According to IMD classification of seasons, Oct-Dec period forms post monsoon or retreating south–west monsoon because it follows the SW-monsoon (June-Sept). SW–Monsoon begins to retreat from north India from the middle (15th) of September and completes withdrawal from whole of India by 15th October. Simultaneously monsoon circulation pattern also changes. Instead of low pressure systems forming over north Bay (during SW monsoon) a low pressure area forms over central and south Bay of Bengal, which further shifts to south Bay during November and south of Lat 5°N during December. An east-west trough extends from this low to south Arabian sea with associated upper air trough extending to 3.0 km asl. In the mid and upper troposphere easterlies prevail over south peninsula but the easterly jet which was present during SW-monsoon disintegrates.

The following are the IMD criteria for declaration of NE-Monsoon over Tamilnadu:

(i) Withdrawal of SW-monsoon upto lat 15°N.

(ii) Onset of persistent surface easterly winds along Tamilnadu coast.

(iii) Depth of easterlies extend upto 1.5 km

(iv) At least fairly widespread rainfall over coastal Tamilnadu and south coastal Andhra Pradesh and neighbourhood.

(v) Onset not to be declared before 10th October even if all above conditions exist.

Rainfall Distribution over Coastal Andhra Pradesh and Rayalseema

These areas get rainfall both in summer monsoon and NE monsoon but receives more in SW-monsoon. As in case of SW-monsoon there is no abrupt increase of rainfall (September to October) with the onset of NE-monsoon. Rainfall in October remains practically same as in September. It decreases rapidly from October to November and by middle of December it becomes insignificant. See Table 26.2.

Table 26.1 Comparison of southwest and northeast monsoon.

	SW-monsoon	NE-monsoon
1.	Rainfall increases abruptly onset of SW-monsoon.	There is no significant change in rainfall on onset of NE-monsoon.
2.	Rainfall increases from the coast to western ghats.	Rainfall decreases from the coast to interior Tamilnadu.
3.	Variation of rainfall is high in regions of poor rainfall.	Variation rainfall is high in the regions of highest rainfall.
4.	Presence of tropical easterly jet in the upper troposphere.	Tropical easterly jet completely disintegrates.
5.	Tropical disturbances assume the form of depression and rarely become cyclones.	Tropical disturbances frequently concentrate into cyclonic storms.
6.	Maximum rainfall occurs south of the track of depression.	Heavy rains occur both north and south of the cyclone / depression track.
7.	Active / vigorous monsoon conditions occur along west coast frequently (about 25% of occasions).	Active / vigorous monsoon conditions are less frequent along east coast.

Rainfall Distribution in Tamil Nadu

Tamilnadu gets more rainfall in October-December as compared to SW-monsoon. Rainfall considerably increases from September to October and continues in November. Coastal districts of Tamilnadu get more rain in November than in October, whereas the interior districts get more rain in October than in November. By December rainfall decreases substantially. Thus NE-monsoon rain mainly occurs in October and November.

In Kerala October is the rainset month of NE-monsoon. Here rainfall decreases in November.

Questions

1. The seasonal reversal of wind with copious rainfall is called monsoon. The main characteristic feature of monsoon air mass is that it is...........and extends vertically up to a depth of........
   (a) Dry continental; 16 km (b) Polar maritime; 2 km
   (c) Moist maritime; 3 to 9 km (d) Maritime; 1 km
2. The global area of monsoon during northern summer are:
   (a) Russia, Europe, Alaska
   (b) Large parts of Indian Ocean, Pacific and Atlantic covering India, southeast Asia, parts of Africa
   (c) North America, Canada, Antarctica
   (d) Area between latitudes equator to 30 °S
3. The global areas of monsoon during northern winter (December-March) are:
   (a) Between latitudes equator to 30 °N
   (b) Brazil (south America), eastern Africa, northwest Australia
   (c) Indian sub-continent, eastern Africa, southeast Asia
   (d) Central America, Europe, north China
4. The global monsoon areas are:
   (a) Between latitude 30 °N to 30 °S
   (b) Indian sub-continent, southeast Asia, Japan
   (c) Western and eastern Africa, north west Australia, Brazil,
   (d) Both (b) and (c)
5. Asian monsoon areas are:
   (a) Iran, Afghanistan, Tibet, Korea
   (b) India, Pakistan, Bangladesh, Myanmar, Thailand, Combodia,
   (c) Laos, Vietnam, Malaysia, Singapore, Japan, China
   (d) Both (b) and (c)
6. In Philippines, summer monsoon extends from...........and winter monsoon from..........
   (a) May to September; November to February
   (b) March to May; July-August
   (c) August-September; January-February
   (d) None above

7. Over west coast of Africa, the ITCZ lies..........throughout year and it oscillates ....
   (a) South of equator; with the march of sun
   (b) North of equator; with seasonal march and retreat
   (c) Equator; doest not oscillate
   (d) None above
8. Australian monsoon extends from..........and experiences ....like the Indian monsoon
   (a) June-September; Tropical disturbances
   (b) November-April; Western disturbances
   (c) November-April; Lows, depression, cyclones called willy-willies
   (d) Both (a) and (b)
9. Indian summer monsoon dynamics is viewed as the cause of......of vast land in............and ocean in.............
   (a) Adiabatic heating; south; north
   (b) Differential heating; the north (Asia); the south (Indian Ocean)
   (c) Winter cooling; east; west
   (d) Adiabatic cooling; east; west
10. The normal date of onset of monsoon in Kerala............and standard deviation...............
   (a) 1 January; 15 days (b) 1 June; 8 days
   (c) 1 March; 10 days (d) Both (a) and (c)
11. At about the time of onset of monsoon over Kerala, the sub-tropical westerly jet stream..........and an easterly jet stream observed along latitude.............
   (a) Located along latitude 25 °N; 5°N
   (b) Moves away north ward of north India; latitude 12-13 °N
   (c) Located along latitude 15 °N; 5 °S
   (d) Both (a) and (c)
12. The semi-permanent features of Indian summer monsoon are:
   (a) Heat low over Pakistan and adjoining Rajasthan, monsoon trough extends from heat low to head Bay of Bengal
   (b) A High over Mascareni islands which is responsible for the cross equatorial current and Tibetan high at 300 h Pa level
   (c) Low level Somali jet, which originates near Mauritius. This provides cross equatorial low level jet stream
   (d) All above (a) to (c)

13. During monsoon period, winds in the lower level......... south of monsoon trough axis and.........north of it
    (a) Easterlies, westerlies (b) Westerlies, easterlies
    (c) Southerlies, northerlies (d) Both (a) and (c)
14. During monsoon period over most part of India upper winds 300 hPa and above will be:
    (a) Westerlies (b) Easterlies
    (c) Northerlies (d) Southerlies
15. During monsoon period, a series of low pressure systems form over...........and move in............direction along..........and weaken over west MP or Rajasthan.
    (a) Arabian Sea; southeasterly;.......ridge
    (b) Head Bay of Bengal; westerly or WNW ly; monsoon trough
    (c) South west Bay of Bengal; northerly; east coast of India
    (d) Both (a) and (c)
16. The position of monsoon trough during active monsoon period lies..............normal position and during break monsoon period lies...........normal position
    (a) North of; south of;
    (b) South; north
    (c) Does not change its position, does not change its position
    (d) Both (a) and (c)
17. During active monsoon period, easterly jet stream located at (pressure level).........and (latitude)..........and (strength)......
    (a) 150-100 h Pa level; latitude 13-15 ^{o}N; increased in strength
    (b) 200 h Pa; latitude 27 ^{o}N; decrease in strength
    (c) 850 h Pa; latitude 8 ^{o}N; decrease in strength
    (d) Both (b) and (c)

Answers

1.	(c)	2.	(b)	3.	(b)	4.	(d)
5.	(d)	6.	(a)	7.	(b)	8.	(c)
9.	(b)	10.	(b)	11.	(b)	12.	(d)
13.	(b)	14.	(b)	15.	(b)	16.	(a)
17.	(a)						

CHAPTER 27

The General Circulation of the Atmosphere

Atmospheric motion is a complex phenomenon, it is very difficult to put forth a theory using hydrodynamics and thermodynamic equations, even though the behaviour of atmosphere is governed by the physical and dynamical laws. Broadly speaking, the general circulation may be described as the average atmospheric flow on the earth. It can be described convincingly by statistical terms. The atmospheric flow statistics are generated from the aggregates of daily patterns. Thus general circulation may be described as the time and space average phenomena of winds, temperatures, pressure, humidity and their variations during seasons, years or months.

By collecting and analyzing statistics of meteorological elements for a particular location / area, climatology of that location / area is built. This type of knowledge of climate of a location / area is very useful to any traveler particularly of trade, industry or commercial interest in travel. In general the pilot and navigator must have a good background knowledge of weather of distant localities which help him in supplementing flight forecast. Aviation route planning depends on climatological data.

In order to suppress the random variation of the phenomena, the average quantities considered over a long period (more than 30 years), zonal (east west direction) averages (of winds etc) and meridional (north south direction) averages of (temperature, water vapour, heat transport etc) air considered well suited for describing many phenomena of atmosphere. General circulation takes into account of energy relations in the atmosphere.

The atmospheric motion is mainly driven by solar radiation and the angular momentum of the earth. The primary force driving the atmosphere is the uneven heating of the globe by insolation. This results in continuous transfer of energy from lower latitudes to higher latitudes. It is estimated that about 1 % of the insolation (about 2 watts / m^2) is absorbed by the earth-atmosphere runs this atmospheric heat engine.

By the law of inertia, earth's atmosphere moves with the earth around its axis of rotation. Angular momentum is proportional to the rate of spin of the earth (angular speed of the earth is about 465 m/s at the equator) and the square of the distance of the air parcel from the axis of rotation.

In eighteenth century, George Hadley, a British meteorologist propounded the general circulation of the atmosphere. According to this model, the heated equatorial air ascends from the surface of the earth and moves pole wards in the upper atmosphere (at about 5 km altitude), where it cools and sinks down to the earth. By this a high pressure area develops at the poles. At the surface again winds move towards equator. Thus circulation completes. This is the simplest treatment of meridional circulation of Hadley cell (See Fig. 27.1). However the atmospheric circulation is not so simple.

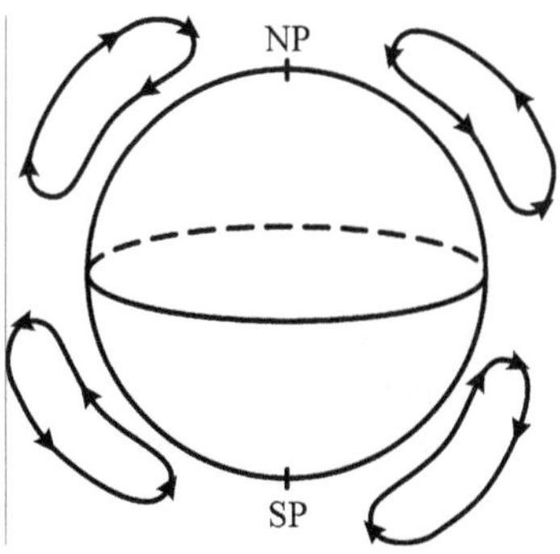

Fig. 27.1 Hadley convective circulation Model.

At any place, close to the surface of the earth, wind is never steady neither in direction nor in speed. It is very difficult to describe the varied atmospheric motions in a simple way. As mentioned earlier, the general circulation of the atmosphere is described as average phenomenon in a lucid patterns by statistical averages.

Trade Winds: The most dominant feature of the general circulation is reflected in trade winds in both hemispheres. Trade winds nearly blow over half of the globe. They originate near sub-tropical high pressure cells and blow from northeast (NE) direction towards equator in the northern hemisphere and from southeast (SE) direction towards equator in the southern hemisphere. In olden days these winds helped the trade vessels constantly blowing in particular direction and thus got the name trade winds. Trade winds are strongest in winter. From the STH belts wind blows north to south (or south to north) direction but deflected by the rotation of the earth (coriolis force) from NE (or SE) direction.

Doldrums/ITCZ: Between the trade winds there is a region near equator where light variable winds accompanied by heavy rain, thunderstorms and squalls are observed. These are traditionally called doldrums. In this zone sailing ships / vessels were be calmed (stranded) for several days or weeks over the sea and thus much feared. This zone lies between the trade winds and in the equatorial trough. Modern meteorologists prefer to call this sector as ITCZ (inter tropical convergence zone). Doldrums region is mostly covered with clouds and rains (murky skies) along with nearly calm or variable winds. It may be noted here that the average high temperature zone lies in this equatorial trough. This called thermal equator (or meteorological equator). It lies about 5 ^{o}N. The equatorial trough migrates with the sun. It is centred about latitude 5 ^{o}S in January and about 12-15 ^{o}N in July. It oscillates about 20 ^{o} latitudes between seasons. The mean position is latitude 5 ^{o}N, which is called meteorological equator.

Horse Latitudes: With the available present knowledge and data, the convective air rising at the equatorial zone does not travel up to poles but descends to a large extent at about 30 ^{o}N (S) where sub-tropical high pressure (STH) belts are located on surface level. The areas of STH belts are dominated by clear skies, light or variable winds. This belt migrates with the sun with a lag of one to two months. Mariners call these areas horse latitudes, because in these latitudes sailing ships / vessels were be calmed and traders sometimes use to throw away horses on board into sea which were being transported from Europe to America. Thus got the name horse latitudes.

The Equatorial Westerlies: In summer hemisphere there is a zone of westerly winds in equatorial region between trade winds, which migrates with the equatorial trough. Over the India Ocean it is located at about latitude 12-15 °N during summer (June-July) and at about 2-3 °S during winter (December-January). However this zone is not observed over Pacific and Atlantic Oceans.

Middle Latitude Westerlies: Between the STH and sub-polar low (SPL) there is a belt of westerlies which are prominent particularly in the southern hemisphere and found around the world. They are called "roaring forties" or "howling fifties" (located at about latitude 40 ° and 50 ° respectively). They are also called ferrel westerlies (See Fig. 27.2).

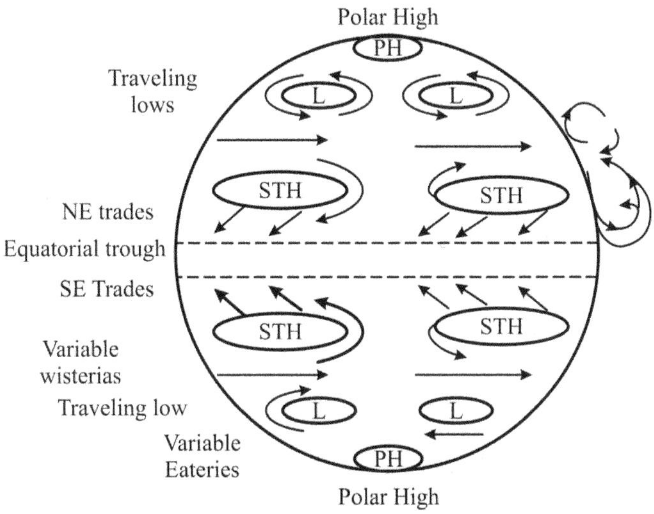

Fig. 27.2 Idealized distribution of surface pressure in general areolation.

Polar Easterlies: Between polar high, sub-polar low there are quasi permanent easterlies (See Fig. 27.2).

Ferrel Cell: In 1856, a British school teacher modified the Hadley cell model to explain the roaring forties. According to him the Hadley cell lies between equator and latitude 30 °, and another cell in opposite direction of wind motion to that of Hadley cell between latitude 30 ° and 60 °. This is called Ferrel Cell or indirect cell. Another weak cell exists between latitude 60 ° and pole which is similar circulation as the Hadley cell with rising winds equator ward and sinking winds at pole. Thus we have three cell model of general circulation of wind (See Fig. 27.3).

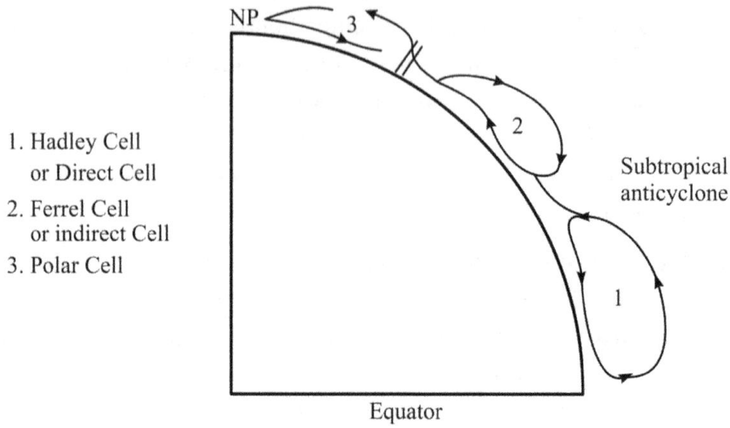

Fig. 27.3 General Circulation three cell model.

In spite of many excellent laboratory studies there is as yet no complete explanation of the general circulation model of the atmosphere. The above described principal wind systems are modified by land and sea surfaces. Land areas heated up quickly during day and cools down quickly during night. In contrast sea surface area heating and cooling goes slowly and hence maintains more even temperature throughout the year. In summer continents behave like heat sources and sea surface areas behave like heat sinks. In winter the reverse is observed. As a result of this temperatures differences, and the atmospheric pressure variations occur. This causes seasonal winds blowing between oceans and continents. This type of seasonal winds observed is case of monsoons.

Atmospheric motions thus differ greatly in their size, duration of life ranging from small turbulent eddies to large cyclones. All these motions are fed by the solar energy and cause circulation of water through the atmosphere (Hydrological cycle). General circulation of the atmosphere may thus be classified as: Planetary winds, terrestrial winds and continental winds.

Planetary winds are further classified as: (i) Equatorial winds or Doldrums, (ii) Trade winds, (iii) Mid-latitude westerlies or horse latitudes, (iv) Prevailing westerlies or roaring forties (or howling fifties) and (v) Polar winds or polar easterlies.

A secondary atmospheric circulation consists of land and sea breeze, valley winds (katabatic and Anabatic winds), mountain waves, tidal and volcanic winds, cyclonic storms.

27.1 Zonal Index

The variation in the mid-latitude westerlies (latitude 35 °N-55 °N) indicate some patterns, particularly in winter months when the zonal westerlies are strong. The variation lasts for about one to two months when the westerlies develop waves, troughs and ridges. They become prominent and finally take cellular shape. The strength of the westerlies between latitude 35 °-55 °N is called zonal index.

High Zonal Index: When westerlies are strong and located a little north of its average position with marked east-west oriented isobars and weak meridional flow is called high zonal index (See Fig. 27.4(a)). In this state the exchange of mass from north and south is not significant.

Zonal Index with Oscillation: When westerlies are strong and located south of its average position with deep trough and ridges with meridional (north-south) mixing of mass is called zonal index with oscillations (See Fig. 27.4 (b) and Fig. 27.4 (c)). In this case westerly jet expands and strengthens with undulation. Meridional heat transfer takes place. Cold air mass from higher latitudes move towards equator simultaneously warm equatorial air mass moves polewards.

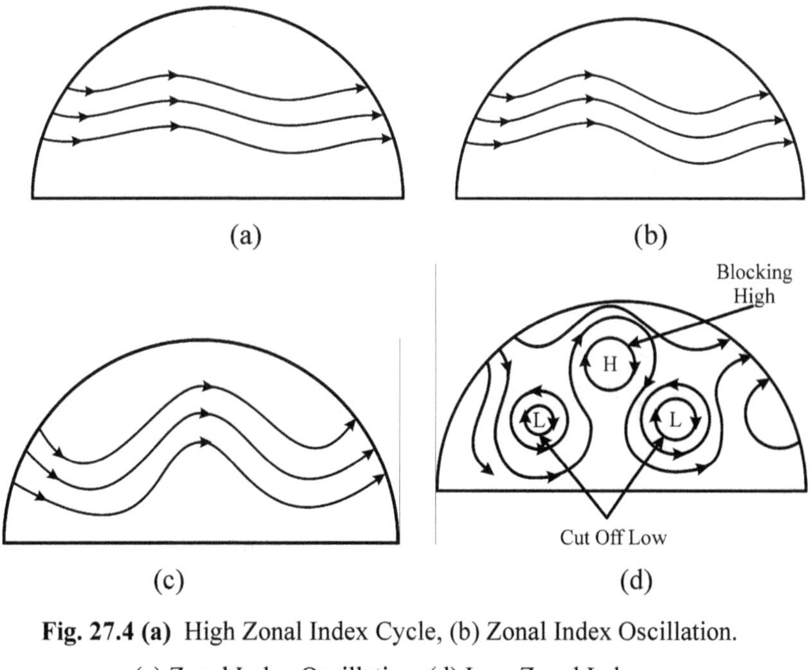

Fig. 27.4 (a) High Zonal Index Cycle, (b) Zonal Index Oscillation. (c) Zonal Index Oscillation, (d) Low Zonal Index.

Low Zonal Index: When the zonal westerlies completely break off with the formation of cut-off lows and blocking highs is called low zonal index (See Fig. 27.4 (d)). In this state westerlies are weak while prominent meridional movement of air is noted. Persistent cloudy weather lies in cut-off lows (in low latitudes) which are blocked by anticyclones (called blocking highs or anticyclones) in higher latitudes having fair weather.

27.2 Global Temperature Distribution

With respect to the idealized earth climate, the position of pressure zones (Equatorial Low, Sub-tropical High, Subpolar Low, Polar High) remain stationary throughout the year.

We have learnt that the rotation of earth about its axis is causing day and night and its revolution in elliptical orbit around the sun (which is at one of the focii of ellipse) and the earth's axis inclination about $23\frac{1}{2}^{o}$ to the elliptical plane causing seasons. Because of the seasonal changes and uneven heating of land and sea, we find lot of deviations from the ideal situations of the earth's climate. In general, temperature decreases with increasing latitudes from equator to pole. In idealized situation we expect maximum temperature at equator and minimum at poles, however in reality this is not the case. In summer hemisphere warmest land belt occurs in equatorial region. In northern hemisphere (NH) summer (in July) the warmest belt is latitude 20-30 °N, while in winter (the southern hemisphere summer) in January the warmest belt is 10-20 °S, this is because the southern hemisphere has more oceanic areas than land. In contrast to this, in January, the coldest region in northern hemisphere is Saiberia instead of the north pole region.

The seasonal variation of mean temperature in Saiberia, Canada, Greenland may be about 60 °C, where as in the equatorial region over the sea area the variation is less than 5 °C.

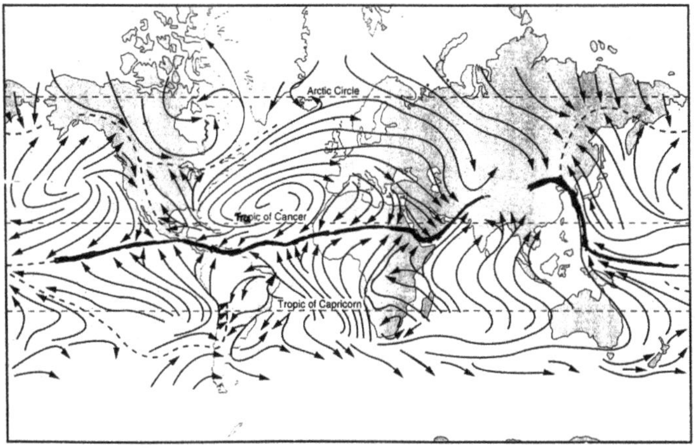

Fig. 27.5 Prevailing surface winds July and location of ITCZ

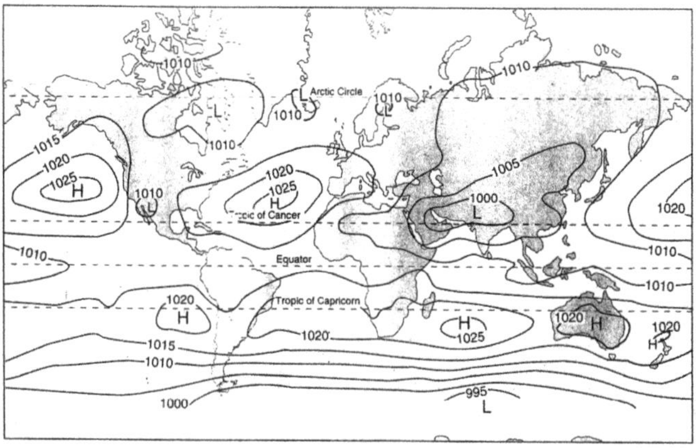

Fig. 27.6 Average pressure (hPa) at mean sea level in July

In Asia, the Himalayas arrest the spread of very cold air from Saiberia to the south into India, Pakistan and Bangladesh. Similarly in North America, the Rockies prevent the land mass from cool moist prevailing wind. In Europe Alps protect the Mediterranean area from the extreme cold weather from the north.

In preparing global map of temperature isotherms distribution, all observations are reduced to sea level using lapse rate 10 °C / km. Changes in global mean temperature is natural due to variations in atmospheric aerosols and gases. As said earlier surface temperature in general decreases from the equator to the poles. However these temperatures are modified by land and sea distribution and ocean currents, and also by seasonal variation. The January and July mean isotherms represent the hottest and coldest period distribution. The highest mean temperature occur in tropics. Northern hemisphere summer is warmer than southern summer due to more land mass in NH.

Global Pressure Distribution: The mean pressure distribution for January and July (which are the coldest and hottest month of NH) are shown in Fig. 27.7, 27.8 (also see fig. 27.9, 27.6) which shows definite pressure belts. As compared to the idealized pattern of pressure the maps show some modifications because of the distribution of land and sea masses, mountain barriers and ocean currents. The pressure belts are equatorial low pressure, the sub-tropical high pressure belt and the sub polar low pressure belt.

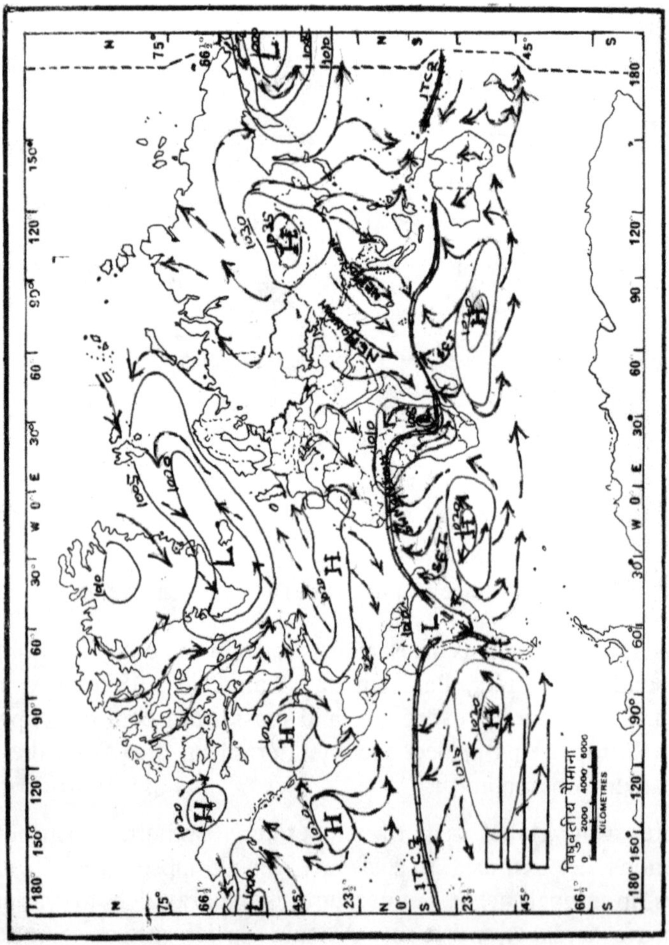

Fig. 27.7 January – Sea level Wind pressure distribution and ITCZ.

In January the axis of equatorial low pressure (ITCZ) belts lies south of the equator and the sub-tropical high pressure belt in NH migrates to the north over continents. In Asia, the STH pressure centre lies close to latitude 30 °N and the most intense cold around the arctic circle. The sub-polar low pressure belts are prominent over ocean area but it is not observed in southern hemisphere. Also the STH pressure belts in southern hemisphere (in January) are over ocean areas in their normal position but not observed over land due to insolation. The sub-polar lows are clear and intense. Pressure tendency shows high over Antarctica.

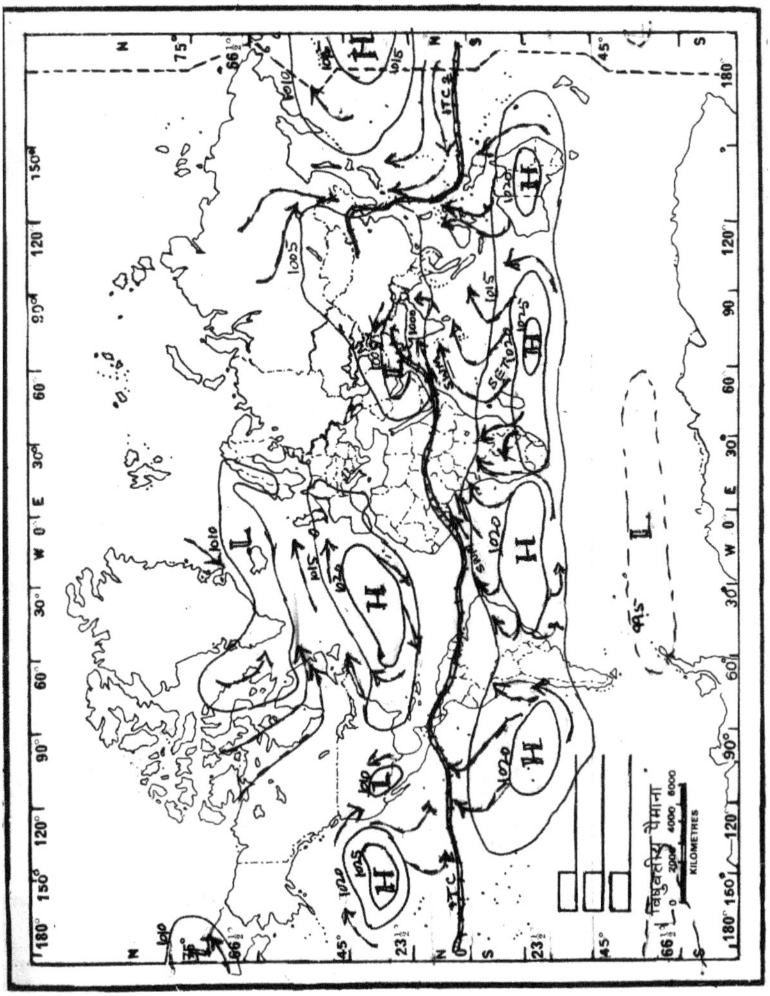

Fig. 27.8 July – Seal level Wind, Pressure Distribution and ITCZ.

In July, the equatorial low (ITCZ) pressure belt migrates to the north with the movement of the sun, while the STH pressure belt intensifies in the SH. The STH pressure belt in the NH extends round the earth with break over eastern land mass (Asia). The development of STHP belt over continental land mass melts due to intense heating, however it is seen clearly on ocean areas. The sub-polar low and equatorial low in NH merge over continents, where as the sub-polar lows can be seen over oceanic areas in NH. The polar High also can be seen.

Global Wind Distribution: In general wind circulation is governed by the pressure pattern, however it is influenced by orography. Like pressure and temperature, wind distribution averages over surface considered for the months January and July – upper wind distribution will be considered separately for India.

January: STH in NH is located near latitude 30 °N and low over Africa and Australia. The NE trades are well developed and steady. Because of the location High and Lows, the NE-trades cross the equator, get deflected to the left by coriolis force in the southern hemisphere. Thus we have cross equatorial flow. At the same time the belt of SE-trades are prominent over ocean areas. Prevailing westerlies are prominent in NH middle latitudes. Polar easterlies are observed on the poleward side of the sub-polar lows. (See Fig. 27.10).

July: The STH in southern hemisphere located about its normal position (latitude 30 °S) and ITCZ or equatorial low shifted to north. The Low lies over India, Pakistan, and trough extends over to Africa. This causes the SE-trades of southern hemisphere crosses the equator and moves towards the Low and deflected to the right by coriolis effect. This cross equatorial flow and right deflection becomes southwest monsoon to Indian sub-continent and southeast monsoon to east Asia. The NE-trades are observed over eastern Atlantic and eastern Pacific. The mid-latitude westerlies are marked close to the western coasts. The polar easterlies are less marked except over Greenland and Iceland where they are slightly stronger (See Fig. 27.5).

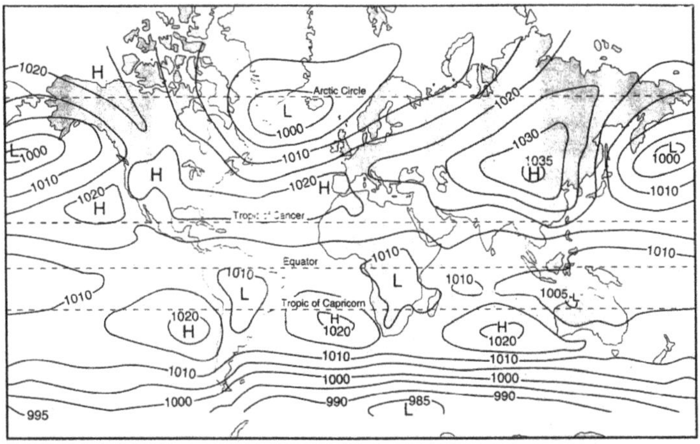

Fig. 27.9 Average pressure (hPa) at mean sea level in January.

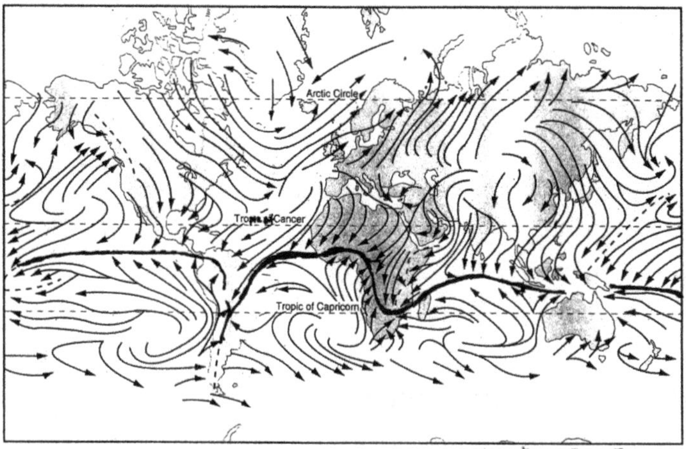

Fig. 27.10 Prevailing surface winds in January and location of the ITCZ.

Questions

1. Atmospheric motion is a complex phenomenon. The general circulation may be described as............ average phenomenon of.........and their variations during seasons, years or months.

   (a) Time and space; winds, temperature, pressure, humidity

   (b) Temperature; winds, precipitation

   (c) Winds, precipitation; globe

   (d) None above

2. Aviation route planning depends on……..:
   (a) Winds (b) Solar radiation
   (c) Temperature (d) Climatological data

3. The atmospheric motion is mainly driven by:
   (a) Pressure
   (b) Solar radiation
   (c) Angular momentum of the earth
   (d) Both (b) and (c)

4. The primary force driving the atmosphere is the ……………...by insolation:
   (a) Pressure gradient (b) Temperature gradient
   (c) Uneven heating of the globe (d) All above

5. It is estimated that about 1% of………… is absorbed by the earth-atmosphere runs the atmosphere heat engine
   (a) Wind energy (b) Insolation
   (c) Atmospheric electrical energy (d) Both (a) and (c)

6. The angular speed of the earth at equator is about…………
   (a) 265 m/s (b) 365 m/s
   (c) 465 m/s (d) Both (a) and (b)

7. At any place, close to the surface of the earth wind is………
   (a) Steady
   (b) Never steady in direction nor speed
   (c) Unsteady
   (d) Both (a) and (c)

8. The most dominant feature of the general circulation is reflected in……………..
   (a) Zonal winds (b) Meridional winds
   (c) Trade winds (d) None above

9. In Northern Hemisphere, trade winds blow from……..
   (a) North (b) East
   (c) Northeast (d) Southeast

10. Trade winds in the Southern Hemisphere blow from….
    (a) North (b) East
    (c) Northeast (d) Southeast

11. In northern hemisphere, trade winds are strongest in .... and blow from...........
   (a) Summer, southeast (b) Winter, northeast
   (c) Rainy season, east (d) Monsoon, southwest

12. Between the trade winds, there is region near equator where winds are light/calm, murky sky, thunderstorms is called..........
   (a) Polar front (b) Warm front
   (c) Cold front (d) Doldrums

13. The average high temperature zone lies in the equatorial trough, which is called.........
   (a) Doldrums, (b) Thermal equator
   (c) Meteorological equator (d) Both (b) and (c)

14. The annual average position of thermal equator is.......
   (a) Lat $23\frac{1}{2}$°N (b) Lat 5 °S
   (c) Lat 5 °N (d) Lat $23\frac{1}{2}$°S

15. The equatorial trough migrates with the sun. In January it is centred about lat.........
   (a) 5 °N (b) 5 °S (c) 15 °N (d) 15 °S

16. The equatorial trough migrates with the sun. In July it is centred about...........
   (a) Lat 5°S (b) Lat $23\frac{1}{2}$°N
   (c) Lat 10-15 °N (d) Lat $23\frac{1}{2}$°S

17. On surface level, the Sub-Tropical High pressure belts are located at about...........
   (a) Lat $23\frac{1}{2}$°N / S (b) Lat 30 °N / S
   (c) Lat $60\frac{1}{2}$°N / S (d) Both a and c

18. The areas of sub-tropical high pressure belt are dominated by......
   (a) Murky skies, light winds
   (b) Clear skies, light or variable winds
   (c) Convective clouds, equally winds
   (d) Clear skies, squally wind

19. In summer hemisphere, there is a zone of westerly winds in equatorial region between trade winds. It migrates with the sun. In Indian ocean it is located about lat.......... during summer (June-July) and at about lat... ....during winter (Dec-Jan).
   (a) 10 ^{O}N; 10 ^{O}S (b) 12-15 ^{O}N; 2-3 ^{O}S
   (c) $23\frac{1}{2}$ °N; $23\frac{1}{2}$ °S (d) Both (a) and (c)

20. Between the Subtropical High and Sub-polar Low, there is a belt of prominent westerlies in southern hemisphere, called.........
   (a) Equatorial westerlies
   (b) Roaring forties or howling fiftees
   (c) Doldrums
   (d) Sub-tropical jet

21. Between Polar High and Sub-polar Low, there is a quasi permanent wind belt called.........
   (a) Equatorial westerlies
   (b) Tropical easterlies
   (c) Polar easterlies
   (d) Sub-tropical westerlies

22. The general circulation wind model consists of ....
   (a) Hadlay cell
   (b) Ferrel cell
   (c) Polar cell
   (d) All above (a), (b) and (c)

23. In High Zonal Index, isobars oriented........, westerly winds are strong, and there is ....
   (a) East-west; little mass exchange from north and south
   (b) North-south; large mass exchange from north and south
   (c) East-west; large mass exchange from north and south
   (d) North-south; little mass exchange from north and south

24. In Low Zonal Index, westerlies completely break off forming.........., and associated with prominent.......

    (a) Cut-off low and blocking highs; meridional movement of air

    (b) Blocking lows and cut-off highs; no meridional movement of air

    (c) Blocking lows and cut-off highs; meridional movement of air

    (d) Cut-off lows and blocking highs; no meridional movement of air

25. The rotation of earth about its axis causes .... and revolution in elliptical orbit around the sun causes...........

    (a) Seasons; day and night

    (b) Day and night; seasons

    (c) Mid night sun; 24 hours darkness

    (d) Both (a) and (c)

26. In preparing global map of temperature isotherms distribution, all observations are reduced to ......using lapse rate........

    (a) Mean seal level (msl); 6.5 °C/km

    (b) 30 ft agl; 10 °C/km

    (c) Msl; 10 °C/km

    (d) 30 ft agl; 6.5 °C/km

27. Annual variation of mean temperature over land may be .... and in equatorial region over the sea......

    (a) 50 °C, 10°C (b) 60 °C, less than 5 °C

    (c) 30 °C, 30 °C (d) 15 °C, 10 °C

28. In general, surface temperature...... with increasing latitudes

    (a) Increases

    (b) Decreases

    (c) Does not change remain constant

    (d) None above

29. Surface temperatures in general, decrease with increasing latitudes. However these temperatures are modified by seasonal variation and ……….
    (a) Soil, vegetation
    (b) Land and sea distribution, Ocean currents
    (c) Soils, ocean currents
    (d) Snow cover, land and sea distribution
30. The highest mean temperature occurs in………
    (a) Sub-tropics (b) Polar region
    (c) Tropics (d) Ocean area
31. Northern hemisphere summer is…….than southern hemisphere summer
    (a) Cooler (b) Warmer
    (c) Same (d) None above
32. In January ITCZ or axis of equatorial low lies………..
    (a) North of the equator (b) South of the equator
    (c) Over the equator (d) None above
33. In January the STH centre in NH lies close to …………
    (a) Equator (b) Lat 30°N
    (c) Lat 30°S (d) Lat 60°N
34. In general, wind circulation is governed by ………..
    (a) Temperature pattern (b) Pressure pattern
    (c) Ice/Snow coverage (d) Ocean currents
35. In January, air flow from Siberian/Asia, High moves over Japan and thence towards ITCZ. It gives … in NH and ………. in SH
    (a) NE monsoon, NW-monsoon
    (b) SW monsoon, NE-monsoon
    (c) SW monsoon, NW-monsoon
    (d) NE monsoon, NE-monsoon
36. In July, air from Australian and Indian Ocean Highs flow towards ITCZ over Asia. It gives………… over SH and …………NH
    (a) SE monsoon; SW-monsoon
    (b) NE monsoon; NW monsoon
    (c) SW monsoon; SE monsoon
    (d) NW monsoon; NE monsoon

Answers

1.	(a)	2.	(c)	3.	(d)	4.	(c)
5.	(b)	6.	(c)	7.	(b)	8.	(c)
9.	(c)	10.	(d)	11.	(b)	12.	(d)
13.	(d)	14.	(c)	15.	(b)	16.	(c)
17.	(b)	18.	(b)	19.	(b)	20.	(b)
21.	(c)	22.	(d)	23.	(a)	24.	(a)
25.	(b)	26.	(c)	27.	(b)	28.	(b)
29.	(b)	30.	(c)	31.	(b)	32.	(b)
33.	(b)	34.	(b)	35.	(a)	36.	(a)

CHAPTER 28

Climatology

Weather and Climate: The physical state of atmosphere at any particular location rarely exhibits a steady state even during short intervals of time. The ever changing physical state of atmosphere constitutes the weather. It is described in terms of instantaneous values or short period mean values of meteorological elements on surface of the earth and in the atmosphere, such as temperature, pressure, humidity, wind, state of sky (cloud), precipitation etc. The average conditions (not only numerical values but available information) of weather at a place over a long period (more than 30 years), together with its extremes and its probabilities constitutes the climate of the place. Weather and climate has been changing right from the origin of atmosphere and life on earth and reached a balanced state that is suitable for life on earth. Climatology may be broadly described as statistical meteorology in relation to geography or natural environment. Climate never fits into rigid demarcation. It changes from one type to another, one generation to the next, from one century to the next and from one ice age to the next. Climatic normals are average values of weather elements over a place for about 10 days or one month (say 1-10 January or January month), averaged over a long period about 30 to 100 years of records. These normals are used as yardstick for describing its behaviour or variation of weather / climate.

Climatic Controls: The factors that control the climate are: (i) Latitude or sun's inclination (ii) Altitude, (iii) Topography, (iv) Land and sea distribution and (v) Ocean currents.

(i) *Latitude*: This determines the amount of heat received from the sun. The sun is the main driving force of the earth-atmospheric systems.

(ii) *Altitude*: The height above sea level generally determines precipitation. With increasing heights pressure and temperature falls and increases precipitation.

(iii) *Topography*: Large mountain barriers stand as climate boundaries. The orientation of the Himalayas is an outstanding example that acts as a climate barrier. The Himalayas prevent the northern cold wind entering over Indian sub-continent during winter. It also acts as a barriers in southwest monsoon that entraps the moist maritime air over India.

(iv) *Land and Sea Distribution*: The amount of heat absorbed or radiated by a surface is a function of nature of surface and its thermal capacity. Because of this there is lot of difference in temperature and pressure over land and sea / ocean at the same latitude. Secondary circulations like land and sea breeze, monsoons are developed due to land-sea distribution.

(v) *Ocean currents*: Ocean currents act as transporters of heat from equator to pole wards. They also greately alter the local coastal climates.

In addition to the above factors, climate is affected by semi-permanent Highs and Lows.

Climate Classification: Climate classification is a complex subject. Most of the classifications are based on the relationship between climate and vegetation or soil. Natural vegetation is the best basis for classification of the world's climate. The following is the simplest which suffices for aviation purposes.

1. **Equatorial Climate:** It is also called humid tropical or rainforest. This zone extends from latitude 10 °N to 10 °S. Over the oceans it is called doldrums. The main features of this zone: light winds, high humidity and temperature. Diurnal range of temperature exceeds the annual range. There is no dry season. It has two rainy

seasons with the movement of thermal equator. It has most unstable atmosphere which is associated with convective clouds and thunderstorms. Annual rainfall is a function of topography.

2. **Savannah or Trade Wind Littoral Climate:** It extends from latitude 10° to about 25° in both hemispheres. These zones have alternate dry, wet seasons. Wet season is associated with the sun overhead. The rainfall amount and duration decreases as one moves from lower to higher latitudes. Similarly annual and diurnal temperature ranges increase with increasing latitudes. During winter dry trade winds prevail. Fairly high uniform temperatures prevail all through the year. In this region flying conditions on land are hot, dusty and turbulent.
3. **Steppe or Arid Sub-tropical Climate:** This zone extends from Savannah to about latitude 30° in both hemispheres. The world's great hot deserts - the Sahara, Arizona, Arabia, Kalahari, South America and Australia deserts are all located in this zone. The main features of this zone is subsidence of air (surface level divergence), very hot summers with large diurnal range of temperature, meagre rain with very short rainy season bordering areas. Bordering Savannah rainfall confined to summer while bordering pole wards winter rainfall. Annual temperature range small. In summer dust/sand storms observed.
4. **Warm temperature zone or Mediterranean type:** This zone extends between latitude $30 - 45^{\circ}$ and includes parts of California, central Chile, extreme South Africa, parts of southern Australia. In summer it comes under the influence of arid-subtropical zone but in winter it experiences unsettled weather of cool temperate zone. The later is due to the shifts of Sub-tropical High towards the equator. Summers are dry and hot but winters may be wet and cool.
5. **Cool temperature zone:** This zone extends between latitude 35-50°. The main features of these zones mobile westerlies, frequent passages of frontal depressions and cold anticyclones or ridges. No dry season and winters can be cold and very clod in areas which are away from oceans. Winds are predominantly westerlies with frequent gales.

6. **Boreal Zone:** This zone roughly extends between latitude 40-60 °N and observed only in northern hemisphere. It is well developed and covers North America, Scandinavia and former USSR. The chief feature of this zone are cool and moist summer, very cold winters and large annual range of temperature. It is dominated by tropical continental air mass in summer and polar continental air mass in winter.
7. **Polar and Tundra Zones:** This zone extends between latitude 55 °N to north pole and latitude 50 °S to south pole. The chief features of these zones is that it has 20 hours of day sunshine for three months in summer and 20 hours per day of darkness in winter. Mean temperature in warmest months is little higher than freezing point (0 °C). It has some tundra vegetation like mosses, lichens and grasses otherwise the surface is continuously covered by snow or ice. The ground is permanently frozen. There is no summer and climate is humid inspite of being very cold. Precipitation is mainly snow or snow grains.

The above 7 zones of climate are as shown in Fig. 28.1.

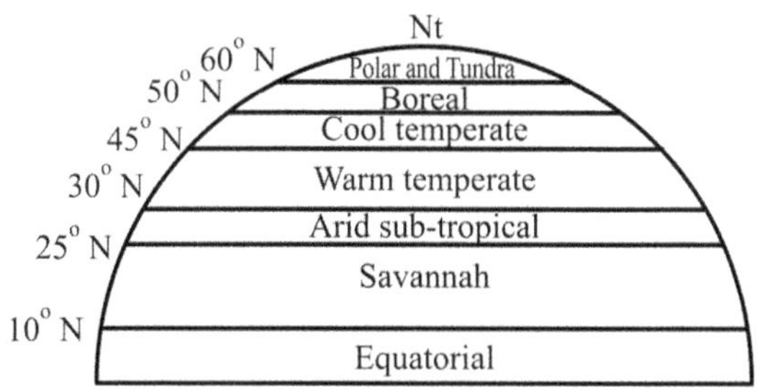

Fig. 28.1 Climatic zones of Northern Hemisphere.

8. **Monsoon Climate:** We have already studied about monsoons. The seasonal reversals of wind direction and temperature cooling is the main feature. The cause of changes is the differential heating of large land mass and sea area altering with the season (like land and sea breeze). Monsoon climate is best explained by summer monsoon over the Indian sub-continent. The onset of southwest monsoon heralds the rainy season over India and the northeast monsoon of winter returns to the dry weather in most of India.

Questions

1. The ever changing physical state atmosphere constitutes..........It is described by the instantaneous values or short period mean values of...... on surface and in atmosphere
   (a) Weather, meteorological elements
   (b) Climate, averages
   (c) Instability, temperature and humidity
   (d) Non above
2. The physical state of atmosphere is described by:
   (a) Temperature, pressure
   (b) Humidity, wind
   (c) State of sky, precipitation
   (d) All above (a) , (b) and (c)
3. The average conditions of weather at a place over a long period together with its extremes and their probabilities constitutes.......of that place
   (a) Weather (b) Climate
   (c) Hazards (d) Both (a) and (c)
4. Climatic normals are used for describing its behaviour or variations of......
   (a) Weather (b) Climate
   (c) Aridity (d) Both (a) and (b)
5. The factors that control the climate are:
   (a) Latitude or suns inclination, altitude
   (b) Topography, land and sea distribution
   (c) Ocean currents
   (d) All above (a) (b) and (c)
6. The climatic zone between latitude 10 oN to latitude 10 oS is called............
   (a) Humid tropical
   (b) Rainforest
   (c) Equatorial
   (d) All above (a), (b) and (c)

7. The equatorial climate area over ocean is called........
   (a) Savannah
   (b) Doldrums
   (c) Horse latitudes
   (d) Roaring forties
8. The main features of equatorial climate:
   (a) Light or variable winds, high humidity and temperature
   (b) No dry season, most unstable atmosphere
   (c) No wet season, cool temperatures all through the year
   (d) Both (a) and (b)
9. In both hemispheres savannah extends between latitude:
   (a) Latitude 5 to 15 ^{o}
   (b) Latitude 10 ^{o} to about latitude 25 ^{o}
   (c) Latitude 5 ^{o} to 23 ½ ^{o}
   (d) Latitude 20 ½ ^{o} to 60 ^{o}
10. The characteristic features of Savannah are:
    (a) Alternate dry and wet seasons with fairly high uniform temperatures
    (b) Rainfall amount and duration decreases as we move from lower to higher latitudes
    (c) During winter dry trade winds prevail
    (d) All above (a) , (b) and (c)
11. The zone of arid sub-tropical or steppe climate zone extends....
    (a) From Savannah to latitude 30 ^{o} in both hemispheres
    (b) From 10 ^{o} to latitude 30 ^{o}
    (c) From equator to latitude 30 in both hemispheres
    (d) Both (a) and (b)
12. The characteristic features of arid-subtropical climate zone are:
    (a) Subsidence of air, very hot summers
    (b) Worlds great hot deserts are located
    (c) Annual temperatures range small and rainfall meagre
    (d) All above (a) , (b) and (c)

13. Mediterranean type or warm temperate zone extends between............and includes parts of............
    (a) Latitude 20-30 °; India, SE Asia and NW Australia
    (b) Latitude 30-45 °; parts of California, central Chile, extreme south Africa and parts of southern Australia
    (c) Latitude 23 ½ -latitude 60 °; India, Sahara
    (d) Both (a) and (c)
14. The characteristic features of Mediterranean climate are:
    (a) Summers are dry and hot, winters are wet and cool
    (b) In summer it comes under the influence of arid-subtropical zone but in winter it experiences unsettled weather of cool temperature zone
    (c) No wet season, high temperature, low humidity
    (d) Both (a) and (b)
15. The climatic zone between latitude 35-50 ° is called:
    (a) Warm temperate zone (b) Cool temperate zone
    (c) Boreal zone (d) Tundra zone
16. The chief features of cool temperature zones are:
    (a) Mobile westerlies, frequent passages of frontal depressions
    (b) Very cold winter and large diurnal temperature range with no wet season
    (c) No dry season, winter can be very cold in areas which are away from oceans
    (d) Both (a) and (c)
17. The climatic zone which is observed only in northern hemisphere between latitude 40-60 °N is called......
    (a) Polar zone (b) Tundra zone
    c) Boreal zone (d) Mediterranean
18. The Boreal zone observed only in NH is well developed and covers.........
    (a) North America, Scandinavia and former USSR
    (b) South America, Brazil, Australia
    (c) India, Pakistan, South China
    (d) Both (b) and (c)

19. The chief features of Boreal zone are:
    (a) Tropical maritime air mass in summer and polar maritime air mass in winter
    (b) Cool and moist summer, very cold winters
    (c) Tropical continental air mass in summer and Polar continental air mass in winter
    (d) Both (b) and (c)

20. Polar and Tundra zones extend.....
    (a) Between latitude 55 ^{o}N to north pole
    (b) Between latitude 50 ^{o}S to south pole
    (c) Between latitude $66\frac{1}{2}$ to pole
    (d) Both (a) and (b)

21. The chief features of Polar and Tundra zones are:
    (a) 20 hrs per day sunshine for three months in summer and
    20 hrs darkness per day for three months in winter
    (b) It has some tundra vegetation like mosses, lichens and grasses otherwise permanently covered by snow or ice
    (c) No dry season and no winter
    (d) Both (a) and (b)

Answers

1. (a)	2. (d)	3. (b)	4. (d)
5. (d)	6. (d)	7. (b)	8. (d)
9. (b)	10. (d)	11. (d)	12. (d)
13. (b)	14. (d)	15. (a)	16. (d)
17. (c)	18. (a)	19. (d)	20. (d)
21. (d)			

CHAPTER 29

Climate of India

Physiography of Indian Sub-continent

In the north of the Indian sub-continent the mighty Himalayan ranges are spread roughly in east-west direction. The ranges are emanating from Pamir knot in the northwest of India, one range Hindukush spread towards west-south west, another range Sulaiman passes southwest wards to Beluchistan from Indus plain. In the northeast India, from Assam to Myanmar Khasi, Jaintia and Garo Hill ranges are spread. South of the peninsular India is occupied by the Indian Ocean. West of the peninsula, the Arabian Sea and east of it the Bay of Bengal. In south India, Western Ghats are spread in a north-south direction along the west coast, while Eastern Ghats are spread along east coast.

The Indian sub-continent is spread over tropical, sub-tropical and temperate zones. The coastal area of peninsular India is influenced by maritime air mass. India has widely contrasting climatic conditions. The annual rainfall varies from less than 22 cm (217 mm Jaisalmer, Rajasthan) to more than 1100 cm (1142 cm Cherrapunji, Assam). Temperatures vary between – 45 °C (Dras, December 1910) in winter to more than 50 °C in summer (Alwar, east Rajasthan 1956).

Seasons

The dominant climate over India is monsoon circulation. In one half of the year wind blows from cooler maritime area to dry land (which is summer monsoon or southwest monsoon). The other half season cold dry wind from Asian anticyclone to warm Indian Ocean (which is winter monsoon or northeast monsoon). Thus there is a reversal of pressure and wind pattern between the halves. The change in pattern takes place gradually. The transition between these two predominant systems lies two intermediate seasons. The seasons in India are thus classified into four seasons. Which are given below.

1. Winter (December-February)
2. Pre-monsoon season (summer season) (March-May)
3. SW-monsoon season or Rainy season (June-September)
4. Post-monsoon season or cool season (October-November)

1. Winter (cold) season: (December-January-February)

During January NE-trades are prominent with Siberian High (msl) or Anticyclone centred at about latitude 45 °N, longitude 105 °E and India lies at its periphery. Continental cold air blows over central parts of Asia and a part of which comes over north India, while a major part of it is prevented by the mighty Himalayas. Intense Siberian high persists during December, January and February. Sub-tropical jet stream core shifts to south of the Himalayas and is located at about latitude 22 °N.

During this period the STH in southern hemisphere is found between equator and latitude 10 °S. It extends from north Australia to eastern parts of Africa. A marked trough on sea level chart runs from Kerala to Gujarat and another Tannasserin coast to northern Myanmar and Assam. A weak ridge runs from NW-India to central Bihar (see Fig. 29.1).

This is the driest season for the country, except Kashmir, most part of India receives an insignificant percentage of rainfall. However this winter rainfall is important for crops in northern India. The highest rainfall of 20 cm is received in Kashmir and at some places in Nicobar islands. About 5 cm of rain received in most of the states in north India including Assam. In south India south of lat 10 °N and east coastal strip south of lat 15 °N and island groups receive rainfall. The chief synoptic features during this season are about 4-5 western disturbances pass over in north India.

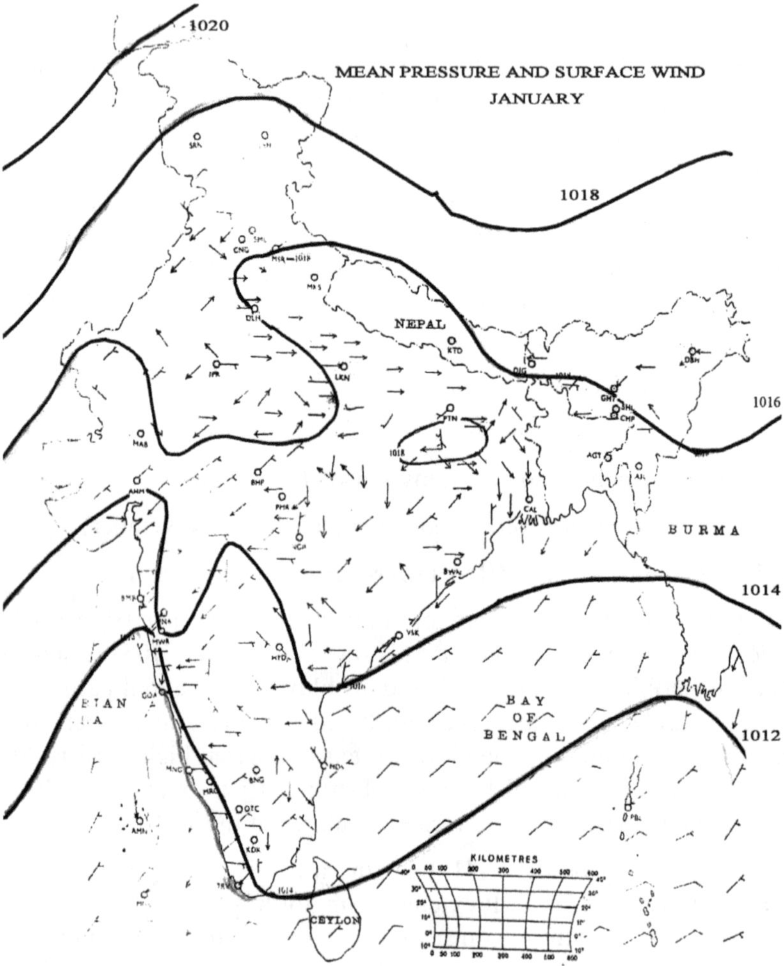

Fig. 29.1 Mean Pressure (hPa) and surface wind (January).

In this season cold waves occur in various parts of the country, particularly north of lat 20 °N. Cold dry air blows from Northwesterly/northerly direction and lowers the day and night temperatures significantly (below normal by 5 to 8 °C). Cold waves do not occur over Bay and Arabian sea Islands, Tamilnadu, Coastal Andhra Pradesh, coastal and south interior Karnataka and Kerala. Jammu and Kashmir is haunted by severe cold waves. Ladakh recorded maximum duration of severe cold wave for 30 days.

Fog and mist are very common in most part of the country affecting visibility particularly in early mornings. Conditions in the rear of the passage of the Western Disturbance in north India thick fog occurs. The most susceptible areas are northwest India, UttarPradesh, Bihar, North Bengal and Assam. Advection fog occurs in the coastal belts. The Brahmaputra valley, Sundarbans, the coastal belts and the hills and valleys of the Peninsula experience this type of fog/mist. The smoke near the cities and towns is being changed into smog and reduces the visibility to about 2 km in the inversion layer. Poor visibility very hazardous for landing of aircraft. Very low thunderstorm activity. The upper air temperature, winds, contour heights are given in the annexture.

2. Summer Season (March to May)

Pressure gradient becomes slack and begins to readjust as the sun moves north (see Fig. 29.2). A feeble High lies over Arabian sea and another over Bay of Bengal. Feeble lows develop over western parts of Rajasthan, Bihar and Myanmar. In this season the Siberian High weakens and shifts to north, while the sub-tropical High in the southern hemisphere shifts to lat 30 ^{O}S. By May the summer heat low extends from north Africa to Asia.

Mainly convective type of rain occurs in various parts of the country. Assam (60–80 cm), Jammu and Kashmir (20 – 40 cm), Kerala and neighborhood (30-40 cm) receives considerable rain due to thunderstorm activity. Hail is also associated with thunderstorms, which decreases with the advance of the season. Cherrapunji receives more than 200 cm of rain. The second important area of rainfall activity extends from west Bengal to North Coastal Andhra Pradesh. Western parts of India, except extreme north and south are driest. Western Rajasthan and adjoining Saurashtra and Kutch are practically dry (about 0.5 cm). Norwesters, Kalbaisakhi and Andhi (local names of severe thunderstorm dealt seperately) are confined to this season only. Tropical cyclones and heat waves are the other synoptic features. During hot weather period (March-June) surface temperatures over many parts of India abnormally shoot up practically over north India where temperatures will be above normal by 5 to 8 °C or more. The incidence of severe heat waves occur mostly in

Uttar Pradesh, but there is no region where recurs successively every year. Bay Islands Lakshadweep, Tamilnadu, Kerala, Coastal and south interior Karnataka are not affected by heat waves. Rest of the country is prone to the incidence of heat waves. Maximum, period of 15 days heat wave lasted over Jammu and Kashmir.

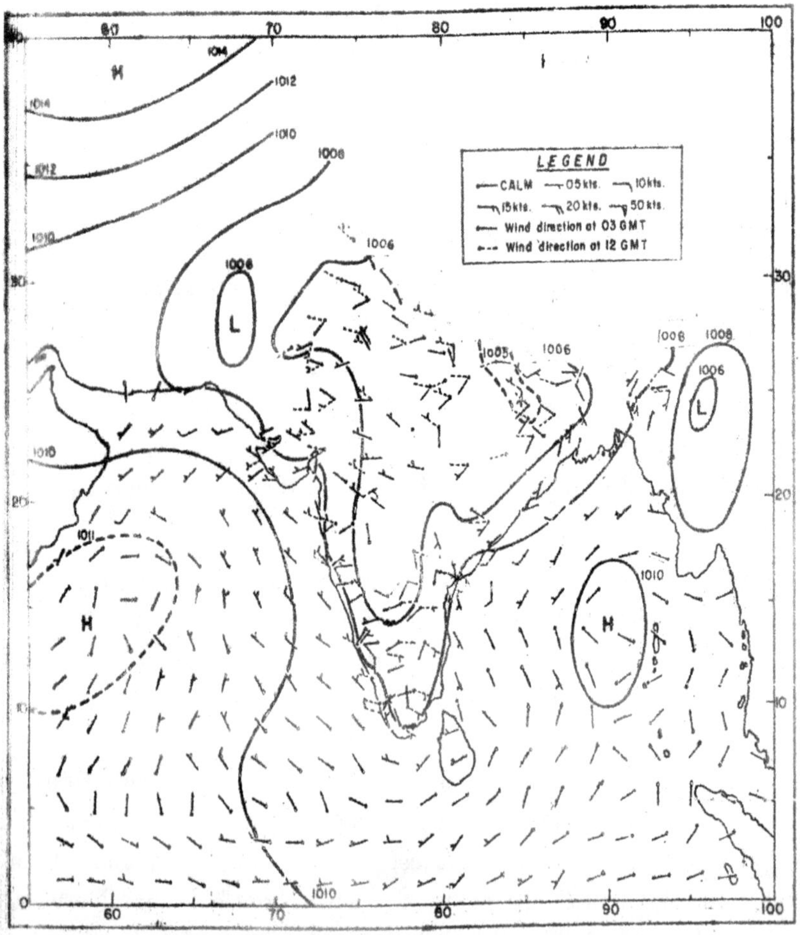

Fig. 29.2 Mean pressure (hPa) and surface winds (April).

Thunderstorm activity is maximum is this season, exceeding 16 thunderstorm days in May over Bengal/Assam and Kerala. Squalls generally accompany thunderstorms in this season. Violent thunder

/ squalls with heavy showers and occasional hail are associated with Norwesters over Bengal, Assam, Orissa and East Bihar. Duststorms (Andhies) are common in north India in this season. In duststorms some times visibility will be less than 2 km and on few occasions dust remains suspended in the atmosphere for a few days.

In coastal areas land and sea breeze affects are pronounced in this season. Upper winds/temperature and contour hights are given in annexture.

3. Monsoon season (June-September)

In this reason low pressure area extends from north Africa to Siberia via Baluchistan and becomes intense. Another trough of low extends from Rajasthan (heat low) to Head Bay of Bengal (which is called monsoon trough in India) which is of great importance to India (see Fig. 29.3). The sub-tropical high in southern hemisphere become more marked and centered around lat 30°S and long 60°E. A weak ridge extends over Arabian sea off west coast of India and another ridge in the Bay of Bengal off Tannasarin coast to Myanmar. By August Afro-Asian low weakens and by September it becomes an east-west trough.

Summer monsoon or southwest monsoon is the main rainy season for India. Except Kashmir and neighborhood and Kerala, the remaining area of the country receives more than 75% of the annual rainfall in this season. Orography and wind direction (wind ward side or lee ward side) plays an important role in distribution of seasonal rainfall as observed in Western ghats and Khasi-jaintia hills, where wind blows at right angles to the ghats/hills. In north Indian plains the minimum rainfall belt runs from northwest Rajasthan to the central parts of west Bengal, roughly along the axis of monsoon trough. Rainfall rapidly decreases along east coast south of latitude 9.5 °N to Kanyakumari. Pamban records only about 4 cm rain during monsoon. In general, July is the rainiest month. Seasons heaviest rainfall zones are Assam (200-400 cm), west coast and adjoining ghats north of latitude 10 °N (200-300 cm). Ladakh valley in the extreme north, western parts of Rajasthan and southernmost part of Tamilnadu gets very small amount (5-10 cm) of rainfall.

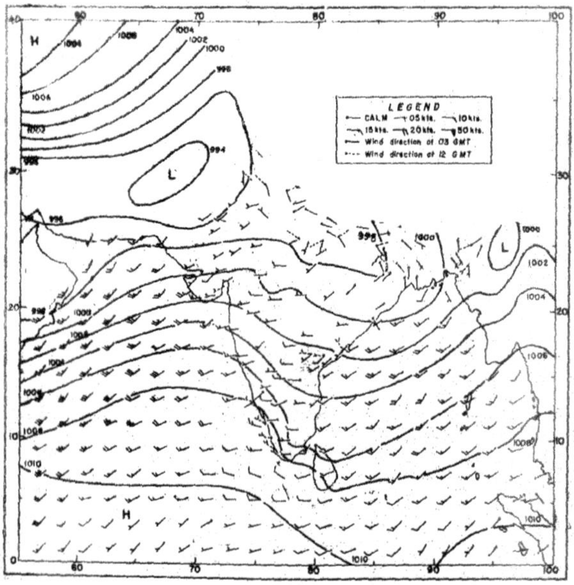

Fig. 29.3 Mean pressure (hPa) and surface winds (July).

The chief synoptic features of southwest monsoon are lows, depressions and their movement along monsoon trough, mid-tropospheric circulations off Gujarat-Konkan coasts and off Andhra, Orissa coasts.

The sub-tropical westerly jet stream migrates to north of Himalayas and tropical easterly jet stream appears over peninsula.

Upper winds / temperature and contour heights are given in annexture.

During monsoon low cloud base comes down to 450 ft to 600 ft agl and during depressions the cloud base may be lower to 100 ft to 200 ft agl. Surface winds of the order 25-30 kt are common and during depressions areas it may be 40-50 kt. Low level jet also appears in low level westerlies. Visibility is good but deteriorates in heavy rain to a few hundred meters.

4. Post monsoon season (October, November)

In this season pressure field becomes slack, by September end monsoon trough shifts to latitude 13 °N (see Fig. 29.4). Low pressure belt extends from Africa to west Pacific through Arabian Sea and Bay of Bengal between equator and latitude 20 °N, with centers over Africa, Bay of Bengal and western Pacific Ocean. The Asian high runs roughly along latitude 50 °N and centered about longitude 9o °E. The Indian Ocean high centered near 30 °S shifts to east and by November it is centered at longitude 80 °E. The trough in the south Bay of Bengal persists till

November end but disappears in December. During this season rainfall activity is mainly confined to south peninsula, the east coast, Assam and parts of Kashmir. It is the wettest season for Tamilnadu (40-80 cm average rain) where NE- monsoon prevails. (In fact the period Oct-Dec is called Northeast Monsoon). Rest of south peninsula gets 5-10 cm of average rain. Western India including Jammu and Kashmir gets rainfall less than 5 cm.

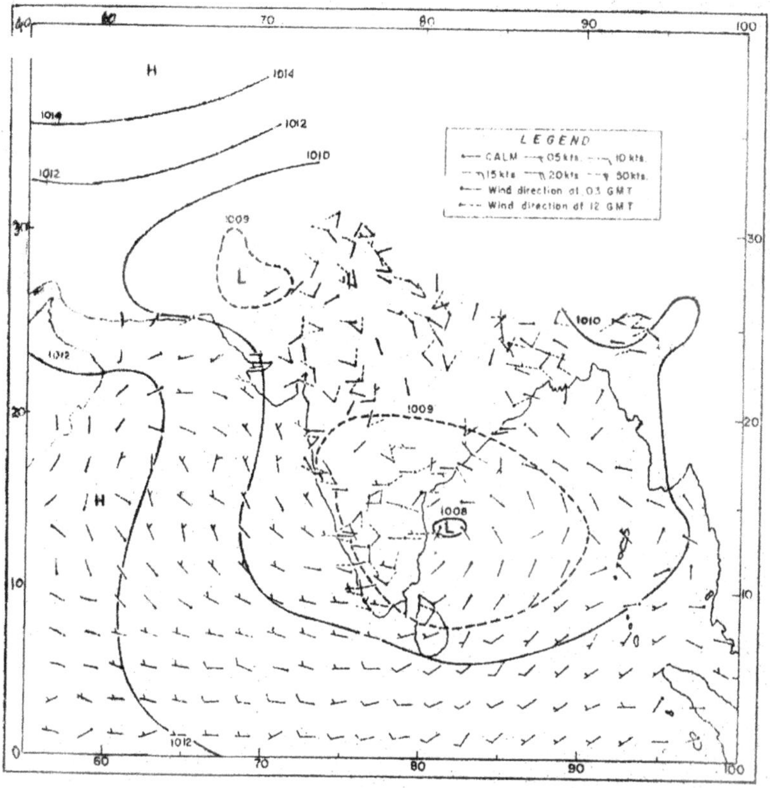

Fig. 29.4 Mean pressure (hPa) and surface winds (October).

The chief synoptic features are tropical cyclones in Bay of Bengal and Arabian Sea. In the north western disturbance start making their appearance and affecting the weather over northwest India. With gradual setting of cold weather, poor visibility due to fog / mist is experienced in the early mornings. Snowfall may be common in the Himalayas in association with Western disturbances. The sub-tropical jet stream core shifts to south and located between latitude 30-35 ^{o}N at 200 hPa level with the average speed 60-70 KT.

Questions

1. The principal climatic types spread over India are:
   (a) Monsoon
   (b) Tropical and Sub-tropical
   (c) Temperate zone
   (d) All above (a), (b) and (c)
2. The annual rainfall over India varies from:
   (a) 200 cm to 1000 cm
   (b) Less than 22 cm to more than 1100 cm
   (c) 50 cm to 500 cm
   (d) None above
3. Temperature over India varies from:
   (a) – 15 °C to + 50 °C (b) – 25 °C to + 45 °C
   (c) – 45 °C to more than 50 °C (d) None above
4. The dominant climate over India is:
   (a) Monsoon circulation (b) Tropical
   (c) Sub-tropical and temperate (d) None above
5. The main seasons in India are:
   (a) Winter (or NE monsoon), summer season (premonsoon)
   (b) Rainy season (or south west monsoon)
   (c) Cool season (or post monsoon)
   (d) All above (a), (b) and (c)
6. During January the centre of Siberian High located at:
   (a) Latitude 30 °N, longitude 90 °E
   (b) Latitude 45 °N, longitude 105 °E
   (c) Latitude 10 °S, longitude 105 °E
   (d) Both (a) and (c)
7. During January the STH in the southern hemisphere lies:
   (a) Between latitude 10 °S and latitude 30 °C
   (b) Between equator and latitude 30 °S
   (c) Between equator and latitude 10 °S
   (d) Between equator and latitude 10 °N

8. During winter, the Himalayas are preventing the spread of cold winds of central Asia over……….

(a) Afghanistan (b) Pakistan

(c) Major part of India (d) Iran and Iraq

9. Winter is the driest season of India. However winter rainfall due to ………… in north India is very important for crops

(a) Thunderstorms (b) Norwesters

(c) Western disturbances (d) Tropical cyclones

10. Cold waves in India do not occur over:

(a) Bay and Arabian Sea Islands

(b) Tamilnadu, coastal Andhra Pradesh

(c) Coastal and south interior Karnataka, Kerala

(d) All above (a), (b) and (c)

11. In winter, most part of the country is a affected by ……………… in the early morning, which is hazardous to aviation

(a) Thunderstorms (b) Hailstorms

(c) Cyclones (d) Fog and mist

12. In the rear of the passages of WD in north India……….occurs which is hazardous to aviation

(a) Thunderstorms (b) Hailstorms

(c) Cyclones (d) Fog and mist

13. During summer (March-May) the Siberian High…., while the STH in southern hemisphere ………..

(a) Strengthens and shifts southwards; shifts to latitude 10 °S

(b) Weakens and shifts to north; shifts to latitude 30 °S

(c) Weakens and shifts to south of Himalayas; shifts to latitude 30 °S

(d) Strengths and shifts to north; shifts to north latitude 10 °N

14. During Indian summer (March-May) rain occurs mainly due to………..

(a) Snowfall (b) Ns, As

(c) Convective clouds (d) Cb

15. During Indian summer (March-May) the main weather systems are:
    (a) Norwesters, Kalbaisakhi and Andhi, cyclones
    (b) Cyclones, western disturbances
    (c) Monsoons, western disturbances
    (d) Monsoons, cyclones
16. During Indian summer (March-May), the maximum thunderstorm activity exceeding 16 days lies over….and minimum over…
    (a) (i) Rajasthan; (ii) Jammu and Kashmir
    (b) (i) Bengal, Assam and Kerala;,(ii) Saurastra Kutch
    (c) (i) Saurastra Kutch; (ii) Assam
    (d) (i) Madhya Pradesh; (ii) Andhra Pradesh
17. During Indian summer (March-May), Norwesters occur over………
    (a) Rajasthan, Saurastra Kutch
    (b) Jammu and Kashmir, Punjab, Haryana
    (c) Bengal, Assam, Orissa and east Bihar
    (d) Uttar Pradesh, Madhya Pradesh, Kerala
18. In Indian summer, Andhi (Dust-storms) occur over ……..
    (a) Kerala, Karnataka, Tamilnadu
    (b) Assam, Bengal, Orissa
    (c) North India, Rajasthan
    (d) Bay Islands
19. In India more than 75% of annual rainfall is received during:
    (a) Winter (b) Summer monsoon
    (c) Northeast monsoon (d) In all above
20. In north Indian plains, the minimum rainfall belt during monsoon runs:
    (a) Northwest Rajasthan to central parts of west Bengal
    (b) Along west coast of peninsular India
    (c) Along the axis of monsoon trough
    (d) Both (a) and (c)

21. During summer monsoon, the minimum rainfall belt (less than 5 cm) runs........
    (a) Along west coast south of latitude 9.5 ^{o}N
    (b) Along east coast south of latitude 9.5 ^{o}N
    (c) Saurastra, Konkan coasts
    (d) Orissa, west Bengal coasts
22. In Southwest monsoon seasons, heaviest rainfall zones are:
    (a) Jammu and Kashmir, Punjab, Haryana
    (b) Tamilnadu coast
    (c) Assam, west coast and adjoining ghats north of latitude 10 ^{o}N
    (d) Along monsoon trough axis
23. During summer monsoon over India, the sub-tropical westerly jet stream located:
    (a) South of the equator
    (b) Migrates to north of the Himalayas
    (c) Along latitude 22 ^{o}N
    (d) Along latitude 10 ^{o}N
24. During summer monsoon over India, the easterly jet stream runs along latitude...........at altitude of..........
    (a) 22 ^{o}N, 200 hPa (or 12 km)
    (b) 13 ^{o}N, 100 hPa (about 16 km)
    (c) 8 ^{o}N, 800 hPa (1.5 km)
    (d) 8 ^{o}S, 100 hPa (5.8 km)
25. During summer monsoon in July, low level westerly jet stream observed along west coast at............level.
    (a) 100 hPa (b) 200 hPa
    (c) 850 hPa (d) 070 hPa
26. During summer monsoon over India, low cloud base comes down to about..........
    (a) 1000 ft agl (b) 2000 ft agl
    (c) 450-600 ft agl (d) 100-200 ft agl
27. During monsoon depression low, cloud base comes down to about.........
    (a) 1000 ft agl, (b) 2000 ft agl
    (c) 450-600 ft agl (d) 100-200 ft agl

28. During summer monsoon heat low, the lowest pressure isobar lies over...........
    (a) 1000 hPa, along west coast (b) 995 hPa, Rajasthan
    (c) 1006 hPa, over Assam (d) Both (a) ft and (c)
29. During post monsoon (October-November) the monsoon trough over India shifts to........
    (a) Latitude 23 °N, (b) Latitude 13 °N
    (c) Latitude 5 °N (d) Latitude 5 °S
30. During post monsoon (October-November), the Asian High would be centred about ....
    (a) Latitude 30 °N, longitude 90 °E
    (b) Latitude 50 °N, longitude 90 °E
    (c) Latitude 30 °S, longitude 80 °E
    (d) Latitude 50 °S, longitude 80 °E
31. During post monsoon (October-November), the southern hemispheric High is located at about........
    (a) Latitude 30 °N, longitude 90 °E
    (b) Latitude 50 °N, longitude 90 °E
    (c) Latitude 30 °S, longitude 80 °E
    (d) Latitude 50 °S, longitude 80 °E
32. The wettest season for Tamilnadu is ....
    (a) Southwest monsoon (June – September)
    (b) Northeast monsoon (October-November)
    (c) Winter
    (d) Summer
33. During winter January, the sub-tropical jet stream core is located at......... /altitude ....with maximum speed........
    (a) Latitude 22 °N / 12 km, 150-200 kt
    (b) Latitude 27 °N / 16 km, 70-100 kt
    (c) Latitude 13 °N / 12 km, 70-100 kt
    (d) Latitude 22 °N / 16 km, 150-200 kt

Answers

1.	(d)	2.	(b)	3.	(c)	4.	(a)
5.	(d)	6.	(b)	7.	(b)	8.	(c)
9.	(c)	10.	(d)	11.	(d)	12.	(d)
13.	(b)	14.	(c)	15.	(a)	16.	(b)
17.	(c)	18.	(c)	19.	(b)	20.	(d)
21.	(b)	22.	(c)	23.	(b)	24.	(b)
25.	(c)	26.	(c)	27.	(d)	28.	(b)
29.	(b)	30.	(b)	31.	(c)	32.	(b)
33.	(a)						

CHAPTER 30

Western Disturbances and Summer Thunderstorms over India

30.1 Western Disturbances (WD)

During November-March, the westerly waves or troughs associated with cyclonic circulations develop in the west Atlantic or in Mediterranean regions. These are extra-tropical frontal type cyclones (ETC). Secondary systems develop over the Persian gulf. Generally, these systems move in an easterly or east-north-easterly directions. During their journey over mountainous terrains of Iran, Afghanistan and Pakistan they modify and may lose their frontal characteristics in the lower levels of atmosphere. These extra-tropical (ET) systems when arrive over India they will be in occluded stage (but sometimes they appear as well-marked low/depression) with characteristics of warm front, warm sector, cold front. Orography plays an important role in breaking up/stagnation and intensification of these disturbances.

A disturbance located in the coastal region, west of the Nile river, may move ENE- wards and effect north Iraq, Iran, Afghanistan, north Pakistan and Jammu and Kashmir. Many of the depressions moving over Iraq bifurcate into two near Iraq-Iran area by high terrain. One cell moves ENE over Iran and the second southern cell travels eastward along the Persian gulf. The later one effects northern India.

During winter (November-March) cloud and precipitation (pptn) belts move from west to east over north India. These are called Western Disturbances (WD). The adjective western signifies that the disturbances approach India from west in contrast to the most of the rain giving systems (monsoon depressions / lows, cyclones etc.,) that approach India from east. As a convention in IMD, the term WD signifies a low or trough either at surface or in upper air in the westerly wind regime (north of latitude 20 °N) that cause clouding and precipitation. In the absence of clouding and precipitation the system is simply referred as trough. When two are more closed, isobars spaced at 2 hPa interval can be drown on sea level chart, the system is described as "western depression". The terms active and weak are associated depending on precipitation amount.

The sequence of precipitation associated with WD is: (i) first day over Jammu and Kashmir (north of latitude 30 ° N) and even Himachal Pradesh, Punjab and Haryana. (ii) The next day over east Rajasthan (north of latitude 26 °N and west of longitude 80 °E) and west Uttar Pradesh (this is primary system). (iii) On many occasions the rainfall may cease at this stage. In some other cases the rainfall now commences over Madhya Pradesh and neighborhood during next two days and then spreads east wards to Assam (this is secondary system). As it spreads east wards, its activity of rainfall stops on west side. (iv) sometimes it happens that the WD may not effect northwest India south of latitude 30 °N but gives rain in north Assam after two to three days (after it caused weather over western Himalayas). (v) After the second rainfall belt moved away ENE-words an extension of this system appears southwards or an independent weak low or circulation may develop over Madhya Pradesh and neighborhood which causes third rainfall belt (tertiary system). A weak low or trough may also form over Gulf of Khambhat (Camby) (as a tertiary system), which while moving in an easterly direction cause weather over Gujarat, north Maharastra. However these tertiary systems are rare. The secondary and tertiary systems that develop over Rajasthan and Madhya Pradesh (under the influence of the primary disturbance that moves across north India and Pakistan) are now referred to as induced lows. The terms primary, secondary and tertiary terms have become obsolete.

As stated earlier, during (December-February) winter, Mediterranean Sea, Black Sea, Caspian Sea, Lake Balkesh form a zone of cyclogenesis for ETC. Of these, Caspian Sea, Aral Sea cyclogenesis effect India. In this season ET depressions / troughs are known to move in a east-north-easterly by direction across southern Russia, Turkey, Iran and sometimes even further south. During January-February ETC originate as wave disturbance and / or cut-off cyclones between latitude 25-40 ^{o}N. Areas lying north of latitude 30 ^{o}N may be at the periphery of these primary western disturbances, which accounts for the first rainfall belt. At this stage a secondary low at latitude 30 ^{o}N is formed or occluded fronts (Nascent depression of ETC) enter this part as WD, while the primary moves away east-north-east wards across the Himalayas.

WDs are in general, lower tropospheric systems. They show variation in strength and vertical extent. These may be super-imposed by mid-and upper tropospheric troughs in westerlies. Movement of induced lows are slow, but they move faster when they are superimposed by mid-and upper tropospheric westerly trough. The life period of a simple sequence of WD is 2 to 4 days with induced lows filling up within that time. Before filling up of induced lows, if a fresh WD arrives from west, the life period prolongs up to 10 days with the formation of complex trough extending to latitude 20 ^{o}N. In such situation extensive rainfall occurs over the country. WDs that cause rainfall over Punjab and Uttar Pradesh arrive in succession of 2 to 5 days interval. At other times these areas are free from weather continuously for one to two weeks. About 50% of the occasions the activity of WDs are confined to Jammu and Kashmir, Himachal Pradesh, Punjab, Haryana and hills of west Uttar Pradesh or Utteranchal. On an average 6 to 7 western disturbances per month move across India in winter. Formation of depression either in the Bay of Bengal or the Arabian Sea inhibits the occurrence of WD over north India.

Weather Sequence

With the approach of a WD (at a station) pressure gradually falls and rise in temperature, minimum temperature and dew point. The winds at surface level and in lower levels back and show southerly component, and sudden increase in clouding. The sequence of clouding is : cirrus, cirrostratus, altostratus. A few cells of stratocumulus, cumulus and cumulonimbus also develop. With the passage of cold front large cumulus and cumulonimbus develops. Associated first with warm front continuous light rain, drizzle occurs, next with warm sector intermittent rain scattered showers occur. Subsequently with the passage of clod front heavy showers, thunderstorms or hailstorms and squalls are observed. In the hills sleet in the beginning, snow fall later then rain or showers.

In the rear of the passage of disturbance weather clears, pressure rises sharply with appreciable fall in temperature that causes cold waves. In the wake of the disturbance radiation fog is a common feature. With the lifting of fog, low stratus appears. Fog may occur both a head and in rear of WD in the Himalayas.

30.2 Summer Thunderstorms over India

During hot weather period (March-May) rapid rise of surface temperature, fall of pressure, intensification of southern Indian ocean anticyclone, Northward movement of equatorial trough, Norwesters, Dust storms, Dust raising winds, Hail storms are common features over India and neighborhood. Thunderstorm activity continues in monsoon season but less marked. The main thunderstorm activity observed during the period are : (i) The area from Northeast India to east Madhya Pradesh, east Vidarbha and adjoining Andhra pradesh, (ii) Southwest Peninsula (Kerala and neighborhood), (iii) northwest India excluding Rajasthan. Squalls associated with summer thunderstorms are very strong/violent. They uproot trees and there were instances in the past that trains being thrown out of tracks. On an average 15 to 20 thunderstorm days recorded in the season. Assam and adjacent states, south Kerala have the highest frequency (30 to 40 days). Entire India is susceptible to thunderstorm activity but Gujarat state has the lowest frequency (about five days in whole of the season).

Thunderstorm activity progressively increases from March. During March the areas of maximum activity (frequency more than 6-days) are Assam and adjacent States and Kerala. The other areas are foot hills of the Himalaya (J & K), Punjab, Himachal Pradesh, east Madhya Pradesh, Bihar, Bengal. The lowest activity lies in Saurastra, Kutch, Gujarat and Konkan coast.

In April thunderstorm activity increases. Maximum frequency (12-14 days) lies over Assam and Kerala. The other areas of high frequency (about 8 days) are east Madhya Pradesh, Vidarbha, Deccan Plateau. The lowest frequency (about one or two days) lies in south coastal Andhra Pradesh, coastal Tamilnadu, Saurastra, Kutch, Gujarat and Konkan coasts.

May is the month of peak thunderstorms activity. The maximum frequency zone runs from Assam to south Kerala. Assam and Bengal have frequency of about 16 days, while Gujarat, Saurastra and Kutch have the least frequency or no activity.

The chief atmospheric features during this season are : (i) a number of meteorological parameters show large diurnal variation, (ii) formation of heat low over central/interior parts of the country, (iii) weak to moderate westerly wind flow in the lower troposphere and (iv) intense convective activity associated with conditional convective instability (steep lapse rate).

Thunderstorm activity continues in monsoon and post monsoon season and becomes least in winter. The highest frequency of thunderstorms (more than 80 days in a year) extends from Assam to east Madhya Pradesh. Assam and neighbouring Bengal have frequency of more than 100 days. Silchar has 101 days, Mohanbari 106 days, Gauhati 102 days and Sibsagar, Krishnanagar have 108 days. The other high frequency zone is central Kerala and hilly regions of extreme north India. The lowest frequency, less than 10 days in a year, are in coastal Saurastra and Kutch.

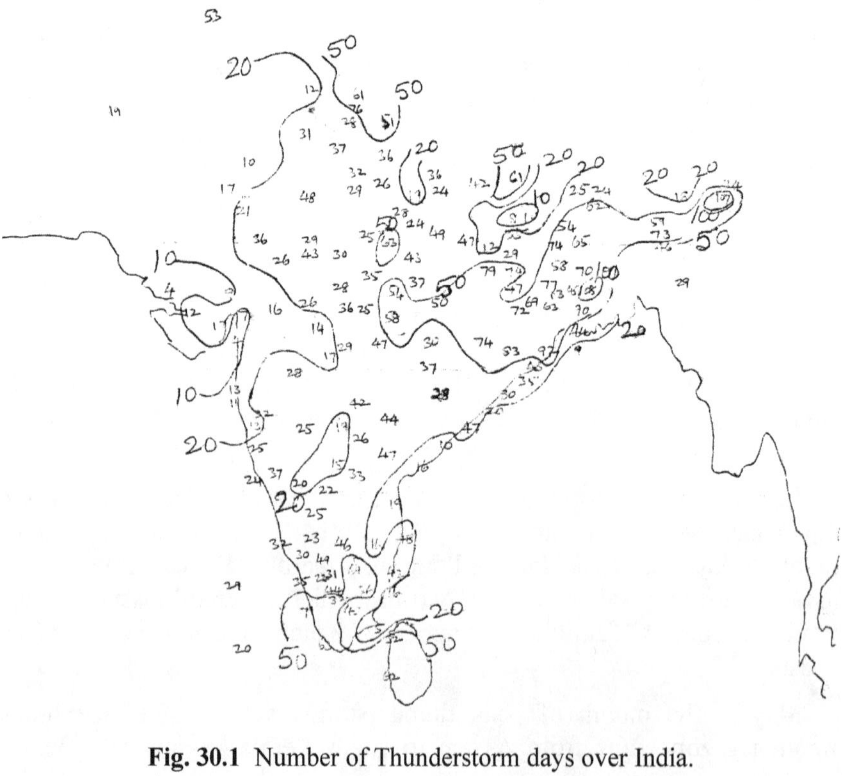

Fig. 30.1 Number of Thunderstorm days over India.

30.2.1 Norwesters

The severe Thunderstorms of hot weather period over Bengal and Orissa are known as Norwesters. They are locally called 'Kal Baisakhi' or the fateful thing of the month Baisakh (15 April to 15 May). These severe thunderstorms usually approach a station from northwest direction (and hence the name Norwester) and burst over a station with great violence raising clouds of dust. The squalls associated with these Norwesters reach tornadic violence (120 to 150 kmph). They cause considerable damage to life and property in their path. Tornadoes rarely occur in north east India. Assam and adjacent states are more susceptible to tornadoes locally called Hatishnura (frequency once in a few years). Generally the direction of the squalls is Northwesterly direction.

Favourable Synoptic Situations

(i) Induced low over Punjab, Haryana, Rajasthan, West Uttar Pradesh or northwest Madhya Pradesh (ii) conditional convective instability in the lower-mid troposphere, (iii) adequate moisture supply in the lower troposphere (Note: Horizontal distribution of moisture at the ground can be ascertained by dew point temperature of surface observations while vertical distribution of moisture is known through Radiosonde ascents) (iv) An extension of trough east-south-east wards from induced low to Bihar Plateau and west wards (v) Penetration of easterlies up to 850 hPa level over Bihar plains, (vi) In mid-and upper troposphere mainly westerly wind flow over north-east India.

Thunderstorms in Gangetic West Bengal, Bihar, Orissa have definite sequence of time. They generally develop over Bihar plateau, south Madhya Pradesh or west Orissa. They travel east or south-east wards towards Gangetic West Bengal or Head Bay of Bengal.

The severe thunderstorms of north-east India are classified into four types.

Type A: They develop over Bihar plateau and adjoining area in the after noon and move with fury in a southeasterly direction.

Type B: They originate in Submontane districts of north Bengal and move south wards. They generally form and move during night or early morning.

Type C: They originate in the hills of Nagaland, Manipur and Mizoram and move west wards.

Type D: They develop near the Khasi hills and move south wards.

In fact Perturbation in the westerlies caused by a trough or Jet stream or a cut-off low. Deep westerly trough causes widespread thunderstorm activity over north-east India.

30.2.2 Thunderstorms and Dust-storms over North-west India

During March-June (in all months), thunderstorms occur in north-west India and Uttar Pradesh. The activity is more in western Himalayas as compared to the plains. Thunderstorm activity progressively increases from March and attains peak in May and June. Dust storms occur in plains of UP, Punjab, Haryana and north Rajasthan. They begin in April and attain peak in June. The dust storms locally called Andhi – meaning blinding.

Duststorms are of two types .(i) Pressure gradient type and (ii) convective type. In pressure gradient type, strong winds raise loose dust from the surface of the earth into the atmosphere. In this case no Cb is involved. In convective type dust is raised into the atmosphere by strong downdrafts from Cb cloud. In convective type dust storm precedes a thunderstorm.

The maximum frequency of dust-storms occurs in Rajasthan. In pressure gradient type strong surface winds are built up due to steep gradient around low. Wind raised dust suspends in the atmosphere and reduces the visibility to below one km. Pressure gradient of 1 to 1.5 hPa or more per degree latitude prevails over a longer period. Winds of the order 30 to 50 Kt prevail up to 1.5 km (850 hpa level). Dust raising winds may commence in the morning and continue for the whole day with maximum intensity in the afternoon when the lapse rate near the ground reaches super-adiabatic. In convective type dust-storms the duration of strong winds is of short duration (few minutes or fraction of an hour). Even after duststorm/dust raising winds subside, the raised dust in the atmosphere may go up to 700 hPa level and remain suspended for days together and reduce the visibility.

Favourable Synoptic Situation

Shallow heat low forms over central and northwest India. Quasi Stationary wind discontinuity in low levels runs in north-south direction to Peninsula and another south-east Madhya Pradesh to Assam. A shallow anticyclone over Bay of Bengal feeds moisture from the south on one side of the wind discontinuity and contrasting dry northerly or north-westerlies prevail on the other side of the discontinuity line. Over north India in the mid-and upper troposphere sub-tropical jet stream runs, which provides upper divergence. Near the surface of the earth and in planetary boundary layer super-adiabatic temperature lapse rate favours convective activity.

30.2.3 Thunderstorms Over South Peninsula

During March-May high frequency of thunderstorm activity prevails over Kerala (about 15 days each month) and neighborhood. Orography plays an important role. Abundant moisture feed comes from the neighbouring sea. Kodaikanal: seasonal 34 days, annual 75 days; Trivendrum: seasonal 36 days, annual 70 days.

30.2.4 Hailstorms

Hailstorms in India occur in northern parts of the country. They are much less common over Peninsula. Hailstorms are always associated with Cb cloud only. Hailstorms are common in winter (Dec-Feb) over northwest India, Uttar Pradesh, Madhya Pradesh, and in pre-monsoon period over north-east India, Peninsula, Vidarbha and Madhya Pradesh. No hailstorm reports from any part of the country during monsoon (June-September) period. During post monsoon (October-November) there is little activity in isolated pockets north of Lat 20 ^{O}N and west of long 90 ^{O}E.

The annual frequency of hailstorms over northeast India is about 10 days, over Uttar Pradesh, east Bihar and Agartala about 6 days, over east Madhya Pradesh, Chattisgarh, Jharkhand is 2 to 3 days. Maximum activity over south Peninsula is 1 to 2 near Madurai. The entire west coast, south coastal Andhra Pradesh are free from hailstorm activity.

Favourable Synoptic Features

Intense convective conditional instability, large Cb cloud extending to upper troposphere (12 to 14 km). In lower and mid-troposphere high moisture content, strong vertical wind shear surface to mid-troposphere, low freezing level, lowering of trough in the jet core.

Questions

1. During November-March, the westerly waves of extra-tropical frontal systems or secondary systems that arrive over India will be in occluded stage. They show the characteristics of sequence:
   (a) Cold front, warm front, cold sector
   (b) Warm front, warm sector, cold front
   (c) Cold sector, cold front, warm sector
   (d) Warm sector, warm front, cold front

2. The extra-tropical depressions moving over Iraq, bifurcate into two near Iraq-Iran area by high terrain. The .... travels........ along the Persian Gulf which effects north India as......
   (a) Northern cell; west wards; tropical depression
   (b) Southern cell; east wards; western disturbance
   (c) Western cell; east ward; ETC
   (d) Western cell; west wards; tropical depression
3. During winter, cloud and precipitation belts move west to east over north India they are called.........
   (a) ETCs (b) ET depressions
   (c) Western Disturbances (d) Tropical disturbances
4. The sequence of precipitation associated with WD is:
   (a) First day over Jammu and Kashmir, north of latitude 30 °N, secondary over area north of latitude 26 °N, subsequently over Madhya Pradesh and then moves eastwards
   (b) First between latitude 30 °N, latitude 26 °N, next over Madhya Pradesh, then induced low over Gulf of Camby, Gujarat, north Maharastra
   (c) First over Gulf of Camby, then Madhya Pradesh and subsequently east wards
   (d) First over Jammu and Kashmir, next over Gulf of Camby and then over Madhya Pradesh
5. With the approach of a WD at a station, the changes that appear in (i) pressure, (ii) temperature, (iii) minimum temperature (iv) and dew point are:
   (a) Rise, fall, fall, rise (b) Fall, rise, rise, rise
   (c) Fall, fall, rise, fall (d) Rise, rise, rise, fall
6. In the rear of the passage of WD, the changes in pressure and temperature are:
   (a) Rise, fall (b) Fall, rise
   (c) Rise, rise (d) Fall, fall
7. In the Himalayas, with the passage of a WD, fog may occur:
   (a) Ahead of WD (b) In the rear of WD
   (c) During passage (d) Both (a) and (b)

8. With the passage of a WD over a station, the sequence of clouding appears:

(a) Cu, Cb, Ns, Ac Ci (b) Ci, Cs, As, Sc, Cu, Cb

(c) As, St, Ns, Cu (d) St, Cu, Cb, Fs

9. With the passage of a WD, the weather sequence at a station:

(a)(i) Warm front weather-continuous light rain, drizzle,

(ii) Warm sector weather-intermittent rain, scattered showers

(iii) Cold front weather-heavy showers thunderstorms, hailstorm, squalls

(b) Warm sector, cold front, warm front

(c) Cold front, warm front, warm sector

(d) Cold front, warm sector, warm front

10. In north India, Duststorms begin in April and attain peak in May, June. They occur in:

(a) Plains of Uttar Pradesh, Punjab

(b) Haryana, north Rajasthan

(c) Gujarat, Madhya Pradesh

(d) Both (a) and (b)

11. In north India, the duststorms are locally called:

(a) Norwesters (b) Andhi

(c) Hathisnura (d) Loo

12. Duststorms are of two types namely:

(a) Pressure type, stable type

(b) Convective type, unstable type

(c) Mechanical type, orographic type

(d) Pressure gradient type, convective type

13. In pressure gradient type duststorms, loose dust is raised by .... from the surface of the earth and.........

(a) Squalls; Cb involved

(b) Strong winds; no Cb involved

(c) Squalls; Cb not involved

(d) Both (a) and (c)

14. The maximum frequency of duststorms occur in:
    (a) Punjab, Haryana
    (b) Madhya Pradesh, Uttar Pradesh
    (c) Rajasthan
    (d) Gujrat, Andhra Pradesh
15. During March-May, thunderstorm activity is maximum over.......... Kodaikanal, Trivandrum. Seasonal thunderstorm days are..........
    (a) Kerala; 10; 20
    (b) Madhya Pradesh; 20; 30
    (c) Kerala; 34; 36
    (d) Both (a) and (b)
16. Hailstorms are associated with:
    (a) Cu (b) Ns (c) Cb (d) Ci
17. Hailstorms are common in winter (December-February) over:
    (a) Northwest India, Uttar Pradesh, Madhya Pradesh
    (b) Rajasthan, Gujarat, Andhra Pradesh
    (c) Assam, gangetic west Bengal, Orissa
    (d) Both (b) and (c)
18. Hailstorms are common in premonsoon (March-May) over:
    (a) Rajasthan, Jammu and Kashmir, Uttar Pradesh
    (b) Northeast India, Peninsula, Vidarbha, Madhya Pradesh
    (c) Northwest India, Uttar Pradesh, Gujarat
    (d) Both (a) and (c)
19. The lowest frequency of occurrence of thunderstorms in India over:
    (a) Assam, Bengal, Kerala
    (b) Gujarat, Saurastra and Kutch
    (c) Jammu and Kashmir, Uttar Pradesh, Bihar
    (d) Andhra Pradesh, Tamilnadu and Karnataka
20. May is the peak month of thunderstorm activity. The maximum frequency in this month about.......over........
    (a) 12 days; saurastra and Kutch
    (b) 16 days; Assam, Bengal
    (c) 14 days; Uttar Pradesh, Bihar
    (d) Both (a) and (c)

21. Thunderstorm activity in India, least during:

(a) Summer (b) Monsoon

(c) Post monsoon (d) Winter

22. The annual highest frequency of thunderstorm days........extends over ....

(a) More than 100 days; Assam and neighbouring Bengal

(b) More than 120 days; Kerala, Karnataka

(c) 110 days; Gangetic west Bengal, Bihar

(d) 106; Uttar Pradesh, Madhya Pradesh and Orissa

23. In India, the highest frequency of thunderstorm days...recorded at...........

(a) 101 days, Gauhati

(b) 106 days, Mohanbari

(c) 108 days; Sibsagar and Krishna nagar

(d) 120 days; Silchar

24. In India, the lowest annual frequency of thunderstorm days.......over.......

(a) 25; Karnataka

(b) 10; Rajasthan

(c) Less than 10 days, coastal Saurastra and Kutch

(d) Less than 10; Kerala

25. The severe thunderstorms over Bengal and Orissa (15 April to 15 May) are called.....,locally called.........

(a) Western disturbances, Hatisnoora

(b) Norwesters, Kalbaisakhi

(c) Norwesters, Andhi

(d) Both (a) and (c)

26. In Baisakh (April-May), Thunderstorms in Gangetic west Bengal, Bihar Orissa travel:

(a) West or west north west (b) East or south-east

(c) South or south-east (d) North or north-east

CHAPTER 31

General Aviation Services

Abbreviations used in Aeronautical information service publications, by Airports Authority of India

(AIP India GEN 2-2-3 01 Jan 2010)

A	Amber
ACT	Active/Activated/Activity
A/A	Air-to-Air
AAL	Above Aerodrome Level
ABT	About
ABV	Above
AC AS	Airborne Collision Avoidance System
ACC	Area Control Centre or simply Area Control
ACCID	Notification of an aircraft accident
ACFT	Aircraft

CHAPTER 31

General Aviation Services

Abbreviations used in Aeronautical information service publications, by Airports Authority of India

(AIP India GEN 2-2-3 01 Jan 2010)

A	Amber
ACT	Active/Activated/Activity
A/A	Air-to-Air
AAL	Above Aerodrome Level
ABT	About
ABV	Above
AC AS	Airborne Collision Avoidance System
ACC	Area Control Centre or simply Area Control
ACCID	Notification of an aircraft accident
ACFT	Aircraft

ACK	Acknowledge
ACT	Active/Activated/Activity
AD	Aerodrome
ADDN	Addition/Additional
ADF	Automatic direction finding equipment
ADIZ	Air Defense identification zone
ADJ	Adjacent
ADS	Automatic dependent surveillance
ADVS	Advisory service
AFIS	Aerodrome flight information service
AFS	Aeronautical fixed service
AGL	Above ground level
AGN	Again
AIC	Aeronautical information circular
AIP	Aeronautical information publication
AIREP	Air-report
ALT	Altitude
ALTN	Alternate (Aerodrome)
AMD	Amend/Amended
AMDT	Amendment
AMSL	Above mean sea level
AMSS	Aeronautical mobile satellite service
AP	Airport
APCH	Approach
APN	apron
APR	April
APRX	Approximate / approximately
AR FOR	Area Forecast
ARNG	Arrange

ARP	Aerodrome reference point
ARR	Arrive / Arrival
ASC	Ascent / ascending to
ATC	Air traffic control
ATS	Air traffic service
AS	Altostratus
AUG	August
AVG	Average
B	Blue
BA	Braking action
BASE	Cloud base
B C	Patches
BCN	Beacon
BCST	Broadcast
BDRY	Boundary
BECMG	Becoming
B FR	before
BKN	broken
B L	Blowing
BLO	Below clouds
BLW	below
B R	mist
BLSN	Blowing Snow
BCFG	Fog patches
BASE	cloud base
BTN	Between
B TL	Between layers
C	Degrees Celsius (centigrade)

CAT	Category / Clear Air Turbulence
CAVOK	Visibility, cloud and present weather better than prescribed value or condition.
C B	Cumulonimbus
C I	cirrus
CC	cirrocumulus
CLD	cloud
CM	centimeter
CMB	climb / climbing to
CNS	Communication Navigation and Surveillance
CNL	Cancelled / cancel
COM	Communication
CONT	Continue/continued
CL	Centre line
CNS	Continuous
COND	Condition
COP	Change over point
COR	Correct / correction / corrected
COT	At the cost
COV	Cover / covered / covering
CPL	Current flight plan
CS	Call sign (used to request a call sign), Cirrostratus
CTA	Control Area
CS	Cirrostratus
CU	Cumulus
CUF	Cumuliform
DA	Decision altitude
DCT	Direct
DEC	December
DEV	Deviation

DEG	Degrees
DIV	Divert
DIF	Diffuse
DEP	Depart / departure
DGCA	Director General of Civil Aviation
DH	Decision height
DISP	Displaced
DS	Dust storm
DIST	Distance
DUR	Duration
DLY	Daily
DUC	Dense upper cloud
DME	Distance measuring instrument
DOC	Document
DRG	During
DPT	Depth
DTRT	Deteriorate / deteriorating
DP	Dew point temperature
DZ	Drizzle
DR	Low drifting
DVOR	Doppler
DU	Widespread dust
EA	Expected approach time
EB	East bound
E	East/Eastern
ELBA	Emergency location beacon aircraft
ELEV	Elevation
ELR	Extra long range
EMBD	Embedded in a layer

ENE	East-north-east
END	Stop-end (related to RVR)
ENRT	En-route
ENR	En-route
ESE	East-south-east
EST	Estimate / estimated
ETA	Estimated time of arrival
ETC	Etcetera
ETD	Estimated time of departure
EV	Every
EXC	Except
EXP	Expected / expect / expecting
EXTD	Extend / extending
F	Fixed
FBL	light, feeble
FAC	Facilities
FC	Funnel cloud
FAP	Final approach point
FAX	Facsimile transmission
FCST	Forecast
FEB	February
FG	Fog
FEW	few
FIR	Flight information Region
FIC	Flight information centre
FIS	Flight information service
FISA	Automated flight information service
FL	Flight level
FM	From

FLUC	Fluctuating
FNA	Final approach
FREQ	Frequency
FZ	Freezing
FZDZ	Freezing drizzle
FRQ	Frequent
FZFG	Freezing fog
FST	First
FZRA	Freezing rain
FT	Feet
FU	Smoke
G	Green
G /A	Ground to air
G / A / G	Ground to air and air to ground
GEN	General
GRADU	Gradual / gradually
GLD	Glider
GND	Ground
GNSS	Global navigation satellite system
GP	Glide path
GR	hail / soft hail
GPS	Global positioning system
GRASS	Grass landing Area
GS	Ground speed
H+	Hours plus
H 24	Continuous day and night service
HDG	Heading
HEL	Helicopter
HEL-L	Light helicopter

HEL-M	Medium helicopter
HEL-H	Heavy helicopter
HF	High frequency (3000 to 30,000 KHz)
HPA	Hecto Pascal
HURCN	Hurricane
HQ	Headquarters
HR	Hours
HVY	heavy
HZ	Haze / Hertz
IAS	Indicated air speed
ID	Identifier / Identify
ICE	Icing
IFR	Instrument flight rules
ILS	Instrument landing system
IM	Inner marker
IMPR	Improve / Improving
IMC	Instrument meteorological conditions
ISA	International standard atmosphere
ISOL	Isolated
INC	In Cloud
INTSF	Intensify / Intensifying
JAN	January
JUL	July
JTST	Jet stream
JUN	June
KG	Kilogram
KHZ	Kilohertz
KM	Kilometers
KMH	Kilometers per hour

KT	knots
KW	Kilowatt
L	Left/Locator
LAT	Latitude
LEN	length
LDG	Landing
LF	Low frequency (30 to 300 kHz)
LNG	Long
LO	Locator/outer
LOC	Local/Located/Location
LONG	Longitude
LORAN	LORAN (Long range navigation system)
LRG	Long range
LSQ	Line squall
LT	Local time
LTD	Limited
LTT	Landline teletypewriter
LV	Light and variable (related to wind)
LVL	Level
LYR	Layer or layered
M	Mach number/meters
MAR	March
MAX	Maximum
MAY	May
MI	Shallow
MCA	Minimum crossing altitude
MDA	Minimum descent altitude
MET	Meteorology/meteorological
METAR	Aviation routine weather report

MHZ	Megahertz
MIN	Minutes
MIFQ	Shallow fog
MLS	Microwave landing system
MNTN	Maintain
MOC	Minimum obstacle clearance
MOD	Moderate
MNM	Minimum
MON	Monday/above mountains
MOV	Move/moving/movement
MPS	Meters per second
MRG	Medium range
MRP	ATS/MET reporting point
MS	Minus
MSA	Minimum sector altitude
MSG	Message
MTW	Mountain waves
MSL	Mean seal level
MWO	Meteorological watch office
N	North/northern latitude
NAV	Navigation
NC	No change
NDB	Non directional radio beacon
NE	Northeast
NM	Nautical mile
NNE	North-north-east
NNW	North-north-west
NOF	International NOTAM office
NOSIG	No significant change

NOTAM	A notice containing information concerning the establishment, condition or change in any aeronautical facility service, procedure or hazard, the timely knowledge of which is essential to personnel concerned with flight operations.
NOV	November
NS	Nimbostratus
NSC	no significant cloud
NSW	no significant whether
NW	Northwest
NXT	Next
OAC	Oceanic area control centre
OAS	Obstacle assessment surface
OBS	Observe/observed/observation
OBST	Obstacle
OBSTR	Obstruction
OCA	Oceanic control area / obstacle clearance altitude
OCH	Obstacle clearance height
OCNL	Occasional
OCT	October
OM	Outer maker
OPMET	Operational Meteorological (information)
OPN	Open / opened / opening
OPR	Operator / operate / operational
OPS	Operations
O/R	On request
OTP	On top
OT LK	Outlook
OUBD	Out of bound
OVC	Overcast

PALS	Precession approach lighting system
PANS	Procedures for air navigation services
PAPI	Precision approach path indicator
PAR	Precision approach radar
PARL	Parallel
PO	Dust/Sand whirls
PAX	Passenger(s)
PCN	Pavement classification number
PER	Performance
PLN	Flight plan
PN	Prior notice required
PNR	Point of no return
POB	Persons on board
PPI	Plan position indicator (in radar)
PPR	Prior permission required
PRI	Primary
PROB	Probability
PROC	Procedure
PS	Plus
PSN	Position
PSR	Primary surveillance Radar
QBI	Compulsory IFR flight
QBM	Magnetic heading (zero wind)
QDR	Magnetic bearing
QFE	Atmospheric pressure at aerodrome elevation (or at runway threshold)
QFU	Magnetic orientation of runway
QNH	Altimeter sub-scale setting to obtain elevation when on runway/ground
QTE	True bearing

QUAD	Quadrant
R	Red/Runway (identification)
RA	Rain
RCA	Reach cruising altitude
RCC	Rescue coordination centre
RCF	Radio communication failure
RCL	Runway centre line
RDL	Radial
RAFC	Regional area forecast centre
RDO	Radio
RAPID	Rapid/rapidly
RASH	Rain showers
RASN	Rain and snow
REC	Receiver/Receive
REDL	Runway edge light(s)
REG	Registration
RE	Recent
RENL	Runway edge light(s)
REP	Reporting point/ report
REQ	Request/requested
RG	Range
REF	Reference to
RH	Rescue helicopter
RMK	Remark
ROBEX	Regional OPMET bulletin exchange
ROC	Rate of Climb
ROD	Rate of descent
ROFOR	Route forecast (in aeronautical meteorological code)
RPT	Repeat

RTE	Route
RQP	Request flight plan
RSC	Rescue sub centre
RSCD	Runway surface condition
RTF	Radio telephone
RTG	Radio telegraph
RTT	Radio teletypewriter
RV	Rescue vessel
RVR	Runway visual range
RWY	Runway
S	South/ southern latitude
SAP	As soon as possible
SAR	Search and rescue
SARPS	Standards and recommended practices (ICAO)
SAT	Saturday
SATCOM	Satellite communication
SB	South bound
SC	Stratocumulus
SCT	scattered
SS	Sand storm
SDBY	Standby
SE	Southeast
SEV	Severe
SEC	Seconds
SFC	Surface
SECT	Sector
SG	Snow grains
SELCAL	Selective calling system
SEP	September

SH	showers
SFC	Surface
SID	Standard instrument departure
SIGMET	Information concerning enroute weather phenomena which may effect the safety of aircraft operations
SIG WX	Significant weather
SPECI	Aviation selected special weather repot (in aeronautical meteorological code)
SPECIAL	Special meteorological report (in abbreviated plain language)
SIMUL	Simultaneous / simultaneously
SKC	Sky clear
SPOT	Spot wind
SQ	Squall
SR	Sunrise
SRG	Short range
SRR	Search and rescue region
SS	Sunset/sand storm
SSE	South-south-east
SSR	Secondary surveillance radar
SST	Supersonic transport
SSW	South-south-west
ST	Stratus
STD	Standard
STF	Stratiform
STN	Station
STNR	Stationary
STOL	Short take-off landing
SUN	Sunday

SW	Southwest
T	Temperature
TC	Tropical Cyclone
TA	Transition altitude
TACA	UHF tactical air navigation
TAF	Aerodrome forecast
TAIL	Tailwind
TAR	Terminal area Surveillance Radar
TAS	True air speed
TAX	Taxing/Taxi
TDO	Tornado
TCU	Towering cumulus
TDZ	Touch down zone
TEMPO	Temporarily/temporary
THU	Thursday
TEND	Trend /trending to
TIL	Until
TOP	Cloud top
TP	Turning point
TR	Track
TROP	Tropopause
TRNG	Training
TRANS	Transmit / transmitter
TRL	Transition level
TS	Thunderstorm
TSGR	Thunderstorm with hail
TSSS	Thunderstorm with duststorm or sandstrom
TT	Teletypewriter
TURB	Turbulence

TYPH	Typhoon
UAC	Upper area control centre
UAR	Upper air route
UHF	Ultra high frequency (300 to 3000 MHz)
ULR	Ultra long range
U/S	Unserviceable
UTC	Coordinated universal time
VA	Volcanic ash
VCY	Vicinity
VER	Vertical
VFR	Visual flight rules
VHF	Very high frequency (30 to 300 MHz)
VIP	Very important person
VIS	Visibility
VLF	Very low frequency (3 to 30 KHz)
VLR	Very long range
VMC	Visual meteorological conditions
VOLMET	Meteorological information for aircraft in flight
VOR	VHF omni directional radio range
VRB	Variable
VSP	Vertical speed
W	West/ western latitude/ white
WDI	Wind direction indicator
WDSPR	Widespread
WAFC	World area forecast centre
WED	Wednesday
WEF	With effect from
WI	Within
WID	Width

WIE	With immediate effect
WKN	Weaken/weakening
WNW	West-north-west
WRNG	Warning
WS	Wind shear
WSW	West-south-west
WT	Weight
WTSPT	Waterspout
WX	Weather
X	Cross
XBAR	Crossbar
XNG	Crossing
Y	Yellow
YCZ	Yellow caution Zone (runway lighting)

31.1 Air Traffic Services (ATS)

Airports Authority of India is the responsible authority for the provision of air traffic service within the area for the entire Indian Territory including territorial waters as well as in airspace over high seas encompassed by Kolkata, Chennai and Mumbai Flight Information Regions, (Exceptions certain military areas, State military and private aerodromes).

Type of services: With the exception of services provided at military air bases, the following types of services are provided.

(a) Air Traffic Control Service

(b) Air Traffic Advisory Services

(c) Flight Information Service and Alerting Service

(d) Radar Service

There is no distinction between upper and lower controlled space. A number of prohibited, restricted and danger areas have been established.

Minimum Flight Altitude: The minimum flight altitude on the ATS routes have been determined so as to ensure at least 1000 ft (300 m) vertical clearance above the highest obstacle within 18 km on each side of the centerline of the route.

31.2 Communication Services

Airports Authority of India is responsible for the overall administration of the Aeronautical Telecommunications, Navigation and Surveillance (CNS) facilities of India.

The Chairman
Airports Authority of India
Rajiv Gandhi Bhavan
Safdarjang Airport
New Delhi – 110003 (India)

Communication services are provided for the entire INDIAN FIRs and the area of responsibility covers all Aeronautical Telecommunications, Navigation and Surveillance Services in India. Responsibility for the day-to-day operations is vested with General Manager (COM) / Airport Director / Controller of Communication / Officer-in-charge at the local Aeronautical Telecommunication unit concerned located at each airport functioning under five Regional Executive Directors with Regional Head Quarters at airports at New Delhi, Mumbai, KolKota, Chennai and Guwahati, who is also responsible for the application of the regulations concerning the design, type and installations of aircraft radio stations.

Types of Service: (1) The following types of Radio Aids to Navigation are available.

(a) Medium Frequency (300-3000 KHZ) Non Directional Beacon (MF NDB)
(b) Instrument Landing System (ILS)
(c) VHF Omini directional Radio Range (VOR)
(d) Distance Measuring Equipment (DME)
(e) Airport Surveillance Radar (ASR) / Monopulse Secondary Surveillance Radar (MSSR)

***Mobile Service*:** The aeronautical stations maintain a continuous watch on the frequencies stated within the published hours of service.

An aircraft shall normally communicate with ATS unit providing service in the area in which the aircraft is flying. Aircraft should maintain a continuous watch on the appropriate frequency of the ATC unit, and shall not abandon watch except in an emergency without informing the ATC unit. Aircraft shall, to the extent possible, also maintain watch on the emergency frequency 121.5 MHz.

All radio telephony communications between aircraft and Indian ATS units are recorded.

Fixed Services: Messages to be transmitted over the AFTN (Aeronautical Fixed Telecommunications Network) will be accepted when they satisfy certain conditions.

Broadcasting Service:

(i) VOLMET BROADCAST

(ii) ATIS (Automatic Terminal Information Service)

Language used: English

31.3 Meteorological Services

The Meteorological Services for civil aviation are provided by the India Meteorological Department, Ministry of Earth Sciences, Government of India.

Director General of Meteorology, India Meteorological Department, Mausam Bhavan, Lodhi Road, New Delhi – 1100 03.

Area of Responsibility: Meteorological services are provided within the Delhi, Kolkata, Chennai and Mumbai FIR.

Types of Services

1. ***Area Forecast***: Under the World Area Forecast System of the International Civil Aviation Organization (ICAO), a regional area forecast centre (RAFC) is located at New Delhi. Forecast charts covering the area bounded by Lat Equator to 40° N, Long 30 °E to 125 °E are prepared and broadcast on facsimile from the centre. Prognostic charts for the surfaces:

   (i) Sea level, (ii) 500 hPa (FL 180) (iii) 300 hPa (FL 300) (iv) 250 hPa (FL 340) and (v) 200 hPa (FL 390) and for significant weather below flight level 400 (150 hPa) and for the tropopause/maximum winds are broadcast according to the schedule.

2. ***Route Forecasts and Aerodrome Forecasts***: Route forecast, aerodrome forecast, local forecasts and trend forecasts are issued by class I Meteorological Offices (i.e., offices having forecasting facilities). During the closed hours of watch of the offices with restricted forecasting hours, aerodrome forecasts and route forecasts, if necessary are issued by the associated Class I Meteorological Offices

3. ***Aerodrome warning and wind shear warnings***: Aerodrome warning for meteorological conditions that adversely affect aircraft and other facilities on the ground and windshear warning for aircraft during landing, take-off, circling approach between runway level and 500 m above that level and aircraft on the runway during landing roll or take-off run. These warnings are issued by Class I Meteorological Office in the airports. The elements for which warnings are issued are – cyclone, thunderstorm, hail, sandstorm/duststorm, strong surface wind, gusts and squalls.

   These warnings are supplied to ATS units of Airports Authority of India for further dissemination over the Aerodrome to operators and other agencies.

   Aerodromes and windshear warnings are not issued in those airports where forecasting offices do not exists.

4. ***Briefing and Display***: In all Class I Meteorological Offices in airports, briefing is available to the pilots and operators representatives about the prevailing and anticipated weather conditions. Latest surface and upper-air synoptic charts, meteorological reports and forecasts of destination and its alternate(s), SIGMET information, AIREPs, prognostic charts, ground based weather radar information and satellite cloud imagery are normally displayed in meteorological offices and are available for consultation and briefing. Depending on the weather situation, the meteorological officer may insist that the pilot-in-command of a flight be present personally to be briefed.

5. ***Documentation***: Documentation consists of: (1) flight forecast in tabular form/pictorial cross sectional form/chart form, (2) terminal aerodrome forecasts of destination and alternates, (3) take-off and

ascent/descent data (wherever necessary) are provided to flights taking off from airports within the country. The chart form of forecast consists of (a) prognostic significant weather charts below flight level 460 and (b) prognostic upper wind and temperature charts for the flight level 300 hPa, 250 hPa and 200 hPa. Chart form of documentation is provided at Mumbai, Kolkata, Delhi and Chennai. Forecasts are normally given up to next halt where a meteorological forecasting office exists, when the duration of the halt is more than 45 minutes. Return forecasts are provided for flights, when the halt for the destination is less than 45 minutes (60 minutes in the case of Goa and Srinagar) and there is no change of crew. If sufficient notice is given, arrangements are also made to provide documentation even at Airports without meteorological forecasting facility. If planned halt(s) at intermediate airports are for long periods, a decision whether fresh forecasts will be required at intermediate halt(s), will have to be taken by mutual consultation between that pilot and the meteorological office at the original departure aerodrome taking into consideration the expected whether developments. A fresh forecast has to be obtained at intermediate halts when the period of validity of the route and of/or aerodrome forecasts supplied has elapsed or expected to elapse before the completion of the flights due to changes in the original plan. In such cases, the operator has to make arrangements to obtain the fresh forecast from the aeronautical meteorological office at the original departure aerodrome or from the nearest aeronautical meteorological office. The responsibility for the collection and transmission of the forecasts to the pilot at intermediate halts rests with the operator.

6. ***In-flight Services***:

   (a) **SIGMETs:** Significant Meteorological information pertaining to the FIRs are issued by the Meteorological watch offices (MWOs) at Mumbai, Kolkata, Delhi and Chennai and are known as SIGMETs. These are supplied to the ATS units at the Flight Information Centers (FICs) for the transmission to aircraft in-flight.

(b) **VOLMET BROADCASTS:** Current weather repots and Aerodrome forecasts of certain stations are broadcast in HF from Mumbai and Kolkata at half hourly intervals.

(c) **ATIS (Automatic Terminal Information Service) BROADCAST:** Latest weather report of the airport, together with trend forecast (valid for next two hours) is included in the Automatic Terminal Information Service (ATIS) Broadcast from Mumbai, Kolkata and Delhi.

Questions

1. Aviation services include:
   (a) Air Traffic Services
   (b) Communication Services
   (c) Meteorological Services
   (d) All above (a), (b), (c)
2. Air Traffic Services include:
   (a) Air Traffic Control Service and Radar service
   (b) Air Traffic Advisory Services
   (c) Flight information service and Alerting service
   (d) All above (a), (b), (c)
3. Minimum flight altitude on the ATS routes determine to ensure atleast...........vertical clearance above the highest obstacle within........on each side of the centerline of the route.
   (a) 500 ft, 10 km   (b) 1000 ft, 18 km
   (c) 1500 ft, 25 km   (d) Both (a) and (c)
4. In India, who is responsible for the overall administration of the Aeronautical Telecommunications, Navigation and Surveillance facilities of India?
   (a) Communication Authority
   (b) Navigation Authority
   (c) Meteorological Authority
   (d) Airport Authority of India

5. Radio Aids to Air Navigation Include:
   (a) MF NDB, ILS (b) VOR, DME
   (c) ASR, MSSR (d) All above
6. The meteorological services for Civil Aviation include:
   (a) Area forecast
   (b) Route forecasts and Aerodrome forecasts
   (c) Aerodrome Warnings and Windshear Warnings
   (d) All above
7. Meteorological flight documentation consists of:
   (a) Flight forecast in tabular form / pictorial cross-section/ chart form
   (b) Terminal aerodrome forecasts of destination and alternates
   (c) Take-off and ascent / descent data whenever necessary
   (d) All above
8. In-flight services to aircraft are:
   (a) SIGMETs
   (b) VOLMET BROADCASTS
   (c) ATIS BROADCASTS
   (d) All above
9. Aerodrome warnings are issued by Class I Meteorological Office in the airports. The elements for which warnings are issued are:
   (a) Cyclones
   (b) Thunderstorm, hail, sandstorm / Dust storm
   (c) Strong surface wind, gusts and squalls
   (d) All above

Answers

1. (d) 2. (d) 3. (b) 4. (d)
5. (d) 6. (d) 7. (d) 8. (d)
9. (d)

CHAPTER 32

Meteorological Services for International Air Navigation

32.1 International Standards Recommended Practices

Definitions: The meaning of the terms used in the Standards and Recommended Practices for Meteorological Service for International Air Navigation are given below.

Aerodrome: A defined area on land or water (including any buildings, installations and equipment) intended to be used either wholly or in part for the arrival, departure and surface movement of aircraft.

Aerodrome Climatological Summary: Concise summary of specified meteorological elements at an aerodrome, based on statistical data.

Aerodrome Control Tower: A unit established to provide air traffic control service to aerodrome traffic.

Aerodrome Elevation: The elevation of the highest point of the landing area.

Aerodrome Meteorological Office: An office located at an aerodrome, designated to provide meteorological service for international air navigation.

Aeronautical fixed service (AFS): A telecommunication service between specified fixed points provided primarily for the safety of air navigation and for the regular, efficient and economical operation of air services.

Aeronautical Fixed Telecommunication Network (AFTN): A world-wide system of aeronautical fixed circuits provided, as part of the aeronautical fixed service, for the exchange of messages and / or digital data between aeronautical fixed stations having the same or compatible communications characteristics.

Aeronautical Meteorological Station: A station designated to make observations and meteorological reports for use in international air navigation.

Aircraft: Any machine that can derive, support in the atmosphere from the reactions of the air other than the reactions of the air against the earth's surface.

Aircraft Observation: The evaluation of one or more meteorological elements made from an aircraft in flight.

AIRMET Information: Information issued by a meteorological watch office concerning the occurrence or expected occurrence of specified en-route weather phenomena which may affect the safety of low-level aircraft operations and which was not already included in the forecast issued for low-level flight in the flight information region concerned or sub-area there of.

Air-report: A report from an aircraft in flight prepared in conformity with requirements for position, and operational and / or meteorological information.

Air Traffic Services Unit: A generic term meaning variously, air traffic control unit, flight information centre or air traffic services reporting office.

Alternate Aerodrome: An aerodrome, to which an aircraft may proceed when it becomes either impossible or inadvisable to proceed to or to land at the aerodrome of intended landing. Alternate aerodromes include the following.

(a) ***Take-off Alternate:*** Alternate aerodrome at which an aircraft can land should this become necessary shortly after take-off and it is not possible to use the aerodrome of departure.

(b) ***En-route Alternate:*** An aerodrome at which an aircraft would be able to land after experiencing an abnormal or emergency condition while en-route.

(c) ***Destination Alternate:*** An alternate aerodrome to which an aircraft may proceed should it become impossible or inadvisable to land at the aerodrome of intended landing.

Note: The aerodrome from which a flight departs may also be an en-route or destination alternate aerodrome for that flight.

Altitude: The vertical distance of a level, a point or an object considered as a point, measured from mean sea level (MSL).

Area of Coverage (World Area Forecast System): A geographical area for which a regional area forecast centre supplies forecasts for flights departing from aerodromes in its service area.

Area of responsibility (World Area Forecast System): A geographical area for which a regional area forecast centre prepares significant weather forecasts.

Briefing: Oral commentary on existing and / or expected meteorological conditions.

Consultation: Discussion with a meteorologist or another qualified person of existing and / or expected meteorological conditions relating to flight operations; a discussion includes answers to questions.

Control Area: A controlled air space extending upwards from a specified limit above the earth.

Cruising level: A level maintained during a significant portion of a flight.

Elevation: The vertical distance of a point or a level, on or affixed to the surface of the earth, measured from mean sea level.

Flight Crew member: A licensed crew member charged with duties essential to the operation of an aircraft during flight time.

Flight documentation: Written or printed document including charts or forms, containing meteorological information for a flight.

Flight information centre (FIC): A unit established to provide flight information service and alerting service.

Flight Information Region (FIR): An airspace of defined dimensions within which flight information service and alerting service are provided.

Flight Level: A surface of constant atmospheric pressure which is related to a specific pressure datum, [1013.2 hPa], and is separated from other such surfaces by specific pressure intervals.

Note: A pressure type altimeter calibrated in accordance with the standard atmosphere:

(a) when set to a QNH altimeter setting, will indicate altitude;

(b) when set to a QFE altimeter setting, will indicate height above the QFE reference datum;

(c) when set to a pressure of 1013.25 hPa may be used to indicate flight levels.

Forecast: A statement of expected meteorological conditions for a specific time or period, and for a specified area or portion of air space.

GAMET Area Forecast: An area forecast in abbreviated plain language for low-level flights for a flight information region or sub-area thereof, prepared by the meteorological office designated by the meteorological authority concerned and exchanged with meteorological offices in adjacent flight information regions, as agreed between the meteorological authorities concerned.

Grid point data in digital form: Computer processed meteorological data for a set of regularly spaced points on a chart, for transmission from a meteorological computer to another computer in a code form suitable for automated use.

Grid point data in numerical form: Processed meteorological data for a set of regularly spaced points on a chart, in a code form suitable for manual use.

Height: The vertical distance of a level, a point or an object considered as a point, measured from specified datum.

Meteorological Authority: The authority providing or arranging for the provision of meteorological service for international air navigation on behalf of Contracting State.

Meteorological Bulletin: A text comprising meteorological information preceded by an appropriate heading.

Meteorological Information: Meteorological report, analysis, forecast, and any other statement relating to existing or expected meteorological conditions.

Meteorological Office: An office designated to provide meteorological service for international air navigation.

Meteorological Report: A statement of observed meteorological conditions related to a specified time and location.

Meteorological Satellite: An artificial earth satellite making meteorological observations and transmitting these observations to earth.

Nephanalysis: The graphical depiction of analysed cloud data on a geographical map.

Observation (Meteorological): The evaluation of one or more meteorological elements.

Operator: A person, organization or enterprise engaged in or offering to engage in an aircraft operation.

Pilot-in-command: The pilot responsible for the operation and safety of the aircraft during flight time.

Prognostic Chart: A forecast of a specified meteorological element(s) for a specified time or period and a specified surface or portion of airspace, depicted graphically on a chart.

Regional Area Forecast Centre (RAFC): A meteorological centre designated to prepare and supply area forecasts for flights departing from aerodromes within its service area and to supply grid point data in digital form for up to world-wide coverage.

Reporting Point: A specified geographical location in relation to which the position of an aircraft can be reported.

Rescue Co-ordination Centre: A unit responsible for promoting efficient organization of search and rescue service and for co-ordinating the conduct of search and rescue operations within a search and rescue region.

Runway: A defined rectangular area on a land aerodrome prepared for the landing and take-off of aircraft.

Runway Visual Range (RVR): The range over which the pilot of an aircraft on the centre line of a runway can see the runway surface markings or the lights delineating the runway or identifying its centre line.

Search and Rescue Services Unit: A generic term meaning, as the case may be, rescue co-ordination centre, rescue sub-centre or alerting post.

SIGMET Information: Information issued by a meteorological watch office concerning the occurrence or expected occurrence of specified en-route weather phenomena which may affect the safety of aircraft operations.

Standard Isobaric Surface: An isobaric surface used on a world-wide basis for representing and analysing the conditions in the atmosphere example msl Chart, upper wind / temperature isobaric surfaces 850, 700, 500, 400, 300, 250, 200, 150, 100 and 070 hPa surfaces.

Threshold: The beginning of that portion of the runway usable for landing.

Touch Down Zone: The portion of a runway beyond the threshold, where it is intended landing aeroplanes first contact the runway.

Upper Air Chart: A meteorological chart relating to a specified upper-air surface or layer of the atmosphere.

VOLMET Broadcast: Routine broadcast of meteorological information for aircraft inflight.

World Area Forecast Centre (WAFC): A meteorological centre designated to prepare and supply upper air forecasts in digital form on a global basis to regional area forecast centres and also issue significant weather forecast.

World Area Forecast System (WAFS): A world-wide system by which world and regional area forecast centres provide aeronautical meteorological en-route forecasts in uniform standardized formats.

Categories of Precision Approach and Landing Operations

Category I (CAT I) Operation: A precision instrument approach and landing with a decision height not lower than 60 m (200 ft) and with either a visibility not less than 800 m or RVR not less than 550 m.

Category II (CAT II) Operation: A precision instrument approach and landing with a decision height lower than 60 m (200 ft), but not lower than 30 m (100 ft), and RVR not less than 350 m.

Category III A (CAT III A) Operation: A precision instrument approach and landing with:

(a) a decision height lower than 30 m (100 ft) or no decision height; and

(b) RVR not less than 200 m.

Category III B (CAT III B) Operation: A precision instrument approach and landing with:

(a) a decision height lower than 15 m (50 ft) or no decision height; and

(b) RVR less than 200 m but not less than 50 m.

Category III C (CAT III C) Operation: A precision instrument approach and landing with no decision height and no RVR limitations.

ASHTAM: A special series of NOTAM notifying by means of a specific format, change in activity of a volcano, a volcanic eruption and / or volcanic ash cloud that is of significance to aircraft operations.

Cloud of Operational Significance: A cloud with the height of cloud base below 1500 m (5000 ft) or below the height minimum sector altitude, whichever is greater.

Heliport: An aerodrome or defined area on a structure intended to be used wholly or in part for the arrival, departure and surface movement of helicopters.

International Airways Volcano Watch (IAVW): International arrangements for monitoring and providing warnings to aircraft of volcanic ash in the atmosphere.

Minimum Sector Altitude: The lowest altitude which may be used which will provide a minimum clearance of 300 m (1000 ft) above all objects located in the area contained within a sector of a circle of 46 km (25 NM) radius centred on a radio aid to navigation.

NOTAM: A notice distributed by means of telecommunication containing information concerning the establishment condition, or change in any aeronautical facility, service, procedure or hazard, the timely knowledge of which is essential to personnel concerned with flight operations.

Tropical Cyclone Advisory Centre (TCAC): A meteorological centre designated by regional air navigation agreement to provide advisory information to meteorological watch offices (MWOs), world area forecast centres and international OPMET data banks regarding the position, forecast direction and speed of movement, central pressure and maximum surface wind of tropical cyclones.

Visibility: It is defined as the greatest distance at which a black object of suitable dimensions, situated near the ground, can be seen and recognised when observed against a bright background **or** the greatest distance at which lights in the vicinity of 1000 candles can be seen and identified against an unlit background.

Volcanic Ash Advisory Centre (VAAC): A meteorological centre designated by regional air navigation agreement to provide advisory information to meteorological watch offices, area control centres, flight information centres, world area forecast centres and international OPMET data banks regarding the lateral and vertical extent and forecast movement of volcanic ash in the atmosphere following volcanic eruptions.

32.2 Objective, Determination and Provision of Meteorological Services

The main objective of meteorological services for air navigation to contribute to the safety, regularity and efficiency of air navigation.

India Meteorological Department (IMD) provides the necessary meteorological service for air navigation, both national and international, operating through civil aerodromes in India. The services are provided through AMO (Aerodrome Meteorological Offices) and AMS (Aeronautical Meteorological Stations) working at various airports in India. The IMD services are in response to International Civil Aviation Organisation's regulatory documents. The meteorological persons are trained at Regional Meteorological Training Centre (RMTC), India.

The meteorological branches of the Indian Air force and Indian Navy also provide meteorological services for national and international air navigation operating through air fields controlled by them.

All AMOs / AMSs maintain close liaison with Operators and ATS personnel on matters, which affect the provision of meteorological service for national and international air navigation.

Notification Required from Operators: An operator requiring meteorological service or changes in existing meteorological service shall have to notify, sufficiently in advance, the meteorological office(s) concerned. The minimum notice required is as follows:

(a) ***For National Flights:***

(i) At AMOs

At least 3 hours in advance of the time of departure of the flight

(ii) At AMSs

At least 18 to 24 hours in advance of the time of departure of the flight.

(b) *For International Flights*:

(i) At aerodrome meteorological offices at Mumbai, Kolkata New Delhi, Chennai, Nagpur, Bangalore, Guwahati, Hyderabad and Tiruvananthapuram.

At least 3 hours in advance of the time of departure of the flight.

(ii) At other aerodrome meteorological offices:

At least 12 hours in advance of the time of departure of the flight

(iii) At Aeronautical Meteorological Stations:

At least 18 to 24 hours in advance of the time of departure of the flight.

The above notice is required is respect of non-scheduled flights.

For scheduled flights sufficiently in advance.

The notification should contain the following information:

(a) aerodrome of departure and estimated time of departure

(b) destination aerodrome and estimated time of arrival

(c) route to be flown and (i) estimated time(s) of arrival at, (ii) and departure(s) from; any intermediate aerodrome(s)

(d) alternate aerodromes needed to complete the operational flight plan

(e) cruising level (FL)

(f) type of flight (weather under VFT or IFR)

(g) type of meteorological information requested for a flight crew-member – namely flight documentation and/or briefing or consultation.

(h) time(s) at which briefing consultation and/or flight documentation required.

32.3 Some Definitions used in Air Traffic Services

Air traffic refers to all aircraft in flight or operating on the manoeuvring area of an aerodrome. The manoeuvring area is the part of an aerodrome used for take-off and landing of aircraft and for the movement of aircraft associated with take-off and landing. Manoeuvring area includes the landing area, taxi ways etc.

Flight Information Region (FIR): An airspace of defined dimensions, within which flight information service and alerting service are provided.

Air Traffic Services Include:

(a) Flight information service

(b) Air traffic advisory service

(c) Air traffic control service (for preventing collisions, maintaining orderly flow of air traffic).

(d) Alerting Service

Air Traffic Control Units: Collective bodies are established to perform the various air traffic services for aircraft. Each collective body is called a unit.

Air Traffic Services Units Include:

(a) Flight information centre

(b) Air traffic control units (which include Area Control Centre, Approach Control Office and Aerodrome Control Tower)

Aerodrome Meteorological Minima

Limiting meteorological conditions are prescribed for the purpose of determining the usability of an aerodrome, either for take-off or for landing. These are known as Aerodrome Meteorological Minima.

The Flight Plan: An operational flight plan has to be completed for every intended flight; indicating that the flight can be conducted with safety and that the prescribed standards for weather conditions and for fuel and oil supply have been complied with. This plan must be approved and signed by the pilot-in-command and, where applicable, the flight operations officer. It is then filed with the operator, his agent or the aerodrome authority.

At least one alternate aerodrome is listed in the flight plan.

The flight plan includes the meteorological minima for the aerodromes to be used including their designated alternatives.

Visual Flight Rules (VFR): The rules which apply, when flying by means of visual reference to the ground.

Instrument Flight Rules (IFR): The rules which apply when flying by means of reference to the instruments in the cockpit. Within control zones and aerodrome traffic zones, under VFR flights there is weather minima (which is lower limit of flying over that aerodrome). Visibility not less than 5 km (or 3 miles). No cloud below 500 ft (vertically) and up to a distance of 1.8 km (or 1 mile) horizontally.

Aborted Landing: When a preceding aircraft is still on the runway, the on coming aircraft is asked to go round by ATC. This is commonly called Aborted landing.

Aborted Take-off: When an aircraft is cleared to take-off and another aircraft is still on the runway or there is some obstacle on the runway, then the aircraft which was cleared to take-off is asked to not to take-off till the runway is cleared. This is called Aborted take-off.

32.4 IMD set up of Meteorological Offices for Aviation

For providing the meteorological service to air navigation, IMD has set up an aviation organisation which has:

(a) 4 Meteorological Watch Offices (at Mumbai, Kolkata, New Delhi and Chennai)

(b) 18 Aerodrome Meteorological Offices (under the 4 MWOs)

  (i) at Chennai, Bangalore, Hyderabad, Tiruvananthapuram,

  (ii) Kolkata, Agartala, Bhubneshwar, Guwahati, Dibrugarh, Patna,

  (iii) Mumbai, Ahmedabad, Bhopal, Nagpur,

  (iv) New Delhi, Jaipur, Luknow, Safdarjang

(c) and 51 Aeronautical Meteorological Stations (AMS)

15 under MWO Chennai,

13 under MWO Kolkata,

14 under MWO Mumbai

9 under MWO New Delhi

A Tropical Cyclone Advisory Centre (TCAC) is functioning at IMD New Delhi.

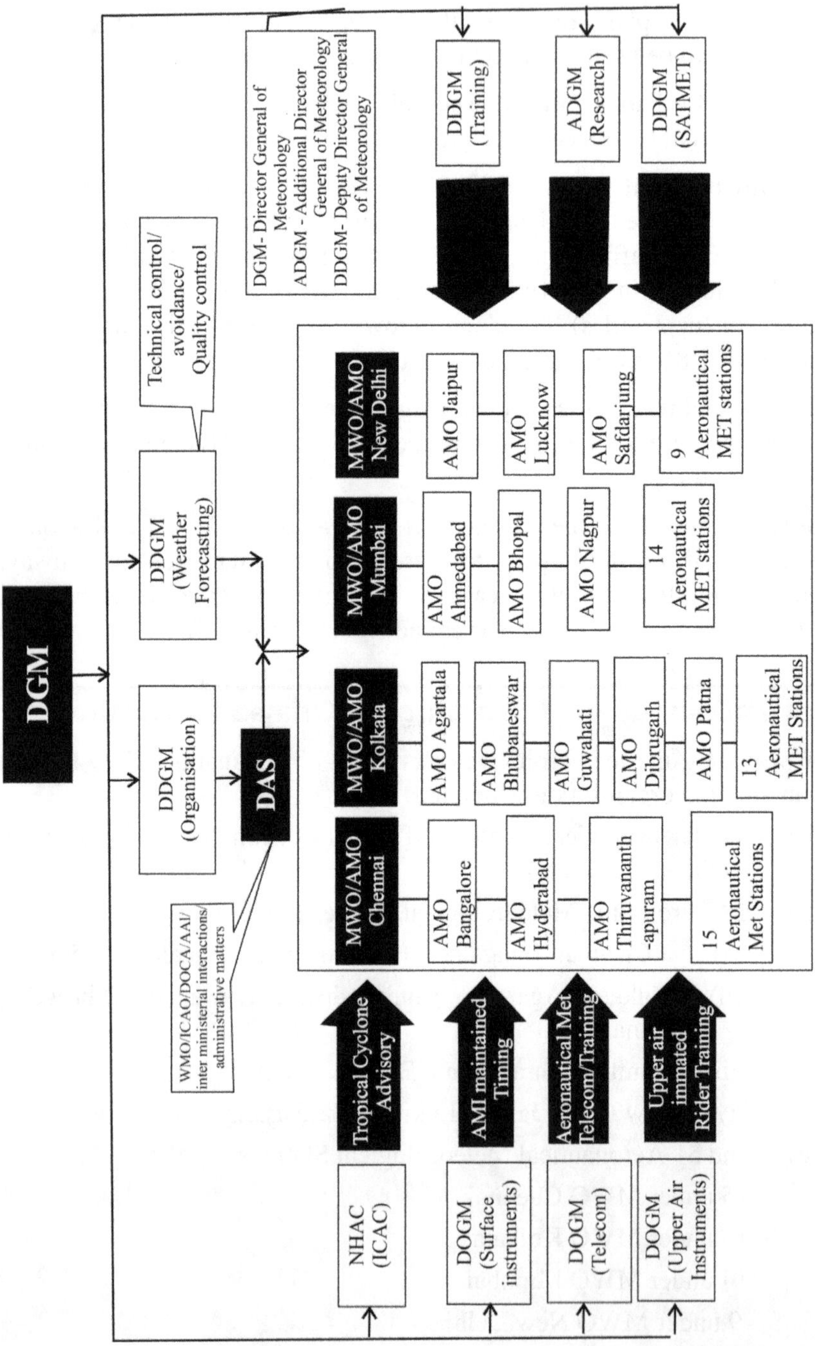

Fig. 32. 1 Aviation Meteorological Organisation of IMD.

32.4.1 Responsibilities of Aerodrome Meteorological Offices (AMOs)

(i) Preparation of forecasts and / or obtaining forecasts and other relevant information for flights operating from their aerodromes.

(ii) Preparation and / or obtaining forecasts for local meteorological conditions

(iii) keeping a continuous meteorological watch over the aerodrome and other aerodromes selected by their associated Aeronautical Meteorological Stations (AMSs)

(iv) Providing briefing, consultation and flight documentation to flight crew members and / or other flight operational persons.

(v) Supply of meteorological information to aeronautical users.

(vi) Display of latest available meteorological information

(vii) Exchange / sharing of meteorological information with other meteorological offices

(viii) Providing landing and take-off forecasts

(ix) Supply of flight planning information

(x) Supply of forecasts to their associated Aeronautical Stations in respect of flights operating from their aerodromes

(xi) Supply of aerodrome forecasts of relevant aerodromes to the pilots/ flight operators

(xii) Supply of all information received in respect of volcanic ash / volcanic eruption / volcanic ash cloud to aeronautical users

(xiii) Issuing of aerodrome warnings for local aerodromes and also for their associated AMS.

Meteorological Informations in (v) includes the following:

(a) Hourly / half hourly METARs / SPECIs and RVR observations (current weather observations)

(b) Landing / take-off reports on request for the required meteorological elements

(c) Meteorological information for VOLMET / ATIS / VOR broadcasts

(d) QNH / QFE (pressure data)

(e) Radar / Satellite observations

(f) SIGMETs (in respect of all FIRs)

(g) AIREPs available

(h) METARs / SPECIs of others stations as required

(i) Low level wind shear and temperature inversions.

32.4.2 Responsibilities of Aeronautical Meteorological Stations (AMSs)

(a) Provide current weather observations (METARs / SPECIs) to aeronautical users in respect of their station and also those stations as required (by obtaining from the stations concerned).

(b) Provide documentation for flights originating from their station after obtaining forecasts from their associated AMOs

(c) Supply of aerodrome warnings to aeronautical users of their aerodrome after getting from their associated AMO

(d) Supply of local aerodrome forecast to aeronautical users, of their aerodrome after getting from their associated AMO

(e) Supply of information received on pre-eruptive volcanic activity, a volcanic eruption / volcanic ash cloud to aeronautical users

32.4.3 Responsibilities of the Meteorological Watch Offices (MWOs)

(a) Maintain a continuous watch of meteorological conditions over their respective FIRs

(b) Prepare SIGMET and other weather information messages for their FIRs

(c) Exchange SIGMET information with other MWOs in the neighbouring countries namely: Bahrain, Iran, Israel, Jordan, Kuwait, Oman, Pakistan, Qatar, Saudi Arabia, Syrian Arab Republic, United Arab Emirates, Yemen, Australia, Bangladesh, Bhutan, Combodia, China (Beijing), Guam (US), Indonesia, Peoples Demoeratic Republic of Lao, Malaysia, Maldives, Myanmar, Nepal, Philippines, Russian Federation, Singapore, Srilanka, Thailand, Uzbekistan, Vietnam.

(d) Supply SIGMET information to their associated ATS (Air Traffic Services) units as well as SIGMET messages received from other MWOs.

(e) Disseminate their SIGMET information to other forecasting offices in India.

(f) Supply of information received on pre-eruptive volcanic activity, a volcanic eruption and volcanic ash cloud for which a SIGMET has not already been issued to its associated FICs (Flight Information Centres) / Area Control Centres (ACC).

(g) Supply information received concerning the accidental release of radio active materials into the atmosphere, in the area for which it maintains watch or adjacent areas, to its associated FIC / ACC. The information shall comprise of location date and time of the accident and forecast trajectories of the radio active materials.

32.4.4 Responsibilities of Tropical Cyclone Advisory Centre (TCAC)

1. Monitor the development of tropical cyclones in its area of responsibility. Using surface / upper air observational data, satellite (geostationary and polar orbiting) data and radar data.
2. Issue of advisory information relating the position of the cyclone Centre (Latitude, longitude and distance from well known station). Its direction of movement and speed, central pressure and maximum, surface wind near the centre (and tendency of development / forecast) in abbreviated plain language to the following interests.
   (a) MWOs in its area of responsibility; and
   (b) other TCACs whose areas of responsibility may be affected; and
   (c) WAFCs, international OPMET data banks, and centres designated for the operation of aeronautical fixed service satellite distribution systems, namely, Mumbai, Kolkata, Colombo, Dhaka, Delhi, Jakarta, Karachi, Kula Lumpur, Chennai, Male, Tehran, Yangon, Abu Dhabi, Jeddah, Kuwait, Muscat, Sanaa, Tehran.
3. Issue of up dated advisory information to MWOs for each tropical cyclone, as necessary, but at least every six hours.

The advisory information on tropical cyclone comprises the following in the order indicated:

(i) **TC** ADVISORY

(ii) **DTG:** Year month data (yyyy mm dd) / time (in UTC) (using "Z") of issue;

(iii) **TCAC:** name of TCAC (location indicator or full name)

(iv) **TC:** name of tropical cyclone

(v) **NR:** advisory number (starting with "01" for each cyclone);

(vi) **PSN:** Position of the centre in degrees and minutes ["N nnnn" or "S nnnn", "W nnnn" or E nnnn]

(vii) **MOV:** direction and speed of movement respectively at least 8 compass points (N, NE, E, SE, S, SW, W, NW) and km/h (or kt)

(viii) **C:** Central pressure (in hPa)

(ix) **MAX WIND:** maximum surface wind near the centre (mean over 10 minutes, in km/h (or kt);

(x) **FCST PSN + 12 HR:** forecast of centre position for fixed valid time of ....UTC (12 hours after time of issuance of the advisory)

(xi) **FCST MAX WIND + 12 HR:** forecast of maximum surface wind for fixed valid time of ... UTC (12 hour after time of issuance of the advisory)

(xii) **FCST PSN + 18 HR:** forecast of centre position for fixed valid time of ....UTC (18 hours after time of issuance of the advisory)

(xiii) **FCST MAX WIND + 18 HR:** forecast of maximum surface wind for fixed valid time of...UTC (18 hours after time of issuance of the advisory);

(xiv) **FCST PSN + 24 HR:** forecast of centre position for fixed valid time of...UTC (24 hours after time of issuance of the advisory);

(xv) **FCST MAX WIND + 24 HR:** forecast of maximum surface wind for fixed valid time of...UTC (24 hours after time of issuance of the advisory);

(xvi) **NXT MSG:** Expected year month date (yyyy mm ss) / time (in UTC using "2" for issuance of next advisory (using "BFR" if applicable or "NO MSG EXP".

Questions

1. A world-wide system of aeronautical fixed circuits provided, as part of the aeronautical fixed service, for the exchange of messages and / or digital data between aeronautical fixed stations having same or compatible communications characteristics is called:

   (a) AIRMET (b) AFS (c) AFTN (d) ATIS

2. Information issued by a Meteorological Watch Office concerning the occurrence or expected occurrence of specified en-route weather phenomena which may affect the safety of aircraft operations is called:

   (a) Flight information (b) Air-report

   (c) SIGMET information (d) None above

3. A precision instrument approach and landing with a decision height lower than 60 m (200 ft) but not lower than 30 m (100 ft) and RVR not less than 350 m is :
   (a) CAT – I operation (b) CAT – II operation
   (c) CAT – III operation (d) None above
4. The lowest altitude which may be used, which will provide a minimum clearance of 300 m (1000 ft) above all objects located in the area contained within a sector of circle of 25 NM radius centered on a radio aid to navigation is called:
   (a) Minimum sector altitude (b) Decision altitude
   (c) Airport meteorological minima (d) None above
5. A cloud with the height of cloud base below 1500 m (5000 ft) or below the height minimum sector altitude, whichever is greater is called:
   (a) Airport meteorological minima
   (b) Minimum sector altitude
   (c) Significant cloud
   (d) Cloud of operational significance
6. An operator requiring meteorological service or changes in existing meteorological service for national flight at AMOs have to notify:
   (a) At least 12 hours in advance of the ETD
   (b) At least 3 hours in advance of the ETD
   (c) At least 18-24 hours in advance of the ETD
   (d) None required
7. An operator requiring meteorological service or changes in existing meteorological service for international flight at AMS have to notify:
   (a) At least 6 hours before ETD
   (b) At least 3 hours before ETD
   (c) At least 18-24 hours before ETD
   (d) None required
8. An airspace of defined dimensions within which flight information service and alerting service are provided is called:
   (a) Flight information centre
   (b) Flight information region
   (c) Flight information service
   (d) None above

9. Limiting meteorological conditions are prescribed for the purpose of determining the usability of an aerodrome, either for take-off or for landing. These are known as:
   (a) Aerodrome flight minima
   (b) Aerodrome Meteorological Minima
   (c) Visual flight rules
   (d) Instrument flight rules
10. In India, Meteorological Watch Offices are located at:
   (a) Mumbai, Kolkata
   (b) Hyderabad, Bangalore
   (c) Delhi, Chennai
   (d) Both (a) and (c)
11. Under MWO Chennai, the AMOs are:
   (a) Chennai, Bangalore
   (b) Hyderabad, Trivandrum
   (c) Visakhapatnam, Minicoy
   (d) Both (a) and (b)
12. In India, a Tropical Cyclone Advisory Centre is located at:
   (a) IMD Mumbai (b) IMD Chennai
   (c) IMD New Delhi (d) IMD Kolkata
13. In India, there are......... Aerodrome Meteorological Offices and .... Aeronautical Meteorological Stations are working
   (a) 4, 18 (b) 18, 51 (c) 1, 4 (d) 51, 1

Answers

1. (c)	2. (c)	3. (b)	4. (a)
5. (d)	6. (b)	7. (c)	8. (b)
9. (b)	10. (d)	11. (d)	12. (c)
13. (b)			

CHAPTER 33

World Area Forecast System (WAFS)

Under the provisions of ICAO World Area Forecast Centers are located at UK Meteorological Office (Bracknell) and at Washington USA (based in Kansas city). Three INTELSAT 604 Geostationary Satellites provide global imagery data. UK Meteorological Office also uses SADIS satellite data, located at 60°E and covers Europe, the Middle East and South Asia. The Washington USA Meteorological Office covers the rest of the globe.

33.1 World Area Forecast System (WAFS)

A world wide system by which world and regional area forecast centers provide aeronautical meteorological en-route forecasts in uniform standard format.

33.2 World Area Forecast Center (WAFC)

A meteorological center designated to prepare and supply upper air forecasts in digital form on a global basis.

Upper wind and temperature charts are prepared for ten global regions, twice a day for nine levels. The objective of WAFS is to provide a uniform system, and forecast of upper air temperature, humidity, wind direction, speed and height of Wind Maxima, Tropopause height and temperature and forecasts of Significant whether phenomena in digital form in a cost effective manner.

Flight documentation is provided as shortly before departure and contains the information (received) within the frame work of WAFS for flight documentation of international flights and national long haul flights.

33.3 Responsibilities of WAFC

1. To prepare global forecasts for grid point in digital form for all required levels and in a standard format. The forecasts shall consist of upper winds, upper air temperatures and humidity, tropopause heights and height of wind maxima with maximum wind speed and direction.
2. To prepare global forecasts of significant whether phenomena in digital form.
3. To issue forecasts referred above (1 & 2) in digital and/ or pictorial form to meteorological authorities and other users in its service area, as approved by the contracting state on advice from the meteorological authorities.
4. To prepare and issue amendments to the forecasts.
5. To receive information about accidental release of radioactive materials into the atmosphere from its associated WMO Regional Specialised Meteorological Center (RSMC) for the provisions of transport model products for the radiological environment emergency response, in order to include the information in significant whether forecasrs, and
6. To establish and maintain contact with VACCs (Volcanic ash coordinating centers) for the exchange of information on volcanic activity to coordinate the inclusion of information on volcanic eruptions in significant weather reports.

33.4 The Criteria for Issuing Amendments to WAFS Products

(a) ***Upper Wind*** Change in direction of 30° or more, provided the wind speed is 30 kt (60 kmph) or more, before or after the change; change in speed of 20 kt (40 kmph) or more.

(b) ***Upper Air Temperatures*** Change of more than 5°C.

(c) ***Significant En-route Weather Forecasts*** In respect of the following three elements

(i) *Aircraft Icing and Turbulence*: Newly expected occurrence or error in expected position of phenomena; intensity increasing; intensity decreasing from severe to light or nil, **or** from moderate to nil.

(ii) *Jetstreams*: Newly expected occurrence of disappearance.

Error in expected position more than 400 km, error in speed more than 20%, error in core height more than 900 m (3000 ft).

(iii) *Other Significant En-route Weather Phenomena*: Newly expected occurrence; no longer expected.

(iv) *Volcanic Eruptions or Accidental Release of Radioactive Material into Atmosphere of Significance to Aircraft Operations*: Inclusion or removal of activity symbol or radiation symbol.

33.5 Regional Area Forecast Centers (RAFC)

RAFC is a Contracting State, having accepted (by regional air navigation agreement) the responsibility for providing regional area forecast within the framework of the area forecast system have the following responsibilities:

(a) To receive global digital grid-point data from a WAFC to meet the needs of meteorological authorities and other users within its service area, including those needs related to centralized flight planning.

(b) To store the digital grid-point data received from a WAFC and to process and supply selectively these data to meteorological authorities and other users in its service area in an agreed format.

(c) To prepare upper wind and temperature charts on the basis of the data received and to supply the relevant charts and abbreviated plain language amendments thereto to users as agreed between the RAFC and the users within its service area.

Note: The upper wind and temperature charts will be produced from grid-point data received from a WAFC, except on those occasions when the RAFC considers it essential to adjust the charts on the basis of recent basic data received.

(d) To notify the relevant WAFC immediately of the content of and reasons for any amendments it has issued to the forecast received from the WAFC.

(e) To prepare significant weather charts, and as required, significant weather forecast messages in abbreviated plain language for its area of responsibility.

Note: In order to prepare these charts and amendments thereto the RAFC will also need to receive basic synoptic and asynoptic data, including satellite data (polar orbiting and geostationary) and aircraft meteorological reports.

(f) To supply significant weather charts, significant weather forecast messages in abbreviated plain language and plain language amendments thereto to the users as agreed between the RAFC and the users within its service area.

(g) To exchange the significant weather charts and abbreviated plain language amendments thereto with other RAFCs as necessary, so as enable each center to provide significant weather charts for its area of coverage.

(h) To prepare WINTEM messages and issue them to users as required.

***Note*:** RAFC products are issued four times a day for fixed valid times 0000, 0600, 1200 and 1800 UTC. Transmission of each forecast should be completed at least 9 hours before the valid time. The flight documentation requirements for inter regional flights and exchange of SIGWX charts between the RAFCs concerned depends on the regional agreement.

The digital data should be transmitted to meteorological authorities and other users with minimum delay after receipt from the WAFC. The significant weather charts should include the phenomena listed in utilization of WAFS products.33.6. (a) Between FL250 and FL 450. (b) FL 100 and FL 250 for limited geographical areas, as determined by regional air navigation agreement. (c) FL450 and FL 600, if so determined by regional air navigation agreement. Forecast covering the layer between FL450 and FL600 should be combined with those at (a) above.

The upper wind and upper air temperature charts should be provided for flight levels as determined by regional air navigation agreement. Amendments to significant weather forecasts should be supplied in the form of abbreviated plain language messages.

RAFCs should apply the criteria for the amendment of significant en-route weather forecasts given in 33.4 (c) i to iv for WAFS products. RAFCs should supply to meteorological authorities and other users within their service areas amendments to upper wind and upper-air temperature forecasts with minimum delay after receipt from a WAFC.

33.5.1 Regional Area Forecast System (RAFS)

Under ICAOs Regional Area Forecasts system 19 RAFCs are working, located at various parts of the world. One RAFC is functioning at Northern Hemispheric Analysis Centre (NHAC)

IMD New Delhi Web sites are:

www. imd. ernet. in and www. imd. gov. in

NHAC cater the needs of aviation forecasts in the form of prognostic charts (PROGs) for the area bounded by latitude Equator to 40 °N and longitude 30 °E to 125° E. The prognostic charts are made Facsimile broadcasts. The prognostic charts broadcasts are:

(i) Surface chart,

(ii) Significant weather chart for below FL 460,

(iii) Tropopause, Maximum wind,

(iv) Winds and temperatures for isobaric levels of 500, 300, 250, and 200 hPa

Area of coverage of charts.

***Map Area A*:** Latitude 25 °S to latitude 45 °N and longitude 30 °E to long 125 °E.

***Map Area B*:** Latitude 0-40 °N and longitude 30 °E to 125 °E

Projection: Mercator, Scale 1: 20,000,000

Upper air PROGs are valid for the fixed time 24 hours after the time of observation.

Map B: Significant weather PROGs
valid 09-21 (time of observation 0000 UTC)
valid 15-03 (time of observation 0600 UTC)
valid 21-09 (time of observation 1200 UTC)
valid 03-15 (time of observation 1800 UTC)

33.6 Utilization of WAFS Products

Use of fixed time prognostic charts, based on raw data from WAFC London. Fixed time prognostic charts are prepared with lead times 06 hr, 12 hr, 24 hr, 30 hr and 36 hr after the time of observations 0000, 0600, 1200 and 1800 UTC. The charts have validity of TT ± 3 hr (TT is the lead time). For example a chart prepared based on 0000 UTC with lead time of 6 hours can be utilized for a flight whose entire flight duration falls within 03-09 hr.

The prognostic chart form of documentation upper wind and upper air temperature is supplied in the form of fixed time prognostic charts for 300, 250, 200 hPa valid for 2100 and 0900 UTC, issued (twice a day) at 1200 UTC and 0000 UTC (based on 0000 UTC and 1200 UTC observation charts respectively).

In WAFCs, the forecasts of upper winds and upper air temperatures, direction, speed and height of Maximum winds and Tropopause heights prepared twice daily and valid for 12, 18, 24 and 30 hr after the time (0000 and 1200 UTC) of the synoptic data on which the forecasts are based and be available not later than 6 hours after standard time of observation.

In WAFCS, the forecasts of Significant Weather phenomena charts prepared and issued four times a day for fixed valid times of 0000, 0600, 1200 and 1800 UTC. The forecasts are made available at least 12 hours before its validity time.

Significant en-rout whether phenomena includes:

(a) Thunderstorms, (b) Tropical Cyclone, (c) severe line sqalls, (d) moderate or severe turbulence (in cloud or in clear air), (e) moderate or severe icing, (f) widespread sandstorm/duststrome, (g) For flights, between FL 100 to FL 250, clouds associated with (a) to (f) above, (h) For flights above FL 250, Cumulonimbus clouds associated with (a) to (f) above, (i) Surface position of well defined convergence zones, (j) surface positions, speed and direction of movement of frontal systems, (k) tropopause heights, (l) jetstreams and (m) information on the location of volcanic eruptions, which are producing ash clouds of significance to aircraft operations, name of volcano and time of first eruption if known.

Note: For aircrafts operating above FL 250 items (a) to (f) are given only when expected to be above that level.

Frontal systems are included only when associated with significant en-route weather phenomena.

Significant weather charts for low level flights operating up to FL100 include the following:

(a) Fronts and convergent zones and their expected movements,

(b) Areas and levels affected by thunderstrom, tropical cyclone, squall line, hail, moderate/severe turbulence (in clouds or in clear air), mountain waves and associated updrafts, aircraft icing, freezing precipitation, widespread sandstrom/duststrom, fog, precipitation and other phenomena causing widespread reduction of visibility to less than 5000 m.

(c) Cloud amount, type and height indications of bases and tops.

(d) Surface visibility if less than 5000 m.

(e) Pressure centers and their expected movement.

(f) Height of $0^{\circ}C$ isotherm level(s), if lower than the top of the air space for which the forecast is provided.

(g) Sea surface temperature and state of the sea if determined by regional air navigation.

(h) Information on the location of volcanic eruptions which are producing ash clouds of significance to aircraft operations, name of volcano, and time of its first eruption if known.

Note: (1) WAFC/RAFC wind and temperature data given for flight levels – FL50 (850 hPa), FL100 (700 hPa), FL180 (500 hPa) , FL240 (400 hPa), FL300 (300 hPa), FL340 (250 hPa), FL390 (200 hPa), FL450 (150 hPa) and where required FL530 (100 hPa), FL600 (070 hPa)

(2) 140 km represents a distance of about 1.25° of latitude.

(3) RAFC products issued four times a day for fixed valid times of 0000, 0600, 1200, 1800 UTC.

(4) SIGWX charts

(a) between FL250-450

(b) between FL100-250

(c) between FL450-600

33.7 Significant Weather Chart

Significant weather forecast is shown with standard abbreviations / symbols. Areas of SIGWX (significant weather) like turbulence (),icing (), thunderstorm (), CB are shown within scalloped boundaries with base and tops marked in flight levels shown in boxes.

(a) *EX* [ISOL EMBD CB 420 XXX]: This indicates the scalloped area forecasted with isolated embedded CB clouds with tops extending to FL 420 (42000 ft) and base not known, which means, it is below the flight levels of the SIGWX chart.

(b) Tropopause heights are shown in horizontal boxes at regular intervals across the chart

EX: [500]---- ---- --[480] : This indicates at the grid point interval heights of tropopause vary from FL 500 to FL 480.

(c) Clear air turbulence shown within pecked lines.

EX: CAT Area [1]  $\frac{400}{300}$

This means, at area [1] marked moderate turbulence between FL 300 and FL 400

(d) Wind maximum shown with arrows and flight level marked on arrow.

EX:  : Maximum wind speed of 115 knots from westerly direction at flight level 340.

(e) The forecast surface position of centres of Low pressure / Depression / Cyclone or high pressure and fronts are shown with likely movement of direction by arrow and speed in knots.

Descriptions of symbols and meaning or legend shown in a box at the corner of the chart.

Spot Winds, Temperature Chart: This chart contains forecast winds and temperature for a particular flight. Boxes refer to spot winds degrees true and knots at intersection of latitude / longitude lines covered by the box; temperature in degrees Celsius, heights above msl (in thousands of feet).

Ex:

24	160	20	– 28
18	180	25	– 15
10	120	25	+ 04
05	140	15	+ 12
02	160	10	+ 16

At 2000 ft wind 160° / 10 kt, temperature + 16 °C
At 5000 ft, wind 140° / 15 kt, temperature + 12 °C
At 10000 ft, wind 120° / 25 kt, temperature + 04 °C
At 18000 ft, wind 180° / 25 kt, temperature – 15 °C
At 24000 ft, wind 160° / 20 kt, temperature – 28 °C

Tabular Forecast of Upper Winds and Upper Air Temperatures or Grid Point Forecast

Data is presented is the following sequence at grid point centre

(i) FL of tropopause
(ii) FL dd fff of maximum wind
(iii) dd, fff, MTT FL 450 / 150 hPa
dd, fff, MTT FL 390 / 200 hPa
dd, fff, MTT FL 340 / 250 hPa
dd, fff, MTT FL 300 / 300 hPa
dd, fff, MTT FL 240 / 400 hPa
dd, fff, MTT FL 180 / 500 hPa

where dd = direction of wind in tens of degrees
fff = wind speed (in knots)
TT = temperature (in °C) preceded by M or P (minus or plus).

Ex: 550
39031020
10010 M 66
31020 M 56
31015 M 44
36005 M 33
33010 M 18
21005 M 06

Decoding:

Tropopause: Level FL 550 or 55000 ft

Wind Maxima: FL 390 and wind direction and speed 310 / 20 kt

Wind and temperature

at FL 450, 100 / 10 kt, – 66 °C
at FL 390, 310 / 20 kt, – 56 °C
at FL 340, 310 / 15 kt, – 44 °C
at FL 300, 360 / 05 kt, – 33 °C
at FL 240, 330 / 10 kt, – 18 °C
at FL 180, 210 / 05 kt, – 06

RAFC fixed time progonstic upper wind and temperature chart for standard isobari surface

Wind directions are shown by arrow and speed by feathers or penants. Temperature shown at the end of arrow.

Units used knots, degrees Celsius

Temperature is negative unless prefixed by + symbol.

Winds are shown at latitude / longitude grid points. Longitudes and latitudes are shown separately.

Upper Wind / Temperature Charts drawn for the levels referred above, at grid points wind arrows flying with the wind speed shown by pennants (50 kts) and feathers (10 kts) or half feathers (3 to 7 kts) indicate wind speed (eg. 

Westerly 75 kts, temperature – 30 °C) Celsius temperature shown on arrow shaft. If it is plus it annoted Ps otherwise it is negative.

Questions

1. The objective of WAFS is to provide a uniform system and forecast of:
   (a) Upper air temperature, humidity and wind
   (b) Height of wind maxima, height of tropopause and temperature
   (c) Forecasts of significant weather phenomena
   (d) All above
2. Under ICAOs, RAFS 19 Regional forecast centres are working and one is located at IMD New Delhi; which uses WAFS products and broadcasts prognostic charts of:
   (a) Surface chart, significant weather chart for below FL 460 including tropopause and maximum wind

(b) Winds and temperatures for isobaric levels 500, 300, 250, and 200 hPa

(c) Rofors, Tafors, Metars

(d) Both (a) and (b)

3. WAFS / RAFS upper air PROGs charts prepared twice a day are valid for:

(a) 36 hrs

(b) 24 hrs, fixed time after the time of observation

(c) 12 hrs, fixed time after the time of observation

(d) None above

4. WAFS / RAFS, significant weather PROGs issued:

(a) Twice a day and valid for 24 hours

(b) Four times a day and valid for 12 hours

(c) Once a day valid for 48 hours

(d) Both (a) and (c)

5. WAFS / RAFS, upper wind and upper air temperature PROGs charts supplied in the form of fixed time Prognostic charts valid for 2100 and 0900 UTC issued respectively at:

(a) 0000, 1200 UTC for isobaric levels 300, 250 200 hPa

(b) 1200, 0000 UTC for isobaric levels 300, 250, 200 hPa

(c) 0600, 1800 UTC for isobaric levels 300, 250 200 hPa

(d) Both (a) and (c)

6. In WAFCS, the significant weather PROGs prepared and issued.......; for fixed valid times.........and made available at least 12 hours before its validity time.

(a) Four times a day; 0000, 0600, 1200 and 1800 UTC

(b) Twice a day; 0000, 1200 UTC

(c) Twice a day; 0600, 1800 UTC

(d) Both (b) and (c)

7. In WAFC, SIGWX chart FL 250 / 450 scalloped area contains box

ISOL EMBD CB 420 XXX

what is the base and top of CB

(a) Base unknown, top 42000 ft
(b) Base below FL 250, top FL 420
(c) Base unknown, top 420 m
(d) Base below FL 420, top FL 420

8. In WAFC, SIGWX chart contain at regular intervals boxes at grid points across chart. What the following indicates?

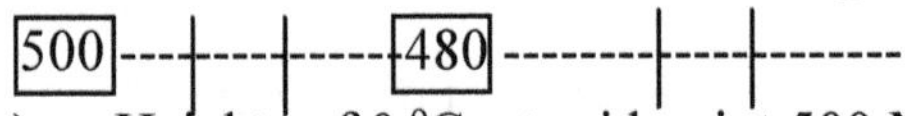

(a) Heights of 0 °C, at grid point 500 M, 480 M
(b) Heights of tropopause, at grid point FL 500 (50000 ft) and FL 480 (48000 ft)
(c) Wind maxima 500 kt, 480 kt, at grid points
(d) None above

9. In WAFC SIGWX chart contain within pecked lines boxes. The following what it indicates?

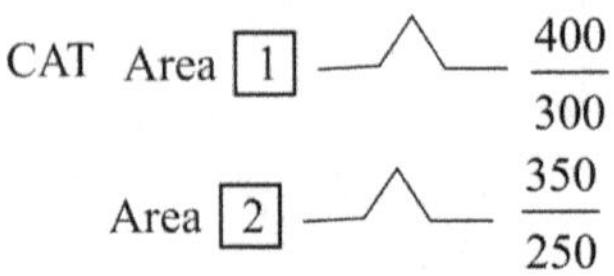

(a) In area 1 moderate clear air turbulence between FL 300 FL 400,
in area 2 mod CAT between FL 250, FL 350
(b) In area 1 wind motion between FL 300, FL 400
in area 2 wind variation between FL 250 FL 350
(c) In area 1 tropopauses at FL 300, and at FL 400
in area 2 tropopauses at FL 250, and at FL 350
(d) None above

10. In WAFC spot winds, temperature chart contains the following box. What is the wind and temperature at FL 180?

24	150	20	– 30
18	180	30	– 20
10	120	15	+ 04
05	140	15	+ 12
02	160	10	+ 16

(a) Wind 160 / 10 kt, temperature + 16 °C
(b) Wind 180 / 30 kt, temperature – 20 °C
(c) Wind 150 / 20 kt, temperature – 30 °C
(d) Wind 120 / 15 kt, temperature + 04 °C

11. In WAFC tabular forecast of upper winds and upper air temperature contains a box at grid point given below

550

39031040

12010 M 65

31035 M 56

31015 M 44

36010 M 33

33010 M 18

21005 M 06

What is the tropopause level, wind maximum, and wind and temperature at FL 340?

(a) Tropopause level 550 hPa
Wind maximum at 390 hPa, 310 / 40 kt
Wind and temperature at FL 340, 310 / 15 kt, – 44 °C

(b) Tropopause FL 550
Wind maximum FL 390, 310 / 40 kt
Wind and temperature at FL 340, 310 / 15 kt – 44 °C

(c) Tropopause FL 550,
Wind maximum FL 390, 310 / 40 kt
Wind and temperature at FL 340, 120 / 10 kt – 65 °C

(d) Both (a) and (c)

Answers

1. (d)	2. (d)	3. (b)	4. (b)
5. (b)	6. (a)	7. (b)	8. (b)
9. (a)	10. (b)	11. (b)	

CHAPTER 34

Meteorological Observations

Observations of various meteorological elements are made with the help of instruments or sensory observations. Surface synoptic observations are made by an observer at or near ground level or sea level using instruments. These observations are made at meteorological observing stations, also called observatories.

Meteorological stations are classified as:

(a) Synoptic stations (land and sea)

(b) Climatological stations

(c) Aeronautical Meteorological stations

(d) Agricultural Meteorological stations

(e) Special stations.

A station at which observations made for synoptic meteorology, like current weather, is called synoptic station. These observations are used especially for prediction of future weather developments. Synoptic stations include both surface and upper air observation stations.

Meteorological data that is used for climatological purposes are obtained from climatological stations. It is concerned with the long term weather data. Synoptic stations data are also used for climatological purposes.

Meteorological stations located at aerodromes to serve the special needs of aviation are called aeronautical meteorological stations. Synoptic and climatological observations are also made at these stations.

Agricultural meteorological stations are maintained to assist agriculture and also horticulture, animal husbandry and forestry.

Special stations are established for evaluating particular meteorological events. These include atmospherics radar detection of hydrometeors, hydrology, radiation measurement, ozone measurements etc.

Networks of Stations: WMO has recommended the density of stations for the basic synoptic purposes to be one station at least for every 150 km apart. In case of upper air stations the density to be one station for every 300 km or less.

Synoptic observations include the following elements.

(a) Wind (direction and speed)

(b) Visibility

(c) Weather (present and past weather)

(d) Cloud amount, type and height of base of cloud

(e) Humidity (Dry and wet bulb temperatures)

(f) Atmospheric pressure

In addition to the above observations of (in respect of land stations):

(a) Pressure tendency characteristics

(b) Extreme temperatures

(c) Amount of precipitation

(d) State of ground

(e) Direction of cloud movement

(f) Special phenomena (hail squall etc)

In respect of ocean weather stations:

(a) Ships course and speed

(b) Sea surface water temperature

(c) Wave observations-direction of movement, period and height of waves

(d) Ice

(e) Special Phenomena

34.1 Synoptic Observations are made at Fixed Universal Time (UTC)

The standard times of surface synoptic observations are 0000, 0600, 1200 and 1800 UTC and intermediate observations are made at 0300, 0900, 1500 and 2100 UTC.

The table 34.1 gives the meteorological elements, its units and instruments of measurement.

Table 34.1

S. No	Elements	Units	Instruments/aids
1	Pressure	Hecto pascals	Barometer, precision aneroid barometer, Barograph
2	Temperature and dew point	Degrees Celsius	Dry bulb, wet bulb, Maximum Minimum, grass minimum thermometers / thermographs, Hygrograph
3	Humidity	Percentage	Dry bulb, wet bulb thermometers and hygrograph
4	(a) wind direction	Degrees from truth north	Cup anemometer, hand anemometer, wind vane, manual estimation, pressure tube anemometer
	(b) wind speed	Knot	Cup anemometer, hand anemometer, wind vane, manual estimation, pressure tube anemometer
5	Cloud		
	(a) type	Oktas (eighths) of sky covered	(a) International cloud Atlas
	(b) amount		(b) Visual

Table *contd...*

S. No	Elements	Units	Instruments/aids
	(c) base height	Feet	(c) Cloud base recorder (ceilometer) (i) Fixed beam ceilometer (ii) Rotating beam ceilometer (iii) laser cloud Cloud search light, balloons, alidade
6	Visibility	Meters/kilometers,	Visual – known distance of fixed object, Transmissometer meter, visibility meter scopograph
7	Weather (present/past)		Visual, radar, rainfall rate recorder
8	Precipitation	Millimeters	Rain gauge, rainfall rate recorder, weather radar
9	Snow depth	Centimeters	ruler
10	State of ground		Visual, representative patches of bare ground
11	Sunshine	Hours	Sunshine recorder

34.2 Units of Measurements

ICAO has recommended the use of decimal metric system of units, with the inclusion of the nautical mile (be adopted for use in air-ground communication). These units are given in table 34.2. In addition to the ICAO unit, British units are given in the last column which is called Blue Table Units.

1 nautical mile = 1852 meters

Table 34.2 Units of measurement

S. No.	Measurement of	Units in ICAO Table	Units in Blue Table
1	Distance used in navigation, position reporting etc., generally more than 2 to 3 nautical miles	Nautical miles and tenths	Nautical miles and tenths
2	Relatively short distances, such as those relating to aerodromes (e.g., runway length)	Meters	Meters

S. No.	Measurement of	Units in ICAO Table	Units in Blue Table
3	Altitudes, elevations and heights	Meters	Feet
4	Horizontal speed including wind speed	Knots	Knots
5	Vertical speed	Meters per second	Feet per minute
6	Wind direction for landing and taking off	Degrees magnetic	Degrees magnetic
7	Wind direction except for landing and taking off	Degrees true	Degrees true
8	Altimeter setting	hecto Pascals	hecto Pascals
9	Temperature	Degrees Celsius (Centigrade)	Degrees Celsius (Centigrade)
10	Weight	Metric tons or Kilograms	Metric tons or Kilograms
11	Time	Hours and minutes, the day of 24 hrs beginning at midnight UTC	Hours and minutes, the day of 24 hrs beginning at midnight UTC

34.3 Observatory Enclosure

Observation of meteorological elements to be comparable for synoptic purpose WMO suggested certain standards and uniformity of methods of observation. The observatory must be located at a site which allows unobstructed wind flow, exposure for instruments. Accordingly the site of observatory enclosure will be a plain level ground free from any obstructions to avoid vitiated exposure of instruments installed inside. A plot of level ground (normally covered with short grass) size 9 m × 6 m for outdoor temperature, humidity sensing instruments and recording / non recording rain gauge, evaporimeter. The site should be away from the influences of trees, tall structures / buildings, and should not be close to steep slopes, ridges cliffs or hollows. Depending on the type of observatory the following instruments are installed in the observatory enclosure.

(a) Standard Stevenson screen (SS screen) to house dry bulb, wet bulb, maximum and minimum thermometers, thermograph and hygrograph.

(b) Wind instruments

(c) Rain gauge (both self recording and non-recording rain gauges)

(d) Sunshine recorder

(e) Snow gauge
(f) Open pan evoprimeter
(g) Pilot Balloon observation pillor

Current weather instruments system will be located in a fenced enclosure as close to take-off / landing surface as permitted. Observatory fencing, gate and light should be maintained by periodic painting and the area be free from tall grass and weeds.

The standard exposure for wind instrument is 4 m above ground on a mast or tower, away from tall buildings or trees. The distance between the anemometer and any obstruction must be 10 times the height of the obstruction above anemometer. If the anemometer and wind vane are fixed on the same platform or mast, they should be 2 m apart.

Standard screen is a double louvered box, which permits free air circulation but prevents sunshine and hydrometers to fall on the instruments inside. In the northern hemisphere, the opening of the screen shutter should face north. The screen is erected on four stout wooden posts (or angle iron stand) with its door opening to the north such that the height of the thermometer bulbs is 1.25 m above the ground. The posts are fixed absolutely vertical in the ground in concrete, such that the bottom of the SS box is perfectly horizontal and about 1.2 m agl (above ground level). The dry bulb, wet bulb, maximum (Max) and minimum (Min) thermometers are installed inside the screen. The water bottle of the wet bulb must be slightly away from it for free evaporation and the reading of the dry bulbs are not vitiated. Dry bulb, wet bulb thermometers be installed absolutely vertical, while maximum and minimum thermometers should be supported in a horizontal position. In case of max thermometer, the bulb must be 0.5 cm lower than the other end. Thermographs, hygrographs, pitched evaporimeter may also be housed inside the SS screen (large / double serene).

34.4 Automatic Weather Observing Systems (AWOS)

This system is intended to provide real time display of all aviation parameters, representative of air field conditions at users end (Met. Section and ATC cabin). The meteorological elements to be sensed, conditioned, transmitted, displayed and printed are given below.

(a) Wind direction and speed (option for two locations)
(b) Visibility and RVR
(c) Cloud base

(d) Temperature

(e) Dew point and RH

(f) Pressure – QFE and QNH

(g) Lightning / Cb location

The proposed AWOS consist of a package of quality meteorological sensors of standard type, signal conditioners at site, data transmission system and a central data acquisitions and processing system. At the users end necessary interface equipment such as data terminals, recorders, printers. VDUs (Visual Display Units) auxiliary displays are envisaged. They are usually located near runway ends (depending on local conditions). Data transmission from site to user's location should be on telephone type cable or preferably VHF link. At the processing end, the data must be checked for accuracy, and completeness and processed according to suitable stored programme. The processor then formats and outputs the data to various display and recording devices that have been selected for the system. Operator interaction with the system is provided by means of standard key board. Remote transmission of data over long distances are envisaged.

34.5 Upper Air Observations

(a) **Pilot Balloon Ascents:** Hydrogen filled balloon with attached standard tail of flags let off into the atmosphere. The movement of the balloon and its flags observed through an optical theodolite. The plot of these readings give the balloon trajectory. From the trajectory, wind direction and wind speed at standard levels are calculated. These are called pilot balloon ascents. This method has limitations during cloudy and rainy weather, when the balloon is obscured and cannot be observed.

(b) **Radiosonde Observation:** In this method a radiosonde instrument is attached to the hydrogen filled balloon and let off into the atmosphere. The radiosonde instrument consists of a barrows switch, thermister and hygrister which is connected to transmitter. As the balloon makes journey into the atmosphere the radio-transmitter sends the signals of pressure, temperature and humidity at various heights in the atmosphere. A radiosonde instrument of common type is a modulated audio frequency type, involving a tiny radio-transmitter which broad-casts a special frequency radio waves. The measuring instrument automatically converts data on the pressure, temperature and humidity into electrical pulses which are broadcast by transmitter to a receiving and recording set on the ground.

(c) **Rawin Observation:** Radiosonde stations use tracking radars, which provides the trajectory of the balloon and thus provides wind direction and speed which is called Rawin. For tracking the radiosonde balloon a passive reflector is attached at the tail of the radiosonde instrument. The combination of passive reflector and radiosonde instrument together provide both upper winds and temperature, humidity at various isobaric levels / altitudes. Electronic instruments are used for meteorological observations include: (a) Digital Aneroid Barometer (DAB), (b) Precision Aneroid Barometer (PAB) and (c) Laser Ranger Finder. These are very sensitive, sophisticated electronic sensors.

34.6 Direct Reception System (DRS)

DRS is an equipment that receives conventional meteorological data, satellite imageries and graphics (facsimile or Fax) data directly from INSAT satellite. India Meteorological Department (IMD), Lodi Road, New Delhi relays meteorological data to INSAT, which in turn transmits the data in S-band to all users having DRS system. The components of DRS are:

(i) Antenna (to receive S-band signals)

(ii) Front End Converter and Helical Feed (this is to amplify the S-band signals and convert into suitable frequency)

(iii) Power supply unit, FDM Demux and TDM Demux

(iv) TP / Fax interface unit (between TDM Dumex on the input side and teleprinter and fax recorder on the output side). Fax interface converts digital signal to analogue signal and provides amplitude – modulated signal for conventional HF fax recorders. TP interface convert TP channel to + 60 V to drive conventional teleprimter machines.

34.7 Satellite Imagery Reception and Video Display System (SIRAVDS)

SIRAVDS is a window based software for receiving, storing, analysing and printing, the satellite picture and Fax data. Two special customize card, namely NITEL interface and ADC card are also installed in the computer as add-on cards. The system works on MS-DOS and

WINDOWS environment. A colour monitor and Laser Jet Printer (LJP) are amongst the main peripherals.

IMD receives satellite pictures in raw form from INSAT and after processing, retransmits to satellite for further inturn transmission to earth. The signal transmitted by satellite is received by DRS at any ground station through MDD link. DRS separates TP data and satellite pictures signals. Analogue satellite signal is converted into digital signal by ADC card. The digitilised signal is recorded by computer and is also displayed simultaneously on the monitor. Instrument receives IR and visible imageries. It has the capability of zooming and density slicing etc.

34.8 Meteorological Message Handling System (MMHS)

The MMHS is a PC based system with software for receiving, sorting, storing and a printer for printing of meteorological data. It is capable of handling eight data channels (five DRS channels, one simplex, one duplex and one for RTP). MMHS is capable of storing the messages according to headers and the station index numbers into Surface / Upper air data of different observation times, data wise, time wise etc. The stored messages can be viewed on monitor and / or printed on Dot Matrix Printer. Supervisor can control the channel selection. System is capable of storing data for 3 days. Previous data is automatically deleted at midnight as the date changes.

34.9 The Effects of Weather on Aircraft

Meteorological observations require in three phases of aircraft flight namely take-off, level flight or cursing and landing phase.

Jet-engine aircraft fly at high altitudes (above 10 km) at high speed above 100 knots using engine efficiency at cold temperatures. Aircrafts land and take-off at high speeds but engine efficiency is low during these phases. Jet aircraft cover long sector routes (more than 1000 km). During cruising of flight they are above most of the bad weather, but they may encounter jetstreams, CAT and Cb tops.

Turbo-propeller aircraft operate in the layer 5-8 km altitude at moderate speed about 300 knots. Take-off and landing speeds are comparatively low.

Piston-engined aircraft (including helicopters) generally fly below 3 km altitude at speed about 100 to 200 knots. Landing and take-off speeds are also low.

The effects of weather on aircraft is very critical in landing and take-off phases in respect of all aircrafts.

Upper Wind: In steady level flight (cruising phase) though aircraft moves in straight line (AB in fig) but it is displaced due to wind (BC in fig), the actual path of aircraft AC.

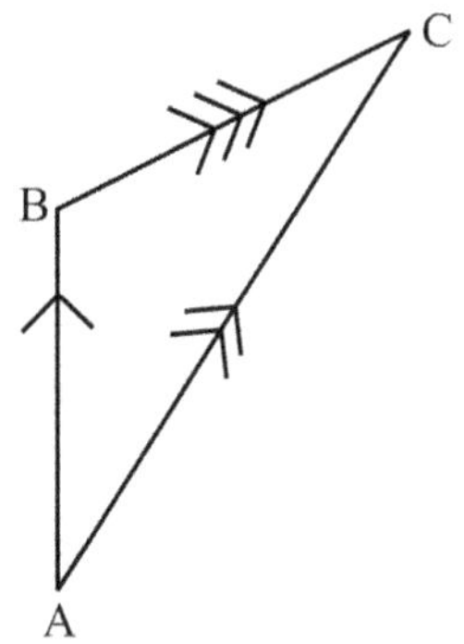

AB = Path of the aircraft in still air

BC = Wind

AC = Actual path of aircraft in relation to the ground.

The information of upper wind speed and direction is important for navigation and planning the fuel load.

Surface Wind: For runway selection, the ATC and the pilots use surface wind observations. The pilots use surface wind observations in planning take-off weight. Strong cross winds are hazardous for landing. Light aircrafts are more susceptible to surface wind fluctuations but respond more quickly to the pilots touch in countering the effects of gustiness.

Vertical Wind Shear: Vertical wind shear is very important in take-off and landing areas.

Descending into decreasing wind speeds the aircraft may undershoot, while descending into increasing wind speeds the aircrafts may overshoot.

Climbing into increasing wind speeds the angle of climb increases, but climbing into decreasing wind speeds the angle of climb of aircraft decreases.

Temperature

(a) **Upper Air Temperature:** The efficiency of an engine is greater with low ambient air temperature. If temperatures are higher than normal, more fuel required to maintain cruising power. Thus the upper air temperatures are required during the planning to determine the fuel load. Upper air temperatures also required for assessing aircraft icing.

(b) **Surface Temperature:** Engure efficiency is lower at higher temperatures. Air density is a function of temperature and pressure. For a given pressure with above normal temperature require higher un-stick speed, thus requires a longer run. However runway lengths are fixed. To adjust it, pay load has to be reduced.

The other observations of atmospheric pressure / air density, precipitation, fog, mist, dust, sand, smoke, squalls, thunderstorms require, which have serious effects on aircraft operations have been discussed in various chapters having these headings.

Questions

1. WMO recommended the density of stations for basic synoptic purposes to be at least one station for every:
   (a) 100 km apart (b) 150 km apart
   (c) 200 km apart (d) 250 km apart
2. WMO recommended the density of upper air stations to be at least one station for every:
   (a) 200 km apart (b) 300 km apart
   (c) 100 km apart (d) 400 km apart
3. The Standard times of surface synoptic observations are:
   (a) 0000, 0300, 1200, 1800 UTC
   (b) 0000, 0600, 1200, 1800 UTC
   (c) 0300, 0900, 1500, 2100 UTC
   (d) 0000, 0900, 1200, 2100 UTC
4. The unit of measurement of pressure for meteorological purpose is:
   (a) Bar (b) Pascal
   (c) Hecto Pascal (d) Atmosphere
5. The unit of temperature and dew point measurement in degrees:
   (a) Kelvins (b) Fahrenheit
   (c) Celsius (d) Knot

6. In Aviation meteorology, wind direction and speed are measured in:
   (a) Degrees magnetic, celsius
   (b) Degrees from true north, knot
   (c) HectoPascals, Celsius
   (d) None above
7. In Aviation meteorology, cloud amount and base height are reported:
   (a) Tenths of sky, meters
   (b) Oktas (eighth's of sky), feet
   (c) Oktas, meters
   (d) Tenth's of sky, feet
8. Pressure and wind speed are measured with:
   (a) Ceilometer , wind vane
   (b) Barometer, anemometer
   (c) Skopograph, alidade
   (d) Rain gauge, thermometer
9. Visibility measured with.......... and reported in ....
   (a) Ceilometer; kilometers
   (b) Skopograph / transmissometer; meters/ kilometers
   (c) Barometer; hectopascals
   (d) Alidade; centimeters
10. Horizontal speed (including wind speed) reported in ... and vertical speed in ....
    (a) Knots, meters per second (b) Feet, feet/second
    (c) Meters, meters/ second (d) Kilometers, km/hr
11. Wind direction for landing and take-off of an aircraft is given in:
    (a) Degrees true north (b) Degrees magnetic
    (c) Degrees Celsius (d) All above
12. The height of bottom of SS screen box is about:
    (a) 3 ft agl (b) 1.25 m agl
    (c) 1.2 m agl (d) 3 m agl

13. The upper air temperatures, humidity and wind measurements are made with:

(a) Pilot balloon (b) Radiosonde

(c) Radar (d) Both b & c

14. DRS (Direct Reception System) is an equipment that receives conventional meteorological data, setillite images and graphics data directly from:

(a) TV (b) Landline

(c) Fax (d) INSAT (satellite)

15. The meteorological message handling system (MMHS) is a PC based system with software for receiving, sorting, storing and a printer for printing meteorological data. It is capable of handling:

(a) Five DRS channel

(b) One duplex and one simplex channel

(c) One RTP channel

(d) All above eight data channels

16. Meteorological observations are required in aircraft flight phase of:

(a) Take-off

(b) Level flight or cruising

(c) Landing phase

(d) All above three phases

17. The information of upper wind speed and direction is important for flight:

(a) Navigation

(b) Planning the fuel load

(c) Terrain clearance

(d) Both (a) and (b)

18. Descending into decreasing wind speeds, the aircraft may……, while descending into increasing wind speeds, the aircraft may………

(a) Undershoot, overshoot

(b) Overshoot, undershoot

(c) Undershoot, stall

(d) Stall, overshoot

19. Climbing into increasing wind speed, the angle of climb………, but climbing into decreasing wind speeds, the angle of climb of aircraft……..
    - (a) Decreases; increases
    - (b) Increases; decreases
    - (c) Does not change; decreases
    - (d) Increases; does not change
20. The efficiency of an engine is greater with ……….. If the temperatures are higher than normal……… to maintain cruising power.
    - (a) High ambient air temperature; less fuel required
    - (b) Low ambient air temperature; more fuel required
    - (c) Low ambient air temperature; less fuel required
    - (d) High ambient air temperature; more fuel required
21. At ground surface, the engine efficiency is lower at……….
    - (a) Lower temperature
    - (b) Higher temperature
    - (c) Does not depend on temperature
    - (d) None above
22. At ground surface, for a given pressure with above normal temperature require…..unstick speed, thus require……..run.
    - (a) Higher; a longer
    - (b) Lower; a shorter
    - (c) No change; no change
    - (d) None above

Answers

1.	(b)	2.	(b)	3.	(b)	4.	(c)
5.	(c)	6.	(b)	7.	(b)	8.	(b)
9.	(b)	10.	(a)	11.	(b)	12.	(b)
13.	(d)	14.	(d)	15.	(d)	16.	(d)
17.	(d)	18.	(a)	19.	(b)	20.	(b)
21.	(b)	22.	(a)				

CHAPTER 35

Satellite Meteorology

35.1 Satellite Meteorology

Introduction

Satellite meteorology is the result of combination of science, satellite technology, computers and communications. Space age has began with the launching of Sputnik-I by the Soviet Union on 4th October 1957. It may be regarded as communication satellite by virtue of radio beacon.

Observing the weather from space began with the launching of TIROS-I (Television Infrared Observational Satellite) by USA on 1st April 1960, since then 27 meteorological satellites have been launched by USA. The first weather satellite Cosmos-122 was launched by Soviet Union in June 1966 and 13 followed thereafter. INSAT-1, the first Indian national satellite was launched on 18th July 1980. INSAT-I series, is a multiple concept consisting of communication, television, radio broadcasting and meteorological services. METSAT launched by India in 2002, later renamed as Kalpana-1, is exclusively dedicated to meteorological services. INSAT-3D launched in 2007 contains a sounder.

All objects in the universe emit radiant energy as long as their temperature is more than 0 ^{o}K, but cease to radiate when it is 0 ^{o}K or less.

The link between the distant stars and the earth is the electromagnetic radiation emitted by them. Radiant energy travels in space / vaccum in the form of electromagnetic waves, which travel in all directions away from the souce with speed of light, 3×10^8 m/s.

The range of all possible radiation wavelengths constitutes electromagnetic spectrum. It consists of gamma rays, X-rays, ultraviolet radiation, visible light radiation, infrared radiation, microwaves and short and long radio waves.

The wavelength (λ) and frequency (f) are related by the formula

$$C = f\lambda$$

where, C is velocity of light and associated energy (E) is given by the formula

$$E = h\nu,$$

where h = Planks constant,

ν = frequency.

The electromagnetic spectrum is given in Table 35.1

Table 35.1 Eelectromagnetic spectrum

Wavelength		Wavelength	
10^{-6} nm		1mm	Millimeter waves (mm)
10^{-5} nm 10^{-4} nm 10^{-3} nm 10^{-2} nm	Gamma rays (MeV)	1 cm 10 cm 1 m	Microwaves (cm, GHz)
10^{-1} nm 1 nm 10 nm	X-rays (A^{0})	10 m 100 m 1 km 10 km	
100 nm	Ultra-violet (nm)	100 km	
1 μm	Visible, near Infrared (μm)	1000 km 1000 km	Radio waves (MHz, kHz)
10 μm	Thermal Infrared (μm)	10000 km	
100 μm	Far Infrared (μm)	100000 km	

nm = nanometer = 10^{-9}m, μm = micometer = 10^{-6} m.

Wavelength range of visible radiation, nomenlecture of microwave and radiowave frequencies, microwave bands are given in Tables 35.2, 35.3 and 35.4.

Table 35.2 Wavelength range of visible colours

Colour	Wavelength	
	nm	μm
Violet	380 - 430	0.38 - 0.43
Indigo	430 - 500	0.43 - 0.50
Blue	500 - 520	0.50 - 0.52
Green	520 - 565	0.52 - 0.565
Yellow	565 - 590	0.565 - 0.59
Orange	590 - 625	0.590 - 0.625
Red	625 - 740	0.625 - 0.740

stand 35.3 Nomenclature of microwave and radiowave frequencies

Abbreviation	Full form	Frequency	Wavelength
EHF	Extremely high frequency (microwaves)	30-300 GHz	1 mm-1 cm
SHF	Super high frequency (microwaves)	30-3 GHz	1 cm-10 cm
UHF	Ultra-high frequency	3 GHz-300 MHz	10 cm-1 m
VHF	Very high frequency	300-30 MHz	1 m-10 m
HF	High frequency	30-3 MHz	10 m-100 m
MF	Medium frequency	3 MHz-300 kHz	100 m-1 km
LF	Low frequency	300-30 kHz	1-10 km
VLF	Very low frequency	30-3 kHz	10-100 km
VF	Voice frequency	3 kHz-300 Hz	100-1000 km
ELF	Extremely low frequency	300-30 Hz	1000-10000 km

Table 35.4 Microwave bands.

Band	λ (Wavelength)	f (frequency)
mm - Band	1-7.5 mm	40-300 GHz
ku - k - ka Band	0.75-2.5 cm	12-40 GHz
X - Band	2.5-4 cm	8-12 GHz
C - Band	4-8 cm	4-8 GHz
S - Band	8-15 cm	2-4 GHz
L - Band	15-30 cm	1-2 GHz

Satellite Orbits

They are two kinds:

(i) Sun-synchronous or near polar orbiting (elliptical or circular orbits), Tropical orbit.

(ii) Geostationary or geosynchronus (circular orbits)

The *polar orbitting satellite* crosses the equator at the same local time and views any given place under comparable lighting conditions. They cover the whole earth and see every location of the earth twice a day at the same approximate time. Generally the altitude of the satellite varies 500-1500 km.

The satellite based sensors can provide:

(i) Atmospheric vertical temperature profile of the globe.

(ii) High resolution picture transmission (HRPT)

(iii) Automatic picture transmission (APT)

(iv) Collection and transmission of data from automatic weather stations to the ground stations.

Geostationary satellites move west to east in a circular orbit in the equatorial plane at an altitude of about 36000 km and have the circular velocity (464 m/s) same as the that of the earth. Consequently they appear sationary to a fixed ground object on equator. These satellites provide continuous weather data round the clock over fixed area from about 60^{0} N to 60^{0} S (but it will not see the polar regions). A constellation of 5 or 6 geostationary satellites spaced around the equator will provide the near global pictures. The satellite based sensors provide vertical temperature profiles, cloud motion wind vectors and sea-surface temperature. They act as a platform for reception of data from automatic weather stations and transmit to the ground stations.

The photographs taken by the satellites are called imageries. They are basically two types. (1) Visual, which are obtained by T.V. cameras, and (2) Infrared (IR), which are obtained by scanning radiometer. The visual photographs taken by T.V. cameras look black and white and represent what we could see with eyes from a satellite. This imageries is a measure of earth's albedo. IR images represent temperatures and temperature contrasts (warm or cold, clouds at different heights).

TV cameras on board satellites measure radiation in visible portion of electromagnetic spectrum about 0.4–0.7 μm. This is possible only from sunlit portion. The IR scanning radiometer measures long wave radiation emitted by the earth, atmosphere (cloud, fog etc) and water bodies in the wave lengths 8 to 13 μm.

Some Characteristics of Satellite Data

Day visual

The brightness of TV images on board satellite is a measure of earth's albedo (reflectivity). White patches indicate the areas of high albedo and black or gray patch shows the areas of lowest albedo. Different gray shades correspond to scenic information. White may represent clouds, snow, ice or sand (as seen by naked human eye). Vegetation, soil and clouds of different heights (high, low, medium clouds) and thickness appear as shades of gray to black to white. Some satellites (SMS/GOES) take visual photographs of the earth about every half an hour and have resolution 1 to 8 km at the satellite sub-point (a point directly below the satellite at the time of photograph). The resolution 2 km simply means that any ground element/cloud as small as 2 km size may be clearly identified from the satellite photographs. The visible channel thus capable of providing a wealth of data such as thunderstorms, cumulus clouds or convective clouds and low level/high level wind flow. With the help of every half hourly data, meteorologist can identify whether the convective cloud is developing or dissipating, cloud motion wind vectors provide wind speed.

Night Visual

Besides day time visual images there is provision in some satellite based instrumentation to take night time visual images with high resolution. It is a low light sensor, able to detect even in moon light. In the absense of moonlight, city lights and aurora phenomena, volcanoes, oil and gas fields, forest fires etc. can be detected.

IR-Images

In this imagery high clouds look white, low/convective clouds as gray and hot land/water surface as black. This scheme may be reversed or colour films may be used to detect various features such as ocean currents. In general black depicts hot or warm and white denotes cool or cold relatively. The thermal IR imagery is viewed in the interval range of wave length 8–13 μm. A VHRR (very high resolution radiometer) provides IR pictures with resolution of 1 km.

The large diurnal variation of land surface produces marked change in appearance of day and night time IR images. During day time the warm land surface appears dark as compared to the cool water bodies which appear gray and mild cold water bodies together with high clouds appear as white. During night time the cold land appears bright shade of gray in the IR images. The land which is very close to the water bodies may acquire the temperature of the water bodies consequently the image contrast between land and water bodies is not marked. This makes coast line and lake shores less obvious. The variation in IR images is also caused due to seasonal and latitudinal temperature variation. In winter, particularly during night, clouds over cold land areas are not easily detectable, However in tropics there is always temperature difference between land surface and middle/high clouds and hence easily detectable.

As pointed out earlier the brightness differences in visible (TV) pictures indicate difference in reflectivity (albedo) while the brightness difference in IR pictures indicate temperature difference of the emitting surfaces.

The Table 35.5 gives the difference of cloud pictures as seen in visible and IR imagery.

Table 35.5

		Visible imagery			
		White	**Gray**	**Dark gray**	**Black**
IR imagery	(a) Black			Fair weather cumulus and desert	Warm Ocean
	(b) Dark gray	St			Cold Ocean
	(c) gray	St, Sc	Thick Ci		High frosted terrain
	(d) light gray	Ci, As, Cb			
	(e) white	Frontal cloud and active snow	Dead Cb anvil	Dense Ci	

Note: All surfaces with temperature above absolute zero emit radiation.

Note: Satellite imagery clouds classified as:

(i) Convective or cumuliform clouds, (ii) Stratiform clouds and (iii) Cirri form clouds.

Before the advent of weather satellites cyclone detection was based on coastal synoptic observations, Radar observations and a few ship observations (which were very sparse or none from the cyclone field). Lack of observations was a great draw back in detection and forecasting of a cyclone, its intensity, movement (direction and speed) and land fall. There might have been occasions of missing detection when cyclones were formed far away from coast (away from radar range of 400 km) and dissipated over the sea itself. After the advent of geostationary weather satellites there was not even a single occasion of missing detection of cyclone over the sea. The geostationary satellites stationed over equatorial circular orbit are GOES – east (long 15 ^{O}W), GOES - west (long 135 ^{O}w) of USA, METEOSAT of Europe at zero meridian, INSAT-ID of India at 85 ^{O}E, GMS of Japan at 140 ^{O}E (see Fig. 35.2). These virtually cover the global detection of cyclones upto 60 ^{O} latitude on either side of the equator. Earlier it was believed that the intensity of cyclone was a function of its size but it is now confirmed by satellite images that it is not so. Dvorak, an American meteorologist propounded an empirical cyclone intensity code called Dvorak T-number (T-stands for tropical cyclone). Even now with some minor modifications Dvorak intensity code is followed (See Fig. 35.3, 35.4, 35.5).

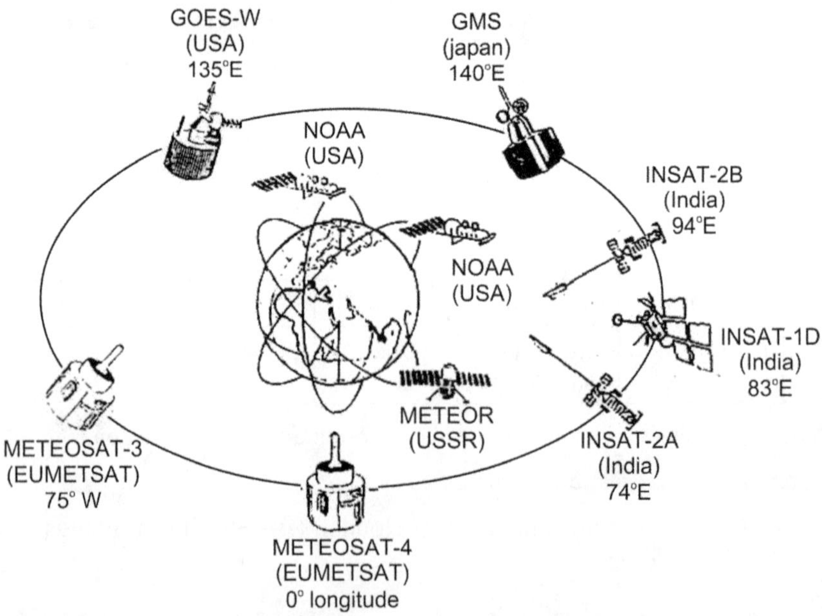

Fig. 35.2 Global system of operaional meteorologocal satellites.

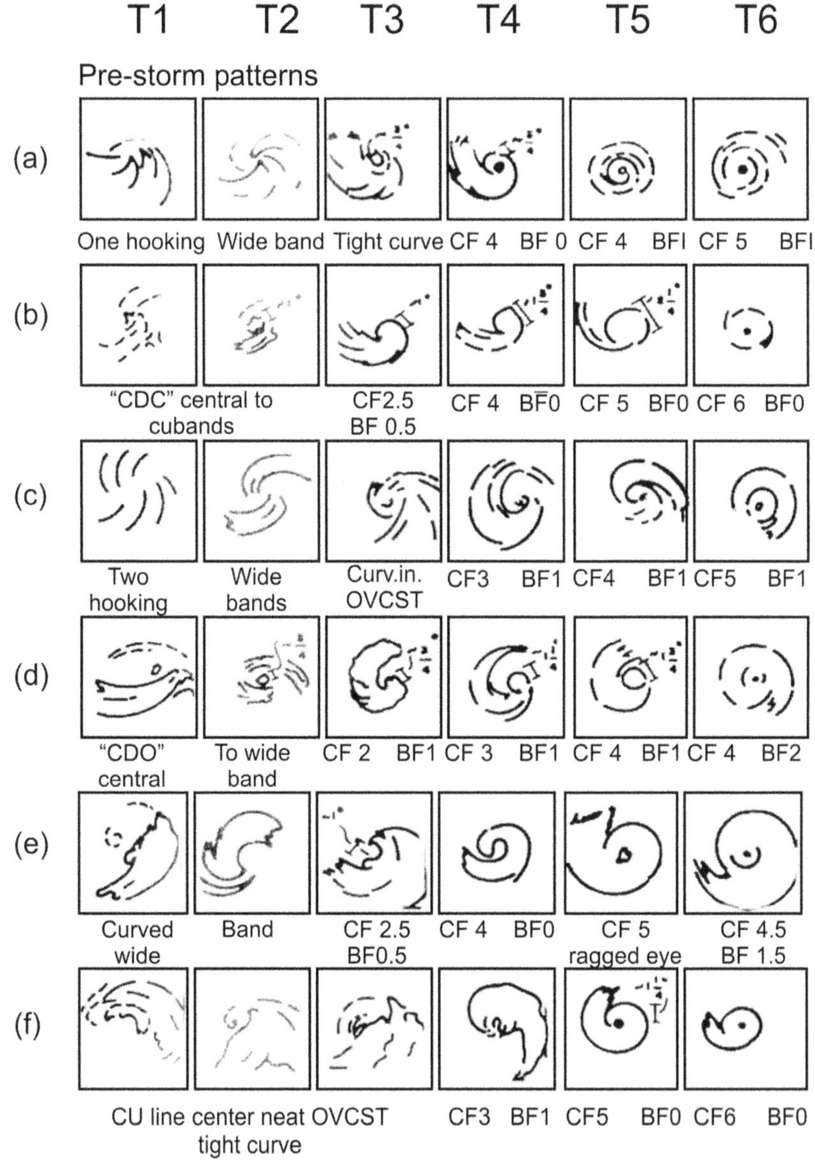

Fig. 35.3 Common tropical cyclone patterns and their corresponding T-numbers. The T-numbr shown must be adjusted for cloud systems displaying unusual size. The patterns may be rotated to fit a particular picture of a cyclone (Dvorak. 1975).

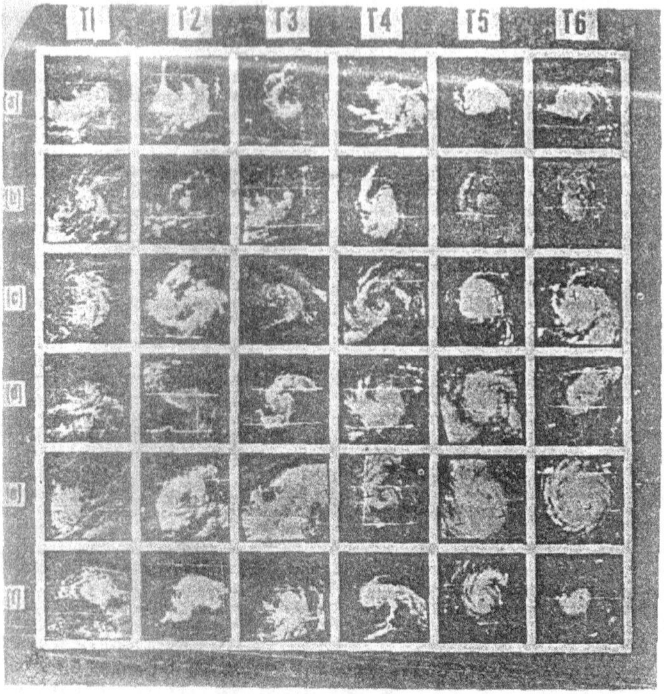

Fig. 35.4 Typical satellite photographs of six stages (T1-T6) of development of tropical vortex from pre-storm stage to intense hurricane stage (Dvorak, 1975).

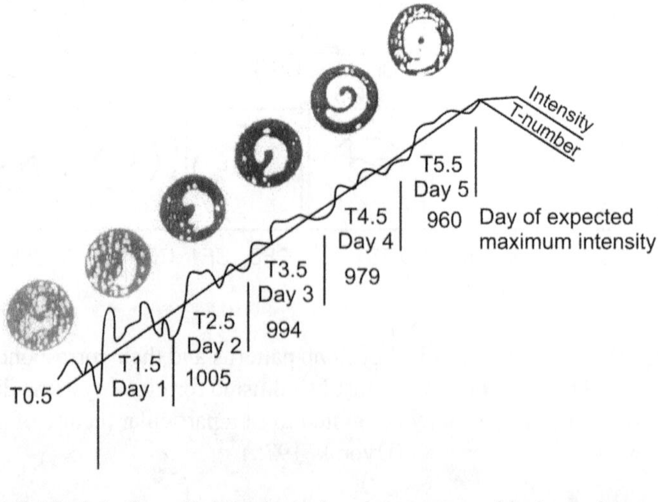

Fig. 35.5 Model of tropical cyclone development used in intensity analysis (curved band pattern type).

T-number is determined on the central features (CF) of Central Dense Over cast (CDO) size, embedded distance of Eye and Banding Features (BF). Digital IR data of cyclone field temperatures help in location of Eye (warmest part) of cyclone and the lowest temperature part indicates the wall cloud region (about 50 km from the centre of Eye).

Note

1. T-number = CF (= size of CDO + embedded Eye distance) + BF (= Amount of banding coil around CDO + vertical depth of wall cloud)
2. Digital IR data of cyclone field provides temperature field of cloud tops. This is determined by gray scale. Warm temperature region (T_W) of central region indicates cyclone Eye and the surrounding coldest (Tc) region indicates wall cloud region. The temperature difference of these two (Tw-Tc) gives an idea of the intensity of the cyclone. The Dvorak scheme categorises the cloud pattern by the CDO, spiral bands and the eye. The CDO features such as round, oval, angular, irregular, ragged edge defines CSC (cyclone system centre). OR cloud line (band features) define CSC.

Note: (Centre of the Eye wall is called CSC which is visible when the cyclone Eye is clearly formed).

The Dvorak's procedure enlists the following steps.

Step 1: Located cloud system centre (CSC).

Step 2: Determine relationship between CSC with CDO. This is done by comparing with standard pattern. If the measurement in this step is clear cut, the intensity is final. If not we go to step 3.

Step 3: Model Expected T-number (MET). In this method current picture is compared with the previous days picture to ascertain whether the storm is intensified or weakened by the past trend.

Step 4 to 6: MET is further refined comparing with the standard patterns corresponding to the intensity and then upgraded or downgraded the intensity. The intensity thus fixed is subject to certain rules to make final adjustments. The rules are–to hold variation of intensity to one T-number per day in the pre-storm stage and to within one number of the MET during the later stage. As an example, let in the prestorm stage two days earlier the T-number T_{-2}, one day earlier T_{-1}, to day T, the next day T_1 and so on.

The main features of the standard patterns (description of photo chart) may be categorized as (i) curved band, (ii) shear, (iii) Eye, (iv) CDO (v) Embedded centre. In addition to this another rare pattern is (vi) central cold cover (CCC) which indicates the halt of further development of the storm. The CCC pattern analysis is taken as Step 3. When the storm centre is masked by cold overcast clouds, the CCC pattern is recognized in which outer curved lines and curved bands appear very weak or not seen. As a result visible pictures are not useful for determining CCC pattern. The persistence of CCC pattern is an indication that storm development is halted. It may be noted that very cold, commapattern contains cirrus streaks/lines on the peripheri of dense overcast, however such cirrus lines are absent in CCC pattern.

Step 7: This covers the rules that are followed in determining T-number. When measurements of cloud features are clear the data T-number (D T- number) is fixed and it is final. When the features are obscure or not clear, the satellite pictures are compared with standard pattern and assign pattern T-number (P – T number). When both D-T number and P – T number are not clear it is then MET (Model Expected T- number) is used. The last technique is used when data are received 3-hourly as in the case of INSAT pictures.

Step 8: This covers assigning final T-number imposing some constraints. The constraints are. "Hold the final T-number close to one number per day" during the pre-storm evolution stages and within one number of the MET during rest of the evolution. It may be noted here that no lowering of T-number is allowed during night in the beginning two days.

Current Intensity (CI)

CI gives directly the cyclone intensity. In development stage the CI is equal to T-number but CI number > T-number when the cyclone is weakening. As a rule CI number is not lowered for about 12 hours after the cloud features show the signs of weakening. At that time CI number = T-number + ½ to 1. If a storm weakens and then intensifies the CI number is not lowered even if the CI number is greater than T-number. The empirical relation between the CI number and storm Maximum wind speed (MWS) is given Table 35.6.

Table 35.6

CI number	MWS (kt)	MSL Pressure Atlantic h Pa	MSL pressure NW- Pacific h Pa	Δ p = pressure drop Indian seas h Pa
1	25			
1.5	25			
2	30	1009	1000	4.5
2.5	35	1005	997	6.1
3	45	1000	991	10.0
3.5	55	994	984	15.0
4	65	987	976	20.9
4.5	77	979	966	29.4
5.0	90	970	954	40.2
5.5	102	960	941	51.6
6	115	948	927	65.6
6.5	127	935	914	80.0
7	140	921	898	97.2
7.5	155	906	879	119.1
8	170	890	858	143.3

MSL : Minimum sea level (pressure)

msl : Mean sea level

IR picture colour code No's with temperatures given in Table 35.7.

Table 35.7

Code No	Colour	Temp ^{0}C
1	OFF white	9 to – 30
2	Dark gray (DG)	– 31 to – 41
3	Medium gray (MG)	– 42 to – 53
4	Light gray (LG)	– 54 to – 63
5	Black	– 64 to – 69
6	White	– 70 to – 75
7	Cold light gray (CLG)	– 76 to – 79
8	Cold Medium gray (CMG)	– 80 to – 84
9	Cold dark gray (CDG)	– 85 to – 89

Naming of Cyclones in Bay of Bengal and Neighourhood

In 2005, a panel of 8 member countries (India, Pakistan, Srilanka, Bangladesh, Myanmar, Oman, Thailand and Maldevis) was set up by WMO for recommending names, which will be adopted and declared by IMD. Of the 40 names proposed by the panel includes:

1. Payaar – Myanmar
2. Baaz – Oman
3. Fanoos – Pakistan
4. Mala – Srilanka
5. Mukda – Thailand

The Indian names include:

1. Akash
2. Bijli,
3. Jal,
4. Lahar, and
5. Megh

The naming of cyclones were introduced in 2005.

35.2 Satellite Pay Loads, TV Camera

The payload carried by the TIROS-1 satellite in 1960 was a TV camera that relayed to the ground, whatever it observed of land, oceans and clouds.

Depending on the application besides TV cameras, the following are the common meteorological payloads.

(i) A scanning radiometer (SR)
(ii) Very high resolution radiometer (VHRR)
(iii) Vertical temperature profile radiometer (VTPR)
(iv) Scanning microwave radiometer (SMR)
(v) Data collection transponder (DCT)
(vi) Tape recorders
(vii) Solar cells

SR : Scanning radiometer has two channels

(i) Visible channel 0.5 to 0.17 μm

(ii) Infrared channel (window) 0.7 to 10.5 μm

Scanning radiometer functions like TV camera, transmits strip image along the orbital track. It takes continuous images of earth's atmosphere and surface. The resolution of visible channel is 4-12 km and IR channel 25 km.

VHRR : It acts like SR with better resolution of visible channel 1 km and IR channel 8 km.

Note: Radiometer is an instrument used to measure electromagnetic radiation energy. Satellite based radiometers can provide radiation budget of surface – air system, sea surface temperature (SST), cloud cover distribution and cloud top temperatures.

Radiometer operates in visible, IR and microwave regions.

Spectrometer: A satellite based spectrometer provides:

(i) Chemical composition of the surface layers in the atmosphere and ocean

(ii) Measures the atmopheric concentrations of water vapour.

Resolution of an instrument is the smallest area on the ground that the given instrument is able to identify.

VTPR: It determines the temperature profile of the atmosphere on realtime operational basis over oceans and other regions (where the ground based upper air sounding station network is scanty or absent).

SMR: It operates on microwave bands and can provide global coverage of surface temperature data with an accuracy of $1^{o}C$.

DCT: It is used for collection of meteorogical and oceanographic data from remote uninhabited places on land and at sea by means of land based, ocean based data collection platforms (DCPs).

Tape Recorder: To store data before transmission to ground station. Such data storage on board makes provision for acquiring the global data at CDA (command data acquisition) station. CDA are managed by satellite launching countries.

CDA: A ground based station (on earth) at which various functions to control the satellite operations and to obtain data from the satellite are performed. The CDA transmits programming signals to the satellite, and commands transmission of data to the ground. Processing of data (by electronic machines and by hand) is accomplished at the CDA station. Raw and processed data are disseminated from CDA stations.

Solar Cells: Used for power. The sides and top of the satellite are studded with more than 9000 of these solar cells.

35.3 Remote Sensing

The process of observing an object with the help of electromagnetic spectra (band of radiation wavelengths) that an human eye cannot see is called remote sensing. In meteorology this term is used for observing meteorological parameters with the aid of radiometers on board satellites. Infact remote sensing is measurement of returned (or back) radiation from the earth-atmosphere system. The satellite based sensors are used for the detection of earth observation (both atmosphere and sea), detection of severe events like cyclones, tornadoes, tsunamises etc, delivary of warnings directly to the people of the affected area and for data relay and communication capabilities.

Microwave remote sensing is a powerful tool in application of meteorology and oceanography. It has the ability to measure water vapour and liquid water even in the presence of most clouds. Land surface emissivity in microwave region is high and is variable. It must be noted that microwave sensors reception signal strength on board geostationary satellites is poor and hence not useful, while its reception on polar low orbit altitudes is fairly good and hence mostly used.

TV Carmeras: TIROS-1 was operational for 78 days but it clearly demonstrated beyond doubt the potential use of satellites for surveying global weather conditions from space.

TIROS-1 has two TV cameras one with low resolution and the other one high resolution. The cameras were operating only in the portion of the earth which was sunlit.

TIROS-8 was launched on 21st December 1963, equipped with APT (Atutomatic picture transmission) capability, which provided satellite pictures world over to meteorological agencies in real time as the satellite passed over their region. There were 50 ground stations including one at Mumbai, India.

The successor TIROS series was ESSA (Environmental Science Services Administration) series initiated in 1966 by U.S. ESSA series transmitted thousands of space based weather system images which were received by more than 300 APT reception stations in 45 countries by 1970.

NASA (National Aeronautical and Space Administration) US launched Nimbus series satellites which had on board very advanced instruments for mapping ozone, sea ice, radiation budget components, coastal zone properties and sea surface temperature. The series carried improved Vidicon Camera system for daylight coverage, which were transmitted through the APT system.

Another APT service was provided parallelly by the Russian METEOR series of satellites.

***NOAA, AVHRR*:** TIROS next generation satellite programme, name TIROS-N launched on 13th October 1978 followed by many satellites in the series. This followed by NOAA (National Oceanic and Atmospheric Administration) series by US. This series till 2005 had NOAA - 18 (series).

All NOAA satellites placed in near circular polar orbit. It carried on board Advanced Very High Resolution Radiometer (AVHRR) which had 4-channel radiometer initially, subsequently by 6 channel radiometer. One in visible (VIS), two in near infrared (NIR), one in mid-wave infrared (MIR) and two in thermal infrared (TIR) wavelength regions (See Table 35.8). All channels had resolution 1.09 km at the sub-satellite point.

Table 35.8 NOAA AVHRR channel specifications

Channel Number	Spectral Band	Wave length Range (um)	Uses
1	VIS	0.58 - 0.68	Daytime cloud and surface mapping
2	NIR	0.725 - 1.00	Land-water boundary delineation
3A	NIR	1.58 - 1.64	Snow and ice detection
3B	MIR	3.55 - 3.93	Night time cloud mapping, SST retrieval
4	TIR	10.3 - 11.3	Night cloud mapping, SST retrieval
5	TIR	11.5 - 12.5	SST retrieval

35.4 Geostationary Satellites

The first US Geostationary Operational Environment Satellite (GOES) was launched on 16th October 1975, the first new generation (GOES-8) launched on 13 April 1994. GOES-12 or GOES-East is located on equatorial orbit at 75^0W and GOES-10 or GOES-west is located at 135^oW.

The current GOES satellites have 5 channel imaging radiometers whose specification are given in the Table 35.9.

Table 35.9 GOES channel specifications.

Channel Number	Spectral Band	Wave length range (μm)	Resolution
1	VIS	0.55 - 0.75	1
2	SWIR	3.80 - 4.00	4
3	WV	6.50 - 7.00	8
4	FIR	10.20 - 11.20	4
5	TIR	11.50 - 12.50	4

(SWIR : shortwave infrared)

INSAT: (Indian National Satellite)

INSAT 1: was launched on 18th July 1980, (India became the seventh nation after Soviet Union, USA, France, Japan, China and UK in satellite launching).

INSAT was a multipurpose satellite used for long distance telecommunication, meterorological earth observation and data relay, and direct TV broadcasting.

INSAT-1 A launched on 10th April 1982 was deactivated on 6th September 1982 due to malfunctioning. INSAT - IB launched on 30th August 1983 was a geostationary satellite, located at 74^oE.

India launched METSAT on 12th September 2002, which was renamed Kalpana - I, is an exclusive meteorological satellite.

INSAT - 3D launched 2007 carried a sounder.

The other geostationary satellites are:

METEOSAT of Europe locate at 0^0 meridian

INSAT - ID of India located at 85^0E

GMS of Japan located at 140^0E

35.5 IMDPS (INSAT Meteorological Data Processing System)

IMDPS is receiving and processing Meteorological data from AWS (automatic weather stations). An AWS station records, stores hourly data and transmits the same to Kalpana - 1 through INSAT Data Relay Transponder (DRT). Cloud imageries are supplied and transmited through INSAT-3C to various IMD forecasting offices and also to other users through MDD (Meteorological Data Dissemination) scheme.

Questions

1. The satellite visual images are obtained by T.V. cameras. They look black and white and represent what we could see with our eyes from a satellite. These images are a measure of:
   (a) Earth's albedo
   (b) Temperature and contrasts
   (c) Earth's absorption insolation
   (d) None above
2. The satellite IR images represent temperature and temperature contrasts (warm or cold, clouds at different heights). The IR scanning radiometer measures long wave radiation emitted by the earth, atmosphere and water bodies in the wave lengths:
   (a) 0.4 to 0.7 μm (b) 1 to 8 μm
   (c) 8 to 13 μm (d) Both (a) and (b)
3. The geostationary satellites move .... in a circular orbit in the equatorial plane at an altitude of about .... and have the same velocity as that of the earth.
   (a) West to east; 36000 km (b) East to west; 3600 km
   (c) East to west; 3600 miles (d) West to east; 36000 miles
4. The geostationary satellites provide continuous weather data over a fixed area round the clock. It provides:
   (a) Vertical temperature profile
   (b) Cloud motion (wind) vectors
   (c) Sea surface temperature (SST)
   (d) All above
5. The brightness difference in visible images indicate .... while the brightness difference in IR pictures indicate ....
   (a) Temperature difference; (albedo) reflectivity difference
   (b) (Albedo) reflectivity difference; temperature difference

(c) Temperature lapse rate; wind shear

(d) None above

6. All objects in the universe emit radiant energy as long as their temperature is more than............ But cease to radiate when it is .... or less.

(a) 0 °C; 0° C (b) 0° K; 0° K

(c) 100° K; 100° k (d) 100 °C; 100 °C

7. Radiant energy travels in space/vaccum in the form of electromagnetic waves, which travel in all directions away from the source with a speed of:

(a) 3×10^{10} ft/s (b) 3×10^{8} m/s

(c) 3×10^{10} m/s (d) 3×10^{8} ft/s

8. The frequency and wavelength of UHF are:

(a) 30 to 3 GHz; 1 cm-10 cm

(b) 3 GHz to 300 MHz; 10 cm-100 cm

(c) 300 to 30 MHz; 1 m-10 m

(d) 30 to 3 MHz; 10 m-100 m

9. X-band frequency, wavelengths are:

(a) 8–12 GHz, 2.5 to 4 cm (b) 4-8 GHz, 4-8 cm

(c) 2-4 GHz, 8-15 cm (d) 1–2 GHz, 15-30 cm

10. S-band frequency, wavelengths are:

(a) 8-12 GHz; 2.5 to 4 cm (b) 4-8 GHz; 4-8 cm

(c) 2-4 GHz; 8-15 cm (d) 1-2 GHz; 15-30 cm

11. The polar orbitting satellite crosses the equator at the same local time and views any given place under comparable lighting conditions. They cover......and every location of the earth ....at the same approximate time, the altitudes of satellite varies 500-1500 km.

(a) Between latitudes 60 °N to 60 °S; once a day

(b) Whole of the earth; twice a day

(c) Half of the globe; four times a day

(d) Both (a) and (c)

Answers

1. (a)	2. (c)	3. (a)	4. (d)
5. (b)	6. (b)	7. (b)	8. (b)
9. (a)	10. (c)	11. (b)	

CHAPTER 36

Climatology for Aviators

36.1 Introduction

Modern radio aids contributed a lot to aviation safety, because of which jet aircrafts fly at higher altitudes and cover long distances. They are equipped with more sophisticated navigation, communication and position fixing systems with automated reporting system on aircraft, transmitted through satellite communications. However the pilot must be thoroughly aware of the atmospheric systems for the safety of the passengers and crew. Aviation climatology helps him in this respect. Climatology of a location is mainly concerned with statistics of average state of the atmosphere together with its extremes at that location. This includes temperature, pressure, humidity, winds (both surface and upper air), precipitation, solar radiation etc. Pilots require weather forecasts and reports for planning the flight. They require information of upper winds, temperatures, tropopause heights, jet streams, freezing level (0°C isotherm) heights, mountain waves, thunderstorms, tropical cyclones, CAT, volcanic activity. They also require terminal weather and the alternate airports nominated, both en-route and destination.

Fig. 36.1 North America and North Atlantic – Typical January Day.

In many parts of the world meteorological services have climatological data of all airfields. The information mainly contain air temperature (including maximum, minimum), surface wind, upper air temperature, humidity and wind, rainfall, sunshine, low clouds visibility, thunderstorm activity with squalls, and hail. In the appendix such data is provided in respect of main airports in India. Upper winds and temperatures together with contour analysis for representative month of four seasons have been presented (in the appendix P 795 – 809). A typical day weather systems in January and July over North America and North Atlantic shown in figs. 36.1, 36.2 respectively.

Brief aviation climatological features in respect of international aviation routes are presented in this chapter.

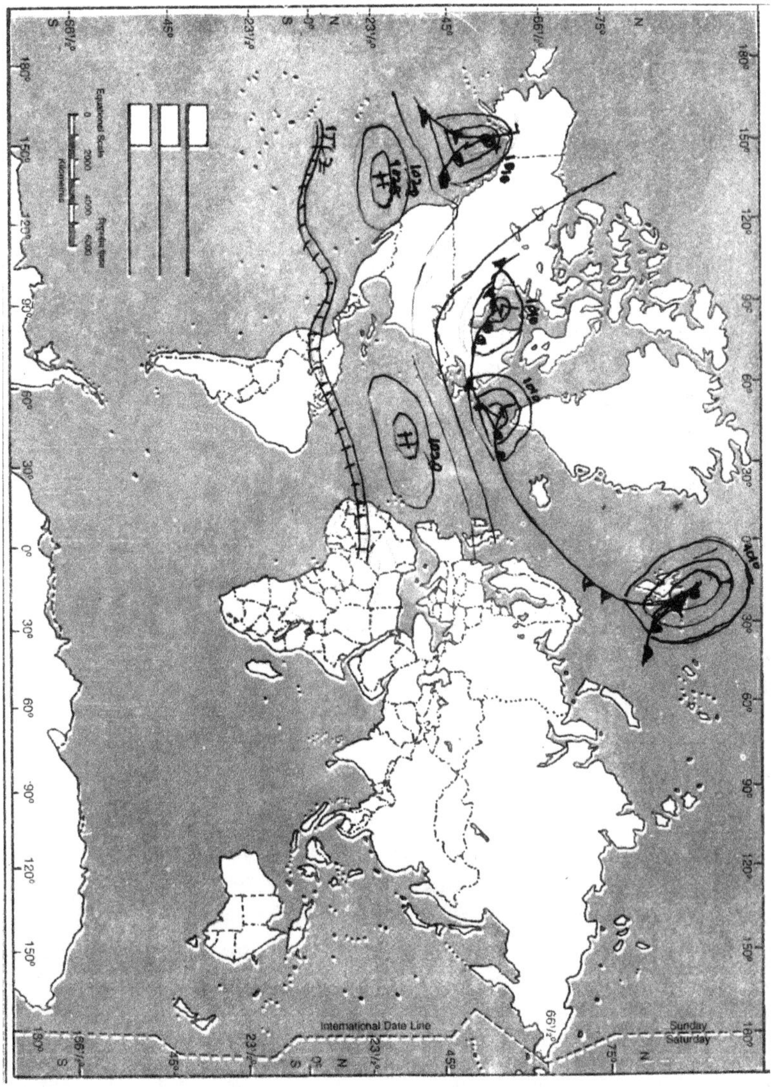

Fig. 36.2 North America and North Atlantic – A typical July day.

36.2 North Polar Regions

Within the Arctic circle there are place to place and year to year changes. In this region winter lasts for five to six months but summer about three months. Around winter solstice, a continuous darkness period prevails, while at around summer solstice a continuous day light. In winter anticyclones are persistent. Polar depressions move northeast wards in continuous series between Greenland and Norway. During summer depressions reach most part of the Arctic from Iceland to Novaya Zemlya. In winter sea remains frozen permanently and in summer air temperature over ice fields are about 0 °C.

In winter Arctic troposphere is colder than the ICAO ISA, while in summer tropospheric temperatures are close to ICAO ISA. All precipitation is in the form of snow thunderstorms and hail are practically abscent. The area will be covered by snow from November to April and near the pole it is all through the year.

The areas of Greenland, Norway, Barents and Kara seas are more cloudy during summer than in winter. Clear days frequency is high during winter but most of sea areas are cloudy with base sometimes on the ground and tops extending to 18000 ft (500 hPa).

Generally visibility is very good except in precipitation. During blizzards and snowfall it is very poor. Radiation fog over land is common during autumn and winter, ice fog is peculiar to the Arctic. During ice fog horizontal visibility is very poor but vertical visibility is good. Ice fog forms over aerodromes with the movement of aircraft and ground vehicles. Ice accretion on aircraft is not great problem as compared to temperate latitudes. Rime is common.

36.3 North Atlantic (lat 30 °N to 60 °N)

The area is mainly covered by Greenland and Iceland (in the north) to Azores and Bermuda (in the south) and air route London to New york (western Europe to North America). This region is under temperate climatic zone. In the south Sub-tropical High pressure belt is a semi-permanent feature. In winter High pressure centre (1020 hPa) located at about 30 °N and in summer it is (High pressure 1025 hPa) located at about 35 °N. In this zone climate is mainly dominated by continental air, either very warm air moving northwards or very cold air moving south wards. Icelandic low moves north-east wards and creates depression on the Atlantic polar front and moves along with north Atlantic drift. In winter, in general low is located between south Greenland and Iceland

(1000 hPa). In January the polar front occupies in its southern limit over the north Atlantic. Depressions move from Florida towards southwest of England. Moving in a east-north-easterly direction they occlude over Bay of Biscy. In addition to the frontal lows/depressions, polar air depressions form over the ocean. The easternside of North America is visited by the polar front lows and also continental lows, which move from west to east. These lows/depressions are associated with bad weather (see Fig. 36.3).

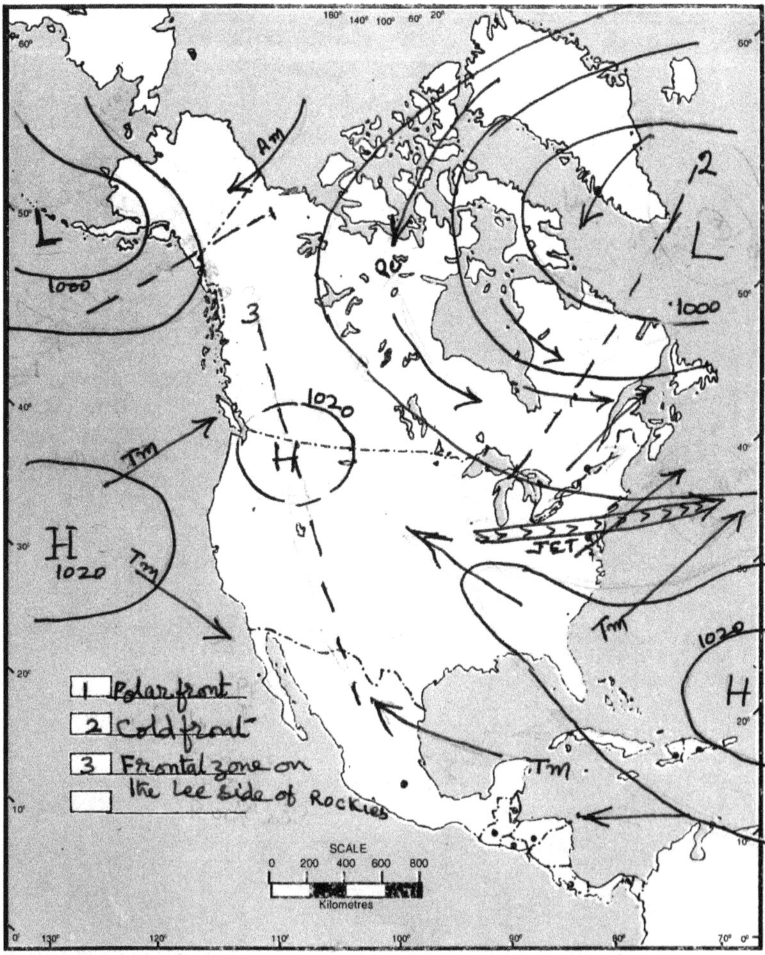

Fig. 36.3 January – Wind, Pressure distribution and Fronts.

During spring sub-tropical high intensifies and moves northward. In July Atlantic Polar front extends from Newfoundland to northwest

Scotland but comparatively weak. In extreme south, north Atlantic and Florida, North America (Florida and Caribbean) lies throughout the year.

Under the influence of north-east trade winds, ITCZ effects the Gulf of Maxico in July. The north American High is prominent only in winter, which will be occupied by low in summer. The New York Boston areas get precipitation under the influence of polar continental outflow and warm maritime air moving north (see Fig. 36.4).

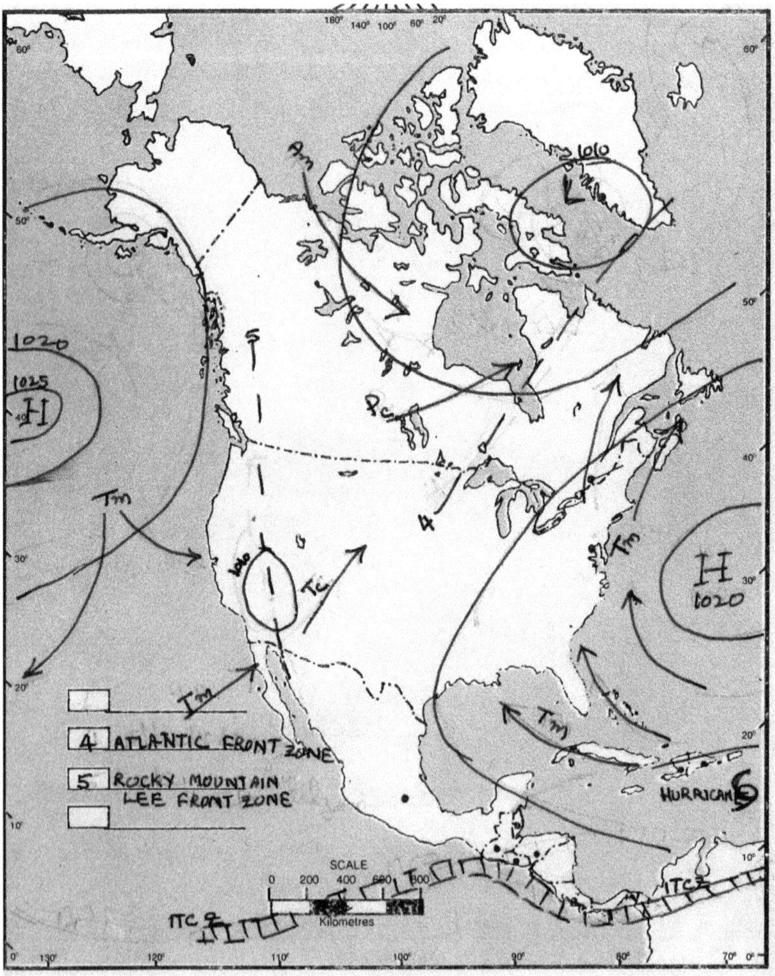

Fig. 36.4 July – Wind, Pressure Distribution and Fronts.

In winter, the extension of Siberian High lies as a Scandinavian High and which affects eastern area with severe cold winds to UK. During

August-October, tropical cyclones (called hurricanes) affect the area around Bermuda.

Surface winds: Mostly northwesterly, westerly or southwesterly. In the north of the region easterlies may be seen.

Upper winds: At 30, 000 ft (FL 300 / 300 hPa) westerly 50 kt, on London-New York route.

Westerly polar jet stream speed 100 kt or more throughout the year. In January the jet stream speed exceeds 200 kt when the sub-tropical jet stream lies between 25 °N and 40 °N, however in summer months it shifts to 40 °N and 45 °N and weakens.

Tropopause height

	At about Arctic circle (65 °N)	At 30 °N
Winter	31000 ft (9.5 km)	45000 ft (13.7 km)
summer	33000 ft (10.1 km)	47000 ft (14.3 km)

Freezing level

	At about Arctic circle (65 °N)	At 30 °N
Winter	Surface level (0 ft)	11000 ft (3350 m)
Summer	6000 ft (1830 m)	16000 ft (4870 m)

Fog

During spring and early summer advection fog occurs over sea and along European coast. The interaction of cold Labrador current and warm Gulf Stream, advection fog occurs over Newfoundland throughout the year.

During autumn and winter radiation fog also occurs along coastline and inland over Europe and America. Radiation fog may also occur at any time of the year. In the vicinity of cities, both in America and Europe smog/fog occurs in association with industrial pollution, particularly in the morning hours.

36.4 North America

North America has mainly continental climate affected by very warm air that moves northwards or very cold air that moves southwards.

In January, polar front depressions that move from Florida to northeast wards. Polar air depressions also form over the oceans. Under the influence of these systems Eastern America experiences bad weather near Great lakes (see Fig. 36.3). On western side, Pacific polar front depressions reach in occlusion state.

In July, Pacific coast of North America lies under the influence of High over Pacific and ITCZ affects the southern parts of Gulf of Maxico (see Fig. 36.4). Pacific High causes little weather on west coast, but the ITCZ causes bad weather in the south.

Surface wind: Mainly northerly in association with High pressure.

In July, southwesterly in eastern parts of North America and Northwesterly on Pacific coast.

Upper winds: Strong westerlies in the east, decreasing to west.

During winter: westerly jetstream appears frequently

Tropopause Height

	January	July
At lat 40° N	37000 ft (11.3 km) drops to 30,000 ft (9.1 km)	41000 ft (12.5 km)

Freezing Levels
January

(i) Over eastern parts and at most places: Surface of the earth

(ii) Near west coast: At 50°N 4000 ft (1220m)

At 35°N 9000 ft (2740m)

July

15000 ft (4570 m)

Tornadoes: Occur mostly during April, May,

Location: Lowland areas of central and upper Mississippi. However almost every state of North America witnesses Tornado. Most frequent in Texas, Oklahoma, Kansas, Iowa, Alabama.

Annual frequency: More than 1000.

36.5 South Atlantic, South America

January

Axis of High pressure belt runs roughly along lat 35°S from South Atlantic to South Pacific across South America.

ITCZ: Over South America central parts (about lat 15 °S, long 60 °W), crosses the equator to over to north Atlantic and then extends to 10°N of African coast, over Africa it moves southeastward to 15°S, 35°E and thence to Madagascar.

Frontal depressions move from west to east across the Mediterranean sea. However they do not move inland over Africa, except over west North Africa over Atlas mountains.

Generally cold fronts are active and are associated with significant clouds. Warm fronts are very weak. Frontal depressions affect the south-western parts of south America (see Fig. 36.5).

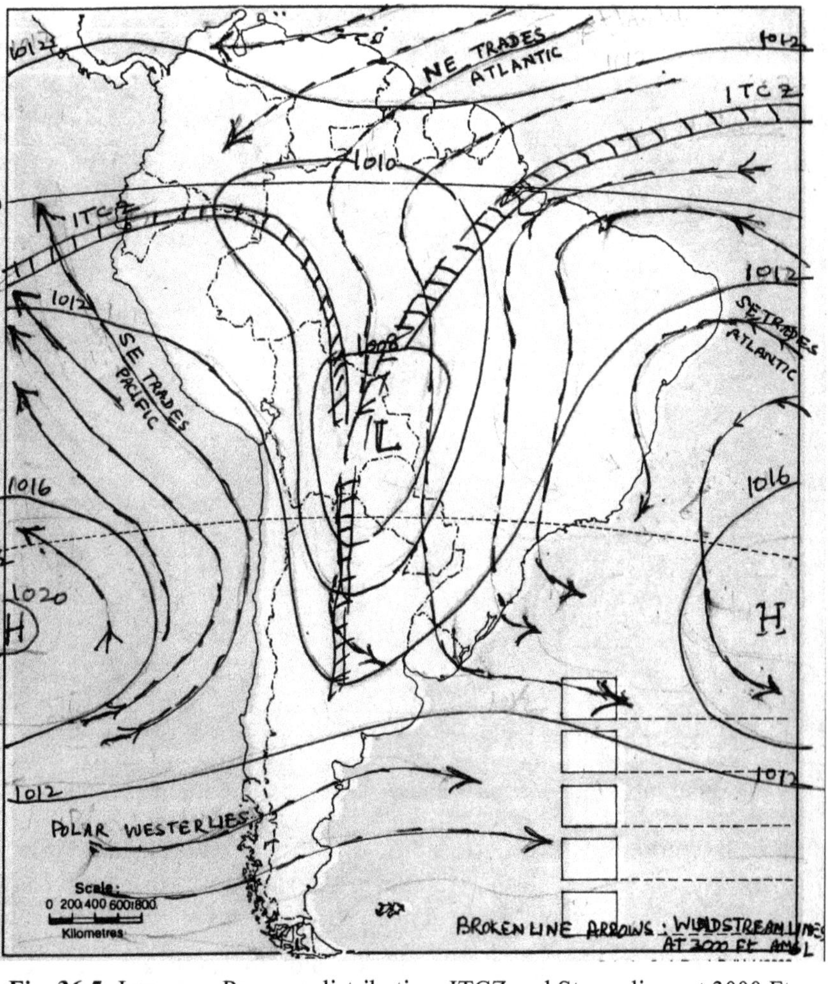

Fig. 36.5 January – Pressure distribution, ITCZ and Streamlines at 3000 Ft.

July:

Axis of High pressure belt runs along Tropic of Capricorn, over South Pacific, South America, South Atlantic and Africa (see Fig. 36.6).

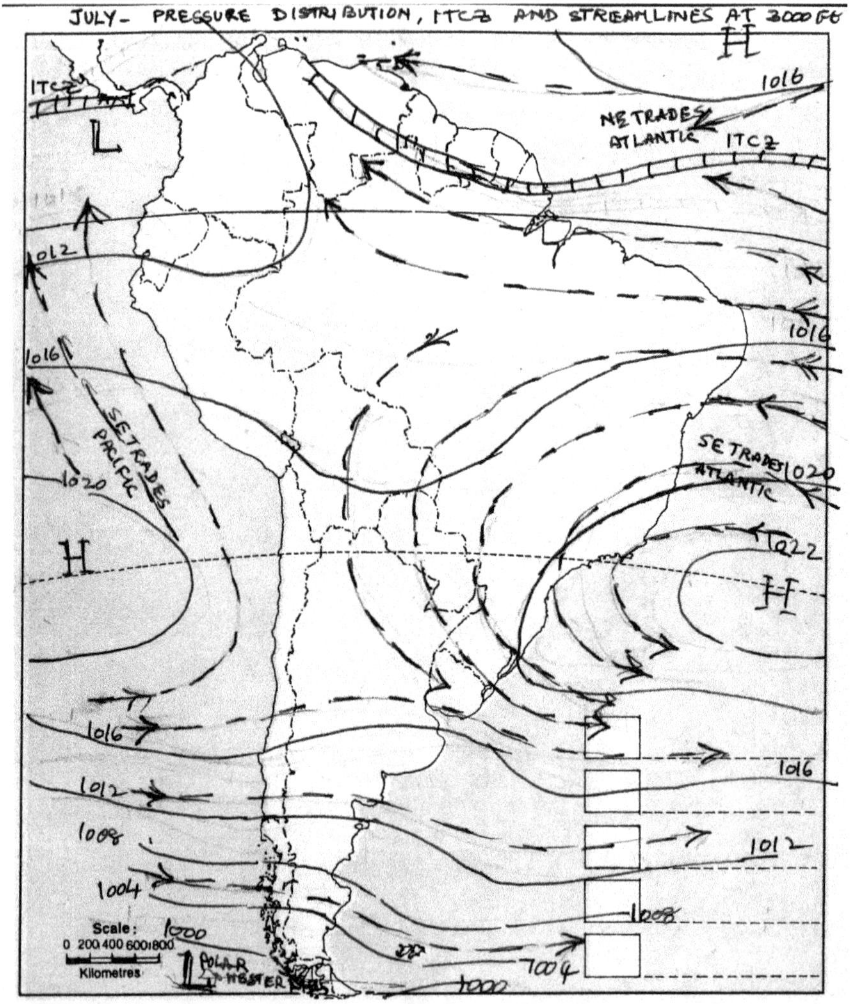

Fig. 36.6 July – Pressure distribution, ITCZ and Streamlines at 3000 ft.

ITCZ: Extends from Pacific to north to northern tip of South America (lat 10 °N) thence to North Atlantic to African coast to about 20 °N. Thence to southern tip of Arabia across Africa.

Both South America, Africa from equator to about 25° - 30 °S lie under the influence of High pressure / Anticyclone. Thus frontal activity does not affect the area except over extreme southerly coastal regions, where frontal activity affects.

Winds

Between equator to 20°–30° S Winds will be mainly Southeasterly Trade winds, and at 40 °S Westerly Roaring forties.

Upper Winds

Between	January	July
10 °N–20 °S	Easterly upto 30 kt	Easterly upto 30 kt
40–50 °S	Westerly 50–60 kt	Westerly 55-65 kt
40–60 ° S	-	Westerly jet stream more than 100 kt

Tropopouse Height: About 56000 ft (17 km) throughout the year.

Freezing level

Between	January	July
10°N to 20°S	16000 ft (4.85 km)	14000 ft (4.3 km)
40°S	10000 ft (3.0 km)	8000 ft (2.5 km)
50°S	less then 8000 ft (2.5 km)	less then 5000 ft (1.5 km)

Note: The Atacama Desert

The Atacama desert is located in South America coast, between lat 18°S (Peruvican border) to lat 30°S (Santiago). The main feature of this desert is that it has medium and low clouds, mist often present but completely bereft of rain. It is the driest place on earth, there was no rain for 400 years. Climate of this place is dominated by the permanent High pressure area over Pacific and cold upwelling of sea water on the west coast of Peru.

36.6 Africa

Africa extends from 30°N to 35°S, and has two deserts, one in the north Sahara and the other Kalahari in the south. These are located in two sub-tropical high pressure regions.

During October to April: The northern east-west region is under the influence of Mediterranean climate, namely dry summer and comparatively humid winter.

The ITCZ/Equatorial Low meanders with the sun by about 15 to 20° latitudes (15° to 6° latitudes in the west and 25° in the east). In July its location is farthest north and in January it is farthest south, particularly inland and on east coarst. Frontal depressions move across the Mediterranean in west-to-east direction and affect the north coastal regions of North Africa.

January: The location of ITCZ over Atlantic 10 °N to inland Africa 10 °N, 20 °E, thence to south-south-east to 20 °S, 20 °E, thence to 20 °S, 30 °E, and into Indian Ocean along 20 °S (see Fig. 36.7 and 36.11).

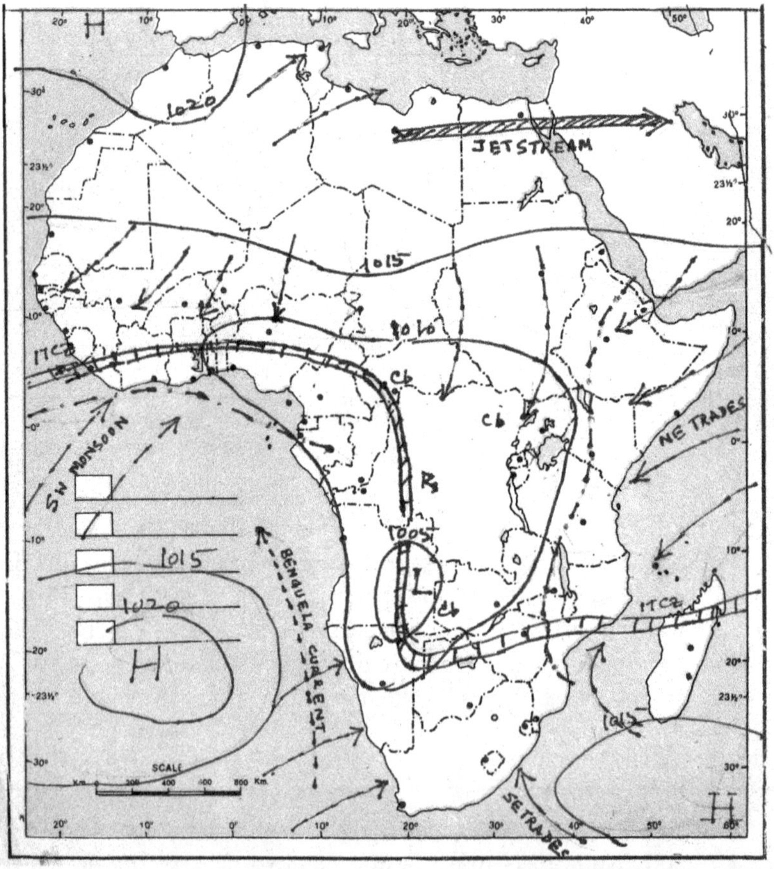

Fig. 36.7 January – Pressure distribution, Wind and ITCZ.

On western coast in South Atlantic southwest monsoon between 10 °S to 10 °N and south of 10 °S Bengula cold current.

On eastern coast Arabian Sea / Indian ocean

North of ITCZ – Northeast Trades and

South of ITCZ – Southeast Trades

Sub tropical jet stream lies along about 30 °N

July: The location of ITCZ on west coast 15 °N over Atlantic to 10 °N Gulf of Aden on east coast (see Fig. 36.8).

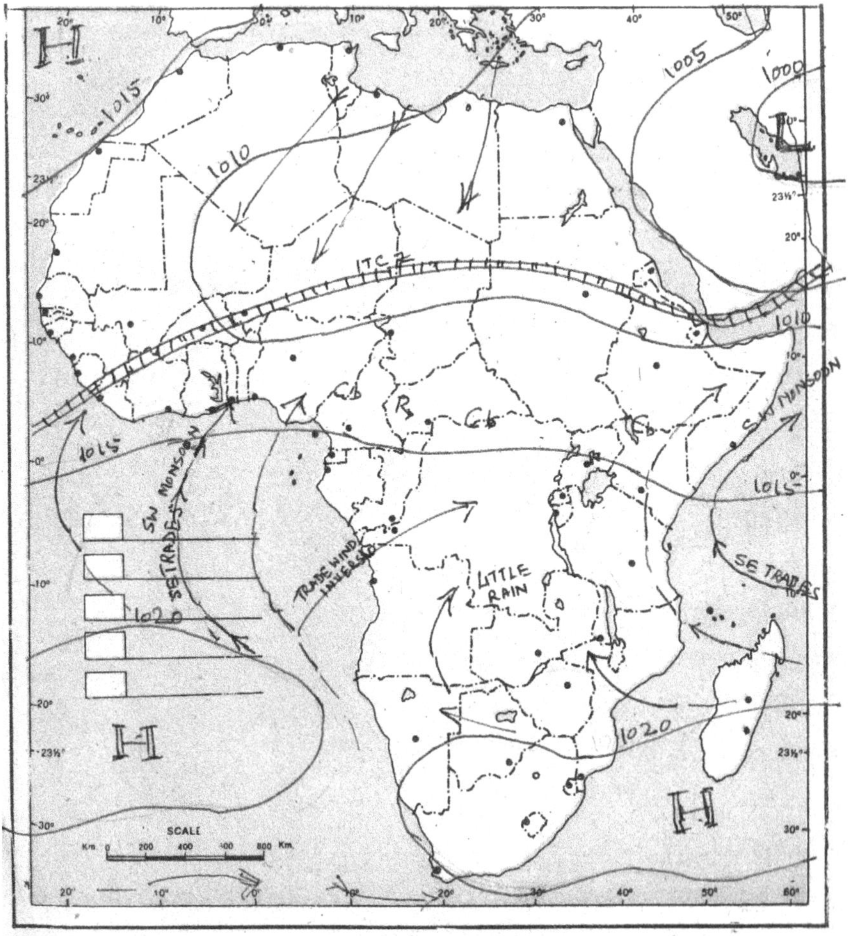

Fig. 36.8 July – Pressure distribution, Wind and ITCZ.

Over Atlantic: Southeast Trades south of 10 °S, and Southwest monsoon north of it. Trade wind inversion along west coast of Africa.

On Eastern Coast of Africa: Subsidence due to sub-tropical anticyclone up to Madagascar and Southeast Trades north of it. These become

Southwest monsoon after crossing equator on Indian Ocean / Arabian Sea.

Over African Land: South of ITCZ to equator will have squall lines, Cb cells and thunderstorm activity.

Upper Winds: Cairo to Johannesburg:

Equatorial region – easterlies. On either side of it light westerlies.

Jet stream:

January: Subtropical westerly jetstream located at 200 hPa level between 25-30° N, speed 100 kt.

July: Easterly jet stream located at 150 hPa level at 15 ^{O}N, speed 80 kt.

In the south of the route equator to Johannesburg

Light westerlies becoming moderate in July.

Tropopause Height

North equator - Cairo to equator - 55000 ft (16.8 km)

Equator to Johannesburg – 52000 ft (15.9 km)

Freezing level:

North equator - Cairo equator – 14000 to 16000 ft (4.3 km to 4.9 km)

Equator to Johannesburg – 16000 to 14000 ft (4.9 km to 4.3 km)

Thunderstorm Annual Frequency:

Kampala (located on equator) in Uganda – 242 days

Buma (Zaire) – 228 days

Kamambe (Rwanda) – 221 days

Calabar (Nigeria) – 215 days

Entebbe (located on equator) (Uganda) – 206 days

Mamfe (Cameroon) – 201 days

36.7 Mediterranean – Europe – Middle East

36.7.1 Mediterranean

Hot dry summer, mild, relatively wet winter. The climate is between temperate maritime and arid sub-tropical desert.

Mediterranean is a landlocked sea, surrounded by mountains baring in the southeast. Thunderstrorms, rainshowers are frequent from August, associated with cold fronts.

Sea surface temperature, generally warmer than air temperature (more than 2 ^{O}C) during winter.

Depressions that form on the lee of mountains Atlas, Alps, Pyrenees cause more rain in winter than Atlantic depressions, All depressions move eastward to Asia. In frontal zone temperature discontinuity in Mediterranean is of the order 12 to 16 ^{o}C in winter.

In winter, a combination of air masses interact namely Arctic (A), Polar continental / maritime (P_{c},/P_{m}) and Tropical continental / maritime (T_c/T_m) air masses interact.

In summer mainly Polarmaritime (P_m) and Tropical continental / maritime (T_c, T_m) air masses interact.

36.7.2 Central Europe

Northwest Europe gets maritime air, while eastern side gets continental Russian air.

In winter temperature decreases as we move eastward rather than northwards. Highest summer temperature occurs in interior Europe. Winds mostly westerlies. Winters are colder, most rivers get frozen during peak winter, when surface temperature in Central Europe records sub-zero (below 0 ^{O}C).

In summer, Central Europe, particularly north of Italy, Bulkan countries are hot. Eastern Scandinavia which is on Arctic Circle records high temperature about 37°C.

There is no dry season. Precipitation is spread out all through the year. February is the driest month under the influence of Siberian High subsidence of air.

In eastern Europe (Finland, Poland & Russia) experiences continental cold air in winter and warm dry air in summer (37°C).

Mediterranean to 60°E, covering Gulf to Iran: The region has temperature transition zone in the north and subtropical climate in the south. Mostly arid area.

In winter, Lows move from Mediterranean to north India across Persian Gulf, Afghanistan. In summer, it is rainless. Over Iraq, northwesterly dusty winds (called Shamal) are persistent.

In winter, westerly jet stream lies over Gulf and Arabia between altitude 35000 to 50000 ft (10.7 to 15.2 km).

In summer, low level northwesterly winds become light easterlies at 12000 ft (3.66 km).

During November to March: Siberian High lies over Asia and sub-tropical depressions (about 4 per month) move from Aegean to Black sea/ Caspian sea and their cold fronts and secondary lows move over to Pakistan and north India (where they are called Western Disturbances). Warm fronts moving from southeast bring widespread precipitation. Cold fronts are associated with thunderstorms, squalls (about three days in a month). Cold fronts sometimes bring in duststorms particularly over Egypt, Iraq (called Shamal) when visibility falls below 1000 m.

In summer (June-September): Generally clear skies prevail except along Black sea coast, where convective clouds, thunderstorms occur and occasionally duststroms. Winds mostly Northerlies.

Upper Winds: During winter westerlies, with subtropical jet stream (westerly) speed 110 to 140 kt.

During summer sub-tropical jet stream not seen over the area. Generally easterlies 15-30 kt observed

In winter Siberian High penetrates into eastern Europe.

The main pressure systems are:

Icelandic Low, Azores High, Seasonal High and Low over Asia and travelling polar front depressions which move west to east without any hindarance of mountain ranges. As said earlier, there will be interaction of Arctic maritime, polar continental, polar maritime, tropical continental and tropical maritime air masses.

36.7.3 Middle East

Tropopause Height

	Europe	Middle east	Mediterranean
January	35000 ft (10 – 7 km)	34000 ft–56000 ft (10.4 - 17.1 km)	40000 ft (12.2 km)
July	39000 ft (11.9 km)	34000 ft–56000 ft (10.4 to 17.1 km)	44000 ft (13.4 km)

Freezing level

	Europe	Middle east	Mediterranean
January (winter)	4000 ft (1220 m)	(8000–12000 ft) (2440–3660 m)	4000–8000 ft (1.2–2.45 km)
July (summer)	12000 ft (3660 m)	16000 ft (4.85 km)	14000–16000 (4.3–4.85 km)

Upper winds

(a) Europe

Generally westerlies but ENE or SE from Siberian High. Westerlies increase with height.

Polar front jetstream between 30,000–35000 ft (9.1 to 10.7 km), westerly 100 kt or more in winter but less speed in summer.

(b) Middle East

***Winter*:** Westerly sub-tropical jet stream – westerly 110–140 kt

***Summer*:** Sub-tropical jet stream not seen

Wind : Easterly/15-30 kt

(c) Mediterranean

***Winter*:** Westerly/80–120 kt during November to March

***Summer*:** Westerly/40 kt–70 kt. No jet stream

36.8 Asia–Indian Ocean

It is a very vast area, we will confine to the area from Arabian Gulf to Singapore. The dominant weather feature is Asian monsoon, ITCZ, tropical cyclones. Much has been described about these features in Monsoons, Cyclones and Indian climatology.

In winter Siberian High is well marked and established. Dry continental cold winds blow over Asian land mass but were prevented entry into India by the mighty Himalayas. In summer the Siberian High weakens and summer (or heat) Low lies over northwest India. This Low draws moist equatorial air from the Indian ocean. Like the land and sea breeze circulations, a reverse circulation to that of winter prevails in summer. This seasonal reversal wind is the main climatic feature. Thus the region has the climate of the combination of sub-tropical and Equatorial climate. As described in Indian climatology, the seasons are described as below.

1. December-February (Representative month January):
   Cold season or winter – Northeast Monsoon.
2. March-May (Representative month - April):
   Hot season or summer - Pre-monsoon.
3. June-September (Representative month-July):
   Rainy season –Southwest Monsoon.
4. October–November (Representative month – October):
   cool season – post-Monsoon.

It must be noted that there is some variation of weather near about Singapore.

Upper winds, temperatures for these representative months are given in appendix (Page 813-841) and also freezing levels over India (Page 810-813). The aviation climatological features briefly given below.

January: Sub-tropical High pressure lies over Siberia (centre on latitude 45 °N) in the northern hemisphere and another over Indian ocean (along latitude 35 °S). Tropical belt is under the influence of NE – monsoon.

ITCZ: Extends from African coast 15 °S, at long 75 °E near equator and near Australia latitude 15 °S. The common features associated with ITCZ are thunderstorms and squalls.

Freezing level:

Height varies between latitudes

30 °S to 18 °N	16000-14000 ft (4.9-4.3 km)
30 °S-45 °S	14000-8000 ft (4.3-2.4 km)
45 °S-60 °S	8000-2000 ft (2.4-0.6 km)
South 60 °S	surface (0 ft)
18-35 °N	14000-1000 ft (4.3-3.1 km)
North of 40 °N	surface

Tropical Tropopause Height

Between 20 °S-20 °N	56000 ft (17.1 km)
20 °N-40 °N	54000 ft (16.5 km)
20 °S-40 °S	51500 ft (15.7 km)

Polar Tropopause Height

Between 40 °N-60 °N	36000-30000 ft (11.0–9.1 km)
north of 60 °N	29000 ft (8.8 km)
40 °S-60 °S	38000 ft (11.6 km)
south of 60 °S	30000 ft (9.1 km)

Subtropical Jet Stream:

Between 25 °N to 40 °N	40000 ft (200 hPa)
	Westerly, speed 100-150 kt

Polar jet stream:

	30000 ft (300 hPa)
	Westerly, speed 60-90 kt

July: Sub-tropical High pressure area lies along latitude 30 °S in the southern hemisphere.

In the northern hemisphere it mergers with the polar High.

ITCZ/Low Pressure: Lies along lat 30 °N over India and adjoining Pakistan, which is monsoon seasonal low.

Tropical Asia and Indian ocean are under the influence of southwest Monsoon.

Freezing level

Between lat 25-30°N		18000 ft (5.5 km)
30-35°N and lat 25°N to equator		16000 ft (4.9 km)
35-60 °N	varies	16000-10000 ft (4.9-3.1 km)
60-75 °N	varies	10000-1000 ft (3.1-0.3 km)
North of 75 °N	on surface	(0 ft)
Between Eq-30 °S		varies16000-8000 ft (4.9-2.4 km)
30-60°S		varies 8000 ft to surface (2.4 km to surface)
South of 60 °S		on surface

Tropical Tropopause Height

55000 ft (16.8 km)

Subtropical Jet Stream

In southern hemisphere 40000 ft (200 hPa), westerly about 80 kt

Winds at 300 hPa or 30000 ft: 090°/15-30 KT

In northern hemisphere – Subtropical westerly jet stream weakens and moves to northern latitudes.

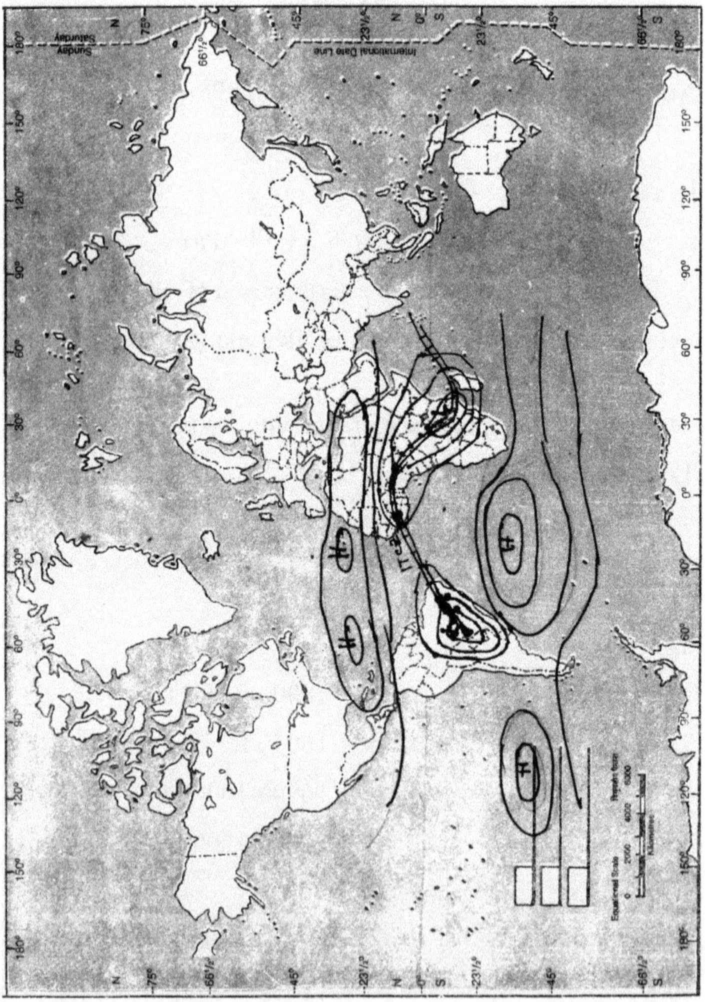

Fig. 36.9 Africa and South America-typical January day

Tropical Easterly Jet Stream: Active, lies along lat 13 °N at 46000 ft to 54600 ft (150 hPa-100 hPa), mean speed about 70 kt with temperature gradient below core is about 1°C/1 lat.

Temperature increases from North to South below the core, while above the core it is from South to North.

Note: Tropopause is the level at which there is an abrupt fall in lapse rate to less than 2 °C / km and this lapse rate maintained for subsequent kilometer.

Polar Tropopause: This occurs north of 70 °N (north of Polar front jet stream). Mean height 8 km. Polar tropopause diffuse with absence of marked inversion.

Extra tropical or Mid-latitude Tropopause: This occurs between 30 °N and 70 °N (that is between sub-tropical jet stream and polar jet stream. Mean height 11 km. In this type there is sudden change in lapse rate, at times becoming isothermal but there is no inversion.

Tropical Tropopause: This occurs between equator and the sub-tropical jet stream. Its mean height varies from 11 to 17 km. Tropopause is sharply defined. There is a marked inversion immediately above it. To the south of latitude 30 °N to equator, the tropical tropopause prevails throughout the year. To the north of 30 °N, the tropical tropopause observed during monsoon months. During winter sub-tropical jet stream appears over north India. At that time tropical tropopause is found in higher levels while sub-tropical tropopause if found in lower levels. Thus in north India we find two tropopauses, extra-tropical at 10 to 11 km and tropical at about 16 km. This is a common feature over northwest India in mid winter months.

36.9 Australia

North Australia is influenced by ITCZ, while the main land by Sub-tropical High pressure belt. Southern parts of Australia influenced by travelling depressions.

Australia is affected by the interaction of different air masses, particularly Polar maritime (modified), Southern maritime, Tropical maritime. Tasman / Pacific, Tropical Indian maritime / Equatorial maritime, Tropical continental and Sub-tropical continental air masses.

In January (southern summer), north Australia affected by ITCZ (see Fig 36.10).

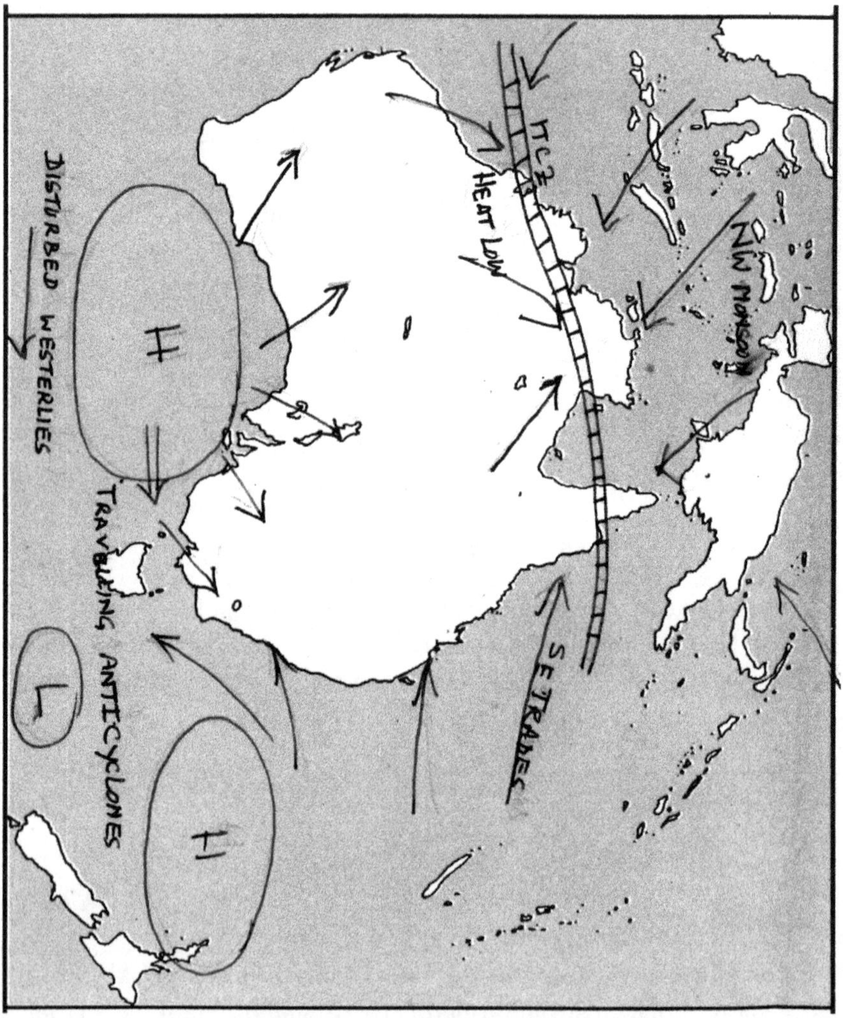

Fig. 36.10 A typical January day (Summer)

During July (winter) the cold fronts affect the southern Australia (see Fig. 36.11).

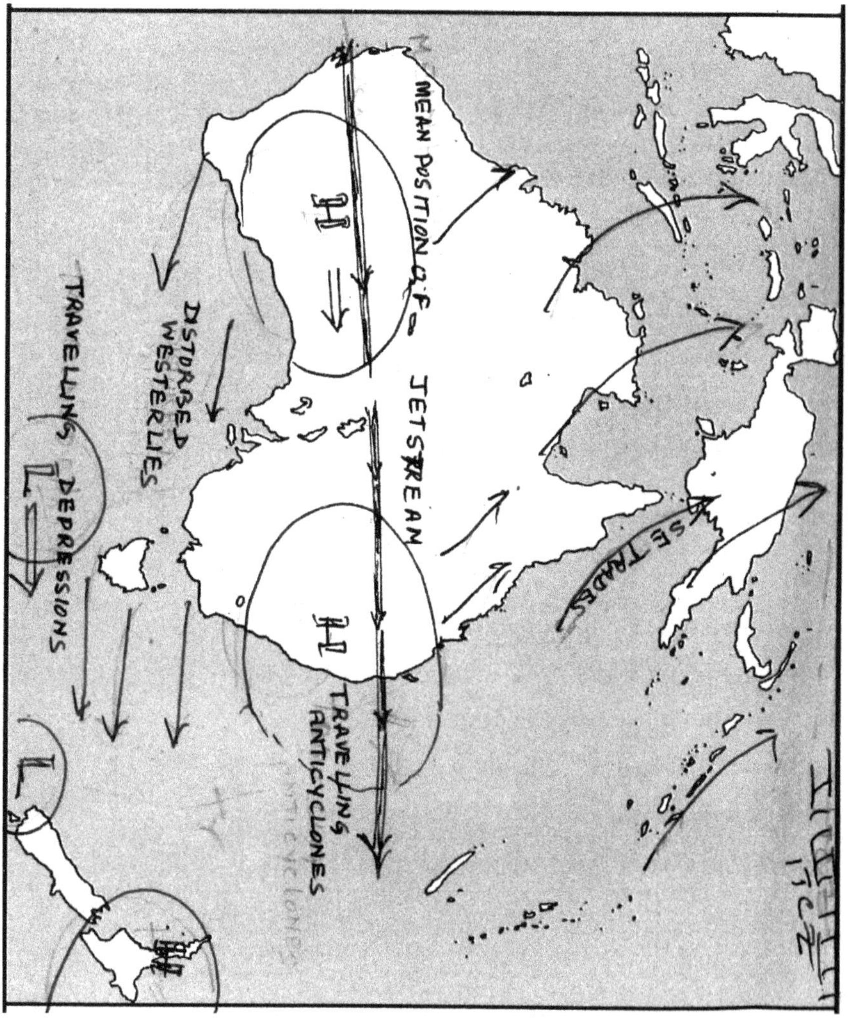

Fig. 36.11 A typical July day (Winter)

Tropical cyclones (locally called Willy-Willies) affect Australia north of 20 °S during December to April, however the peak period is January – February – March.

Accompanied by thunderstorms, line squalls gusting to 70 kt occur in southern Australia. In association with line squalls temperature drops by about 10-15 °C in a span of few minutes.

During November–March

(a) ***Near Darwin***: Surface wind northwesterly. Upper wind at Darwin mostly westerly. Axis of sub-tropical jet stream at 40000 ft (200 hPa) westerly speed 60-70 Kt. ITCZ moves up to 15 °S. Surface temperature 20-30 °C all through the year.

(b) ***Surface Wind at Sydney***: Variable

Upper wind at Sydney – Mostly westerlies.

Surface temperatures 19–29 °C.

During April-October

Surface wind at Darwin: light easterlies

Upper wind at Darwin: Mostly westerlies

Axis of sub-tropical jet stream at 40000 ft (200 hPa); position 22 °S, speed 100 kt.

Location of ITCZ–north of Singapore.

Surface temperature 20-32°C all through the year

Surface wind at Sydney: variable

Upper wind at Sydney: Mostly westerlies

Surface temperature: 9-15°C

Freezing level over Australia and New Zea land

	January (Summer)	**July (Winter)**
At 20°S	16000 ft (4.9 km)	1000 ft (3.1 km)
At 40°S	12000 ft (3.7 km)	6000 ft or less (1830 m or less)

Height of Tropopause over Australia and New Zealand

	January (Summer)	**July (Winter)**
At 20°S	55000 ft (16.8 km)	53000 ft (16.2 km)
At 40°S	38000-51000 ft (11.6-15.5 km)	34000-49000 ft (10.4-14.9 km)

36.10 Pacific

It is a very vast area and it will be discussed in important segments.

During October to March, depressions form off Chinese coast (while on land area winter anticyclones dominate) and along polar fronts depressions form to the east. The later affect the Hawaiian Islands (20°N). The ITCZ runs north of Australia, extends eastwards into ocean, thence northeastwards to NE of Fiji Islands and then runs west-east at about 5°N to South America. Subtropical High pressure belt lies along lat 25°S between Australia and South America.

From April to September the ITCZ lies entirely in the northern hemisphere. It emerges into sea in the vicinity of Korea, moves southeast wards to the Philippines and then eastwards across north Pacific.

36.10 (a) Singapore – Japan

This area is under the influence of Asiatic monsoons and the ITCZ, Typhoons, Siberian High / Anti-cyclone cold front depressions.

In winter (November-March): it is under the influence of NE monsoons

In April–August: lull in monsoon

In January–April: Crachin

The strongest mean wind speeds over the globe are found between Singapore – Japan

During winter, the sub-tropical jet stream (westerly) attains the speed of 250 kt, but 150 kt are frequent at latitude 30 °N. Westerly direction varies southwest to northwest.

During June–July months, it is on the periphery of Southwest monsoon. Monsoon depressions develop and move across Japan.

During September–October months northerly winds blow and northeast monsoon slowly ushers in.

Northern China and Japan experience cold-front weather.

In late summer and autumn, this area experiences typhoons from Pacific which develop in ITCZ.

Freezing Level

	Singapore	Japan
January	16000 ft (4.85 km)	3000 ft (915 m)
July	17000 ft (5.2 km)	10000 ft (3.1 km)

Upper winds
January: Over Japan: westerly /150 kt to 200 kt at 40000 ft (200 hPa)
Over Singapore: Light westerlies and light easterlies.
July: Subtropical Jetstream disappears; light westerlies in the north (Japan side) and light easterlies in the south (Singapore side).
Tropopause height:
Singapore: 56000 ft (17.1 km), throughout the year
Japan: Winter 35000 ft (10.7 km)
Summer 42000 ft (12.8 km)

36.10 (b) Singapore – Australia

Weather in this area is governed by equatorial trough/ ITCZ, Doldrums, Subtropical High pressure.

In this area ITCZ moves from about 15 °S in January to about 20 °N in July.

Northern coast of Australia at times affected by tropical cyclones. Thunderstorms, squalls, convective clouds are frequent.

January: Northeast Monsoon prevails near Singapore and Northwest Monsoon over northern Australia.

July: Under the influence of Sub-tropical High in the southern hemisphere southeast Trades prevail in the south, which become southwest monsoon north of the equator. Land, sea breezes are prominent.

Freezing level: At about 16000 ft (4.85 km) throughout the year.

Tropopause height: About 52000–56000 ft (15.9–17.1 km) throughout the year.

Upper winds: Mostly light easterlies (090°/15-30 kt) throughout the year except near Darwin.

Near Darwin: January - light waterlies

July – waterlies up to 60kt but no jet streams

36.10(c) Southwest Pacific (New Zealand)

January (summer): In Pacific ITCZ always lies north of the equator. However, there is a secondary convergence zone called South Pacific Convergence Zone (SPCZ) which extends from 20°S, 180°W to 25°S 145°W

Travelling anticyclones move eastward along lat 40°S. Thunderstorms are associated with convergence zone

July (winter): Northern hemisphere ITCZ is prominent, while the southern hemisphere SPCZ is very weak or absent

Traveling anticyclones observed along lat 35 °S.

SE-Trades extend up to equator in the southern hemisphere

Anticyclone lies over New Zealand, Tasman sea.

South of lat 45°S, disturbed westerlies. Tropical Cyclones and Tornadoes also occur in the area.

Upper winds:

In January

I at 18000 ft (500 hPa or FL 180) -Cyclonic westerly or westerly, strong between 40-50°S

II at 40000 ft (200 hPa or FL390)

(a) In the equatorial region: Easterlies in the west of date line and Westerlies in the east of date line

(b) In middle latitudes: Westerlies, between 45-50°S: Westerlies/ 50 kt

In July: Sub-tropical westerly jet stream at 200 hPa level, speed 150 kt, may go up to 250 kt.

Tropopause Height:

In January (summer) – 43000-48000 ft (13.1-14 km)

In July (winter) – 35000-33000 ft (10.7-10.1 km)

36.10 (d) North Pacific

(a) **January (Winter):** North Pacific and North Atlantic have similar weather. The area is affected by travelling depressions along polar front. However winds at higher levels are stronger than over Atlantic (Westerly more than 50 kt).

Polar jet stream mixes up in sub-tropical jet stream and have very strong winds. Polar front moves to lower latitudes and also polar front depressions.

Aleutian low is marked during October–March. Some polar front depressions from Mongolia and Siberia enter Pacific near Japan. At the rear of these depressions very cold continental air prevails.

Thus winter is dominated by Siberian anticyclone, polar cold winds. Siberian High is a source of NE-monsoon near Japan.

(b) **July (Summer):** In summer and rarely in winter, typhoons occasionally move into higher latitudes which affect North Pacific north of Aleutia.

Subtropical High pressure centre will be located between 25–35 °N.

Upper winds: Mostly westerly airflow. Westerly sub-tropical Jetstream is strongest in winter and located in a southerly position.

Polar front jetstream is found at 30000–40000 ft (9.1-12.2 km 300-200 hPa)

Polar front jet stream is found about 350-700 km behind coldfront and 700-1400 km ahead of warmfront between latitudes 30°–60°N and at right angles to the freshly occluded fronts. Speed 100–200 kt. Direction of jetstream is mostly westerly but frequently takes wave shape.

Sub-tropical jetstream is located between 20–35 °N in winter and 40–50 °N in summer, height 40,000 ft (200 hPa), speed about 150 kt.

When sub-tropical jetstream mixes up with polar front jetstream speed may reach to 300 kt

Height of Freezing Level:

	Winter (January)	Summer (July)
at 60 °N	surface (0 ft)	6000 ft (1830 m)
at 30 °N	8000 ft (2440 m)	16000 ft (4.9 km)

Tropopause height

	Winter (January)	Summer (July)
At 60 °N	29000 ft (8840 m)	36000 ft (11.0 km)
At 30 °N	37000–54000 ft (11.3–16.5 km)	43000–51000 ft (13.1–15.5 km)

Questions

1. Within the Arctic circle, winter lasts for five to six months and summer about three months. In this region around the winter solstice ... prevails, while at around summer solstice .... prevails.
   (a) 12 hrs darkness; 12 hrs day light
   (b) 18 hrs darkness; 18 hrs day light
   (c) 24 hrs darkness; 24 hrs day light
   (d) None above

2. In winter, polar depressions in north polar regions move:
   (a) Eastwards (b) Northeast wards
   (c) Westwards (d) Southwest wards
3. In winter, within polar circle, sea remains........
   (a) Liquid state (above 0 °C)
   (b) Partly liquid partly frogen
   (c) Permanently frogen
   (d) None above
4. In winter, Arctic troposphere is ... than ICAO ISA and in summer .... than CAO ISA.
   (a) Warmer; cooler
   (b) Cooler; close to ISA temperature
   (c) Cooler; warmer
   (d) Nearly ISA; cooler
5. In north polar region, all precipitation is in the form of :
   (a) Rain (b) Snow
   (c) Rain and snow (d) Virga
6. The areas of Greenland, Norway, Barents and Kara seas are ..... during summer than in winter
   (a) More clear skies (b) More cloudy
   (c) Completely clear sky (d) Over cast
7. Radiation fog over land in north polar region during .... and .... is peculiar to the Arctic.
   (a) Spring and summer; Ice fog
   (b) Autumn and winter; Ice fog
   (c) Summer and winter; snow
   (d) Spring and Autum; snow
8. In North Atlantic, in summer, High pressure centre is located about ... and in winter at about.........
   (a) 25°N; 20°N (b) 35°N; 30°N
   (c) 30°N; 25°N (d) 35°N; 25°N
9. In July, the NewYork, Boston areas get precipitation under the influence of:
   (a) Polar continental outflow and warm maritime air moving north
   (b) Polar front depressions, polar air depressions
   (c) ITCZ, polar air depressions
   (d) ITCZ, polar continental out flow

10. In winter, the extension of Siberian High lies over Scandinavia which affects the eastern areas of UK with...........
   (a) Warm maritime air   (b) Tropical cyclones
   (c) Severe cold winds   (d) Both (a) and (b)
11. Under the interaction of Cold Labrador current and warm Gulf Stream .... occurs over Newfoundland.......
   (a) Radiation fog; in summer
   (b) Advection fog; in winter
   (c) Advection fog; throughout the year
   (d) Radiation fog; in winter
12. In January, over north America, Polar front depressions move from ....to northeast wards and also polar air depressions form ....
   (a) NewYork; over central America
   (b) Canada; over Pacific
   (c) Florida; over oceans
   (d) Both (a) and (b)
13. In July, pacific coast of North America lies under the influence of ....and .... affects southern parts of Gulf of Maxico
   (a) Aleutian Low; ITCZ
   (b) Pacific High; ITCZ
   (c) Travelling depressions; Atlantic High
   (d) Both (a) and (c)
14. Tornadoes mostly occur in north America, during........... They are most frequent in...........
   (a) April-May; Texas, Okalhama, Kansas, Iowa, Alabama
   (b) Nov-Feb; Along west coast
   (c) July-Oct; along east coast
   (d) Jan; Mexico
15. Annual frequency of Tornadoes over North America is:
   (a) More than 100   (b) More than 500
   (c) Less than 1000   (d) More than 1000
16. In January, frontal depressions affect.....
   (a) North east South America
   (b) Central parts of South America
   (c) Southeastern parts of South America
   (d) Southwestern parts of South America

17. In July ITCZ extends over South America.........(to).........
    (a) 10°S to equator
    (b) Northern tip of South America (10°N) to North in Atlantic ocean to African coast 20°N
    (c) 10°N of South America to 20°S Africa
    (d) ITCZ is abscent
18. In July, over Atlantic off African coast, Southeast Trades extend up to .... and .... north of it
    (a) 10°S; Southwest monsoon
    (b) 20°N; Southeast monsoon
    (c) 20°S; Northeast monsoon
    (d) None above
19. Annual maximum frequency 242 days of thunderstorms occur at ....
    (a) Cherrafunji (India)
    (b) Mamfe, (Cameron) Africa
    (c) Enteble (Uganda) Africa
    (d) Kampala (Uganda) Africa
20. The characteristics feature of Mediterranean is ....
    (a) Wet summer, dry winter
    (b) Wet all through the year
    (c) Dry all through the year
    (d) Dry summer, relatively wet winter
21. In central Europe, winters are ... and most rivers ..... during peak winter
    (a) Milder; flow (b) Colder; frogen
    (c) Warmer; frogen (d) None above
22. In summer, central Europe is .... and eastern Scandinavia (which on Artic circle) records temperature about.........
    (a) Cool; – 5°C (b) Warm; + 20°C
    (c) Hot; + 37°C (d) None above
23. In winter, Lows/depressions from Mediterranean move to.... across....
    (a) Iran; Arabia
    (b) Iraq; Europe
    (c) North India; Persian gulf, Afghanistan
    (d) Arabia; Persian gulf

24. High pressure exists over all the worlds continents in ....

    (a) Summer (b) Winter

    (c) Spring (d) Autumn

25. In northern hemisphere Sub-tropical High pressure exists over Asia, centred about......... and another over Indian ocean at (latitude)...... in the southern hemisphere

    (a) Lat 45°N, Lat 35°S (b) Lat 30°N, Lat 45°S

    (c) Lat 60°N, Lat 45°S (d) Lat 30°N, Lat 30°S

26. The approximate mean heights of tropical, mid-latitude and polar tropopauses in summer in northern hemisphere are:

    (a) 56000 ft (17.1 km); 36000 ft (11.0 km); 26000 ft (8.0 km)

    (b) 60000 ft (18.3 km); 36000 ft (11.0 km); 20000 ft (6.0 km)

    (c) 46000 ft (14.0 km); 30000 ft (9.1 km); 20000 ft (6.0 km)

    (d) 26000 ft (8.0 km); 26000 ft (8.0 km); 26000 ft (8.0 km)

27. Over Asia, the Subtropical westerly jet steam in winter lies between lat.......... at altitude.......... speed.......

    (a) 30°N-50°N; 200 hPa (40000 ft); 80-120 kt

    (b) 25 N-40 °N; 200 hPa (40000 ft); 100-150 kt

    (c) equator-15°N; 100 hPa (54000 ft); 80-100 kt

    (d) 40°N-60°N; 300 hPa (3000 ft); 60-90 kt

28. Over Asia, polar jet stream exists at..........and speed............

    (a) 200 hPa; westerly 100–150 kt

    (b) 100 hPa; easterly 80–120 kt

    (c) 300 hPa; westerly 60–90 kt

    (d) 500 hPa; easterly 80 kt

29. Tropical easterly jet exists during June–September between Africa-Singapore at latitude...........height.......and speed..........

    (a) 13 °N; 54000 ft (100 hPa); 70 kt

    (b) 25 °N; 46000 ft (200 hPa); 80-120 kt

    (c) 5 °N; 3000 ft (300 hPa); 50 kt

    (d) Both (b) and (c) above

30. Polar tropopause occurs north of latitude......and its mean height..........It is diffused with absence of marked inversion
    (a) 60 °N; 46000 ft (200 hPa)
    (b) 40 °N; 46000 ft (200 hPa)
    (c) 70 °N; 26000 ft (8 km)
    (d) 40 °N; 54000 ft (100 hPa)
31. Extra–tropical or mid-latitude tropopause occurs between latitude...or between.......(jetstreams), mean height........
    (a) 30 °N and 70 °N; subtropical and polar jetseteams; 11 km
    (b) Equator and 25 °N; equatorial, and sub-tropical jetstreams; 16 km
    (c) North of 70 °N; Arctic circle and N-pole; 8 km
    (d) Both (b) and (c) above
32. In northern hemisphere, tropical tropopause occurs between....its mean height....
    (a) Tropic of Cancer and Arctic circle; 11 km
    (b) Arctic circle and North pole; 8 km
    (c) Equator and sub-tropical jetstream; 17-11 km
    (d) Both (a) and (b) above
33. Tropical cyclones that affect north Australia (north of 20 °S) are called...........
    (a) Cyclones (b) Typhoons
    (c) Baguios (d) Willy-willies
34. In winter (November-March), Singapore-Japan area is affected by............
    (a) North–east monsoons (b) South-west monsoons
    (c) Doldrums (d) SE-Trades
35. The strongest mean wind speeds over the globe are found.......
    (a) Over Australia
    (b) Over north Indian Ocean
    (c) Between Singapore-Japan
    (d) Between South America-Africa
36. During winter, the sub-tropical westerly jetsteam attains speed of 250 kt or more near.......
    (a) Sydney (b) Darwin
    (c) Japan (d) Colombo

CHAPTER 37

Aviation Weather Reports and Codes

Routine Observations and Reports

At aerodromes, routine meteorological observations are made throughout the 24 hours each day except otherwise agreed between meteorological authority, the appropriate Air Traffic Services (ATS) authority and the operator concerned. Such observations are made at intervals of one hour or at intervals of one half-hour as determined by regional air navigation agreement. At other aeronautical meteorological stations, such observations are made depending on the requirements of air traffic services units and aircraft operations.

Reports of routine observations are issued as routine reports to local air traffic services units as required and made available to the operators and other users at the aerodrome.

Routine reports are disseminated beyond the aerodrome of origin in accordance with regional air navigation agreement.

CHAPTER 37

Aviation Weather Reports and Codes

Routine Observations and Reports

At aerodromes, routine meteorological observations are made throughout the 24 hours each day except otherwise agreed between meteorological authority, the appropriate Air Traffic Services (ATS) authority and the operator concerned. Such observations are made at intervals of one hour or at intervals of one half-hour as determined by regional air navigation agreement. At other aeronautical meteorological stations, such observations are made depending on the requirements of air traffic services units and aircraft operations.

Reports of routine observations are issued as routine reports to local air traffic services units as required and made available to the operators and other users at the aerodrome.

Routine reports are disseminated beyond the aerodrome of origin in accordance with regional air navigation agreement.

37.1 Aviation Weather Codes

Aviation weather codes are used for exchange of observed and forecast weather data / information for aviation safety and efficiency. The codes have been evolved by WMO in consultation with ICAO. The present codes which have come into effect from 1 November 2001 (resembles that given in the Manual on codes – WMO No.306) and 9th edition of IMD codes (2004).

In order to ensure ready identification in transmission, in some cases, code name is included as a symbolic prefix in the code form. The following are the main code forms.

METAR	-	Aviation routine weather report (with or without trend forecast)
SPECI	-	Aviation selected special weather report (with or without trend forecast)
TAF	-	Aerodrome forecast
ARFOR	-	Area forecast for aviation
ROFOR	-	Route forecast for aviation
CODAR	-	Upper air report from an aircraft (other than weather reconnaissance aircraft)
WINTEM	-	Forecast upper wind and temperature for aviation

37.2 METAR/SPECI Code

We shall now give the code forms and explain the each term

METAR or SPECI CCCC YYGGggz (Auto)

ddd ff Gf$_m$ f$_m$ [KT or KMH or MPS] $d_n\, d_n\, d_n\, V\, d_x\, d_x\, d_x$

$$\left[\begin{array}{l} VVVV\,D_v \\ \quad \text{or} \quad V_x V_x V_x V_x D_v \\ CAVOK \end{array} \left[\begin{array}{c} R D_R D_R / V_R V_R V_R V_R \\ \text{or} \\ R D_R D_R / V_R V_R V_R V_R V V_R V_R V_R V_R \end{array}\right]\right]$$

$$W'W'\begin{bmatrix} N_sN_sN_s\,h_sh_sh_s \\ \text{or VV///} \\ \text{or SKC} \\ \text{or NSC} \end{bmatrix} T'T'/T'_dT'_d \quad Q\,P_HP_HP_HP_H$$

$$RE\,W'W'\begin{bmatrix} WSRWY\,D_RD_R \\ \text{or} \\ WSALLRWY \end{bmatrix}\left(WT_sT_s/SS\right)$$

$(R_R\,R_R\,E_R\,C_R\,e_R\,e_R\,B_R\,B_R)$

$$\begin{bmatrix} \begin{matrix} TTTTT \\ \text{or} \\ NOSIG \end{matrix} & TT\,GG\,gg & ddd\,ff\,G\,f_m\,f_m \begin{bmatrix} \text{KT or} \\ \text{KMH or} \\ \text{MPS} \end{bmatrix}\begin{bmatrix} VVVV \\ \text{or} \\ CAVOK \end{bmatrix} \end{bmatrix}$$

$$\begin{bmatrix} W'W' \\ \text{or} \\ NSW \end{bmatrix}\begin{bmatrix} N_sN_sN_sh_sh_sh_s \\ \text{or VV///} \\ \text{or SKC} \\ \text{or NSC} \end{bmatrix}$$

(RMK …)

Note: METAR is the name of the code for an aviation routine weather report.

SPECI is the name of the code for an aviation selected special weather report

A *METAR* report and a SPECI report may have a trend forecast appended.

The groups enclosed in brackets are used in accordance with regional or national decisions.

Trend forecast identified by a change indicator (TTTTT = BECMG or TEMPO as the case may be) or by the code word NOSIG.

General Regulations: The sequence of weather reported elements in METAR, SPECI are:

CCCC	:	Identification of station
YY GG gg Z	:	Time of observation including date in UTC

ddd ff G $f_m f_m$	:	Surface wind, including gusting
VVVV D_V	:	Visibility
R D_R D_R / $V_R V_R V_R V_R$	:	Runway visibility
$W' W'$	:	Present weather
$N_S N_S N_S h_S h_S h_S$	:	Clouds
CAVOK	:	Cloud visibility OK
$T'T' / T'_d T'_d$	:	Temperature and dew point (°C)
Q $P_H P_H P_H P_H$	:	Pressure (hPa)

Supplementary information (wind shear)

TTTTT	:	Change indicator for trend forecast.

Symbolic Letters of Code

METAR or SPECI – Code name will be included

CCCC – ICAO international four letter location indicator

Example

VOHY	:	Hyderabad;	VOTV: Trivandrum
VOMM	:	Chennai;	VOTP: Tirupathi
VOBG	:	Bangalore;	VABB: Mumbai
VOBP	:	Bhopal;	VAAH: Ahmadabad
VAID	:	Indore;	VANP: Nagpur
VILK	:	Lucknow;	VIDP: Delhi
VECC	:	Kolkata;	VEBS: Bhubneshwar
VEGT	:	Guwahati;	VEPT: Patna etc.
VGKK	:	London/Gatwick	OBBI: Bahrain
EGLL	:	London/Heathrow	SBGL: Riode Janeiro
ASSY	:	Sydney Australia (NSW)	ZBAA: Beijing/Capital
EHAM	:	Amsterdam	NZAA: Aukland
KJFK	:	Newyork / Kennedy	RPMM: Manila
RJTT	:	Tokyo (Honchu I.Japan)	EDDW: Breman
YSSY	:	Sydney	

CYMX : Montreal / Mirabel

EDDF : Frankfurt

HKNA : Nairobi/Jomo Kenyatta

LFPO : Paris/Orly

LFPG : Pairs/Charles de-Gaulle

Group YYGGggZ

YY Date of the month (UTC)

GG Valid time to the nearest whole hour UTC of the beginning of the forecast

GGggZ Time of observation in hours, minutes UTC followed by the letter Z

AUTO Inserted before wind group, indicating a report containing fully automated observations without human intervention.

Group dddffG$f_m f_m$

ddd: True wind direction, in degrees, rounded off to the nearest 10°, from which wind is blowing

ff : Wind speed in KT, KMH or MPS

$f_m f_m$: Maximum wind speed in KT, KMH or MPS

G : gusting (wind)

Group d_n d_n d_n v$d_x d_x d_x$

$d_n d_n d_n$: Extreme counter-clockwise direction of variable wind, reported wrt true north (wrt : with respect to)

V: Variable

$d_x d_x d_x$: Extreme clockwise direction of variable wind reported wrt true north

Group VVVVD_V

VVVV: Horizontal visibility at surface in meters

D_v: Direction of observation given by one or two letters indicators of the eight points of the compass (N, NE, E, SE, S, SW, W, NW)

CAVOK: Both visibility and cloud OK

Group $V_XV_XV_XV_XD_V$

$V_XV_XV_XV_X$: Maximum horizontal visibility at surface, in meters

D_v : Direction of maximum visibility

Group $RD_RD_R/V_RV_RV_RV_R$

R: Runway (number)

D_RD_R: Number of the runway to which the runway visible range given by $V_RV_RV_RV_R$ refers

$V_RV_RV_RV_R$: Runway visual range, in meters

Group $R\ D_RD_R/V_RV_RV_RV_R\ V\ V_RV_RV_RV_R$

R: Runway

D_RD_R: Runway number

$V_RV_RV_RV_R\ V\ V_RV_RV_RV_R$: Variable visibility; one minute mean maximum and mean minimum to be given instead of 10 minute mean.

Group $W'W'$: significant present weather

Group $N_SN_SN_Sh_sh_sh_s$: Category of cloud amount and base

$N_SN_SN_S$: Category of cloud amount - Few, Scattered, Broken or Overcast.

FEW : 1 to 2 oktas

SCT : 3 to 4 oktas

BKN : 5 to 7 oktas

OVC : 8 oktas

$h_sh_sh_s$: height of base of cloud layer or mass

000 less than 30 meters

001 30 m

002 60 m

....

....

008 240 m

010 300 m

etc.

098	2940 m	
099	2970 m	
100	3000 m	etc.
150	4500 m	
……………		
……………		
990	29700 m	
999	$\geq$ 30000 m	

VV/// Vertical visibility

When sky is obscure, vertical visibility to be reported.

Group $T'T'/T'_d T'_d$ Air temperature and dew point in whole degrees Celsius. 0.5 °C rounded up to next higher

eg. 9.5 → 10
3.3 → 03

Group Q $P_H P_H P_H P_H$: QNH value, in whole hectoPasclas

QNH observed value rounded down to the nearest whole hectoPascal.

QNH 995.6 → reported Q 0995

Supplementary phenomena

REW'W' : Recent significant present weather.

WSRWY $D_R D_R$:

WS: Wind shear

RWY: Runway

$D_R D_R$: Direction of runway / Runway number on which wind reported / expected

WSALL RWY: Wind shear reported / expected on all runways.

(W $T_S T_S$ / SS): Sea surface temperature and state of the sea.

$T_S T_S$: Sea surface temperature (°C)

SS: State of sea (code given-below)

SS code fig	Description	Height (meters)
0	Calm (glassy)	0
1	Calm rippled	0 – 0.1
2	Smooth (wave lets)	0.1 – 0.5
3	Slight	0.5 – 1.25
4	Moderate	1.25 – 2.5
5	Rough	2.5 – 4.0
6	Very rough	4.0 – 6.0
7	High	6.0 – 9.0
8	Very high	9.0 – 14.0
9	Phenomenal	more than 14.0

(R_RR_R E_RC_R e_Re_R B_RB_R): state of runway

R_RR_R	Runway designator
E_R	Runway deposits
C_R	Runway contamination extent
e_Re_R	Depth of the contamination
B_RB_R	The frictional coefficient/Braking

Code E_R (Runway Deposits)

E_R Code Fig.	Description
0	Clear or dry
1	Damp
2	Wet and water patches
3	Rime and frost covered depth normally less than 1 mm
4	Dry snow
5	Wet snow
6	Slush
7	Ice
8	Compacted or rolled snow
9	Frozen ruts or ridges
/	Type deposit not reported (i.e., due to runway clearance in progress)

C_R Code: Extent of runway contamination

C_R Code Fig.	Description
1	Less than 10% of runway contaminated (or covered)
2	11-25% of runway contaminated (or covered)
3, 4	Reserved
5	26-50% of runway contaminated (or covered)
6, 7, 8	Reserved
9	51-100% of runway contaminated (or covered)
/	Not reported (e.g. due to runway clearance in progress)

e_Re_R: Depth of deposit code.

e_Re_R code fig	Description
00	less than 1 mm
01	1 mm
02	2 mm
....	....
....	....
89	89 mm
90	90 mm
91	Reserved
92	10 cm
93	15 cm
94	20 cm
95	25 cm
96	30 cm
97	35 cm
98	40 cm or more
99	Runway or Runways non operational due to snow, slush, ice, large drifts or runway clearance but depth not reported.
//	Depth deposit operationally not significant or not measurable

B_RB_R: Friction coefficient / Braking action

B_RB_R code fig	Description
00	Friction coefficient 0.00
01	Friction coefficient 0.01
02	Friction coefficient 0.02
…………………… ……………………	
88	Friction coefficient 0.88
89	Friction coefficient 0.89
90	Friction coefficient 0.90
91	Braking action poor, frication coefficient $\leq$ 0.25
92	Braking action medium/poor, frication coefficient 0.26 – 0.29
93	Braking action medium, friction coefficient 0.3 – 0.35
94	Braking action medium/good, friction coefficient 0.36 – 0.39
95	Braking action good, friction coefficient $\geq$ 0.4
96 – 98	Reserved
99	Unreliable
//	Braking conditions not reported and / or runway not operational.

First three numbers, mean true wind direction over the last ten minutes.

ddd: True direction, in degrees, rounded off to the nearest 10°, from which wind is blowing (or will blow).

ddd: Shall normally be encoded as VRB only when the mean wind speed is 3 knot (2 MPS or 6 KMH) or less.

Gf_mf_m: When the maximum wind speed is more than 10 KT (5 MPS or 20 KMH) the maximum wind speed (f_mf_m) will be indicated immediately after dddff.

If the highest wind speed during ten minutes exceeds the mean speed by 10 KT or more then this is reported in G f_mf_m.

Ex: 25015 G 28 KT

Meteorological actual reports contain the information in the following sequence:

(i) Type of report
(ii) ICAO international four letter location indicator
(iii) Date and time of observation in UTC
(iv) Surface wind direction and speed
(v) Visibility
(vi) RVR (runway visual range) when applicable
(vii) Present (significant) weather
(viii) Cloud (amount, type only for CB, TCU at or near the aerodrome) and height of base
(ix) Temperature, dew point
(x) QNH (in hPa) (QFE when applicable).
Supplementary information
(xi) Recent significant weather (since last observation)
(xii) Wind shear
(xiii) Trend forecast for subsequent two hours
(xiv) State of runway.

37.3 Notes on Reporting

1. *Some of these items are conditional*:

If the wind direction during 10 minutes before the report changed by 60° or more and mean wind speed was 4 KT or more, then the two extreme directions are reported inserting V inside.

e.g., 090 18 G 30 KT 060 V 140, this means mean wind 090° (T), speed 18 KT, gusting to 30 KT and direction changed between 060° (T) and 140° (T) during 10 minutes before the report.

Incase of variable wind, direction *ddd* is encoded as VRB when the wind speed is 3 knots or less. Available wind at higher speeds will be reported if the variation of wind direction is 180° or more.

e.g., VRB 02 KT (variable with mean wind speed 2 knots); VRB 12 KT, (imples mean speed 12 knots) but the wind direction change is 180° or more). Incase of direction change is more than 60° but less that 180° and direction change is clockwise, this is reported as 300 V 080, meaning change of wind direction from 300° to 080°. Calm wind is reported as 00000 KT (or MPS, KMH).

dddff – is the mean wind direction and speed taken over a period of 10 minutes immediately preceeding the report prepared. The direction of the wind is reported in terms of degrees. During 10 minute period preceding the observation, if the maximum wind in gust exceeds the mean speed by 10 knots or more, this will be reported as G f_m f_m.

Ex: 27015 G 28 KT

2. As per ICAO, visibility is coded in meters and kilometers. When there is no marked variation in horizontal visibility it is reported VVVV and D_V is ignored.

Ex: 4000

Visibility variation is considered when the difference is 50% or more and visibility is less than 5000 m.

When horizontal visibility is different in different directions the minimum visibility is reported followed by D_v (in one or two letters like N, NE, S, W, NW etc)

Ex: 4000 NE (Visibility is 4000 m to the northeast)

When minimum visibility reported is less than 1500 m, while visibility in another direction is more than 5000 m, the group V_x V_x V_x V_x D_v is used.

Ex: 1400 S 6000 NE (Visibility is 1400 m in the south and 6 kilometers in the northeast)

Horizontal visibility is reported in the following steps:

(i) upto 800 meters, rounded down to the nearest 50 meters e.g. 680 m rounded 650 m.

(ii) between 800 and 5000 meters, rounded down to the nearest 100 meters (e.g 1250 m reported as 1200 m)

(iii) between 5000 meters upto 9999 meters, rounded down to the nearest 1000 meters [7660 m reported as 7000 m]

(iv) 9999 indicates visibility 10 km or more

CAVOK stands for simultaneous occurrence of:

(i) Visibility 10 km or more

(ii) No cloud below 1500 meters or below the highest minimum sector altitude,

(iii) no Cb

(iv) No significant weather phenomena.

3. Significant present and forecast weather W′ W′

Qualifier		Weather Phenomena		
Intensity or proximity (1)	Descriptor (2)	Precipitation (3)	Obscuration (4)	Other (5)
- light Moderate (No qualifier)	MI Shallow BC Patches	DZ Drizzle RA Rain	BR Mist FG Fog	PO Dust / sand whirls (dust devils)
+ Heavy (Well developed in the case of dust/sand whirls and funnel clouds) VC in the vicinity Ex: RA, light rain	PR Partial (covering part of aerodrome) DR Low drifting BL Blowing	SN Snow SG Snow grains IC Ice crystals diamond dust	FU Smoke VA Volcanic Ash DU Dust Wide spread	SQ Squalls Fc Funnel clouds Tornado or Water spout
	SH showers	PL Ice pellets	SA Sand	
RA, moderate rain	TS Thunderstorm	GR Hail	HZ Haze	SS Sand storm
+ RA, heavy rain	FZ Freezing (or super cooled)	GS Small hail and / or Snow pellets		DS Dust storm

4. RVR (Runway visual range) is reported only when visibility is less than 1500 meters.

Group R D_R D_R – Runway designator

ex: R 09 means Runway 09

R 27 / 1200 means RVR on runway 27 is 1200 m

Runway Designator **D_R D_R:** In case of parallel runways letters L, C or R used to indicate Left, Central or Right respectively.

The runway visual range (RVR) is a mean value and tendency (i) over a period of 10 minutes just before time of observation.

i = U for increasing

i = D for decreasing

i = N for no distinct change in RVR observed

Ex: R 27 / M 0150 N, meaning RVR on runway 27 is less than 150 meters and tendency no change.

R 09 / P 1500 U, meaning, on runway 09, the RVR is greater than 1500 m and tendency increasing.

Abbreviations: P = Positive side or greater than

M = Minus side or less than.

R D_R D_R / V_R V_R V_R V_R V V_R V_R V_R V_R i: Significant variations of RVR.

When RVR at a runway varies significantly during 10 minute preceding normal observation time, the one-minute mean extreme value (vary by 50 m or more or more than 20% of the mean value whichever is greater) the one minute mean minimum, mean maximum will be reported in that order in the form RD_R D_R/V_R V_R V_A V_R V V_R V_R V_R V_R i

5. W′ W′ present significant weather phenomena.

The abbreviations used in combinations

+SHRA	heavy shower rain
MIFG	shallow fog (visibility at 2 meters agl ≥ 1000 m) (fog: horizontal visibility < 1000 m)
DZFG	light drizzle and fog
SNRA	moderate snow and rain
FZFG	freezing fog
BLDU	blowing dust
BLSA	blowing sand
BLSN	blowing snow
TSGR	thunderstorm along with hail
TSGS	thunderstorm with small hail or snow pellets
DRSA	drifting sand (visibility less than 5000 m)
BR	mist (Visibility more than 1000 m but less than 2000 m)
FG	fog (visibility less than 1000 m)
VCFG	fog in the vicinity of aerodrome
BCFG	patches of fog in the vicinity of aerodrome (visibility less than 1000 m upto 2 meters height above ground level)
SQ	squall

6. N_s N_s N_s h_s h_s h_s cloud group

N_s N_s N_s	cloud amount
FEW	1 – 2 oktas
SCT	3 – 4 oktas
BKN	5 – 7 oktas
OVC	8 oktas
h_s h_s h_s	Height of the cloud base in flight level
SKC	clear sky
NSC	No significant cloud

Significant convective clouds to be reported are CB (cumulonimbus) and TCU (Cumulus congestus or Towering cumulus)

***Clouds Reported with the flowing Criteria*:**

1st group	–	lowest layer. FEW, SCT, BKN or OVC
2nd group	–	SCT, BKN or OVC
3rd group	–	BKN or OVC
Additional groups	–	Significant convective clouds CB/TCU

Ex: 1okta stratus base 500 ft

2 oktas cumulonimbus base 1000 ft

3 oktas cumulus base 1800 ft

6 oktas stratocumulus base 2500 ft

These respectively reported as

FEW 005, FEW010CB, SCT 018, BKN 025

The height of the base of cloud layer reported in steps of 100 ft (30 meters) upto 10000 ft (3000 meters). Above 10000 ft (3000 meters) in steps of 1000 ft (300 meters) in the form of h_s h_s h_s.

At mountain station, when cloud base is below station level (that is on ground and above) the cloud group reported as N_s N_s N_s ///

Ex: SCT ///

When sky is obscured vertical visibility VV /// reported in lieu of N_s N_s N_s h_s h_s h_s

7. $T'T' / T_d'T_d'$ (Temperature, dew point)

Rounded to the nearest whole degree Celsius. 0.5 °C rounded up to next higher Celsius degree

Ex: $T'T' = 9.5^{\circ}C$; $T_d'T_d' = 3.3^{\circ}C$ reported as 10/03

–9 °C reported as M09

–0.5 °C reported as M00

8. $QP_H P_H P_H P_H$ (pressure QNH)

QNH rounded down to the nearest whole hPa

Ex: QNH996.8 hPa reported as Q 0996

QNH 1002.3 hPa reported as Q 1002

9. Supplementary information groups

$$R\,E\,W'\,W' \begin{bmatrix} W\,S \;\; R\,W\,Y\,D_R\,D_R \\ \text{or} \\ W\,S \;\; \text{ALL RWY} \end{bmatrix} \qquad (W\,T_S\,T_S\,/\,S\,S)$$

$(R_R\,R_R\,E_R\,C_R\,e_R\,e_R\,B_R\,B_R)$

For international dissemination, recent significant weather phenomena of operational significance reported, like wind shear in the lower layers.

Recent weather phenomena (RE WW) means, weather phenomena observed since last routine report or last hour but not at the time of observation, up to three groups of recent weather phenomena can be given.

Recent weather phenomena includes:

- Freezing precipitation
- Moderate or heavy drizzle, rain or snow
- Moderate or heavy ice pellets, hail, small hail and / or snow pellets
- Moderate or heavy blowing snow (including snow storm)
- Sandstorm or duststorm
- Thunderstorm
- Funnel clouds (Tornado or Water spout)
- Volcanic ash

Wind Shear: In the lower layer $\left[\begin{array}{l} \text{W S RWY } D_R D_R \\ \text{or} \\ \text{W S ALL RWY} \end{array}\right.$

Wind shear along the take-off or approach path between runway level and 500 m (1600 ft) agl.

($W\ T_sT_s/SS$) Sea surface temperature and the state of the sea.

State of Runway ($R_R R_R E_R C_R e_R e_R B_R B_R$)

$R_R R_R$ runway designator

E_R runway deposits

C_R extent of runway contamination

$e_R e_R$ depth of deposits

$B_R B_R$ the friction coefficient / braking action

Instead of state of runway group, abbreviation SNOCLO used when aerodrome is closed due to extreme deposit of snow.

37.4 Landing Forecast

Landing forecasts are prepared either in the form of self contained forecast or in the form of trend type forecast.

A *self-contained landing forecast* consists of a concise statement of the expected meteorological conditions at the aerodrome (concerned) and contains any or all of the elements – surface wind, visibility, significant weather and cloud. The period of validity is two hours from the time of issue.

Trend Forecasts

Trend-type landing forecasts issued in India.

(i) Trend forecast given in a routine METAR or SPECI, consists of expected significant changes in Wind, Visibility, Present Weather and Clouds.

(ii) The period of validity of trend forecast appended to routine report is two hours from the time of METAR or SPECI.

(iii) *Change indicators*: BECMG or TEMPO.

These are used with abbreviations FM, TL or AT followed by a time group. AT is not used with TEMPO.

(a) BECMG (becoming) describes expected changes in meteorological parameters to reach or pass through specified values.

(b) TEMPO (temporary) describes anticipated changes in meteorological parameters to reach or pass specified values which last for less than one hour in each instances and in totality cover less than one-half of the period of the forecast

(c) NOSIG is given when no significant change is anticipated in all the elements (viz wind, visibility, present weather and clouds) during the next two hours.

(d) If trend is attached to METAR or SPECI of 0900Z the validity period is 0900Z to 1100Z.

The use of FM(from), TL(till) and AT (particular time).

AT (particular time) when used in the trend have the following meanings:

Change Groups	Time of commence of change	Time of completion of change
BECMG FM0930 TL 1030	0930	1030
BECMG FM 0930	0930	1130
BECMG TL 1030	0900	1030
BECMG AT 1030	1030	1030
BECMG (without any time group)	0900	1100
	(or any time between 0900 and 1100)	

TEMPO FM 0930 TL 1030: Temporary fluctuations during the period 0930 to 1030Z.

TEMPO FM 0930: Temporary fluctuations during the period 0930 to 1100Z

TEMPO TL 1030: Temporary fluctuations during the period 0900 to 1030Z

TEMPO (without any time group): Temporary fluctuations during the period 0900 to 1100Z

The "PROB" group is not used in the trend forecast.

EX: Expected surface wind 250° at 35 KT with maximum speed (gusts) to 50 KT throughout the period of trend forecast is given as TEMPO 25035 G 50 KT

37.5 Significant changes that form the Criteria for Inclusion in the Trend Forecast

(a) ***Wind*:**

(i) Change in mean wind speed by 10 KT or more

(ii) Change in wind direction by 60° or more with mean speed remaining 10 KT or more either before and / or after the change

(b) ***Visibility*:** When visibility is expected to reach or pass through 5000 M, 3000 M, 1500 M, 800 M, 600 M, 350 M, or 150 M.

In case of significant change (deterioration in visibility or improvement) the causing phenomena is indicated.

(c) ***Present Weather*:** Onset, cessation or change in the intensity of the following significant weather phenomena or combinations.

(i) Freezing Precipitation

(ii) Freezing fog

(iii) Moderate or heavy precipitation (including showers)

(iv) Low drifting dust, sand or snow

(v) Blowing dust, sand or snow (including snowstorm)

(vi) Duststorm

(vii) Sandstorm

(viii) Thunderstorm (with or without precipitation)

(ix) Squall

(x) Funnel cloud (Tornado or Waterspout)

(xi) Other significant weather phenomena if they are likely to cause significant change in visibility.

End of significant weather is indicated by NSW (No significant weather).

Ex: Cessation at 1630 UTC of significant weather, such as thunderstorm indicated in the form

"BECMG AT 1630 NSW"

Ex: Temporary moderate freezing rain between 0300 and 0430 UTC indicated as

"TEMPO FM 0300 TL0430 FZRA"

(d) ***Clouds*:**

(i) When the height of the base of lowest cloud layer of BKN (5-7 oktas) or OVC (8 oktas) reach or pass through 450 M (1500 ft), 300 M (1000 ft), 150 M (500 ft), 90 M (300 ft), 60 M (200 ft), 30 M (100 ft).

(ii) When the amount of cloud layer below 450 M (1500 ft) changes from to or reverse as shown below.

$$\left.\begin{matrix}\text{BKN}\\ \text{or}\\ \text{OVC}\end{matrix}\right] \begin{matrix}\rightarrow\\ \leftarrow\end{matrix} \left[\begin{matrix}\text{SCT}\\ \text{FEW}\\ \text{SKC}\end{matrix}\right.$$

A change to clear sky indicated by SKC

Ex: A forecast rapid increase in stratus cloud at 1130 UTC from SCT to OVC with base 1000 ft indicated as BECM AT 1130 OVC010

Ex: When no CB and no cloud below 1500M (or the highest minimum sector altitude whichever is greater) are forecast and CAVOK and SKC are not appropriate the abbreviation "NSC" (no significant cloud) is used.

(e) In trend forecasts, which are appended to METAR or SPECI, it should be in the coded form.

Significant changes when expected in any of the elements as per above criteria, change indicators used for TTTTT are BECMG or TEMPO.

The time group GGgg is preceded by one of the letters TT by FM (from), TL (till) or AT (at).

The change indicator BECMG is used to describe expected changes in meteorological elements which reach or pass specified threshold criteria. All these will be clear with the model (fictitious) examples

Ex:

METAR	YUDO	221530Z	27010KT	0800
R09/1000U	FGDZ	SCT008	OVC020	18/17
Q1016	BECMG	TL1600	1000 FG	BECMG
AT1700	9999	NSW		

Decoded

Routine meteorological report for Doulon/international (fictitious location), issued on 22nd of the month at 1530 UTC.

Surface wind (w/v): 270° / 10 KT

Visibility: 800 M

RVR for runway 09 = 1000 M with upward tendency

Present weather: Fog and drizzle

Cloud: (SCT) 2-3 oktas base 800 ft

(OVC) 8 oktas base 2000 ft

Temperature / Dew point: 18°C/17°C

QNH: 1016 hPa

Trend forecast for next two hours (upto 1730 UTC)

Becoming by 1600 UTC visibility 1000 M in fog and at 1700 UTC becoming 10 km or more.

Significant weather forecast: Nil

Ex: METAR EGHI 271250Z 23015G28KT 200V270 6000 SCT010 BKN020 18/16 Q1008

Decoded:

Routine weather report.

Station ICAO indicator: EGHI

Date/time of observation (T00): 27th day of the month; T00: 1250 UTC

Wind/Velocity (W/V): 230° (T) / 15KT gusting to 28 KT

Wind direction varied between 1240-1250 UTC from 200°(T) to 270°(T)

Visibility: 6000 M

Clouds: 3 – 4 oktas base 1000 ft above aerodrome elevation

5 – 7 oktas base 2000 ft above aerodrome elevation

Temperature / Dew point: 18 °C / 16 °C

QNH: 1008 hPa

Ex: METAR EGPD 200950 Z 24015 KT 9999
FEW020 SCT040 BKN090 16/13
Q 0996 TEMPO 6000 BKN 015

Decoded:

Routine Weather Report

ICAO station indicator:	EGPD
DATE:	20th day of the month, time of observation 0950 UTC
W / V:	240° (T) / 15 KT
Visibility:	10 km or more
Cloud:	1 – 2 oktas, base 2000 ft above aerodrome elevation
	3 – 4 oktas, base 4000 ft above aerodrome elevation
	5 – 7 oktas, base 9000 ft above aerodrome elevation
Temperature / dew point:	16 °C / 13 °C
QNH	996 hPa

***Trend forecast*:** Temporary phenomena valid for next two hours covering less than one hour.

Visibility deteriorate to 6000 M and

Cloud will be 5 – 7 oktas base 1500 ft above aerodrome elevation.

Ex: METAR EDDF 260850Z 26010 KT CAVOK 18/12 Q 1012 NOSIG.

Decoded: ICAO station indicator EDDF

Date / time of observation: 26th day of the month at 0850 UTC

W/V : 260° / 10 knots

CAVOK: Visibility 10 km or more

No cloud below 1500 M (5000 ft) above aerodrome elevation

No cumulonimbus cloud and no significant present weather (i.e., no precipitation, thunderstorm, shallow fog or drafting snow).

Temperature Dew point: 18 °C/12 °C

QNH: 1012 hPa

Trend forecast: No significant change expected during next two hours (upto 1050 UTC)

Ex : METAR LIPZ 101050 Z 00000 KT 1300 R09 / P1500 U

BR FEW 005 SCT 008 BKN 012 16/15 Q 1020

Decoded: Routine weather report

ICAO station indicator LIPZ

Date/time of observation: 10th day of the month at 1050 UTC

W/V: Calm

Visibility: 1300 Meters

RVR: Runway 09 RVR greater than 1500 M and increasing tendency

Present weather: Mist

Cloud: 1-2 oktas base 500 ft

3-4 oktas base 800 ft

5-7 oktas base 1200 ft

Temperature/Dewpoint: 16 °C/15 °C

QNH: 1020 hPa

(No trend forecast given)

Ex: METAR CYWG 251800Z 27006G13MPS 1000 R36/1200D SNBLSN BKN 009 OVC050 MO4/MO6 Q 1006 REFZRA WS RWY 36

Decoded:

Routine weather report

ICAO station indicator CYWG (Winni peg)

Date/time of observation : 25th day of the month at 1800 UTC

Surface wind (w/v): 270°/06 MPS gusting to 13MPS

Visibility: 1000 Meters

RVR: Runway 36 RVR 1200 Meters, decreasing

Present weather: Moderate snow, snow blowing

Cloud: 5 – 7 oktas base 900 ft

8 oktas base 5000 ft

Temperature/Dew point: Minus 04 ^{O}C/Minus 6 ^{O}C

QNH 1006 hPa

Recent weather Freezing rain

Wind shear reported on runway 36

Ex: METAR ULLI 060650Z 12012KT 4000 + RA BKN008 OVC 012 06/05 Q0996 NOSIG RMK ALL RWYS CLEAR AND DRY 52-95 PCT LESS THAN1MM BA72

Decoded

Routine weather report

1CAO station indicator: ULLI

Date/time of observation: 6th day of the month, at 0650 UTC

Wind (w/v): 120° / 12 knots

Visibility: 4000 Meters

Present weather: Heavy rain

Cloud: 5-7 oktas base 800 ft

8 oktas base 1200 ft

Temperature/Dew point: 06°C / 05°C`

QNH: 996 hPa

Trend forecast for next two hours (up to 0850 UTC): No significant change

Remarks: All runways clear and dry, contamination with coefficient of friction between 0.52 and 0.95.

Braking action good.

Less than 1 mm of water standing on runway.

37.6 Criteria for Issuing SPECI

Special observations and special reports:

A list of criteria for special observations are established by Meteorological Authority in consultation with the appropriate Air Traffic Services Authority, Operators and others concerned. The list includes:

(i) those values which correspond with Operating Minima of the operators using the aerodrome.

(ii) those values which satisfy other local requirements of Air Traffic Services Units and of the Operators.

(iii) an increase in air temperature of 2 °C or more that given in the latest report.

(iv) the available supplementary information concerning the occurrence of significant meteorological conditions in the approach and climb-out areas.

[*Significant meteorological conditions*:

Cumulonimbus cloud, thunderstorm, moderate or severe turbulence, windshear, hail, severe squall line, moderate or severe icing, freezing precipitation, severe mountain waves, sandstorm, duststorm, blowing snow or funnel cloud (Tornado or Water spout)]

(v) those values which constitute criteria for selected special reports.

Special reports are made available to ATS Units, Operators and other users at the aerodrome.

SPECI – is a selected special report prepared in accordance with criteria (which is given below), and intended primarily for dissemination beyond the aerodrome of origin.

SPECI *Criteria*: SPECI issued for any of the following significant changes.

(a) *Surface wind*:

(i) When the mean surface wind direction has changed by 60° or more with respect to (w.r.t) its latest report, the mean wind speed before and/or after the change being 10 knots or more.

(ii) When the mean surface wind speed has changed by 10 knots or more from that given in the latest report.

(iii) When the variation from the mean surface wind speed (gusts) has increased by 10 knots or more from that given in the last (latest) report, the mean speed before and/or after the change being 15 knots or more.

EX: METAR VOMM 060610Z 24010 KT 7000 SCTO20 SCT100 33/14 Q1008 BECMG TL 0630 29025 KT

SPECI VOMM 060635Z 27025 G 40 KT 7000 SCT020 SCT100 34/14 Q1008 BECMG FM0740 23010 KT

(b) ***Visibility*:** When visibility changes to or passes : 5000 M, 3000 M, 1500 M or 800 M

(c) RVR (Runway visual range)

When RVR changes to or passes 800 M or 600 M.

In addition, SPECI is issued when RVR changes to or passes 350 M 150 M if the minimum recordable value of RVR is less than 350 M.

These SPECIs are issued where instrumental recording facilities of RVR available

***EX*:**

(i) METAR VABB 200040Z 00000KT 6000
SCT012 25 / 23 Q1009 TEMPO TL 0210 2000 BR

(ii) SPECI VABB 200110Z 00000KT 3000 HZ
SCT012 25/23 Q1010 TEMPO TL0210 1200 BR

(iii) SPECI VABB 200125Z 00000KT 1200 R 27/P1500
BR SCT012 25/23 Q1010 BECMG 0230 4000

(d) ***Present weather*:** when the onset, cessation or change in intensity of any of the following weather phenomena or combination thereof occurs:

(i) freezing precipitation
(ii) freezing fog
(iii) moderate or heavy precipitation (including showers thereof)
(iv) low drifting dust, sand or snow
(v) blowing dust, sand, or snow, (including snowstorm)
(vi) duststorm
(vii) sandstorm

(viiii) thunderstorm (with or without precipitation)
(ix) squall
(x) funnel cloud (tornado or waterspout)

EX: SPECI VIDP 250945Z 27015KT 3000 TSRA SCT025 SCT030 CB 28/27 Q 1005 BECMG NSW

METAR VECC 221050Z 18010KT 6000 SCT025 SCT030CB 26/25 Q1008 TEMPO TL1200 15030 KT 4000 + TSRA

SPECI VECC 221115Z 16025G35 3000 + TSRA SCT025 SCT030CB 26/25 Q1008 BECMG FM1150 18010KT 6000 SCT025 SCT040

(e) Cloud

(i) when the height of base of the lowest cloud layer of BKN or OVC extent changes to or passes 30, 60, 150, 300 or 450M (100, 200, 500, 1000 or 1500 ft)

(ii) When the amount of cloud layer below 450 M (1500 ft) changes:

- from SKC, FEW or SCT to BKN or OVC or
- from BKN or OVC to SCT, FEW or SKC

(iii) when the sky is obscured and the vertical visibility changes to or passes through 30, 60,150 or 300 M (100, 200, 500 or 1000 ft)

Ex: METAR VOTP 250240Z 27005 KT 6000

BKN 012 OVC 100 24/22 Q1008

TEMPO RA BKN008 OVC 080

SPECI VOTP 250300Z 27005 KT 6000

RA BKN008 OVC100 24/22 Q1008 TEMPO + RASH

SPECI VOTP 250340Z 27005 KT 6000 BKN 015 OVC120 24/21

Q 1009 BECMG NSC

Note: ***RVR related abbreviations***: Runway visual range is observed from more than one location along the runway, the value representative of the touchdownzone be given first followed by the values representative of the mid-point and stop-end. The locations for which these values are representative should be indicated as "TDZ" (touch down) "MID" (middle), and "END" (stop-end) respectively.

Ex: "RVR RWY 09 TDZ 600 M MID 500 M END 400 M"

CHAPTER 38

Aviation Weather Forecasts and Codes

Owing to the variability of meteorological elements in space and time, to limitations of forecasting techniques and to limitations caused by the definitions of some of the elements, the specific value any of the elements given in a forecast should be treated as the most probable value which the element is likely to assume during the period of the forecast. Similarly, when the time of the occurrence or change of an element is given in a forecast, this time should be treated as the most probable time.

The issue of a new forecast by a meteorological office, such as routine aerodrome forecast (TAFOR) shall be understood to cancel automatically any earlier forecast of the same type given to the same period and for the same time (period of validity) or part thereof.

38.1 Aerodrome Forecast

An aerodrome forecast is prepared by the meteorological office designated by the meteorological authority concerned.

An aerodrome forecast, in general, consists of a concise statement of the expected meteorological conditions at an aerodrome during a specified period. It includes surface wind, visibility, weather and cloud, and expected significant changes to one or more of these elements during the period. TAF is a forecast of terminal weather conditions encoded (in IMD written on T-10 form).

Code form

TAF CCCC YYGGggZ $Y_1Y_1G_1G_1G_2G_2$

ddd ffG $f_m f_m$ (KT or KMH or MPS) [VVVV or CAVOK

$$\begin{bmatrix} \text{W'W'} \\ \text{or NSW} \end{bmatrix} \begin{bmatrix} N_sN_sN_sh_sh_sh_s \\ \text{or VV ///} \\ \text{or SKC (or NSC)} \end{bmatrix} \quad \text{PROB } C_2\,C_2 \qquad G\,G\,G_e\,G_e$$

$$\begin{bmatrix} \text{TTTTT } GGG_eG_e \\ \text{or} \\ \text{TT GG gg} \end{bmatrix} \quad \left(\text{TXT}_F\text{T}_F / \text{G}_F\text{G}_F\text{Z} \quad \text{TN T}_F\text{T}_F / \text{G}_F\text{G}_F\text{Z}\right)$$

General Regulations (Code Symbolic letters)

TAF is the code name of an aerodrome forecast. It is to be included at the beginning of an individual aerodrome forecast. In case of meteorological bulletin, which may consists of more than one aerodrome forecast the code name TAF is included at the beginning of the text of the bulletin. However in a bulletin the code name TAF is omitted.

CCCC : ICAO international four-letter location indicator

YYGGggz (group)

YY : Day of the month (UTC) with 01 indicating first day, 02 the second day etc.

GG gg Z: Time of forecast in hours and minutes UTC followed by the letter Z as an abbreviated indicator of UTC.

$Y_1Y_1G_1G_1G_2G_2$ (group)

Y_1 Y_1 : Day of the month of the beginning of the period of validity

G_1 G_1 : Time of commencement of period of forecast in whole hours UTC

$G_2 G_2$: Time of ending of period of forecast in whole hours UTC.

(i) When the period of forecast ends at midnight $G_2 G_2$ will be encoded 24.

(ii) When the period is between 25 to 48 hours after $G_1 G_1$, $G_2 G_2$ encoded by adding 50 to the time of ending of the period of forecasts

dddffG $f_m f_m$ (group) [Also see in METAR code]

ddd : surface wind true direction, in degrees, rounded off to the nearest 10°, from which wind will blow.

If the wind speed in less than 3 KT, ddd is encoded as 'VRB'

ff : wind speed in KT, KMH or MPS

For wind speed of 100 units or more, the exact number of wind speed units will be given in lieu of the two figure code ff or $f_m f_m$.

When the wind speed is 100 Knots or more the groups ff and $f_m f_m$ are preceded by the letter indicator P and reported as p99 KT (or P 49 MPS or P 199 KMH)

$Gf_m f_m$: Maximum wind speed in gusts in KT, KMH or MPS

Wind speed less than 1 Knot indicated by "(CALM)" 00000 KT

VVVV or CAVOK Group

VVVV : Horizontal visibility at surface in meters, in increments (or steps) of 50 meters up to 500 M, in increments of 100 M between 500 to 5000 M, in increments of 1000 M between 5000 upto 9999 M with 9999 indicating visibility of 10 Km or more.

If the value is between two increments, it is rounded off downward to the lower of two increments

Ex: A visibility of 370 M is reported as 0350,

a visibility of 580 M is reported as 0500,

a visibility of 5700 M is reported as 5000,

CAVOK: When it is expected that the following conditions will apply simultaneously.

(a) Visibility 10 km or more,

(b) no cloud below 1500 M (5000 ft) or below the highest minimum sector altitude, whichever is greater,

(c) no Cumulonimbus and

(d) no significant weather phenomena.

W′W′ *or NSW (group)*: W′W′ significant forecast weather. This is same as given in METAR code.

NSW : No significant weather.

N_s N_s N_s h_s h_s h_s or V V /// or SKC (or NSC) (cloud group):

N_s N_s N_s h_s h_s h_s: The order of inclusion of the groups will be lower to higher levels. The height of the base of forecast cloud layer (mass) will be in units of 30 meters (100 ft) in the form h_s h_s h_s.

Types of forecast clouds except CB/TCU not given. When clear sky is forecast, the cloud group will be replaced by SKC.

V V ///: Vertical visibility. When the sky is expected to be obscured and clouds cannot be forecasted, it is indicated by V V /// in lieu of N_s N_s N_s h_s h_s h_s. When CAVOK or SKC are not appropriate the abbreviation NSC is used.

PROB C_2 C_2 GGG_eG_e (Group)

The probability occurrence of alternate value(s) of forecast element(s), during a defined period of time the group PROBC_2C_2 GGG_eG_e is replaced directly before the alternate value(s).

C_2C_2 : Probability in percent rounded to whole tens.

C_2C_2 cannot exceed 50 (50%)

If the probability of occurrence of an element exceeds 50%, then that occurrence will be predominant feature of the forecast.

For C_2C_2 only values 30 to 40 are used to indicate probabilities of 30% and 40% respectively.

GG : The forecast beginning valid time (to the nearest whole hour) UTC

G_eG_e : The forecast end time to the nearest whole hour UTC

When probability 50% or more, this is indicated by BECMG, TEMPO or FM as appropriate. The group PROBC_2C_2 is not used in combination with the change indicator group BECMG or the time indicator group FMGGgg.

TTTTT GGG_eG_e

or **(group)**

TTGGgg

TTTTT Change indicator (BECMG or TEMPO). When during the period G_1G_1 to G_2G_2, a change in some or all of the elements forecast is expected to occur at some intermediate time GGgg or during the period GG to G_eG_e. Such groups are introduced untill all the data groups necessary to describe the elements forecast in the period G_1 G_1 to G G or GG gg have been given.

The change groups TTTTT

GGG_eG_e in the form of BECMG GGG_eG_e will indicate a change to forecast meteorological condition expected to occur at either a regular or irregular rate at an unspecified time within the period GG to G_eG_e. The duration of the period GG to G_e G_e normally will not exceed two hours and in any case will not exceed four hours. The change groups are followed by a description of all the elements for which a change is forecast. When an element is not described in data groups which follow the change groups, the description of this element for the period between G_1 G_1 to G_2 G_2 is considered to remain (valid) the same.

The conditions described after the groups, BECMG GGG_eG_e are those expected to prevail from G_eG_e until G_2G_2 unless a further change is expected, in which case a further set of change groups BECMG GGG_eG_e or FM GG gg is necessarily used.

The change groups TTTTT GGG_eG_e in the form of TEMPO GGG_eG_e indicate frequent or infrequent temporary fluctuations to forecast meteorological conditions which are expected to last less than one hour in each instance and in the aggregate cover less than half the period indicated by GGG_e G_e

(TX T_F T_F / G_F G_F Z TN T_F T_F / G_F G_F Z) (group)

TX : Forecast maximum temperature

TN : Forecast minimum temperature

T_F T_F : Forecast temperature in whole degrees Celsius

G_F G_F : (Forecast temperature) valid time to the nearest whole hour UTC.

This group is used to indicate forecast maximum and minimum temperatures expected to occur at the time indicated by G_F G_F Z.

Temperature between – 9 °C and + 9 °C are to be preceded by '0' (zero), temperatures below 0 °C are to be preceded by the letter 'M' (that is minus).

38.1.1 Criteria for inclusion of change group in Aerodrome Forecast

The criteria used for the inclusion of change groups in aerodrome forecasts or for the amendment of aerodrome forecasts are based on the following.

(a) when the surface wind is forecast to change through values of operational significance the threshold values should be established by meteorological authority in consultation with the appropriate ATS authority and operators concerned, taking into account changes in the wind which would:

  (i) require a change in runway(s) in use; and

  (ii) indicate that the runway tailwind and crosswind components will change through values representing the main operating limits for typical aircraft operating at the aerodrome.

(b) When the visibility is forecast to change to or pass through 5000 M, 3000 M, 1500 M, 800 M, 600 M, 350 M and 150.

(c) When any of the following weather phenomena or combinations thereof are forecast to begin or end or change in intensity.

  - freezing precipitation
  - freezing fog
  - moderate or heavy precipitation (including showers thereof)
  - low drifting dust, sand or snow
  - blowing dust, sand or snow (including snowstorm)
  - duststorm
  - sandstorm
  - squall
  - funnel cloud (tornado or waterspout)
  - other significant weather phenomena (given in SPECI criteria)

(d) when the height of base of the lowest layer or mass of cloud of BKN or OVC extent is forecast to change to or pass through 450 M (1500 ft), 300 M (1000 ft), 150 M (500 ft), 60 M (200 ft), 30 M (100 ft)

(e) When the amount of a layer or mass of cloud below 450 M (1500 ft) is forecast to change

(i) from SKC, FEW or SCT to BKN or OVC or (ii) from BKN or OVC to SCT, FEW or SKC

(f) When cumulonimbus clouds are forecast to develop or dissipate.

(g) When the vertical visibility is forecast to change or pass through 30, 60, 150 or 300 M (100, 200, 500 or 1000 ft).

(i) Any other criteria based on local aerodrome operating minima, as agreed between meteorological authority and the operators, including those concerned which should be included in aerodrome forecasts for local dissemination.

38.1.2 Period of Validity of Routine Aerodrome Forecasts

- Not less than a hour and not more than 30 hours
- Routine aerodrome forecasts valid for less than 12 hours and issued every 3 hours and those valid for 12 to 24/30 hours issued every 6 hours.
- Aerodrome forecast is meant for an aerodrome and its surrounding 8 km (5 miles) radius.

In India TAFs are issued at least three hours in advance of valid period. Generally for domestic flights the validity period is 9 hours and issued every three hours; for international flights the validity period is 24/30 hours and issued every six hours.

In India all Aerodrome Meteorological Offices issue aerodrome forecasts for their own aerodromes and their associated Meteorological Stations.

In case of restricted watch of Aerodrome Meteorological Offices, the responsibility is taken by the Concerned Regional Centre or State Capital Centre for the closed watch period.

Forecasts of all other aerodromes are to be obtained from the meteorological office preparing the same.

As said earlier, an aerodrome fore cast consists of a concise statement of the expected meteorological conditions at an aerodrome during the specified period. They (TAF_s) include the following in the order indicated.

(a) Identification of the type of forecast,
(b) (LCAD) location indicator,
(c) Time of issue of the forecast,
(d) Identification of a missing forecast (Nil may be used) when applicable,
(e) Date and period of validity of forecast,
(f) Identification of cancelled forecast (CNL used) when applicable,
(g) Surface wind,
(h) Visibility,
(i) Weather,
(j) Cloud,
(k) Expected Significant Change to one or more of these elements during the period of validity.

Aerodrome forecasts issued by a meteorological office are to be kept under continuous, review in order to issue Amendment when required. TAF that cannot be kept under continuous review shall be cancelled.

38.1.3 IMD Routine Aerodrome Forecasts

Routine aerodrome forecasts issued by IMD are valid for 9 and 24 hours and are issued as below.

Based on chart (UTC)	Time of issue (UTC)	Period of validity UTC	
		9 hours	24 hours
1800	2100	00 – 09	00 – 24
2100	0000	03 – 12	-
0000	0300	06 – 15	06 – 06
0300	0600	09 – 18	-
0600	0900	12 – 21	12 – 12
0900	1200	15 – 24	-
1200	1500	18 – 03	18 – 18
1500	1800	21 – 03	-

9 hourly TAFs in respect of Aerodrome Meteorological Stations are issued by the concerned Aerodrome Meteorological Offices/Meteorological Centre to meet specific requirement of national and international flights.

In India Meteorological Offices at Mumbai, Kolkata, Delhi, Chennai and Nagpur aerodromes issue 9 hourly TAFs 8 times a day at every three hours with TAF 03 – 12 UTC issued at 00 UTC and exchange these amongst themselves throughout the 24 hours. In addition, Meteorological offices at Mumbai, Kolkata, Delhi and Chennai aerodromes issue 18 hourly TAFs four times a day at six hourly intervals, beginning with 00 UTC for exchange amongst themselves.

38.2 Dissemination of TAF

TAF and amendments thereto are disseminated to international OPMET data banks and the centres designated by regional air navigation agreement for the operation of aeronautical fixed Service Satellite distribution System.

38. 2.1 ROBEX Collection and Dissemination of TAF Bulletins

ROBEX Centre: Mumbai

Aerodromes: Ahmedabad, Mumbai, Nagpur, Kolkata, Patna, Delhi, Luknow, Amritsar, Varnasi, Jaipur, Colombo, Kathmandu, Cochin, Calicut, Hyderabad, Chennai, Tiruchurapally, Tiruvananthapuram.

Examples

Ex 1: TAF YUDO 150000 Z 150624 12010KT 8000 BKN025 BECMG 0609 SCT015CB BKN020 TEMPO 0912 16015G30KT 1000 TSRA SCT010CB BKN020 FM1230 14010KT 9999 BKN025 OVC 090

Decoded : Aerodrome forecast for Donlon/international (fictitious location), issued on 15th day of the month at 0000 UTC, valid from 0600-2400 UTC on the 15th of the month

Surface wind (W/V): 120°/10KT

(W/V) : 120°/10 KT

Visibility : 8 km

Cloud : 5-7 oktas base height 2500 ft aae
(aae = above aerodrome elevation)

Change : Becoming between 0600 to 0900 UTC

Cloud : 3-4 oktas cumulonimbus cloud base height 1500 ft aae

5-7 oktas base height 2000 ft aae

Temporary change between 0900-1200 UTC (less than half the period)

surface wind (W/V) : 160 ° (T) / 15 knots gusting to 30 knots

Visibility : 1000 meters

Weather : Moderate thunderstorm with rain

Cloud : 3-4 oktas cumulonimbus base height 1000 ft aae

5-7 oktas base heigh 2000 ft aae

Change from 1230 UTC onwards to

Surface wind (W/V) : 140^{O}(T)/10 knots

Visibility : 10 km or more

Cloud : 5-7 oktas base height 2500 ft aae

8 oktas, base height 9000 ft aae

***Ex* 2**: TAF VIDP 201500Z 201818 05008KT 2500DU SKC BECMG 0306 09010G20KT 4000 BLDU FEW030 SCT100 BECMG 1214 05006KT 2500DU TEMPO0918 27030KT 0800 DS/TSRA FEW030CB

Decoded: Aerodrome forecast for Delhi, issued on 20th day of the month at 1500 UTC, valid from 1800 UTC of 20th to 1800 UTC of 21st day of the month

Surface wind (W/V): 050^{O}(T)/08 knots

Visibility : 2500 meters in widespread dust

Cloud : Clear sky

Change : Becoming between 0300–0600 UTC of 20th

Surface wind (W/V): 90^{O}(T)/10 knots gusting to 20 knots

Visibility : 4000 meters

Weather : Blowing dust

Cloud : 1-2 oktas, base height 3000 ft aae

3-4 oktas base height 10000 ft aae

Change : Becoming between 1200-1400 UTC of 20th

Surface wind (W/V): 050^{O}(T)/06 knots

Visibility : 2500 meters in widespread dust

Temporary change between 0900-1800 UTC of 20th

Surface wind (W/V): 270° / 30 knots

Visibility : 800 meters

Weather : Dust storm or thunderstorm with rain

Cloud : 1-2 oktas cumulonimbus with base height 3000 ft aae

***Ex* 3:** TAF VABB 182100Z 190024 27005KT 3000 HZ SCT020 BKN100 BECMG 0406 26010KT 6000 BECMG 1214 27005KT 4000 HZ TEMPO 2000 SHRA SCT010 FEW025TCU OVC 090

Decoded: Aerodrome forecast for ICAO station indicator VABB (Mumbai) issued on 18th day of the month at 2100 UTC, valid from 19th day 00-24 UTC

Surface wind (W/V) : 270° / 05 knots

Visibility : 3000 meters in haze

Weather : Haze

Cloud : 3-4 oktas base height 2000 ft aae (above aerodrome elevation)

5-7 oktas base height 10000 ft aae

Change : Becoming between 0400 to 0600 UTC

Surface wind (W/V) : 260° / 10 knots

Visibility : 6000 meters / 6 km

Change : Becoming between 1200 to 1400 UTC

Surface wind (W/V) : 270 ° / 05 knots

Visibility : 4000 meters in Haze

Weather : Haze

Temporary change : Less than half the period

Visibility : 2000 meters in shower rain

Weather : Shower rain

During shower rain cloud:		3-4 oktas base height 1000 ft aae
		1-2 oktas CU congestns aae or towering CU base height 2500 ft aae
		8 oktas base height 9000 ft aae

***Ex* 4:** TAF VOHY 121500Z 121818 05005KT 8000 SCT020 SCT100 BECMG 2403 07010KT 5000 –RA SCT018 BKN090 TEMPO 1013 06010 G25KT 3000 TSRA/SHRA FEW020/ SCT025CB BKN090

Decoded: Aerodrome forecast for ICAO station indicator VOHY (Hyderabad), issued on 12th day of the month at 1500 UTC, valid from 12th day 1800 UTC to 13th day 1800 UTC.

Surface wind (W/V)	:	050°/05 knots
Visibility	:	8000 meters
Cloud	:	3-4 oktas base height 2000 ft aae
		3-4 oktas base height 1000 ft aae
Change	:	Becoming between 2400 UTC of 12th and 0300 UTC of 13th
Surface wind (W/V)	:	070°/10 knots
Visibility	:	5000 meters in light rain
Weather	:	light rain
Cloud in light rain	:	3-4 oktas base height 1800 ft aae
		5-7 oktas base height 9000 ft aae

Temporary change between 1000-1300 UTC of 13th day

Surface wind (W/V)	:	060°/10 knots gusting to 25 knots in TS RA/SHRA
Visibility	:	3000 meters in TS RA/SHRA
Weather	:	TSRA/SHRA (thundershower rain or shower rain)
In TSRA/SHRA cloud	:	1-2 oktas base height 2000 ft
		3-4 oktas cumulonimbus base height 2500 ft
		5-7 oktas base height 9000 ft

Ex 5: TAF EGPD 201506Z 201623 27015G30KT 9999 SCT018 BKN030 TEMPO 1623 6000 -RA BKN015 PROB 40 + RA BKN007 BECMG 1820 25008 KT

Decoded : Aerodrome forecast for ICAO station indicator. EGPD (fictious location) issued or transmitted on 20th day of month at 1506 UTC, valid from 20th day 1600-2300 UTC

Surface wind (W/V)	:	270°/15 knots gusting to 30 knots
Visibility	:	10 km or more
Cloud	:	3-4 oktas base height 1800 ft aae
		5-7 oktas base height 3000 ft aae

Temporary change between 1600-2300 UTC of 20th

Visibility	:	6000 meters in light rain
Weather	:	temporary light rain
Cloud in light rain	:	5-7 oktas base height 1500 ft aae
Probability or chance	:	There is a 40% probability less than half the period between 1600-2300 UTC that there will be heavy rain.
During heavy rain	:	Cloud, 5-7 oktas base height 700 ft.
Change	:	Be coming between 1800-2000 UTC
Surface wind	:	250°/08 knots for the rest of the forecast period

Ex.6 TAF VOMM 251000Z 251812 24005KT CAVOK TEMPO 0712 24012G25KT

Decoded: Aerodrome forecast for ICAO station indicator VOMM (Chennai) issued or transmitted on 25th day of the month at 1000 UTC, valid from 25th 1800 UTC to 26th 1200 UTC

Surface wind (w/v)	:	240°/05 knots
Visibility	:	10 km or more
Cloud	:	No cloud below 1500 m (5000 ft) (above aerodrome elevation) or MSA (minimum safe altitude) whichever is higher, no thunderstorm or significant weather.
Change	:	Temporary Change between 0700-1200 UTC (for less than half the period) on 26th day of the month
Surface wind	:	240°/12 knots gusting to 25 knots

*Ex.*7 TAF LOWG 201000Z 201812 VRB 03 KT 4000 BR NSC TEMPO 2022 3000 MIFG BECMG 2224 0300 BCFG PROB30 TEMPO 0006 0100 FG VV001 BECMG 0710 3000 BR FEW040 BECMG 1012 24008KT 9999

Decoded: Aerodrome forecast for ICAO station indicator LOWG issued (or transmitted) on 20th at 1000 UTC, valid from 1800 UTC of 20th to 1200 UTC of 21st.

Surface wind	:	Wind direction variable and speed 3 knots
Visibility	:	4000 meters in mist
Weather	:	Mist
Cloud	:	No significant cloud or no cloud below 1500 M (5000 ft) aae or MSA whichever is the higher. No thunderstorm (MSA = Minimum safe altitude)

Temporary change between 2000-2200 UTC of 20th,

Visibility	:	3000 meters in shallow fog
Weather	:	Shallow fog
Change	:	Becoming between 2200-2400 UTC of 20th
Visibility	:	300 meters in banks or patches of fog
Weather	:	Patches of fog
Probability or Chance	:	There is 30% probability of occurrence temporarily between 0000-0600 UTC of 21st day (less than half the period).
Visibility	:	deteriorate to 100 meters in fog with a vertical visibility of 30 meters (100 ft).
Change	:	Becoming between 0700-1000 UTC of 21st day
Visibility	:	3000 meters in mist
Weather	:	Mist
Cloud	:	1-2 oktas, base height 4000 ft a a e or MSA whichever is higher
Change	:	Becoming between 1000-1200 UTC of 21st
Surface wind	:	240° / 08 knots

Visibility : 10 km or more this improvement assume it to occur at 1200 UTC of 21st.

Ex 8: TAF VECC 251000Z 251212 14010G25KT 5000 BRRA BKN010 TEMPO 1220 1500 SHRA BKN008 BKN012CB BECMG 2224 VRB03KT TEMPO 2006 1000 SHRA BKN008 BKN012CB BECMG 0608 24015G25KT TEMPO 0612 1500 SHRA BKN006 BKN010CB

Decoded: Aerodrome forecast for ICAO station indicator VECC (Kolkata), issued (or transmitted) on 25th day of the month at 1000 UTC, valid from 1200 UTC of 25th to 1200 UTC of 26th day.

Surface wind (W/V) : 140°/10 knots gusting (maximum) to 25 knots

Visibility : 5000 meters in mist with moderate rain

Weather : Mist with moderate rain

Cloud : 5-7 oktas base height 1000 ft aae

Temporary change between 1200-2000 UTC of 25th (less than half the period)

Visibility : 1500 meters in shower rain

Weather : shower rain

Cloud during shower rain : 5-7 oktas base height 800 ft aae

5-7 oktas cumulonimbus base height 1200 ft aae

Change : Becoming between 2200-2400 UTC of 25th

Surface wind : Variable in direction with 3 knots (speed)

Temporary change (less than half the period) between 2000 UTC of 25th and 0600 UTC of 26th

Visibility : 1000 meters in shower rain

Weather : Shower rain

Cloud during shower rain : 5-7 oktas base height 800 ft aae

5-7 oktas cumulonimbus base height 1200 ft aae

Change : Becoming between 0600-0800 UTC of 26th

Surface wind : 240°/15 knots maximum gusting to 25 knots

Temporary change (less than half the period) between 0600-1200 UTC of 26th

Visibility	:	1500 meters in shower rain
Weather	:	Shower rain
Cloud in shower rain	:	5-7 oktas base height 600 ft aae
		5-7 oktas cumulonimbus base height 1000 ft aae

38.3 Landing Forecast

This is already discussed in METAR code, however briefly described again

A landing forecast is prepared by the Meteorological office designated by the meteorological authority concerned. These forecasts are intended to meet the requirements of local users and of aircraft within about one hour's flying time from the aerodrome. Landing forecasts prepared either in the form of a self-contained forecast or in the form of a trend-type forecast, as determined by regional air navigation agreement.

Landing forecast contains the expected meteorological conditions at an airfield for an aircraft intending to make a landing. It is required in the flight stage and is prepared by meteorological office of the airfield concerned. It provides a more up-to-date assessment of expected weather (during next two hours) that is already given in the relevant TAF.

The period of validity of TREND forecast is two hours from the time of the report or to which it is appended.

When a change indicator (BECMG or TEMPO) appears in a trend forecast, the meteorological elements which normally given are: (i) wind and its variation (ii) visibility (iii) weather and (iv) cloud conditions

38.4 Take–off Forecast

A forecast for take–off is prepared by the meteorological office designated by the meteorological authority concerned. In IMD (in India) take-off forecast is prepared by Aerodrome Meteorological Offices.

Take-off forecast is prepared for specified period of time and contain the information on expected conditions over the runway complex in respect of the following elements.

(a) Surface wind direction, speed and any variation thereof

(b) Temperature

(c) Pressure (QNH)

(d) Visibility

(e) Significant weather and

(f) any other element as required.

Take-off forecast is supplied to operators and flight crew members on request within three hours before the expected time of departure (ETD).

The order of the elements and terminology, units and scales used in forecasts for take-off will be the same as those used in METARs. When necessary, amendment forecast is issued promptly. The criteria for amendment shall be consistent with the corresponding criteria for special report.

TTTT indicator group for TREND which is indicated by BECMG, TEMPO or NOSIG

WX NIL : indicate the expected end occurrence of significant weather

SKC : indicate clear sky

ALLOK : indicate visibility 10 km or more, clear sky conditions, temperature less than 40 °C.

CAVOK : No cloud below 1500 meters, no precipitation no thunderstorm, no fog expected.

BECMG/TEMPO followed by wind, visibility, weather and cloud.

NOSIG : When none of the elements is expected to change in such a way which require to indicate the change, this is indicated by the term NOSIG. NOSIG appended to a METAR or SPECI is a complete landing forecast.

38.5 Operationally Desirable Accuracy of Forecasts

Element	Desirable accuracy	Minimum % of cases within range
1	2	3
A: For aerodrome forecasts		
1. Wind direction	$\pm$ 30°	80% of cases
2. Wind speed	$\pm$ 5kT up to 25 kT $\pm$ 20% above 25 kT	80% of cases
3. Visibility	$\pm$ 200 M up to 700 M $\pm$ 30% between 700 M to 10 Km	80% of cases
4. Precipitation	Occrrence or non-occurrence	80% of cases
5. Cloud amount	$\pm$ 2 oktas	70% of cases
6. Cloud height	30 M up to 120 M 100 ft up to 400 ft 30% between 120 M and 3000 M	70% of cases
7. Air temperature	$\pm$ 1°C	70% of cases
B: For Take-off Forecasts		
1. Wind direction	$\pm$ 30°	90% of cases
2. Wind speed	$\pm$ 5 kT up to 25 kT $\pm$ 20% above 25 kT	90% of cases
3. Air temperature	$\pm$ 1 °C	90% of cases
4. Pressure value (QNH)	$\pm$ 1 hPa	90% of cases
C: For Landing Forecasts		
1. Wind direction	± 30°	90% of cases
2. Wind speed	± 5 KT up to 25 KT ± 20% above 25 KT	90% of cases
3. Visibility	± 200 M up to 700 M ± 30% between 700 M to 10 KM	90% of cases
4. Precipitation	occurrence or non-occurrence	90% of cases
5. Cloud amount	± 2 oktas	90% of cases
6. Cloud height	± 30 M up to 120 M 30% between 120 M – 3000 M	90% of cases

D: For Area, Flight and Route Forecasts

1. Upper air temperature	± 3°C (mean for 900 km/500 NM)	90% of cases
2. Upper wind	± 15 KT up to flight level 250 ± 20 KT above FL 250	90% of cases
3. Significant en-route weather	occurrence or non-occurrence	80% of cases
4. Phenomena location	± 100 km/60NM	70% of cases
5. Cloud	vertical extent ± 600 M/2000 ft	70% of cases

38.6 Local and Area Forecast

As said earlier, a weather forecast is an expected meteorological conditions over a specified area and time.

Local and Area forecasts are statements about the expected meteorological conditions over a given airfield, flying sector or range for a specified period for operational use, and for Air Traffic Control (ATC) authorities. In India, Aerodrome Meteorological offices (AMOs) issue aerodrome forecast, Area forecasts and local forecast for their own aerodromes and for their associated Aeronautical Meteorological stations (AMSs).These forecasts are issued for the use of local flights including helicopters.

In India IMD ***Area forecast*** covers 100 NM (nautical miles) or more as per operational requirement. ***Local forecast*** covers the aerodrome and 50 NM around. They are valid for 6-12 hours and may have "outlook" for a further period of 6 hours. These forecasts normally issued thrice a day. The time of origin (T 00) and period of forecast given below (Met T-2 form used).

T00 of Local forecast		**Period of validity 12 hours**	
UTC	(IST)	UTC	IST
0130	(0700)	0200 – 1400	(0730 – 1930)
0730	(1300)	0800 – 2000	(1330 – 0130)
1230	(1800)	1300 – 0400	(1830 – 0930)

Format

Local/Area forecast issued in Met. T-2 form (specimen given appendix Page 774). It contains area covered, period of validity T00, salient synoptic situations and evaluation briefly, weather warnings if any, and outlook for a further period of 6 to 12 hours. Freezing level and Mintra level, expected weather hazard elements like CAT, windshear, cloud cover, visibility, weather warning if any will be underlined in Red.

It is disseminated manually and LAN (Local Area Network) to: ATC, Chief Operation Officer (COO) and Base Operations Centre (BOC) (see model example).

38.7 Area Forecast Qode Form

Section 1 ARFOR (YYGGggZ)
$Y_1Y_1G_1G_1G_2G_2$ (KT or KMH or MPS)

AAAAA (VVVV) $(W_1W_1W_1)$

$$\left(\begin{bmatrix} N_sCCh_sh_sh_s \\ \text{or (VV///)} \\ \text{or SKC (or NSC)} \end{bmatrix}\right) \quad 7\, h_t h_t h_t h_f h_f h_f$$

$6\, I_c h_i h_i h_i t_L$ $\quad$ $5B\, h_B h_B h_B t_L$

$(4\, h_x h_x T_h T_h$ $\quad$ $d_h d_h f_h f_h f_h)$ $\quad$ $(2h'_p h'_p T_p T_p)$

Section 2 (11111 $\quad$ $QL_a L_a L_o L_o$ $\quad$ $h'_j h'_j f_j f_j f_j)$

Section 3 (22222 $\quad$ $h'_m h'_m f_m f_m f_m$ $(d_m d_m VV))$

Section 4 $9i_3$ nnn

Notes:

(i) ARFOR is the name of the code for an aviation forecast in figure code prepared for a specific area.

(ii) As said in TAF code, due to time space variability and limitation of forecasting the specific value of any of the forecast elements is to be taken as the most probable value during the forecast period. Similarly time of occurrence and change of an element is given in forecast this time is to be taken as the most probable time.

(iii) The code form is divided into four sections as under.

Section number	Symbolic figure groups	Contents
1	–	Code identification and time groups; Area forecast.
2	11111	Jet-stream data (optional)
3	22222	Data of maximum wind and vertical wind shear (optional)
4	–	Supplementary phenomena

Regulations (description of code symbolic letters) Group

Group

ARFOR : It is code name for area forecast for aviation, prefixed to individual coded area forecast.

YYGG gg Z : YY-day of the month (UTC)

GGggZ: Time of forecast in hours (GG), minutes (gg) UTC followed by the letter Z as an abbreviated indicator.

$Y_1Y_1G_1\ G_1G_2G_2$:

Y_1Y_1 - Day of the month of forecast valid

G_1G_1 - Time of commencement of period of forecast in whole hours UTC.

$G_2\ G_2$ - Time of ending of period of forecast in whole hours UTC

When the period of forecast ends at midnight G_2G_2 will be encoded 24.

When the period is between 25 and 48 hours after $G_1G_{1,}$ then G_2G_2 will be encoded by adding 50 to the time of ending period of forecast.

KT = knots, KMH = kilometers per hour, MPS = Meters per second

AAAAA : Description of area in plain language

***Ex*:** VOHY AND 50 NM AROUND

(VVVV) : Visibility group, omitted when visibility is not forecast

VVVV is horizontal visibility at surface given in meters in increments of 50 M up to 500 M, in increments of 100 M between 500 M to 5000 M and in increments of 1000 M between 5000 M to 9999 M and 9999 indicate visibility of 10 km or more.

If the value lies between two increments or steps it will be rounded down.

Ex: 370 M Reported as 0350

580 M reported as 0500

3560 M reported as 3500

5600 M reported as 5000

(W_1 W_1 W_1) : Forecast weather group

W_1 W_1 W_1 – forecast weather

This group is used when any of the following phenomena are forecast

Tropical cyclone, severe line squall, hail, thunderstorm, marked mountain waves, widespread sandstorm or duststorm or freezing rain.

Weather abbreviation codes given below.

W_1 W_1 W_1 forecast weather

Code figure	Abbreviation	
111	TS	Thunderstorm
222	TRS	Tropical cyclone
333	LSQ	Severe line squall
444	HAIL	Hail
555	MTW	Marked mountain waves
666	SAND	Widespread sandstorm
777	DUST	Widespread duststorm
888	FZR	Freezing rain

$$\left[\begin{array}{l} N_S CC\ h_s h_s h_s \\ \text{or VV ///} \\ \text{or SKC (or NSC)} \end{array}\right] \text{Cloud group}$$

N_s : Amount of individual cloud layer

CC : Genus of cloud in two letters, code given below

Cloud	CC	Cloud	CC
Cirrus	CI	Nimbostratus	NS
Cirrocumulus	CC	Stratocumulus	SC
Cirrostratus	CS	Stratus	ST
Altocumulus	AC	Cumulus	CU
Alto stratus	AS	Cumulonimbus	CB

Cloud not visible due to darkness, fog, duststorm, sandstorm or other analogues phenomena indicated 'CC' by "//"

$h_s h_s h_s$: height of base of cloud mass

Code for $h_sh_sh_s/h_th_th_t/h_fh_fh_f/h_Bh_Bh_B/h_ih_ih_i/h_xh_xh_x$

Code figure	Meters
000	Less than 30
001	30
002	60
003	90
⋮	⋮
009	270
010	300
011	330
…	…
099	2970
100	3000
110	3300
…	…
990	29700
999	$\geq$ 30000

The above code is direct reading in units of 30 meters.

VV///: Vertical visibility, when sky is expected to be obscured and clouds cannot be forecast, it will be indicated as VV/// in lieu of cloud group.

7 $h_th_th_th_fh_fh_f$: When the heights above msl (mean sea level) of both the base and top of a number of layers are forecast, the cloud and 7 groups will be used in pairs for each layer.

When the 0 °C isotherm is forecast but no forecast is made for top of clouds, the 7-group will take the form 7/// $h_fh_fh_f$.

6 I_c $h_ih_ih_it_L$: Icing group.

I_c: Ice accretion on the external parts of aircraft.

I_c code

Code Fig.	Meaning
0	No icing
1	Light icing
2	Light icing in cloud
3	Light icing in precipitation
4	Moderate icing
5	Moderate icing in cloud
6	Moderate icing in precipitation
7	Severe icing
8	Severe icing in cloud
9	Severe icing in precipitation

$h_i h_i h_i$: Height of lowest level of icing

t_L = thickness of layer.

Code for t_L

Code fig	Meaning	Code fig	meaning
0	Up to top of cloud	5	1500 M
1	300 M	6	1800 M
2	600 M	7	2100 M
3	900 M	8	2400 M
4	1200 M	9	2700 M

5B h_B h_B $h_B t_L$: Turbulence group

B : Turbulence Code

Code fig	Meaning
0	None (non turbulence)
1	Light turbulence
2	Moderate turbulence in clear air, occasional
3	Moderate turbulence in clear air, frequent
4	Moderate turbulence in cloud, occasional
5	Moderate turbulence in cloud, frequent
6	Severe turbulence in clear air, occasional
7	Severe turbulence in clear air, frequent
8	Severe turbulence in cloud, occasional
9	Severe turbulence in cloud, frequent

h_B h_B h_B : Height of lowest level of turbulence (code given above)

t_L : Thickness of layer (code given above)

(4 h_x h_x h_x T_h T_h d_h d_h f_h f_h f_h): Temperature and wind at upper levels.

These groups will be used together and repeated for each level for which temperature and wind are forecast.

h_x h_x h_x : Altitude to which temperature and wind refer (height code given above $h_s h_s h_s / h_f h_f h_f / h_p h_p h_p / h_t h_t h_t / h_i h_i h_i / h_x h_x h_x$)

T_h T_h: Air temperature in whole degrees Celsius, at the height indicated by h_x h_x h_x

d_h d_h: True direction of wind in tens of degrees from which wind will blow at the altitude h_x h_x h_x

$f_h f_h f_h$: Forecast wind speed (in KMH, KT or MPS) at the level given by $h_x h_x h_x$

($2 h'_P h'_P T_P T_P$) : Tropopause group.

This group will be omitted when tropopause data are not forecast.

$h'_P h'_P$: Height of tropopause level.

The heights are indicated in ICAO flight level (FL) numbers, last figure omitted.

ICAO flight levels datum pressure 1013.2 hPa and are separated by a nominal distance of 500 ft [or FL 05]

Scheme of Coding

$h'_P h'_P$ code fig	ICAO FL number	Meters (approx)	Feet
20	200	6000	20000
20	205	6150	20500
21	210	6300	21000
21	215	6450	21500
⋮	⋮	⋮	⋮
30	300	9000	30000

$T_P T_P$: Tropopause temperature in degrees Celsius.

***Section* 2**

$\left(11111 \;\; Q L_a L_a L_o L_o h'_j h'_j f_j f_j f_j\right)$: Jet stream group

This group will be omitted when Jet stream data are not forecast.

Q: Octant of the globe (see Fig. on page 735)

Q code fig	Longitude	Hemisphere	Q code fig	Longitude	Hemisphere
0	0° - 90° W		5	0° - 90° W	
1	90° - 180° W	Northern hemispher	6	90° - 180° W	Southern hemispher
2	180° - 90° E		7	180° - 90° E	
3	90° E - 0° E		8	90° E - 0° E	

$L_a L_a$: Latitude in whole degrees

$L_o L_o$: Longitude in whole degrees

$h'_j h'_j$: Height of jet stream core level, given in ICAO flight levels as given above for tropopause height

$f_j f_j f_j$: Wind speed (in KMH, KT or MPS) in the jet core

***section* 3:** $\left(22222 \quad h'_m h'_m f_m f_m f_m (d_m d_m v v)\right)$ wind maximum group

when the maximum wind is forecast but no forecast is made for the vertical wind shear, the last group of the section will be of the form $d_m d_m$ //.

When only information of vertical wind shear is made, the group $h'_m h'_m f_m f_m f_m$ is omitted and the group $d_m d_m v v$ will be of the form // v v

$h'_m h'_m$: Height of the maximum wind level, which is coded in ICAO flight levels as given earlier above for tropopause height.

$f_m f_m f_m$: Maximum wind speed (in KMH, KT or MPS) at flight level given by $h'_m h'_m$

$d_m d_m$: direction of wind in tens of degrees from which maximum wind will blow at the height $h'_m h'_m$

VV : Vertical windshear in knots per 300 meters

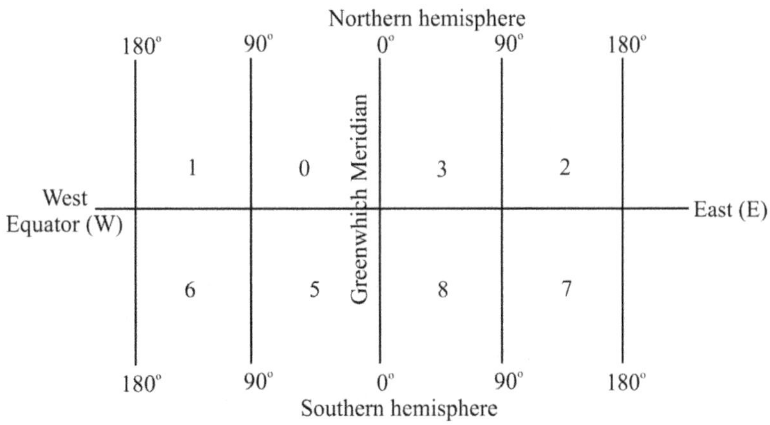

Fig.

Section **4**

9$i_3$nnn Frontal activity group.

The groups 91 P_2 P_2 P_2, 92 F_t L_a L_a 93 F_t L_oL_o, 94 F_t GG if required shall always be placed at the end of the relevant part of the message.

The groups 92 F_t L_a L_a, 93 F_t L_oL_o, 94 F_t GG shall only be used to indicate the type front, together with the position or time of passage. The type of weather during the frontal passage shall be indicated separately. Eg., by separating the forecasts into different periods., or by the groups 96 GGG_P and 97 GGG_P or by a combination of both methods.

$P_2P_2P_2$: Pressure reduced to msl in whole hecto Pascals

F_t : Type of front = code given below

F_t Code fig	Description
0	Quasi stationary front at the surface
1	Quasi stationary front above the surface
2	Warm front at the surface
3	Warm front above the surface
4	Cold front at the surface
5	Cold front above the surface
6	Occlusion
7	Instability line
8	Intertropical front
9	Convergence line

GG : Valid time (to the whole hour) UTC of the beginning of the forecast

G_P : Period, to the nearest whole hour

i_3 : Indicator for supplementary phenonena

nnn : Specifications related to supplementary phenomena.

9$i_3$nnn

91 $P_2P_2P_2$: Forecast lowest mean sea level pressure

92 F_t L_aL_a : Type of front and its position (track of aircraft approximately N-S)

93 F_t L_oL_o : Type of front and its position (track of aircraft approximately E-W)

94 F_t GG : Type of front and time of passage

Used only in ROFOR

951 //	:	Gradual change along the route
952 L_aL_o	:	Change at latitude L_aL_a north along the route
953 L_aL_a	:	Change at latitude L_aL_a south along the route
954 L_oL_o	:	Change at longitude L_oL_o east along the route
955 L_oL_o	:	Change at longitude L_oL_o west along the route
96 GGG_p	:	(a) When $G_p = 0$, a self contained part of the forecast beginning at GG. All prior forecast conditions are superseded. (b) When $G_p = 1$ to 4, change at either a regular or irregular rate at an unspecified time within the period beginning at GG and indicated by G_p.
97GGG_p	:	Frequent or infrequent temporary fluctuations taking place within the period indicated by G_p.
9999 C_2	:	(a) When used in combination with 99GG_P probability C_2 occurrence of alternate value of a forecast element indicated in tens of percent. (b) When used in combination with 97 GG_P probability C_2 of occurrence of temporary fluctuation, indicated in tens of percent.
99 GGG_p	:	used in combination with 9999 C_2, time period G_P beginning at GG that the alternative of a forecast element may occur.

***Note*:** Local variations in ARFOR and ROFOR may be described if required with the following expressions:

LOC – Locally

LAN- Inland

COT- At the coast

MAR- At sea (Maritime)

VAL- In valley

CIT – Near or over towns

MON – Above high ground or mountains

SCT – Scattered (phenomena scattered both in space and time)

Ex: ARFOR 150230 Z 150309KT VOHY AND
50 NM AROUND 4000 HZ 2ST010 5STSC015
7///130 611506 50//// 407008 14005
410002 19010 4150M02 27015 4185M07
29025 250M75 11111 32580 22080
22222 24100 2705

***Decoded*:** Area forecast for Hyderabad and 50 nautical miles around, issued on 15th day of the month at 0230 UTC, valid for 15th day between 0300-0900 UTC, wind reported in knots.

Surface visibility : 4000 M in haze.

Cloud : 2 okta stratus base 1000 ft (300 M)

5 okta stratus / stratocumulus base 1500 ft (450 M).

Height of freezing level: 3900 meters (13000 ft)

Icing : light icing above 4500 M (15000 ft), thickness 1800 meters,

No Turbulence;

Upper wind Height FL	Temperature	Direction/speed
FL 070	+ 08 °C	140°/ 05 kt
FL 100	+ 02 °C	190°/10 kt
FL 150	– 02 °C	270°/15 kt
FL 185	– 07 °C	290° /25 kt

Tropopause: height FL 500, temperature – 75° C

Jetstream: located at latitude 25 °N, long 80 E; FL 220, Jet core speed 80 kt

Wind maxima: FL 240, speed 100 kt direction 270° (T)

Vertical wind shear 5 kt/300 meters

38.8 ROFOR

***Route / Flight Forecasts*:** A route forecast is the forecast for the route of a planned flight valid for the entire period of the forecast for the whole route, whereas a flight forecast is valid only for the time the aircraft is expected to be over the specific portion of the route. For the en-route phase of the of the flight, the forecast shall cover the route from the aerodrome of departure to the final aerodrome. A statement or other indication of conditions expected between the final aerodrome and one alternate aerodrome will be provided when requested as per details provided by the captain of the aircraft.

In IMD, Route forecast (ROFOR) is written on T-3 or T- 4 or (Met T-3 page 775, or Met. T-4 page 780) forms. T-3 form is used for flights covering less than 500 NM and T- 4 form is used for flights covering 500 NM or more.

ROFOR Code form

Section 1 ROFOR (YYGGggZ) $Y_1Y_1G_1G_1$ G_2G_2 [KT or KMH or MPS]

CCCC Q LaLaLoLo CCCC Oi_2 ZZZ

(VVVV) $(w_1w_1w_1)$ $N_sCCh_sh_sh_s$ $7h_th_th_th_fh_fh_f$

$6I_ch_ih_ih_it_L$ $5Bh_Bh_Bh_B$ t_L

$(4h_xh_xh_xT_hT_h$ $d_hd_hf_hf_hf_h)$ $\left(2\, h'_p\ h'_p\ T_p\ T_p\right)$

Section 2 (1111 Q $L_aL_aL_oL_o$ h'_j h'_j $f_jf_jf_j$)

Section 3 (22222 $(h'_mh'_mf_mf_mf_m(d_md_mVV))$

Section 4 $9i_3$ nnn

***Notes*:**

1. ROFOR is the name of the code for an aviation forecast in figure code for the route between two specified aerodromes.
2. Forecast elements are most probable values likely to assume during the period of forecast. Time of occurrence or change of an element given in a forecast are also to be the most probable time.
3. Code form is divided into four sections. Sections 2, 3, & 4 are same as in ARFOR.

Section 1 contains code identification and time groups, route forecast

Section 2 indicated by symbolic figure group 11111 and contains jet stream data (optional)

Section 3 indicated by symbolic figure group 22222 and contains data of Maximum wind and Vertical wind shear (optional)

Section 4 contains supplementary phenomena

Groups (symbolic letters of code)

YYGGggZ : Date and time of origin of forecast

YY- Day of the month (UTC).

GGggZ – Time of forecast in hours and minutes UTC followed by Z as an abbreviated indicator.

$Y_1Y_1G_1G_1G_2G_2$ - Date and time of validity of forecast

Y_1Y_1 – Day of the month (UTC)

G_1G_1 – Beginning of the period of forecast in UTC

G_2G_2 – End period of the forecast in UTC

[KT or KMH or MPS] – Unit of wind speed used indicated by KT or KMH or MPS

Forecast description contain atleast information about wind, visibility, weather and cloud. Forecast period G_1 G_1 to G_2 G_2 may be divided into two or more self contained groups in the form FM GG gg.

CCCC : Internatioanal ICAO four letter location indicator of the aerodrome at either end of the route (ex: VOHY to VIDP)

$QL_aL_aL_oL_o$: Q : Octant of the globe

L_a L_a : Latitude in whole degrees

L_o L_o: Longitude in whole degrees

Code figure octant of the globe same as given for ARFOR

Qi_2 ZZZ : This group should be used at the beginning of the forecast for each section or point.

i_2 = Zone type indicator

ZZZ = Zone specification

0000	:	up to the turning point indicated by the first group Q L_aL_a L_oL_o which appears between the index numbers at the beginning of the message.
01 Q L_a L_a		up to latitude L_a L_a
02 Q L_o L_o		up to longitude L_o L_o
03 Q CCCC		up to the aerodrome identified by CCCC
04 nnn	:	up to the point at a distance of nnn kilometers from the preceding point
050 ZZ	:	For the area indicated in 5° zone numbers
06 Q L_a L_a	:	At latitude L_a L_a
07 Q L_o L_o	:	At longitude L_o L_o
08 CCCC		At the aerodrome identified by CCCC
09 nnn		At a point at a distance from nnn kilometers from preceding point
(VVVV)	:	Horizontal visibility at surface, coded as given in ARFOR
($W_1W_1W_1$)	:	Forecast weather coded as given in ARFOR, mainly
Code	:	Phenomena
111 TS	:	Thunderstorm
222 TRS	:	Tropical cyclone
333 LSQ	:	Severe line squall
444 HAIL	:	Hail
555 MTW	:	Marked mountain wave
666 SAND	:	Widespread sandstorm
777 DUST	:	Widespread duststorm
888 FZR	:	Freezing of rain
N_s CC $h_sh_sh_s$	:	Amount, genus and height of base of cloud. Coded as given in ARFOR
7 $h_th_th_t$ $h_fh_fh_f$	:	Height of 0 °C isotherm ($h_fh_fh_f$), coded as given in ARFOR
6 I_C $h_ih_ih_i$ t_L	:	Type of icing, height of lowest level of icing

I_C – type of ice accretion on the external parts of aircraft

$h_ih_ih_i$ – lowest level of icing height

t_L: thickness layer of icing. Coded as given in ARFOR

$5\ B\ h_Bh_Bh_B\ t_L$: Turbulence group

B: turbulence

$h_Bh_Bh_B$: height of lowest level of turbulence. Coded as given in ARFOR

t_L : thickness layer, coded as given in ARFOR

$(4\ h_Xh_Xh_X\ T_hT_h$: temperature and winds

$d_hd_h\ f_hf_hf_h)$ $h_Xh_Xh_X$- altitude to which temperature and winds refer

$T_h\ T_h$ – temperature in whole degrees Celsius corresponding to the height $h_Xh_Xh_X$

d_hd_h – direction of wind in tens of degrees at the height $h_Xh_Xh_X$

$f_hf_hf_h$ – wind speed in KT at the level given by $h_Xh_Xh_X$

$(2\ h'_Ph'_P\ T_PT_P)$: Tropopause information. Omitted when tropopause data not given

$h'_Ph'_P$- height of tropopause level given in ICAO flight levels last figure omit

T_PT_P - temperature at tropopause, in whole degrees

Celsius, M for negative. Coded as given in ARFOR.

Section 2 Jetstream information. This will be omitted if jetstream data not forecast.

11111- Jetstream indicator group

$Q\ L_a\ L_a\ L_o\ L_o$: Q-octant of the globe

$L_a\ L_a$ – latitude in whole degrees

Lo Lo – longitude in whole degrees

$h'_j h'_j f_j f_j f_j$: $h'_j h'_j$ - Height of jetstream in tens of Flight level

$f_j f_j f_j$ – wind speed in knots in jetcore

Section 3 Wind Maxima information and vertical wind shear

22222 – Wind Maxima indicator group symbolic

$h'_m h'_m$ - height of wind maxima in tens of FL

$f_m f_m f_m$ – wind speed in Wind Maxima in KT

$d_m d_m$ – direction of wind in tens of degrees in Wind Maxima

VV - Vertical wind shear in Wind Maxima

d_m dm // or // VV indicate only direction or only wind shear, and wind shear not known or direction of wind in shear not known

VV expressed in knots per 300 meters

Section 4

Supplementary phenomena

9 i_3 nnn : i – indicator for supplementary phenomena

nnn – Specification of supplementary phenomena

Coded as given in ARFOR

Ex **1:**	ROFOR	060300Z	060612 KT VIDP		32580	
	403028	27010	405020	28015	407015	30020
	410010	33025	611409	523306	11111	32775
	40120	22222	38150	2715 =		

Decoded: Route forecast from (VIDP) Delhi to lat 25 °N, long 80° E, issued on 6th at 0300 UTC, valid for 6th between 0600–1200 UTC, windspeed measurement in knots.

Upper winds and temperature.

	FL/ft(M)	Temp	direction/speed
At	030/3000 ft (900 M)	28 °C	270°/10 kt
	050/5000 ft (1500 M)	20 °C	280°/15 kt
	070/7000 ft (2100 M)	15 °C	300°/20 kt
	100/10000 ft (3000 M)	10 °C	330°/25 kt

Icing : light icing in air above FL 140/14000 ft, thickness layer 270 M (900 ft)

Turbulence : Moderate turbulence in clear air above FL 330/33000 ft, thickness layer 180 meters (600 ft)

Jetstream : Core location lat 27°N, long 75 °E ICAO 3rd octant Jetcore height FL 400/40000 ft (12000 M)

Jetcore wind speed 120 kt

Wind Maxima : FL 380/38000 ft (11400 M)

Maximum wind speed 150 knots direction 270° Wind shear 15 knot/300 meters

***Ex* 2** ROFOR 070530Z 070915 KT VECC VANP

4000 444 3CB020 7///170

641809 541609 405016 16012 410006 24015

[415003 28025 4200M0 427045 4300M25

28080 11111 32881 40150 22222

40160 2825 =

***Decoded*:** Route forecast from (VECC) Kolkata to (VANP) Nagpur, issued on 7th at 0530 UTC, valid on 7th between 0900–1500 UTC

Visibility : 4000 Meters

Forecast weather : Hail

Cloud : 3 oktas CB (cumulonimbus) base 2000 ft (agl)

Freezing level : FL 170 (17000 ft)

Icing : Moderate icing above FL 180 (18000 ft), thickness layer 270 meters (900 ft)

Turbulence : Moderate turbulence above FL 160/(16000 ft), thickness layer 270 meters (900 ft)

Upper winds and temperatures at levels given below:

FL	Temperature	Wind direction/speed
FL050/5000 ft (1500 M)	16°C	160°/12 kt
FL100/1000 ft (3000 M)	06 °C	240°/15 kt
FL150/15000 ft (4500 M)	03 °C	280°/25 kt
FL200/20000 ft (6000 M)	– 04 °C	270° /45 kt
FL300/30000 ft (9000 M)	– 25 °C	280/80 kt

Jetstream: Location lat 28 ^{0}N, long 81 ^{o}E, ICAO 3rd octant

Jet core level FL 400 (40000 ft)

Jet core mean speed 150 kt

Wind Maxima: FL 400/(40000 ft)

Wind speed 160 kt, direction 280°

Wind shear 25 kt / 300 meters.

***Ex* 3:**

ROFOR	01000Z	010610KT	VECC	VILK	3SC030
2CBO30	3AC100	5CI300	7///170	621809	
541508	405022	28015	407010	28020	
410005	29030	4200M05	27045	4340M41	
27105	11111	32870	380120	22222	
36140	2825 =				

Decoded: Route forecast form (VECC) Kolkata to (VILK) Luknow, issued on 1st at 0000 UTC, valid for 1st between 0600–1000 UTC, winds speed reported in knots

Cloud : 3 okta SC (Stratocumulus), base height 3000 ft (900 meters)

2 okta CB (Cumulonimbus),base height 3000 ft (900 meters)

3 okta AC (Altocumulus), base height 10000 ft (3000 meters)

5 okta CI (Cirrus), base height 30000 ft (9000 meters)

Freezing level: Height FL170 (17000 ft)

Icing : Light icing in cloud above flight level 180 (18000 ft), thickness layer 270 meters (900 ft)

Turbulence : Moderate turbulence in cloud above FL 150 (15000 ft), thickness layer 240 meter (800 ft)

Upper winds and temperature: At levels given below:

FL	Temperature	Wind direction/speed
050 (5000 ft/ 1500 meters)	22 °C	280°/15 kt
070 (7000 ft/ 2100 meters)	10 °C	280°/20 kt
100 (10000 ft/ 3000 meters)	05°C	290°/30 kt
200 (20000 ft/ 6000 meters)	– 05°C	270°/45 kt
340 (340000 ft/ 10200 meters)	– 41°C	270°/105 kt

Jet stream : Location 28°N, 70°E, ICAO 3rd octant

Height of Jet core: FL 380 (38000 ft), wind speed jet core 120 KT.

Maximum wind : Height FL 360 (36000ft), wind speed 140 kt, wind direction 280°, wind shear 25kt/300 meters.

Ex 4 : ROFOR 150400Z 150813KT VOTV VECC 02380
6000 SCT 025CB 7160140 BKN 100 TEMPO 0912
1500 111 TS BKN012 SCT 020CB 7260 /// 651504
551103 410015 27025 415003 30025 4200M 05
30035 03VECC 6000 SCT025 SCT030CB BKN090
410014 24025 415M03 28030 4200M07 23030

38.9 WINTEM

(Forecast upper wind and temperature for aviation)

CODE FORM

Section 0

WINTEM $Y_F Y_F G_F G_F$ g_F g_F [KMH or KT or MPS]

Section 1

L_a^1 L_a^1 I_a^1 A L_o^1 L_o^1 L_o^1 L_o^1 B L_o^2 L_o^2 L_o^2 I_o^2 B ...... L_o^j L_o^j L_o^j I_o^j B

(TROP n_t n_t n_t n_t n_t n_t .... n_t n_t n_t)

(MAXW n_m n_m n_m d_m d_m f_m f_m f_m n_m n_m n_m d_m d_m f_m f_m f_m …… n_m n_m n_m d_m d_m f_m f_m f_m)

F n_1 n_1 n_1 dd fff STT dd fff STT ….. dd fff STT

F n_2 n_2 n_2 dd fff STT dd fff STT …… dd fff STT

: : : : : : …… : :

F n_k n_k n_k dd fff STT dd fff STT …… dd fff STT

L_a^2 L_a^2 I_a^2 A

(TROP n_t n_t n_t n_t n_t n_t n_t n_t n_t)

(MAXW n_m n_m n_m d_m d_m f_m f_m f_m n_m n_m n_m d_m d_m f_m f_m f_m ….. n_m n_m n_m d_m d_m f_m f_m f_m)

F n_1 n_1 n_1 dd fff STT dd fff STT ….. dd fff STT

F n_2 n_2 n_2 dd fff STT dd fff STT …… dd fff STT

….. ….. ….. ….. …..

….. ….. ….. ….. …..

L_a^j L_a^j I_a^j A

(TROP n_t n_t n_t n_t n_t n_t ……. n_t n_t n_t)

(MAXW n_m n_m n_m d_m d_m f_m f_m f_m …… n_m n_m n_m d_m d_m f_m f_m f_m……
n_m n_m n_m d_m d_m f_m f_m f_m)

F n_1 n_1 n_1 dd fff STT dd fff STT ….. dd fff STT

F n_2 n_2 n_2 dd fff STT dd fff STT …… dd fff STT

: : : : : : …… : :

F n_k n_k n_k dd fff STT dd fff STT ….. dd fff STT

Notes:

1. WINTEM is the name of the code used to provide forecast upper wind and temperature for aviation,
2. The forecast data are valid at the points of a rectangular geographical grid.
3. A WINTEM message is identified by the word WINTEM
4. The code form is divided into two sections

Section-0: Contains identification and time of validity forecast.

Section-1: Contains grid point coordinates and data groups for tropopause height, level of maximum wind with specified flight levels.

Regulations (code symbolic letters)

Y_FY_F – Day of the month (UTC) on which the WINTEM message is valid

$G_FG_Fg_Fg_F$ – Time in whole hours UTC, at which the WINTEM message is valid (g_Fg_F will be equal to 00).

KMH or KT or MPS: wind speed scale KMH or KT or MPS

A – Direction of latitude (N = North, S = South)

B – Direction of longitude (E = East, W = West)

$L_aL_aL_a$ – Latitude in terms of a degree, tenths are obtained by dividing the number of minutes by six (disregarding remainder)

$L_oL_oL_oL_o$ – Longitude in terms of a degree, tenths are obtained by dividing the number of minutes by six (disregarding the remainder).

$$\left.\begin{matrix} L_a^1 & L_a^1 & I_a^1 \\ L_a^2 & L_a^2 & I_a^2 \\ . & . & . \\ L_a^j & L_a^j & I_a^j \end{matrix}\right] \begin{matrix} \text{Latitude, coordinates of grid points, where} \\ I_a^j \text{ is tenths of a degree of latitude, } I_a^j = 0 \text{ or } 5 \end{matrix}$$

$$\left.\begin{matrix} L_o^1 & L_o^1 & L_o^1 I_0^{'} \\ L_o^2 & L_o^2 & L_o^2 I_0^2 \\ . & . & . \\ L_o^j & L_o^j & L_o^j I_0^j \end{matrix}\right] \begin{matrix} \text{Longitude coordinates of grid points, where} \\ I_o^j \text{ is tenths of a degree of longitude, } I_o^j = 0 \text{ or } 5 \end{matrix}$$

TROP – Tropopause

MAXW – Maximum wind

In print form, the format of the WINTEM is a direct reading data table.

The geographical grid will be rectangular (its boundaries are delineated by means of two meridians and two parallel circles).

The data associated with a given grid point will be included in the following order

(a) Tropopause height

(b) Maximum wind level

(c) Specified flight levels arranged in decreasing order.

Tropopause height and/or Maximum wind-level data will be omitted whenever these data are not required for operational purposes.

Example:

	110° E		115° E	120° E
15° S	Forecast values apply to the centre point of 5° squares of superimposed grid		**Data presentation** FL of tropopause FL dd fff of Maximum wind dd fff MTT FL 450 (150 hPa) dd fff MTT FL 390 (200 hPa) dd fff MTT FL 340 (250 hPa) dd fff MTT FL 300 (300 hPa) dd fff MTT FL 240 (400 hPa) dd fff MTT FL 180 (500 hPa)	
20° S	460 36029100 31060 M61 31080 M54 29095 M41 30095 M34 30065 M20 310 45 M 00	480 38032065 31040M64 32060M57 30055M45 28060M34 30040M19 31030M08		

Between 20–25° S and 110-115° E

Tropopause height FL 460 (46000 ft)

Max wind at FL 360 (36000 ft), direction 290°, speed 100 knots.

Level	Wind direction/speed	Temperature
At FL 450	310°/60 kt	– 61°C
At FL 390	310°/80 kt	– 54°C
At FL 340	290°/95 kt	– 41°C
At FL 300	300°/95 kt	– 34°C
At FL 240	300°/65 kt	– 20°C
At FL 180	310°/45 kt	– 00°C

Note: VOLMET is meteorological information for aircraft inflight.

The contents of VOLMET Broadcast contain current METAR, SPECI, TAF and SIGMET by means of continuous and repetitive voice broadcasts.

In India, VOLMET broadcasts are made at Mumbai and Kolkata and contain current aerodrome reports with appended trends, aerodrome forecasts and SIGMETs.

Mumbai Broadcast contain

SIGMETs of – Mumbai, Chennai

SPECI / METARs of – Mumbai, Ahmedabad, Chennai, Colombo, Karachi, Male.

TAFs of – Mumbai, Colombo, Male.

Kolkata Broadcast Contains

SIGMENTs of – Kolkata, Delhi,

SPECI / METAR of – Kolkata, Delhi, Dhaka, Yangon, Kathmandu

TAFs of – Kolkata, Delhi, Ho-chi-Minh

METARs include QNH/TEMP/DP/Trend forecasts.

(Trend broadcast between 0110 to 1610 only).

CHAPTER 39

SIGMET and AIRMET Information and Aerodrome Warnings

39.1 SIGMET

***Definition: SIGMET Information*:**

Information issued by a Meteorological Watch Office (MWO) concerning the occurrence or expected occurrence of special en-route weather phenomena which may affect the safety of aircraft operations.

SIGMET denotes significant meteorological information. It is issued by MWOs for their respective FIRs.

IMD MWOs are located at Mumbai, Kolkata, New Delhi, and Chennai. They issue SIGMET messages for their respective FIRs (Flight Information Regions) to the concerned FICs (Flight Information Centres), who are responsible to communicate the information to aircraft inflight.

SIGMET is valid for 4 hours (or 6 hours). It is issued every three hours or more frequently. In case of Tropical cyclones (TCs), Volcanic Ash (VA) an outlook is included providing information up to 12 hours beyond the period of validity-concerning the trajectory of volcanic ash cloud and position of tropical cyclone. SIGMET messages for VA and TC are updated at least every six hours.

SIGMET is issued for one or more of the following weather phenomena and aircrafts inflight are warned when these phenomena are occurring or expected to occur on the route ahead up to 500 NM or two hours flying time.

Weather Phenomena

At subsonic cruising level	At Transonic and supersonic level. This is identified by SEGMET SS
Active thunderstorm area	Moderate or severe turbulence
Tropical cyclone / tropical storm	Cumulonimbus clouds
Severe line squall	Hail
Heavy hail	
Severe turbulence	
Severe icing	
Marked mountain waves	
Widespread sandstorm / dust storm	
Volcanic ash	

SIGMET messages will contain the following information in the order indicated below

- Identification of the information / Location indicator
- Period of validity
- Phenomena
- Description
- Types of information (observed or forecast)
- Location
- Movement
- Expected development

SIGMET information can be obtained from Approach Area Control (AAC). Before crossing a FIR boundary the crew should obtain the current SIGMET for the next FIR to update or forewarned of the significant weather. The sequence of SIGMET messages are indicated by figures or combination of figures and letters.

Ex: "SIGMET 4"

Example 1: VECC SIGMET 3 VALID 251600/252200 VOMM FIR TC BIJLI OBS 15.5 N 83.2 E AT 1600 UTC FRQTS TOPS FL 450 WI 100 NM OF CENTRE MOV NW 10 KT INTSF OTLK TC CENTRE 260300 16.5 N 82.0 E 260600 16.5 N 81.5 E ON /LAND

Ex: YUCC SIGMET 5 VALID 221215/221600

YUDO – AMSWELL FIR SEV TURB OBS AT 1210 YUSB FL250 MOVE 40 KMH WKN

Example 2: VECF SIGMET2 VALID 200700/201100 VECC

CALCUTTA FIR FRQTS FCST TOPS FL450 23N 86E 26N 90E 50NM MOV SE 20KT INTSF(.)

Example 3: VABF SIGMET1 VALID 12 1715/12 2100 VABB

BOMBAY FIR MOD MTW OBS AT 1710 VAAS FL210 MOVE WKN(.)

Example 4: VIDF SIGMET3 VALID 161805/162200 VIDP

DELHI FIR SEV TURB OBS AT 1802 UT VIJP FL 350 MOV E 20 KT WKN(.)

Explanation: The fifth SIGMET massage issued for the AMSWELL (fictitious location) flight information region (identified by in abbreviated plain language and by YUCC AMSWELL area control centre) by the Donlon/ International (fictitious location) meteorological watch office (YUDO) since 0001 UTC; the message is valid from 1215 UTC to 1600 UTC on 22nd of the month; severe turbulence was observed at 1210 UTC over YUSB (Siby / Bistock –a fictitious location) at flight level 250, the turbulence is expected to move east wards at 40 kilometers per hour and to weaken in intensity.

Note: The elements turbulence and icing are normally assumed to be associated with an active thunderstorm area, tropical revolving storm and severe line squall. They are not specifically mentioned when SIGMET information is issued for active thunderstorm area, tropical revolving storm and severe line squall. However, they are mentioned when actually observed or reported by aircraft.

Note: The weather phenomena for sub-sonic cruising levels will be indicated using abbreviation as under.

1. Thunderstorm
   - obscured OBSC TS
   - embedded EMBD TS
   - frequent FRQ TS
   - squall line SQL TS
   - obscured with heavy hail OBSC TS HVYGR
   - embedded with heavy hail EMBD TS HVYGR
   - frequent with heavy hail FRQ TS HVYGR
   - squall line with heavy hail SQL TS HVYGR
2. Tropical Cyclone
   - tropical cyclone with 10 minute mean surface wind speed of 63 KMH/34 KT or more TC (+ cyclone name)
3. Turbulence
   - Sever turbulence SEV TURB
4. Icing
   - severe icing SEV ICE
   - severe icing due to freezing rain SEV ICE (FZRA)
5. Mountain wave
   - Severe mountain wave SEV MTW
6. duststorm
   - heavy sandstorm HVY DS
7. Sandstorm
   - heavy sandstorm HVY SS
8. Volcanic ash
   - Volcanic ash VA(+ Volcano name if known)

For transonic and supersonic cruising levels:

1. turbulence
   - moderate turbulence MOD TURB
   - severe turbulence SEV TURB
2. Cumulonimbus
   - Isolated cumulonimbus ISOL CB
   - Occasional cumulonimbus OCNL CB
   - Frequent cumulonimbus FRQ CB
3. hail
   - hail GR
4. Volcanic ash
   - Volcanic ash VA(+ Volcano name if known)

39.2 AIRMET Information

Definition: Information issued by a meteorological watch office concerning the occurrence or expected occurrence of specified en-route weather phenomena which may affect the safety of low level aircraft operations and which was not already included in the forecast issued for low level flights in the FIR concerned or sub-area thereof.

AIRMET information issued by a meteorological watch office in accordance with regional air navigation agreement, taking into account the density of air traffic operating below flight level 100 (FL 100).

AIRMET information provides a concise description in abbreviated plain language concerning the occurrence and / or expected occurrence of specified en-route weather phenomena, which have not been included in the area forecast for low level flights and which may affect the safety of low-level flights, and of the development of those phenomena in time and space. The information is indicated using one of the following appropriate abbreviations.

At cruising levels below F100 (or below flight level 150 (FL 150) in mountainous areas):

- Wide spread mean surface wind speed above 30 kt	SFC WSPD (+ wind speed and units)
- Wide spread areas affected by reduction of visibility to less than 5000 meters	SFC VIS (+ visibility)
- Isolated thunderstorms without hail	ISOL TS
- Occasional thunderstorms without hail	OCNL TS
- Isolated thunderstorms with hail	ISOL TSGR
- Occasional thunderstorms with hail	OCNL TSGR
- Mountains obscured	MT OBSC

- Widespread areas of broken or overcast cloud with base height less than 300 meters (1000 ft) agl

broken BKN CLD (+ height of the base and unit)

overcast OVC CLD (+ height of the base and unit)

- isolated CB without thunderstorm ISOL CB
- occasional CB without thunderstorm OCNL CB
- frequency CB without thunderstorm FRQ CB
- moderate icing
  (except for icing in convective clouds) MOD ICE
- moderate turbulence
  (except for turbulence in convective clouds) MOD TURB
- moderate mountain waves MOD MTW

Note: The specification for SIGMET information which is also applicable to low level flights.

AIRMET information shall be cancelled when the phenomena are no longer occurring or no longer expected to occur in the area.

Example: "YUDD AIRMET 1 VALID 151520 / 151800 YUSOSHANLON FIR TS OBS TOPS ABV FL100 N OF 50 DEGS STNR WKN"

Cancellation of AIRMET: "YUDD AIRMET 2 VALID 151650 / 151800 YUSOSHANLON FIR CNL AIRMET 1 151520 / 151800"

39.3 Aerodrome Warnings

Aerodrome warnings are issued by Aerodrome Meteorological Offices (AMOs) during forecasting watch hours, contain the concise information of meteorological conditions (in plain language), which could adversely affect aircraft on the ground including parked aircraft and aerodrome facilities and services. In respect of Aeronautical Meteorological Stations (AMS) these warnings will be issued by the associated AMOs at the state capital or regional centre.

Aerodrome warnings shall be cancelled when the conditions no longer expected to occur at the aerodrome. Aerodrome warnings are related to occurrence or expected occurrence of adverse weather phenomenon single or in combination.

The criteria for issue of Aerodrome warnings are given below.

1. *Wind*:
   (a) When wind speed expected to reach 30 knots or more even in gusts.

(b) When rapid direction change by 45° or more with wind speed before and after expected to be 20 knots or more.

2. *Squall*: Whenever squall expected, the likely direction and speed shall be indicated.
3. *Tropical cyclone*: This is to be included if the 10 minute mean surface wind speed at the aerodrome expected to be 34 knots or more.

   When affected by low pressure system with intensity classification cyclonic storm and above, the position of centre, expected direction of movement and speed of movement to be indicated. Also associated adverse phenomena for which warnings are issued.
4. Duststorm
5. Sandstorm
6. Thunderstorm
7. Hail
8. Frost
9. Hoar Frost or rime
10. Snow (including the expected or observed snow accumulation)
11. Freezing precipitation
12. Rising sand or dust
13. Volcanic ash

Warnings for gliders and light aircraft and helicopters are issued and appended to local forecast when wind speed is expected to reach 17 knots or more and issued with heading “Warnings for light aircraft”.

Reduction in visibility and lowering of cloud base associated with warning elements are mentioned in warnings.

Aerodrome warnings are issued at all aerodromes where AMOs are functioning during their watch hours.

Aerodrome warnings are passed on to local ATS units, who will further disseminate to the operators (over the aerodrome).

Aerodrome warnings are issued at least half to one hour earlier to the expected weather. The validity period in general should not exceed 4 hours. If the weather phenomena (like cyclone) expected to continue for a longer time, the warning extended suitably. Amendments to the warnings are also issued when necessary.

For example: If a warning for thunderstorm issued valid for 4 hours, but thunderstorm passed over the station quickly and no longer expected, a cancellation is issued. If warning was issued only for thunderstorm but if it is realized it will accompany with a squall or hail, then an amended warning will have to be issued accordingly.

Apart from this weather warnings are also included in local forecasts, which are anticipated over the aerodrome and 50 NM around during the validity period of local forecast.

39.4 Low Level Wind Shear Warnings

Wind shear warnings contain concise information of the observed or expected wind shear which could adversely affect aircraft on the approach path or take-off path or during circling approach between runway level and 500 meters (1600 ft) above that level and the aircraft on the runway during the landing roll or take-off run. Wind shear warnings are prepared by Aerodrome Meteorological Offices. The warnings are prepared and disseminated for aerodromes where wind shear is considered a factor for aircraft safety in accordance with local arrangements with the ATS authority and operators concerned. Where local topography shown to produce significant wind shears at heights in excess of 500 meters (1600 ft) above runway level, then 500 meters (1600 ft) above runway should not be considered restrictive.

Wind shear warnings are cancelled when aircraft reports indicate that wind shear no longer exists or alternately after an elapsed time of two hours.

The following weather phenomena generally associated with wind shear.

- thunderstorms, microbursts, funnel cloud (tornado or waterspout), gust fronts
- frontal surfaces
- strong surface winds coupled with local topography
- sea breeze fronts
- mountain waves (including low level rotors in the terminal area)
- low level temperature inversion

39.4.1 Evidence of Existence of Wind shear shall be Derived From

(a) ground-based wind shear remote-sensing equipment (for example-Doppler radar),

(b) ground based wind shear detection equipment (for example, a system of surface wind and/or pressure sensors located in an array monitoring a specific runway or runways and associated approach and departure paths),

(c) aircraft observations during the climb-out or approach phases of flight, or

(d) other meteorological information (for example, from appropriate sensors located on existing masts or towers in the vicinity of the aerodrome or nearby areas of high ground)

39.4.2 Format Dissemination of Wind shear Warning

1. wind shear warnings are prepared in abbreviated plain language
2. wind shear in approach area are to be reported.

   Ex: "WS WRNG SURFACE WIND 320/10KT

   WIND AT 60 M 360/25 APCH"

3. When micro-bursts are observed, reported by pilots or detected by ground-based wind shear detection or remote-sensing equipment, then the wind shear warning must include a specific reference to micro-burst

   Ex: "WS WRNG MBST APCH RWY 27"

4. when information got form ground-based wind shear detection or remote-sensing equipment, the warning includes specific sections of the runway and distances along the approach path or take-of path

   Ex: "WS WRNG 30KT AIRSPEED LOSS 4KM FINAL RWY 13"

5. when an aircraft report is used to prepare a wind shear warning or to confirm a warning previously issued, the aircraft report, including the type of aircraft, should be given unchanged in the warning.

   Ex: "WS WRNG B747 REPORTED MOD/SEV WS IN APCH RWY 27 AT 1510"

Ex: YUDD SIGMET 2 VALID 101200 / 101600 YUSOSHANLON FIR / UIR OBSC TS FCST TOPS FL390 S OF 54 DEG N MOVE WKN

Cancellation SIGMET

YUDD SIGMET 3 VALID 101345 / 101600 YUSOSHANLON FIR/UIR CNL SIGMET 2 101200 / 101600

Dissemination: The meteorological section will inform the weather warnings and cautionary meteorological reports issued immediately to the following authorities.

1. DATCO
2. Base operations officer (BOO)
3. Chief operations officer (COO)

It is the responsibility of above officers for further dissemination to the concerned.

Note: UIR – Upper Flight Information Region

39.4.3 Information for Parked and Moored Aircraft

Warnings for the protection of parked and moored aircraft relating, the following phenomena as determined by local agreement.

- Gale, squall, thunderstorm, sandstorm, duststorm, rising sand or dust, hail, frost, rime, snow, freezing precipitation, rough sea, swell.
- Unless otherwise agreed locally, gale warning will be issued when mean surface wind speed is expected to exceed 34 knots.

Note:

1. Aerodrome warning does not include poor visibility or low cloud base associated with the warning elements.
2. Aerodrome warnings issued only for the local aerodrome. It is meant for the protection of the aircraft parked over the aerodrome. These warnings are not disseminated beyond the aerodrome.

39.5 RAREP

Weather rader reports code is named RAREP, which is described below as used by IMD.

Code:	RAREP	FFBB iiiii YYGGg
	CHARACTER	$(b_1\ b_1\ b_1/r_1r_1r_1.....b_n\ b_n\ b_n/r_n\ r_n\ r_n)$
	INTENSITY	
	TENDENCY	$d_sd_sf_sf_s$ ALTD $(bbb/H_t\ H_t/rrr)$

Note: Rereps are normally reported at synoptic hours, depending on weather situations additional messages (even hourly) may be transmitted.

Regulations

***CHARACTER*:** It includes the following features.

(i) *EYE (eye)*: An echo identified definitely as the eye wall of a tropical cyclone.

(ii) *SPRL BND (spiral bond)*: A continuous or broken curved line of echoes recognizable as a spiral band associated with a cyclonic system or cyclone.

(iii) *SQL LN (squall line)*: This pattern normally will have a length to width ratio about 10 : 1 and about 60 km or more.

(iv) *BRKN LN (broken line)*: A broken line of echoes

(v) *SLD (solid)*: An area fully covered by echoes.

(vi) *BRKN (Broken)*: An area 4/8 to 7/8 covered with echoes

(vii) *SCT (scattered)*: An area 1/8 to 4/8 covered with echoes

(viii) *WDLY SCT (widely scattered)*: An area less than 1/8 covered with echoes

(ix) *ISLTD (isolated)*: Isolated solid mass of echo

(x) *ECHO ALOFT (echo aloft)*: Echo seen only at elevations higher than half the beam width

(xi) *bbb*: Azimuth in three digits (degrees) of points on the periphery of an echo area

(xii) *rrr*: Range (three digits) in units of kilometers

INTENSITY code

Code	For radars having Quantitative Measurement Facility		Other radars
	dBZ	Approximate rainfall rate mm/hr	
WK	23 to 32	Less than 4	Quantitatively determined as in draft weather Radar Manual
MDT	33 to 42	4 to 15	
STG	43 to 52	16 to 63	
VRY STG	53 or more	64 and above	

Note:

1. The rainfall rates are based on the relationship $Z = 200\ R^{1.6}$ and may be taken only as a rough guide.
2. Intensity is to be reported only of echoes within 200 km range.
3. The intensity of the strongest echo is the group is to be reported.

TENDENCY

Code	Meaning
INCG	Increasing
DCG	Decreasing
NO CHG	No change

Note: Tendency to be reported only in case of isolated cells or group of cells or a line mainly for aviation purpose.

d_sd_s - direction in tens of degrees towards when the echo or group of echoes is moving

f_sf_s - speed in kilometers per hour of the echo or group of echoes

In case of group of echoes or of a line only the overall movement of the group of echoes to be reported. Movement to be observed over a period of about 30 to 60 minutes (or with respect to previous observation).

ALTD (Altitude): Indicator for echo height information

H_t H_t: Height of top of echo above msl in kilometers.

Special phenomena will be reported at the discretion of radar meteorologist, such as Bright Band, Anomalous propagation, in plain language at the end of the message.

Notes regarding Character Report:

1. The groups within brackets ( ) may be repeated as many times as necessary.
2. In case of line echoes, the spiral band and eye wall, as many bbb/rrr points along the line as necessary be given to define the shape. The points may be given along the line in anticlockwise direction.
3. The case of areas, the point should begin from the northernmost point of the shape and continued in the anticlockwise order as in case 2 above.

4. The character of echoes to be reported in the order given the group description above.
5. In case of embedded echoes, for example SPRL BND or BRKN LN in a large area of echoes it should be reported as SPRL BND or BRKN LN in addition to the area reported

39.6 CODAR (Upper air report from an aircraft)

CODAR is the name of the code for an upper-air report from aircraft (other than weather reconnaissance aircraft) in figure code.

CODE Form

$M_i\, M_i\, M_j\, M_j$ YYGGg 99 $L_a\, L_a\, L_a$ $QL_oL_o\, L_o\, L_o$ $P_aP_aP_aB_zS_h$ $TTT_a\, n_s\, n_m$

(40 $L_a\, L_a\, L_a$ $Q_C\, L_o\, L_o\, L_o\, L_o$) ddfff

(41 $L_a\, L_a\, L_a$ $Q_C\, L_o\, L_o\, L_o\, L_o$ dd fff)

………………………………..

……………………………..

(49 $L_a\, L_a\, L_a$ $Q_C\, L_o\, L_o\, L_o\, L_o$ ddfff)

(6 HHHH)

***Note*:** CODAR report is identified by $M_i\, M_i\, M_j\, M_j$ = LLXX

Regulations General

(i) The code name CODAR is not included in the report

(ii) The identification group $M_i\, M_i\, M_j\, M_j$ included in the first line of the text of meteorological bulletin of CODAR reports.

Individual reports in the bulletin will not contain the group $M_i\, M_i\, M_j\, M_j$.

Wind data: If both spot and mean wind are available, spot winds are reported first.

If one spot wind only is reported, it refers to the position given at the beginning of the report.

In case of more than one spot wind, the position of the points where the second and the following spot winds were measured are included immediately before the relevant dd fff group(s).

Symbolic Letters of Code

M_iM_i – identification letters of the report

M_jM_j – identification letters of the part of the report or version of code

M_iM_i – LL For aircraft report

M_jM_j – XX No distinction

$L_aL_aL_a$ – latitude in tenths of a degree. Tenths be obtained by dividing the number of minutes by 6, the disregarding the remainder.

Q_C – Quadrant of the globe

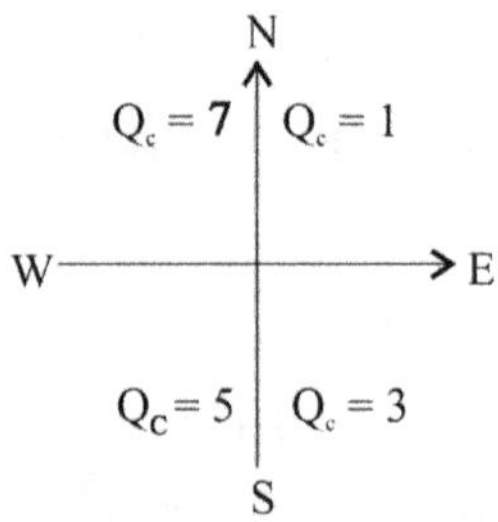

Code Fig	Latitude longitude
1	North East
3	South East
5	South West
7	North West

$L_oL_oL_oL_o$ – Longitude in tenths of a degree

$P_aP_aP_a$ – Pressure at the level at which the aircraft is flying (in hPa)

Where pressure corresponds to ICAO standard atmosphere (flight level indicated report received from the aircraft)

B_Z – High level turbulence

Code Fig. B_Z	
0	None
1	Moderate
2	Severe

S_h – Type of temperature and height data

Code Fig. S_h	
0	Observed air temperature-D-value positive
2	Observed air temperature-D-value negative
4	Observed air temperature- no D-value reported
6	Air temperature reduced to the nearest standard isobaric surface. Height reduced to nearest isobaric surface.

TT – Temperature (tens and unit digits) in degrees Celsius

T_a – approximate tenths value and sign (plus or minus) of the air temperature at the level given by $P_aP_aP_a$

dd – True direction, in tens of degrees from which wind is blowing

fff – wind speed in KMH or KT or MPS

HHHH – D-value or height reduced to the nearest standard isobaric surface in tens of meters.

CHAPTER 40

Meteorology for Helicopter Operations

Unlike fixed wing aircraft helicopter operations require some different considerations, which are briefly described in relation to different meteorological elements.

1. **Effect of surface wind**

   (a) *On ground operations:* For handling helicopter on ground some cautions have to be taken. When surface wind is 25 knots or more or gusty, on such occasions rotor blades may under go marked vertical oscillations (called blade sailing) when rotating slowly during starting up and stopping and to flap up and down at rest. Both blade sailing and flapping are serious hazards to personnel and equipment within the rotor diameter and also to the aircraft fuselage. Both these phenomena are likely to the lee of hills, large buildings (hangers etc.), large trees and other big structures. Some helicopters are fitted with stops to restrict these oscillations within safe limits in wind speeds up to about 40 knots. As wind speed increases, taxing the helicopter across wind becomes gradually more difficult.

(b) *During take-off and landing:* The direction and speed of wind are very important during take-off and landing of helicopter in very restricted areas. During night marked cooling may occur at surface and backing of wind as compared to a few hundred feet above ground. As a result turbulent conditions may develop on the last stage of descent of helicopter. Before landing is made in a valley, the wind direction must be checked because valley-winds direction may change considerably as compared to the general wind.

(c) *Running snow:* When a strong wind blows over frozen (or partly frozen snow) and grains of snow are set in motion across the surface, running snow is encountered. A layer of moving snow is formed and the pilot may be under the impression that he is moving forward, particularly when he is touching down over a large flat snowfield. Snow may become a frozen surface layer due to melting in the sun followed by refreezing. Several landings will break up this surface and in strong winds the snow may 'run' on the landing area.

2. **Effect of Turbulence:** Turbulence is hazardous in helicopter operations. In mountainous terrain turbulence rules in landing. It is likely that even entry into autorotation will not stop a climb induced by a strong updrafts or application of full power may not stop a helicopter from descending marked downdraft. The most serious effect of turbulence is the effect on retreating-blade stall speed. An up-gust increases the blades angle of attack and can cause a blade which is operating at a high angle of attack to stall. It is recorded that a helicopter with computed stall speed of 80 knots stalled at 40 knots because of turbulence.

   In mountainous regions, standing waves or rotor-streaming turbulence may be dangerous or even prohibitive for helicopter operation. There is a probability of sharp changes in wind speed and direction in rotor zones close to cliff-edges, which is a dangerous site for landing in strong winds.

3. **Effect of Temperature:** For a given throttle setting, an increased air-intake temperatures decrease the power output of a piston engine. For a gas turbine engine same applies for a given compressor speed. This will limit the power available and affect the performance of the helicopter. This has to be taken into account on performance of helicopter during planning operation. Low air-intake temperature may cause engine icing.

4. **Effect of Icing:** Horizontal visibility less than 1500 meters for safety reasons necessitates to reduce cruising speed to below

economic speed, which results in increasing fuel consumption and loss of endurance. If there is smoke or haze beneath an inversion, the haze is thicker at inversion than that at surface. An helicopter flying just below the inversion encounter poor visibility as compared at ground level.

5. **Effect of Cloud:** In general, low cloud base is not a problem to helicopter pilot, however flying close to the ground may create problem due to the turbulence. Flying in cloud, in general, may be avoided, although there may be limited flying in stable stratified cloud. Helicopter flying should not be planned in very turbulence cloud or in any type of cloud at sub-zero temperature. Diffused light conditions are very hazardous when flight is near snow covered mountain because in such conditions it is very difficult for the pilot to recognize relief (of orography) or distance to a point on the ground.

6. ***Helicopter Icing:*** In comparison to fixed wing aircraft, helicopters are more sensitive to the effect of ice-accretion. A small ice deposit on the blades may have very hazardous effect, as it destroys the aerodynamic lift, causes vibrations and increases the centrifugal stress on the rotor head. The rotor blades are finely balanced to avoid vibrations and the rotor head experiences large centrifugal stress. Ice accretion on the head may interfere with the control of the blades. Consequently, within minutes of entry into icing conditions dangerous situation may develop. Even on ground running, ice formation on the blades cause vibrations which are liable to transmit to the springs of the under-carriage, where sympathetic vibrations may become sufficiently large and overturn the helicopter. At high speeds there is a risk of ice pieces leaving the blades which may act as missiles. Problems also develop like injection of snow by engines, icing of windows and aerials.

   Normally helicopters are operated in clear skies, even which may effect the flight when ambient air temperature is sub-zero. Consequently icing protection systems are being developed, to enable helicopter operations in adverse winter conditions. The deposition of ice on helicopter blades on ground is expected to occur as for a fixed wing aircraft.

7. ***Precipitation:*** If visibility falls below 1500 meters in precipitation, there may be loss of endurance as expected above.

   When a flight is made at low speed in moderate/heavy snow or when helicopter is hovering, appreciable snow may collect on nose and may cover air-intake grills. This causes air starvation, loss of power and possibly engine failure. If this snow becomes compacted (by turning to ice) and brakes off and enters the engine turbine damage may occur. When the ground is covered with powdery snow, helicopter's taking-off and landing or flying close to the ground often experiences extremely poor visibility (white-out) because of re-circulation of the snow.

40.1 Weather Forecast for Gliders and Ballooning

For the purpose of gliding, Area/Local forecast in Met T-2 form is useful. In addition to the usual Area/Local forecast, special emphasis is given to the thermal activity, surface and low level inversion. In coastal areas sea breeze affects are emphasized, also mountain lee waves depending on the location. Period of validity mostly after sunrise to sufficiently before sunset. Forecast is always in conjunction with AIRMET.

We have already studied about thermal wind, however a few salient features are given here.

A thermal wind is caused by a temperature difference between two adjacent air masses. If the surface pressure is the same at two adjacent air masses but if there is temperature difference then surface wind will be calm but wind increases aloft. Wind aloft will be blowing parallel to the isotherms with low isotherm on the left in the Northern hemisphere (and on the right in the Southern hemisphere). Thermal turbulence results from convection currents set up by surface heating. Speed of thermal wind is proportional to the temperature gradient (closer the isotherms stronger the thermal wind).

In general low level (surface level) warm temperature associated with low pressure area, while at the same location in the upper level high pressure area and cold temperatures are associated, that is, low level convergence, rising of winds or cyclonic winds in the low level and anticyclonic winds in the upper level.

The model forecast for Gliding and Ballooning are given below.

A. GLIDING FORECAST

For---------- day----------month----------year

Met. Forecast No……….. Prepared by--------

Recipient site

LONDON GLIDING CLUB CHILTERNS

Period: DAWN TO DUSK

ALL HEIGHTS ABOVE MEAN SEA LEVEL

ALL TIMES UTC

NOTE: THIS FORECAST TO BE USED IN CONJUNCTION WITH AIRMET

THERMAL ACTIVITY: MOD TO AROUND 4500 FT LATE MORNING AND AFTERNOON;

TRIGGER TEMP PS15C. DYING AFTER ABOUT 1400Z

INVERSIONS: SLIGHT AT 5000 FT AND 8000 FT WITH SHALLOW SURFACE INVERSION INLAND AT FIRST

SURFACE WIND: SOUTHERLY 05 KT, BUT VARIABLE OR CALM AT TIMES

LEE WAVES: NIL

SEA BREEZE EFFECT: WEAK TO MODERATE/SEA BREEZE AFTER 0700Z

CONFINED TO SOUTH COAST AND NOT PENETRATING FAR INLAND

LOWEST FORECAST PRESSURE (QNH): 1016 HPA

LONDON GLIDING CLUB

B. BALLOONING FORECAST

NOTE: THIS FORECAST IS PROVIDED AS A SUPPLEMENT TO, AND NOT A SUBSTITUTE FOR "AIRMET" OR CHART FORMS 214/215

AREA	:	SOUTHEAST AND CENTRAL SOUTHERN ENGLAND
PERIOD	:	DAWN TO MIDDAY, FRIDAY 8TH MAY 2010
WINDS	:	(DEGREES TRUE AND KNOTS)
SURFACE	:	VRB 05 KT, BUT CALM AT TIMES
THERMALS	:	WEAk MOD to 5000 FT, TRIGGER TEMP PS14C
INVERSIONS	:	SLIGHT INVERSION AT 5000FT AND 8000 FT
SEA BREEZES	:	WEAK/MOD SEA BREEZE AFTER 0700Z CONFINED TO SOUTH COAST

LEE WAVES : NIL
LOWEST PRESSURE (QNH) : 1016 HPA
SURFACE AIR TEMPERATURE:
0600 : PS07C
0900 : PS10C
1200 : PS15C
OUTLOOK UNTIL DUSK: THERMALS BECOMING MODERATE

ISSUED AT 080400Z

TIME	UTC	AREA OF INTEREST

Question Bank

1. Hazardus (aviation) weather that is expected at Begumpet Hyderabad above a squall line:
   (a) Poor visibility (b) Severe updrafts
   (c) Severe down drafts (d) extreme turbulence
2. The most severe icing expected over mountains:
   (a) to the windward side
   (b) to the lee (ward) side
   (c) above the ridge to the windward side
   (d) just above the highest ridge
3. Flying at FL 50 and lowering stratus clouds encountering the associated weather expected with
   (a) a line squall (b) a warmfront
   (c) colder weather (d) anti-cylone
4. In flight maximum icing expected:
   (a) during falling pressure (b) in a high pressure area
   (c) with slowly rising barometer (d) in a frontal zone
5. Which front is steeper?
   (a) warm front, (b) cold front
   (c) occluded front (d) retarded front
6. In a Cb cloud where hail is formed, the vertical currents likely to be:
   (a) 80 kmph (b) 16 kmph
   (c) 80 mph (d) more than 160 kmph

7. To what height does a "Squall line" extend?

(a) 20,000 ft (FL 200) (b) 4000 ft (FL 40)

(c) 12000 ft (FL 120) (d) 8000 ft (FL 80)

8. Ice Seldom forms on a plane with in temperature zero (^{O}C) because:

(a) the moisture freezes with sleet before striking the plane

(b) the moist air forms snow crystals

(c) there is very little water present

(d) moisture forms line frost crystals that do not cling (stick) to plane at FL 50

9. If a pilot were flying at FL050 through a warmfront from the warm air side, to the cold air side he would expect to find his wind shift:

(a) roughly over the location of the surface front

(b) about 240 km (150 miles) beyond the position of surface front

(c) about 240 km (150 miles) before he reached the surface front

(d) at no place as he would be flying too high to encounter the front

10. If a pilot is flying from an airport where the pressure is high to another airport where the pressure is much lower, and leaves his altimeter set to the altimeter setting of his departure airport, on his arrival at his destination, he would find his altimeter would read:

(a) his exact elevation above the field

(b) higher than his true altitude

(c) lower than his true altitude

(d) exactly 400 ft higher than at departure

11. If a certain station should report a surface temperature of 8 ^{O}C, a pilot would normally expect to find icing conditions at an altitude of about:

(a) 2000 ft (b) 4000 ft

(c) 6000 ft (d) 10000 ft

12. In summer warm tropical air from the tropical Atlantic moves in over the coast where the temperature is much higher. The most likely result would be:
    (a) good visibility, smooth air except for sporading thunderstorms and sea fog
    (b) Cirrus cloud, smooth air and high winds
    (c) Stratus clouds, severe turbulence, and thunderstorms
    (d) Strato-cumulus clouds, low ceilings, and moderate turbulence up to about 10,000 ft

13. A fog formed by warm moist air flowing over a colder surface is called:
    (a) frontal fog (b) upslope fog
    (c) radiation fog (d) advection fog

14. If a pilot set his altimeter to read zero in the evening when he left his airplane, and in the morning he found it read 500 ft. The barometric pressure in the morning would be:
    (a) 15 hPa higher than before
    (b) 15 hPa lower than before
    (c) 33 hPa higher than before
    (d) 33 hPa lower than before

15. In the NH (lat 30-45 N) at altitude 2000 ft or 3000 ft the air flows generally:
    (a) parallel to the isobars and westwards
    (b) parallel to the isobars and east wards
    (c) across the isobars towards the south
    (d) across the isobars towards north

16. If you are flying at an indicated altitude of 1000 ft and the outside temperature is – 10 °C, your true altitude would be:
    (a) the same, since the temperature is standard for that level
    (b) higher than your indicated altitude
    (c) lower than your indicated altitude
    (d) about 500 ft higher than the altimeter indicate

17. If you are flying into an area marked on the weather map as "low" and holding a constant altitude, your actual altitude would be:
    (a) the same
    (b) lower than your indicated altitude
    (c) higher than your indicated altitude
    (d) increase at the rate of about 200 ft for each 3 hPa difference in pressure

18. If you are flying into an area marked on the weather map as 'High' and holding a constant altitude, your actual altitude would be:
(a) the same
(b) lower than your indicated altitude
(c) higher than your indicated altitude
(d) decrease at the rate of about 200 ft for each 3 hPa difference in pressure

19. In approaching a rapidly moving cold front a pilot would expect to find, (at a point) 320 km (200 miles) in advance of the front:
(a) snow storms mixed with sleet
(b) pre-frontal storms
(c) hail and rain
(d) clear weather with stratiform clouds visible near the front

20. On the lee side of mountain and buildings, you would normally expect to find:
(a) downdrafts (b) updrafts
(c) eddies and downdrafts (d) eddies and updrafts

21. If you have to fly through an occlusion running north-south, you should, in relation to the warm sector of the occlusion, fly to the:
(a) northeast of it (b) northwest of it
(c) south of it (d) west of it

22. The clouds which are mainly characteristic of a warm front are known as:
(a) alto stratus (b) nimbo stratus
(c) cumulus (d) cumulonimbus

23. The word "Frontolysis" printed on a weather map means:
(a) the generation of, or intensification of a front
(b) the dissipation of a front
(c) the result of a warm and cold front meeting
(d) the state of a cold front overtakes a warm front and under runs it

24. The word 'Frontogenisis' printed on a weather maps means:
(a) the generation of, or intensification of a front
(b) the dissipation of a front
(c) the result of a warmfront and cold front meeting
(d) the state of a cold front overtakes a warmfront and underruns it

25. If you were flying in an area about 80 km (50 miles) in advance of a typical warm front, the type of clouds which you could expect to encounter would be:
(a) High clouds (b) Cb
(c) Ac (d) Ns

26. On a clear, calm night which of the following fogs do you expect?
(a) steam (b) advection
(c) radiation (d) frontal

27. The typical weather associated with a warm front would be characterized by:
(a) low ceiling and good surface visibility
(b) low ceiling, poor surface visibility and widespread showers
(c) scattered to broken cloudiness with occasional showers
(d) low ceilings, poor surface visibility and steady precipitation over a large area

28. A certain area station reports a surface temperature of 7 °C and surface dew-point 5 °C. A steady drizzle is falling at the station. A pilot flying over the station would expect to encounter icing conditions at an altitude of about:
(a) 2000 ft (b) 3500 ft
(c) 4500 ft (d) 10,000 ft

29. A certain station reports surface temperature, dew point 7 °C, 7 °C respectively. A steady rainfall occurring at the station. A pilot flying over the station would expect to encounter icing conditions at about an altitude of:
(a) 2000 ft (b) 3500 ft
(c) 4500 ft (d) 10,000 ft

30. A certain station reports surface temperature, dew point 9 °C, 0 °C respectively. A pilot flying over the station would expect to encounter icing conditions at about an altitude of:
(a) 4500 ft (b) 3000 ft
(c) 12080 ft (d) 5000 ft

31. If the wind above certain station is reported to be from westerly at an altitude of about 3000 ft, what would you normally expect to be surface wind from:
(a) NW (b) W
(c) SW (d) E

32. During a cross-country flight, you find that the ceilings are considerably lower than were forecast. In making a radio report to the weather Bureau, you would report the height of the clouds:
   (a) above ground level
   (b) above your flight level
   (c) on the basis of pressure altitude
   (d) above sea level after calling for a current altimeter setting
33. A pilot planning a long distance flight in a westerly direction would find the most favourable winds:
   (a) immediately behind a cold front
   (b) north of a High pressure area
   (c) south of a High pressure area
   (d) south of a low pressure area
34. You are flying a constant heading and observe a sequence of Ci, Cs and Ac clouds. This would indicate:
   (a) a pre-frontal thunderstorm (b) lowering temperature
   (c) lowering pressure (d) steady rain
35. Presence of hail in a cloud always indicate:
   (a) Ns
   (b) high winds
   (c) strong verical currents
   (d) stable weather conditions
36. An airplane should land at a higher than normal speed when:
   (a) there is a calm wind conditions
   (b) landing on a snow covered run ways
   (c) the wind velocity is greater than 50% of the normal stalling speed
   (d) the wind is gusty
37. A pilot is enroute on an instrument flight. ATC requests that he makes a report of current weather conditions in his vicinity. In so doing the pilot should report the base of the clouds:
   (a) relative to his pressure altitude reading
   (b) from his altimeter reading corrected for temperature
   (c) above the terrain
   (d) relative to their altitude above sea level

38. You are planning a flight in the vicinity of a cold front. The weather map indicated "Frontogenesis" in close proximity. This should indicate you that:
   (a) the cold front is dissipating and your flight planning can proceed with no apprehensions about the weather
   (b) this particular front will remain stationary for next 24 hours
   (c) the cold front is intensifying and you should plan your flight accordingly
   (d) this front may veer to the north west and remain clear of your proposed route

39. On an instrument flight a pilot noted that each station on his route reported a lower pressure than the preceding station. He should know that if he does not keep his altimeter corrected his actual altitude will be:
   (a) higher than his indicated altitude
   (b) remain the same
   (c) lower than his indicated altitude
   (d) be erratic and constantly change to higher and lower readings

40. A pilot on an instrument flight requests a weather report. He is given the report and in addition is advised that a squall line lies in the path of his flight. He would be justified in assuming that the area of squall line will:
   (a) have thick stratus clouds
   (b) have reduced visibility due to dense fog
   (c) consists of a series of thunderstorms
   (d) have exceeding high temperature

41. Thunderstorm activity on the windward side of mountain ranges is caused by:
   (a) mechanical lifting of the air, resulting in an increase in the air temperature
   (b) mechanical lifting of the air, resulting in decrease in the air temperature
   (c) thermal lifting of the air, resulting in an increase in temperature
   (d) thermal lifting of the air, resulting in violent eddies

42. An instrument pilot is informed that there may be low clouds of bad weather on his proposed flight path. This type of cloud is described as:
   (a) Sc (b) Cs (c) Ns (d) Cu

43. A smooth local flight may be expected if the radiosonde observation shows a:
(a) DALR
(b) SALR
(c) temperature inversion
(d) Super adiabatic lapse rate

44. An instrument pilot is warned of a possible icing area along his proposed route. He should know that ice generally forms on aircraft when:
(a) the aircraft is being flown in cirrus clouds
(b) ice crystal are present in the area of flight
(c) super cooled water droplets are present
(d) the outside air temperature is below 0 °C

45. The station circle (model) on a weather map indicates changes that have occurred over a six-hour period with exception of:
(a) visibility (b) temperature
(c) dew point (d) pressure tendency

46. Icing on aircraft leading edges is generally determined by the super cooled water content of the air through which the aircraft passes. Super cooled water exist when liquid water is cooled below:
(a) 5 °C (b) 10 °C (c) 35 °C (d) 32 °F

47. Weather may decrease in intensity if there is:
(a) an increase in the speed of a cold front
(b) an increase in the slope of a cold front
(c) a decrease in the speed of a cold or warm front
(d) a frontogensis reaction

48. The region in the atmosphere between ground and stratosphere is known as:
(a) Thermosphere (b) lithosphere
(c) troposphere (d) ionosphere

49. The change in temperature of the atmosphere with height is called:
(a) adiabatic change (b) isothermal change
(c) lapse rate (d) SALR

50. Increase in temperature with height in the atmosphere is called:
(a) lapse rate (b) adiabatic change
(c) inversion (d) isothermal

51. Lines of equal pressure are called:
(a) isobars (b) isollobars
(c) isotachs (d) isotopes

52. Lines of constant wind speed on weather charts are called:
(a) isobars (b) isollobars
(c) isotachs (d) isotopes

53. Lines of equal pressure tendency are called:
(a) isobars, (b) isollobars
(c) isotachs (d) isotopes

54. When a parcel of air is subjected to changes in pressure, temperature and volume without loss of heat (which remains constant), the system process is called:
(a) adiabatic (b) isothermal
(c) expansion (d) contraction

55. Norwesters occur in .... and are frequent in the month of........
(a) North west India, July
(b) Jammu and Kashmir, December
(c) Bengal, May
(d) Kerala, August

56. Westerly jet streams appear over north India during:
(a) January (b) April
(c) August (d) November

57. The east coast of Tamilnadu gets its bulk rainfall during:
(a) South west monsoon (b) North east monsoon
(c) premonsoon (d) Western disturbances

58. Instrument for measuring humidity is called:
(a) thermograph (b) hygrograph
(c) hytograph (d) pyromograph

59. The conditions that exists when the distribution of winds within a given area is such that there is a net horizontal out flow of air outwards from the region is called:
(a) updrafts, (b) downdrafts
(c) convergence (d) divergence

60. The distribution of winds within a given area is such that there is net horizontal in flow of air inside the region is called:
(a) updrafts (b) downdrafts
(c) convergence (d) divergence

61. The tropical tropopause level during winter season:
(a) comes down (b) goes up
(c) oscillates (d) does not change

62. The average height of tropopause over equator and over poles are:
(a) 30 km, 15 km (b) 16 km, 8 km
(c) 10 km, 5 km (d) 16 km, 16 km

63. The only tropical ocean in which no tropical revolving storms, occurs is:
(a) Indian ocean (b) North Atlantic
(c) South Atlantic (d) South Pacific

64. Tropical revolving storms in Bay of Bengal/Arabian sea are called:
(a) Hurricanes (b) Typhoons
(c) Cyclones (d) Willy willies

65. Tropical revolving storms in China sea are called:
(a) Hurricanes (b) Typhoons
(c) Cyclones (d) Willy willies

66. Tropical revolving storms in West Indies are called:
(a) Hurricanes (b) Typhoons
(c) Cyclones (d) Willy Willes

67. Tropical revolving storms in the vicinity of Australia are called:
(a) Hurricanes (b) Typhoons
(c) Cyclones (d) Willy Willies

68. The steady winds which flow on the equatorial side of the sub tropical High pressure regions in both Hemisphere are called:
(a) Katabatic winds (b) Anabatic winds
(c) Fohn winds (d) Trade winds

69. If the pressure is falling all around the centre of a Tropical depression, it is called:
(a) Frontolysis (b) Frontogenisis
(c) deepening (d) Filling

70. If the pressure is rising all around the centre of tropical depression, it is called:
(a) Frontolysis (b) Frontogenisis
(c) deepening (d) filling

71. The speed of a warmfront is generally considerably .... to the geostrophic component
(a) less than (b) more than
(c) equal (d) no relation

72. The speed of a cold front is generally .... to the geostrophic component
(a) less than (b) more than
(c) equal (d) none above

73. A front always tends to move towards the direction of:
(a) falling pressure (b) rising pressure
(c) steady pressure (d) none above

74. A warm anticyclone is likely to collapse if invaded by a:
(a) warm front (b) cold front
(c) cold occlusion (d) both (b) and (c)

75. The temperature of the air is measured in Meteorological observatories in a well ventilated shade at a height above ground level of:
(a) 2 ft (b) 4 ft
(c) 30 ft (d) zero ft (ground)

76. The boiling point of water at sea level corresponds to:
(a) 373 °K (b) 273 °K
(c) 212 °K (d) 100 °K

77. The freezing point of water or melting point of ice at sea level corresponds to:
(a) 0 °F (b) +32 °F
(c) 100 °F (d) 273 °F

78. The boiling point of water at sea level corresponds to:
(a) 100 °F (b) 212 °F
(c) 273 °F (d) 373 °F

79. The diurnal variation of surface pressure in middle latitudes is about:
(a) less than 1 hPa (b) 2 hPa
(c) 4 hPa (d) 10 hPa

80. The diurnal variation of surface pressure in equatorial region is about:
(a) less than 1 hPa (b) 2 hPa
(c) 4 hPa (d) 10 hPa

81. The diurnal variation of surface pressure in polar regions is about:
(a) less than 1 hPa (b) 2 hPa
(c) 4 hPa (d) 10 hPa

82. 'D' – value is calculated by subtracting:
(a) Pressure altitude from true altitude
(b) Pressure altitude from flight level
(c) true altitude from height above msl
(d) true altitude from flight level

83. While flying over the sea, the difference between simultaneous reading of radio altimeter and pressure altimeter, the sub-scale of the later being set to 1013.2 hPa gives:
(a) QNH
(b) QFE
(c) true altitude amsl
(d) the altimeter correction

84. 'D' value (altimeter correction) is:
(a) always positive (b) always negative
(c) either positive or negative (d) always zero

85. Surfaces of constant atmospheric pressure which are related to a specific pressure datum 1013.2 hPa, and separated by specific pressure intervals are reffered to as:
(a) freezing level (b) icing level
(c) condensation level (d) flight level

86. The height indicated by an ICAN altimeter when the sub scale is set to 1013.2 hPa is the:
(a) Pressure altitude (b) freezing level
(c) Condensation level (d) icing level

87. The height of the aerodrome in standard atmosphere correspond to airport surface pressure is reffered to as:
(a) QFE (b) QNH (c) QNE (d) QFF

88. The pressure altitude of the aerodrome is:
(a) QFE (b) QNH (c) QNE (d) QFF

89. The temperature attained by a parcel of air when it is compressed (or expanded) adiabatically to a pressure of 1000 hPa is called:
(a) Drybulb temperature (b) dew point temperature
(c) virtual temperature (d) potential temperature

90. The geostrophic force (or coriolis force) results from:
(a) Solar heating
(b) temperature variation
(c) rotation of the earth on its axis
(d) revolution of the earth around the sun

91. Clear Air Turbulence is observed near:
(a) jet streams
(b) Cold fronts
(c) Cumulonimbus cloud
(d) tropical revolving storm

92. Flying through a thunderstorm the region between freezing level and – 10 °C is:
(a) the most hazardous for flight
(b) the least hazardous for flight
(c) the least favourable for icing
(d) both (b) and (c)

93. The vector difference between the geostrophic wind at a higher level and the geostrophic wind at a lower level is called:
(a) gradient wind (b) Cyclostropic wind
(c) thermal wind (d) both (a) and (b) above

94. The DALR of the atmosphere is about:
(a) 5 °C/km
(b) 10 °C/km(or 3 °C/1000 ft)
(c) 15 °C/km
(d) 6.5 °C/km

95. For a particular parcel of air its virtual temperature is
(a) variable (b) increases
(c) decreases (d) constant

96. At sea level the SALR is about
(a) 1.8 °C/1000 ft (b) 3 °C/1000 ft
(c) 5 °C/1000 ft (d) 6 °C/1000 ft

97. In the atmosphere where temperature is less than – 40 °C
(a) SALR > DALR
(b) SALR < DALR
(c) SALR = 6 °C/km
(d) SALR = DALR = 3 °C/1000 ft

98. At msl, the dew point lapse rate (DPLR) is about:
(a) 0.5 °C/1000 ft (b) 1 °C/1000 ft
(c) 1.8 °C/1000 ft (d) 3 °C/1000 ft

99. The cloud base height is (T = surface temperature, T_d = surface dew pont)
(a) $(T - T_d)100$ (b) $(T - T_d)\ 200$
(c) $(T - T_d)\ 300$ (d) $(T - T_d)\ 400$

100. When ELR is greater than DALR then the air is said to be:
(a) stable (b) conditionally stable
(c) unstable (d) neutral

101. When ELR is less than SALR, then the air is said to be:
(a) stable (b) conditionally stable
(c) unstable (d) neutral

102. On a tephigram, the dry adiabat from surface dry bulb temperature, saturate adiabat from wet bulb temperature and isolygric from dew point temperature meet at a point, which is called
(a) Normand point (b) freezing point
(c) turbulence level (d) entropy

103. Normand point on a tephigram indicates:
(a) turbulence level, (b) freezing level
(c) lowest condensation level (d) jetstream level

104. The appearance of lenticular clouds indicate the presence of
(a) cold front
(b) warm front
(c) occluded front
(d) standing waves in the lee ranges of hills

105. A wind which flows under the influence of only pressure gradient and coriolis force and which is balanced, is called:
(a) gradient wind (b) cyclostrophic wind
(c) thermal wind (d) geostrophic wind

106. A wind which flows in a curved path under the influence of pressure gradient, coriolis force and centripetal force and which is balanced is called:
   (a) gradient wind (b) cyclostrophic wind
   (c) thermal wind (d) geostrophic wind

107. The temperature at which the air would be cooled to saturate it at constant pressure, is called:
   (a) dry bulb temperature (b) wet bulb temperature
   (c) dew point temperature (d) potential temperature

108. If the temperature of a sample of air is higher than the dew point temperature, the sample of air is called:
   (a) saturated (b) unsaturated
   (c) super saturated (d) stable

109. If the temperature and dew point of a sample of air are equal, then the sample of air is called:
   (a) saturated (b) unsaturated
   (c) supersaturated (d) stable

110. As compared to the visibility in mist, the visibility in fog is:
   (a) less (b) more
   (c) equal (d) not comparable

111. For the formation of dew and fog, clear skies:
   (a) assist (b) hinder
   (c) no effect (d) cant say

112. IN the formation of radiation fog, a light wind (2 to 4 knots) is:
   (a) necessary (b) harmful
   (c) no effect (d) not related

113. In a Cb cloud the maximum values of both updrafts and downdrafts occur in:
   (a) the base of the cloud
   (b) the top of the cloud
   (c) the middle and higher levels of the cloud
   (d) none above

114. If lightning flash is seen but thunder is not heard then it can be inferred that:
(a) the Cb cloud is very distant
(b) the thunder did not occur at the time of lightning
(c) the Cb cloud was not fully developed
(d) the Cb cloud was dissipated

115. In developed Cb the height band between freezing level and FL 250 is the worst:
(a) in regard to bumpiness (b) icing
(c) hail formation (d) all above

116. The procedure to avoid hazard due to hail while flying is to fly:
(a) just above freezing level
(b) between freezing level and F 250
(c) as far below freezing level as practicable
(d) just below F 250

117. In the night, wind blowing from land towards sea is called:
(a) Katabatic wind (b) Anabatic wind
(c) land breeze (d) sea breeze

118. The major hazard in Altocumulus lenticularis is:
(a) updrafts (b) downdrafts
(c) poor visibility (d) icing

119. Heading north in NH you experience starboard drift when:
(a) flying towards high pressure
(b) flying towards low pressure
(c) flying away from high pressure
(d) flying away from low pressure

120. In NH, you experience port drift when:
(a) flying towards high pressure
(b) flying towards low pressure
(c) flying away from high pressure
(d) flying away from low pressure

121. In India, the mean position of the sub-tropical jet stream found roughly at lat .... at a height of ...... and with mean wind speed of.....
(a) 15 °N; 16 km; 90 kt (b) 20 °N; 10 km; 150 kt
(c) 27 °N, 12 km; 100 kt (d) 35 °N; 8 km; 150 kt

122. Cloud that is associated with a high degree of turbulence is:
(a) Cu (b) Sc (c) St (d) As

123. Showery type of precipitation is associated with cloud:
(a) Cu type (b) St type
(c) Ci type (d) As type

124. In a squall the minimum wind speed should be:
(a) 6 BF (b) 9 BF
(c) 12 BF (d) 15 BF

125. Haboob is the name given to:
(a) Tropical rotating storms
(b) extra tropical depressions
(c) severe local thunderstorms in West Bengal
(d) Summer dust storms originating with Cb cloud near Khartoum

126. The critical temperature at which contrails possible are marked by a line on a tephigram and this line is called:
(a) freezing level (b) MINTRA level
(c) Drytra level (d) MAXTRA level

127. Nephoscope is an instrument used for:
(a) observing wind speed
(b) observing cloud height
(c) observing wind direction on runway
(d) observing the movement of clouds

128. A transmissometer is used to measure:
(a) wind velocity (b) wind direction
(c) visibility (d) cloud ceiling

129. RVR observations are taken when horizontal visibility becomes:
(a) less than 10000 meters (b) less than 6500 meters
(c) less than 4000 meters (d) less than 1500 meters

130. The surface discontinuity between the polar easterlies and the temperate westerlies is called:
(a) the polar front (b) the equatorial front
(c) tropopause (d) polar tropopause

131. Depressions move towards the area of .... and anticyclones move towards area of ....
(a) rising pressure; falling pressure
(b) no change in pressure; no change in pressure
(c) falling in pressure; rising in pressure
(d) constant; falling in pressure

132. In the Bay of Bengal tropical depressions in the month of July form:
(a) at the head of the Bay (b) Central Bay
(c) South Bay (d) South west Bay

133. In the Bay of Bengal the favourable area of formation of tropical depressions in the month of December:
(a) North Bay (b) Central Bay
(c) Head of the Bay (d) South Bay

134. The altimeter of an aircraft standing at an airfield indicated a height of 450 feet when the sub-scale of the instrument was set to 1013 hPa. If the QNH at the airfield was 1000 hPa, what was the elevation of the airfield? (use 1 hPa pressure change represents a height change of 27 feet)
(a) 99 ft (b) 549 ft (c) 351 ft (d) 801 ft

135. If the geostrophic wind at 8000 ft is 270°/15 kt and the geostrophic wind at 2000 ft is 090/10 kt, the thermal wind between 2000 ft and 8000 ft is:
(a) 270°/25 kt (b) 090°/5 kt
(c) 270°/5 kt (d) 090°/25 kt

136. If the geostrophic wind at 8000 ft is 360°/10 kt and the geostrophic wind at 2000 ft is 180°/15 kt, the thermal wind between 2000 ft and 8000 ft is:
(a) 360°/25 kt (b) 360°/5 kt
(c) 180°/25 kt (d) 180°/5 kt

137. If the geostrophic wind at 8000 ft is 045°/20 kt and the geostrophic wind at 2000 ft is 225°/10 kt, the thermal wind between 2000 ft and 8000 ft is:
(a) 045°/30 kt (b) 045°/10 kt
(c) 225°/10 kt (d) 225°/30 kt

138. If the geostrophic wind at FL80 is 270°/25 kt and the geostrophic wind at FL20 is 270°/15 kt, the thermal wind between FL 020 and FL080 is:
(a) 270°/40 kt (b) 270°/10 kt
(c) 090/40 kt (d) 090/10 kt

139. When the eye of a tropical cyclone crosses a station, the wind changes direction through:
(a) 360° (b) 180° (c) 090° (d) O°

140. In general pressure gradients are steeper in .... hemisphere
(a) Summer (b) Winter
(c) northern (d) southern

141. A vector difference between the real wind and the geostrophic wind is called:
(a) thermal wind (b) gradient wind
(c) ageostrophic wind (d) cyclostrophic wind

142. A front at which the warm air ascending the frontal surface to higher altitudes is termed:
(a) warm front (b) cold front
(c) occluded front (d) ana front

143. In free air temperature below – 30 °C, mist consisting of ice crystals or very light ice fog is called:
(a) Polar mist (b) diamond dust
(c) Arctic mist (d) blizzard

144. Within troposphere if the direction and wind speed remains the same in a vertical plane, it is called:
(a) homogeneous atmosphere (b) barotropic atmosphere
(c) baroclinic atmosphere (d) Isothermal atmosphere

145. Precipitation of very small ice crystals that are super saturated at temperature below – 30 °C in interior Antarctica is called:
(a) Arctic mist (b) Polar mist
(c) diamond dust (d) blizzards

146. A violent squall of wind from mountain ravines on the Pacific side of Central America is called:
(a) Hurricane (b) line squall
(c) gully squall (d) microburst

147. The ratio of the mass of water vapour to the mass of dry air, expressed in grams of water vapour per kilogram of dry air is called:
(a) relative humidity (b) humidity mixing ratio
(c) isohygric (d) specific humidity

148. A line drawn on chart joining places of equal rainfall amount is called:
(a) isotherm (b) isobar
(c) isopleth (d) isohyet

149. A meteorological centre designated to prepare and supply upper air forecasts in digital form on a global basis to RAFC (regional area forecast centres) is called:

(a) WMO (b) WAFC
(c) WAFS (d) MWO

150. A system by which world and regional area forecast centres provide aeronautical meteorological en-route forecasts in a uniform standard format is called:

(a) WMO (b) WAFC
(c) WAFS (d) RAFC

151. Dry, dusty, hot (very hot by day) continental tropical air flow from equator side of the Sahara Desert is called:

(a) Bora (b) Mistral
(c) Cohan (d) Hermattan

152. The hot and dusty winds (or dust storms) occur in spring or summer over Arabia and north African desert that last about 20 minutes are called:

(a) Bora (b) Mistral
(c) Simoon (d) Hermattan

Fill in the Blanks

153. Aviation weather codes are used for exchange of .... and ....weather data/information for aviation safety and efficiency.
(a) climatological data; extreme data
(b) observed; forecast
(c) location; hazardous
(d) wind maxima; jet stream

154. In METAR/TAF wind direction always indicates .... in degrees, rounded off to the nearest .... from which wind is flowing.
(a) true direction; 10°
(b) magnetic direction; North or South
(c) 8 points of compass; $22\frac{1}{2}^{\circ}$
(d) 16 points of compass; 12°

155. In METAR/TAF wind speed is given in:
(a) knots (b) KMH
(c) MPS (d) all above

156. VRB stands for direction of wind as variable only when the speed is ....
(a) 3 kt or less (b) 2 MPS or less
(c) 6 KMH or less (d) all above

157. VRB 12 KT means, wind speed 12 knots but wind direction change is:
(a) 60° or more (b) 90° or more
(c) 180° or more (d) none above

158. VRB 12 KT 300V080 means wind speed 12 knots, but wind direction change is:
(a) 300° to 080° veering (b) 300° to 080 backing
(c) 080° to 300° veering (d) 080° to 308 backing

159. In a METAR, wind reported as 09018 G 30 KT 060 V 140. This stands for:
(a) mean wind 090° and speed 18 kt, gusting to 30 KT
(b) direction changed between 060° to 140° during 10 minutes before the report
(c) mean wind speed 18 KT, direction 060° changes to 30 KT with direction 140°
(d) both (a) and (b)

160. In a METAR, calm wind is reported as:
(a) CALM (b) 99999
(c) 00000 (d) 00000 KT

161. In a METAR, during 10 minutes period preceding the observation if the maximum wind speed in gusts 28 KT exceeds the mean 15 kt by 10 KT and direction 270° will be reported as:
(a) 27015 KT (b) 270 15 G 28 KT
(c) 27028 KT (d) 270 10 G 25 KT

162. In a METAR, when horizontal visibility is different in different directions the minimum visibility is reported followed by direction code when it is less than 5000 m. The code of horizontal visibility 1400 S 6000 NE is:
(a) Visibility is 1400 meters in the south
(b) Visibility in 6000 meters in the northeast
(c) lowest visibility is 1400 meters in the south but 6000 meters (6 km) the northeast.
(d) visibility 1400 meters

163. What will be reported in METAR when visibility is 10 km or more, no cloud with base below 1500 m and no Cb and no significant weather phenomena.

(a) NOSIG (b) NSC

(c) SKC (d) CAVOK

164. What does it indicate if the wind group in METAR reported 270 p99 KT?

(a) Variable wind speed direction 270°
(b) direction 270°, wind speed 99 KT
(c) direction 270°, wind speed more than 100 KT
(d) none above

165. What does it indicate the visibility group in METAR reported as 9999?

(a) Visibility unknown
(b) Visibility variable
(c) Visibility 9999 meters
(d) Visibility 10 km or more

166. In METAR runway visual range (RVR) is reported only when visibility is less than 1500 meters. What does the group R09/P 1500 U in METAR indicates?

(a) RVR on runway 09 is less than 1500 meters
(b) RVR on runway 09 is 1500 meters
(c) RVR on runway 09 is more than 1500 meters and tendency over a period of 10 minutes just before time of observation is increasing
(d) None above

167. In a METAR what does the group R27/MO150 N indicate?

(a) RVR on runway 27 is 150 meters
(b) RVR on runway 27 is more than 150 meters
(c) RVR on runway 27 is less than 1500 meters with tendency of improvement
(d) RVR on runway 27 is less than 150 meters and tendency over a period of 10 minutes before time of observation no change

168. What does it indicate FEW 010 CB in a METAR?
(a) cloud amount 1 to 2 oktas CB cloud with base 100 meters
(b) cloud amount 2 to 3 oktas, CB cloud with base 1000 ft (300 m)
(c) cloud amount 1 to 2 oktas CB cloud with base 1000 ft (300 meters)
(d) cloud amount 4 to 5 oktas of CB cloud with base 1000 feet (300 m)

169. What does it indicate OVC020 in a METAR?
(a) 3 to 4 oktas of cloud with base 120 meters
(b) 4 to 5 oktas of cloud with base 2000 feet
(c) 8 oktas (over cast) cloud with base 2000 feet
(d) 1 to 2 oktas of cloud with base 120 feet

170. A mountain station reported SCT /// in a METAR. What does it indicate?
(a) 3 to 4 oktas of low cloud whose base is below the station level
(b) 4 to 5 oktas of cloud whose base is obscure
(c) 6 to 7 oktas of cloud whose base is more than 10 km
(d) overcast cloud with base on ground

171. In a METAR, 1 oktas stratus base 500 ft, 2 oktas cumulonimbus base 1000 ft and 5 oktas stratocumulus base 2500 ft will be reported as:
(a) IST005, 2CB010, 5SC025
(b) FEW 005, FEW 010CB, BKN 025
(c) FEW 500, SCT 1000CB, BKN 2500
(d) IST 150, 2CB 300, 5SC 750

172. In a METAR the groups 3000 + TSRA SCT025 CB indicate:
(a) Visibility 3000 meters, present weather heavy thunderstorm rain, 3 to 4 oktas CB with base 2500 ft
(b) Visibility 3000 ft, present weather thunderstorm with hail and 5 oktas CB base 250 meters
(c) Visibility 3000 ft, present weather light thunderstorm with rain and 1 oktas CB base 750 ft
(d) Visibility 3000 ft, present weather thunderstorm and 2 oktas CB base 2500 ft

173. In a METAR the groups 28/27 Q 1014 indicate:
(a) Wind variation 280° – 270°, QNH 1014 INCHES
(b) Drybulb/dewpoint temperatures 28 °C/27 °C, QNH 1014 hPa
(c) Drybulb temperature/RH 28 °C/27% QFE 1014 hPa
(d) Drybulb/wetbulb temperature 28 °C/27 °C QNE 1014 hPa

174. In a METAR, TEMPO TL 1200 15030 G 45 KT indicate:
(a) Trend forecast valid till 1200 UTC (1000-1200 UTC) temporary change (not exceeding 1 hour) in wind 150° / 30 KT gusting to 45 KT
(b) Temporary change in next 12 hours in wind 150°/45 KT
(c) Temporary change observed in 10 minutes just before time of observation 1200 UTC, wind 150°/30 KT gusting to 45 KT
(d) Temporary change between 1200-1800 UTC, wind variation 150°/30 KT to 150°/45 KT

175. Trend type landing forecast issued in India. Trend forecast given in a routine METAR or SPECI. The period of validity of trend forecast is:
(a) 4 hours from the time of METAR/SPECI
(b) 6 hours from the time of METAR/SPECI
(c) 12 hours from the time of METAR/SPECI
(d) 2 hours from the time of METAR/SPECI

176. Trend forecast appended to METAR/SPECI consists of expected change in:
(a) wind, visibility
(b) present weather and clouds
(c) RH and dew point
(d) both (a) and (b) above

Given:

METAR EDDB 260850Z 26010KT CAVOK 18/12 Q 1012 NOSIG.

177. In the above METAR what is the visibility trend forecast?
(a) Not given (b) 10 km
(c) more than 10 km (d) 10 nm

178. In the given METAR above, what is validity time period of trend forecast?
(a) Not given
(b) 08 50 UTC to 1050 UTC of 26th day of the month
(c) 0850 UTC to 1250 UTC of 26th day of the month
(d) 0850 UTC to 1450 UTC of 26th day of the month

179. In the given METAR above what is the trend forecast given and its meaning?
(a) NOSIG; meaning no significant cloud
(b) NOSIG; meaning no significant weather
(c) NOSIG; no significant change expected between 0850 UTC to 1050 UTC of 26th day of the month
(d) both (a) and (b) above

Given:

METAR LIPZ 101050Z 00000 KT 1500 R09/P1500U BR FEW 005. SCR008 BKN012 16/15 Q1020 NOSIG

180. In the above METAR what is the wind and its trend forecast?
(a) Variable; no significant change
(b) calm; no significant change meaning remains calm
(c) northerly speed variable; likely to remain as it is
(d) both (a) and (c) above

181. In the METAR given above, what is the value of RVR?
(a) 1500 meters
(b) visibility on runway 09 more than 1500 meters and trend is increasing. Trend forecast no significant change expected 1050 – 1250 UTC
(c) Visibility on runway 09 less than 1500 ft and trend is decreasing
(d) Visibility on runway 09, 1500 ft and likely to continue for next 2 hours

182. In the METAR above (LIPZ) what is the cloud reported
(a) 1-2 oktas base 500 ft, 3 – 4 oktas base 800 ft, 5-7 oktas base 1200 ft
(b) 1-2 oktas base 500 M, 3-4 oktas base 800 M 5-7 oktas base 1200 M
(c) 3-4 oktas bas 5 nm, 3-4 oktas base 8 nm, 5-7 oktas base 12 nm
(d) both (b) and (c) above

183. In the METAR LIPZ above what is the present weather QNH reported?
(a) Fog; 1020 inches HG
(b) Mist (more than 1000 m but less than 2000 m), 1020 hPa
(c) Rain; 1500 hPa
(d) Thunderstorm; 1020 cm Hg

184. In the METAR LIPZ above what is the temperature/dew point and pressure trend forecast?
(a) 15 °C/16 °C; rise
(b) 16 °C/15 °C; no change
(c) 16 °C/15 °C pressure trend will not be given
(d) 16 °F/15 °F; pressure fall

185. The validity of a TAF is:
(a) 6 hours (b) 12 hours
(c) 18 hours (d) specified in the TAF

186. It is given, temperature –1.5 °C and dew point – 2.5 °C. How this will be reported in a METAR?
(a) M01/M02 (b) M02/M03
(c) M 1.5/M 2.5 (d) M 02 M02

187. It is given surface wind over an airfield 270°/15 kt. What would be the most likely wind at 2000 ft?
(a) 300°/30 kt (b) 330°/15 kt
(c) 280°/20 kt (d) 260°/10 kt

188. The approximate height of 300 hPa level is:
(a) 30000 ft (b) 34000 ft
(c) 18000 ft (d) 38000 ft

189. The approximate flight level corresponding to a pressure level of 250 hPa is:
(a) FL 180 (b) FL 300
(c) FL 340 (d) FL 390

190. The approximate height of 150 hPa level is:
(a) FL 600 (60000 ft) (b) FL 530 (53000 ft)
(c) FL 450 (45000 ft) (d) FL 390 (39000 ft)

The grid point forecast for a particular point is given as below:

550
34031020
10010 M 65
31015 M 57
31020 M 45
36005 M 32
32010 M 18
27005 M 07

For the above data answer the following questions (191 – 196)

191. The tropopause level is:
(a) 50000 ft (b) 60000 ft
(c) 55000 ft (d) 40000 ft

192. The wind maxima level and wind is:
(a) FL 300; 360°/05 kt (b) FL 340; 310°/20 kt
(c) FL 450; 100°/10 kt (d) FL 180; 270°/05 kt

193. The temperature and wind for FL 180 is:
(a) – 07 °C, 270°/05 kt (b) – 18 °C, 320°/10 kt
(c) – 32 °C, 360°/05 kt (d) – 45 °C, 310°/20 kt

194. The wind and temperature for FL 390 is:
(a) 100°/10 kt, – 65 °C (b) 310°/15 kt, – 57 °C
(c) 320/20 kt, – 45 °C (d) 360°/05 kt, – 32 °C

195. The wind and temperature for FL 450 is:
(a) 360°/05 kt, – 32 °C (b) 310°/20 kt, – 45 °C
(c) 310°/15 kt, – 57 °C (d) 100°/10 kt, – 65 °C

196. What is the wind and temperature at wind maxima level?
(a) 310°/20 kt, – 45 °C (b) 310°/15 kt, – 57 °C
(c) 100°/10 kt, – 65 °C (d) 270°/05 kt, – 7 °C

For a particular flight spot winds and temperature chart is given below:

24	18015	– 30
18	18020	– 18
10	14025	+02
05	12015	+12
02	16010	+16

From the above data answer the following questions (197 – 200)

197. What is the wind and temperature at 10000 ft?
(a) 120°/15 kt, + 12 °C (b) 140°/25 kt, + 02 °C
(c) 180°/20 kt, – 18 °C (d) 180°/15 kt, – 30 °C

198. What is the approximate height of freezing level?
(a) 10800 ft (b) 11600 ft
(c) 10000 ft (d) 9200 ft

199. What is the wind and temperature at about FL 180?
(a) 180°/15 kt, – 30 °C (b) 180°/20 kt, – 18 °C
(c) 140°/25 kt, + 02 °C (d) 120°/15 kt, + 12 °C

200. What is the wind and temperature at 400 hPa level?
(a) 180°/15 kt, – 30 °C (b) 180°/20 kt, – 18 °C
(c) 140°/25 kt, + 02 °C (d) 120°/15 kt, + 12 °C

201. In India, generally for routine domestic flights the period of validity of TAFOR is............. and issued every ...... hours.
(a) 9 hours; 3 hours (b) 12 hours; 6 hours
(c) 18 hours; 6 hours (d) 24 hours; 12 hours

202. In India, generally for routine international flights the period of validity of TAFOR is .... and issued every..........hours.
(a) 24 hours; 6 hours (b) 30 hours; 12 hours
(c) 18 hours; 3 hours (d) 12 hours; 3 hours

203. In general, the period of validity of routine aerodrome forecasts is:
(a) not less than 9 hours and not more than 30 hours
(b) not less than 12 hours and not more than 24 hours
(c) not less than 18 hours and not more than 30 hours
(d) 24 hours

204. Landing forecasts are intended to meet the requirements of local users and of aircraft within about .... flying time from aerodrome.
(a) 6 hours (b) 4 hours
(c) 2 hours (d) 1 hour

205. The trend type of landing forecast period of validity is .... hours from the time of report to which it is appended.
(a) 6 hours (b) 4 hours
(c) 2 hours (d) 1 hour

206. Operationally desirable accuracy of aerodrome forecast wind (direction/speed)
(a) $\pm 30^{\circ}$; $\pm$ 5 kt upto 25 kt 80% of occasions and $\pm$ 20% above 25 kt
(b) $\pm 10^{\circ}$; $\pm$ 10 kt 90% of occasions
(c) $\pm$ 20%; 90% of occasions
(d) $\pm$ 10%; $\pm$ 10% 70% of occasions

207. A Col (region) in winter may produce:
(a) thunderstorm (b) fog/low stratus
(c) frost (d) both (b) and (c)

208. A Col (region) in summer may produce:
(a) thunderstorm (b) fog/low stratus
(c) frost (d) both (b) and (c)

209. ITCZ generally moves along the sun. In central north Pacific ITCZ is:
(a) meanders between 5 °N, 5 °S in a year
(b) stationary roughly throughout the year
(c) not seen
(d) always south of the equator

210. Cold anticyclones are ……… on upper level chart. Siberian High is an example of …………
(a) not present; warm anticyclone
(b) not present; cold anticyclone
(c) present; warm anticyclone
(d) present; cold anticylone

211. At a given level in the atmosphere, a low characterized by colder air near its centre than around its periphery is called:
(a) cold-core low (b) cold-core-cyclone
(c) cold-core high (d) both (a) and (b)

212. At latitudes 25-40° over oceanic area are found ………. The area experiences no surface cooling or haustable layer
(a) warm anticylone (b) cold anticyclone
(c) occluded anticyclone (d) both (b) and (c)

213. The Azores High is an example of ……..
(a) warm core anticyclone (b) cold core anticyclone
(c) occluded anticyclone (d) none above

214. All through the year over all the worlds oceans (North Pacific, South Pacific, North Atlantic, South Atlantic) exists:
(a) High pressure (b) low pressure
(c) Col (d) both (b) and (c)

215. The Polar front jet is found in northern hemisphere all through the year. It follows the front in summer from Newfoundland to Scandinavia. In Winter it follows the front …….. from Florida to South west England
(a) north; south (b) south; north
(c) north; north (d) south; south

216. Light upper easterly winds extend from10 °S to 5 °N only over Africa and Asia during:
(a) summer (b) winter
(c) all the year (d) none above

217. In NH summer, upper easterly winds extend from 5 °S to 20 °N over Africa and Asia, They are .... along the equator, becoming .... by 20 °N.
(a) light; strong (b) strong; strong
(c) strong; light (d) light; light

Answers

1.	(d)	2.	(c)	3.	(b)	4.	(d)	5.	(b)
6.	(d)	7.	(c)	8.	(c)	9.	(b)	10.	(b)
11.	(b)	12.	(d)	13.	(d)	14.	(b)	15.	(b)
16.	(c)	17.	(b)	18.	(c)	19.	(b)	20.	(c)
21.	(b)	22.	(b)	23.	(b)	24.	(a)	25.	(d)
26.	(c)	27.	(d)	28.	(b)	29.	(b)	30.	(b)
31.	(c)	32.	(d)	33.	(c)	34.	(d)	35.	(c)
36.	(d)	37.	(d)	38.	(c)	39.	(c)	40.	(c)
41.	(b)	42.	(c)	43.	(c)	44.	(c)	45.	(d)
46.	(d)	47.	(c)	48.	(c)	49.	(c)	50.	(c)
51.	(a)	52.	(c)	53.	(b)	54.	(a)	55.	(c)
56.	(a)	57.	(b)	58.	(b)	59.	(d)	60.	(c)
61.	(a)	62.	(b)	63.	(c)	64.	(c)	65.	(b)
66.	(a)	67.	(d)	68.	(d)	69.	(c)	70.	(d)
71.	(a)	72.	(c)	73.	(a)	74.	(a)	75.	(b)
76.	(a)	77.	(b)	78.	(b)	79.	(b)	80.	(c)
81.	(a)	82.	(a)	83.	(d)	84.	(c)	85.	(d)
86.	(a)	87.	(b)	88.	(c)	89.	(d)	90.	(c)
91.	(a)	92.	(a)	93.	(c)	94.	(b)	95.	(d)
96.	(a)	97.	(d)	98.	(a)	99.	(d)	100.	(c)
101.	(a)	102.	(a)	103.	(c)	104.	(d)	105.	(d)
106.	(a)	107.	(c)	108.	(b)	109.	(a)	110.	(a)
111.	(a)	112.	(a)	113.	(c)	114.	(a)	115.	(d)

116.	(c)	117.	(c)	118.	(d)	119.	(b)	120.	(a)
121.	(c)	122.	(a)	123.	(a)	124.	(a)	125.	(d)
126.	(b)	127.	(d)	128.	(c)	129.	(d)	130.	(a)
131.	(c)	132.	(a)	133.	(d)	134.	(a)	135.	(a)
136.	(a)	137.	(a)	138.	(b)	139.	(b)	140.	(b)
141.	(c)	142.	(d)	143.	(c)	144.	(b)	145.	(c)
146.	(c)	147.	(b)	148.	(d)	149.	(b)	150.	(c)
151.	(d)	152.	(c)	153.	(b)	154.	(a)	155.	(d)
156.	(d)	157.	(c)	158.	(a)	159.	(d)	160.	(d)
161.	(b)	162.	(c)	163.	(d)	164.	(c)	165.	(d)
166.	(c)	167.	(d)	168.	(c)	169.	(c)	170.	(a)
171.	(b)	172.	(a)	173.	(b)	174.	(a)	175.	(d)
176.	(d)	177.	(c)	178.	(b)	179.	(c)	180.	(b)
181.	(b)	182.	(a)	183.	(b)	184.	(c)	185.	(d)
186.	(a)	187.	(a)	188.	(a)	189.	(c)	190.	(c)
191.	(c)	192.	(b)	193.	(a)	194.	(b)	195.	(d)
196.	(a)	197.	(b)	198.	(a)	199.	(b)	200.	(a)
201.	(a)	202.	(a)	203.	(a)	204.	(d)	205.	(c)
206.	(a)	207.	(d)	208.	(a)	209.	(b)	210.	(b)
211.	(d)	212.	(a)	213.	(a)	214.	(a)	215.	(a)
216.	(b)	217.	(c)						

Appendix

FIR (Flight Information Region)

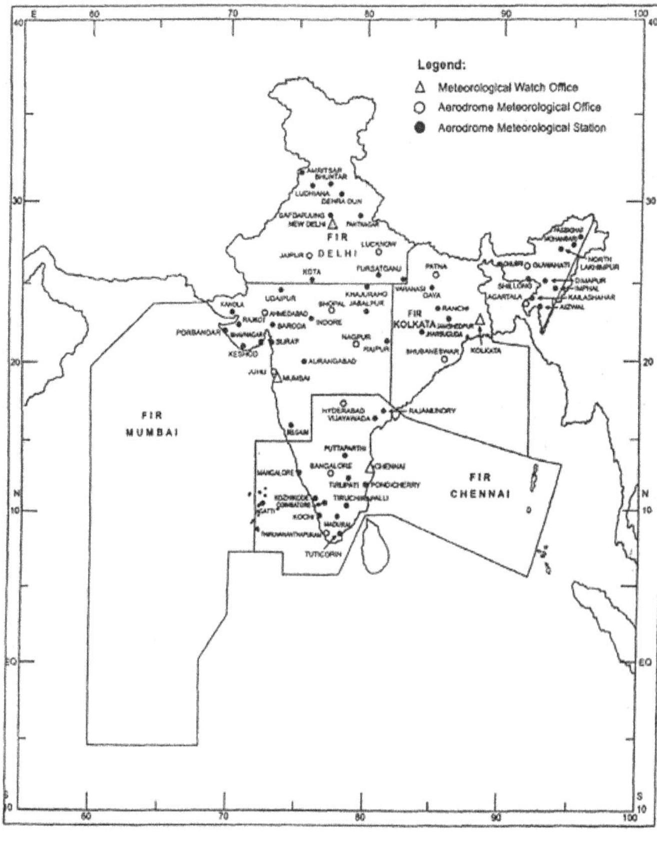

FORMAT OF LOCAL FORECAST

मेट.टी–2
MET.T-2

भारत मौसम विभाग

INDIA METEOROLOGICAL DEPARTMENT

मौसम कार्यालय
METERORОLOGICAL OFFICE *XXXX AIRPORT*

स्थानीय पूर्वानुमान
LOCAL FORECAST *FOR VXXX AND 50 NM AROUND*

तारीख –से — यू.टी.सी
from *0600* hours UTC *1998-05-15* date

तारीख से यू.टी.सी
till *1400* hours UTC *1998-05-15* date

पूर्वानुमान भेजने का समय —— यू.टी.सी

Time of Origin : *15/0530* hours UTC

सतह पवन
SURFACE WIND : *090 / 10 KT MAX20 BECMG 11 / 13 VRB / 03 KT*

उपरी पवन
UPPER WIND :

9000M	300 / 55	-25	3000M	290 / 30	+06
7500M	300 / 45	-15	2100M	270 / 25	+09
6000M	280 / 35	-02	1500M	250 / 10	+12
4500M	290 / 30	+01	900M	120 / 05	+20

मौसम
WEATHER: *HZ TEMPO 06 / 10 MOD TSRA (.) TEMPO 06 / 10 MOD/SEV TURB AND ICING IN CB*

दृश्यता
VISIBILITY: *4000M IN HZ TEMPO 06 / 10 1200M IN MOD TSRA*

मेघ
CLOUD : *SCT SC 600M/750M SCT CU 600M/5000M TEMPO BKN ST 06 / 10 240M ISOL CB 900M/8000M OVC AS 2400M*

हिम स्तर
FREEZING LEVEL: *4800M*

अतिरिक्त टिप्पणियाँ
ADDITIONAL NOTES: *TEMPO 06 / 10 MOD / SEV TURB AND ICING IN CB*

चेतावनी
WARNING: *VISIBILITY LIKELY 4000M IN HZ AND 1200M IN MOD TSRA BETWEEN 06 / 10 MOD TSRA LIKELY BETWEEN 06 / 10 WHEN LOW CLOUD BASE 240M WITH 5 OKTA OR MORE LIKELY WARNING FOR LIGHT AIRCRAFT – WIND SPEED MAY REACH 20 KTS IN GUST WITH DIRECTION FROM 090*

सूर्योदय
Sunrise *16 / 0643EF*

सूर्यास्त
Sunset *15 / 1729EF*

चन्द्रोदय
Moonrise *16 / 0834EF*

चन्द्रास्त
Moonset *15 / 1841EF*

चंद्रकला
Phase of Moon

तारीख *19* को बजे जारी किया गया

Issued at *0530* hrs. UTC *1998-05-15*

Met T-3

उडान पूर्वानुमान / FLIGHT FORECAST

मार्ग Route: VXXX से To VYYY; बरास्ता Via: DIRECT; वायु की अनुमानित वास्तविक गति Assumed True Air speed________ नाट Kts

छुटने के लिए मान्य Valid for departure: 0800/1998-05-25 (समय यु.टी.सी. और तारीख) (Time U.T.C. and Date); पहुँचने के लिए मान्य Valid for arrival: 1200/1998-05-25 (समय यु.टी.सी. और दिनांक) (Tme U.T.C. and Date)________

मौसम स्थिति के विशेष लक्षण / Special Features of Meteorological Situation.

उपरी वायु/ Upper wind (वास्तविक डिग्री व नाट) और तापमान (से) (Degree true and knots) and Temperatures (°C)		मार्ग के सेक्शन; (क्षेत्र सं अक्षांश और देशान्तर या भौगोलिक सूचक) SECTION OF ROUTE (ZONE Nos. LATITUDE AND LONGITUDE OR GEOGRAPHICAL INDICATORS)				
		VXXX	VZZZ			VYYY
उडान स्तर FLIGHT LEVELS	200	270 / 25	- 3	28[illegible] / 20		- 4
	150	260 / 15	+ 2	290 / 20		+ 2
	100	250 / 15	+ 8	290 / 15		+ 7
निम्नतम परत / LOWEST LAYER मात्रा और किस्म / AMOUNT AND TYPE		SCT CU		SCT ST	ISOL CB	SCT CU
उडान स्तर FLIGHT LEVELS	आधार पर. / of Base	030		010	020	025
	शिखर पर/ of Top	120		--	300	150
बादल / CLOUD उच्चतर परत / HIGHER LAYER मात्रा और किस्म / Amount and Type				BKN AS		
उडान स्तर / FLIGHT LEVELS	आधार पर./ of Base			100		
	शिखर पर / of Top			140		
सतह दृश्यता / SURFACE VISIBILITY		10 KM		6 KM REDUCING TO 2000M IN TS		
सार्थक मौसम / SIGNIFICANT WEATHER		NOSIG WX		ISOL TS (.) MOD / SEV TURB IN CB		
0° से.की समतापीय उडान स्तर FLIGHT LEVELS OF 0°C ISOTHERM		160		160		
न्यूनतम माध्य समुद्रतल का दाब LOWEST MSL PRESSURE						
टिप्पणियाँ / REMARKS						

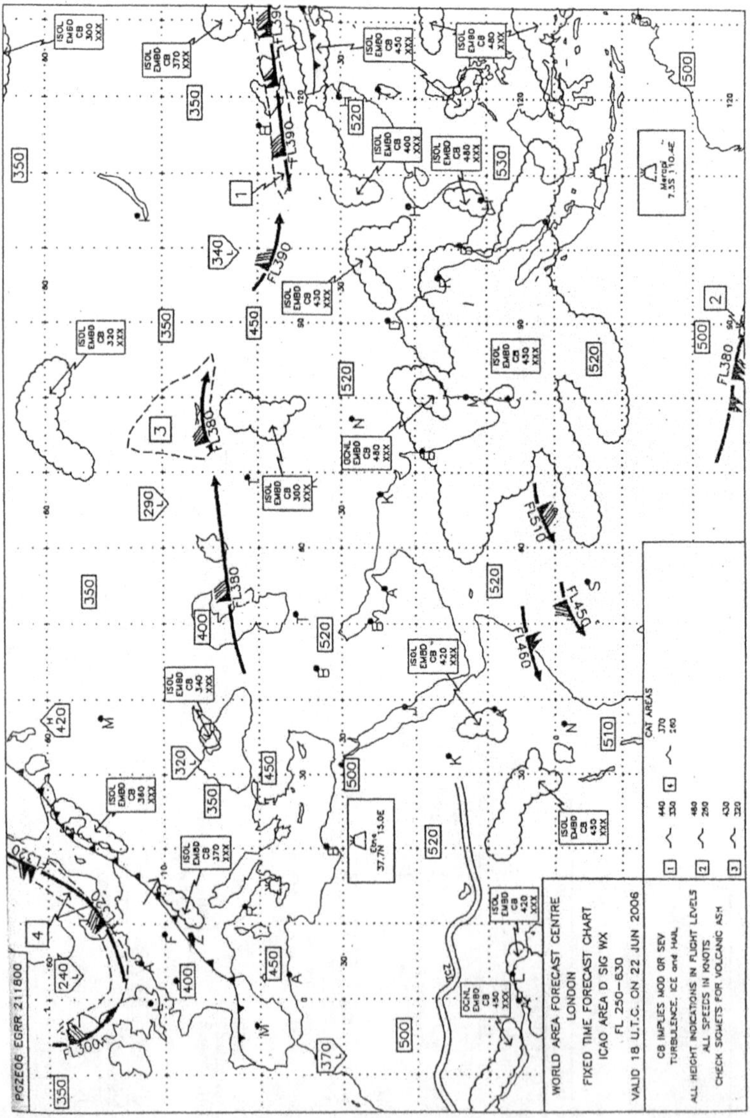

Example of WAFC SIGWX CHART

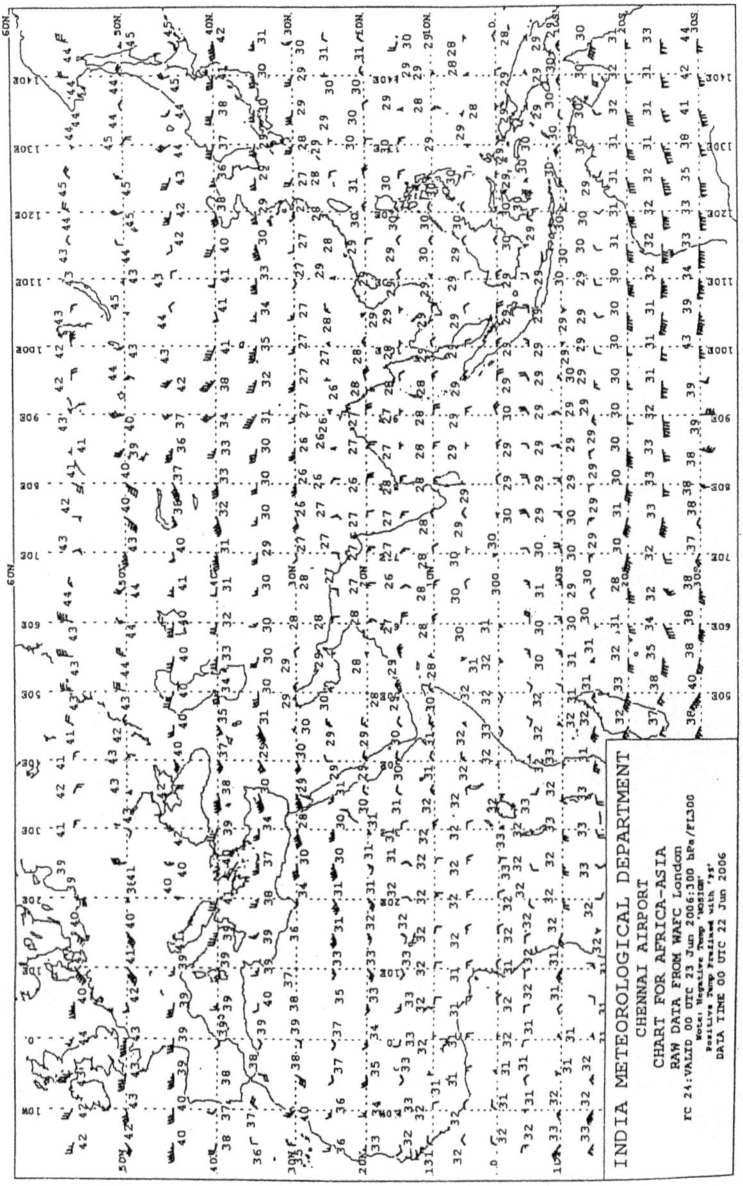

Examples of WAFC wind and Temperature chart.

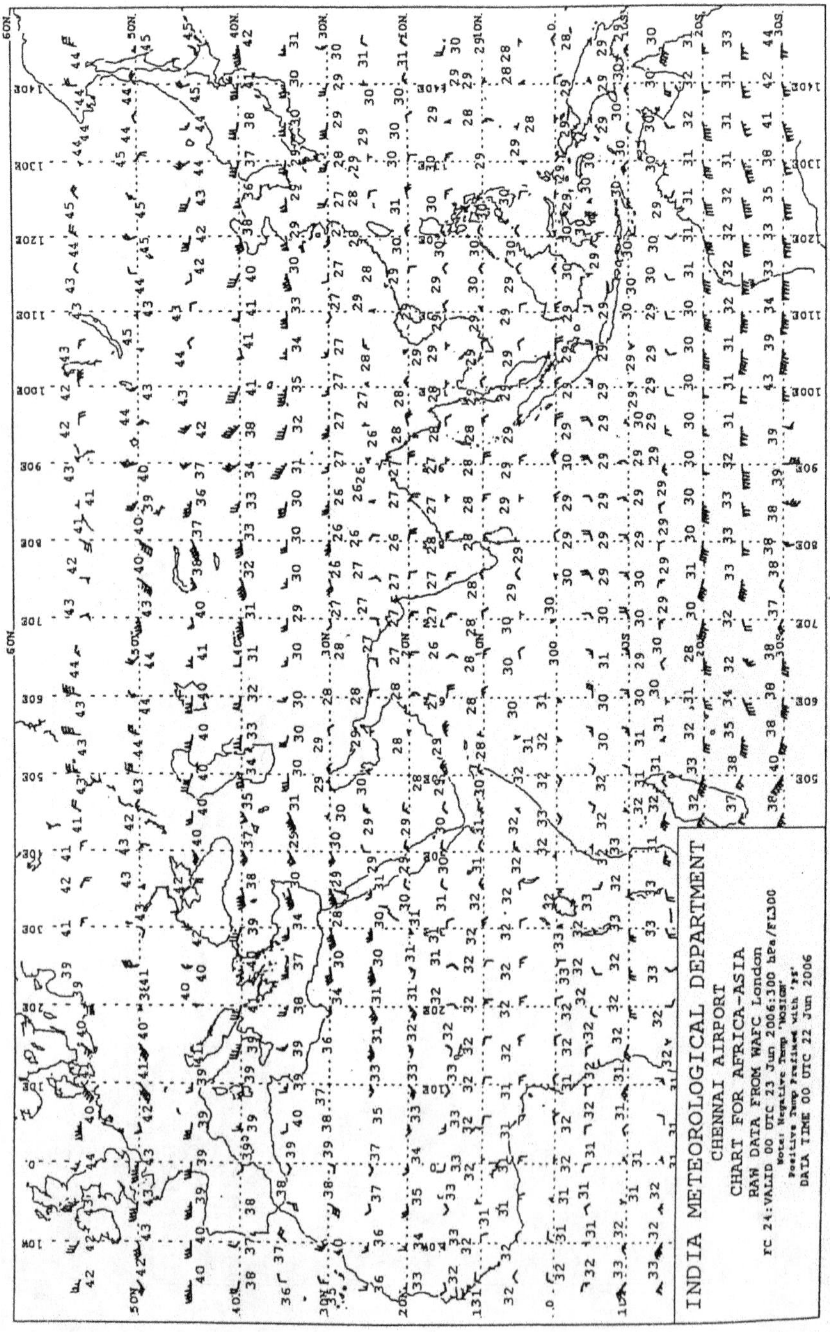
INDIA METEOROLOGICAL DEPARTMENT
CHENNAI AIRPORT
CHART FOR AFRICA-ASIA
RAW DATA FROM WAFC London
FC 24:VALID 00 UTC 23 Jun 2006:300 hPa/FL300
DATA TIME 00 UTC 22 Jun 2006

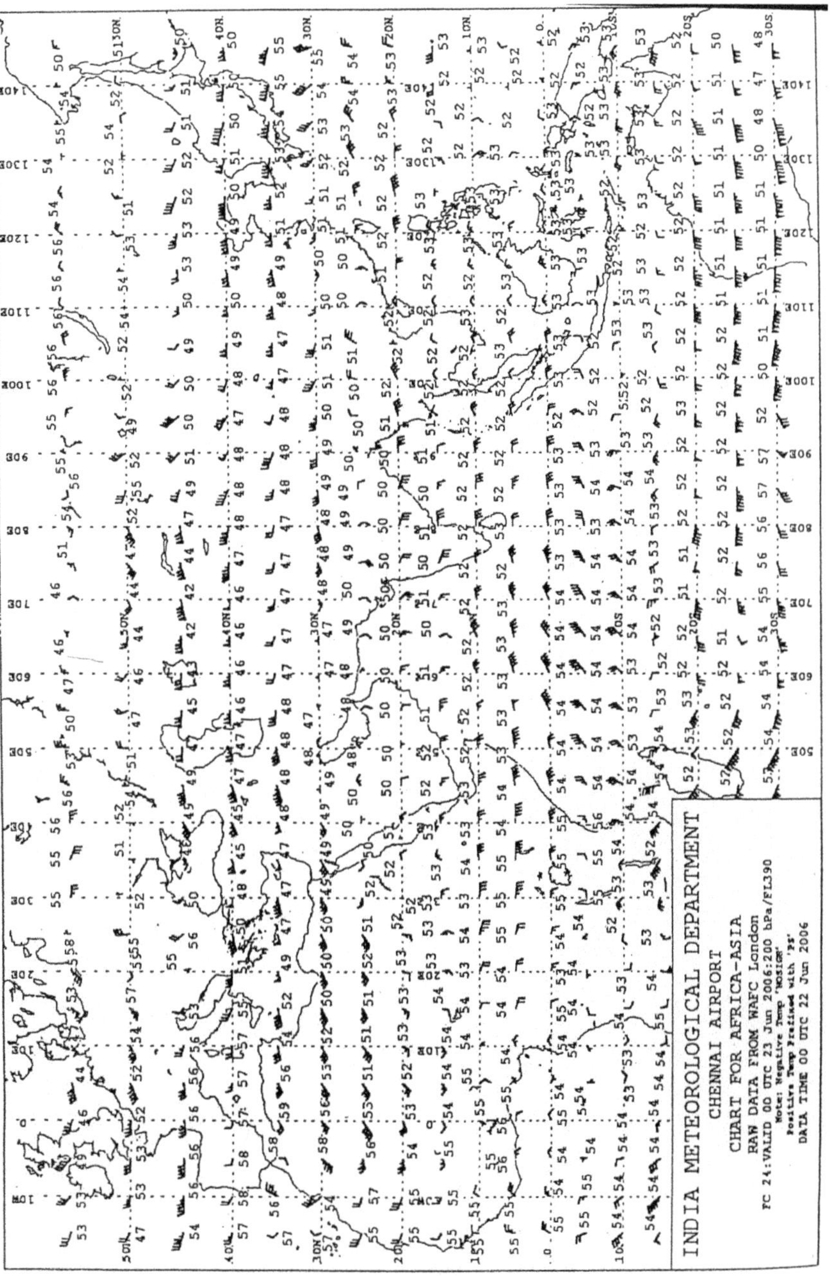
INDIA METEOROLOGICAL DEPARTMENT
CHENNAI AIRPORT
CHART FOR AFRICA–ASIA
RAW DATA FROM WAFC London
FC 24:VALID 00 UTC 23 Jun 2006:200 hPa/FL390
DATA TIME 00 UTC 22 Jun 2006

Met. T-4

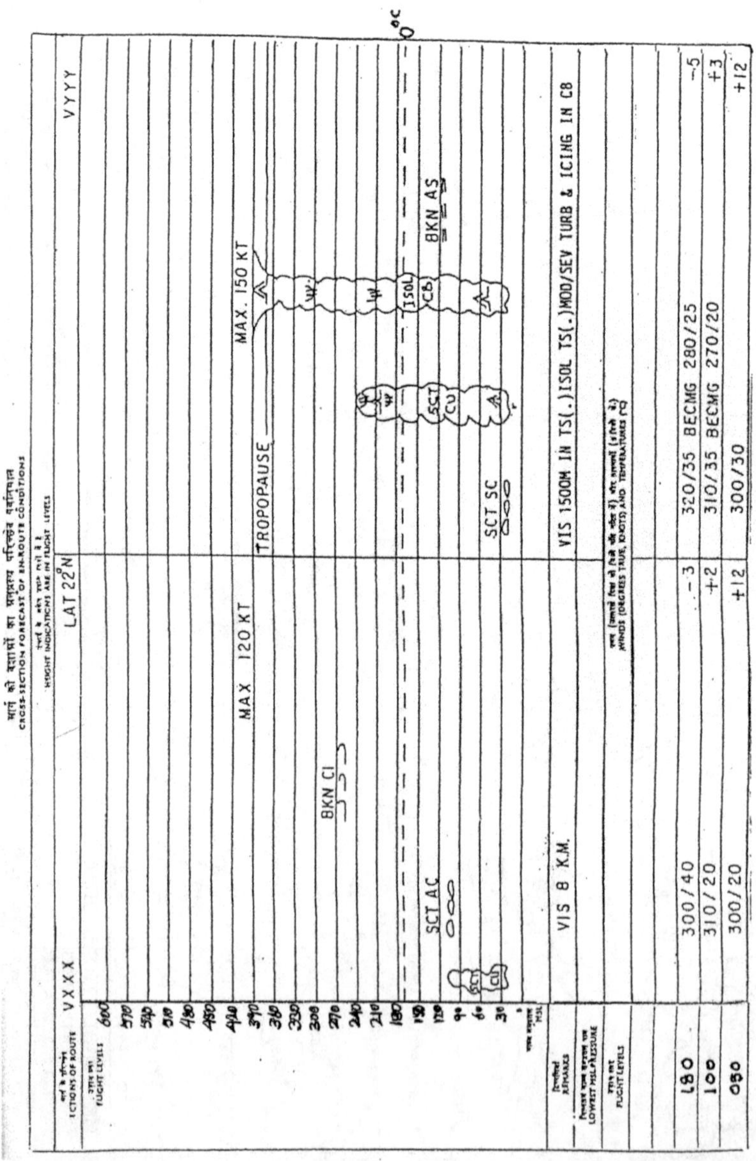

Example of Route Forecast

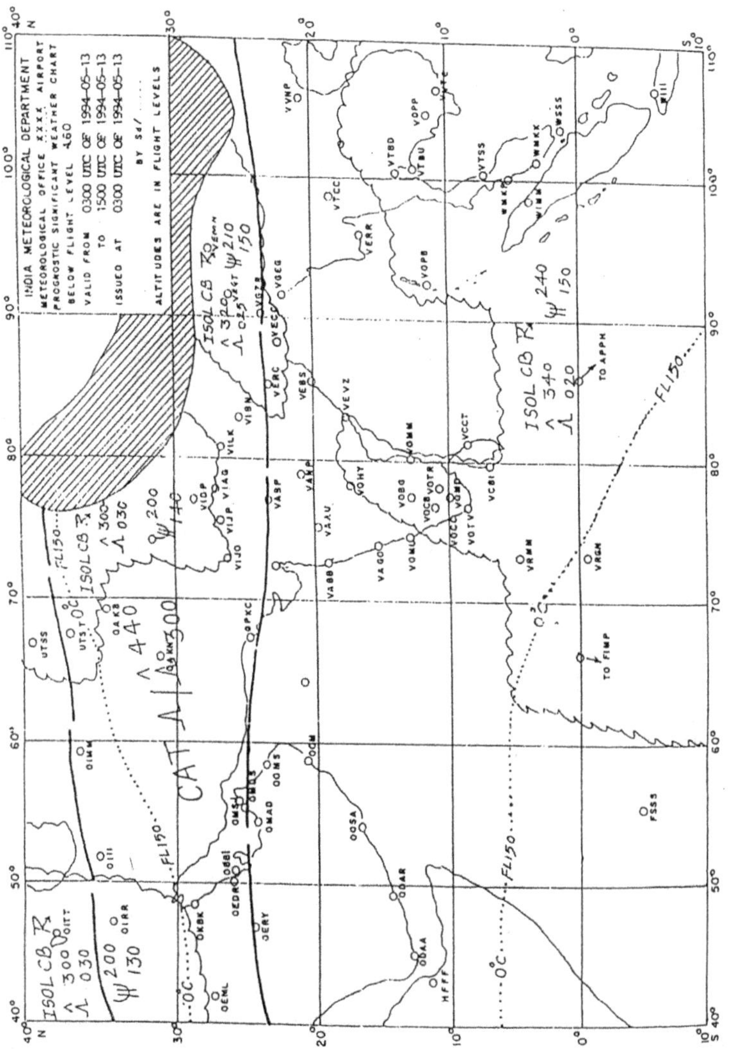

National SIGWX chart

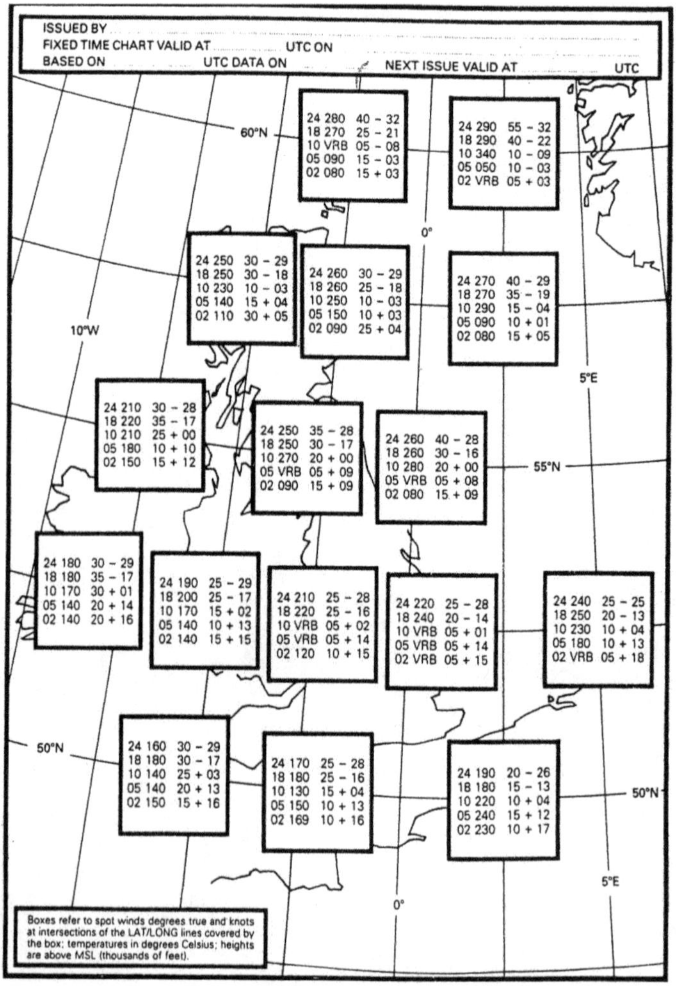

Model TB. Tabular forecast of upper winds and upper-air temperatures
Example 1 – Spot locations

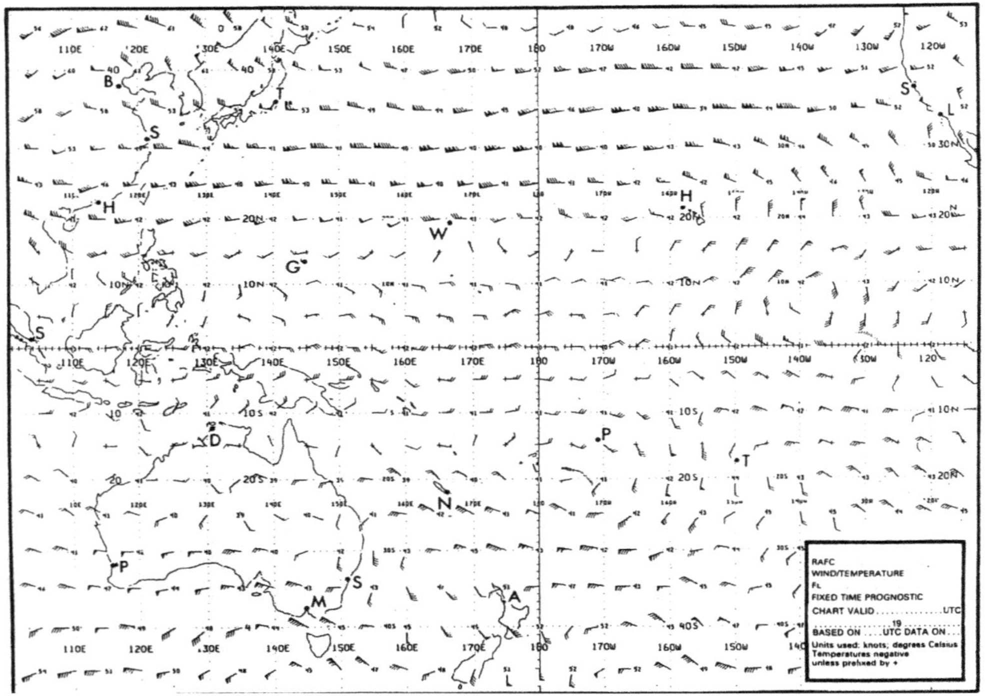

Model IS. Upper wind and temperature chart for standard Isobaric surface
Example 1 – Arrows and feathers (Mercator projection)

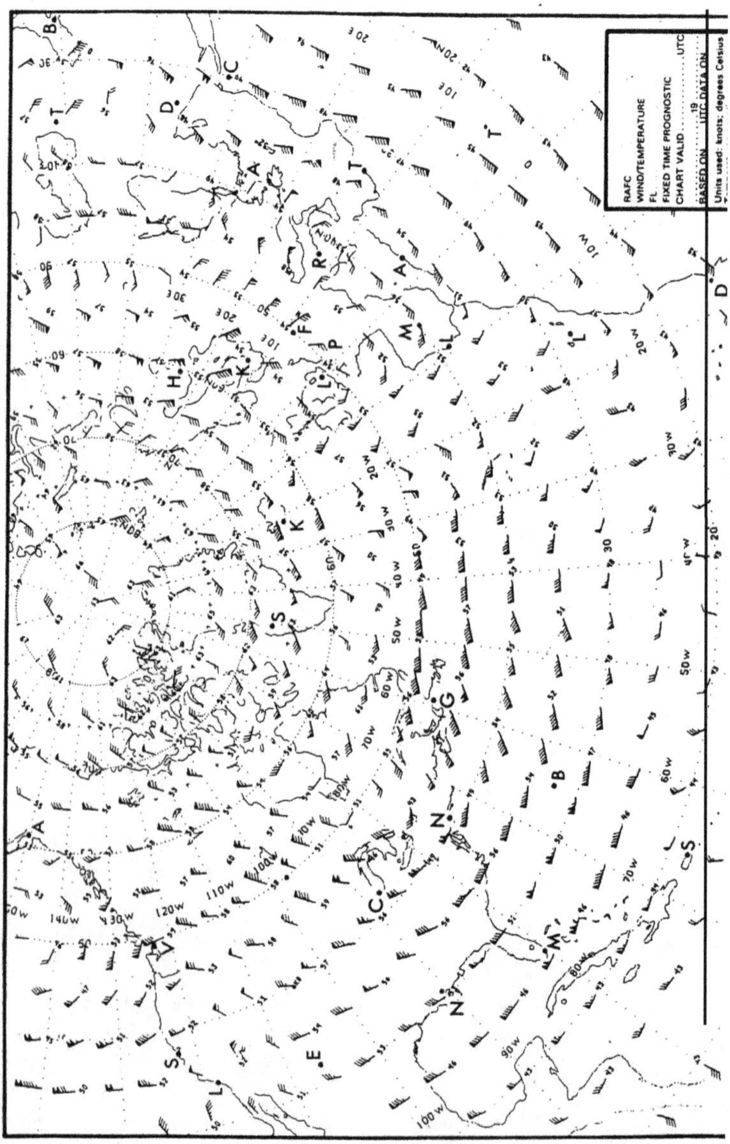

Model IS. Upper wind and temperature chart for standard isobaric surface
Example 2 – Arrows and feathers (Polar stereographic projection)

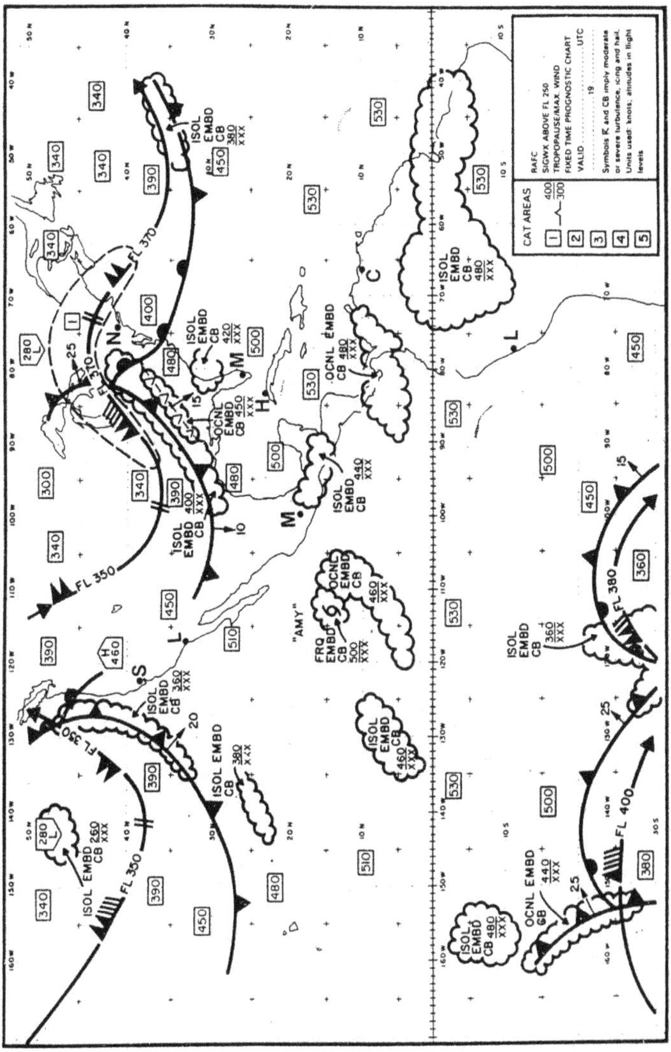

Model SWH. Significant weather chart (high level)
Example 1 – Mercator projection.

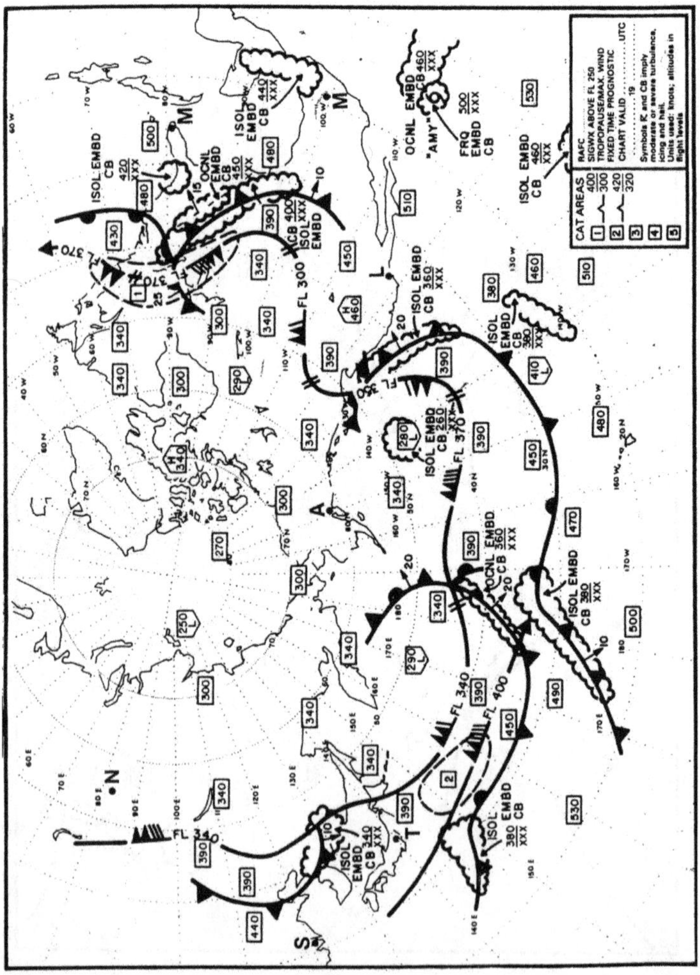

Model SWH. Significant weather chart (high level)
Example 2 – Polar stereographic projection.

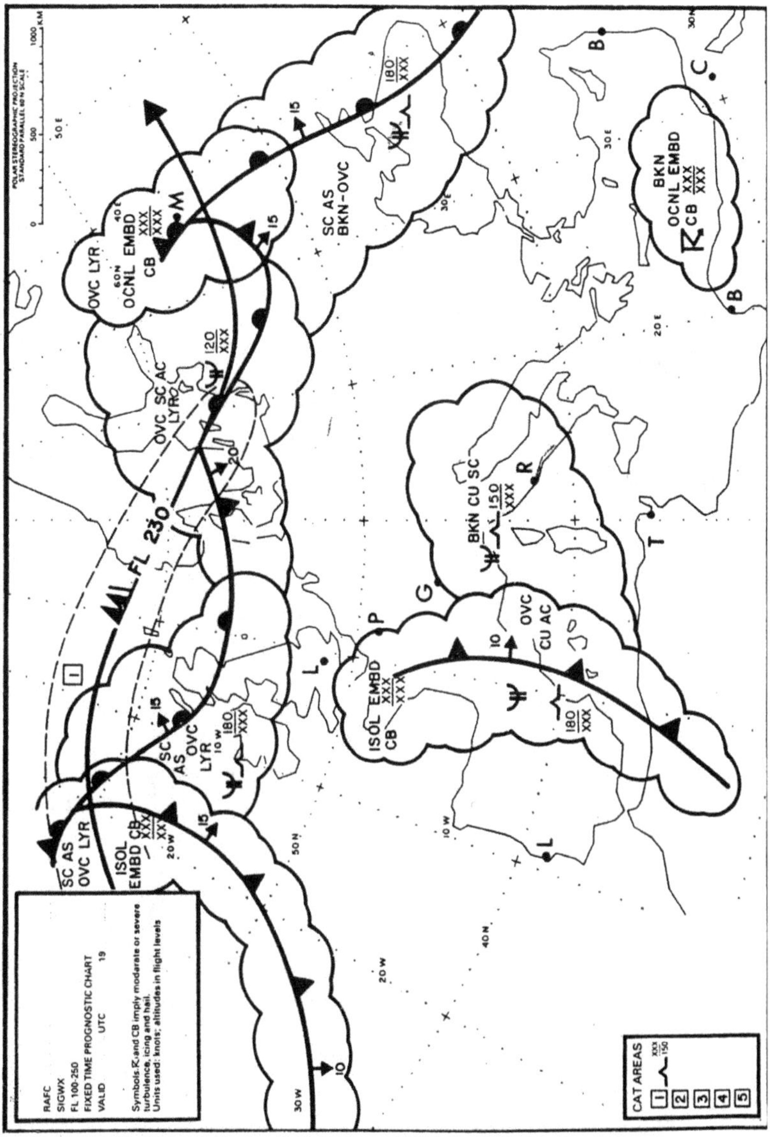

Model SWM. Significant weather chart (medium level)

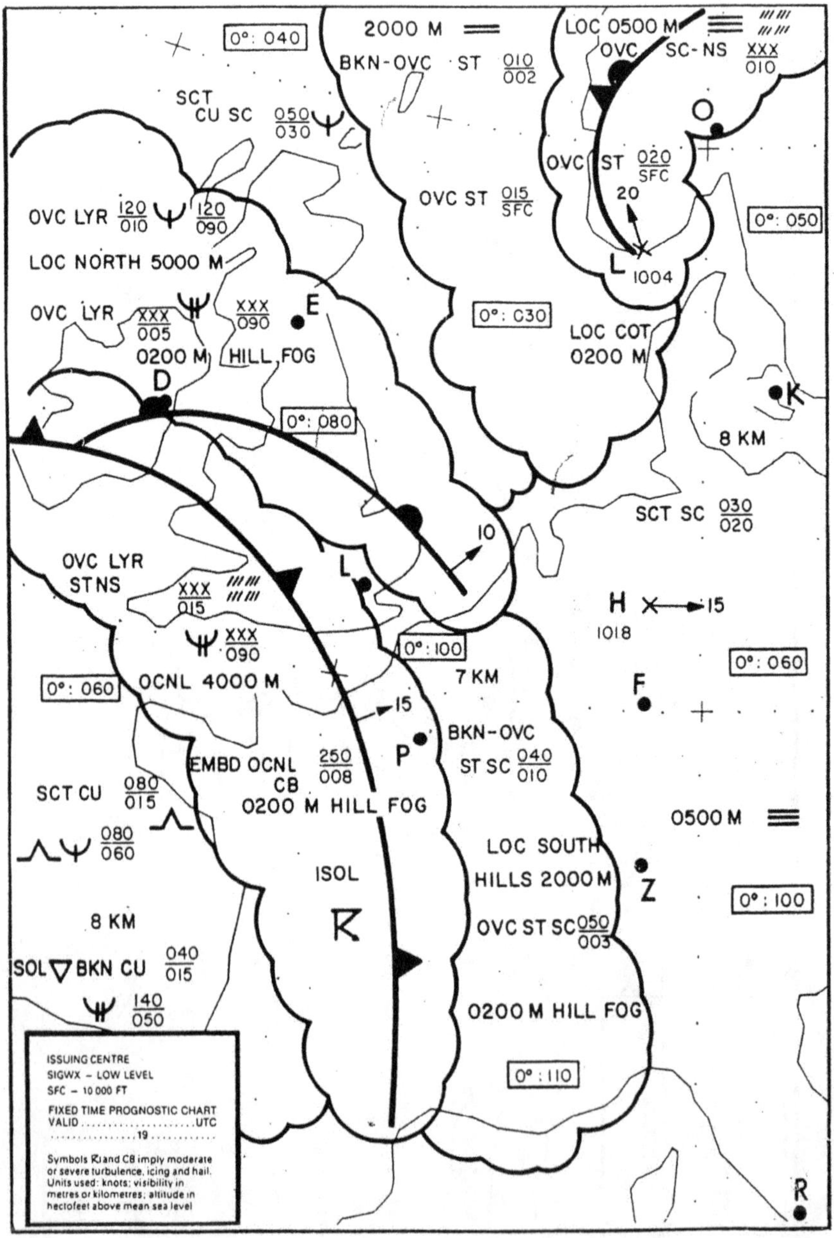

Model SWL. Significant weather chart (low level)
Example 1

FIXED TIME CHART VALID. . . . . UTC. . . . . 19. . . BASED ON. . . . . .UTC DATA ON . . . .

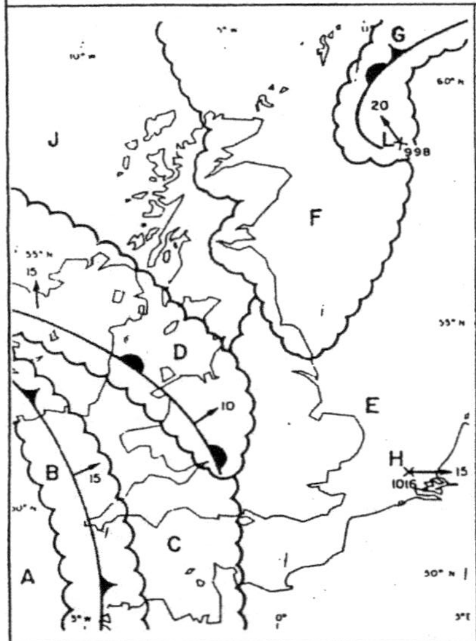

VARIANT	VIS	WEATHER	CLOUD, TURBULENCE, ICING	0°C
AREA A	10		SCT CU 025/080 050/080	50
ISOL	8	SHOWERS	BKN CU 015/XXX 050/XXX	
AREA B	10	RAIN	OVC LYR ST NS 015/XXX 050/XXX	50
OCNL	4000	HEAVY RAIN	EMBD CB 008/XXX HILL FOG	
ISOL	1000	THUNDERSTORM		
AREA C	7		BKN to OVC ST SC 010/040	100
LOC SOUTH COT HILLS	2000	DRIZZLE	OVC ST SC 003/050 HILL FOG	
AREA D	10	RAIN	OVC LYR SC NS 010/XXX O90/XXX	90
LOC NORTH	5000	RAIN	OVC LYR ST NS 005/XXX O90/XXX HILL FOG	
AREA E	8		SCT SC 020/030	40
LOC LAND	0500	FOG		
AREA F	2000	MIST	BKN to OVC ST 020/010	30
LOC COT HILLS	0200	FOG	OVC ST SFC/015	
AREA G	5000	RAIN	OVC CU SC NS 010/XXX 030/150	30
LOC NORTH	0500	FOG	OVC ST SFC/010	
AREA J	10		SCT CU SC 030/050 040/050	40
LOC HILLS NORTH			MOD CAT BLW 070	

SIGWX BELOW 10 000 FT
ISSUED BY. . . . . .AT . . . . .UTC

Note: 1. Pressure in hPa and speeds in knots.
2. Vis in m or km. Hill fog implies vis 200M or less.
3. Altitude in hectofeet above MSL XXX = above 10 000 ft.
4. ⋜ and CB imply MOD/SEV icing and turbulence.

WARNING AND/OR REMARKS:
EAST TO NE GALES SHETLAND TO HEBRIDES - SEVERE MOUNTAIN WAVES NW SCOTLAND - FOG PATCHES EAST ANGLIA - WDSPR FOG OVER NORTH FRANCE, BELGIUM AND HOLLAND

Model SWL.. Significant weather chart (low level)
Example 2

National TAFs Bulletin Generated on 04-01-2011 at 11:04 UTC

TAF VOMM 080600Z 0809/0818 36010KT 6000 SCT020=

TAF VOTR 080600Z 0809/0818 05010KT 6000 FEW020 SCT250=

TAF VOMD 080600Z 0809/0818 07010KT 6000 FEW020 SCT100=

TAF VOCB 080600Z 0809/0818 09010KT 6000 FEW015 BECMG 0812/0813 00000KT 4000 HZ=

TAF VOSM 080300Z 0806/0815 05005KT 6000 SCT020 SCT100=

TAF VOBL 080600Z 0809/0818 09005KT 8000 NSC=

TAF VOBG NO DATA AVAILABLE=

TAF VOML 080600Z 0809/0818 27005KT 8000 NSC=

TAF VOTV 080600Z 0809/0818 25010KT 6000 SCT015 SCT100 TEMPO 0809/0815 3000 TSRA/RA SCT012 FEW025CB BKN080 BECMG 0815/0816 00000KT 4000 HZ=

TAF VOCI 080600Z 0809/0818 27010KT 6000 FEW020 SCT100 TEMPO 0809/0813 SCT012 FEW025CB BKN100 BECMG 0816/0817 00000KT 4000 HZ FEW015 SCT100=

TAF VOCC 072100Z 0800/0809 00000KT 3000 HZ FEW015 SCT100 TEMPO 0800/0803 1200 MIFG/BR FEW005 SCT250 BECMG 0805/0806 25010KT 6000 =

TAF VOCL 080600Z 0809/0818 29010KT 7000 FEW020 SCT100 BECMG 815/0816 00000KT 5000 HZ FEW015 BKN100=

TAF VOHS 080600Z 0809/0818 05010KT 5000 HZ FEW020 BKN200=

TAF VOHY 080600Z 0809/0818 05010KT 5000 HZ FEW020 BKN200=

TAF VOBZ 080600Z 0809/0818 07010KT 6000 FEW015 SCT200 BECMG 0815/0816 4000 HZ=

TAF VOTP 080600Z 0809/0818 03010KT 6000 SCT015 SCT100 BECMG 0814/0816 4000 HZ=

TAF VOTK 080300Z 0806/0815 03010KT 6000 FEW020 SCT100=

TAF VORY 080600Z 0809/0818 07010KT 5000 HZ FEW015 SCT200 BECMG 0815/0816 3000 HZ=

TAF VANP 080300Z 0806/0815 VRB03KT 6000 NSC BEC 0810/12 09005KT 4000 HZ NSC=

TAF VARP 080000Z 0803/0812 VRB03KT 2500 HZ NSC BECMG 0804/0806 05005KT 5000 HZ BECMG 0807/0809 6000=

TAF VECC 080600Z 0809/0818 34005KT 3000 HZ BECMG 0812/0813

VRB03KT 2000 HZ BECMG 0816/0817 00000KT 1500 BR=

TAF VEBS 080600Z 0809/0818 36005KT 3000 HZ NSC BECMG 0814/0816 2500 HZ=

TAF VEVZ 080300Z 080615 09008KT 4000 HZ NSC BECMG 0709 13012KT 5000 HZ BECMG 1315 3000 HZ FEW 250 =

TAF VEPT 080300Z 0806/0815 29004KT 1600 BR NSC BECMG 0808/0810 29006KT 2000 HZ BECMG 0810/0812 32006KT 25004KT BECMG 0812/0814 00000KT 2000 HZ=

TAF VERC 080300Z 0806/0815 32004KT 2000 HZ NSC BECMG 0808/0810 32006KT 2500 HZ BECMG 0812/0814 00000KT 2000 HZ=

TAF VOPB 072100Z 0800/0812 14006KT 4500 HZ SCT012 SCT090 BKN220 BECMG 0801/ 0802 15010KT 6000 TEMPO 0802/0806 17010G20KT 2000 RA/TSRA FEW008 SCT010 FEW018TCU/CB BKN090=

TAF VABB 080600Z 0809/0818 03008KT 5000 FU NSC BECMG 0810/0812 31008KT 4500 FU BECMG 0813/0815 33007KT 4000 FU BECMG 0816/0818 05005KT 3500 FU =

TAF VAAH 080600Z 0809/0818 10008KT 6000 NSC BECMG 0813/0815 00000KT 4000 FU/NSC=

TAF VAPO 080300Z 0806/0815 09005KT 6000 NSC TEMPO 0812/0814 5000 FU=

TAF VAGO 080600Z 0809/0818 29012KT 6000 FEW018 SCT250 BECMG 0816/0818 VRB03KT=

TAF VIDP 080600Z 0809/0818 11006KT 2500 FU FEW030 BECMG 0811/0813 1200 FU/BR BECMG 0815/0817 VRB02KT 0700 MIFG=

TAF VILK 080600Z 0809/0818 VRB02KT 0800 MIFG VV/// BECMG 0810/0812 30003KT 1500 BR SCT010 SCT015=

TAF VIBN 080600Z 0809/0818 VRB02KT 0800 MIFG VV/// BECMG 0810/0812 30003KT 1500 BR SCT010 SCT015=

TAF VIJP 080600Z 0809/0818 06003KT 5000 HZ NSC BECMG 0813/0815 VRB02KT 3000 HZ BECMG 0816/0817 00000KT 1500 BR=

Model TA1. Tabular forecast of en-route conditions

DATE	HEIGHTS IN FEET ABOVE MSL	
ROUTE FROM ....BIGGIN HILL.... TO ....AMSTERDAM.... VIA ....AIRWAYS....		
VALID FOR DEPARTURE BETWEEN ......1500...... UTC AND ......1700..... UTC AND FOR ARRIVAL BETWEEN ....1700.......... UTC AND ......2100..... UTC		
SPECIAL FEATURES OF THE METEOROLOGICAL SITUATION: ACTIVE COLD FRONT FROM HUMBER TO CHANNEL ISLES AT 1000 UTC MOVING EAST AT 20 KNOTS TO LIE NORTH/SOUTH ACROSS TRACK ABOUT 40 NM WEST OF AMSTERDAM BY 1900 UTC.		
ZONE	**LONDON**	**02° E AMSTERDAM**
UPPER WINDS (DEGREES TRUE AND KNOTS) TEMPERATURES (DEGREES CELSIUS) 10 000 ft 5000 ft 2000 ft	280/30 MS 12 290/25 MS 03 290/20 PS 03	250/45 MS 09 240/35 00 230/30 PS 06
CLOUD	SCT CU 18 000/1500 BKN SC 10 000/2500 BKN AC LYR 18 000/12 000	ISOL EMBD CB 30 000/1000 BKN ST 800/500 OVC SC AS LYR 24 000/2000
SURFACE VISIBILITY	1500 M IN SHOWERS	5000 M IN RAIN AND 1000 M IN THUNDERSTORMS
SIGNIFICANT WEATHER	OCNL RAIN SHOWERS MODERATE OCNL SEVERE ICING 10 000/3500 MODERATE TURBULENCE IN CU 18 000/1500	MODERATE/HEAVY RAIN ISOL THUNDERSTORMS MODERATE OCNL SEVERE ICING 13 000/5000 MODERATE OCNL SEVERE TUBULENCE IN CB AND FRONTAL ZONE 30 000/1000
HEIGHT OF 0°C ISOTHERM	3500	5000
FORECAST LOWEST MSL PRESSURE (hPa)	1008	1004
SUPPLEMENTARY INFORMATION		

Issued by .... at .... UTC on .... 19 ... by .... Forecaster.

Notes: 1. Positive and negative values are indicated by the prefix "PS" (plus) and "MS" (minus) respectively.
2. When a single numerical value of an element is given in a forecast it is to be interpreted as representing the most probable mean of a range of values which the element may assume during the period of the forecast.

Abbreviations: SKC—O oktas, FEW—1 to 2 oktas, SCT—3 to 4 oktas, BKN—5 to 7 oktas, OVC—8 oktas, LYR—Layered, LOC—Locally, ISOL—Isolated, OCNL—Occasional, FRQ—Frequent, EMBD—Embedded.

Model TA2. Tabular forecast of en-route conditions

DATE .... HEIGHTS IN PRESSURE ALTITUDE IN HUNDREDS OF FEET		
ROUTE FROM ...... BIGGIN HILL ..... TO ...... AMSTERDAM ...... VIA ...... AIRWAYS		
VALID FOR DEPARTURE BETWEEN ...... 1500 ...... UTC AND ...... 1700 ..... UTC AND FOR ARRIVAL BETWEEN .... 1700 .... UTC AND ...... 2100 ..... UTC		
SPECIAL FEATURES OF THE METEOROLOGICAL SITUATION (SURFACE CENTRES AND FRONTS): ACTIVE COLD FRONT FROM HUMBER TO CHANNEL ISLES AT 1000 UTC MOVING EAST AT 20 KNOTS TO LIE NORTH/SOUTH ACROSS TRACK ABOUT 40 NM WEST OF AMSTERDAM BY 1900 UTC.		
ZONE	LONDON 02 ° E	AMSTERDAM
UPPER WINDS FL 300 (DEGREES TRUE AND KNOTS) FL 240 TEMPERATURES FL 180 (DEGREES CELSIUS) FL 100	250/50 MS 52 260/40 MS 40 270/35 MS 26 280/30 MS 12	230/65 MS 50 240/60 MS 36 240/50 MS 24 250/45 MS 09
SIGNIFICANT WEATHER AND ASSOCIATED CLOUD	MODERATE TURBULENCE IN SCT CU 180/015	MODERATE TO SEVERE ICING AND TURBULENCE IN ISOL EMBD CB 300/010
HEIGHT OF O°C ISOTHERM	035	050
* TROPOPAUSE HEIGHT	———	———
* JET STREAM	———	———
SUPPLEMENTARY INFORMATION		

* Above planned cruise level if not specified

Issued by .... at .... 1400 .... UTC on .... 19 ... by .... Forecaster.

Notes : 1. Pressure altitude is the height in feet of a level in the standard atmosphere above the datum level corresponding to a pressure of 1013.2 hPa

2. Positive and negative values are indicated by the prefix "PS" (plus) and "MS" (minus) respectively.

3. Only cloud associated with significant weather is shown. Low stratus and fog, when expected, will be shown for terminal areas in appropriate aerodrome forecasts.

4. When a single numerical value of an element is given in a forecast, it is to be interpreted as representing the most probable mean of a range of values which the element may assume during the period of the forecast.

Abbreviations : SKC—O oktas, FEW—1 to 2 oktas, SCT—3 to 4 oktas, BKN—5 to 7 oktas, OVC—8 oktas, LYR—Layered, LOC—Locally, ISOL—Isolated, OCNL—Occasional, FRQ—Frequent, EMBD—Embedded.

GRID POINT FORECAST

ISSUED BY

At UTC

Valid UTC

Based on UTC data on

DATA PRESENTATION

FL of tropopause
FL, dd, fff of maximum wind
dd, fff, MTT FL450/150 hPa
dd, fff, MTT FL390/200 hPa
dd, fff, MTT FL340/250 hPa
dd, fff, MTT FL300/300 hPa
dd, fff, MTT FL240/400 hPa
dd, fff, MTT FL180/500 hPa

Legend
FL = Flight level
dd = wind direction (tens of degrees)
fff = wind speed (knots)
TT = temperature (°C) preceded by M or P as appropriate

Forecast values apply to centre points of 5°C squares of superimposed grid

10°S

540	540	550	550	550	540	540	530	530
43006020	42002015	39031020	37030030	39030030	37030030	36032030	37033025	37035020
06015M66	08010M66	10010M66	35005M66	31010M66	18010M66	16015M66	15015M66	12015M67
35010M56	35010M55	31020M56	30025M57	30030M57	29020M58	32020M58	34020M59	36015M59
02010M44	01010M44	31015M44	31025M45	31025M45	31020M45	33020M45	33020M45	32010M45
33005M34	30005M34	36005M33	36010M34	29020M34	32025M34	34025M34	30020M33	28015M33
35005M18	35005M18	33010M18	34010M18	28020M19	28020M18	29015M19	31015M18	31010M18
20005M06	23005M06	21005M06	34005M07	04005M07	34005M07	32005M07	35010M06	04015M06

480	500	540	550	550	540	540	520	520
40032025	39032020	39024020	37038045	38027050	39027050	38027060	38030060	38029050
32020M64	29015M65	25005M66	21005M66	25010M66	26020M85	25025M65	26025M65	24015M65
33025M56	32020M56	29020M56	28040M57	27045M58	27050M58	28050M59	30055M58	30040M58
30020M44	29020M45	28020M45	28040M45	28040M45	27040M45	29050M46	30055M45	29045M42
29025M33	29015M33	28015M33	28050M34	28045M34	28040M34	29050M35	29055M34	28060M32
29020M18	28015M18	31010M18	29030M19	27025M19	27020M20	26035M19	28040M19	30040M17
29015M07	28010M06	35010M06	32015M07	23010M08	23005M09	23030M08	26035M07	33035M06

20°S

460	480	500	520	520	500	480	460	460
36029100	38032065	38030055	38028040	38026060	38026055	38025110	37027130	32027150
31060M61	31040M64	29040M65	26030M66	26045M66	26040M65	27050M64	27070M62	27075M61
31080M54	32060M57	30050M57	27035M57	26055M58	26050M58	28075M57	28095M53	27110M53
29095M41	30055M45	30045M46	28030M45	26050M44	27045M47	26090M43	28100M38	28100M40
30095M34	28060M34	29050M34	28035M34	26050M34	27050M35	26100M33	27100M31	30095M34
30065M20	30040M19	30035M19	29025M19	25035M20	25035M21	24080M21	27075M21	31080M20
31045M00	31030M08	31025M07	29020M08	25020M10	22030M10	23070M11	26055M13	30080M13

390	410	430	450	410	400	390	320	300
37031170	39031125	39031090	38028070	39026080	38024115	38022140	31026100	29030150
29110M56	30095M60	30065M63	27060M64	26050M63	25065M63	25080M62	25065M55	28070M51
29050M52	31125M54	31090M56	28065M58	26080M57	24110M58	25115M56	26085M48	30140M47
30155M42	30120M45	31080M47	30055M48	25070M49	24110M50	24120M43	25080M41	30145M45
30145M37	30115M37	30080M37	30060M36	25065M38	24105M38	24110M33	26080M37	30135M42
30115M26	31090M24	30060M23	29045M22	25040M24	23075M24	22085M24	24050M29	29080M31
30090M17	32075M15	31045M12	28035M11	25020M12	22050M14	21070M18	20040M22	28040M24

30°S

320	350	400	430	400	390	360	280	260
31030160	30031150	37031170	39029110	37024120	37022145	31022115	31023070	26031075
29105M50	30110M57	30105M61	28090M62	27080M62	26085M63	23070M60	24035M54	29035M60
30150M49	31140M55	30130M54	29110M57	24115M60	22140M61	23105M57	23060M48	32070M48
30135M44	31150M47	30135M48	29100M52	24115M54	22140M53	23110M47	23065M45	31060M49
30120M41	31135M41	30095M40	30070M41	23080M43	20130M41	22100M39	22065M46	31050M48
29085M32	32100M29	31080M27	31045M26	23040M28	20080M28	20085M29	20050M33	27015M34
29060M25	33080M20	31065M18	32025M15	25010M17	20045M19	18060M22	18030M24	19025M23

290	340	360	380	370	350	300	280	280
28029105	31032140	35031155	27029130	36027120	35022120	28018120	26019060	27020040
27045M50	30075M56	30095M59	30090M61	28085M62	26070M61	23040M59	23035M54	27025M51
29105M50	31135M57	31135M56	29120M60	26105M64	22105M62	22095M58	21045M52	27020M49
29100M51	31135M54	31120M56	31105M59	26095M60	21110M58	22105M54	21045M53	27025M49
29080M48	32135M47	32105M48	31080M47	25075M47	20105M47	21090M45	19045M49	22025M50
31035M36	33095M33	33085M33	32065M31	26040M32	20065M32	19080M32	18056M35	20030M36
02015M26	34065M23	34070M21	33055M20	30020M21	21035M21	19065M23	18040M25	19035M25

40°S

				370	350	320	300	300
				36027100	33024070	31020050	29016050	29014040
				28080M60	26065M60	25045M59	26025M56	24020M53
				28090M66	24065M64	22045M61	19020M56	19015M53
				27070M65	23055M63	20040M59	17035M56	16020M54
				28060M52	23045M52	20040M50	16040M52	14030M53
				29050M35	25035M35	20035M35	16045M37	14035M38
				31045M22	25030M23	20035M24	16045M26	15035M27

110°E 120°E 130°E 140°E 150°E

Climate Data of Selected IMD Stations (page. 827-841)

VOMM Chennai (Minambakam) lat 13° 00′ N, long 80°, 11′ E

Height amsl 16 m

Mean	January	February	March	April	May	June	July	August	Sept.	October	Nov.	Dec.	Annual
DB I	23.5	24.7	27.1	29.9	31.5	31.0	29.3	23.8	23.6	27.6	27.6	24.1	27.7
II	26.2	27.5	29.4	31.0	32.4	32.8	31.5	31.1	30.3	28.6	28.6	25.7	29.4
WB I	21.3	22.1	23.8	25.8	25.9	24.4	24.1	24.3	24.5	24.7	23.5	22.0	23.9
II	21.9	22.5	24.3	26.4	27.2	26.3	25.5	25.6	25.8	25.3	23.5	22.1	24.7
Daily Max.	28.8	30.6	32.7	34.9	37.6	37.3	35.2	34.5	33.9	31.8	29.2	28.2	32.9
Daily Min.	20.3	21.1	23.1	26.0	27.8	27.6	26.3	25.8	25.4	24.4	22.5	21.0	24.3
RH% I	83	80	77	72	63	58	65	69	73	81	83	84	74
II	67	63	64	68	66	59	61	64	69	76	76	71	67
C	3.6	3.0	2.7	4.2	4.5	6.4	7.1	6.2	5.8	5.6	5.9	4.7	4.8
Cl	1.9	1.3	1.4	2.1	1.1	2.7	3.0	2.8	2.1	1.9	2.5	2.5	1.8
Mean wind speed kmph	9.0	9.2	10.2	10.5	13.0	16.4	14.6	13.6	11.1	9.2	11.7	12.6	11.8
No. of days GR	0	0	0	0	0	0	0	0	0	0	0	0	0
No. of days Ts	0.2	0	0.7	2	3	5	5	7	8	9	3	0.5	44
No. of days FG	0.8	0	0.3	0.1	0	0	0	0	0	0.4	0.1	0.4	3
No. of days DS	0	0	0	0	0.2	0.1	0	0	0	0	0	0	0.3
No. of days SQ	0.1	0	0.3	0.6	1.2	1.0	3	1.5	1.2	1.5	0.6	0.1	11
Visibility less then Km	0	0.1	0	0	0.1	0	0	0	0	0	0.2	0.1	0.5
Visibility 1 -4 Km	2	1.5	1.0	0.2	0.3	0.2	0.3	0.3	0.3	0.8	1.6	1.5	10

Abbreviations

DB = Dryball Temperature ^{o}C   I = 0300 UTC

WB = Wetball Temperature ^{o}C II – 1200 UTC

Daily Max = Daily Maximum temperature ^{o}C

Daily Min = Daily Minimum temperature ^{o}C

C = Total cloud amount in oktas

C_L = Low cloud amount in oktas

GR = Hail days

TS = Thunderstorm days

FG = Fog days

DS = Duststorm days

SQ = squalls days

(WDD) New Delhi (Safalarjung) lat 23o 35 N long 77o 12′ 5 5 Height above Mean sea level 216 m

Mean	January	February	March	April	May	June	July	August	Sept.	October	Nov.	Dec.	Annual
DB I	10.0	13.5	19.8	27.0	31.5	32.5	30.1	28.9	27.9	23.8	16.6	11.0	22.7
II	19.2	23.5	29.2	35.2	35.2	38.0	33.2	31.7	32.1	30.4	25.5	20.8	29.8
WB I	7.6	9.8	13.7	17.0	20.52	24.1	25.9	25.4	21.3	18.0	11.1	7.8	16.9
II	12.8	14.1	16.6	18.9	2.0	24.6	26.6	26.4	24.8	20.3	16.0	13.5	19.7
Daily Max.	21.3	23.8	30.2	36.2	40.5	39.9	35.3	33.7	34.1	33.1	28.7	23.4	31.7
Daily Min	7.3	10.1	15.1	21.0	26.6	28.7	27.2	26.1	24.6	18.7	11.8	8.0	18.8
RH% I	72	59	47	32	31	48	73	77	70	54	48	63	56
II	41	28	21	16	18	32	60	65	54	35	31	38	37
C	3.0	2.6	2.7	2.3	1.8	2.8	5.3	5.5	3.1	1.0	1.0	0.3	2.8
Cl	0.9	0.6	0.7	0.7	1.1	1.6	2.9	3.0	1.9	0.6	0.1	0.3	1.2
Mean wind speed kmph	8.4	10.1	10.8	10.9	12.9	14.6	10.3	9.1	9.6	6.3	6.7	7.0	9.8
No. of days GR	0.1	0.1	0.4	0.1	0.1	0	0	0	0	0	0	0	0.8
No. of days Ts	1.5	1.7	3	3	5	5	7	8	4	1.3	0.4	0.7	41
No. of days FG	3	0.5	0.1	0	0.7	0.2	0.1	0	0	0.1	0.6	1.7	7
No. of days DS	0	0	0.5	1.0	3	3	0.5	0	0.1	0.3	0	0	8
No. of days SQ	0.3	0.4	1.0	1.6	3	2	1.5	1.0	0.6	0.1	0.1	0	12
Visibility less then 1 Km	2	0.4	0.2	0.2	0.5	0.7	0	0	0	0	0.2	0.8	5
Visibility 1- 4 Km	7	4	2	3	4	5	1.7	1.2	1.1	2	4	7	41

43003 Mumbai lat 19° 07′ N long 72o 51′ E Height AMSL 15 M (Santacruz) VABB

Mean	January	February	March	April	May	June	July	August	Sept.	October	Nov.	Dec.	Annual
DB I	20.3	21.5	25.3	28.5	29.8	28.4	27.0	26.3	26.5	26.7	25.0	22.2	25.6
II	27.2	28.3	29.8	30.9	31.1	21.8	27.9	27.3	28.2	29.7	29.3	28.1	29.0
WB I	16.0	16.8	20.7	23.8	25.5	258.6	25.3	24.8	24.7	23.6	19.9	17.5	22.0
II	19.9	20.2	22.2	24.4	26.0	26.0	25.5	25.0	24.8	24.7	22.4	21.0	23.5
Daily Max	30.6	31.6	32.8	33.3	33.2	31.7	29.6	29.1	29.9	32.4	33.1	32.1	31.6
Daily Min	16.3	16.9	20.2	23.8	26.2	25.4	24.8	24.4	23.9	229	19.8	17.7	21.9
RH% I	63	61	65	67	69	80	86	87	86	75	61	62	72
II	48	48	50	57	65	74	82	82	75	65	53	50	62
C	1.4	1.2	1.5	2.7	4.2	6.3	7.4	7.4	6.0	3.8	2.1	1.5	3.8
Cl	0.3	0.2	0.6	1.4	3.2	4.2	4.4	4.6	3.4	1.6	0.5	0.2	2.0
Mean wind speed kmph	8.3	9.2	10.4	11.6	12.8	15.1	19.0	16.1	10.8	7.6	7.1	7.6	11.3
No. of days GR	0	0	0	0	0	0	0	0	0	0	0	0	0
No. of days Ts	0	0.2	0.2	0.6	2	6	1.8	0.8	4	3	0.1	0	19
No. of days FG	0	0.6	0.4	0	0	0	0	0	0	0.1	0	0.1	1.2
No. of days DS	0	0	0	0	0	0	0	0	0	0	0	0	0
No. of days SQ	0	0	0	0.1	0.3	2	4	3	0.8	0.5	0	0	11
Visibility less then 1 Km	0	0.3	0.1	0	0	0.2	0.1	0.1	0	0.1	0	0	0.5
Visibility 1- 4 Km I	9	10	7	1.6	0.3	3	6	5	2	1.3	2	6	52
II	0.5	0											

42809 VECC Kolkata (Dum Dum) lat 22° 3′ 9 N, long 88° 27′ E Height amsl 6m

Mean	January	February	March	April	May	June	July	August	Sept.	October	Nov.	Dec.	Annual
DB I	17.6	20.3	25.6	29.1	30.6	30.0	28.8	28.9	29.0	27.5	22.6	18.1	25.7
II	22.4	26.1	30.5	32.2	31.9	30.4	29.2	29.8	28.8	27.7	24.5	22.0	27.9
WB I	14.7	17.4	22.0	25.3	27.0	27.1	26.6	26.5	26.5	24.7	19.6	15.5	22.8
II	17.1	19.0	22.1	25.2	27.0	27.1	26.7	26.7	26.4	24.6	20.7	17.5	23.3
Daily Max.	26.4	29.0	33.8	36.0	35.7	33.8	31.8	31.8	32.0	31.4	29.0	26.5	31.4
Daily Min.	12.3	15.1	20.4	24.3	26.0	26.2	26.0	26.0	25.9	23.6	17.6	13.0	21.4
RH% I	74	72	70	70	75	79	83	82	81	77	72	75	76
II	57	49	47	55	68	77	81	82	81	77	68	63	67
C	1.6	2.3	2.5	3.1	4.5	6.5	7.1	6.9	6.5	4.4	2.3	1.5	4.0
Cl	0.5	0.7	1.2	1.9	3.1	3.6	4.5	4.3	3.7	2.3	0.7	0.3	2.2
Mean wind speed Kmph	4.1	5.1	8.4	13.0	16.2	12.4	11.5	10.0	8.4	5.0	3.9	3.5	8.5
No. of days GR	0	0	0.2	0.1	0	0	0	0	0	0	0	0	0.3
No. of days TS	0.5	0.6	4	5	7	11	7	9	10	7	0.5	0.2	62
No. of days FG	5	7	5	0.8	0	0.1	0.3	0.1	0.3	1.6	1.8	5	27
No. of days DS	0	0	0	0.1	0.1	0	0	0	0	0	0	0	0.2
No. of days SQ	0.1	0.1	0.5	3	3	3	0.1	0.2	0.5	0.1	0	0	11
Visibility less than 1 km	1.4	1.7	0.8	0.1	0	0	0.1	0	0.1	0.1	0	1.1	5
1-4 km	15	13	9	4	0.9	1.3	2	1.8	2	5	8	14	76

42867 VANP Nagpur (Sonegaon) lat 21°. 06′ N long 79° 03′ E Height ams 310 m

Mean	January	February	March	April	May	June	July	August	Sept.	October	Nov.	Dec.	Annual
DB I	16.9	20.3	25.5	30.5	34.3	30.4	26.1	25.7	26.1	24.9	20.4	17.3	24.9
II	26.5	31.1	34.5	37.8	40.9	34.3	28.7	28.2	28.5	28.7	26.3	25.3	30.9
WB I	13.3	14.3	16.9	20.2	22.3	23.8	24.0	23.8	23.7	21.1	15.7	13.7	19.4
II	17.7	18.7	20.1	22.1	23.3	24.3	24.7	24.6	24.2	22.0	18.3	17.4	21.5
Daily Max	28.6	32.5	36.4	39.7	42.8	38.4	31.2	30.4	31.5	31.9	29.9	28.7	33.5
Daily Min	12.7	15.1	19.1	23.9	28.4	26.9	24.0	23.7	23.1	20.0	14.1	12.1	20.3
RH% I	65	49	40	37	34	59	83	85	81	71	59	64	61
II	38	26	24	23	19	46	72	74	70	54	43	42	44
C	2.1	1.7	2.4	3.7	4.2	6.2	7.3	7.3	6.4	4.1	2.2	2.0	4.1
CL	1.1	0.6	1.2	2.2	2.6	3.8	3.7	3.7	3.5	1.7	0.3	0.3	2.1
Wind speed Kmph	-	-	-	-	-	-	-	-	-	-	-	-	-
No. of days GR	0	0.1	0.3	0.1	0	0.1	0	0.3	0	0	0.2	0	1.1
No. of days TS	0.9	1.9	4	5	5	12	8	7	7	2	0.6	0.5	54
No. of days FG	0.2	0	0	0	0	0	0	0.1	0.2	0.2	0	0.1	0.8
No. of days DS	0	0	0	0.3	0.2	0	0	0	0	0	0	0	0.5
No. of days SQ	0.1	0	0	1.3	1.3	3	0.6	0.8	0.8	0.5	0	0	8
Visibility less than 1 km	0.2	0	0	0	0	0	0	0	0	0	0	0.1	0.3
1 – 4 km	3	1.5	1.8	0.3	0.1	1.1	2	1.5	0.9	0.9	0.6	2	15

43296 VOBG Bangalore Aerodrome lat 12° 57′ N long 77° 38′ E Height amsl 897 meters

Mean	January	February	March	April	May	June	July	August	Sept.	October	Nov.	Dec.	Annual
DB I	18.8	20.7	23.4	25.0	24.5	22.6	21.5	21.5	21.7	22.3	21.0	18.6	21.8
II	26.0	28.8	31.3	31.4	29.9	26.8	25.2	25.5	26.2	25.4	24.5	24.3	27.1
WB I	16.3	16.3	18.4	21.1	21.3	20.2	19.7	19.6	19.7	20.1	18.3	16.4	18.9
II	18.0	17.5	18.2	20.3	21.5	21.1	20.5	20.8	20.9	20.8	19.1	17.9	19.7
Daily Max.	28.0	30.6	33.2	34.0	32.9	29.7	27.8	27.9	28.6	28.3	27.5	26.9	29.6
Daily Min.	15.3	16.4	18.7	21.3	21.2	19.9	19.3	19.4	19.3	19.3	17.4	15.4	18.6
RH% I	78	64	61	69	75	80	84	85	83	80	77	80	76
II	45	30	25	36	49	61	66	66	62	67	60	53	52
C	3.7	3.0	2.7	5.3	6.1	6.8	7.0	7.3	6.5	6.6	5.1	4.2	5.2
CL	2.3	0.9	1.4	2.9	3.5	3.8	4.9	4.8	3.9	3.4	2.5	2.1	2..7
Mean wind speed Kmph	-	-	-	-	-	-	-	-	-	-	-	-	-
No. of days GR	0	0	0	0	0	0	0	0	0	0	0	0	0
No. of days TS	0	0.4	0.7	6	12	4	3	3	4	7	0.8	0.4	41
No. of days FG	3	0.9	0.3	0.1	0	0.1	0	0	0.3	0.6	1.2	3	9
No. of days DS	0	0	0.2	0.1	0.2	0.1	0	0	0	0	0	0	0.6
No. of days SQ	0	0.1	0.3	1.7	3	1.7	0.7	1.2	0.4	0.2	0.2	0.1	10
Visibility less than 1 km	0.9	0.1	0	0	0.2	0	0.3	0	0.1	0.2	0	0.9	2
1 – 4 km	3	2	1.5	0.8	0.7	0.7	1.3	0.8	0.6	1.1	0.9	3	16

43128 VOHY Hyderabad (Begumpet) lat 17° 27′ N, long 78o 28′ E Height above msl 545 m

Mean	January	February	March	April	May	June	July	August	Sept.	October	Nov.	Dec.	Annual
DB I	19.6	21.5	25.4	28.8	30.4	27.0	24.7	24.6	24.7	25.1	22.7	19.3	24.5
II	26.9	30.0	33.2	34.9	35.7	31.7	27.7	27.5	27.1	27.7	26.3	25.7	29.5
WB I	16.3	17.0	18.8	21.4	22.2	22.7	22.2	22.0	22.2	21.3	18.4	15.7	20.0
II	18.0	18.9	20.1	21.8	23.1	23.7	23.2	23.2	23.0	21.7	18.5	17.2	21.0
Daily Max.	28.6	31.2	34.8	36.9	38.7	34.1	29.8	29.5	29.7	30.3	28.7	27.8	31.7
Daily Min.	14.5	16.7	20.0	28.7	26.2	24.1	22.3	22.1	21.6	19.8	16.0	13.4	20.0
RH% I	79	64	54	51	50	71	83	82	82	73	68	71	69
II	36	35	30	31	33	54	69	70	71	58	48	42	48
C	2.4	2.2	2.1	3.7	4.4	6.2	7.0	6.7	6.4	4.8	3.7	2.5	4.3
CL	1.4	1.0	1.3	2.1	2.4	3.2	3.4	3.1	3.0	2.1	1.2	0.8	2.0
Mean wind speed Kmph	8.1	8.9	9.6	10.9	12.4	23.8	22.1	18.3	12.6	8.9	8.0	7.4	12.6
No. of days GR	0	0	0.1	0	0	0	0	0	0	0	0	0	0.1
No. of days TS	0.1	0.5	1.9	5	6	5	1.5	2	4	3	0.3	0	29
No. of days FG	0.8	0.2	0	0	0	0	0	0	0	0.1	0.2	0.8	2
No. of days DS	0.2	0.1	0.2	0	0.3	0	0	0	0	0	0.2	0.1	1.1
No. of days SQ	0	0	0.6	1.7	2	3	2	0.9	1.0	0.1	0	0	11
Visibility less than 1 km	0.9	0.4	0.3	0.2	0.0	0	0	0	0.2	0	0.3	0.8	3
1 – 4 km	6	5	4	3	0.8	1.4	1.2	0.8	0.6	1.1	5	7	36

43371 VOTV Trivandrum lat 29′ N long 76° 57′ E Height amsl 64 m

Mean	January	February	March	April	May	June	July	August	Sept.	October	Nov.	Dec.	Annual
DB I	25.1	25.1	26.7	27.8	27.3	25.8	25.5	25.8	25.8	25.9	25.7	25.2	25.9
II	28.7	29.1	29.7	29.3	28.8	27.0	26.7	27.0	27.2	27.0	27.3	28.1	28.0
WB I	22.2	22.6	24.0	25.2	25.4	24.2	23.8	23.8	23.8	24.1	23.8	22.6	23.8
II	23.6	24.0	25.0	25.6	25.7	24.8	24.4	24.2	24.3	24.5	24.4	23.9	24.5
Daily Max	31.3	31.7	32.5	32.4	31.6	29.4	29.1	29.4	29.9	29.9	30.1	30.9	30.7
Daily Min	22.3	22.9	24.2	25.1	25.0	23.6	23.2	23.3	23.3	23.4	23.1	22.5	23.5
RH % I	77	79	80	81	84	90	89	88	86	87	87	80	84
II	63	63	66	73	77	82	81	78	77	80	78	69	74
C	3.5	3.6	4.3	5.7	6.0	6.6	6.7	6.1	5.6	6.4	5.9	4.3	5.4
CL	1.4	1.9	2.8	3.4	3.3	3.6	3.3	3.0	2.5	3.4	3.3	2.0	2.8
Mean wind speed Kmph	5.1	5.9	6.6	7.8	9.2	9.6	10.9	11.2	10.4	7.3	5.5	4.8	7.9
No. of days GR	0	0	0	0	0	0	0	0	0	0	0	0	0
No. of days TS	1.5	3	8	16	12	3	1.2	1.4	3	9	9	3	70
No. of days FG	0	0	0	0	0	0	0.1	0	0	0.1	0.6	0	0.8
No. of days DS	0	0	0	0	0	0	0	0	0	0	0	0	0
No. of days SQ	0	0	0	0	0	0.1	0.2	0	0.1	0	0	0	0.4
Visibility less than 1 km	0	0.1	0.1	0.2	0.5	0.5	0.6	0.3	0.3	0.5	0.3	0.1	3
1 – 4 km	13	13	8	3	3	9	8	7	4	7	7	10	92

43540 Srinagar lat 34° 05′ N, long 74° 50′ E Height amsl 1586 m′

Mean	January	February	March	April	May	June	July	August	Sept.	October	Nov.	Dec.	Annual
DB I	-0.9	1.1	5.9	11.3	16.1	19.9	22.4	21.1	16.8	9.6	2.9	0.1	10.5
II	2.4	6.2	11.7	17.6	22.4	26.6	28.8	27.6	24.9	18.0	11.5	6.0	17.0
WB I	-1.4	0.4	4.7	9.1	12.9	15.7	18.9	18.5	14.3	7.8	1.9	-0.7	8.5
II	1.0	3.9	7.8	11.8	15.3	17.9	20.8	20.6	17.3	12.1	7.3	3.7	11.6
Daily Max.	4.4	7.9	13.4	19.3	24.6	29.0	30.8	29.9	28.3	22.6	15.5	8.8	19.5
Daily Min.	-2.3	-0.8	3.5	7.4	11.2	14.4	18.4	17.9	12.7	5.7	-0.1	-1.8	7.2
RH% I	88	87	84	7.7	71	75	73	79	77	82	85	88	81
II	78	69	60	53	49	45	49	53	47	51	54	67	56
C	7.1	6.3	5.5	5.2	4.8	3.9	4.0	4.4	2.9	2.6	2.9	5.2	4.3
CL	5.9	4.9	3.7	3.5	3.4	2.6	2.7	3.1	2.1	1.6	1.6	3.7	2.9
Mean wind speed Kmph	3.6	4.3	5.6	5.3	4.3	4.1	4.0	3.6	3.7	3.3	3.0	3.4	4.0
No. of days GR	0	0	0.2	0.1	0.2	0	0	0	0	0	0	0	0.5
No. of days IS	0.4	0.2	3	7	10	7	7	6	4	2	0	0.2	47
No. of days FG	1.1	0.7	0.1	0	0	0	0	0	0	0	0.4	1.1	3
No. of days DS	0	0	0.1	0.1	0.3	0.6	0.2	0.2	0.3	0.1	0	0	1.9
No. of days SQ	0.1	0	0.1	0.2	0.2	0.3	0.1	0.1	0.2	0.1	0	0	1.4
Visibility less than 1 km	1.3	0.9	0.3	0.2	0.1	0	0	0	0	0.1	0.3	0.6	4
Visibility less than 1-4 km	9	7	5	3	1.9	2	2	2	1.1	1.7	1.7	5	41

42056 Jammu lat 32° 40′ N long 74° 50′ E Height above msl 366 m

Mean	January	February	March	April	May	June	July	August	Sept.	October	Nov.	Dec.	Annual
DB I	10.5	13.1	18.0	24.7	30.4	32.0	29.0	27.6	26.4	21.9	15.9	11.5	21.7
WB I	7.7	9.5	13.1	16.4	19.4	22.1	24.6	24.7	22.5	16.7	11.2	8.2	16.3
Daily Max	18.4	21.1	26.4	33.1	39.0	40.4	35.4	33.2	33.3	31.4	26.2	21.1	30.0
Daily Min	8.3	10.6	14.8	20.5	25.8	27.7	26.0	25.1	23.9	19.4	13.4	9.3	18.7
RH% I	68	64	55	42	32	41	71	81	71	57	48	64	58
C	3.8	3.3	3.3	2.5	2.3	2.3	4.9	4.9	2.5	1.1	1.5	3.1	3.0
CL	1.8	1.5	1.4	0.8	0.8	1.0	3.1	3.3	1.4	0.4	0.3	0.9	1.4
Mean wind speed Kmph	-	-	-	-	-	-	-	-	-	-	-	-	-
No. of days GR	0.1	0.3	0.5	0.3	0.1	0.1	0	0.1	0	0.1	0	0.1	1.7
No. of days TS	0.9	0.8	2	3	2	4	5	4	3	1.2	0.4	0.4	27
No. of days FG	0.1	0.2	0.1	0	0	0	0.1	0.3	0	0.2	0	0	1.0
No.of days DS	0	0.1	0.2	0.4	0.9	1.4	0.2	0.1	0.4	0.1	0	0	4
No. of days SQ	0.3	0.2	0.2	0.1	0.4	0.2	0.2	0.1	0	0	0	0.1	1.8
Visibility less than 1 km	0.3	0.2	0.1	0	0.2	0.2	0.4	0.6	0.3	0	0	0.1	2
1 – 4 km	3	1.5	1.9	0.7	1.1	1.6	4	4	2	0.4	0.4	1.2	22

42647 VAAH Ahmedabad lat 23o 04′ N, long 72o 38′ E Height amsl 55 m

Mean	January	February	March	April	May	June	July	August	Sept.	October	Nov.	Dec.	Annual
DB I	14.8	17.2	22.9	27.6	29.7	29.8	27.6	26.8	26.8	25.3	20.8	19.1	24.0
II	26.8	29.9	34.7	38.5	40.3	36.8	31.3	30.2	31.5	33.3	30.3	27.8	32.6
WB I	10.4	12.1	16.2	20.3	25.0	26.3	25.4	24.5	24.1	20.6	15.3	11.8	19.3
II	16.2	17.6	19.6	21.1	23.7	26.1	26.3	25.7	25.2	21.9	18.8	17.1	21.6
Daily Max.	28.7	31.10	35.7	39.7	40.7	38.0	33.2	31.8	33.1	35.6	33.0	29.6	34.2
Daily Min.	11.9	14.5	18.6	23.0	26.3	27.4	25.7	24.6	24.2	21.2	16.1	12.6	20.5
RH% I	55	52	47	49	68	77	86	84	80	64	52	56	64
II	28	24	20	18	21	41	68	69	60	35	29	29	37
C	1.5	1.3	1.4	1.5	1.8	4.8	6.9	6.9	4.6	1.8	1.3	1.4	2.9
CL	0.4	0.2	0.3	0.4	1.2	2.9	4.4	4.5	2.5	0.8	0.3	0.2	1.4
Mean wind speed Kmph	5.4	5.1	7.3	7.8	9.5	10.8	10.8	8.3	7.0	4.6	4.1	2.6	6.9
No. of days GR	0	0	0	0	0	0	0	0	0	0	0	0	0
No. of days TS	0.4	0.1	0.3	0.8	0.7	4	3	1.8	3	1.0	0.4	0.1	16
No. of days FG	0	0	0	0	0.1	0	0	0	0	0	0	0.2	0.3
No. of days DS	0	0	0	0.4	0.2	0.5	0	0	0	0	0	0	1.1
No. of days SQ	0	0	0	0.1	0.1	0.6	0	0.3	0.1	0.1	0.1	0	1.4
Visibility less than 1 km	0.1	0	0	0	0.1	0	0.1	0.1	0	0	0	0.1	0.5
1 – 4 km	10	7	4	0.9	1.1	1.4	1.9	1.9	1.4	4	5	7	46

42348 VIJP Jaipur lat 26° 49′ N, long 75o 48′ E Height amsl 390 m

Mean	January	February	March	April	May	June	July	August	Sept.	October	Nov.	Dec.	Annual
DB I	12.1	14.8	20.9	28.2	32.4	31.9	28.7	27.0	27.2	25.6	19.5	13.5	23.5
II	20.3	23.8	29.8	35.2	38.8	36.5	31.4	29.6	31.8	30.4	25.9	21.4	29.6
WB I	8.3	9.9	13.6	17.4	20.9	23.3	25.1	24.6	23.5	18.4	13.1	9.3	17.3
II	12.6	13.9	16.8	19.7	22.2	23.6	25.4	25.5	24.5	19.6	15.6	13.6	19.4
Daily Max.	22.0	25.4	30.9	36.5	40.6	39.2	34.1	31.9	33.2	33.2	29.0	24.4	31.7
Daily Min.	8.3	10.7	15.5	210	25.8	27.3	25.6	24.3	23.0	18.3	12.0	9.1	18.4
RH% I	60	49	38	29	33	47	75	82	74	51	48	56	53
II	35	28	22	18	19	33	62	71	57	32	30	35	37
C	2.4	2.2	2.2	2.1	1.7	3.1	5.9	6.3	3.9	1.2	1.0	2.4	2.9
CL	1.0	0.6	0.7	0.9	1.3	2.2	3.6	3.8	2.8	1.0	0.3	0.4	1.5
Mean wind speed Kmph	11.0	11.5	12.0	13.1	16.5	17.8	16.2	12.4	12.3	10.0	7.6	8.1	12.4
No. of days GR	0.2	0.4	0.1	0.1	0	0.5	0.5	0	0.1	0	0	0.1	2
No. of days TS	1.7	1.4	2	3	5	9	9	9	6	1.9	0.5	0.5	49
No. of days FG	0	0.1	0	0.2	0	0.1	0	0.8	0	0.1	0	0	1.8
No. of days DS	0	0	0.3	0.7	1.7	3	0.5	0.1	0	0	0.1	0	6
No. of days SQ	0.1	0	0.1	0.5	0.3	1.0	0.7	0.5	0.5	0.3	0.1	0.1	4
Visibility less than 1 km	0.3	0	0.1	0.1	0.5	0.4	0.1	0.1	0	0	0	0	1.4
1 – 4 km	1.9	1.4	1.7	1.6	5	3	1.2	1.9	0.6	0.8	1.1	1.5	22

43 333 Portblair lat 11° 40′ N long 92° 45 E Height amsl 79 m

Mean	January	February	March	April	May	June	July	August	Sept.	October	Nov.	Dec.	Annual
DB I	27.9	27.8	28.4	30.1	29.2	27.9	27.6	27.5	27.3	28.1	28.4	28.3	28.2
II	25.6	26.4	27.2	28.1	27.3	26.4	26.3	26.2	25.9	25.7	25.8	25.4	26.3
WB I	23.6	23.6	24.5	25.7	25.9	25.4	25.3	25.1	24.9	25.1	25.0	24.0	24.8
II	22.8	23.2	23.9	25.1	25.3	25.1	24.9	24.8	24.5	24.4	24.0	23.1	24.3
Daily Max.	29.2	29.8	31.1	31.9	30.9	29.1	28.9	28.8	28.7	29.0	29.2	29.0	29.6
Daily Min.	22.7	21.8	22.5	24.2	24.4	24.2	24.1	24.1	23.6	23.6	23.6	23.1	23.5
RH% I	70	71	69	70	77	83	89	84	84	81	76	72	77
II	77	76	74	77	84	88	88	88	89	89	85	80	83
C	5.4	3.9	3.9	5.3	6.6	7.4	7.3	7.9	6.9	6.7	5.7	5.3	6.0
CL	3.7	2.9	2.4	3.1	4.7	5.4	5.3	5.5	5.1	4.5	4.8	4.7	4.3
Mean wind speed Kmph	13.4	9.5	8.0	8.7	12.1	19.8	19.9	18.9	15.9	10.7	11.7	13.9	13.5
No. of days GR	0	0	0	0	0	0	0	0	0	0	0	0	0
No. of days TS	0.6	0.8	3	9	14	9	7	4	4	4	3	1.6	60
No. of days FG	1.0	2	3	1.2	0.4	0.1	0.3	0.5	1.4	3	1.5	0.9	15
No. of days DS	0	0	0	0.1	0	0	0	0	0	0	0	0	0.1
No. of days SQ	0	0	0	0.2	0.2	0.7	0.6	0.2	0.2	0.1	0	0	2
Visibility less then 1 km	0	0.1	0	0	0.2	0..5	0.2	0.4	0.3	0./2	0.2	0	2
1 – 4 km	0.7	0.8	0.9	1.3	5	7	6	5	5	3	2	1.3	38

42410 VEGT Gauhati (Bhorjar) lat 26o 05′ N, long 91° 43 E Height amsl 54 m

Mean	January	February	March	April	May	June	July	August	Sept.	October	Nov.	Dec.	Annual
DB I	15.7	18.5	22.6	25.8	26.7	27.8	28.4	28.8	28.3	26.2	21.8	17.5	24.0
II	18.2	22.2	26.0	28.9	29.1	29.0	29.3	29.1	28.3	26.3	22.2	18.7	25.6
WB I	14.2	15.6	18.3	21.6	23.9	25.7	26.3	26.6	26.0	24.0	19.8	16.1	21.5
II	15.7	17.0	19.6	22.5	25.0	26.1	26.7	26.6	26.1	24.5	20.5	17.2	22.3
Daily Max	23.4	26.7	30.0	31.9	31.1	31.2	31.7	32.1	31.7	30.1	27.5	24.8	29.3
Daily Min	9.8	11.6	16.0	20.1	22.8	24.9	25.7	25.8	24.7	22.1	17.3	11.9	19.4
RH% I	86	73	66	68	79	84	85	85	83	83	83	86	80
II	74	58	53	57	72	80	81	82	84	85	85	83	75
C	2.7	2.8	3.5	3.2	5.5	6.7	7.0	6.6	5.9	4.5	2.8	2.7	4.3
CL	1.4	1.3	1.9	2.0	4.0	4.8	4.9	4.0	3.9	3.2	1.8	1.5	2.8
Mean wind speed	-	-	-	-	-	-	-	-	-	-	-	-	-
No. of days GR	0	0.2	0.8	1.1	0	0	0	0	0	0	0	0	2
No. of days TS	0.7	2	8	13	20	16	14	17	15	7	5	0.6	118
No. of days FG	18	3	1.4	0.9	0	0	0	0	0.5	5	11	22	62
No. of days DS	0	0	0.1	0.2	0	0	0	0	0	0	0	0	0.3
No. of days SQ	0	0.2	0.9	2	2	0.4	0.1	0.1	0	0	0	0	6
Visibility less than 1 km	3	0.4	0.2	0.1	0	0	0.1	0	0.1	0	0.6	3	7
1-4 km	8	2	5	3	0.8	1.5	1.3	0.5	1.1	2	3	5	33

Heights 0°C isotherm (Freezing level) over India (Page 842-845)

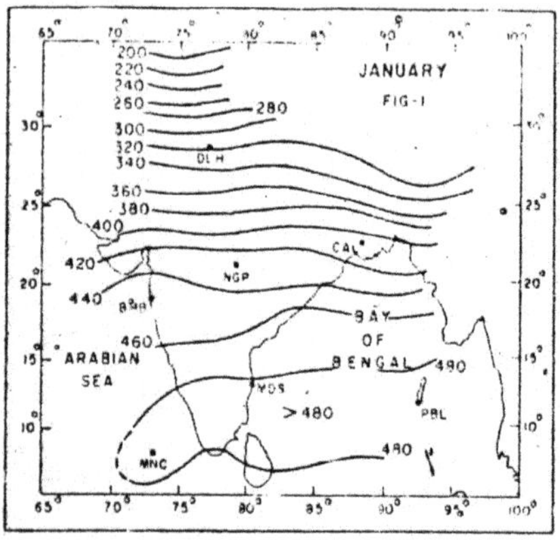

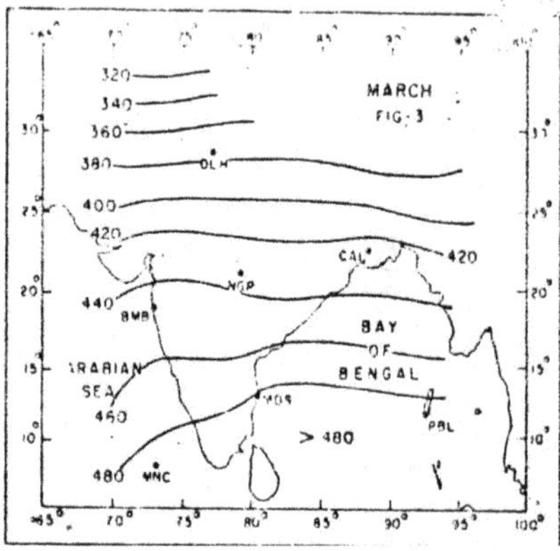

Mean Heights in (Gpdm) of 0 °C Isotherm (Based on 00-12 GMT Data)

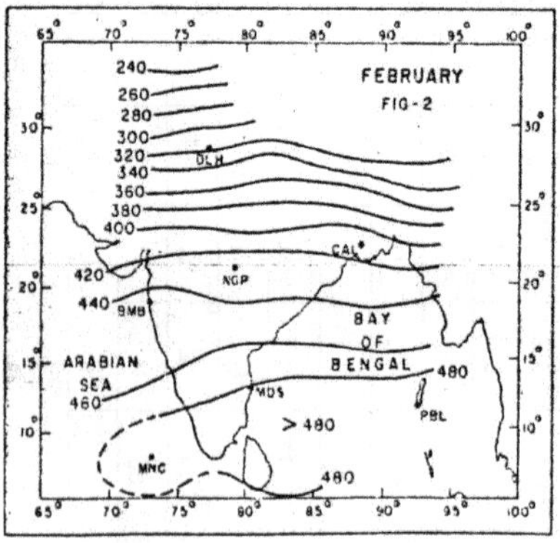

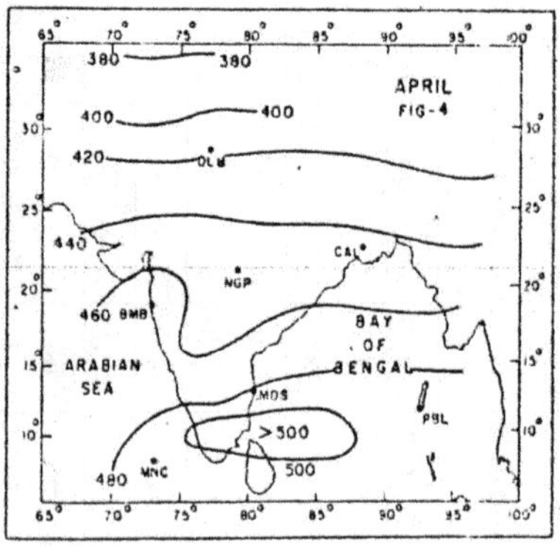

(Based on 1968 – 78 Data)

Mean Heights in (Gpdm) of 0 °C Isotherm (Based on 00-12 GMT Data)

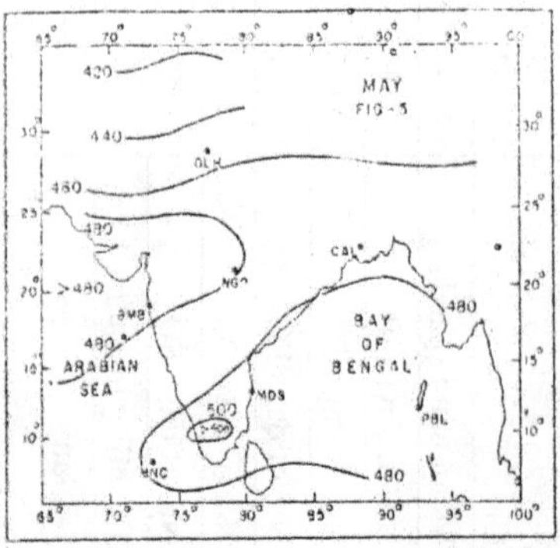

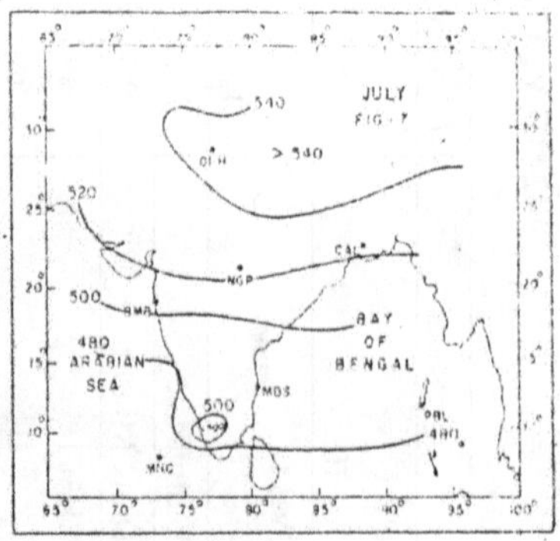

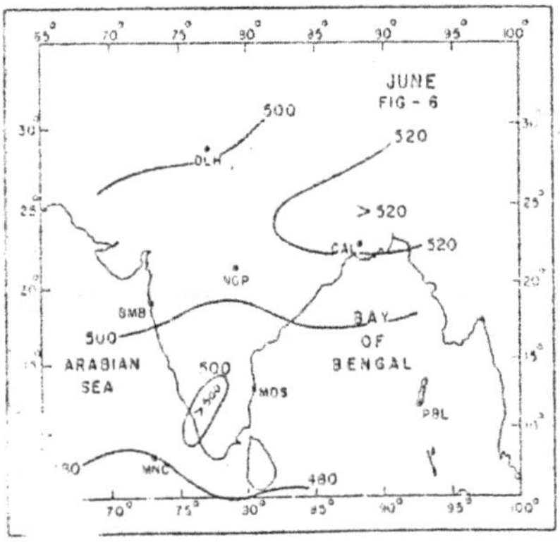

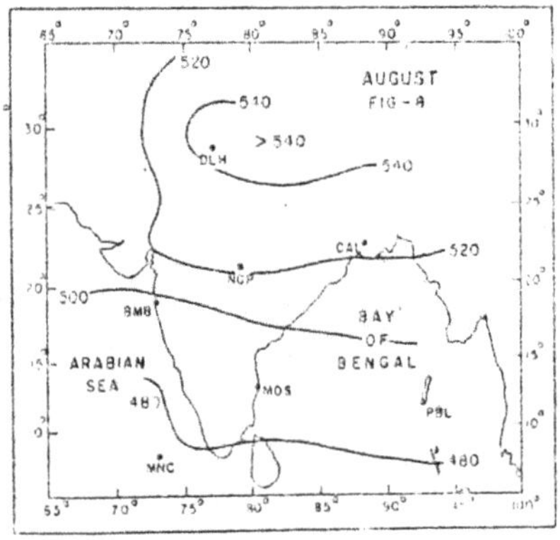

Based on 1968 – 78 Data

Mean Heights in (Gpdm) of 0 °C Isotherm (Based on 00-12 GMT Data)

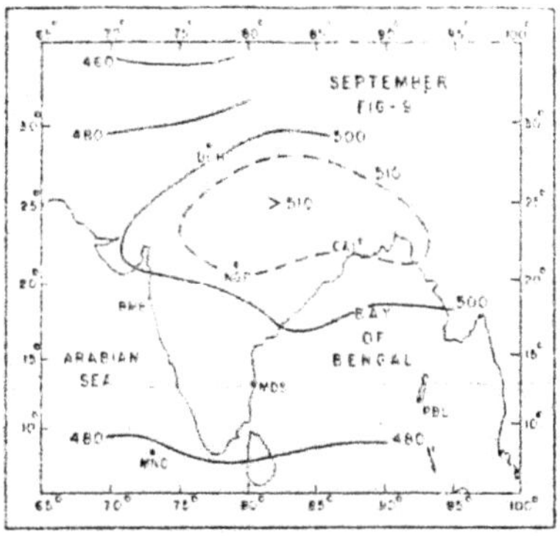

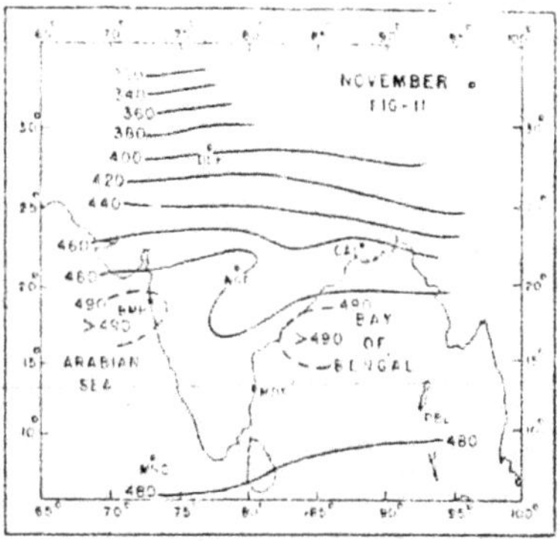

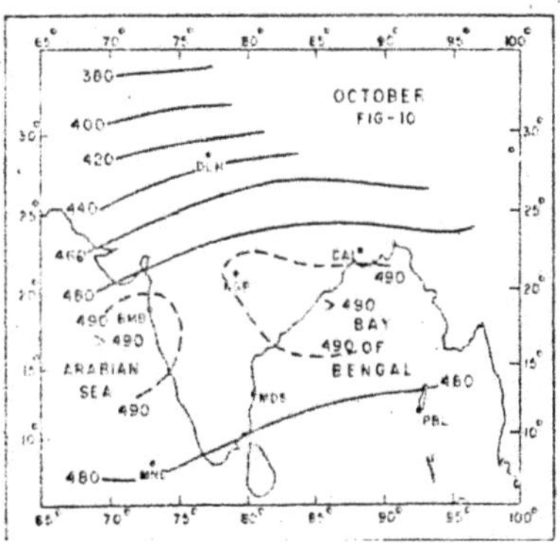

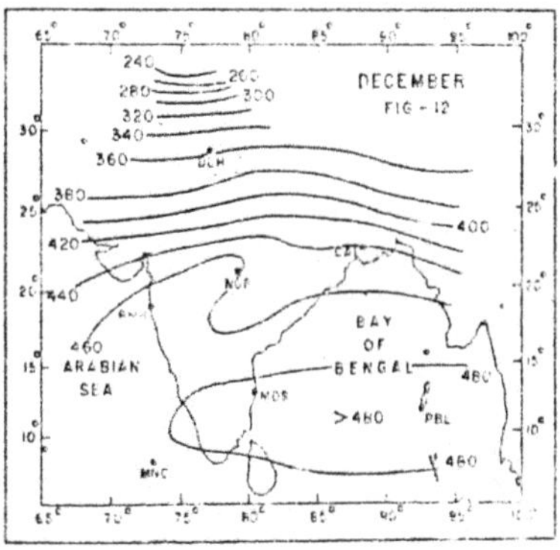

Based on 1968 – 78 Data

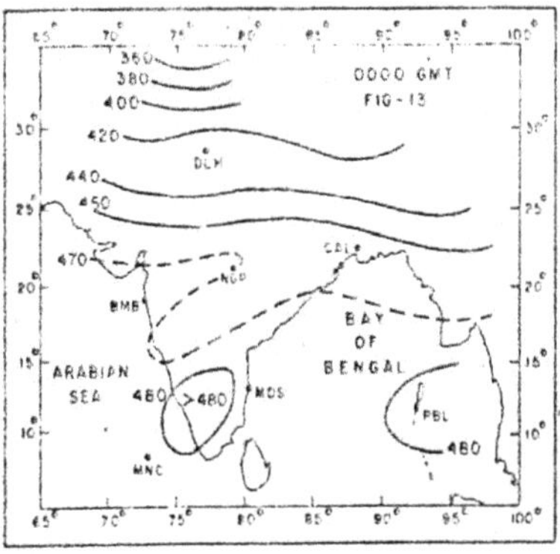

Annual Mean Heights in (gpdm) of 0 °C Isotherm

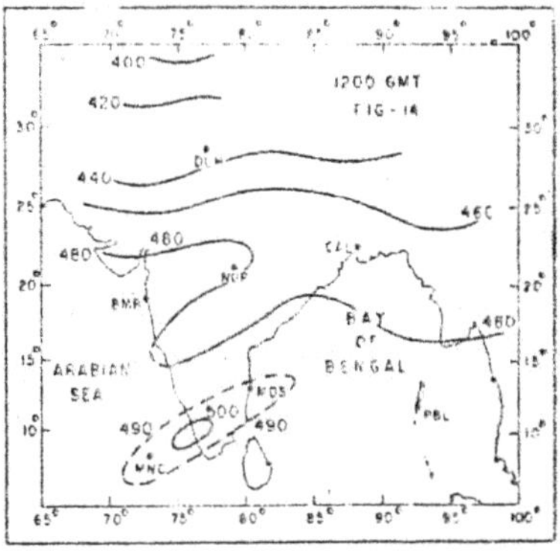

Standard Deviations of Mean Monthly Heights in (gpdm) of 0 °C Isotherm

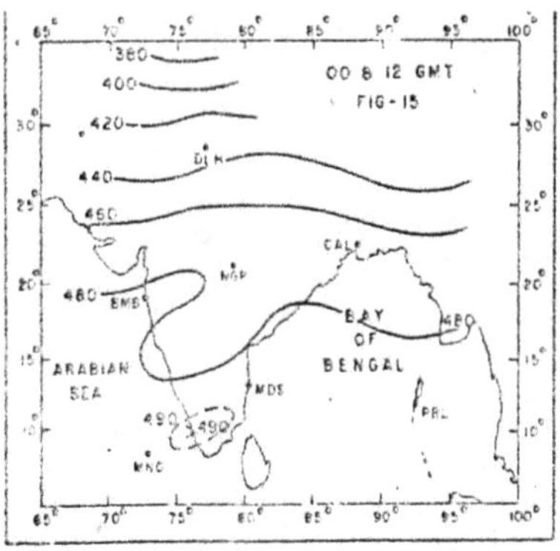

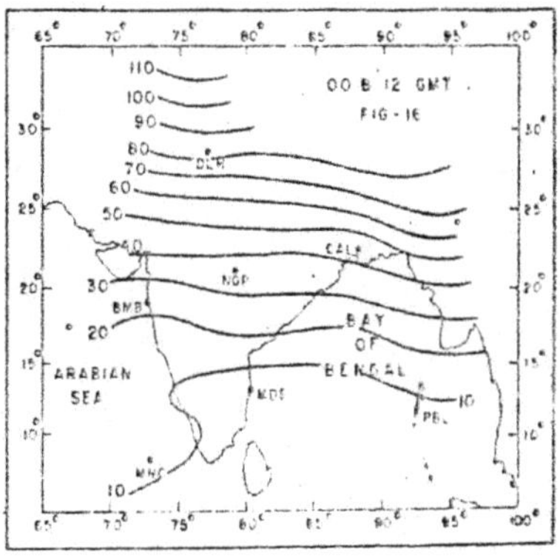

(Based on 1968 – 78 Data)

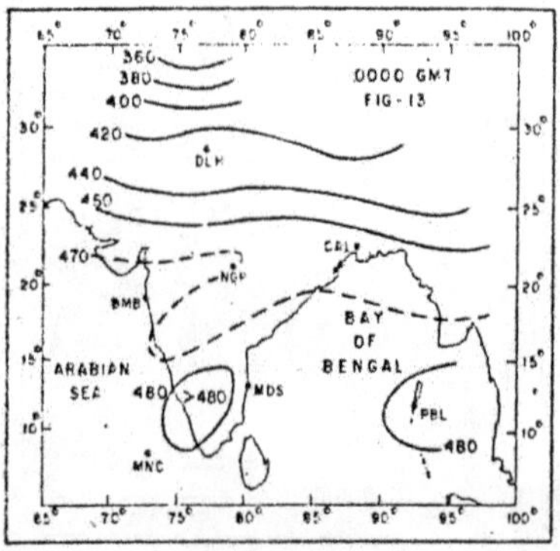

Annual Mean Heights in (gpdm) of 0 °C Isotherm

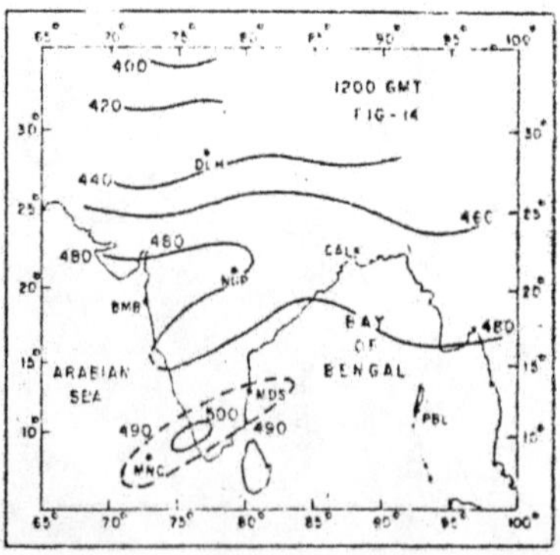

Standard Deviations of Mean Monthly Heights in (gpdm) of 0 °C Isotherm

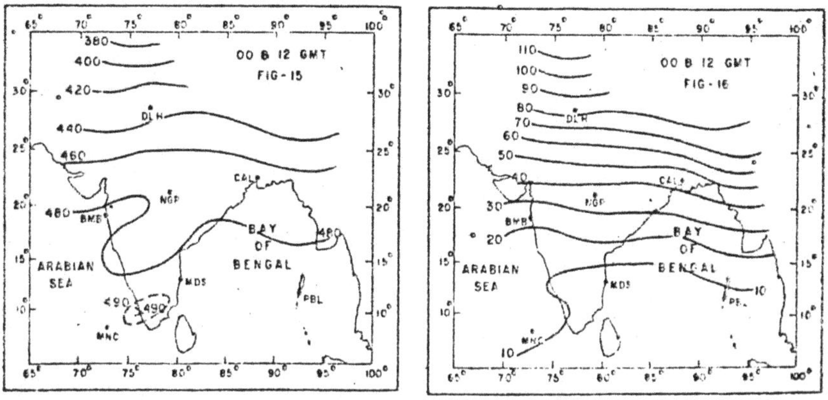

Annual Mean Heights in (gpdm) of 0 °C Isotherm
Based on 1968 – 78 Data

Appendix

Normal upper winds and temperature patterns over India (Page 845-874)

Normal upper air wind and temperature patterns over India and neighbourhood at levels 850 hPa (FL 50), 700 hPa (FL 100), 500 hPa (FL 180), 300 hPa (FL 300), 200 hPa (FL 390), 150 hPa (FL 450) and 100 hPa (FL 530) for representative months January, April, July and October.

The contour lines (continuous lines) (________) are drawn at intervals of 20 gpm (geopotential meters) for levels 850, 700 and 500 hPa for all months except January, where the interval for 500 hPa is 40 gpm. For levels 500, 300, 200, 150, 100 hPa the interval is 80 gpm for all months except July where the interval for 300 hPa is 40 gpm.

The isotherms are shown with hetched line (┼┼┼) are shown at an interval of 2 °C for all months for levels 850 and 700 hPa. For other levels the interval is 5 °C except 500 hPa level for January and October, where the interval is 2 °C. High and low are indicated by the symbols 'H' and 'L' and cold and warm regions by symbols 'K' and 'W' respectively.

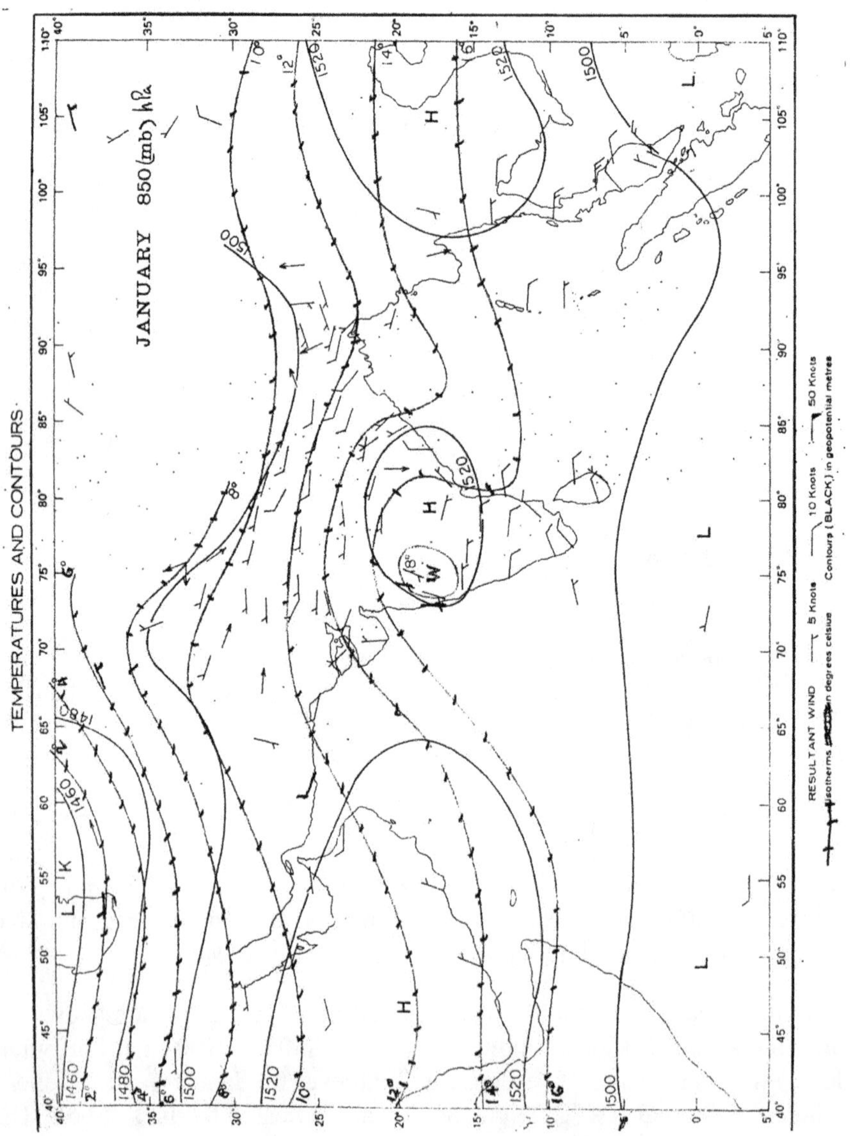
TEMPERATURES AND CONTOURS
JANUARY 850 (mb) hPa
RESULTANT WIND 5 Knots 10 Knots 50 Knots
Isotherms in degrees celsius Contours (BLACK) in geopotential metres

TEMPERATURES AND CONTOURS

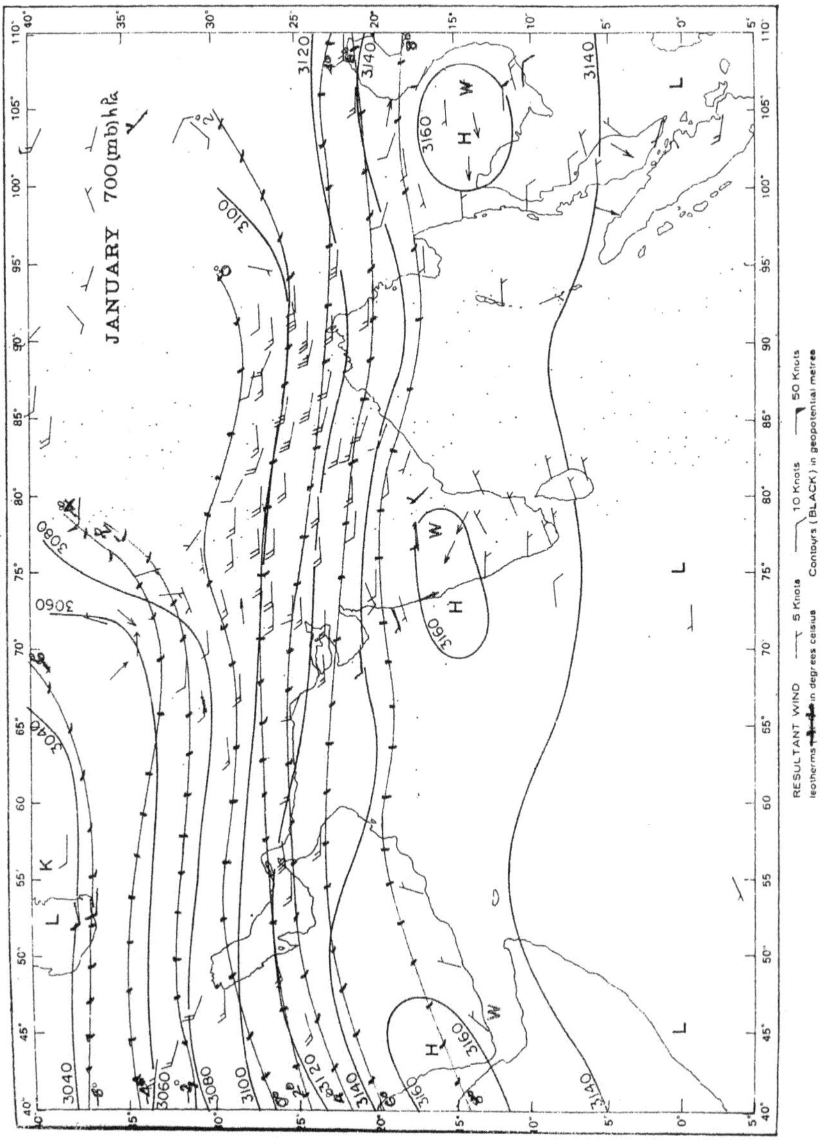

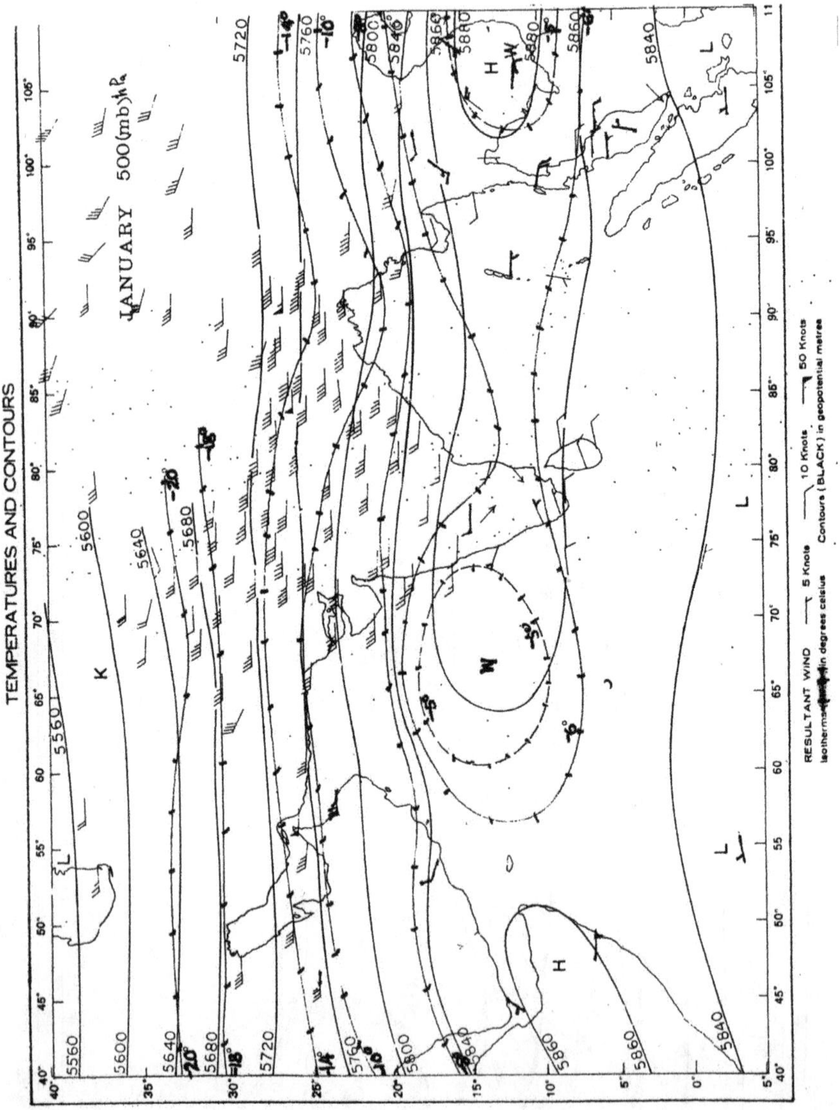
TEMPERATURES AND CONTOURS
JANUARY 500(mb)hPa
RESULTANT WIND 5 Knots 10 Knots 50 Knots
Isotherms in degrees celsius Contours (BLACK) in geopotential metres

TEMPERATURES AND CONTOURS

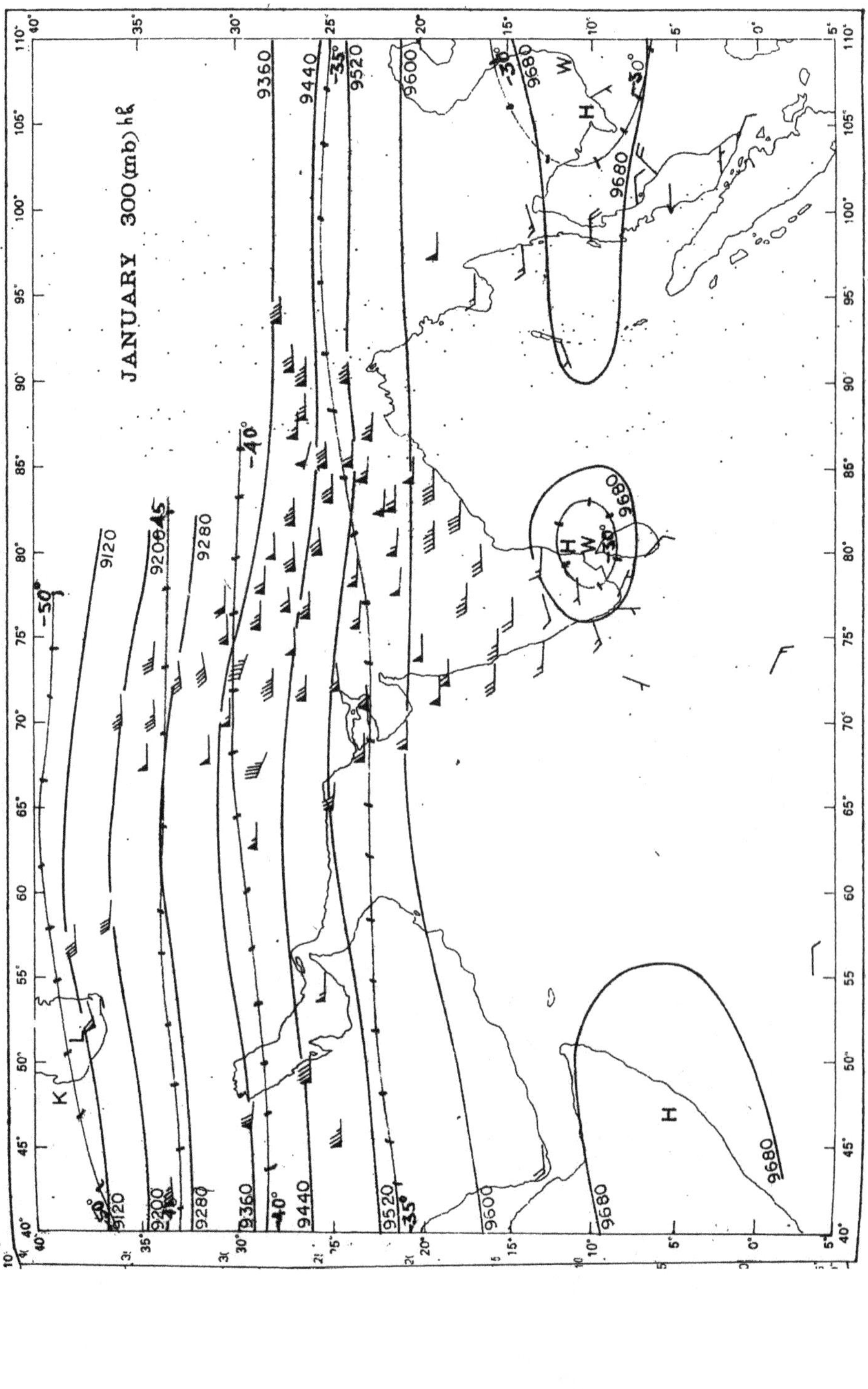

RESULTANT WIND — 5 Knots — 10 Knots — 50 Knots

Contours (BLACK) in geopotential metres

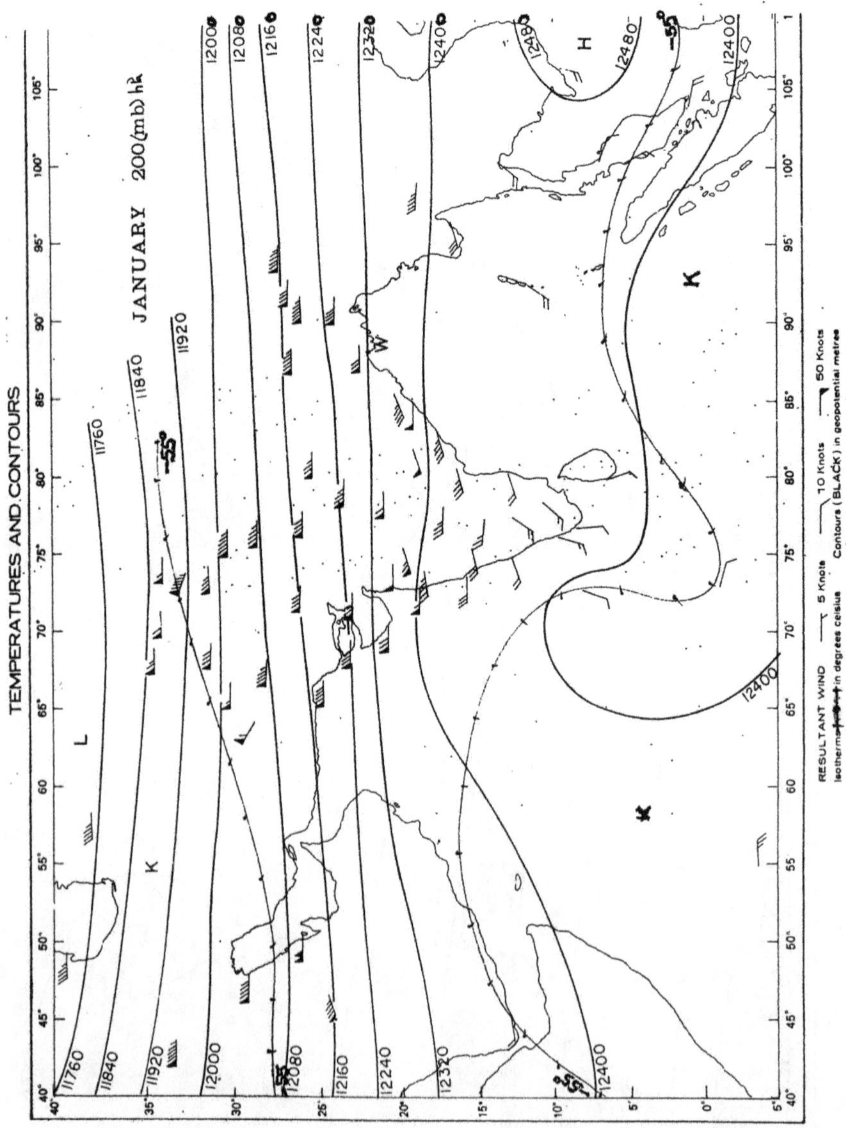
TEMPERATURES AND CONTOURS
JANUARY 200(mb) hk
RESULTANT WIND 5 Knots 10 Knots 50 Knots
Isotherms in degrees celsius Contours (BLACK) in geopotential metres
11760
11840
11920
12000
12080
12160
12240
12320
12400
12480
L
K
W
H
-55°

TEMPERATURES AND CONTOURS

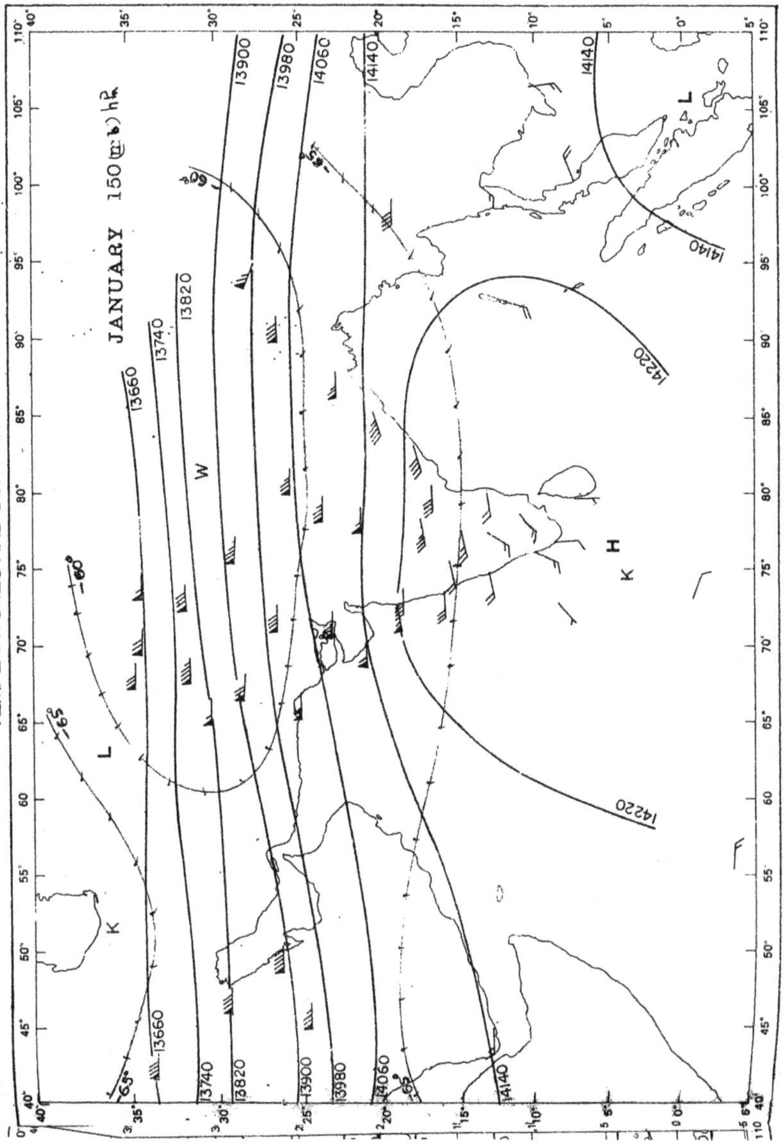

RESULTANT WIND 5 Knots 10 Knots 50 Knots
Isotherms in degrees celsius Contours (BLACK) in geopotential metres

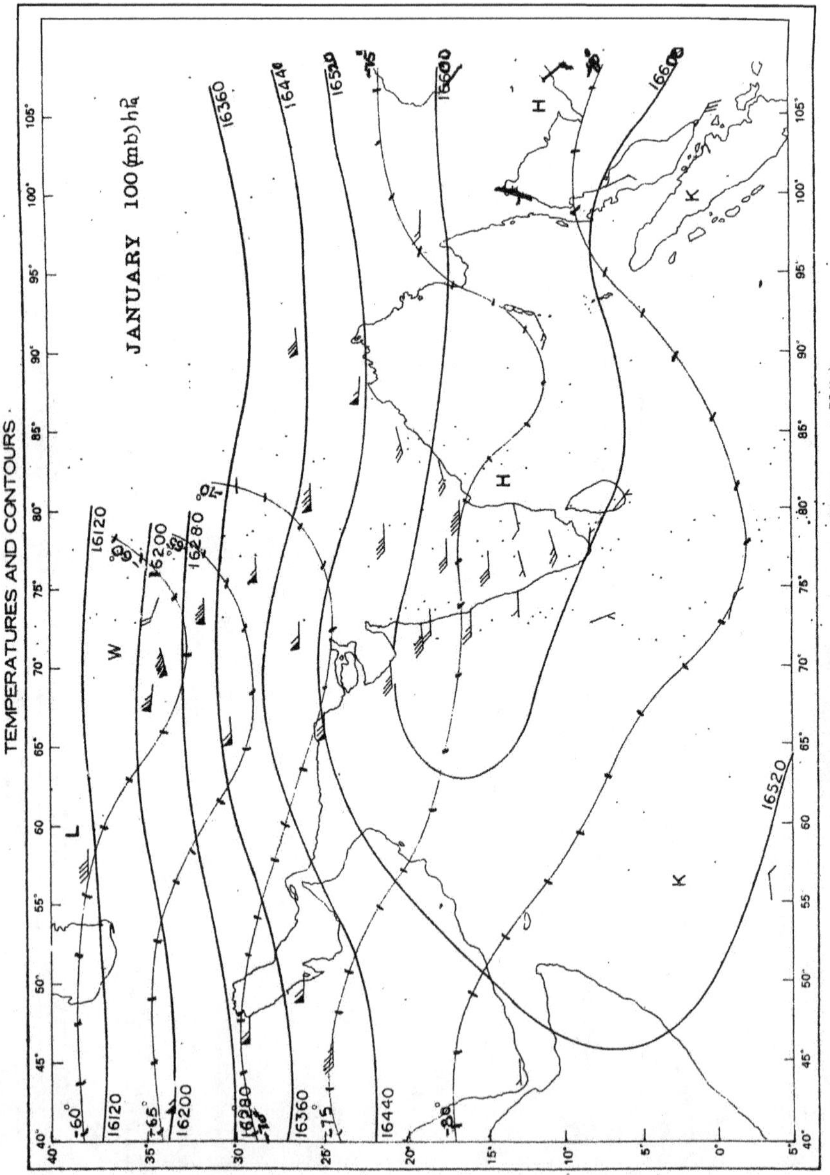
TEMPERATURES AND CONTOURS
JANUARY 100 (mb) hPa
16120
16200
16280
16360
16440
16520
16600
H
K
L
W
RESULTANT WIND 5 Knots 10 Knots 50 Knots
Isotherms in degrees celsius Contours (BLACK) in geopotential metres

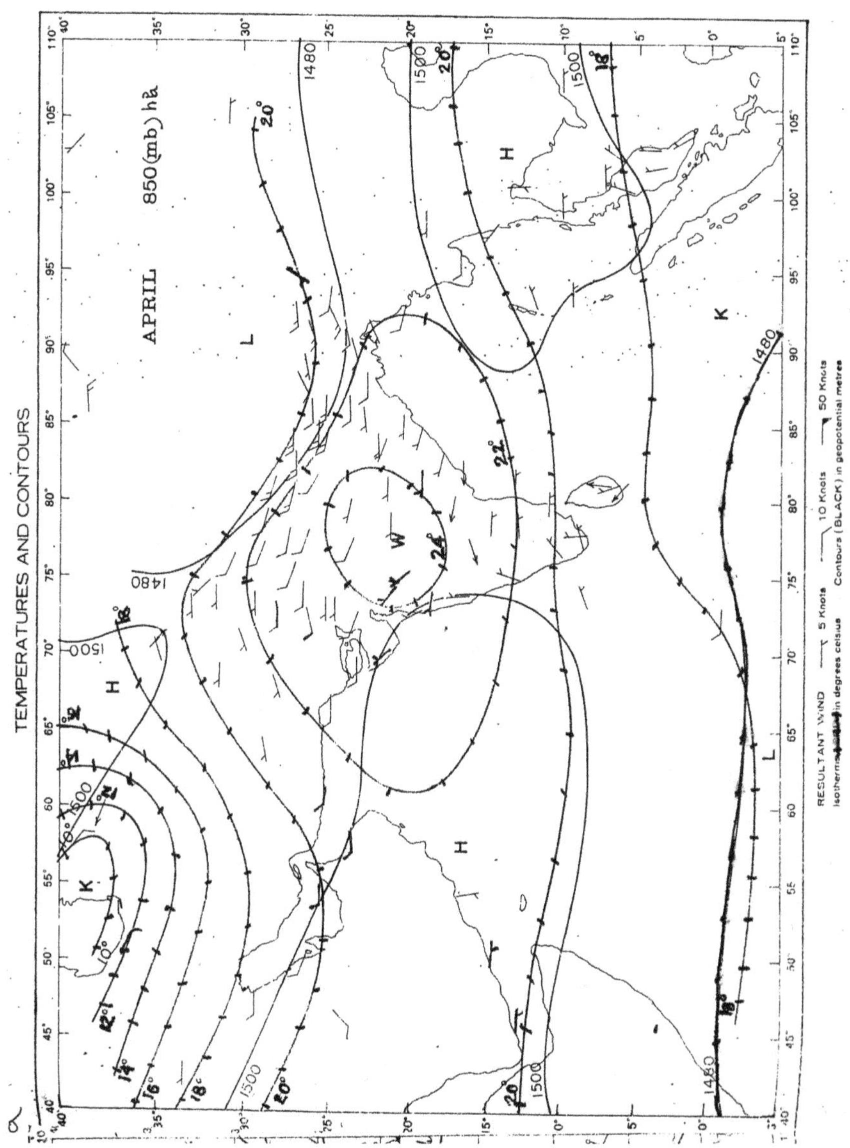
TEMPERATURES AND CONTOURS
APRIL
850(mb) hPa
RESULTANT WIND
5 Knots
10 Knots
50 Knots
Isotherms in degrees celsius
Contours (BLACK) in geopotential metres

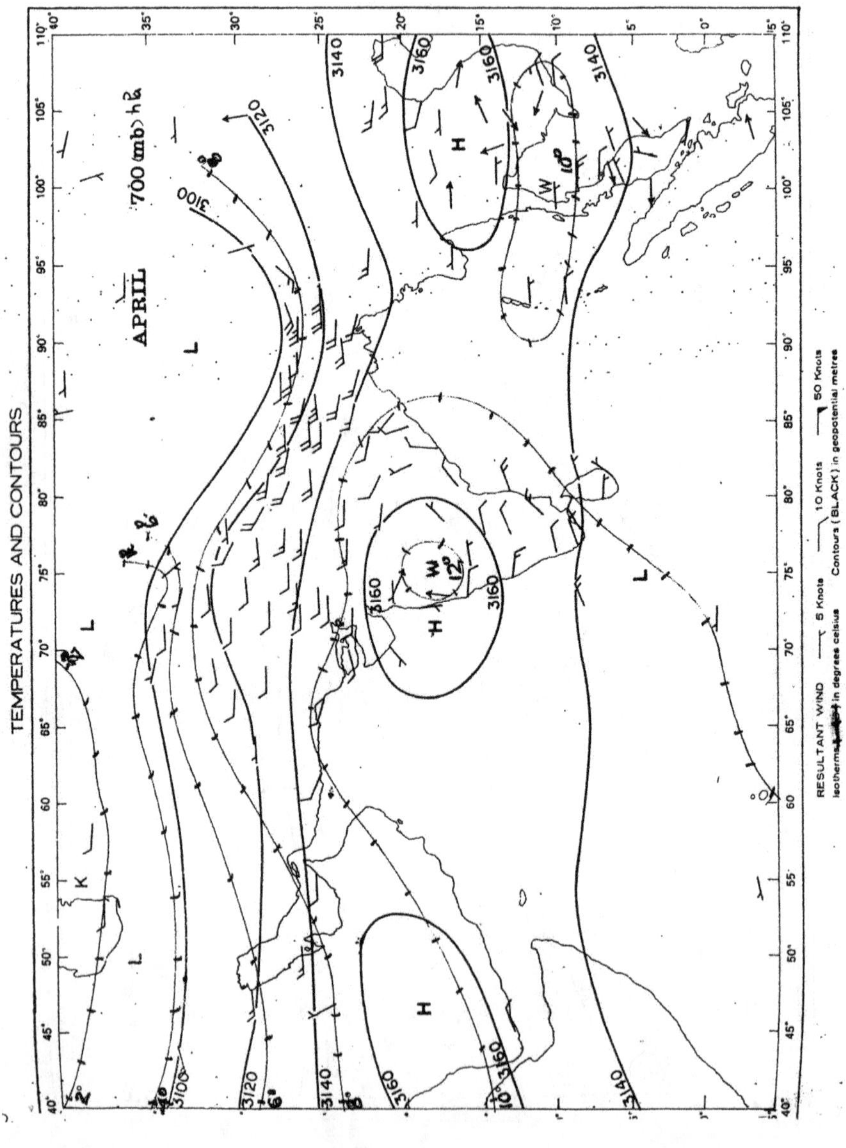
TEMPERATURES AND CONTOURS
APRIL
700 mb
3100
3120
3140
3160
H
L
W
RESULTANT WIND 5 Knots 10 Knots 50 Knots
Isotherms in degrees celsius Contours (BLACK) in geopotential metres

TEMPERATURES AND CONTOURS

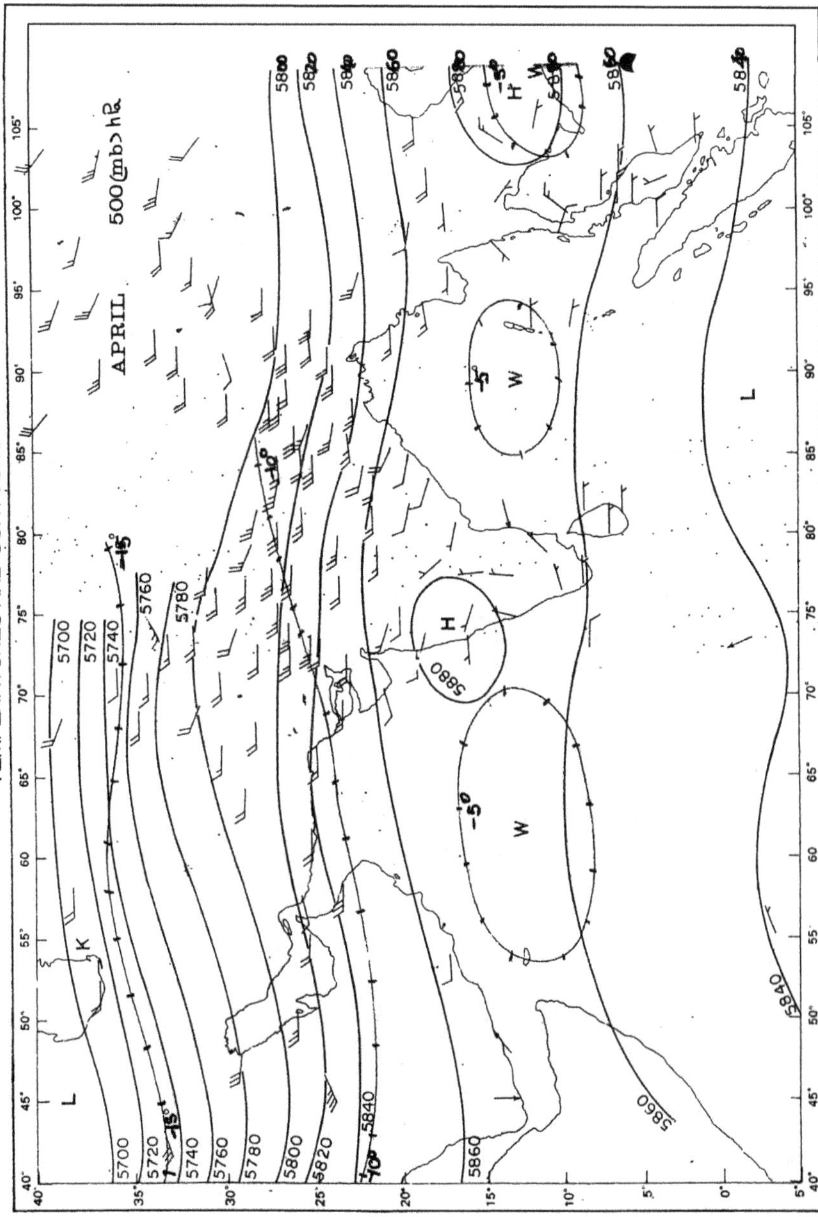

RESULTANT WIND — 5 Knots — 10 Knots — 50 Knots
Isotherms ([illegible]) in degrees celsius Contours (BLACK) in geopotential metres

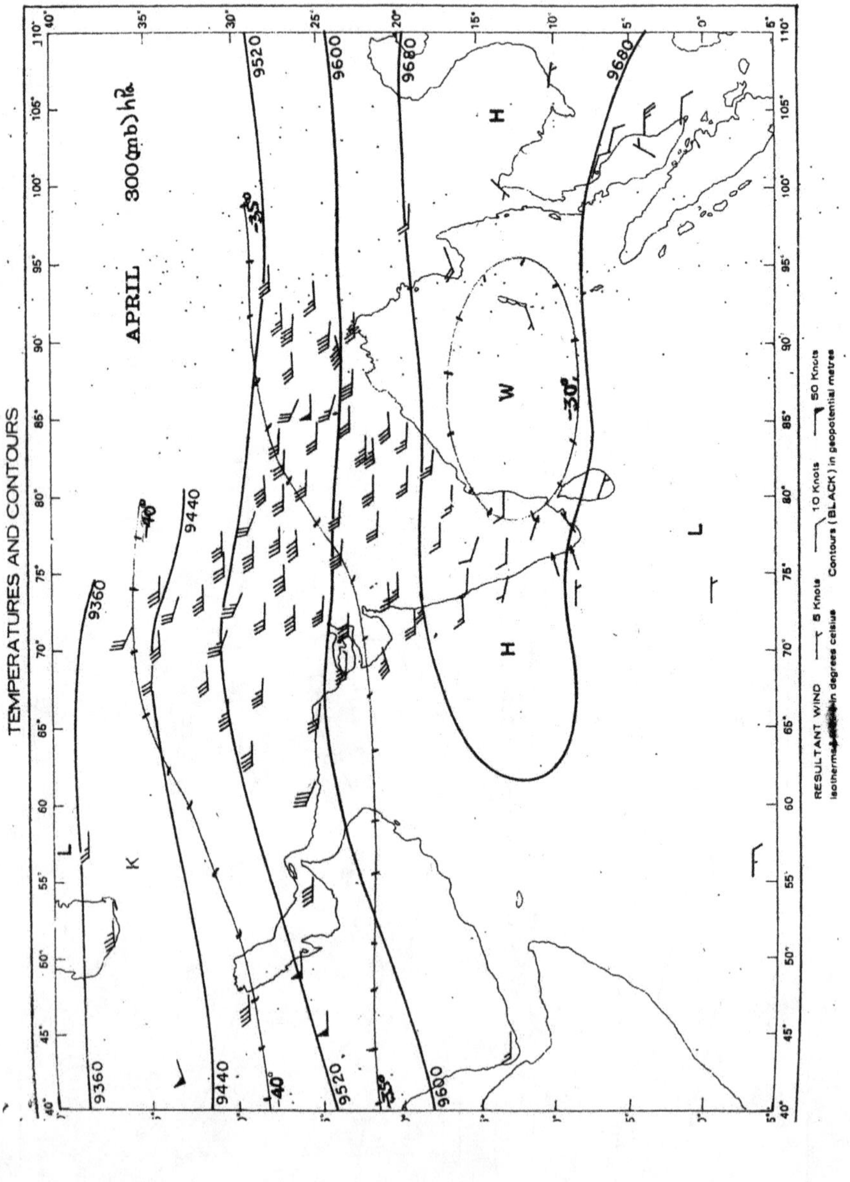
TEMPERATURES AND CONTOURS
APRIL
300(mb) hPa
9360
9440
9520
9600
9680
-40°
-35°
-30°
H
L
W
K
RESULTANT WIND — 5 Knots — 10 Knots — 50 Knots
Isotherms in degrees celsius — Contours (BLACK) in geopotential metres

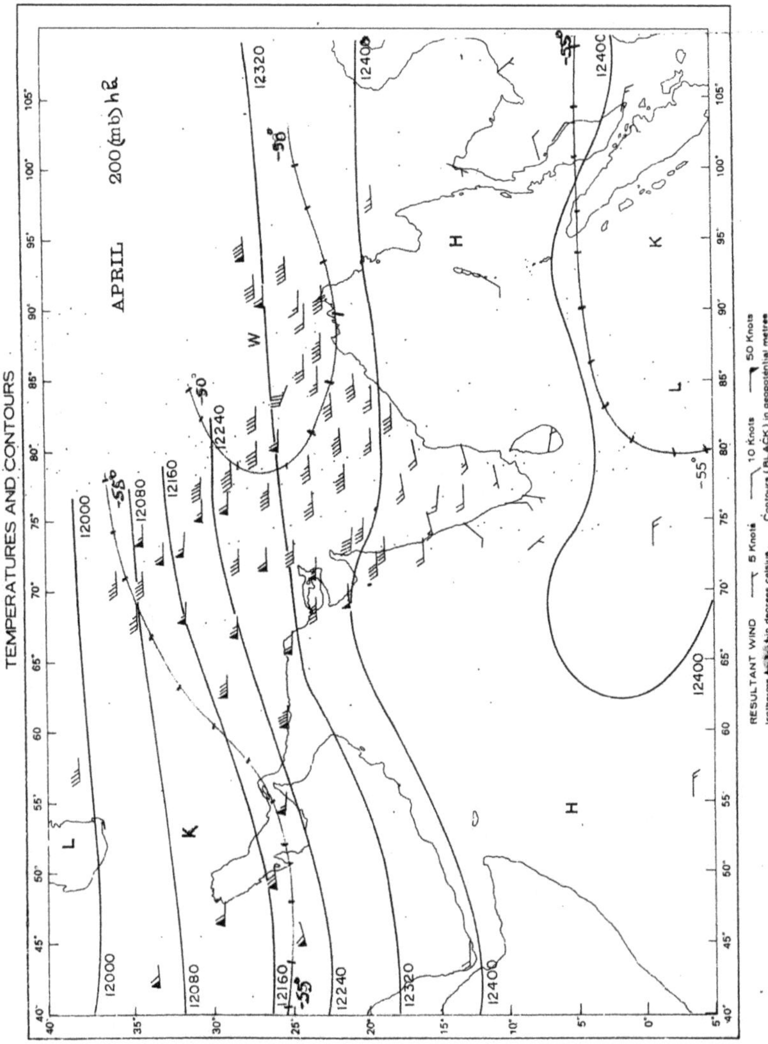
TEMPERATURES AND CONTOURS
APRIL
200 (mb) hPa
12000
12080
12160
12240
12320
12400
RESULTANT WIND
5 Knots
10 Knots
50 Knots
Contours (BLACK) in geopotential metres

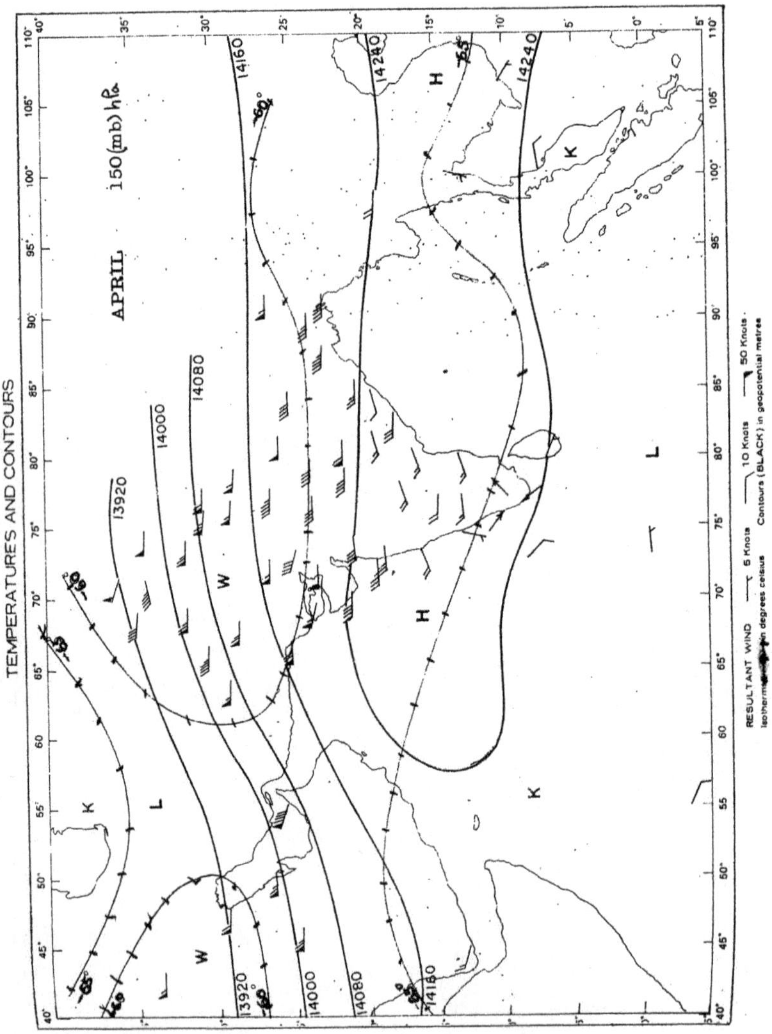
TEMPERATURES AND CONTOURS
APRIL
150(mb) hPa
13920
14000
14080
14160
14240
RESULTANT WIND
5 Knots
10 Knots
50 Knots
Isotherms in degrees celsius
Contours (BLACK) in geopotential metres

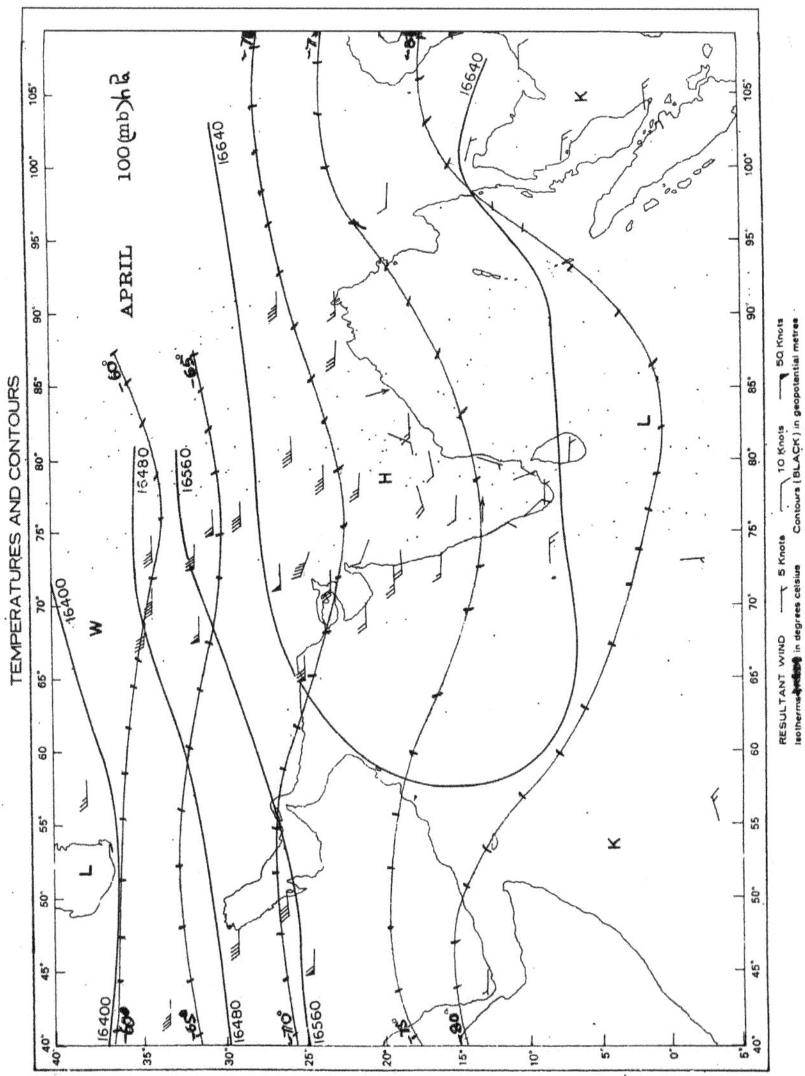
TEMPERATURES AND CONTOURS
APRIL 100(mb)hPa
16400
16480
16560
16640
W
L
H
K
RESULTANT WIND 5 Knots 10 Knots 50 Knots
Isotherms in degrees celsius Contours (BLACK) in geopotential metres

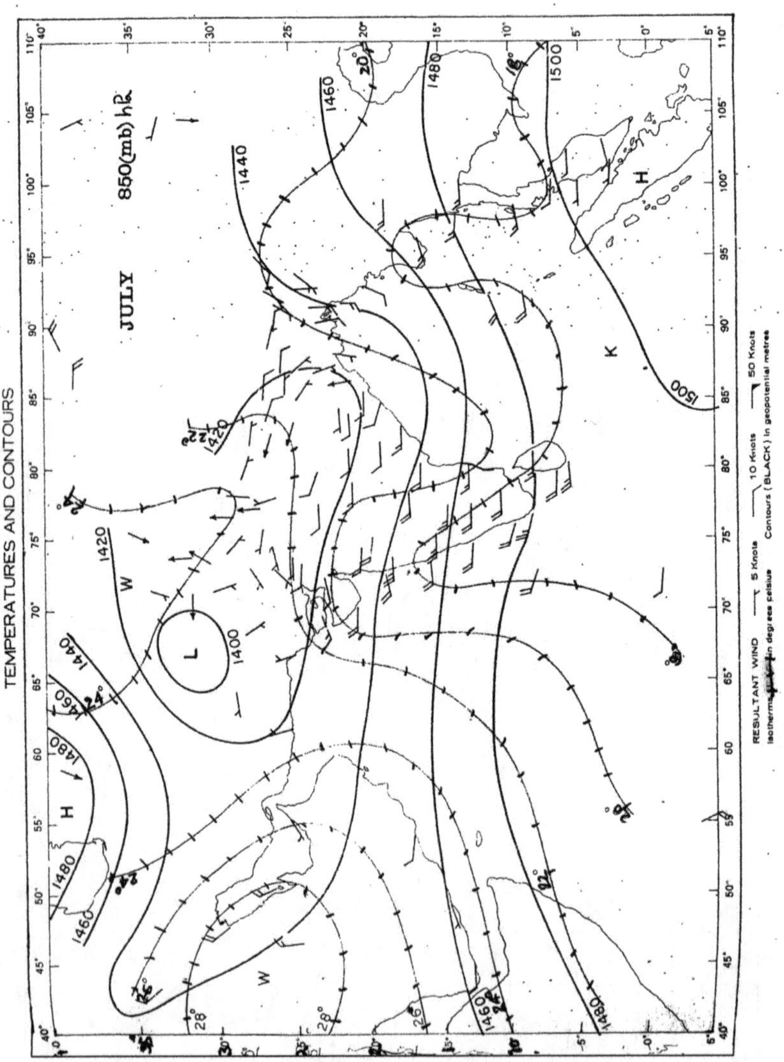
TEMPERATURES AND CONTOURS
JULY
850(mb) hPa
RESULTANT WIND
5 Knots
10 Knots
50 Knots
Contours (BLACK) in geopotential metres

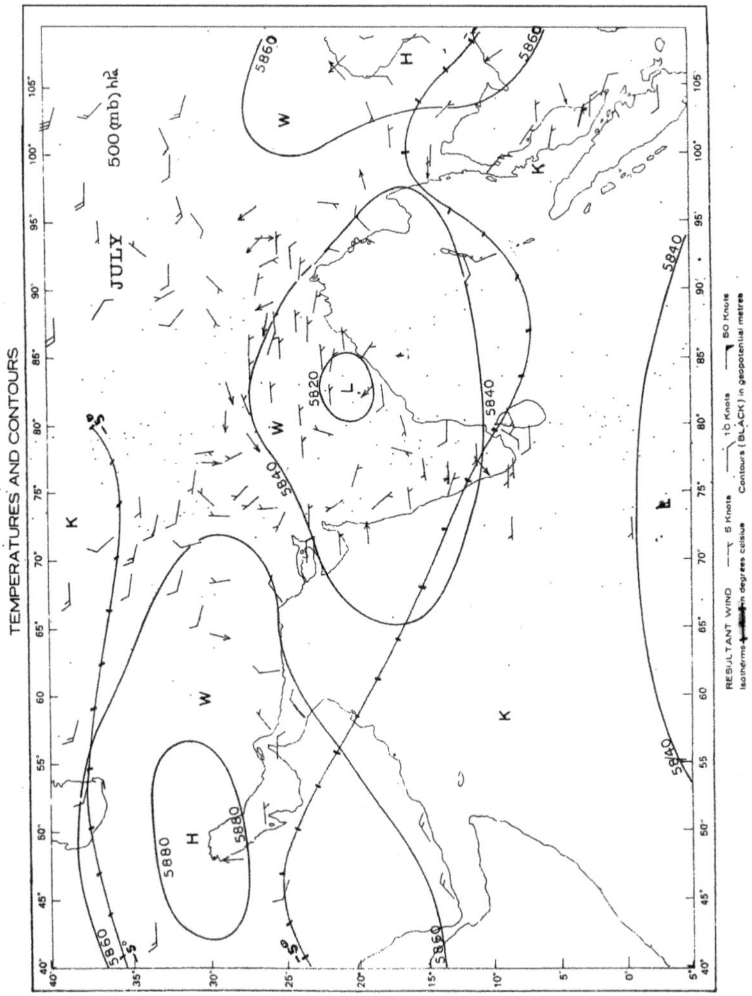
TEMPERATURES AND CONTOURS
JULY
500 (mb) hPa
5880
5860
5840
5820
RESULTANT WIND
5 Knots
10 Knots
50 Knots
Contours (BLACK) in geopotential metres

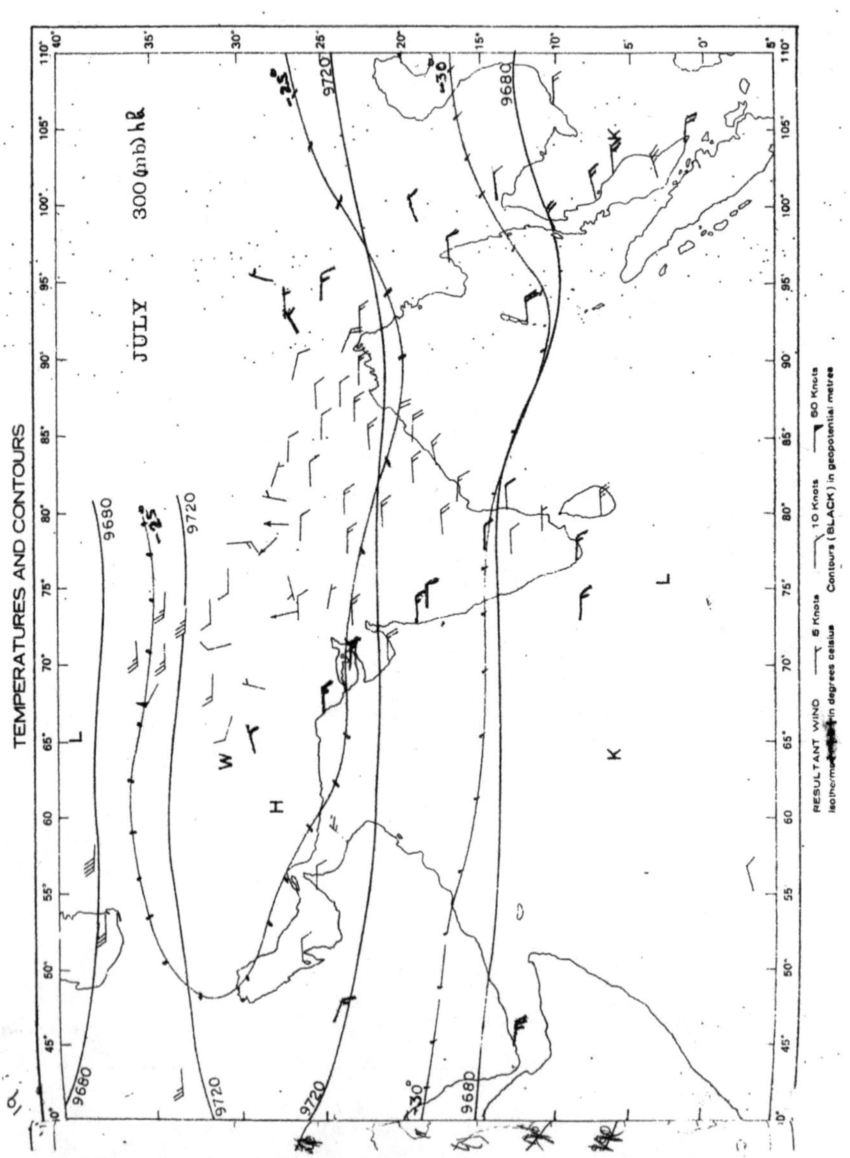
TEMPERATURES AND CONTOURS
JULY
300 (mb) hP
9680
9720
-25°
-30°
H
W
L
K
RESULTANT WIND 5 Knots 10 Knots 50 Knots
Contours (BLACK) in geopotential metres

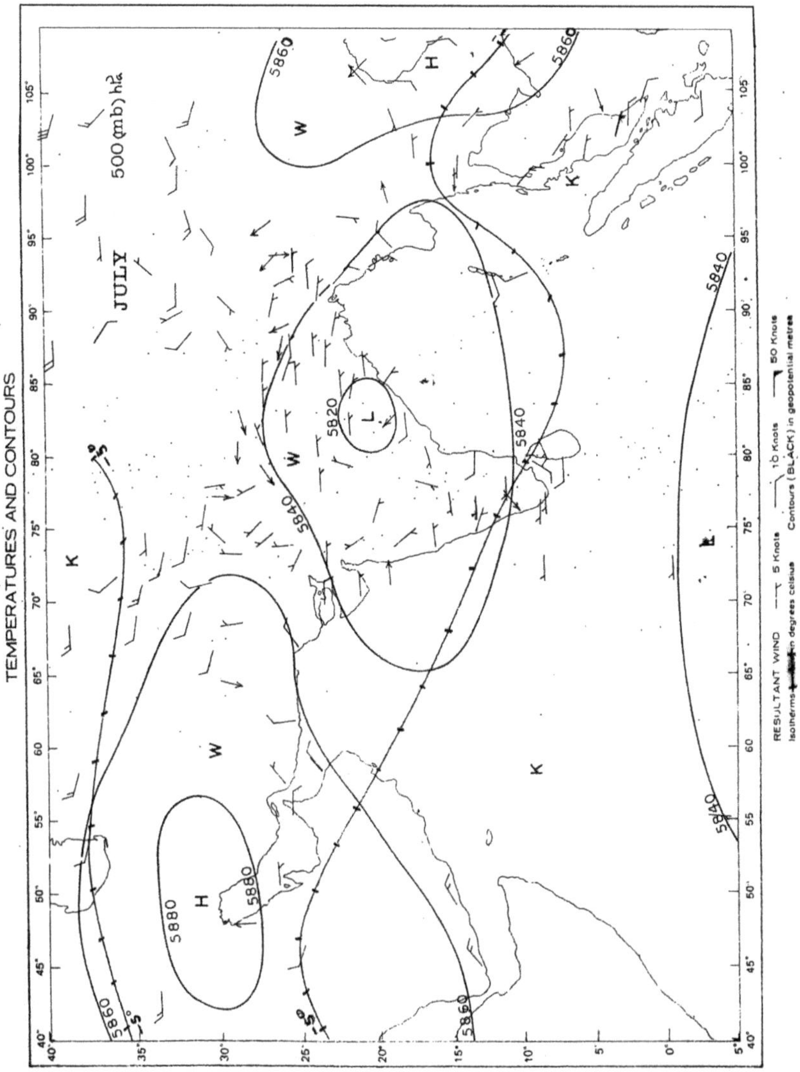
TEMPERATURES AND CONTOURS
JULY
500 (mb) hPa
5880
5860
5840
5820
H
L
W
K
-5°
RESULTANT WIND — 5 Knots — 10 Knots — 50 Knots
Isotherms in degrees celsius
Contours (BLACK) in geopotential metres

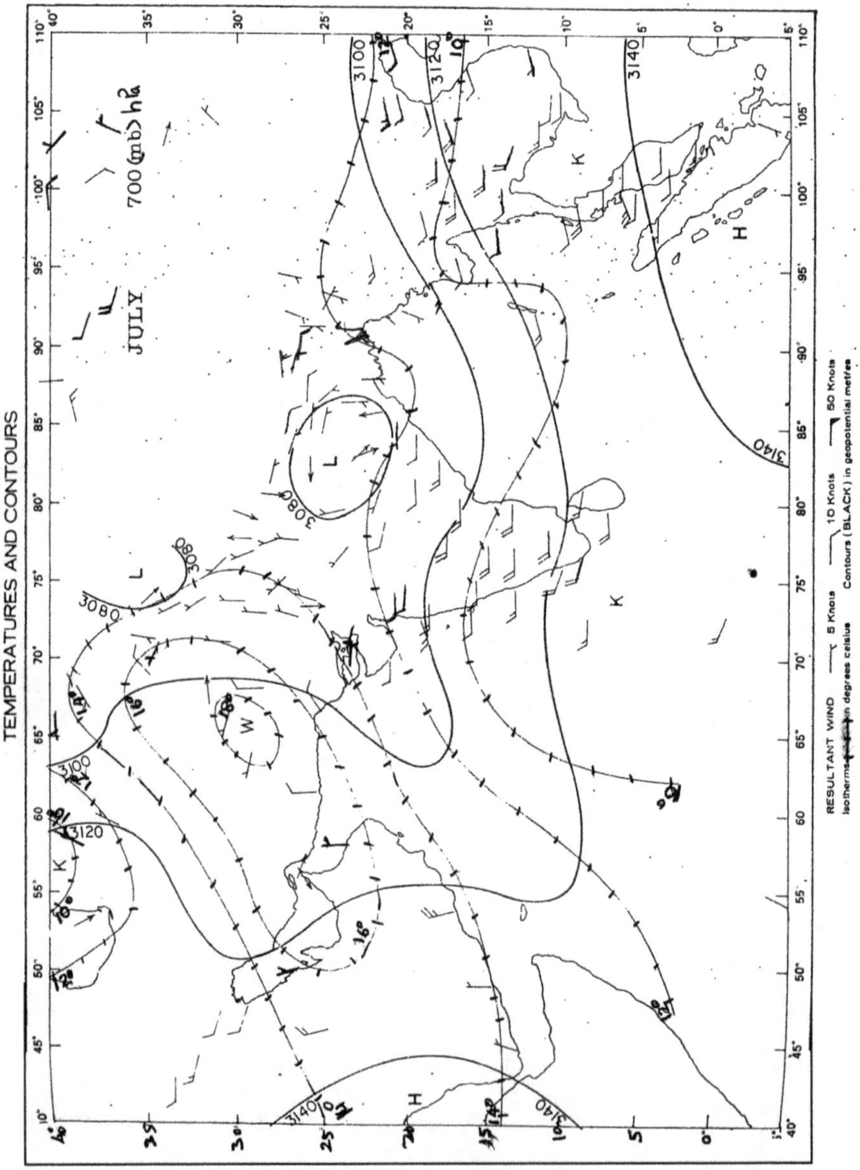
TEMPERATURES AND CONTOURS
JULY
700 (mb) hPa
RESULTANT WIND
5 Knots
10 Knots
50 Knots
Isotherms in degrees celsius
Contours (BLACK) in geopotential metres
3080
3100
3120
3140
L
H
K
W

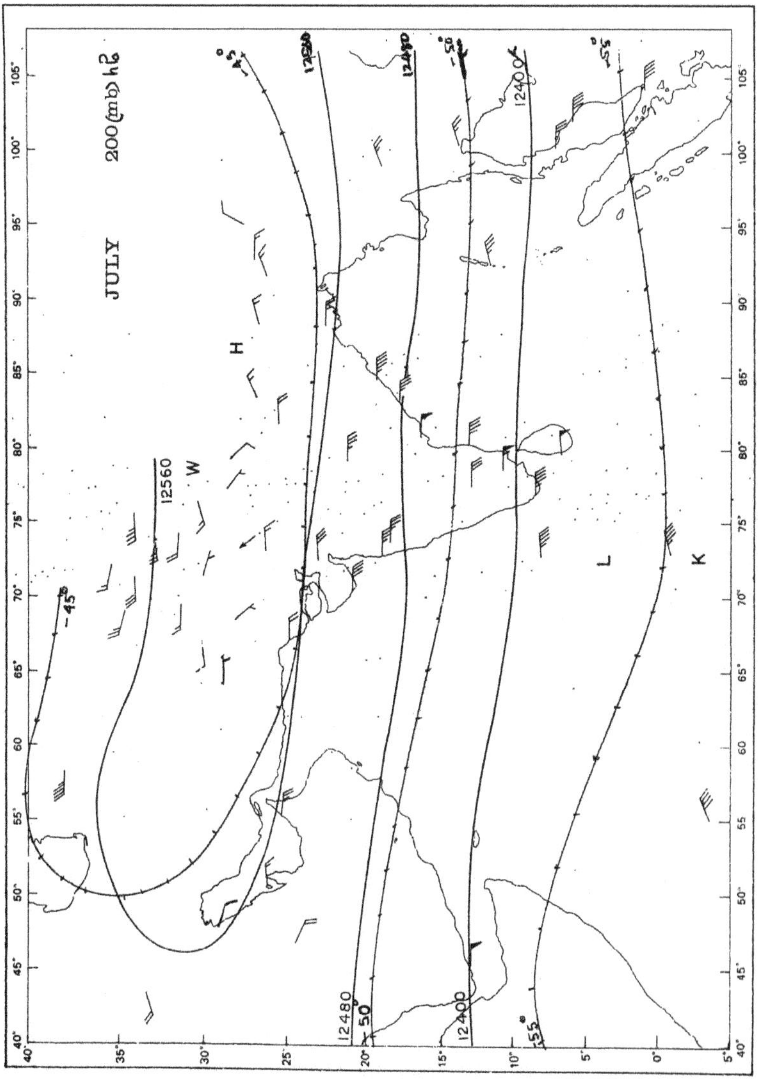
TEMPERATURES AND CONTOURS
JULY
200 (mb) hPa
12560
12480
12400
H
W
L
K
RESULTANT WIND
5 Knots
10 Knots
50 Knots
Isotherms in degrees celsius
Contours (BLACK) in geopotential metres

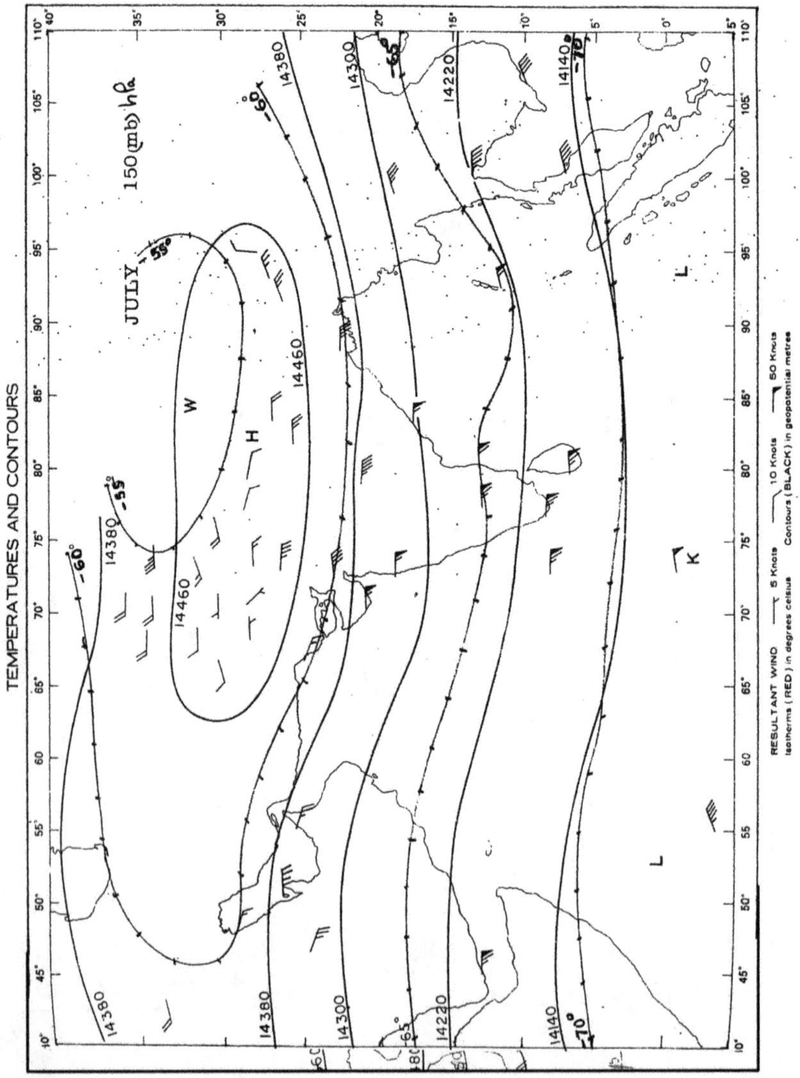
TEMPERATURES AND CONTOURS
JULY
150(mb) hPa
RESULTANT WIND 5 Knots 10 Knots 50 Knots
Isotherms (RED) in degrees celsius Contours (BLACK) in geopotential metres

TEMPERATURES AND CONTOURS

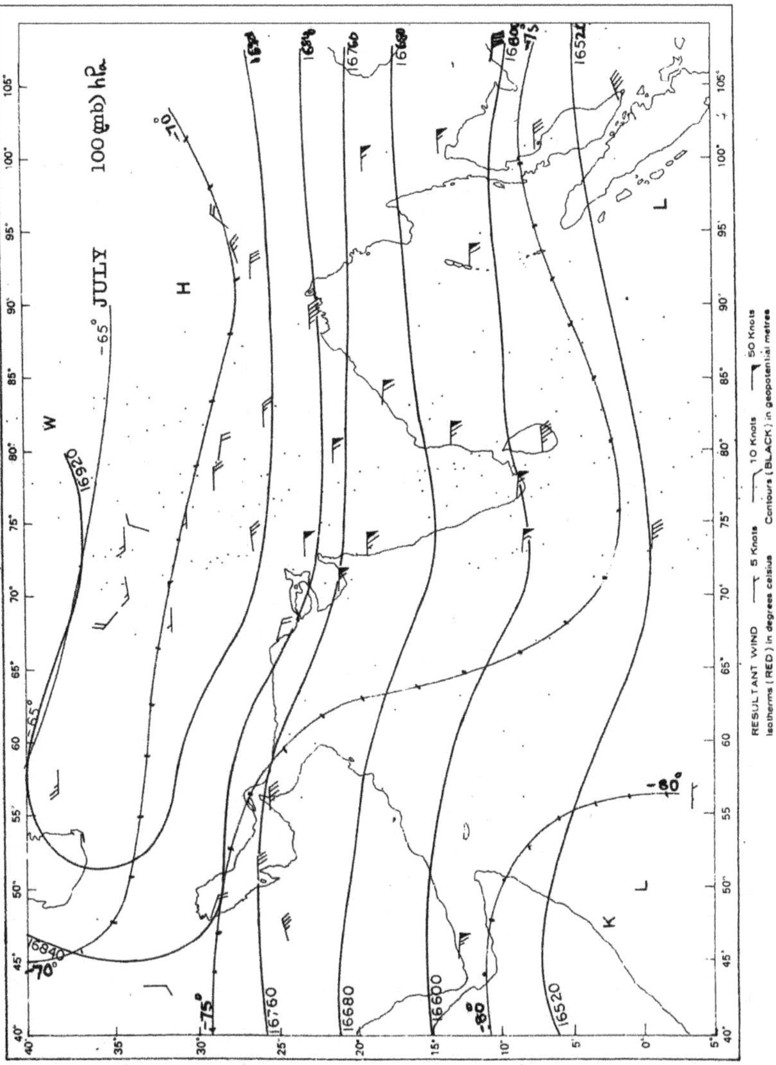

TEMPERATURES AND CONTOURS

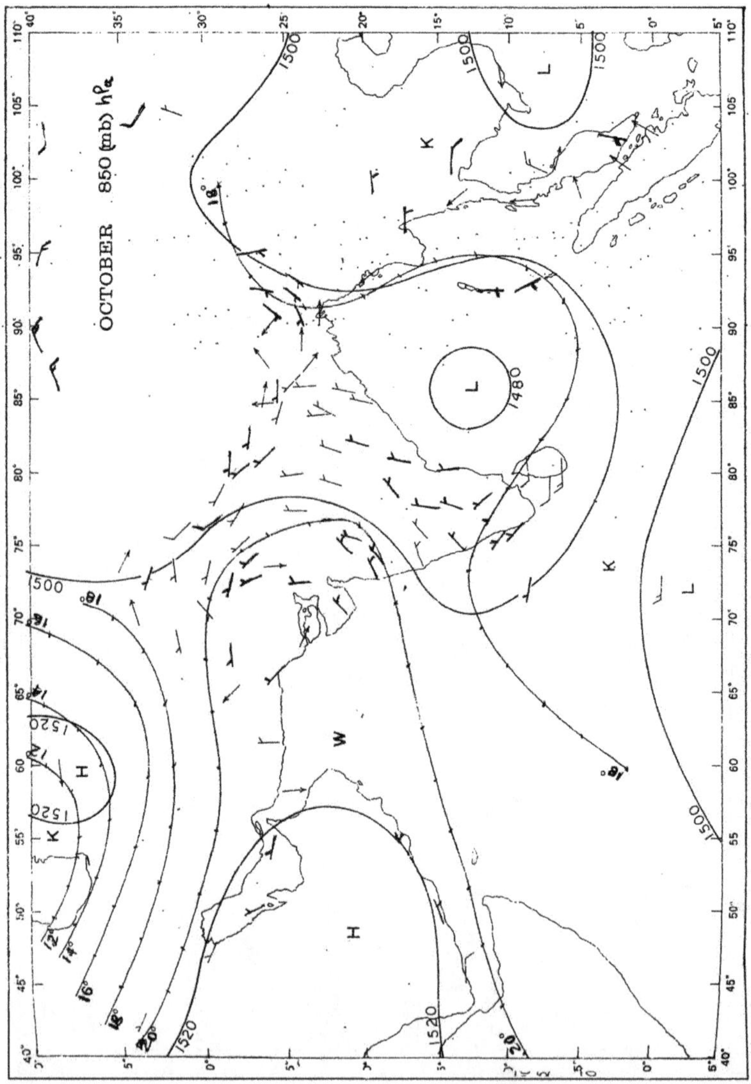

RESULTANT WIND 5 Knots 10 Knots 50 Knots
Isotherms in degrees celsius Contours (BLACK) in geopotential metres

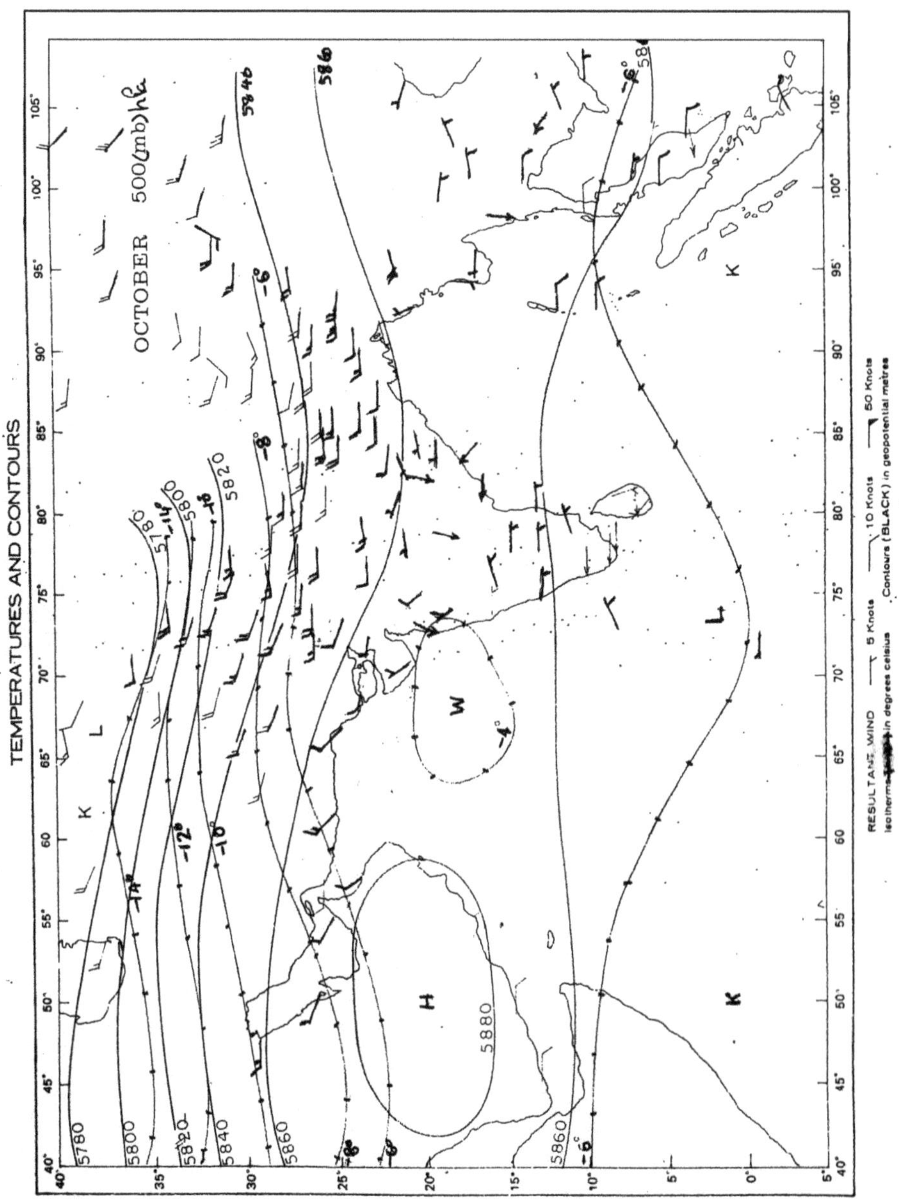
TEMPERATURES AND CONTOURS
OCTOBER 500(mb)hPa
RESULTANT WIND — 5 Knots — 10 Knots — 50 Knots
Isotherms in degrees celsius
Contours (BLACK) in geopotential metres

TEMPERATURES AND CONTOURS

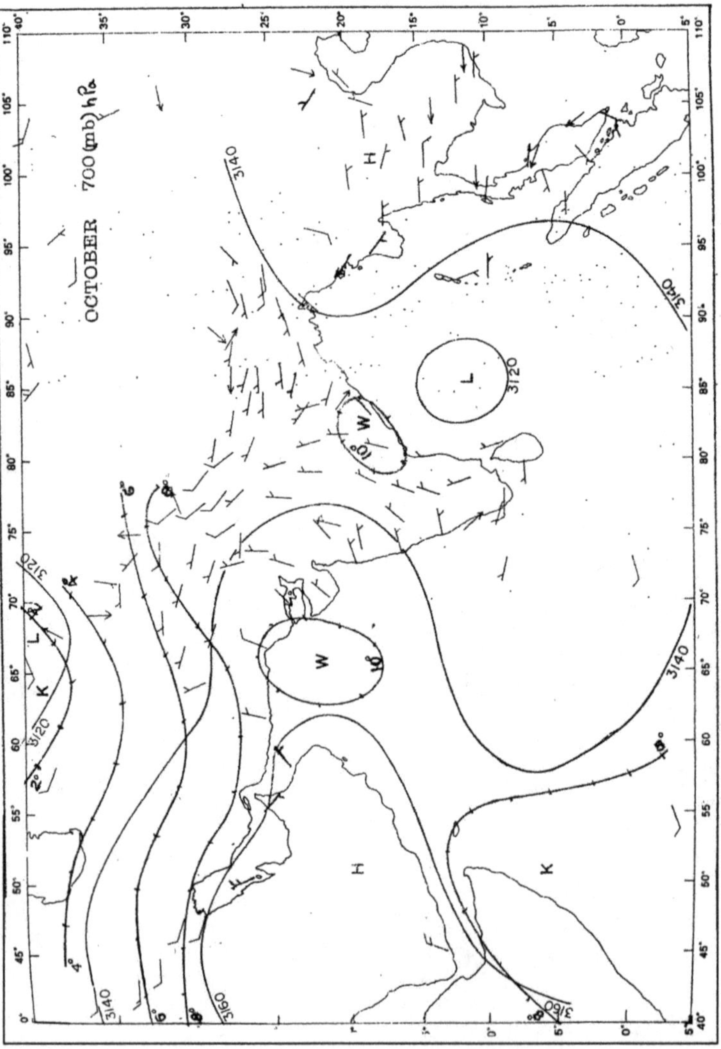

RESULTANT WIND 5 Knots 10 Knots 50 Knots

Isotherms in degrees celsius Contours (BLACK) in geopotential metres

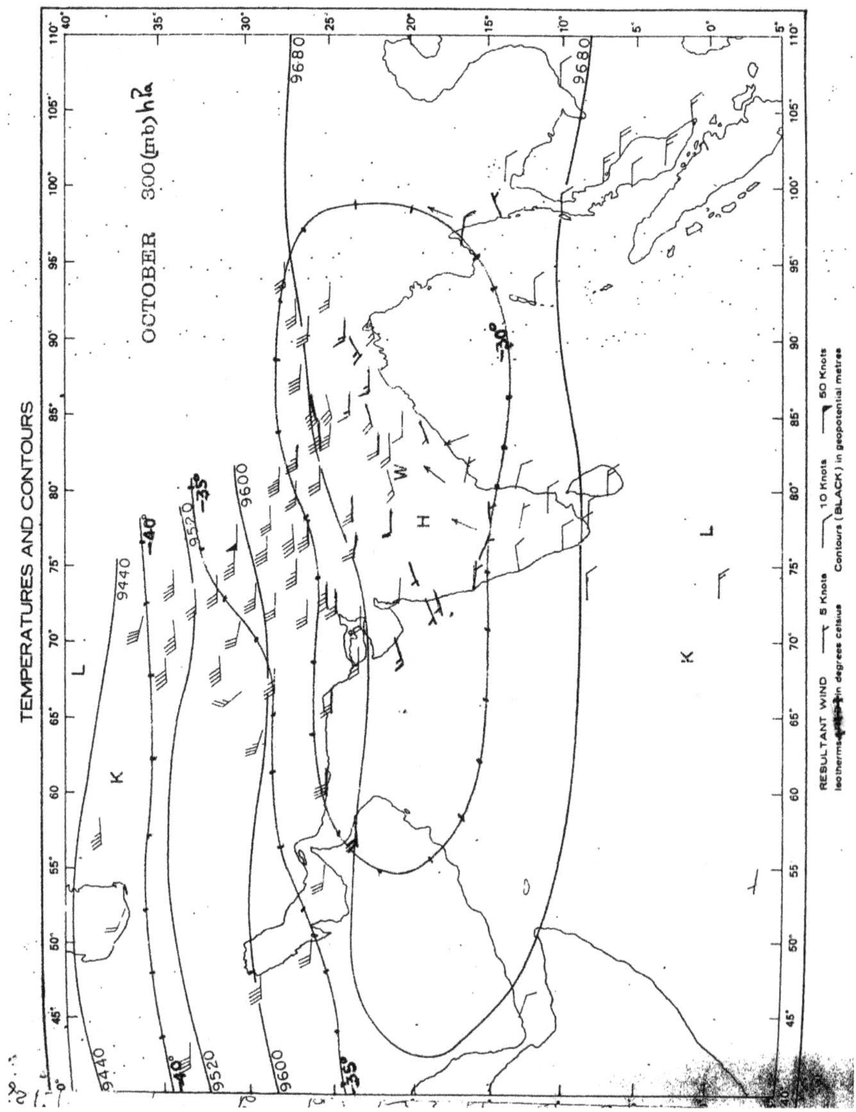
TEMPERATURES AND CONTOURS
OCTOBER 300(mb) hPa
RESULTANT WIND — 5 Knots — 10 Knots — 50 Knots
Isotherms in degrees celsius
Contours (BLACK) in geopotential metres
9440
9520
9600
9680
-40°
-35°
-30°
H
W
K
L

TEMPERATURES AND CONTOURS

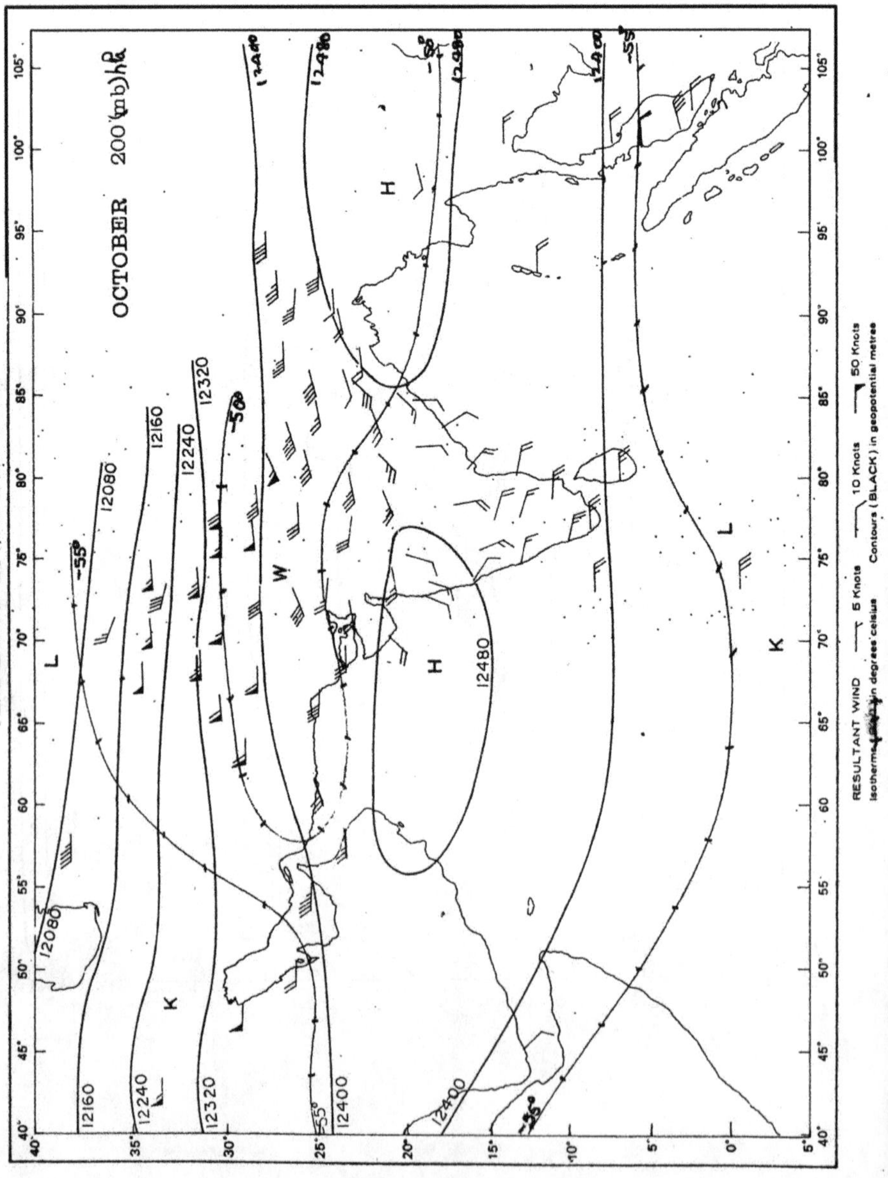

TEMPERATURES AND CONTOURS

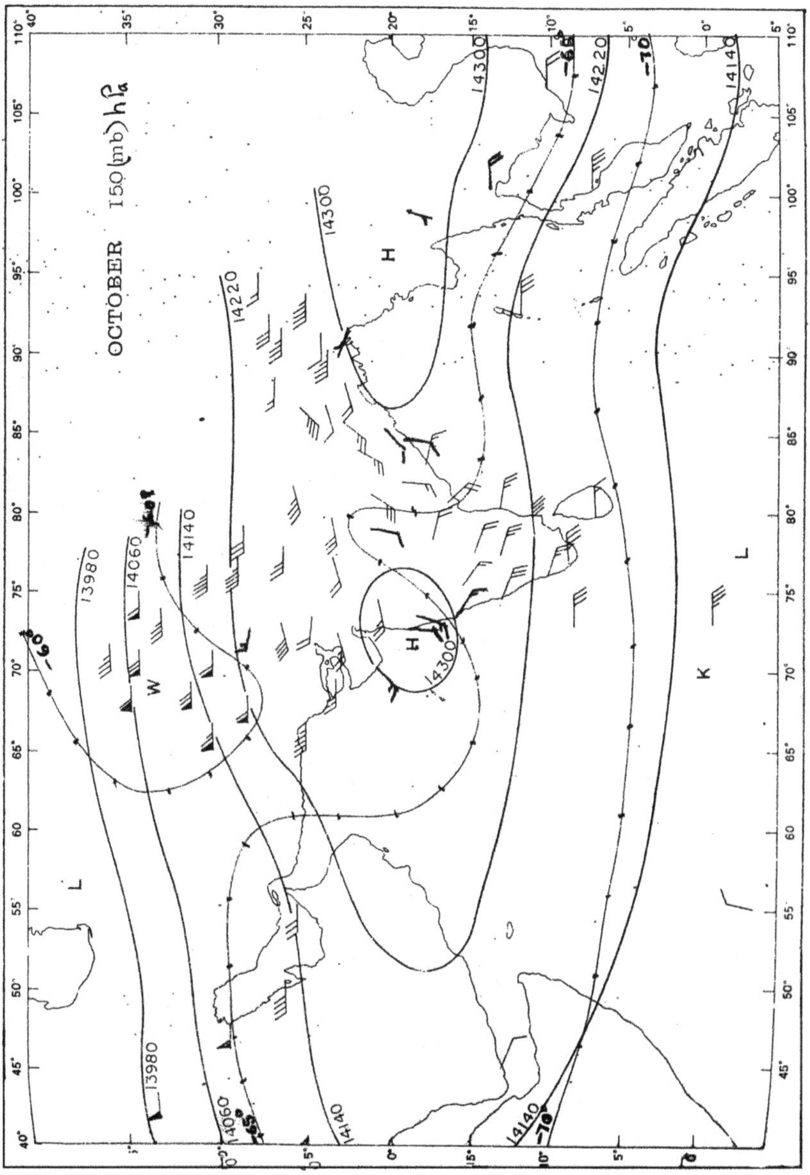

RESULTANT WIND — 5 Knots — 10 Knots — 50 Knots
Isotherms in degrees celsius Contours (BLACK) in geopotential metres

TEMPERATURES AND CONTOURS

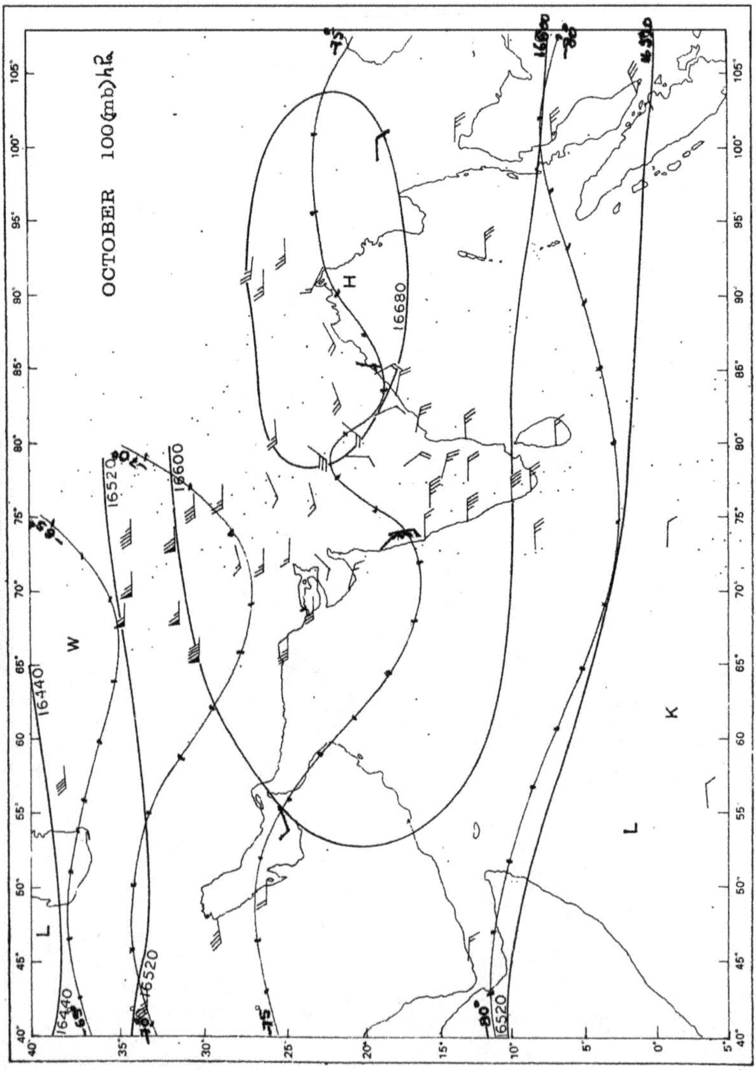

RESULTANT WIND — 5 Knots — 10 Knots — 50 Knots
Isotherms in degrees celsius Contours (BLACK) in geopotential metres

Anticyclones

Subtropical high pressure belts both in north and southern hemisphere have a number are permanent high pressure cells. Some of these are persisting called permanent or semi-permanent Anticyclones. Siberian High is an example of this kind.

Anticyclones (Highs) are associated with cold air (high density) at surface level and surface inversion of temperature. Anticyclones extend to some height in low troposphere and are places of air mass origin with calm or light wind at the centre. Pressure rising at the centre due to upper level convergence/subsidence of air, divergence at surface. ELR becomes stable. If air is dry the area will be cloudless and dew or frost is common during winter. In general, anticyclone areas (Highs) have stable atmosphere, smooth flying conditions and fair weather.

Anticyclones are of two kinds, cold anticyclones warm anticyclone.

Cold anticyclone: Siberrian High in winter is an example of Cold Anticyclone (permanent) and to some extent the High over north American continent in winter. Because of very low surface temperature in winter, the subsidence warming of air is negligible. Surface inversion of temperature and ground frost are common features over land area due to large diurnal variation of temperature. Cold Anticyclones are not possible in the lower and upper layers of the atmosphere. When Anticyclone areas are affected by travelling frontal depressions (lows), in the ridge area temporary cold anticyclone may form which will be associated with fog, low stratus.

Warm anticyclone: The High over Azores sea area is an example of this kind. In this case the subsiding air do not cool down appreciably over sea area, where diurnal variation of temperature is very small. This kind of anticyclones are associated with fine weather, good visibility and smooth flying conditions. When they are effected by travelling frontal depressions (or lows) fog, clouds and drizzle may occur.

In conclusion, a clod anticyclone, in winter may produce fog, surface frost and in summer thunderstorm. A stationary anticyclone in summer is cloud free because of subsidence surface, divergence but in winter fog, low stratus may form. Subsidence is vertical down movement of air.

A warm High is associated with air bulging upwards. In an anticyclone or ridge at surface centre air movement will be subsiding (descending) and diverging. In a cold anticyclone in winter may produce very low temperature (over land) at surface and frost.

Ridges: Air in ridges generally stable and general character of anticyclones are observed.

Cols: In a Col region wind is very light and stagnant at the centre. Weather depends on the air mass.

When a frontal system travels into out-flowing currents of Col it becomes more or less stationary and the system may degenerate. If the (frontal) system enters into inflowing currents (of Col) a weak system may intensity. In autmn and winter over land the Col is associated with fog and low cloud. In summer thunderstorms may occur under favourable conditions.

Clod Anafront: In Greek Anabatic means 'go up' and Katabatic means 'go down'. In an cold anafront warm air ascending and moving forward but less rapidly than the frontal surface. In this type of front heavy rain is accompanied at the frontal passage and slight rain behind the front. Large temperature fall, wind veering sharply with decrease in wind speed observed.

Cold Katafront: In this case comparatively warm air descending and moving forward faster than the frontal surface. It is accompanied by very slight rain at the front but more rain ahead of surface front about 100 km a head. Slight temperature fall. Wind veers gradually with little change in speed.

Surface Synoptic Weather Report Code

After recording the meteorological observations the weather reports are transmitted in a suitable coded form which was developed for easy, rapid and efficient transmission. There has been international agreement concerning the codes for exchange of weather information between national meteorological services within a region. Each code form bears a number preceded by the letters FM (Forecasting Manual of WMO). Here we shall present a SYNOP code form which is used by IMD.

SYNOP

Section 0 Mi Mi Mj Mj YY GG i_w

Section 1 $I_R i_x h$ VV I I i i I N dd ff 1 s_n TTT 2 s_n T_d T_d T_d

	$3\,P_o\,P_o\,P_o\,P_o$	$4\,P\,PPP$	$7\,WW\,W_1W_2$	$8\,N_h\,C_LC_MC_H$
Section 3	333	$1s_n\,T_x\,T_x\,T_x$	$2\,s_n\,T_n\,T_n\,T_n$	$56\,D_L\,D_M\,D_H$
	$57\,C\,D_a\,e_c$	$58\,P_{24}\,P_{24}\,P_{24}$	$59\,P_{24}\,P_{24}\,P_{24}$	$6\,R\,R\,R\,t_R$
	$8\,N_s\,C\,h_s\,h_s$	$9\,S_p\,S_p\,s_p\,s_p$		
Section 5	555	$O\,R_{24}\,R_{24}\,R_{24}\,R_{24}$	$1\,R_T\,R_T\,R_T\,R_T$	$2\,D_q\,F_q\,q\,q_t$

Explanation

Section	Number	
0	–	Data for identifying type, date and time of observation.
1	–	Data for international exchange.
2	222	Maritime data pertaining to a coastal station.
3	333	Data for Regional exchange
5	555	Data for National exchange

Note : All stations in India with elevation between 500 to 800 m will report 4 RRRR only.

Stations with elevations 800 m or more will report 4 PPPP and report 4 a_3 hhh giving the geopotential height of standard isobaric level.

The group 7 WW W_1W_2 will be omitted if both present weather and past weather are insignificant. The codes 00, 01, 02, 03 for WW and 00, 01, 02, for W_1W_2 are considered insignificant.

The meaning of the symbols in the above code :

a_3 : Indicator giving the standard constant pressure level, of which the geopotential is reported.

C : Type of cloud

C_H : Type of High cloud

C_M : Type of Medium cloud

C_L : Type of Low cloud

D_a : Direction in which orographic clouds or C_b, C_u (with vertical development) are observed.

D_L : Direction of Low cloud movement

D_M	:	Direction of Medium cloud movement
D_H	:	Direction of High cloud movement
D_q	:	Direction of squall
dd	:	Direction of surface wind
e_c	:	Angle of elevation of the top of the cloud reported by C
F_q	:	Maximum of squall wind force in Beaufort scale
ff	:	Wind speed in knots
GG	:	Actual time of observation to the nearest whole hour in UTC or barometer to be read at exact time.
h	:	Height agl of the base of the lowest cloud
hhh	:	Geopotential of the standard pressure level given by a_3 in standard geopotential meters.
$h_s h_s$	:	Height agl of the cloud base reported by C.
i_R	:	Indicator for inclusion or omission of precipitation data.
i_w	:	Wind indicator.
i_x	:	Indicator for type of station operation (manned or automatic).
$M_i M_i$	:	Identification letters of the report e.g., a SYNOP report.
$M_j M_j$	:	Identification letters of the part of the report.
		SYNOP reports from land stations is identified by AAXX.
N	:	Amount of cloud coverage (oktas).
N_h	:	Amount of low cloud coverage (Oktas).
N_s	:	Amount of individual cloud layers of type C.
$P_{24} P_{24} P_{24}$	:	Change (fall or rise) in station level pressure in last 24 hours in tenths of hPa.
PPPP	:	Pressure reduced to msl in tenths of a hPa omitting thousandth digit.
$P_o P_o P_o P_o$	:	Station level pressure in tenths of a hPa.

q	:	Nature of squall.
q_o	:	Time of occurrence of squall.
RRR	:	Amount of precipitation in mm since 0300 UTC.
$R_{24} R_{24} R_{24} R_{24}$	:	Amount of precipitation since 0300 UTC in the tenths of mm.
$R_T R_T R_T R_T$	:	Progressive seasonal total precipitation in full mm.
S_n	:	Sign of temperature. '0' stands for positive temperature and '1' stands for negative temperature.
$S_p S_p s_p s_p$	:	special phenomena.
T T T	:	Screen air temperature in tenths of a degree Celsius.
$T_d T_d T_d$	:	Dew point temperature in tenths of a degree Celsius.
$T_n T_n T_n$	:	Minimum temperature in tenths of a degree Celsius.
$T_x T_x T_x$	:	Maximum temperature in tenths of a degree Celsius.
t_R	:	Duration of precipitation.
VV	:	Horizontal visibility
$W_1 W_2$	:	Past weather
WW	:	Present weather
YY	:	Day of the month (Date)

Example of SYNOP

AA xx	:	SYNOP of Inland station
180 34	:	Date 18, time 03 UTC, wind indicator anemometer instrument
43 128	:	Station index 43 block international station number 128 (Hyderabad - India)
214 96 $i_R = 2$	:	Rainfall inclusion, $i_x = 1$ manned, h = 4 (300 – 600 m) VV = 96, visibility coded 96
727 10	:	Total cloud coverage 7 oktas, wind direction, speed 270 deg. 10 kt.
10 248	:	$s_n = 0$ positive temperature, D.B. temperature 24.8 °C

2 0 2 1 0	:	Dew point temperature 21.0 °C
3 9448	:	Station level pressure in first decimal 944.8 hPa
400 47	:	MSL pressure in tenths of hPa 1004.7 hPa
7 0 5 2 2	:	Present weather 05 Haze, past weather 22 no significant weather.
8 5 7 3 0	:	Amount of low cloud 5 oktas, low cloud type 7 Stratus, medium cloud type 3 Altocumulus.
3 3 3	:	Section three
2 0 2 2 7	:	$s_n = 0$ positive temperature, minimum temperature 22.7 °C
5 6 6 9 0	:	Direction of movement of low cloud 6 = west, medium cloud 9 = not known
5 8 0 3 2	:	24 hr pressure rise 03.2 hPa
6 00 3	:	Rainfall since 0300 UTC 003 mm, duration not reported.
8 4 7 1 0	:	Cloud amount 4 oktas, type of cloud Stratus (7) or genus, height of base of cloud 10 = 300 m
8 3 6 1 5	:	Cloud amount 3 oktas, type of cloud stratocumulus (6), height of base of cloud 15 = 450 m.
8 5 3 60	:	Cloud amount 5 oktas, type of cloud Altocumulus (3) or genus, height of base of cloud 60 = 3000 m
555	:	Section five
00026	:	24 hr rainfall (intenths of mm) 2.6 mm
1 0 4 6 2	:	Seasonal total rainfall (in full mm) 462 mm

Example 2

AA XX	:	Synop of inland station.
16 124	:	Date 16, time 1200 UTC, wind indicator instrument.
4 3 1 2 8	:	Station index Hyderabad. (Block 4 3, station number 128 for Hyderabad.
3 1 4 9 6	:	$i_R = 3$ rainfall included.

		i_x = 1 manned.
		h = 4 height of base of lowest cloud 300 to 600 m.
		VV = 96, visibility coded 96.
7 2710	:	Total cloud coverage 7 oktas, wind : 270°/10kt.
10 294	:	'0' positive temperature' DB 29.4 °C.
20 222	:	'0' positive temperature' dew point 22.2 °C.
3 9381	:	station level pressure 938.1 hPa
499 66	:	MSL pressure 996.6 hPa
705 22	:	Present weather 05 = haze, past weather 22 = not significant.
8 59 30	:	Amount of low cloud 5 oktas Type of low cloud (9) cumulonimbus. Type of medium cloud (3) Altocumulus
3 3 3	:	Section 3
10 304	:	'0' positive temperature, maximum temperature 30.4 °C
5 66 90	:	Direction of low cloud (6) westerly, Direction of medium cloud (9) not known
5 79 73	:	9-Cb, direction of CB(7) northwesterly, angle of elevation (3) 3°
590 15	:	24 hr pressure (0) rise 1.5 hPa
8 3615	:	3 oktas low cloud (type 6), Stratus, base height (15) = 450 m
8 3825	:	3 oktas low cloud (type 8) Cumulus, base height (25) = 750 m
8 1 9 3 0	:	okta low cloud (type 9) Cumulonimbus, base height (30) 900 m
8 5 3 6 0	:	5 oktas medium cloud (type 3) Altocumulus, base height (60) = 3000 m

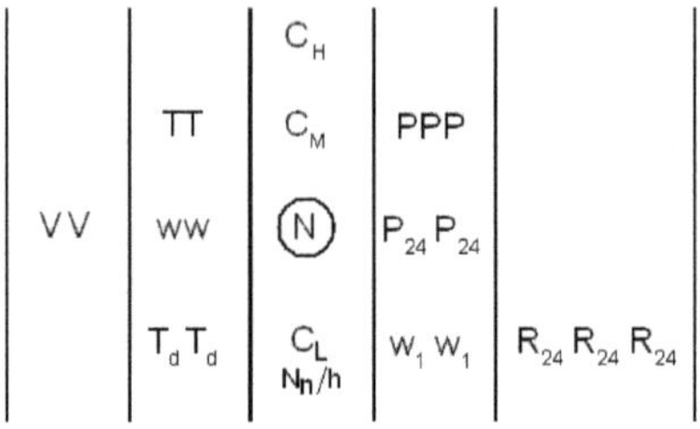

SYNOP Station Plotting Model

Above two examples plotted below.

Example 1

Date 18, time 0300 U T C

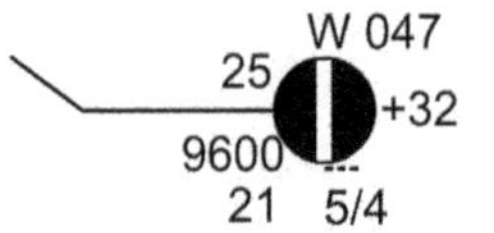

Mini temperature 22.7

Station : Block 43 St.no. 128 Hyderabad

Example 2

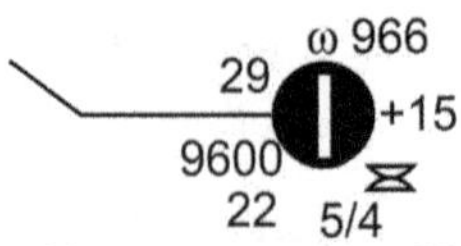

Max temperature 30.4 °C

Date 16, time 1200 UTC

Station Hyderabad

N	0	1	2	3	4	5	6	7	8	9	/
Plotting symbol	○	⦶	◔	◔	◑	⊖	◕	◕	●	⊗	⊠

Important SYNOP Plotting Symbols are given below.

1. N – Total cloud amount

   N = 0 represents clearsky

   N = 9 Sky obscure (with fog etc)

   N = / Sky cloud coverage not known.

2. ddff – wind direction and wind speed in kt.

   Direction is plotted by shaft of an arrow showing the direction of wind blowing.

   Wind speed is plotted by barbs – half barb 5 kt, full barb 10 kt, pentad 50 kt

Example

In northern hemisphere easterly 15 kt

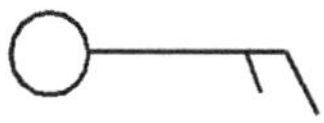

Southerly 20 kt

Southwesterly 45 kt

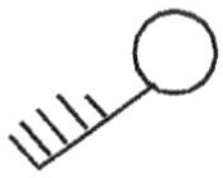

Westerly 60 kt

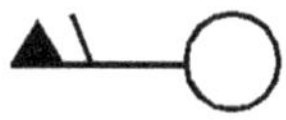

Northwesterly 6 kt

(code 99 =) variable 04 kt

Southeasterly 10 kt

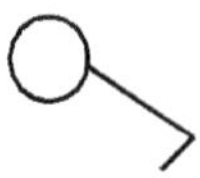

Southerly 20 kt

3. Plotting cloud symbols for C_L, C_M, C_H (8 N_h, C_L C_M C_H)

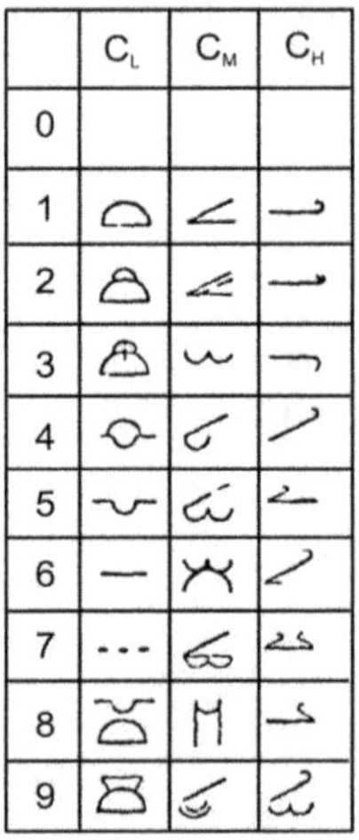

4. Plotting symbols of past weather W

Code	0	1	2	3	4	5	6	7	8	9
Symbol	not	not	not	S→ / ⇸	≡	,	•	*	▽	R
	used	used	used	Sand/ dust storm	fog	Drizzle	Rain	Snow	Show- ers	Thunder storm

5. Plotting symbols for present weather ww

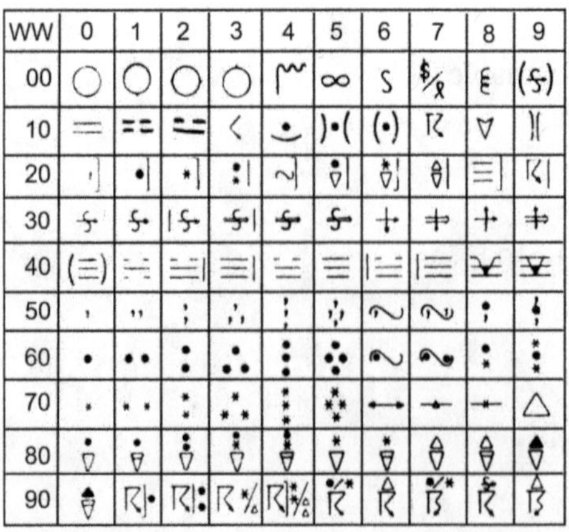

Code ww (00 - 99)

00 – 49 : When there is no pptn at the station at the time of observation.

50 – 99 : There is pptn at the station at the time of observation.

Details

01, 02, 03, No change in the state of sky.

04 – Visibility reduced due to smoke

05 – Haze

06 – Widespread dust suspended in the air

07 – For land station : Dust/Sand raised by wind at the time of observation.

For Ship : Blowing spray at the observation site.

08 – Well developed dust/sand whirl(s) observed at the station or nearby.

09 – Dust/sandstorm at the station or nearby at the time of observation or during the preceding hour.

10 – Mist

11 – Patches of shallow fog

12 – Roughly continuous shallow fog

13 – Fog only lightning visible

Code 14 – 16 pptn within sight, but not at the station

17 – Thunderstorm without pptn at the time of observation

18 – Squall at/within sight at the time of observation or during preceding hour.

19 – Funnel cloud at/within sight of the station at the time of observation or during the preceding hour.

Code 20 – 29 Refers to pptn, fog, ice fog or thunderstorm at the station during preceding hour.

These codes are not used where there is pptn at the time of observation.

28 – Fog or ice fog.

29 – Thunderstorm without pptn.

Code 30 – 39 Refers to duststorm, sand storm, drifting or blowing snow at the station at the time of observation.

30 – 35	Dust and sand storm.
36 – 39	Drifting or blowing snow.
Code 40 – 49	Refers to fog or ice fog at the time of observation.
40 –	Fog or ice fog (not at the station).
41 – 47	Fog or ice fog.
48 – 49	Fog depositing rime.
Code 50 – 59	Refers to drizzle.
Code 60 – 69	Rain (at the station).
Code 70 – 79	Refers to solid precipitation not in showers.
Code 80 – 89	Showery pptn.
Code 90 – 99	Showery pptn with or without thunderstorm .
91 – 94	Thunderstorm during preceding hour.
95 – 99	Thunderstorm at the time of observation.

6. Past weather

0 – 2	No significant change.
3	Sand/dust storm or blowing snow.
4	Fog/ice fog or thick haze.
5	Drizzle.
6	Rain.
7	Snow or rain and snow mixed.
8	Showers.
9	Thunderstorm with or without pptn.

7. N_h All low clouds (C_L) sky coverage in oktas.

   N_h = 9 when the sky is not discernible

   If there are no low clouds, N_h should be reported by medium cloud coverage irrespective of their height.

   $N_h \leq N$, C_L – Clouds are Sc, St, Cu, Cb

   C_M – Clouds are Ac, As, Ns

   C_H – Clouds are Ci, Cc, Cs

8. Symbols for pressure change 'a'.

	0	1	2	3	4	5	6	7	8
a	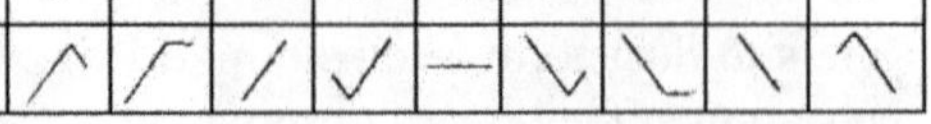								

9. N

0	Clear sky
1	1 Okta or less
2	2 oktas
3	3 oktas
4	4 oktas
5	5 oktas
6	6 oktas
7	7 oktas or more but not overcast
8	8 oktas overcast
9	Sky obscured/sky not discernible.

10. $h_s h_s$ Height of base of cloud layer indicated by C

The code is direct multiple of 30 for 0 – 50

hs hs	
00	< 30 m
01	30 m
..	..
..	..
..	..
20	600 m
..	..
..	..
..	..
50	1500 m

51 – 55 not used

56 – 80 Increments of 300 m after substracling 50

$h_s h_s = 60\ (60 - 50) \times 300 = 3000$ m

$= 70\,(70 - 50) \times 300 = 6000$ m

$= 80\,(80 - 50) \times 300 = 9000$ m

81 – 88 Increments of 1500 m from code 80

$h_s h_s = 81\,(80 - 50) \times 300 + (81 - 80)\ 1500 = 9000 + 1500$

$= 10500$ m

$= 82\ 9000 + (82 - 80)\ 1500 = 12000$ m

$= 86 \quad 9000 + 8 \times 1500 = 21000$ m

$= 89$ is used if the height is greater than 21000 m

Code figures 90 – 99 used under regional instructions.

Tables of meteors and their symbols (same as on Forecast form)

11. C Genus of cloud/type of cloud.

Code fig.	Genus	Symbols
0	Cirrus	
1	Cirricumulus	
2	Cirrostratus	
3	Altocumulus	
4	Altostratus	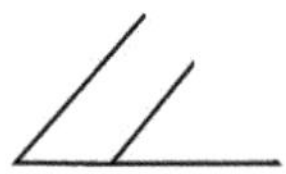
5	Nimbostratus	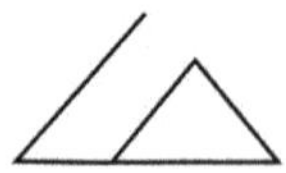
6	Stratocumulus	
7	Stratus	
8	Cumulus	
9	Cumulonimbus	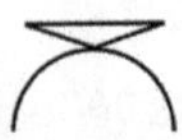
/	Cloud not discernible due to darkness, fog, dust/sand storm etc.	

12. Horizontal Visibility VV

Code 00 – 50 Direct reading in units of 100 meters

e.g., 05 stands for 5 × 100 = 500 m

12 stands for 12 × 100 = 1200 m

50 stands for 50 × 100 = 5000 m

51 – 55 not used.

56 – 80 Substract 50, the remaining figure direct reading in Km

e.g., 65 stands for 65 – 50 = 15 Km

70 stands for 70 – 50 = 20 Km

81 – 89 Upto 80 is 30 Km, thereafter increments in 5 Km

e.g., 81 stands for 30 + (81 – 80) × 5

= 30 + 5 = 35 Km

85 stands for 30 + (85 – 80) × 5 = 55 Km

89 stands for 30 + 9 × 5 = 75 Km

13. Height of base of lowest clouds 'h' (agl = above ground level)

Code fig.	Height in meters
0	0 – 50
1	5 – 100
2	100 – 200
3	200 – 300
4	300 – 600
5	600 – 1000
6	1000 – 1500
7	1500 – 2000
8	2000 – 2500
9	2500 or more or no clouds
/	Height of cloud base not known.

14.i_3

Code fig.	Specification
0	Wind speed estimated
1	Wind speed taken from anemometer, recorded in meters per second.
3	Wind speed estimated, recorded in knots.
4	Wind speed taken from anemometer and recorded in knots.

Reference

1. Handbook of Aviation Meteorology, HMSO. London.
2. Ground Studies for Pilots-Meteorology, R.B. Underdown and John Standen Blackwell Publishing.
3. Compendium lecture notes for class III meteorological persons, WMO publication.
4. Compendium of Meteorology Vol, I WMO No. 364.
5. Atmosphere, Weather and Climate, R.G. Barry, R.J. Chorley 1976 ELBS and Methuen and Co. Ltd.
6. The science of weather and Environment, Navale Pandarinath, B.S. Publications, Hyderabad.
7. IMD – Tracks of Cyclonic Storms 1891 – 1970.
8. IMD – Aviation Weather Codes.
9. IMD – Aviation Meteorology.
10. Climatology for Aviation pilots, H.R. Quantick, Blackwell Science – 2001.
11. International Standards and Recommended Practices –

    Anmex 3 – Meteorological Services for International Air Navigation, 1-1-1996.
12. Airport Authority of India, AIP India 01 January 2001.
13. DGCA – Civil Aviation Requirement Rev. 1, 26 June 2006. Series B. Part II Section 7.
14. Climatology of India and neighbourhood, Y.P. Rao, K.S. Ramanamurthy, IMD FMU No.12.
15. Satellite Meteorology, R.R. Kelkar, B.S. Publication, Hyderabad.

Index

A

Index

A

D

E

F

G

H

I

J

K

L

M

N

O

P

T

Z

www.ingramcontent.com/pod-product-compliance
Lightning Source LLC
Chambersburg PA
CBHW071234230926
51849CB00001B/100
9789385433054